Libra

MASS
TRANSFER

INTERNATIONAL STUDENT EDITION

**McGRAW-HILL KOGAKUSHA, LTD.**
*Tokyo*
Auckland
Düsseldorf
Johannesburg
London
Mexico
New Delhi
Panama
São Paulo
Singapore
Sydney

**THOMAS K. SHERWOOD**
**ROBERT L. PIGFORD**
**CHARLES R. WILKE**
*Professors of Chemical Engineering*
*University of California, Berkeley*

# Mass
# Transfer

This book was set in Times New Roman.
The editors were B. J. Clark and J. W. Maisel;
the production supervisor was Dennis J. Conroy.
The drawings were done by ECL Art Associates, Inc.

Library of Congress Cataloging in Publication Data

Sherwood, Thomas Kilgore, date
    Mass transfer.

    (McGraw-Hill chemical engineering series)
    Previous editions published under title:  Absorption
and extraction.
    Includes bibliographies.
    1.  Mass transfer.  I.  Pigford, Robert Lamar,
date, joint author.  II.  Wilke, Charles R.,
date, joint author.  III.  Title.
TP156.M3S53    1975    660.2'842    75-1295
ISBN 0-07-056692-5

**MASS
TRANSFER**

**INTERNATIONAL STUDENT EDITION**

KOSAIDO PRINTING CO., LTD. TOKYO JAPAN

# CONTENTS

**MASS
TRANSFER**

1

# INTRODUCTION

Mass-transfer phenomena are to be found everywhere in nature, and are important in all branches of science and engineering. The phrase "mass transfer," which has come into common use only in recent years, refers to the motion of molecules or fluid elements caused by some form of potential or "driving force." It includes not only molecular diffusion but also transport by convection and sometimes simple mixing—not the conveyance of a material, as in the flow of a fluid in a pipe.

Mass transfer is involved wherever a chemical reaction takes place, whether in an industrial reactor, a biological system, or a research laboratory. As pointed out by Weisz [1], the reacting substances must come together if the reaction is to proceed; and in many cases, the reaction slows or stops if the reaction products are not removed. The reactants have little difficulty in coming together in the case of homogeneous reactions in a single well-mixed liquid or gas phase, but the rate of mass transfer may completely determine the chemical conversion when reactants must move from one phase to another in order that reaction may occur. This is the case, for example, when reaction occurs at the surface of a very active catalyst in contact with a fluid which carries the reactants and removes the reaction products. In the case of a reversible reaction, the conversion is improved if the desired product is continuously removed by mass transfer to a second phase in which no reaction takes place. Furthermore, the relative rates of mass transfer of the several reacting and product species can greatly affect the selectivity when competing reactions are involved.

The general subject of mass transfer may be divided into four broad areas of particular interest and importance: molecular diffusion in stagnant media, molecular diffusion in fluids in laminar flow, eddy diffusion or mixing in a free turbulent stream, and mass transfer between two phases.

The first has been the subject of much study by scientists for more than a century, and the theory is in good shape for diffusion in gases, though not for diffusion in dense fluids. The second is an application of the first, and is treated by a mathematical manipulation, often difficult, of what is known about molecular diffusion in situations where the flow field can be described or calculated.

Eddy diffusion in a free stream, away from a phase boundary, is the process by which gases leaving a stack are dispersed into the atmosphere, and by which mixing occurs in many situations, as in turbulent jets. Transfer between two phases, across an interface, is of particular importance in engineering, largely because it is involved in most separation processes, as in the recovery of a pure product from a mixture. Evaporation from a reservoir, oxygenation of blood, removal of pollutants from the atmosphere by rain, chemical reaction at the surface of a solid catalyst or within its porous structure, deposition by electrolysis or electrophoresis, drying of wood, and removal of carbon from steel by blowing with air or oxygen are all examples of mass transfer between phases.

The chemical engineer's interest in mass transfer stems primarily from his traditional role as a specialist in the design of separation processes. The materials fed to a chemical process are purified by separation or concentration of the reactants, and the valuable products must be separated from the stream leaving the reactor. Though the separation equipment is ancillary to the reactor, its cost is often the major part of the investment in the plant. This point is illustrated by Fig. 1.1, which is a flow diagram of an industrial process for the production of formaldehyde by the oxidation of methanol. The small reactor contains several layers of a silver gauze catalyst. The rest of the plant involves typical mass-transfer equipment for purification of feed, recovery and purification of the product, and separation and recycling of unreacted methanol. Mass transfer is of concern in the reactor itself, since the production of formaldehyde cannot proceed faster than the rate of transport of methanol and oxygen from the gas stream to the surface of the silver.

Separation of a product from a very dilute stream is usually expensive. This largely explains the high prices of antibiotics, for example, which are produced in very dilute fermentation broths. The difficulty is partly explained by thermodynamics, since the isothermal reversible work required for recovery of a pure material from a mixture is roughly proportional to the logarithm of the reciprocal of the concentration in the mixture in which the substance is found or produced. Figure 1.2 indicates the general relation between the selling price of a pure product and the concentration of the mixture from which it is obtained.

The work or free energy needed for separation is not the only production

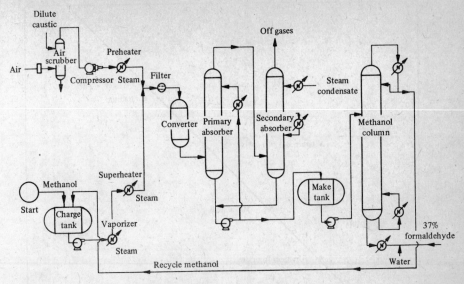

**FIGURE 1.1**
Process flow diagram for the manufacture of formaldehyde from methanol (from R. N. Shreve, "Chemical Process Industries," 3d ed., 1967. McGraw-Hill Book Co., N.Y.).

cost; Fig. 1.2 should thus not be taken too seriously. The free energy of separation of copper from the chemical compound in the natural ore, for example, is arbitrarily excluded in the energy represented by the abscissa. The point for sulfur from stack gases represents the *cost* of separation, not the selling price of sulfur. Mined sulfur, selling at roughly a cent per pound, requires very little purification. The point for gold from sea water, at the far right, illustrates that gold from this source cannot compete with gold from mined ores; the free energy required for the separation is enormously greater than that for separation from available ores. Though the selling prices are influenced by many things, the rough correlation suggested by Fig. 1.2 indicates that separation cost is usually a dominant factor.

Most practical separation schemes are based on the fact that concentrations of two phases in equilibrium with each other are usually quite different. For example, an equimolal mixture of benzene and toluene is in equilibrium with a vapor containing 70 mole percent benzene. Phases in contact tend to equilibrate by mass transfer from one to the other. Subsequent separation of the phases by mechanical means is usually simple and inexpensive. Any method of contacting two phases which results in the selective interphase transport of one of the constituents can form the basis of a separation process. The selectivity is ordinarily the result of different equilibrium relations for the different species; in a very few cases, it may be due to different rates of transport of the several constituents.

Considerations of phase equilibria are essential to an understanding or analysis

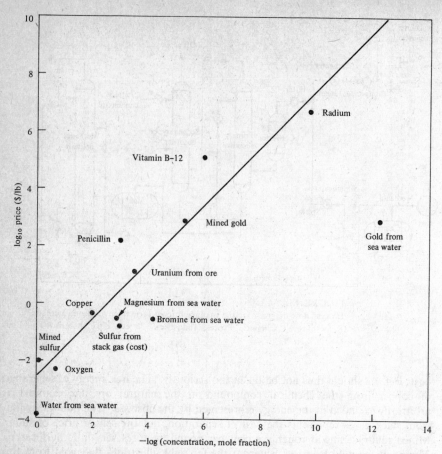

FIGURE 1.2
Relation between the values of pure products and the concentrations in the mixtures in which they are found or produced.

of any mass-transfer process. Here it is that thermodynamics has much to offer, since the second law states the conditions of equilibrium. The thermodynamic theory is well developed, though in many practical cases it does not provide the necessary connection between activities and concentrations. In a few situations, this relation is well known. In others, the empirical Lewis and Randall rule may prove adequate, although in some cases more sophisticated analyses based on complicated equations of state or on molecular theory may be required. In any event, the student or engineer dealing with problems of mass transfer must be familiar with both the theory and the empirical facts of the thermodynamics of phase equilibria. Furthermore, the first law of thermodynamics provides the basis for the calculation of the necessary enthalpy balances, including the heat effects often associated with the mass-transfer phenomena. The important role of thermodynamics in connection with equilibria and enthalpy changes is presented in numerous texts; it is only touched on in the later chapters.

Since the subjects of molecular diffusion and thermodynamics are in relatively good shape, it is the factors determining the *rate* of interphase transfer that have been the particular concern of many chemical engineers. The size and cost of mass-transfer equipment of a given type is nearly inversely proportional to the mass flux, which plays a major role in process design. Despite its crucial importance, information on rates of transfer for various important applications is often unavailable and must be estimated from fragmentary data on the basis of the principles discussed in the chapters which follow.

Most industrial processes depending on mass transfer involve one or more fluids in turbulent flow, and the existing theory of turbulence is quite inadequate as a basis for the development of a practically useful theory of mass transfer at a phase boundary. This lack of understanding of turbulence presents a major stumbling block to the development of a theoretical basis for mass transfer between phases. G. K. Batchelor, a well-known authority on fluid mechanics, writes that "modern technology needs help in describing and analysing turbulent flow, and cannot wait for scientists to understand its mysteries" [2]. Written in 1957, this is evidently still true. Of necessity, therefore, the existing correlations of data on transport rates are largely empirical. These have proved extremely useful in the design of process equipment, although the needed data and correlations are often missing or provide only approximate estimates of the size of the mass-transfer devices and of their performance. Nevertheless, the design engineer must use the tools available, within the constraints of both equilibrium limitations and economics.

Thermodynamic potential gradients of several kinds may serve to implement mass transfer. In electrophoresis and in electrochemical processes, an electrical potential provides the motive force; in the Clusius and Dickel column for the separation of gaseous isotopes, it is a thermal gradient. In a majority of applications in the process industries, however, activity or concentration gradient is employed to effect mass transfer between two phases. Several inconclusive studies have been made to determine if transport rates in such cases are not more nearly proportional to the activity than to concentration gradients. Regardless of the outcome of such studies, however, the use of concentration has enormous practical appeal, since it makes it so much simpler to tie the rate process to the stoichiometry of the process streams. This has proved generally successful; it has a good theoretical basis in some situations, though in others it is a forced simplification employed with little reference to the mechanism by which mass transfer takes place.

In living tissue, diffusion may take place in the direction of a negative concentration gradient. This phenomenon of "active transport" is presumed to be due to an input of free energy or work needed for concentration by diffusion, causing a solute to diffuse "uphill." If this process were understood it might find application in industry.

The exposition which follows is intended to provide the chemical engineer with an understanding of the phenomena of mass transfer and of the existing theory, where the latter may be available and helpful. Though intended primarily for the student, it will hopefully be found interesting and useful by the practicing engineer. Satisfying the needs of both may not be practicable because the former is usually more interested in the theory and the latter in design information. The literature on the subject is extremely voluminous, so that it is no longer possible to collect and correlate experimental data with the completeness that was attempted in the earlier versions of the book ("Absorption and Extraction," 1937 and 1952).

The authors have attempted to emphasize the *engineering* aspects of mass transfer to a greater extent than is done in many of the existing excellent texts on the subject. It has been said that "scientists tackle those problems which can be solved; engineers are faced with problems which *must* be solved." The design engineer must come up with a design. Of necessity he must rely largely on empirical correlations and incomplete theories. He develops a skill in the use of the *form* of a theory for extrapolation purposes, where the theory does not provide the absolute numbers needed for prediction. "Engineering estimates," often crude, are of great value as guides in the selection of feasible designs. It is important to be able to judge critically the validity of experimental data and the conditions under which existing correlations and theories may be employed with confidence. On such judgment hinges the decision about whether or not new laboratory or expensive pilot-plant data must be obtained before the design project should be pursued further. Clear understanding of elementary theory is often essential for making such decisions.

Only the last three chapters of the book deal specifically with design, and here the reader is referred to Perry's "Chemical Engineers' Handbook" and other sources for most of the needed design data. Chapters 2 through 8 attempt to lay a foundation for understanding the phenomena, theories, basic principles, and nature of existing data on mass transfer. Without this understanding, the use of handbook data can lead to serious design mistakes and costly failures of process equipment.

Although the text may appear to emphasize mass transfer in gas absorption, the phenomena, theory, and design principles are equally relevant to distillation, solvent extraction, crystallization, leaching, and other mass-transfer operations of industrial importance. These are treated in various other books, notably King's "Separation Processes" [3], which deals primarily with distillation, and the books on solvent extraction by Treybal [4] and Hanson [5].

An essential aspect of engineering design is optimization based on economic considerations. Alternative conceptual designs are proposed and the most promising selected for detailed analysis. This involves an economic balance of operating and investment charges and estimates of return. The techniques and cost information employed in these steps constitute a large additional body of knowledge, often more readily available to those in industry, which is beyond the scope of this book.

It seems inevitable that the United States will adopt the metric system of units, termed SI, for Système International d'Unités. The change has been promoted off and on since John Quincy Adams's 1821 report, and the U.S. Congress is currently debating various bills proposing studies of how it might be implemented. All other major countries have either gone metric or are committed to metrication. However, it would appear to be a reasonable guess that the United States will change but gradually to SI during the useful life of this book.

Chemical engineers deal with plant and equipment—areas of engineering where the English system is almost universally employed. But the fundamental science is chemistry, and the literature of chemistry is based wholly on the metric system. Of necessity, the chemical engineer must become facile in the use of both systems. The use of both in this book (sometimes in a single calculation) is a deliberate attempt to encourage students to develop this useful skill.

A number of worked examples are included in the text. Whether a university student or a practicing engineer, the reader should consider these to be as important as the text material. Understanding the units to be employed and skill in simple calculation with a slide rule or desk computer are essential to the engineer. The use of numbers helps one to understand equations. In general, we have attempted to minimize the use of advanced mathematics and of machine computation. The problems appearing at the ends of the chapters, for which solutions are not given, are taken mostly from notes on courses taught at the University of California at Berkeley, the University of Delaware, and the Massachusetts Institute of Technology. Their shortcomings should stimulate instructors to develop new and better problems for their own use in class.

We acknowledge with thanks the advice and assistance of numerous colleagues and friends. We are grateful to Dr. Urs von Stockar for help in preparing the illustrative problems of Chap. 9. We owe a special debt to Professors K. A. Smith of Massachusetts Institute of Technology and T. Vermeulen of Berkeley, to our secretary, Miss Myra Baker, and to a number of teaching and research assistants with whom we have worked.

# REFERENCES

1  WEISZ, P. B.: *Sci.*, **179**: 433 (1973).
2  BATCHELOR, G. K.: *J. Fluid Mech.*, **2**: 204 (1957).
3  KING, C. J.: "Separation Processes," McGraw-Hill, New York, 1971.
4  TREYBAL, R. E.: "Liquid Extraction," 2d ed., McGraw-Hill, New York, 1963.
5  HANSON, C. (ed.): "Recent Advances in Liquid-Liquid Extraction," Pergamon, New York, 1971.

# 2

# MOLECULAR DIFFUSION

## 2.0 SCOPE

The chapter begins with definitions of diffusion coefficients. The manner in which mutual diffusion coefficients may be expected to vary with the properties of a binary gas system is illustrated by reference to classical kinetic theory. Methods of estimating binary diffusion coefficients in gases and liquids are described. A short treatment of diffusion in porous solids is included. Basic rate equations for diffusion in stagnant and moving systems are developed in Chap. 3.

## 2.1 PRINCIPAL SYMBOLS

| | |
|---|---|
| $a$ | Activity of component of mixture |
| $c$ | Concentration, g moles/cm$^3$ |
| $c_+, c_-$ | Concentrations of cation, anion, g equiv/cm$^3$ |
| $D$ | Diffusion coefficient, cm$^2$/s |
| $D^*$ | Tracer diffusion coefficient, cm$^2$/s |
| $D_{AB}$ | Mutual diffusion coefficient species $A$ in binary of $A$ and $B$, cm$^2$/s |
| $D_{AB}{}^0$ | Mutual diffusion coefficient of $A$ in binary at infinite dilution of $A$ in $B$, cm$^2$/s |

| | |
|---|---|
| $D_{ABP}$ | Mutual diffusion coefficient of $A$ in a gas binary in porous media, $cm^2/s$ |
| $D_{Am}$ | Mutual diffusion coefficient species $A$ in mixture, $cm^2/s$ |
| $D_{ii}$ | Self-diffusion coefficient, $cm^2/s$ |
| $D_{KP}$ | Knudsen diffusion coefficient in porous media, $cm^2/s$ |
| $D_P$ | Effective diffusion coefficient in porous media, $cm^2/s$ |
| $D_{SP}$ | Surface diffusion coefficient, $cm^2/s$ |
| $E$ | Electrical potential, V |
| Fa | Faraday, 96,488 C/g equiv |
| $G_+, G_-$ | Cation, anion, concentration gradients, g equiv/$(cm^3)$(cm) |
| $H'$ | Solubility coefficient; Henry's-law constant, g moles/$(cm^3)$(atm); *note:* values of the Henry constant $H$ tabulated by Perry have the units atmosphere per mole fraction |
| $J$ | Molal flux density relative to a plane of no net volume transport, g moles/$(s)(cm^2)$ |
| $M$ | Molecular weight |
| $n$ | Molecular number density, molecules/$cm^3$ |
| $n_+, n_-$ | Valences of cation, anion |
| $N_{Avo}$ | Avogadro's number, molecules/g mole |
| $p$ | Partial pressure, atm |
| $P$ | Total pressure, atm |
| $\bar{P}$ | Permeability |
| $R$ | Gas constant, 82.06 $(cm^3)$(atm)/(g mole)$(°K)$; 8.314 J/$(°K)$(g mole) |
| $r_0$ | Radius of sphere, cm |
| $S$ | Cross section normal to diffusion flux, $cm^2$ |
| $S_g$ | BET surface, $cm^2/g$ |
| Sc | Schmidt number $= \mu/\rho D$ |
| $T$ | Temperature, $°K$ |
| $T_c$ | Critical temperature, $°K$ |
| $U$ | Velocity, cm/s |
| $\bar{U}$ | Mean speed of molecule, cm/s |
| $v$ | "Atomic diffusion volume" (see Table 2.1) |
| $V$ | Molal volume, $cm^3/g$ mole |
| $\bar{V}$ | Partial molal volume, $cm^3/g$ mole |
| $V_c$ | Critical volume, $cm^3/g$ mole |
| $x$ | Coordinate |
| $y$ | Coordinate, distance in direction of diffusion, cm |
| $X, Y$ | Mole fractions, liquid, gas |
| $\Delta H$ | Enthalpy of vaporization at the normal boiling temperature, g cal/g mole |
| $\gamma$ | Activity coefficient $= a/c$ |
| $\varepsilon$ | Lennard-Jones potential parameter $(\varepsilon/k = °K)$ |
| $\theta$ | Void fraction of porous media |
| $\lambda$ | Molecular mean free path, cm |
| $\lambda_+^0, \lambda_-^0$ | Limiting (zero concentration) ionic conductances of cation and anion, respectively, (amp)/$(cm^2)(V)/(cm)(g$ equiv/$cm^3)$ |
| $\mu$ | Viscosity, poises; chemical potential, $(cm^3)$(atm)/g mole |
| $\mu_{AB}$ | Viscosity of binary liquid mixture |
| $\mu°$ | Chemical potential of pure component, $(cm^3)$(atm)/g mole |
| $\rho$ | Density, $g/cm^3$ |
| $\rho_p$ | Density of porous media, $g/cm^3$ |
| $\sigma$ | Lennard-Jones potential parameter, Å |
| $\tau$ | Empirical constant ("tortuosity") |

$\phi$        Association parameter of solvent

$\phi(r)$      Intermolecular potential

$\Omega_D$      Diffusion collision integral

$\Omega_v$      Viscosity collision integral

## 2.2 MOLECULAR DIFFUSION

The part played by mass transfer in a variety of technical processes has been outlined in Chap. 1. Separation processes usually involve mass transfer of constituents of a mixture from one phase to a second phase readily separated from the first by mechanical methods. In other cases, as in heterogeneous catalysis, mass transfer occurs within a single phase, simultaneously with transport to and from the solid-phase boundary. The constituents of the reacting mixture must reach a surface, whether or not they cross it. Within a single phase this transport is effected by *molecular diffusion*, *eddy diffusion*, or both. This chapter will be devoted to molecular diffusion; eddy diffusion is discussed in Chap. 4.

In most situations of technical importance, the molecular diffusion is of the type known as *ordinary diffusion*, which results from the thermal motion of the molecules. Molecules move at high speeds but travel extremely short distances before colliding with other molecules and being deflected in random directions. The migration of individual molecules, therefore, is slow except at quite low molecular densities. This is in contrast with the flow due to the rapid pressure equalization in a fluid, which is the result of momentum transport due to molecular collisions. However, if there is a region $H$ in which the concentration of species $A$ is high, surrounded by a region $L$ having a low concentration of the same species, then the molecular flux out of $H$ will carry more molecules of $A$ than the flux into it from $L$. The flux of total molecules in and out of the two regions may be the same, but there is a net outward flux of $A$ due simply to the fact that the concentrations are different in the two regions. If $Y_H$ and $Y_L$ are the mole fractions of $A$, and if $n$ total molecules of mixture are interchanged between the two regions, then $Y_H n$ molecules of $A$ will leave $H$ and $Y_L n$ will leave $L$. The net flux of $A$ from $H$ to $L$ is proportional to the difference $(Y_H - Y_L)$ and to the rate of exchange of total molecules between the two regions. In this way, concentration gradients tend to be eliminated by diffusion.

Molecular diffusion may occur as a result of concentration, temperature, or pressure gradients, or because a directed external electrical or other potential is applied to a mixture. A completely quiescent gas or liquid mixture will develop a concentration gradient in the direction of an imposed temperature gradient (the "Soret effect"). For example, two connected bulbs containing a mixture of 35.6 percent hydrogen and 64.4 percent neon will develop a concentration difference of 6.9 percent if one bulb is held at 290.4°K and the other at 90.2°K. Conversely, a temperature gradient can lead to diffusion of a species in a mixture even in the absence of a concentration gradient. This process, known as *thermal diffusion*, has been employed for the separation of uranium and other isotopes. It is reported [96] that the gradient of viscous shear stresses in a fluid mixture adjacent to a rotating cylinder establishes a steady-state concentration gradient, a phenomenon somewhat analogous to the Soret effect.

The theory describing thermal diffusion is summarized by Hirschfelder, Curtiss, and Bird [47] and by Bird [11]. The latter reference includes a simplified theory of the Clusius and Dickel column, which employs staged thermal diffusion for the separation of gaseous isotopes. Benedict [9] describes the processes developed during World War II for the separation of uranium isotopes by thermal diffusion. Powers and Wilke [79] describe an experimental study of continuous thermal diffusion in liquids and provide design calculations relative to the possibility of commercial hydrocarbon separations.

Molecular transport by thermal, pressure, or forced diffusion have found limited technical application, however, and the discussion which follows in this chapter and in Chaps. 3 and 4 will deal only with "ordinary" molecular diffusion and with eddy diffusion.

## 2.3  SIMPLE MOLECULAR MODEL OF DIFFUSION IN A GAS MIXTURE

The elementary concepts of the classical kinetic theory provide a basis for the development of an approximate theory of diffusion in ideal-gas mixtures. Figure 2.1 represents an element of volume of such a mixture in which a concentration gradient exists in the $y$ direction. This gradient is assumed not to vary with time. Diffusion occurs normal to the plane $RR'$, where the cross section is $S$ square centimeters. At this plane the molecular number density of species $A$ is $n_A$ molecules per cubic centimeters and the concentration gradient is $dn_A/dy$. The total pressure is constant, so the total number density is constant.

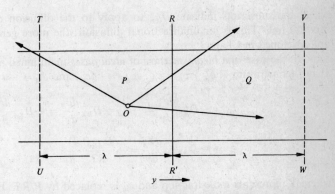

FIGURE 2.1
Diagram illustrating interdiffusion of gases.

The mean free path of $A$ in the mixture is $\lambda$ centimeters, which is the mean distance a molecule of $A$ travels before colliding with another molecule in the mixture. This distance is covered in time $\lambda/\overline{U}_A$, where $\overline{U}_A$ is the mean molecular speed. Planes $TU$ and $VW$ are at distances $\lambda$ removed from $RR'$. The gases in $P$ and $Q$ have mean concentrations

$$n_A - \frac{(\lambda/2)\, dn_A}{dy} \quad \text{and} \quad n_A + \frac{(\lambda/2)\, dn_A}{dy}$$

respectively, and the volume of each is $\lambda S$. Molecules at $O$ move the distance $\lambda$ in time $\lambda/\overline{U}_A$ with equal probability of going in any direction. Of all the molecules in $P$, a fraction cross $RR'$ from $P$ to $Q$, and the same fraction of the molecules in $Q$ cross $RR'$ to $P$. It can be shown by multiple integration that this fraction is $\frac{1}{3}$ [52]. The net transport of molecules of species $A$ from $P$ to $Q$ in time $\lambda/\overline{U}_A$ is then given by

$$J_A N_{\text{Avo}} S \frac{\lambda}{\overline{U}_A} = \frac{1}{3}(S\lambda)\left(n_A - \frac{\lambda}{2}\frac{dn_A}{dy}\right) - \frac{1}{3}(S\lambda)\left(n_A + \frac{\lambda}{2}\frac{dn_A}{dy}\right) \tag{2.1}$$

or

$$J_A = -\frac{\lambda \overline{U}_A}{3 N_{\text{Avo}}}\frac{dn_A}{dy} = -\frac{\lambda \overline{U}_A}{3}\frac{dc_A}{dy} \tag{2.2}$$

where the flux $J_A$ is expressed as gram moles per second per square centimeter and the concentration $c_A$ as gram moles per cubic centimeter; $N_{\text{Avo}}$ is Avogadro's number.

If the *diffusion coefficient* (often called "diffusivity") is defined as the ratio of the flux density to the negative of the concentration gradient in the direction of diffusion, then

$$J_A = -\frac{\lambda \overline{U}_A}{3}\frac{dc_A}{dy} = -D_{Am}\frac{dc_A}{dy} \tag{2.3}$$

where the subscripts indicate $D_{Am}$ to apply to the diffusion of species $A$ in the mixture $(m)$. This is for unidirectional diffusion; the more general rate equations are developed in Chap. 3.

In the case of a *binary mixture* of *ideal gases* maintained at constant pressure $P$ and temperature $T$, $c_A + c_B = c_m$, $dc_A = -dc_B$, and $J_A = -J_B$, whence

$$J_A = -D_{AB}\frac{dc_A}{dy} = -J_B = D_{BA}\frac{dc_B}{dy} = -D_{BA}\frac{dc_A}{dy}$$

$$= -D_{AB}c_m\frac{dY_A}{dy} = -\frac{D_{AB}P}{RT}\frac{dY_A}{dy} \tag{2.4}$$

where $Y$ represents mole fraction and $c_m$ is replaced by $P/RT$. From this it follows that $D_{AB} = D_{BA}$, that is, that the *diffusion coefficient is independent of concentration* and a *property* of the binary ideal-gas pair.

Visualizing the random motion of the molecules, it is clear that the mean free path $\lambda$ will vary inversely as both the mean molecular cross-sectional area $s$ and the number density $n_m$ of all molecules present. The latter is inversely proportional to the total volume of the mixture, which in turn is proportional to $T/P$. Simple kinetic theory shows the mean speed $\overline{U}_A$ to be proportional to $(T/M_A)^{1/2}$. Thus if both species have the same molecular weights, velocity, and cross section $s$,

$$D \propto \overline{U}\lambda \propto \frac{1}{n_m s}\sqrt{\frac{T}{M}} \propto \frac{T^{3/2}\sqrt{1/M}}{Ps} \tag{2.5}$$

This derivation is clearly overly simplified: the mean free path in a mixture is not the same for all molecules and the appropriate molecular speed is neither $\overline{U}_A$ nor $\overline{U}_B$ but the relative velocity of the diffusing clouds of the two species. Thus if the molecular weights are $M_A$ and $M_B$, kinetic theory leads to the relation

$$D_{AB} = \frac{(\text{constant})T^{3/2}[(1/M_A) + (1/M_B)]^{1/2}}{Ps_{av}} \tag{2.6}$$

where $s_{av}$ is an average of the two cross sections. A numerical value of the constant is obtained from classical kinetic theory where the molecular cross section is that of a sphere. As will be shown in Sec. 2.5, modern kinetic theory leads to a more sophisticated form of Eq. (2.6), which is useful in calculating values of diffusion coefficients.

To complete the definition of the diffusion coefficient, it is necessary to choose a plane of reference, i.e., a position, fixed or moving, in relation to which the flux is defined. Though various authors select different reference planes, it has become common in chemical engineering and in much of physics to choose the plane of *no net volume flux*, for which $\sum_i J_i \overline{V}_i = 0$, where $\overline{V}_i$ is the partial molal volume of constituent $i$. If the molecular weights of the diffusing constituents are not equal, there will be a net transport of *mass* across such a plane. In a binary at fixed temperature and pressure, for example, $J_A \overline{V}_A = -J_B \overline{V}_B$, but $J_A M_A \neq -J_B M_B$. If the entire system is moving with respect to the earth, the planes of no net volume transport, no net *mass* transport, and the fixed apparatus present three choices as possible reference planes. The planes of no net volume transport and that of the apparatus are identical choices in the common case of an ideal mixture in a fixed container. Furthermore, in an ideal-gas mixture, the plane of no net volume transport is the same as the plane of no net molal transport.

The treatment of diffusion and mass transfer which follows is based on the choice of the plane of no net volume transport as the reference plane required to specify the flux and to define $D_{Am}$. Equations for diffusion with reference to the other reference planes are given by Hartley [43] and Bird [11, 12].

The literature pertaining to molecular diffusion employs diffusion coefficients defined in several ways. The coefficients $D_{AB}$ and $D_{Am}$ employed in the foregoing are the common *mutual diffusion* coefficients, applying to the diffusion of one constituent of a binary or of a mixture. *Self-diffusion* refers to the diffusion of a

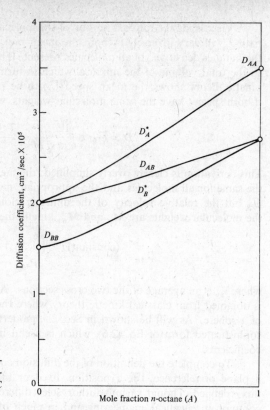

FIGURE 2.2
Mutual, self-, and tracer diffusion coefficients in binary liquid mixtures of *n*-octane (*A*) and *n*-dodecane (*B*) at 60°C [98].

constituent through itself; values for the self-diffusion coefficient $D_{AA}$ cannot be easily measured, since analysis to determine diffusion rates is impractical if all the molecules are alike. Approximate values of $D_{AA}$ may be obtained by measuring the diffusion of a small amount of an isotope of the substance through which diffusion occurs. Though these isotopes are used as "tracers," the process is not to be confused with *tracer diffusion*, which is the diffusion of a small amount of tracer in a *mixture* of two or more substances; the rates and the tracer-diffusion coefficient $D_A^*$ then depend on the composition of the mixture.

Figure 2.2 illustrates the relation [98] between $D_{AB}$, $D_{AA}$, and $D_A^*$ for the diffusion of *n*-octane in liquid *n*-dodecane at 60°C. The mutual diffusion coefficient $D_{AB}$ increases somewhat with increase in *n*-octane concentration. $D_A^*$ is the coefficient of diffusion of an *n*-octane isotope in a mixture of ordinary *n*-dodecane and *n*-octane. In pure *n*-octane, $D_A^*$ becomes equal to the self-diffusion coefficient $D_{AA}$. Similarly, the coefficient $D_B^*$, which relates to the diffusion of an isotope of *n*-dodecane in a mixture of the ordinary substances, becomes equal to the self-diffusion coefficient $D_{BB}$ in pure *n*-dodecane.

The coefficient of greatest practical utility is the mutual-diffusion coefficient $D_{AB}$.

Most of the literature having to do with diffusion employs Fick's law, as applied to unidirectional binary diffusion, in the form

$$J_A = -D_{AB} \frac{dc_A}{dy} \tag{2.7}$$

A basically more fundamental expression for the same case, however, is

$$J_A = -c_m D_{AB} \frac{dY_A}{dy} \tag{2.8}$$

The second is used by Bird, Stewart, and Lightfoot [12] and is not uncommon in the literature of chemical engineering. The forms are evidently identical if the molal density $c_m$ is independent of $y$, as in Eq. (2.4) for a binary mixture of ideal gases at constant temperature and pressure.

The first of the two forms is clearly incorrect in certain applications. For example, a concentration gradient of oxygen exists in air at 1 atm in the immediate vicinity of a hot radiator, since the air density varies with the temperature. Yet there is no gradient of mole fraction, and no diffusion except a trivial flux due to thermal diffusion. Equation (2.7) indicates a finite diffusion flux, but Eq. (2.8) does not.

Diffusion does not occur in a system at equilibrium, and diffusion fluxes increase as the departure from equilibrium is increased, i.e., as regions of different chemical potentials develop. Near equilibrium in an isothermal system, the small diffusion flux might be expected to be proportional to the gradient of the chemical potential. In its most general form, the diffusion coefficient might then be regarded as the ratio of the diffusion flux to the negative of the gradient of chemical potential.

Equation (2.8) can be derived from the exact kinetic theory of ideal-gas mixtures in which temperature and concentration gradients but no thermal diffusion exist. It also follows from the thermodynamics of irreversible processes and the Onsager principle. The derivations, which are complex, are given in Hirschfelder, Curtiss, and Bird [47]. See also Ref. 26.

From a practical point of view it is desirable to employ the form defining a $D_{AB}$ which does not vary with concentration. Neither form does this. However, if the gradient of chemical potential is used, the theory suggests an activity correction term which partially describes the variation of $D_{AB}$ with concentration (see Secs. 2.4 and 2.6). This is an important correction in liquid systems, but not in gases.

There have been only a few published attempts to test the relative merits of concentration and activity as the potential by comparison with experimental data. In one such, Fallat [32] sublimed naphthalene into air and helium at 1.0 and 38 atm in a flow system. In this case a much better correlation of the data with flow parameters was obtained by the use of concentration than was the situation when activity was employed. See also Ref. 46. In high-pressure gas systems, the wide variations of $D_{AB}$ with pressure are not explained by the activity correction term, and the activity-corrected $D_{AB}$ varies widely with concentration in most liquid systems.

Almost all the published values of $D_{AB}$ are based on the use of Eq. (2.7), or an equivalent form, to interpret the experimental data. For engineering calculations it is nearly always acceptable to employ the first form [Eq. (2.7)] together with the activity correction term to make partial allowance for the variation of $D_{AB}$ with concentration. This procedure will be followed in subsequent chapters.

## 2.4 PHENOMENOLOGICAL THEORY OF MOLECULAR DIFFUSION

The cloud of molecules diffusing in a mixture may be said to have a statistical velocity (relative to the velocity $U_m$ of the plane of no net volume flux) given by $\overline{U}_A - U_m = J_A/c_A$. Intuitively, one might expect the flux of species $A$ to be resisted by a frictional force resulting from molecular collisions and proportional both to the number density of the interfering species and to the speed of the cloud of species $A$ relative to the other molecules present. The work energy required to overcome this friction may be assumed to be provided by decreases in the chemical potential of the diffusing species. These assumptions provide the basis for the derivation of an interesting theory relating diffusion coefficients to composition.

The derivation which follows is made simpler by restricting it to the case of a binary mixture of $A$ and $B$ diffusing only in the $y$ direction. The mixture is not necessarily ideal, and may be a gas, liquid, or even a solid solution.

Since $J$ is defined in relation to a plane of no net volume flux,

$$J_A \overline{V}_A = -J_B \overline{V}_B = c_A(\overline{U}_A - U_m)\overline{V}_A = -c_B(\overline{U}_B - U_m)\overline{V}_B \tag{2.9}$$

where $\overline{V}_A$ and $\overline{V}_B$ are the partial molal volumes, and $U_m$ is the velocity (relative to the earth) of the plane of no net volume transport.

The assumption that the chemical potential $\mu_A$ provides the work to overcome friction leads to the relation

$$\frac{d\mu_A}{dy} = -bc_B(\overline{U}_A - \overline{U}_B) \tag{2.10}$$

where $b$ is a constant of proportionality. The concentration $c_B$ is included, since the resistance to diffusion is clearly proportional to the number density of molecules of species $B$ impeding the flux $J_A$. Chemical potential is expressed as a function of temperature, pressure, and composition:

$$\mu_A = \mu_A{}^0(T, P) + RT \ln a_A \tag{2.11}$$

with the activity given by $a_A = \gamma_A c_A$.

From Eq. (2.9),

$$(\overline{U}_A - U_m)c_A \overline{V}_A + (\overline{U}_B - U_m)c_B \overline{V}_B = 0 \tag{2.12}$$

and from the definition of partial molal volume,

$$c_A \overline{V}_A + c_B \overline{V}_B = 1 \tag{2.13}$$

Combination of Eqs. (2.9), (2.10), (2.11), (2.12), and (2.13) gives

$$J_A = -\frac{c_A \overline{V}_B RT}{b} \frac{\partial \ln a_A}{\partial y} = -\frac{RT\overline{V}_B}{b} \frac{\partial \ln a_A}{\partial \ln c_A} \frac{dc_A}{dy}$$

$$= -\frac{RT\overline{V}_B}{b}\left(1 + \frac{\partial \ln \gamma_A}{\partial \ln c_A}\right)\frac{dc_A}{dy} = -D_{AB}\frac{dc_A}{dy} \tag{2.14}$$

A similar equation is obtained for $J_B$ in terms of $b$, $\overline{V}_A$, $a_B$, $\gamma_B$, $c_B$, and $D_{BA}$.

The Gibbs-Duhem equation of thermodynamics may be written for a binary mixture at constant temperature and pressure as

$$c_A d(\ln a_A) + c_B d(\ln a_B) = 0 \tag{2.15}$$

whence

$$\frac{\partial \ln a_A}{\partial \ln c_A} dc_A + \frac{\partial \ln a_B}{\partial \ln c_B} dc_B = 0$$

and

$$J_B = \frac{RT\overline{V}_A}{b} \frac{\partial \ln a_A}{\partial \ln c_A} \frac{dc_A}{dy} = -D_{BA} \frac{dc_B}{dy} \tag{2.16}$$

Division of Eq. (2.14) by Eq. (2.16) leads to

$$\frac{D_{AB}}{D_{BA}} = -\frac{\overline{V}_B}{\overline{V}_A} \frac{dc_B}{dc_A} \tag{2.17}$$

If the system is one for which there is *no volume change on mixing*, then $\overline{V}_A$ and $\overline{V}_B$ are constants, and $\overline{V}_A dc_A = -\overline{V}_B dc_B$. In this case $D_{AB} = D_{BA}$. Furthermore, if $V_A = V_B$,

$$J_A = -D_{AB} \frac{dc_A}{dy} = D_{BA} \frac{dc_B}{dy} = -J_B \tag{2.18}$$

This is perhaps obvious for the case of an ideal-gas mixture at constant temperature and pressure.

Equation (2.14) relates $D_{AB}$ to mixture composition in a binary:

$$D_{AB} = \frac{RT\overline{V}_B}{b}\left(\frac{\partial \ln a_A}{\partial \ln c_A}\right) = \frac{RT\overline{V}_B}{b}\left(1 + \frac{\partial \ln \gamma_A}{\partial \ln c_A}\right) \tag{2.19}$$

This will be used in Sec. 2.6 dealing with the variation of the diffusion coefficient with concentration in a binary liquid system.

## 2.5 ESTIMATION OF DIFFUSION COEFFICIENTS IN BINARY GAS MIXTURES

Molecular diffusion in binary gas systems has been the subject of theoretical and experimental investigations for nearly a century. The definitive review of the modern theory, measurement techniques, and published experimental data by Marrero and Mason [63] is strongly recommended—it is a major contribution to the current literature dealing with the subject. The present section outlines the theory briefly, provides a limited tabulation of experimental values, and describes several semi-empirical correlations useful for engineering estimates of $D_{AB}$ in low-pressure binaries.

Experimental values are available for many gas pairs (see Refs. 108, 35, 63, and 84) and should be employed if available. Quite reliable values may be estimated, however, by the use of any one of a number of semiempirical correlations of experimental data. These include equations proposed by Arnold [4], Gilliland [38], Andrussow [3], Wilke and Lee [108], Slattery and Bird [93], Chen and Othmer [21], Othmer and Chen [74], and Fuller, Schettler, and Giddings [35]. Most of these are empirical modifications of Eq. (2.6). Several of them are reviewed and compared with experimental data in Ref. 84.[1]

Perhaps the best of these empirical correlations is that due to Fuller, Schettler, and Giddings [35], which was developed by machine computation to give the best fit of the form of Eq. (2.6) with some 340 experimental data points. The equation is

$$D_{AB} = \frac{0.00100 T^{1.75}(1/M_A + 1/M_B)^{1/2}}{P[(\Sigma v)_A^{1/3} + (\Sigma v)_B^{1/3}]^2} \qquad (2.20)$$

The quantities $(\Sigma v)_A$ and $(\Sigma v)_B$ are obtained by summing atomic-diffusion volumes for each constituent of the binary. Values of $v$ (and of $\Sigma v$ for simple molecules) are listed in Table 2.1. The average discrepancy between calculated and experimental values for binary gas systems at low pressures is 4 to 7 percent.

The correlation represented by Eq. (2.20) is based very largely on data obtained near room temperature, and the exponent 1.75 on temperature appears to be too large. Perhaps the best estimate of $D_{AB}$ at elevated temperatures is obtained by calculating its value at, say, 300°K, by the use of Eq. (2.20), and then correcting to the higher temperature $T$ by ratio, using the temperature function $T^{3/2}/\Omega_D$ appearing in Eq. (2.22) below.

**Table 2.1   ATOMIC AND MOLECULAR DIFFUSION VOLUMES FOR THE ESTIMATION OF $D_{AB}$ BY THE METHOD OF FULLER, SCHETTLER, AND GIDDINGS [35]***

A   Atomic and structural diffusion volume increments, $v$

| C | 16.5 | (Cl) | 19.5 |
|---|---|---|---|
| H | 1.98 | (S) | 17.0 |
| O | 5.48 | Aromatic ring | −20.2 |
| (N) | 5.69 | Heterocyclic ring | −20.2 |

B   Diffusion volumes for simple molecules, $\Sigma v$

| $H_2$ | 7.07 | CO | 18.9 |
|---|---|---|---|
| $D_2$ | 6.70 | $CO_2$ | 26.9 |
| He | 2.88 | $N_2O$ | 35.9 |
| $N_2$ | 17.9 | $NH_3$ | 14.9 |
| $O_2$ | 16.6 | $H_2O$ | 12.7 |
| Air | 20.1 | $(CCl_2F_2)$ | 114.8 |
| A | 16.1 | $(SF_6)$ | 69.7 |
| Kr | 22.8 | $(Cl_2)$ | 37.7 |
| (Xe) | 37.9 | $(Br_2)$ | 67.2 |
|  |  | $(SO_2)$ | 41.1 |

* Parentheses indicate that the value listed is based on only a few data points.

[1] Scott [92] has called attention to the fact that many of the data points used in developing these correlations are actually numbers calculated in various ways by earlier writers, and not experimental values. Marrero and Mason [63] have recently published an extensive survey of the literature, which provides a basis for the collection of hundreds of experimentally measured values of diffusion coefficients in binary gas pairs.

Nain and Ferron [67] find that values of $D_{AB}$ calculated by Eq. (2.20) are 20 to 35 percent high in the case of three binary systems of quite polar molecules.

Variation of $D_{AB}$ with mixture composition in dilute (low-pressure) binary gas systems is small and is usually ignored. The tracer diffusion coefficient $D_A^*$, however, may vary severalfold with composition in such systems. It has been shown [5, 65] that $1/D^*$ varies linearly with $Y_A$ ($D_A^* = D_{AB}$ at $Y_A = 0$; $D_A^* = D_{AA}$ at $Y_A = 1$).

The modern mathematical theory of nonuniform gases [47, 63] extends the classical kinetic theory by introducing allowances for the mutual interaction of the molecules of a gas. Molecules repel each other when close together and attract when separated. The intermolecular potentials are expressed empirically by any of a number of "potential functions," of which the one most widely used is the Lennard-Jones 6-12 potential:

$$\phi(r) = 4\varepsilon \left[ \left( \frac{\sigma}{r} \right)^{12} - \left( \frac{\sigma}{r} \right)^{6} \right] \tag{2.21}$$

where $\phi(r)$ = potential energy
$r$ = distance between centers
$\varepsilon, \sigma$ = Lennard-Jones potential parameters

Starting with this relation, a theory of diffusion is developed, leading to expressions for $D_{AB}$ and $D_{Am}$ in binary and multicomponent dilute (low-pressure) gas mixtures. Several assumptions are employed: (1) only binary collisions occur, (2) the motions of the colliding molecules can be described by classical mechanics, (3) only elastic collisions occur, (4) there are no quantum effects, and (5) molecular forces operate only through fixed centers of molecules. Furthermore, semiempirical combining rules are adopted for the estimation of $\sigma_{AB}$ and $\varepsilon_{AB}$ from values for pure components to permit the extension of the resulting equations for self-diffusion to systems involving mixtures. The same theoretical development yields equations for the viscosity and other properties of gases, and it is by comparing the equations for viscosity with experimental data on the variation of viscosity with temperature of pure dilute gases that values of $\varepsilon$ and $\sigma$ are normally obtained. Conversely, the viscosity, needed in evaluating the Schmidt number, may be calculated from known or estimated values of the potential parameters, as explained by Bromley and Wilke [15]. In spite of the assumptions and empiricism involved, the theory provides an excellent basis for the estimation of diffusion coefficients in dilute gases.

The theoretical equation for the mutual diffusion coefficient in a low-pressure binary gas mixture is [47]:

$$D_{AB} = \frac{0.001858 T^{3/2} (1/M_A + 1/M_B)^{1/2}}{P \sigma_{AB}^2 \Omega_D} \quad \text{cm}^2/\text{sec} \tag{2.22}$$

where $T$ = temperature, °K
$M_A, M_B$ = molecular weights of constituents $A$ and $B$
$P$ = absolute pressure, atm
$\Omega_D$ = collision integral, $f(kT/\varepsilon_{AB})$; see Table 2.2
$\varepsilon_{AB}, \sigma_{AB}$ = Lennard-Jones force constants for the binary
$k$ = Boltzmann's constant

The constants $\varepsilon_{AB}$ and $\sigma_{AB}$ are obtained from the corresponding values for the pure substances by use of the combining rules

$$\frac{\varepsilon_{AB}}{k} = \left(\frac{\varepsilon_A}{k}\frac{\varepsilon_B}{k}\right)^{1/2} \tag{2.23}$$

and

$$\sigma_{AB} = \tfrac{1}{2}(\sigma_A + \sigma_B) \tag{2.24}$$

Values of $\varepsilon/k$ and $\sigma$ for a limited number of pure substances are given in Refs. 47 and 95; values for common substances are given in Table 2.3. Where tabulated values of $\varepsilon/k$ and $\sigma$ are used they should be employed together; different sources often give quite different values of each, but differences in $\varepsilon/k$ tend to be offset by differences in $\sigma$, and the calculated diffusion coefficients may be nearly the same.

**Table 2.2** VALUES OF THE COLLISION INTEGRAL $\Omega_D$ BASED ON THE LENNARD-JONES POTENTIAL†

| $kT/\varepsilon$‡ | $\Omega_D$‡ | $kT/\varepsilon$ | $\Omega_D$ | $kT/\varepsilon$ | $\Omega_D$ |
|---|---|---|---|---|---|
| 0.30 | 2.662 | 1.65 | 1.153 | 4.0 | 0.8836 |
| 0.35 | 2.476 | 1.70 | 1.140 | 4.1 | 0.8788 |
| 0.40 | 2.318 | 1.75 | 1.128 | 4.2 | 0.8740 |
| 0.45 | 2.184 | 1.80 | 1.116 | 4.3 | 0.8694 |
| 0.50 | 2.066 | 1.85 | 1.105 | 4.4 | 0.8652 |
| 0.55 | 1.966 | 1.90 | 1.094 | 4.5 | 0.8610 |
| 0.60 | 1.877 | 1.95 | 1.084 | 4.6 | 0.8568 |
| 0.65 | 1.798 | 2.00 | 1.075 | 4.7 | 0.8530 |
| 0.70 | 1.729 | 2.1 | 1.057 | 4.8 | 0.8492 |
| 0.75 | 1.667 | 2.2 | 1.041 | 4.9 | 0.8456 |
| 0.80 | 1.612 | 2.3 | 1.026 | 5.0 | 0.8422 |
| 0.85 | 1.562 | 2.4 | 1.012 | 6 | 0.8124 |
| 0.90 | 1.517 | 2.5 | 0.9996 | 7 | 0.7896 |
| 0.95 | 1.476 | 2.6 | 0.9878 | 8 | 0.7712 |
| 1.00 | 1.439 | 2.7 | 0.9770 | 9 | 0.7556 |
| 1.05 | 1.406 | 2.8 | 0.9672 | 10 | 0.7424 |
| 1.10 | 1.375 | 2.9 | 0.9576 | 20 | 0.6640 |
| 1.15 | 1.346 | 3.0 | 0.9490 | 30 | 0.6232 |
| 1.20 | 1.320 | 3.1 | 0.9406 | 40 | 0.5960 |
| 1.25 | 1.296 | 3.2 | 0.9328 | 50 | 0.5756 |
| 1.30 | 1.273 | 3.3 | 0.9256 | 60 | 0.5596 |
| 1.35 | 1.253 | 3.4 | 0.9186 | 70 | 0.5464 |
| 1.40 | 1.233 | 3.5 | 0.9120 | 80 | 0.5352 |
| 1.45 | 1.215 | 3.6 | 0.9058 | 90 | 0.5256 |
| 1.50 | 1.198 | 3.7 | 0.8998 | 100 | 0.5130 |
| 1.55 | 1.182 | 3.8 | 0.8942 | 200 | 0.4644 |
| 1.60 | 1.167 | 3.9 | 0.8888 | 400 | 0.4170 |

† From J. O. Hirschfelder, C. F. Curtiss, and R. B. Bird, "Molecular Theory of Gases and Liquids," John Wiley & Sons, Inc., New York, 1954.

‡ Hirschfelder uses the symbols $T^*$ for $kT/\varepsilon$ and $\Omega^{(1.1)*}$ in place of $\Omega_D$.

If values of $\varepsilon/k$ and $\sigma$ are not available, they may be estimated by the following very approximate rules:

$$\frac{\varepsilon}{k} = 0.75 T_c$$

$$\sigma = \tfrac{5}{6} V_c^{1/3}$$

(2.25)

where $T_c$ = critical temperature, °K

$V_c$ = critical volume, cm³/g mole

$\sigma$ = Lennard-Jones potential parameter, Å

Reference 84 suggests ways of estimating $T_c$ and $V_c$ when these are not available.

Table 2.3  LENNARD-JONES POTENTIAL PARAMETERS [95]*

| Molecule | Compound | $\sigma$, Å | $\varepsilon/k$, °K |
|---|---|---|---|
| A | Argon | 3.542 | 93.3 |
| He | Helium† | 2.551 | 10.22 |
| Kr | Krypton | 3.655 | 178.9 |
| Ne | Neon | 2.820 | 32.8 |
| Xe | Xenon | 4.082 | 206.9 |
| Air | Air | 3.711 | 78.6 |
| $Br_2$ | Bromine | 4.296 | 507.9 |
| $CCl_4$ | Carbon tetrachloride | 5.947 | 322.7 |
| $CF_4$ | Carbon tetrafluoride | 4.662 | 134.0 |
| $CHCl_3$ | Chloroform | 5.389 | 340.2 |
| $CH_2Cl_2$ | Methylene chloride | 4.898 | 356.3 |
| $CH_3Br$ | Methyl bromide | 4.118 | 449.2 |
| $CH_3Cl$ | Methyl chloride | 4.182 | 350.0 |
| $CH_3OH$ | Methanol | 3.626 | 481.8 |
| $CH_4$ | Methane | 3.758 | 148.6 |
| CO | Carbon monoxide | 3.690 | 91.7 |
| COS | Carbonyl sulfide | 4.130 | 336.0 |
| $CO_2$ | Carbon dioxide | 3.941 | 195.2 |
| $CS_2$ | Carbon disulfide | 4.483 | 467.0 |
| $C_2H_2$ | Acetylene | 4.033 | 231.8 |
| $C_2H_4$ | Ethylene | 4.163 | 224.7 |
| $C_2H_6$ | Ethane | 4.443 | 215.7 |
| $C_2H_5Cl$ | Ethyl chloride | 4.898 | 300.0 |
| $C_2H_5OH$ | Ethanol | 4.530 | 362.6 |
| $C_2N_2$ | Cyanogen | 4.361 | 348.6 |
| $CH_3OCH_3$ | Methyl ether | 4.307 | 395.0 |
| $CH_2CHCH_3$ | Propylene | 4.678 | 298.9 |
| $CH_3CCH$ | Methylacetylene | 4.761 | 251.8 |
| $C_3H_6$ | Cyclopropane | 4.807 | 248.9 |
| $C_3H_8$ | Propane | 5.118 | 237.1 |
| $n\text{-}C_3H_7OH$ | n-Propyl alcohol | 4.549 | 576.7 |
| $CH_3COCH_3$ | Acetone | 4.600 | 560.2 |
| $CH_3COOCH_3$ | Methyl acetate | 4.936 | 469.8 |
| $n\text{-}C_4H_{10}$ | n-Butane | 4.687 | 531.4 |
| $iso\text{-}C_4H_{10}$ | Isobutane | 5.278 | 330.1 |
| $C_2H_5OC_2H_5$ | Ethyl ether | 5.678 | 313.8 |

*Continued*

Table 2.3—*Continued*

| Molecule | Compound | $\sigma$, Å | $\varepsilon/k$, K |
|---|---|---|---|
| $CH_3COOC_2H_5$ | Ethyl acetate | 5.205 | 521.3 |
| $n\text{-}C_5H_{12}$ | $n$-Pentane | 5.784 | 341.1 |
| $C(CH_3)_4$ | 2,2-Dimethylpropane | 6.464 | 193.4 |
| $C_6H_6$ | Benzene | 5.349 | 412.3 |
| $C_6H_{12}$ | Cyclohexane | 6.182 | 297.1 |
| $n\text{-}C_6H_{14}$ | $n$-Hexane | 5.949 | 399.3 |
| $Cl_2$ | Chlorine | 4.217 | 316.0 |
| $F_2$ | Fluorine | 3.357 | 112.6 |
| HBr | Hydrogen bromide | 3.353 | 449.0 |
| HCN | Hydrogen cyanide | 3.630 | 569.1 |
| HCl | Hydrogen chloride | 3.339 | 344.7 |
| HF | Hydrogen fluoride | 3.148 | 330.0 |
| HI | Hydrogen iodide | 4.211 | 288.7 |
| $H_2$ | Hydrogen | 2.827 | 59.7 |
| $H_2O$ | Water | 2.641 | 809.1 |
| $H_2O_2$ | Hydrogen peroxide | 4.196 | 289.3 |
| $H_2S$ | Hydrogen sulfide | 3.623 | 301.1 |
| Hg | Mercury | 2.969 | 750.0 |
| $I_2$ | Iodine | 5.160 | 474.2 |
| $NH_3$ | Ammonia | 2.900 | 558.3 |
| NO | Nitric oxide | 3.492 | 116.7 |
| NOCl | Nitrosyl chloride | 4.112 | 395.3 |
| $N_2$ | Nitrogen | 3.798 | 71.4 |
| $N_2O$ | Nitrous oxide | 3.828 | 232.4 |
| $O_2$ | Oxygen | 3.467 | 106.7 |
| $PH_3$ | Phosphine | 3.981 | 251.5 |
| $SF_6$ | Sulfur hexafluoride | 5.128 | 222.1 |
| $SO_2$ | Sulfur dioxide | 4.112 | 335.4 |
| $SnBr_4$ | Stannic bromide | 6.388 | 563.7 |
| $UF_6$ | Uranium hexafluoride | 5.967 | 236.8 |

* Values of $\sigma$ and $\varepsilon/k$ for additional substances are given in Ref. 95.
† Calculated from quantum-mechanical formulas.

The collision integral $\Omega_D$ decreases slowly with increase in temperature, as indicated by Table 2.2, so $D_{AB}$ varies with $T$ as a power slightly greater than 1.5. The integral $\Omega_D$ is not very sensitive to errors in $\varepsilon/k$, but $D_{AB}$ is obviously sensitive to errors in $\sigma$. The activation energy increases with temperature, but for a typical binary for which $\varepsilon_{AB}/k$ is 150, it is approximately 3.3 kcal/g mole over the temperature range 300 to 700°K.

Equation (2.22) suggests that $D_{AB}$ (or $D_{AA}$) should be inversely proportional to total pressure. This appears to be true at pressures from a few millimeters of mercury to 10 to 20 atm, but $DP$ may either increase or decrease as the pressure is increased further. There appears to be neither an adequate theory nor a useful empirical correlation for the prediction of gas diffusion coefficients at high pressures [11, 47, 93].

The theory leading to Eq. (2.22) is being extended to provide a reliable basis for the treatment of diffusion in gases at high pressures, for polar molecules, and for diffusion in gases at high temperatures. The possible advantages of replacing the Lennard-Jones potential function with others, such as that proposed by Kihara, is being investigated [71]. For the present, however, Eq. (2.22) can be used with confidence for technical calculations to obtain $D_{AB}$ over a wide range of pressures and temperatures, even for mixtures containing a polar constituent.

The following example illustrates the use of Eqs. (2.20) and (2.22) for the prediction of diffusion coefficients in binary gas mixtures. Table 2.4 lists a limited number of experimental values of $D_{AB}$ for low-pressure gas binaries.

**EXAMPLE 2.1**  Estimate $D_{AB}$ for the diffusion of methane in hydrogen at 15°C and 1.0 atm.

SOLUTION  $T = 288°K$; $M_A = 16$; $M_B = 2$. From Table 2.3, $\varepsilon_A/k = 148.6$, $\sigma_A = 3.758$, $\varepsilon_B/k = 59.7$; $\sigma_B = 2.827$. From Eqs. (2.23) and (2.24), $\varepsilon_{AB/k} = 94$, $\sigma_{AB} = 3.292$; and from Table 2.2, $\Omega_D = 0.944$. Then from Eq. (2.22),

$$D_{AB} = \frac{0.001858 \times 288^{3/2} \left(\frac{18}{32}\right)^{1/2}}{1 \times 3.292^2 \times 0.944} = 0.66 \text{ cm}^2/\text{s}$$

**Table 2.4  EXPERIMENTAL VALUES OF MUTUAL DIFFUSION COEFFICIENTS FOR COMMON BINARY GAS PAIRS AT LOW PRESSURES $D_{AB}P$, cm²/(s)(atm)***

| System | $T$, °K | $D_{AB}P$ | System | $T$, °K | $D_{AB}P$ |
|---|---|---|---|---|---|
| Air–carbon dioxide | 317.2 | 0.177 | Helium–oxygen | 298 | 0.729 |
| Air–ethanol | 313 | 0.145 | Helium–i-propanol | 423 | 0.677 |
| Air–helium | 317.2 | 0.765 | Helium–water | 307.1 | 0.902 |
| Air–n-hexane | 328 | 0.093 | | | |
| Air–n-pentane | 294 | 0.071 | Hydrogen–acetone | 296 | 0.424 |
| Air–water | 313 | 0.288 | Hydrogen–ammonia | 298 | 0.783 |
| | | | | 358 | 1.093 |
| Argon–ammonia | 333 | 0.253 | | 473 | 1.86 |
| Argon–carbon dioxide | 276.2 | 0.133 | | 533 | 2.149 |
| Argon–helium | 298 | 0.729 | Hydrogen–benzene | 311.3 | 0.404 |
| Argon–hydrogen | 242.2 | 0.562 | Hydrogen–cyclohexane | 288.6 | 0.319 |
| | 448 | 1.76 | Hydrogen–methane | 288 | 0.694 |
| | 806 | 4.86 | Hydrogen–nitrogen | 298 | 0.784 |
| | 1069 | 8.10 | | 573 | 2.147 |
| Argon–methane | 298 | 0.202 | Hydrogen–sulfur dioxide | 473 | 1.23 |
| Argon–sulfur dioxide | 263 | 0.077 | Hydrogen–thiophene | 302 | 0.400 |
| | | | Hydrogen–water | 328.5 | 1.121 |
| Carbon dioxide–helium | 298 | 0.612 | | | |
| Carbon dioxide–nitrogen | 298 | 0.167 | Methane–water | 352.3 | 0.356 |
| Carbon dioxide–nitrous oxide | 312.8 | 0.128 | | | |
| Carbon dioxide–oxygen | 293.2 | 0.153 | Nitrogen–ammonia | 298 | 0.230 |
| Carbon dioxide–sulfur dioxide | 263 | 0.064 | | 358 | 0.328 |
| Carbon dioxide–water | 307.2 | 0.198 | Nitrogen–benzene | 311.3 | 0.102 |
| | 352.3 | 0.245 | Nitrogen–cyclohexane | 288.6 | 0.0731 |
| | | | Nitrogen–sulfur dioxide | 263 | 0.104 |
| Carbon monoxide–nitrogen | 373 | 0.318 | Nitrogen–water | 307.5 | 0.256 |
| | | | | 352.1 | 0.256 |
| Helium–benzene | 423 | 0.610 | | | |
| Helium–ethanol | 423 | 0.821 | Oxygen–benzene | 311.3 | 0.101 |
| Helium–methane | 298 | 0.675 | Oxygen–carbon tetrachloride | 296 | 0.0749 |
| Helium–methanol | 423 | 1.032 | Oxygen–cyclohexane | 288.6 | 0.0746 |
| Helium–nitrogen | 298 | 0.687 | Oxygen–water | 352.3 | 0.352 |

* Marrero and Mason [63] present a thorough review of the literature, with tables of values for gas pairs not included here and references to reported experimental values of $D_{AB}$ for several hundred systems.

Applying the method of Fuller, Schettler, and Giddings [35], $\Sigma v_A$ and $\Sigma v_B$ are found from Table 2.1 to be $16.5 + 4 \times 1.98 = 24.4$ and 7.07, respectively. Substituting in Eq. (2.20),

$$D_{AB} = \frac{0.00100 \times 288^{1.75}\left(\frac{18}{32}\right)^{1/2}}{1 \times (24.4^{1/3} + 7.07^{1/3})^2} = 0.65 \text{ cm}^2/\text{s}$$

An experimental value of 0.694 cm²/s is reported in the literature, as shown in Table 2.4.  ////

The dimensionless Schmidt number $\mu/\rho D_{AB}$ is widely employed in the correlation and representation of mass-transfer data. Its value for a gas system at low pressure (ideal-gas mixture) may be estimated by the use of Eq. (2.22) together with the corresponding equation for gas viscosity [47]. For diffusion of $A$ at low concentration in $B$,

$$\text{Sc}_{AB} = \frac{\mu_B}{\rho_B D_{AB}} = 1.18 \frac{\Omega_D}{\Omega_v} \left(\frac{M_A}{M_A + M_B}\right)^{1/2} \left(\frac{\sigma_{AB}}{\sigma_B}\right)^2 \tag{2.26}$$

$\Omega_v$ is the collision integral for viscosity; the other symbols are defined in the text, following Eq. (2.22). Values of the ratio $\Omega_D/\Omega_v$ needed to calculate Sc are given in Table 2.5. This ratio should properly involve $\Omega_D$ based on $\varepsilon_{AB}$ and $\Omega_v$ based on $\varepsilon_B$, but it varies little with $kT/\varepsilon$, so it is sufficient for most purposes to employ $kT/\varepsilon_B$ for use with Table 2.5.

This method of estimating Sc gives only approximate values in many cases; it is preferable to use experimental values of $\mu$ and $D_{AB}$.

**Table 2.5  VALUES OF $\Omega_D/\Omega_v$ FOR USE IN ESTIMATING THE SCHMIDT NUMBER (Sc) IN GAS BINARIES AT LOW PRESSURE**

| $kT/\varepsilon$ | $\Omega_D/\Omega_v$ |
|---|---|
| 0.50 | 0.9154 |
| 0.75 | 0.9055 |
| 1.00 | 0.9067 |
| 1.50 | 0.9117 |
| 2.00 | 0.9149 |
| 2.50 | 0.9145 |
| 3.0 | 0.9134 |
| 3.5 | 0.9121 |
| 4.0 | 0.9109 |
| 5 | 0.9086 |
| 7 | 0.9048 |
| 10 | 0.9007 |
| 40 | 0.8872 |
| 100 | 0.8722 |

## 2.6  DIFFUSION IN LIQUIDS

The diffusion coefficient $D_{AB}$ for binary liquid systems is very much smaller than in gases at atmospheric pressure, usually falling in the range 0.5 to $2 \times 10^{-5}$ cm²/s in nonviscous liquids at 25°C (compared with typical values of 0.1 to 1.0 cm²/s for common gas pairs at atmospheric pressure). This does not necessarily mean that diffusion is slower in liquids, since the molar densities and concentration gradients are usually much greater. In contrast with the situation in gases, $D_{AB}$ often shows a substantial variation with concentration.

Several thousand published experimental values of $D_{AB}$ in liquid binaries have been collected and tabulated by Bailey, Parlin, and Beckman of the Department of Chemical Engineering at the University of Maryland. The arduous work of assembling this mass of data was supported by a NASA grant. Data on the variation of $D_{AB}$ with liquid composition for many systems are included. A second extensive tabulation of diffusion coefficients in liquid nonelectrolytes has been prepared by Ertl, Ghai, and Dullien [31a]. An earlier review by Himmelblau [46b] deals primarily with gases dissolved in liquids.

Tables 2.6, 2.7, and 2.8 list values of $D_{AB}{}^0$ for a limited number of liquid systems at infinite dilution of $A$ in $B$. Values of $D_{AB}$ for a large number of amino acids, peptides, and sugars at low concentrations in water have been collected as Table 5.2 of Ref. 97. Additional values of $D_{AB}{}^0$ for organic solutes in several aliphatic alcohols are given in Ref. 90. Blander [14] tabulates values of self- and mutual diffusion coefficients for a variety of molten salt systems.

Considerable progress has been made in understanding diffusion in liquids, notably by Hildebrand and coworkers [46a]. However, the present state of development of the molecular theory of the liquid state does not provide a basis for the prediction of diffusion coefficients with the degree of confidence possible for dilute-gas systems. The early Stokes-Einstein equation, based on a model in which a solute sphere is considered to move through a continuum of the solvent, is

$$D_{AB} = \frac{kT}{6\pi r_0 \mu} \tag{2.27}$$

where $r_0$ is the sphere radius. This has since been modified by Wilke and Chang [107] to provide a useful estimation procedure. Their equation, represented graphically by Fig. 2.3, is

$$D_{AB}{}^0 = 7.4 \times 10^{-8} \left[ (\phi M_B)^{1/2} \frac{T}{\mu_B V_A{}^{0.6}} \right] \tag{2.28}$$

where $D_{AB}{}^0$ = mutual diffusion coefficient of solute $A$ at very low concentrations in solvent $B$, cm²/s
$\phi$ = association parameter of solvent $B$
$M_B$ = molecular weight of $B$
$T$ = temperature, °K
$\mu_B$ = viscosity of $B$, cP
$V_A$ = molal volume of the solute at its normal boiling point, cm³/g mole

It is to be noted that Eq. (2.27) is not dimensionally consistent and that the viscosity $\mu_B$ is expressed in centipoises. The molal volume $V_A$ is obtained by addition of the LeBas atomic and group volumes listed in Table 2.9, or by the use of the molal volumes for common compounds given in the footnote to the same table.

The introduction of an association parameter suggests one of the main difficulties in attempts to develop correlations of diffusion coefficients for liquids. Associated molecules behave as large molecules and diffuse more slowly; the degree

**Table 2.6 EXPERIMENTAL VALUES OF DIFFUSION COEFFICIENTS OF VARIOUS GASES AND NONELECTROLYTES AT VERY LOW CONCENTRATIONS IN WATER; VALUES ARE $D_{AB}^0$, $cm^2/s \times 10^5$**

| Solute | $T$, °K | $D_{AB}^0 \times 10^5$ | Solute | $T$, °K | $D_{AB}^0 \times 10^5$ |
|---|---|---|---|---|---|
| Argon [110] | 298 | 2.5 | Methanol | 283 | 0.84 |
| Argon [29] | 298 | 2.00 | Ethanol | 283 | 0.84 |
| Air [110] | 293 | 2.5 | n-Propanol | 288 | 0.87 |
| n-Butane [111] | 293 | 0.89[b] | i-Propanol | 288 | 0.87 |
| Carbon dioxide[a] | 298 | 1.92[c] | n-Butanol | 288 | 0.77 |
| Carbon monoxide [109] | 293 | 2.03 | i-Butanol | 288 | 0.77 |
| Chlorine | 298 | 1.25[d] | i-Pentanol | 288 | 0.69 |
| Ethane [111] | 293 | 1.20[b] | Ethylene glycol [18] | 298 | 1.16[e] |
| Ethylene [29] | 298 | 1.87 | 1,2-Propylene glycol | 293 | 0.88 |
| Helium [34] | 298 | 6.28 | Glycerol | 293 | 0.82 |
| Helium [110] | 293 | 6.8 | Benzyl alcohol | 293 | 0.82 |
| Hydrogen [34] | 298 | 4.50 | Acetic acid | 293 | 1.19 |
| Hydrogen [110] | 293 | 5.0 | Oxalic acid | 293 | 1.53 |
| Krypton [109] | 293 | 1.68 | Benzoic acid | 298 | 1.00[f] |
| Methane [111] | 293 | 1.49[b] | Salicylic acid [62] | 298 | 1.06 |
| Neon [110] | 293 | 3.00 | Glycine | 298 | 1.06 |
| Nitric oxide [109] | 293 | 2.07 | Ethyl acetate | 293 | 1.00 |
| Nitrogen [110] | 293 | 2.6 | Acetone | 293 | 1.16 |
| Nitrous oxide [29] | 298 | 1.69 | Furfural | 293 | 1.04 |
| Oxygen | 298 | 2.10[c] | Urea | 293 | 1.20 |
| Propane [111] | 293 | 0.97[b] | Urethane | 288 | 0.80 |
| Propylene [29] | 298 | 1.1 | Diethylamine | 293 | 0.97 |
| Propylene [101] | 298 | 1.44 | Aniline | 293 | 0.92 |
| Xenon [109] | 293 | 0.60 | Acetonitrile | 288 | 1.26 |
| Ammonia | 285 | 1.64 | Pyridine | 288 | 0.58 |
| Benzene [111] | 293 | 1.02 | Water | 298 | 2.44[g] |

[a] NOTE: Where references are not given, the values are from collected data given in Refs. 107 and 106.
[b] Witherspoon and Bonoli [111] report data on normal paraffin hydrocarbons methane through pentane, benzene, toluene, ethyl benzene, cyclopentane, methylcyclopentane, and cyclohexane for the temperature range 2° to 60°C. Their values for methane, ethane, propane and n-butane are little more than half those reported in Ref. 110.
[c] Estimate from three literature reviews [34, 29, 90].
[d] This value for chlorine is for an equilibrium mixture of hydrolyzed and unhydrolyzed chlorine; Peaceman [76] reports $1.48 \times 10^{-5}$ for molecular chlorine, $1.54 \times 10^{-5}$ for hypochlorous acid, and $1.51 \times 10^{-5}$ for total chlorine.
[e] Reference 18 gives data on glycol-water from 25° to 70°C from 0.7 to 99.2 weight percent glycol.
[f] An average of 1.05 [20] and 0.95 [62].
[g] Self-diffusion coefficient for water [88]; King, Hsueh, and Mao [53] give a table of self-diffusion coefficients for 23 other liquids; Dullien [31] also gives a table of published values.

of association varies with mixture composition and molecular type in a manner which is not well understood. Wilke and Chang recommend the following values of $\phi$ for several solvents: water, 2.6; methanol, 1.9; ethanol, 1.5; benzene, 1.0; ether, 1.0; heptane, 1.0; unassociated solvents generally, 1.0. The solid lines of Fig. 2.3 may be expected to approach tangentially but not cross the dotted line representing Eq. (2.27).

Lusis and Ratcliff [60] recommend $\phi = 3.3$ for methanol and ethanol, and $\phi = 5.1$ for higher alcohols. Akgerman and Gainer [1] report data on $H_2S$ and $C_2H_2$ in water and several gases in a number of organic liquids; they claim their correlation based on absolute rate theory to be superior to that of Wilke and Chang, particularly for small solute gas molecules such as $H_2$ or He, and for liquids having viscosities greater than 3 to 5 cP. The effect of solute-solvent complex formation is discussed in Ref. 81.

A second difficulty encountered in attempts to develop a general correlation is that $D_{AB}$ evidently depends in part on the shape of the molecule. Hayduk and Buckley [44] report $D_{AB}$ for linear molecules to be, in general, some 30 percent greater than for substances composed of essentially spherical molecules having the same molal volume at the normal boiling point.

Table 2.7  MUTUAL DIFFUSION COEFFICIENTS FOR $H_2$, $CH_4$, AND $CO_2$ IN AQUEOUS SALT SOLUTIONS [39, 82]; VALUES ARE $D_{AB}$, $(cm^2/s) \times 10^5$

| Gas | Electrolyte | Electrolyte concentration g moles/l | $D_{AB} \times 10^{-5}$ | | |
|---|---|---|---|---|---|
| | | | 25°C | 45°C | 65°C |
| $H_2$ | KCl | 1.0 | 3.79 | 5.97 | |
| $H_2$ | KCl | 3.0 | 3.34 | 5.27 | 8.30 |
| $H_2$ | $MgCl_2$ | 1.5 | 3.09 | 4.13 | 6.04 |
| $H_2$ | $MgCl_2$ | 3.0 | 1.45 | 2.71 | 3.47 |
| $H_2$ | $MgSO_4$ | 0.5 | 3.90 | 5.29 | 7.18 |
| $H_2$ | $MgSO_4$ | 2.0 | 1.57 | 2.90 | 4.27 |
| $CH_4$ | KCl | 2.4 | ... | ... | 3.86 |
| $CH_4$ | KCl | 3.65 | ... | 2.52 | 3.29 |
| $CH_4$ | $MgCl_2$ | 1.03 | 1.60 | 2.44 | 3.36 |
| $CH_4$ | $MgCl_2$ | 2.95 | 0.79 | 1.42 | 1.70 |
| $CH_4$ | $MgSO_4$ | 0.5 | 1.88 | 2.58 | 3.56 |
| $CH_4$ | $MgSO_4$ | 2.0 | 0.73 | 1.27 | 1.97 |
| $CO_2$ | NaCl | 1.041 | 1.73 | | |
| $CO_2$ | NaCl | 3.776 | 1.30 | | |
| $CO_2$ | $NaNO_3$ | 1.076 | 1.76 | | |
| $CO_2$ | $NaNO_3$ | 3.602 | 1.34 | | |
| $CO_2$ | $Na_2SO_4$ | 0.318 | 1.74 | | |
| $CO_2$ | $Na_2SO_4$ | 0.898 | 1.50 | | |
| $CO_2$ | $MgCl_2$ | 0.377 | 1.80 | | |
| $CO_2$ | $MgCl_2$ | 1.262 | 1.43 | | |
| $CO_2$ | $MgSO_4$ | 0.195 | 1.89 | | |
| $CO_2$ | $MgSO_4$ | 0.969 | 1.28 | | |
| $CO_2$ | $Mg(NO_3)_2$ | 0.215 | 1.85 | | |
| $CO_2$ | $Mg(NO_3)_2$ | 1.219 | 1.64 | | |

**Table 2.8 EXPERIMENTAL VALUES OF DIFFUSION COEFFICIENTS IN NON-AQUEOUS SOLVENTS AT INFINITE DILUTION\*; VALUES ARE $D_{AB}^0$, $(cm^2/s) \times 10^5$**

| Solute $A$ | Solvent $B$ | $T, °K$ | $D_{AB}^0 \times 10^5$ |
|---|---|---|---|
| Acetic acid | Acetone | 298 | 3.31 |
| Benzoic acid | Acetone | 298 | 2.62 |
| Carbon dioxide | Amyl alcohol | 298 | 1.91 |
| Water | Aniline | 293 | 0.70 |
| Acetic acid | Benzene | 298 | 2.09 |
| Carbon tetrachloride | Benzene | 298 | 1.92 |
| Cinnamic acid | Benzene | 298 | 1.12 |
| Ethylene chloride | Benzene | 280.6 | 1.77 |
| Ethanol | Benzene | 288 | 2.25 |
| Methanol | Benzene | 298 | 3.82 |
| Naphthalene | Benzene | 280.6 | 1.19 |
| Carbon dioxide | $i$-Butanol | 298 | 2.20 |
| Acetone | Carbon tetrachloride | 293 | 1.86 |
| Benzene | Chlorobenzene | 293 | 1.25 |
| Acetone | Chloroform | 288 | 2.36 |
| Benzene | Chloroform | 288 | 2.51 |
| Ethanol | Chloroform | 288 | 2.20 |
| Carbon tetrachloride | Cyclohexane | 298 | 1.49 |
| Azobenzene | Ethanol | 293 | 0.74 |
| Camphor | Ethanol | 293 | 0.70 |
| Carbon dioxide | Ethanol | 290 | 3.20 |
| Carbon dioxide | Ethanol | 298 | 3.42 |
| Glycerol | Ethanol | 293 | 0.51 |
| Pyridine | Ethanol | 293 | 1.10 |
| Urea | Ethanol | 285 | 0.54 |
| Water | Ethanol | 298 | 1.132 |
| Water | Ethylene glycol | 293 | 0.18 |
| Carbon dioxide | Gas oil, boiling point 200–300° | 298 | 1.95 |
| Carbon dioxide | Gas oil, boiling point 300–400° | 298 | 0.73 |
| Water | Glycerol | 293 | 0.0083 |
| Carbon dioxide | Heptane | 298 | 6.03 |
| Carbon tetrachloride | $n$-Hexane | 298 | 3.70 |
| Toluene | $n$-Hexane | 298 | 4.21 |
| Carbon dioxide | Kerosine | 298 | 2.50 |
| Tin | Mercury | 303 | 1.60 |
| Water | $n$-Propanol | 288 | 0.87 |
| Water | 1,2-Propylene glycol | 293 | 0.075 |
| Acetic acid | Toluene | 298 | 2.26 |
| Acetone | Toluene | 293 | 2.93 |
| Benzoic acid | Toluene | 293 | 1.74 |
| Chlorobenzene | Toluene | 293 | 2.06 |
| Ethanol | Toluene | 288 | 3.00 |
| Carbon dioxide | White spirit | 298 | 2.11 |

\* This is a condensation of a table of collected values given in Ref. 16; the added values for carbon dioxide are from Ref. 25. *Note:* Data on diffusion in inorganic solids and their melts are found in Refs. 27 and 14. A large collection of values of $D_{AB}^0$ in nonaqueous systems is to be found in the appendix of Ref. 97.

$T = °K$; $D_{AB} = cm^2/sec$; $\mu$ = viscosity of solvent, centipoises;
$M_B$ = molecular weight of solvent; association parameter $\phi$ is 2.6
for water, 1.9 for methanol, 1.5 for ethanol, 1.0 for benzene,
ether, heptane, and unassociated solvents generally; solute
molal volumes are LeBas values (Table 2.9).

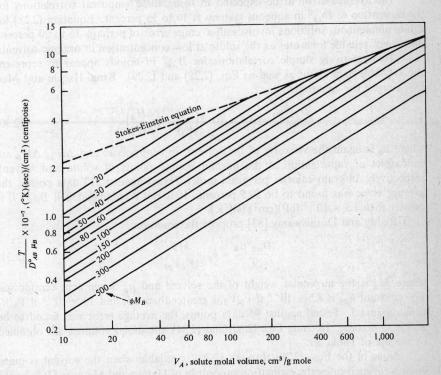

FIGURE 2.3
Wilke-Chang estimation procedure for molecular diffusion coefficients in binary
liquid systems at low concentration of the diffusing solute.

For dilute *aqueous systems* the equation developed by Othmer and Thakar [73]
is somewhat easier to use and gives values which check the experimental data
about equally well. This is

$$D_{AB}{}^0 = 14.0 \times 10^{-5} \mu_B{}^{-1.1} V_A{}^{-0.6} \qquad (2.29)$$

where $V_A$ is the molal volume of the solute ($B$ is water), cubic centimeters per
gram mole; and $\mu_B$ is the viscosity of water, in centipoises, at the temperature in
question. Witherspoon and Bonoli [111], however, found that their data on 11
hydrocarbons in water agreed well with the Wilke-Chang correlation [Eq. (2.28)]
if the parameter $\phi$ is taken to be 2.6.

Using data on diffusivities of 87 substances in dilute *aqueous solutions*, Hayduk
and Laudie [45a] conclude that the Wilke-Chang and Othmer-Thakar correlations

can be improved if $\phi$ for water is taken to be 2.26 in Eq. (2.28), and the constant and exponents in Eq. (2.29) changed from 14.0 to 13.26, $-1.1$ to $-1.4$, and $-0.6$ to $-0.589$.

The average error to be expected in using these empirical correlations for the estimation of $D_{AB}{}^0$ in aqueous systems is 10 to 15 percent. Equation (2.28) for dilute nonaqueous solutions involves an average error of perhaps 15 to 20 percent, and is not reliable for water as the solute at low concentration in organic solvents.

Two relatively simple correlations for $D_{AB}{}^0$ in liquids appear to represent the available data about as well as Eqs. (2.28) and (2.29). King, Hsueh, and Mao [53] suggest the equation

$$\frac{D_{AB}{}^0 \mu_B}{T} = 4.4 \times 10^{-8} \left(\frac{V_B}{V_A}\right)^{1/6} \left(\frac{\Delta H_B}{\Delta H_A}\right)^{1/2} \tag{2.30}$$

where $\mu_B$ is again the viscosity of the solvent, in centipoises; and $\Delta H_A$, $\Delta H_B$ are enthalpies of vaporization at the normal boiling points of solute and solvent, respectively, in gram calories per gram mole. Tested against 213 data points, the average error was found to be 19.5 percent. It is not recommended if $D\mu_B{}^0/T$ is greater than $1.5 \times 10^{-7}$ (cP)(cm$^2$)/(s)($^\circ$K).

Reddy and Doraiswamy [83] propose the equation

$$\frac{D_{AB}{}^0 \mu_B}{T} = k_{RS} \frac{M_B{}^{1/2}}{(V_A V_B)^{1/3}} \tag{2.31}$$

where $M_B$ is the molecular weight of the solvent and $\mu_B$ is again in centipoises. The constant $k_{RS}$ is $8.5 \times 10^{-8}$ if $V_B/V_A$ is greater than 1.5, and $10 \times 10^{-8}$ if $V_B/V_A$ is less than 1.5. Tested against 96 data points, the average error was found to be about 15 percent. In using Eqs. (2.30) and (2.31), the molal volumes are obtained by the use of Table 2.9.

None of the Eqs. (2.28) through (2.30) is reliable when the solvent is quite viscous; in such cases the estimation procedure of Gainer and Metzner [36] should be used. See also the recent paper by Hiss and Cussler [47a]. Equation (2.30) or (2.31) is to be preferred to Eq. (2.28) for water as the solute in low concentrations in organic solvents.

Furthermore, it should be noted that very few data have been reported for diffusion coefficients in liquids at temperatures above 30$^\circ$C. Way [102], who made a study of the limited data on the effect of temperature (all in hydrocarbon systems), reports activation energies of 3 to 6.7 kcal/g mole, increasing with the molecular weight of the solvent. The effect of temperature suggested by Eq. (2.28) corresponds to activation energies of 2.5 to 5 kcal/g mole.

Hayduk and Cheng [45] report the interesting empirical observation that for any one solute, $D_{AB}{}^0$ is a power function of the viscosity of the solvent, that is, $D_{AB}{}^0 = A\mu_B{}^n$, where $n$ varies from $-0.44$ to $-1.15$ with different solutes. This is shown to hold quite well for a variety of systems where $0.25 < \mu_B < 5$ cP.

Limited experimental data on self-diffusion in liquids have been reported. Dullien [31] gives a table of published values and suggests an equation which provides an excellent correlation of the data. This is in the form of a simple expression for $D_{AA}$ in terms of temperature, viscosity, and critical volume. See also Ref. 97a.

As noted above, $D_{AB}$ usually varies considerably with concentration, often passing through a maximum or minimum, with the two extreme values for infinite dilution sometimes differing severalfold. Figure 2.2 shows a substantial variation of $D_{AB}$ with concentration for the rather ideal binary $n$-octane–$n$-dodecane.

The phenomenological theory developed in Sec. 2.4 suggests that this variation might be represented by the following expression, providing $V_B$ is assumed to be independent of concentration:

$$D_{AB} = D_{AB}{}^0 \frac{\partial \ln a_A}{\partial \ln c_A} \tag{2.32}$$

In the case of a binary system for which there is no volume change on mixing ($V_A$ and $V_B$ independent of concentration but not necessarily equal), $\ln c_A$ may be expressed in terms of mole fraction $X_A$ to give

$$D_{AB} = D_{AB}{}^0 \left[ \frac{X_A V_A + X_B V_B}{V_B} \right] \left( \frac{\partial \ln a_A}{\partial \ln X_A} \right)_{T,P} \tag{2.33}$$

**Table 2.9 ADDITIVE-VOLUME INCREMENTS\* FOR THE CALCULATION OF MOLAL VOLUMES AT THE NORMAL BOILING POINT (LEBAS)**

| | Increment, cm³/g mole |
|---|---|
| Carbon | 14.8 |
| Hydrogen | 3.7 |
| Oxygen (except as noted below) | 7.4 |
| In methyl esters and ethers | 9.1 |
| In ethyl esters and ethers | 9.9 |
| In higher esters and ethers | 11.0 |
| In acids | 12.0 |
| Joined to S, P, N | 8.3 |
| Nitrogen | |
| Doubly bonded | 15.6 |
| In primary amines | 10.5 |
| In secondary amines | 12.0 |
| Bromine | 27 |
| Chlorine | 24.6 |
| Fluorine | 8.7 |
| Iodine | 37 |
| Sulfur | 25.6 |
| Ring, three-membered | −6.0 |
| Four-membered | −8.5 |
| Five-membered | −11.5 |
| Six-membered | −15.0 |
| Naphthalene | −30.0 |
| Anthracene | −47.5 |

\* The additive-volume procedure should not be used for simple molecules. The following approximate values are employed in estimating diffusion coefficients: $H_2$, 14.3; $O_2$, 25.6; $N_2$, 31.2; air, 29.9; CO, 30.7; $CO_2$, 34.0; $SO_2$, 44.8; NO, 23.6; $N_2O$, 36.4; $NH_3$, 25.8; $H_2O$, 18.9; $H_2S$, 32.9; COS, 51.5; $Cl_2$, 48.4; $Br_2$, 53.2; $I_2$, 71.5.

Furthermore, since in this case Eq. (2.17) yields $D_{AB} = D_{BA}$, Eq. (2.32) may be written for each of these to relate $D_{AB}{}^0$ to $D_{BA}{}^0$. Equating $D_{AB}$ and $D_{BA}$ in this way, introducing Eq. (2.15), and noting that $dc_A/dc_B = -V_B/V_A$, one obtains the equality $D_{AB}{}^0/D_{BA}{}^0 = V_B/V_A$. Substitution of this in Eq. (2.33) yields

$$D_{AB} = (D_{BA}{}^0 X_A + D_{AB}{}^0 X_B)\left(\frac{\partial \ln a_A}{\partial \ln X_A}\right)_{T,P} \tag{2.34}$$

which indicates that the ratio of $D_{AB}$ to the activity term should be linear in mole fraction. The superscript zero denotes infinite dilution of the substance indicated by the first of the two-letter subscripts.

Vignes [99] has pointed out that the logarithm of the ratio $D_{AB}/(\partial \ln a_A/\partial \ln X_A)$ is linear in $X_A$ over the entire concentration range for most binary liquid systems, including some where association is important. This linearity is represented by the following:

$$D_{AB} = (D_{AB}{}^0)^{X_B}(D_{BA}{}^0)^{X_A}\left(\frac{\partial \ln a_A}{\partial \ln X_A}\right)_{T,P} \tag{2.35}$$

Vignes' empirical observation is explained in terms of the Eyring "hole" model of liquids by Cullinan [23].

Vignes has shown Eq. (2.35) to agree well with data for a number of systems; Dullien [30] found that it gave excellent results for a large number of ideal binary systems but gave an average error of some 16 percent for several nonideal systems. The articles by Vignes [99] and Dullien [30] provide a good collection of published data on $D_{AB}$ over the entire concentration range in binary liquid systems, together with references and data on the activity correction term.

More recently, data on aniline-benzene and ethanol-benzene [80] have appeared. Halaska and Colver [41] report data on three additional nonideal systems: toluene-methylcyclohexane, toluene-aniline, and methylcyclohexane-aniline. A semi-theoretical equation involving the viscosities of the pure components was found to agree with the data on these three systems somewhat better than did Vignes' Eq. (2.35).

**EXAMPLE 2.2** Estimate the mutual diffusion coefficient of acetone at infinite dilution in water and for water at infinite dilution in acetone at 25°C.

SOLUTION Let $A$ represent acetone and $B$ water. Using Eq. (2.28), with $\mu_B = 0.891$ cP, $\phi_B = 2.6$, $V_A = 74$ cm³/g mole (from Table 2.9), and $T = 298°$K,

$$D_{AB}{}^0 = 7.4 \times 10^{-8}\left[(2.6 \times 18)^{1/2} \times \frac{298}{0.891 \times 13.2}\right]$$

$$= 1.28 \times 10^{-5} \text{ cm}^2/\text{s}$$

The experimental value [2] is $1.28 \times 10^{-5}$.

Similarly, using Eq. (2.29),

$$D_{AB}{}^0 = 14.0 \times 10^{-5}\frac{0.891^{-1.1}}{13.2} = 1.20 \times 10^{-5} \text{ cm}^2/\text{s}$$

Using Eq. (2.28) for water in acetone, with $\mu_A = 0.305$ cP and $V_B = 18.9$ (water is now the solute and acetone the solvent, so the subscripts in the equation as written must be reversed),

$$D_{BA}{}^0 = 7.4 \times 10^{-8} \left[ (1 \times 58)^{1/2} \times \frac{298}{0.305 \times 5.83} \right]$$

$$= 9.4 \times 10^{-5} \text{ cm}^2/\text{s}$$

This value is suspect since, as noted in the text, Eq. (2.28) is not reliable for water as the solute in organic solvents. Using Eq. (2.31),

$$D_{BA}{}^0 = \frac{298 \times 10 \times 10^{-8}(58)^{1/2}}{0.305(74 \times 18.9)^{1/3}} = 6.7 \times 10^{-5} \text{ cm}^2/\text{s}$$

and using Eq. (2.30), with $\Delta H_B/\Delta H_A = 9{,}717/6{,}952$,

$$D_{BA}{}^0 = 4 \times 10^{-5} \text{ cm}^2/\text{s}$$

This last is the closest to the experimental value [2] of $5.3 \times 10^{-5}$ cm for a very dilute solution of water in acetone.    ////

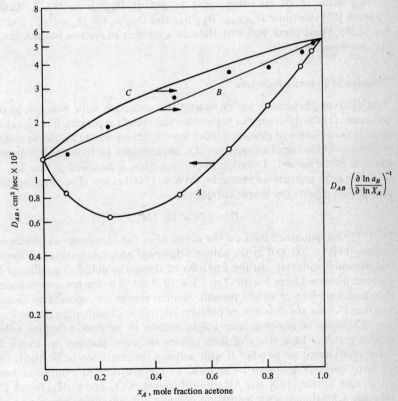

FIGURE 2.4
Mutual diffusion coefficients for the system acetone-water at 25°C. Points are experimental [2]. Curves $B$ and $C$ represent Eqs. (2.33) and (2.34), respectively, using the experimental values of $D_{AB}{}^0$ and $D_{BA}{}^0$.

Example 2.2 shows exceptionally good agreement between the experimental value of $D_{AB}{}^0$ and that calculated by the use of Eq. (2.28). Equation (2.30) gives fair agreement for $D_{BA}{}^0$ in the acetone-water system at 25°C. Figure 2.4, line $A$, shows the experimental values [2] of $D_{AB}$ as a function of composition, with a sharp minimum at about 0.3 mole fraction acetone. This system, which shows positive deviation from ideality, gives a curve of $D_{AB}$ vs. $X_A$ which is concave upwards; systems showing negative deviations give curves concave downwards, which may pass through a maximum. Note that the ordinate scale is logarithmic. The experimental data are seen to follow closely the linear relation between the log of the activity-corrected values and mole fraction, as suggested by Vignes, and represented by the straight line $B$ drawn to connect the two limiting experimental values for infinite dilution. Line $C$, representing Eq. (2.34) and drawn through the experimental values of $D_{AB}{}^0$ and $D_{BA}{}^0$, shows somewhat poorer agreement.

Bearman [8] and Gainer [37] suggest equations relating $D_{AB}$ to the composition of a binary which introduce solution viscosity as a variable; both require but one limiting value of $D^0$ (i.e., they serve to predict $D_{BA}{}^0$ from $D_{AB}{}^0$). Leffler and Cullinan [55] substitute $D_{AB}\mu_{AB}$, $D_{BA}{}^0\mu_A$, and $D_{AB}{}^0\mu_B$ for $D_{AB}$, $D_{BA}{}^0$, and $D_{AB}{}^0$ in Eq. (2.34). These check well with data on a number of systems but not for that on acetone-water.

## Diffusion in Polymer Solutions

Two different phenomena are of interest in connection with diffusion in polymer solutions: (1) the diffusion of a high-molecular-weight substance in a liquid solvent, and (2) the diffusion of gases and other low-molecular-weight solutes in solutions of polymers. Diffusion of oxygen or $CO_2$ in solutions of biological proteins is an example of the second. Diffusion in polymer films is discussed in Sec. 2.9.

For large unhydrated molecules (MW > 1,000) at low concentrations in water, Polson [78] reports the simple correlation

$$D = 2.74 \times 10^{-5} M^{-1/3} \tag{2.36}$$

Paul [75] has published data on the diffusion of the copolymer acrylonitrile–vinyl acetate (MW $\approx$ 200,000) in the solvents dimethyl acetamide, dimethyl formamide, and dimethyl sulfoxide. In the first two of these, the diffusion coefficient for the polymer increased from $1 \times 10^{-6}$ to $1.5 \times 10^{-6}$ cm$^2$/s as the polymer concentration increased from 0 to 25 weight percent. Similar results are reported by Osmers and Metzner [72] for the diffusion of polyacrylonitrile in dimethylformamide.

Diffusion of low-molecular-weight solutes in polymer solutions follows no simple pattern. In a study of four solutes in seven polymer solutions, Li and Gainer [58] found the trend of $D$ with polymer concentration to be highly variable. In some cases $D$ appeared to increase as the polymer concentration increased. Zandi and Turner [112] and Astarita [6] (solutes $CO_2$ and $C_2H_4$) found $D$ to go through a maximum with increasing polymer concentration, which also involves large increases in viscosity. The validity of these maxima is questioned by Navari, Gainer, and Hall [68]. See also Ref. 28. These authors assert that the ratio of the diffusion coefficient in the polymer solution to that in the pure solvent "appears to be completely independent of the solute," and derive a theory for use in the prediction of the ratio of the two coefficients. They also report data on diffusion of oxygen, $CO_2$, and several other solutes in solutions of albumin, gamma globulin, and fibrinogen in human plasma.

Diffusion coefficients of small nonreacting solutes are not much smaller in polymer solutions than in the pure solvent, at least over a modest range of polymer concentrations. Hoshino [48], for example, found $D$ in solutions of polyvinyl pyrrolidone or polyvinyl alcohol to agree within 35 percent with $D$ for the same solute in water. See also Ref. 72. The solutes were urea, sucrose, and sodium chloride; and the polymer concentrations ranged from 30 to 175 g/l. In polymer solutions, and perhaps also in slurries, the probable effect of the additive is to reduce the cross section of the diffusional path in the solvent. In any case, it is clear that the viscosity of the *polymer solution* is not to be employed in equations based on the Stokes-Einstein relation.

## 2.7  DIFFUSION IN ELECTROLYTE SOLUTIONS

When a salt dissociates in solution, it is the ions rather than the molecules which diffuse. In the absence of an applied electrical potential, the diffusion of a single salt may be treated as molecular diffusion, however, since the requirement of zero current causes anions and cations to attract each other strongly and, in the presence of only two ions of opposite charge, to diffuse at the same rate.

The theory of diffusion of salts at low concentrations in aqueous solutions is well developed, though it provides but limited guide about the behavior of the more concentrated solutions of technical interest. (See, however, the comprehensive review of transport in electrolytic solutions by Newman [70].) The diffusion coefficient $D_{AB}{}^0$ for a single salt at infinite dilution is given by the Nernst-Haskell equation:

$$D_{AB}{}^0 = \frac{RT}{(\text{Fa})^2} \frac{(1/n_+ + 1/n_-)}{(1/\lambda_+{}^0 + 1/\lambda_-{}^0)} \tag{2.37}$$

where $D_{AB}{}^0$ = diffusion coefficient at infinite dilution, relating molecular flux to the gradient of molecular concentration, $cm^2/s$

$\quad\quad T$ = temperature, °K

$\lambda_+{}^0, \lambda_-{}^0$ = limiting (zero concentration) ionic conductances at $T$, of cation and anion, respectively, $amp/(cm^2)$ $(V/cm)$ $(g\ equiv/cm^3)$

$\quad\quad$ Fa = Faraday = 96,488 C/g equiv

$\quad\quad R$ = gas constant = 8.315 J/(°K)(g mole)

$\quad n_+, n_-$ = valences of cation and anion, respectively

Values of $\lambda_+{}^0$ and $\lambda_-{}^0$ at 25°C are given in standard texts [42, 66, 88]; Table 2.10 lists values for common ions. Table 2.11 provides a limited collection of experimental values of $D_{AB}{}^0$ in aqueous salt solutions.

As the salt concentration is increased, $D_{AB}$ first decreases (seldom to less than 70 percent of $D_{AB}{}^0$) and then increases, sometimes passing through a maximum (see

Fig. 2.5). Thus, $D_{AB}$ for calcium chloride in water at 25°C decreases from $D_{AB}{}^0 = 1.35 \times 10^{-5}$ to $D_{AB} = 1.11 \times 10^{-5}$ at $0.16N$ and then increases and passes through a maximum of $1.32 \times 10^{-5}$ at $2.25N$ [66]. There appears to be no generally reliable theory or empirical procedure to be used in predicting the variation with concentration in strong solutions; see, however, Ref. 69.

In a system of *mixed electrolytes*, as in the simultaneous diffusion of HCl and NaCl in water, the faster moving $H^+$ ion may move ahead of its $Cl^-$ partner, the electrical current being maintained at zero by the lagging behind of the slower moving $Na^+$ ions. In such systems, the unidirectional diffusion of each ion species results from a combination of electrical and concentration gradients:

$$J_+ = \frac{\lambda_+}{(Fa)^2}\left[-RT\frac{dc_+}{dy} - (Fa)c_+\frac{dE}{dy}\right] \tag{2.38}$$

$$J_- = \frac{\lambda_-}{(Fa)^2}\left[-RT\frac{dc_-}{dy} + (Fa)c_-\frac{dE}{dy}\right] \tag{2.39}$$

where $J_+$, $J_-$ = diffusion flux densities of cation and anion, respectively,
  g equiv/(cm$^2$)(s)
  $c_+$ and $c_-$ = corresponding ion concentrations, g equiv/cm$^3$
  $dE/dy$ = gradient of the electrical potential

Table 2.10 LIMITING IONIC CONDUCTANCES IN WATER AT 25°C;* amp/(cm$^2$) ($V$/cm) (g equiv/cm$^3$)

| Cation | $\lambda_+{}^0$ | Anion | $\lambda_-{}^0$ |
|---|---|---|---|
| $H^+$ | 349.8 | $OH^-$ | 197.6 |
| $Li^+$ | 38.7 | $Cl^-$ | 76.3 |
| $Na^+$ | 50.1 | $Br^-$ | 78.3 |
| $K^+$ | 73.5 | $I^-$ | 76.8 |
| $NH_4^+$ | 73.4 | $NO_3^-$ | 71.4 |
| $Ag^+$ | 61.9 | $ClO_4^-$ | 68.0 |
| $Tl^+$ | 74.7 | $HCO_3^-$ | 44.5 |
| $\frac{1}{2}Mg^{++}$ | 53.1 | $HCO_2^-$ | 54.6 |
| $\frac{1}{2}Ca^{++}$ | 59.5 | $CH_3CO_2^-$ | 40.9 |
| $\frac{1}{2}Sr^{++}$ | 50.5 | $ClCH_2CO_2^-$ | 39.8 |
| $\frac{1}{2}Ba^{++}$ | 63.6 | $CNCH_2CO_2^-$ | 41.8 |
| $\frac{1}{2}Cu^{++}$ | 54 | $CH_3CH_2CO_2^-$ | 35.8 |
| $\frac{1}{2}Zn^{++}$ | 53 | $CH_3(CH_2)_2CO_2^-$ | 32.6 |
| $\frac{1}{3}La^{3+}$ | 69.5 | $C_6H_5CO_2^-$ | 32.3 |
| $\frac{1}{3}Co(NH_3)_6^{3+}$ | 102 | $HC_2O_4^-$ | 40.2 |
| | | $\frac{1}{2}C_2O_4^-$ | 74.2 |
| | | $\frac{1}{2}SO_4^-$ | 80 |
| | | $\frac{1}{3}Fe(CN)_6^{3-}$ | 101 |
| | | $\frac{1}{4}Fe(CN)_6^{4-}$ | 111 |

* From H. S. Harned and B. B. Owen, "The Physical Chemistry of Electrolytic Solutions," ACS Monograph 95, Reinhold Publishing Corporation, New York, 1950. A somewhat more complete table is given by Robinson and Stokes [88], with values at other temperatures.

This gradient may be imposed externally, but it is present in the ionic solution even if there is no external electrostatic field, owing to the small separation of charges which result from diffusion itself. Collision effects, ion complexes, and activity corrections are neglected.

One equation for each cation and one for each anion may be combined with the requirement of zero current at any $y$: $\Sigma J_+ = \Sigma J_-$. Solving for the unidirectional flux densities [100],

$$n_+ J_+ = -\frac{RT\lambda_+}{(Fa)^2 n_+}\left[G_+ - n_+ c_+ \frac{\Sigma\lambda_+ G_+/n_+ - \Sigma\lambda_- G_-/n_-}{\Sigma\lambda_+ c_+ + \Sigma\lambda_- c_-}\right] \quad (2.40)$$

$$n_- J_- = \frac{RT\lambda_-}{(Fa)^2 n_-}\left[G_- + n_- c_- \frac{\Sigma\lambda_+ G_+/n_+ - \Sigma\lambda_- G_-/n_-}{\Sigma\lambda_+ c_+ + \Sigma\lambda_- c_-}\right] \quad (2.41)$$

where $G_+$, $G_-$ are the concentration gradients, $dc/dy$, in the direction of diffusion.

Table 2.11 MUTUAL DIFFUSION COEFFICIENTS OF INORGANIC SALTS IN AQUEOUS SOLUTIONS*

| Solute | $T$, °C | Concentration, g moles/l | $D_{AB} \times 10^5$, (cm²/s) × 10⁵ |
|---|---|---|---|
| HCl | 12 | 0.1 | 2.29 |
| $H_2SO_4$ | 20 | 0.25 | 1.63 |
| $HNO_3$ | 20 | 0.05 | 2.62 |
|  |  | 0.25 | 2.59 |
| $NH_4Cl$ | 20 | 1.0 | 1.64 |
| $H_3PO_4$ | 20 | 0.25 | 0.89 |
| $HgCl_2$ | 18 | 0.25 | 0.92 |
| $CuSO_4$ | 14 | 0.4 | 0.39 |
| $AgNO_3$ | 15 | 0.17 | 1.28 |
| $CoCl_2$ | 20 | 0.1 | 1.0 |
| $MgSO_4$ | 10 | 0.4 | 0.39 |
| $Ca(OH)_2$ | 20 | 0.2 | 1.6 |
| $Ca(NO_3)_2$ | 14 | 0.14 | 0.85 |
| LiCl | 18 | 0.05 | 1.12 |
| NaOH | 15 | 0.05 | 1.49 |
| NaCl | 18 | 0.4 | 1.17 |
|  |  | 0.8 | 1.19 |
|  |  | 2.0 | 1.23 |
| KOH | 18 | 0.01 | 2.20 |
|  |  | 0.1 | 2.15 |
|  |  | 1.8 | 2.19 |
| KCl | 18 | 0.4 | 1.46 |
|  |  | 0.8 | 1.49 |
|  |  | 2.0 | 1.58 |
| KBr | 15 | 0.046 | 1.49 |
| $KNO_3$ | 18 | 0.2 | 1.43 |

Values from International Critical Tables, Vol. V.
* A more complete table for common electrolytes is given by Robinson and Stokes [88].

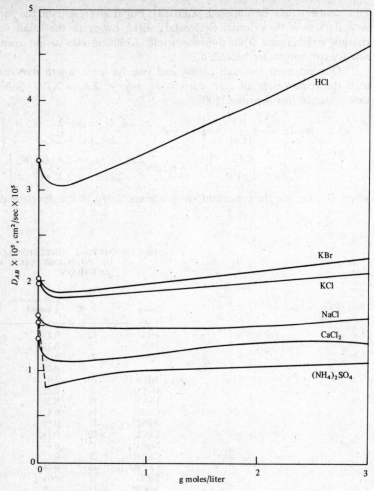

**FIGURE 2.5**
Diffusion coefficients for several common electrolytes in water at 25°C [88].

Vinograd and McBain [100] have shown these relations to represent their data on diffusion in multi-ion solutions. $D_{AB}$ for the hydrogen ion was found to decrease from 12.2 to $4.9 \times 10^{-5}$ cm²/s in a solution of HCl and BaCl₂ when the ratio of H⁺ to Ba⁺⁺ was increased from zero to 1.3; $D_{AB}$ at the same temperature is $9.03 \times 10^{-5}$ for the free H⁺ ion, and $3.3 \times 10^{-5}$ for HCl in water. The presence of the slow-moving Ba⁺⁺ accelerates the H⁺ ion, the potential existing with zero current causing it to move in dilute solution even faster than it would as a free ion with no other cation present, i.e., electrical neutrality is maintained by the

Table 2.12 MUTUAL DIFFUSION COEFFICIENTS FOR $H_2$ AND $CH_4$ IN AQUEOUS SALT SOLUTIONS [39]; VALUES ARE $D_{AB}$, $(cm^2/s) \times 10^5$

| Gas | Electrolyte | Electrolyte concentration g moles/l | $D_{AB} \times 10^5$ | | |
|-----|-----------|------|------|------|------|
| | | | 25°C | 45°C | 65°C |
| $H_2$ | Water | ... | 4.10 | 6.30 | 9.44 |
| $H_2$ | KCl | 1.0 | 3.79 | 5.97 | |
| $H_2$ | KCl | 3.0 | 3.34 | 5.27 | 8.30 |
| $H_2$ | $MgCl_2$ | 1.5 | 3.09 | 4.13 | 6.04 |
| $H_2$ | $MgCl_2$ | 3.0 | 1.45 | 2.71 | 3.47 |
| $H_2$ | $MgSO_4$ | 0.5 | 3.90 | 5.29 | 7.18 |
| $H_2$ | $MgSO_4$ | 2.0 | 1.57 | 2.90 | 4.27 |
| $CH_4$ | Water | ... | 1.81 | 2.78 | 3.84 |
| $CH_4$ | KCl | 2.4 | ... | ... | 3.86 |
| $CH_4$ | KCl | 3.65 | ... | 2.52 | 3.29 |
| $CH_4$ | $MgCl_2$ | 1.03 | 1.60 | 2.44 | 3.36 |
| $CH_4$ | $MgCl_2$ | 2.95 | 0.79 | 1.42 | 1.70 |
| $CH_4$ | $MgSO_4$ | 0.5 | 1.88 | 2.58 | 3.56 |
| $CH_4$ | $MgSO_4$ | 2.0 | 0.73 | 1.27 | 1.97 |

hydrogen ions moving ahead of the chlorine, faster than they would as free ions, while the barium diffuses more slowly than as a free ion.

The interaction of ions in a multi-ion system is important when the several ion conductances differ greatly, as they do when $H^+$ or $OH^-$ is diffusing (see Table 2.10). Where the diffusion of these two ions is not involved, no great error is introduced by the use of "molecular" diffusion coefficients for the salt species present.

Relatively few data on $D_{AB}$ for diffusion of gases dissolved in salt solutions are available. Typical values are those for hydrogen and methane given in Table 2.12.

## 2.8 DIFFUSION IN POROUS MATERIALS

Diffusion in porous materials is of particular importance in heterogeneous catalysis, where porous pellets or extrudates of solid catalysts are commonly employed. Knowledge of diffusion coefficients in such materials is required to make full use of the Thiele theory of diffusion and reaction within the porous mass. The discussion which follows is a brief presentation of the main concepts; for a more detailed treatment see the book by Satterfield [91].

Diffusion in porous solids may occur by one or more of three mechanisms: ordinary ("bulk"), Knudsen, and surface diffusion. In the absence of surface diffusion, the proximity of the pore walls is not important if the pores are large in relation to the mean free path of the molecules in the free gas, and the process is that of ordinary molecular diffusion within the gas or liquid contained in the pores. The diffusion coefficients discussed in earlier sections and the rate equations presented in Chap. 3 then apply.

Knudsen diffusion is encountered in pores containing a gas if the pressure is low or the pore size small. If the mean free path of the gas molecules is large compared with the pore diameter, then the molecules collide much more frequently with the pore walls than with each other. Reflection from the wall is not specular but essentially diffuse; the molecules rebound in nearly random directions. The resistance to diffusion along the pore is then due primarily to molecular collisions with the wall rather than with each other, as in ordinary diffusion. In some ranges of gas densities and pore sizes both types of collisions are important.

Surface diffusion is encountered when the diffusing species is adsorbed by the solid. The equilibrium surface concentration increases with concentration in the gas, so the surface layer tends to develop a gradient of the surface concentration of the same sign and in the same direction as the concentration gradient of the gas in the pore. With some types of adsorption, the adsorbed layer is mobile and tends to move along the solid surface, so that surface diffusion occurs in parallel with diffusion in the gas. Total flux of the adsorbed species is then greater than if adsorption did not occur.

In the case of ordinary molecular diffusion, the flux density based on the total or face area of the porous mass is obviously reduced because the free or open cross section is but a fraction of the total. If the concentration gradient is expressed in terms of the distance normal to the face, then the diffusion coefficient is also reduced because the diffusion distance along the tortuous pore is greater than that normal to the face, and because the plane normal to diffusion at a point has a larger area than that of the plane at the same point, parallel with the face, which is used to base the flux density. An effective diffusion coefficient may be defined by

$$D_{ABP} = \frac{D_{AB}\theta}{\tau} \qquad (2.42)$$

where $D_{AB}$ is the usual diffusion coefficient for ordinary diffusion in a binary system, $\theta$ is the fraction free cross section, and $D_{ABP}$ is the coefficient, corresponding to $D_{AB}$, employed to describe the flux density per unit total face area, per unit concentration gradient normal to the face. The factor $\tau$ is introduced to allow for the fact that the diffusion path is greater than the distance traveled normal to the face, and for the varying cross sections of the pores, which are not straight round tubes. It is sometimes called a "tortuosity factor," but it is required to correct not only for the greater path length but also for the possible existence of tiny orifice restrictions in solids of complex structure. This correction factor must be obtained experimentally except for solids of exceedingly uniform structure and pore size. Note that the free cross section $\theta$ is identical with the fraction void volume, which can be measured without great difficulty; $\theta$ obtained in this way is the average free cross section in a plane parallel to the face—not necessarily normal to the direction of diffusion at any point.

The theory of Knudsen diffusion in straight nonadsorbing round pores is based on the kinetic theory, from which one obtains [52] the following expression for transport with diffuse wall reflection in a pore of radius $r_e$:

$$J = -\frac{2r_e}{3}\left(\frac{8RT}{\pi M}\right)^{1/2}\frac{dc}{dy} \tag{2.43}$$

where $J$ is based on the pore cross section, and $M$ is the molecular weight of the diffusing species.

Equation (2.43) states that the flux is proportional to the concentration gradient, to the free molecular velocity, and to the ratio of pore volume to pore surface. The volume-to-surface ratio of a fine-pore structure may be approximated by dividing the void fraction $\theta$ by either the Brunauer-Emmett-Teller (BET) surface per unit volume, or by the product of the BET surface per gram, $S_g$, and the bulk density $\rho_p$ of the porous mass, grams per cubic centimeter. Since the volume-to-surface ratio of a straight round pore is $r_e/2$, one may define an equivalent or average radius of the pores as $2\theta/S_g\rho_p$. Substituting this in Eq. (2.43) and dividing the flux by the concentration gradient to obtain a diffusion coefficient, one obtains

$$D_{KP} = \frac{8\theta^2}{3\tau S_g \rho_p}\left(\frac{2RT}{\pi M}\right)^{1/2} = 19,400\,\frac{\theta^2}{\tau S_g \rho_p}\left(\frac{T}{M}\right)^{1/2} \tag{2.44}$$

where $D_{KP}$ is the Knudsen diffusion coefficient, which is the ratio of the flux per unit of *total face area* to the concentration gradient normal to the face. Since only collisions with the wall are involved, and not those with other molecules, $D_{KP}$ does not depend on the nature of other species present.

The practical value of Eq. (2.44) depends on knowledge of $\tau$, which is known to vary widely with the structure of the porous mass, the particle or pore-size distribution, and the shapes of the passages; this structure cannot be described by a single number. Measurements of diffusion in porous Vycor glass, which has a relatively uniform pore size, give values of $\tau$ close to 5.9. Porous materials having a bimodal pore-size distribution (distributions concentrated around one large and one small pore size) may give values of $\tau$ less than unity; i.e., diffusion is more rapid than it would be in a straight round pore having a radius calculated from the BET surface and void fraction.

A limited collection of experimental data on diffusion in both unconsolidated and consolidated porous media are quoted by Satterfield [91], but it appears that no reliable correlation is available for the prediction of diffusion coefficients. Perhaps none can be expected until better ways of characterizing the structure of porous media are developed. See, however, Ref. 33.

A common method [103, 105] of measuring the coefficient involves steady-state, constant-pressure counterdiffusion of two gases through a pellet or flat disk of the porous substance. The result depends primarily on the larger pores which extend through the sample; the small side dead-end pores contribute to the measured $S_g$ and $\theta$ but not to the measured flux. Values of the coefficient perhaps more relevant to catalysis are obtained by unsteady-state measurements, as by the use of a chromatograph [40, 54], or by the use of the Thiele theory to calculate effective diffusion coefficients from measurements of chemical reaction rates with pellets of several sizes.

It is clear that there must be a range of pressures and pore sizes in which both ordinary and Knudsen diffusion contribute to the diffusion of a gas in a pore. In this "transition region" the effective diffusion coefficient is given approximately by [77]

$$D_P = \left( \frac{1}{D_{ABP}} + \frac{1}{D_{KP}} \right)^{-1} \tag{2.45}$$

Ordinary or Knudsen diffusion will be "controlling" if one or the other of the reciprocal terms is negligible. This is a better indication than the empirical rule that Knudsen diffusion is controlling if the mean free path of the molecules is greater than the pore diameter. Experimental data on diffusion in small capillaries are reported by Remick and Geankoplis [86].

Multicomponent diffusion in porous materials has received little attention, but a recent article [33a] reports studies of the diffusion of $He-N_2-CH_4$ in gamma-alumina at 1 to 70 atm. Several models are described, each involving a surface diffusion term.

The extent to which surface diffusion becomes important varies widely with the nature of the solid-gas system, since it depends on the existence of an adsorbed layer. Though capillary condensation may cause the pores to fill with liquid if the temperature is near the dew point, it is usually acceptable to assume that the adsorbed layer is so thin that the cross section of the pore available for gas diffusion is not greatly reduced. One may then assume that surface and gas diffusion occur in parallel, and that the fluxes are additive.

If it is also assumed that the surface diffusion flux is proportional to the gradient of the surface concentration, then

$$J = - \left( \frac{1}{D_{ABP}} + \frac{1}{D_{KP}} \right)^{-1} \frac{dc}{dy} - D_{SP} \frac{d(S_g \rho_p c_s)}{dy} \tag{2.46}$$

where $D_{SP}$ is a surface diffusion coefficient, square centimeter per second, and $c_s$ is the surface concentration of the adsorbed diffusing species, gram moles per square centimeter. The product $S_g \rho_p c_s$ is the amount adsorbed expressed as gram moles per cubic centimeter of total volume of the porous mass.

If it is further assumed that the adsorbed layer and gas phase in the pore are in equilibrium, and that the adsorption isotherm is linear ($S_g \rho_p c_s = Kc$), then

$$J = - \left[ \left( \frac{1}{D_{ABP}} + \frac{1}{D_{KP}} \right)^{-1} + K D_{SP} \right] \frac{dc}{dy} \tag{2.47}$$

This suggests that the various solutions to the simple diffusion equation may be employed in approximate analyses of situations where ordinary, Knudsen, and surface diffusion are all involved.

Since helium adsorption on solids at ordinary temperatures is very small, a diffusion test with helium in the Knudsen range will give a value from which the gas-diffusion contribution to the total flux can be estimated $(D_{KP} \propto 1/M)$ and subtracted from the total flux to obtain the flux due to surface diffusion. Employing this technique, Kammermeyer and Rutz [51] find that 50 to 70 percent of the total flux is due to surface diffusion of ethylene, propylene, and propane in porous Vycor glass $(r_e \approx 29 \text{ Å})$ at 25°C. The ratio of surface flux to Knudsen diffusion in Vycor was found in flow experiments to increase as a smooth function of the normal boiling point of the diffusing pure substance. Though helium adsorption is small, a small amount of surface diffusion of helium on Vycor glass has been reported by Kammermeyer and Hwang [50].

Two recent papers by Gilliland, Baddour, Perkinson, and Sladek [38a] provide a basis for the estimation of surface diffusion coefficients. Values of $D_{SP}$ decreased from $1.6 \times 10^{-2}$ to $10^{-13}$ cm$^2$/s as $q/mRT$ ranged from 0 to 55, for various gases and solid surfaces, the decrease being exponential. Here $q$ is the heat of adsorption, and $m$ an integer, 1, 2, or 3, determined by the nature of the adsorption bond. The correlation gives only approximate values of $D_{SP}$ but is useful where an estimate is required. The data on which the study is based were obtained at temperatures below 100°C.

There is presently no way to predict surface-diffusion coefficients in commercial catalysts at the elevated temperatures commonly employed in industry. That surface diffusion may be important in such cases is suggested by the data of Barrer [7], who finds that the surface flux may be as much as 15 to 30 percent of the total in silica-alumina catalysts for ethane and propane at 180°C. If the gas is strongly adsorbed, the adsorbed layer may be thick enough to reduce the gas diffusion appreciably, or even fill the pore. Thus Barrer [7] found SO$_2$ to be so strongly adsorbed by porous carbon that hydrogen could not enter the pores from a gas mixture of SO$_2$ and hydrogen: the carbon served as a semipermeable membrane permitting the transport of SO$_2$ but not hydrogen. This suggests the basis for a possible separation process.

## 2.9 DIFFUSION IN POLYMERS

Diffusion in polymers is of interest to chemical engineers because thin "permselective" polymer membranes may be employed in separation processes, and because processes for the manufacture of polymers often involve diffusion of reactants or products to or from the site of the polymerization reaction. The separation processes include hydrocarbon separation, dialysis, reverse osmosis, blood oxygenators, and artificial kidneys. Diffusion of water and other solvents presents problems in the manufacture of commercial polymers and the spinning of these to produce textile fibers; low water permeability is required in polymer films used to package food. Separation by electrodialysis or by ion-exchange equipment depends on ion diffusion in polymers.

Nonporous polymer membranes permit the passage of most gases and liquids soluble in the polymeric material. The substance diffusing dissolves in the membrane at one face and is desorbed at the other face. The overall potential or driving force is the difference in activity of the solute at the two faces, each essentially equal to the activity of the solute in the ambient media.

Distinction is made between the permeability $\bar{P}$ and the mean coefficient of diffusion in the polymer matrix. The former is the product of the flux density and the membrane thickness, divided by the overall difference of chemical potential between the two external media. The latter is the familiar Fick's-law coefficient $D$ based on the gradient of the solute volumetric concentration in the polymer. The relation between the two is given by the following equation for gas transport across a membrane of thickness $\Delta y$:

$$J_A = \frac{\bar{P}(p_1 - p_2)}{\Delta y} = \frac{\bar{D}(c_1 - c_2)}{\Delta y} = \frac{\bar{D}H'(p_1 - p_2)}{\Delta y} \tag{2.48}$$

where  $J_A$ = solute flux density
$p_1$ and $p_2$ = partial pressures of the solute gas in the upstream and downstream fluid media
$H'$ = Henry's-law solubility coefficient $c/p$
$c_1, c_2$ = volumetric concentrations in the polymer at the two faces

Since $D$ varies considerably with concentration, the symbols $\bar{P}$ and $\bar{D}$ are used to represent averages over the concentration range $c_1$ to $c_2$. The resistance of the membrane is usually large compared with the surface resistances, so equilibrium solubility data may be employed to relate $c_1$ to $p_1$, and $c_2$ to $p_2$. With liquids, $p_1$ and $p_2$ are replaced by the concentrations of the ambient liquid in contact with the two faces.

The physics of diffusion in polymers is exceedingly complicated; it is presently not possible to predict permeabilities from known properties of the polymer and of the diffusing substance. Important variables include the chemical nature of the polymer, its molecular weight distribution, the degree of cross-linking, the nature of plasticizer, if any, the glass transition temperature, and the manner in which it is manufactured and annealed. The properties of the diffusing "solute" or penetrant are evidently relevant—not only its molecular weight but its molecular shape. Of perhaps greatest importance is the interaction of the solute and polymer molecules; the transport is enormously enhanced if solution of the diffusing substance in the polymer causes the latter to swell.

The current status of the theory of solution and diffusion in polymers is described in the excellent book "Diffusion in Polymers," edited by Crank and Park [22], which includes much of the published data on permeabilities for various systems. The shorter review by Rickles [87] describes some of the main features of the theory, with numerous references. The present discussion will be limited to a presentation of typical data.

The complex phenomena can be divided by considering two types of systems: the transport of "gases" which interact little with polymer, and the transfer of vapors and liquids which dissolve appreciably and cause the polymer to swell. The solubility of the more permanent gases generally follows Henry's law, and diffusion coefficients are independent of pressure and solute concentration in the polymer.

Organic compounds which swell the polymer have diffusion coefficients strongly dependent on solute concentration.

Michaels and Bixler [64] report data on both gas solubility and diffusion coefficients of He, $N_2$, CO, $O_2$, A, $CH_4$, $CO_2$, $SF_6$, $C_2H_6$, $C_3H_6$, $C_3H_8$, $C_3H_4$, and $CH_3Cl$ at 25°C in natural rubber and in three different polyethylenes. Apparent heats of solution in polyethylene varied from $+2.4$ kcal/g mole for helium to $-3.5$ kcal/g mole for $C_3H_4$; the solubility of the larger molecules increases with temperature. The gases appear to be insoluble in the polymer crystallites; $H'$ was expressed as $H^*\alpha$, where $\alpha$ is the fraction amorphous polymer. $H^*$ varied widely in the three polyethylenes, being very small for helium but increasing progressively with increasing values of the Lennard-Jones force constant $\varepsilon/k$ for the solute gas. (A similar relation holds for the solubility of gases in nonpolar liquids [49].) Since $\varepsilon/k$ is roughly proportional to the normal boiling temperature, this result parallels the observation of Kammermeyer and Rutz (see Sec. 2.8) that surface diffusion in Vycor glass increases with the normal boiling temperature of the adsorbed gas.

In the same series of gases and polymers, $D$ decreased with increase in molecular complexity, from $31 - 151 \times 10^{-7}$ for helium to $0.016 - 0.56 \times 10^{-7}$ $cm^2/s$ for $SF_6$ in the three polyethylenes at 25°C. Apparent activation energies for diffusion varied from about 6 kcal/g mole for helium to 15 kcal/g mole for $SF_6$. The permeability is represented by the relation

$$\bar{P} = \frac{D^*H^*\alpha}{\tau\beta} \tag{2.49}$$

where $D^*$ is the diffusion coefficient for the gas in the amorphous polymer constituting the fraction $\alpha$ of the membrane, $\tau$ is a "geometric impedance factor" (corresponding to the tortuosity correction factor of Sec. 2.8), and $\beta$ is a "chain immobilization factor" reflecting the cross-linking of the crystallites. The factor $\tau$ increased from 1.0 to 10 as $\alpha$ decreased from 1.0 to about 0.2; $\beta$ evidently varies both with the size of the gas molecule and with the polymer structure.

Stannett, in Chap. 2 of Ref. 22, has collected the known data (through 1966) on diffusion coefficients for simple gases in various polymers. The effect of temperature over a limited range is indicated by tabulated values of $D_0$ and $E_d$ for use in the Arrhenius form

$$D = D_0 \exp \frac{-E_d}{RT} \tag{2.50}$$

Typical values for $D$ at 25°C as quoted by Stannett are shown in Table 2.13.

Brubaker and Kammermeyer [17] report values of the permeability $\bar{D}H'$ for He, $H_2$, $CO_2$, $N_2$, and $O_2$ in membranes of 54 commercial polymers. For several hydrocarbons, Li and Henley [57] found the gas permeability in polyethylene to increase monotonically with the normal boiling point. Li and Long [56] report data on permeation of various liquids and vapor through polyolefin films.

In relation to transport through polymers, the distinction between gases and liquids, suggested above, is by no means sharp. At a given temperature the solubility coefficient $H'$ increases roughly as the boiling point (or critical temperature) of the penetrant. The diffusion coefficient $D$, however, is much greater for the smaller molecules (see Table 2.13). These effects tend to offset, so the permeability $\bar{D}H'$ varies much less than either $\bar{D}$ or $H'$ alone in going from one penetrant to another in a given polymer.

For any one system, as for a hydrocarbon vapor in polyethylene, the solubility coefficient $H'$ is not constant but increases with the concentration of the penetrant, as does $D$. The permeability then increases rapidly with the partial pressure or activity of the vapor. Since both $D$ and $H'$ vary nearly exponentially with concentration, one may write [89]

$$\bar{P} = D_0 H_0' e^{\beta c} \tag{2.51}$$

where $\beta$ is a constant for the system, and $D_0$ and $H_0'$ represent $D$ and $H'$ for $c = 0$. In contrast with the situation with the simple gases, the permeability of a vapor is pressure-dependent because $H'$ varies.

Rogers, Stannett, and Szwarc [89] report data on transport of a variety of hydrocarbons and chlorinated hydrocarbons in polyethylene. The downstream pressure was held near zero, so the variable studied was the upstream vapor activity $a_1$ (the ratio of the partial pressure to the vapor pressure). Over the range of $a_1$ from 0 to 0.8, $\bar{D}$ increased five- to tenfold, while $\bar{P}$ increased considerably more.

Stern, Mullhaupt, and Gareis [94], in a study of the effect of pressure on permeability of vapors in polyethylene, propose an interesting correlation of $H_0'$ with the reduced temperature $T/T_c$, $T_c$ being the critical temperature of the vapor. They also suggest a method of estimating the pressure above which the Henry-law deviation is more than 5 percent ($H'/H_0' > 1.05$).

**Table 2.13 SELECTED VALUES OF DIFFUSION COEFFICIENTS OF SIMPLE GASES IN VARIOUS POLYMERS AT 25°C. (FROM THE LARGE COLLECTION OF DATA IN CHAPTER 2 OF REF. (22); VALUES ARE $D \times 10^6$ ($D$ in cm$^2$/s) AT 25°C AND MODEST PRESSURES**

| | Polymer | He | H$_2$ | O$_2$ | CO$_2$ | CH$_4$ |
|---|---|---|---|---|---|---|
| 1 | Polyethylene terephthalate (glassy crystalline) | 1.7 | $\cdots$ | 0.0036 | 0.00054 | 0.00017 |
| 2 | Polycarbonate (Lexan) | $\cdots$ | 0.64 | 0.021 | 0.0048 | |
| 3 | Polyethylene, density 0.964 | 3.07 | $\cdots$ | 0.170 | 0.124 | 0.057 |
| 4 | Polyethylene, density 0.914 | 6.8 | $\cdots$ | 0.46 | 0.372 | 0.193 |
| 5 | Polystyrene | 10.4 | 4.36 | 0.11 | 0.058 | |
| 6 | Butyl rubber | 5.93 | 1.52 | 0.81 | 0.058 | |
| 7 | Polychloroprene (neoprene) | $\cdots$ | 4.31 | 0.43 | 0.27 | |
| 8 | Natural rubber | 21.6 | 10.2 | 1.58 | 1.10 | 0.89 |
| 9 | Silicone rubber, 10 percent filler (extrapolated) | 53.4 | 67.1 | 17.0 | | |
| 10 | Polypropylene, isotactic | 19.5 | 2.12 | | | |
| 11 | Polypropylene, atactic | 41.6 | 5.7 | | | |
| 12 | Polyethyl methacrylate | 44.1 | $\cdots$ | 0.11 | 0.030 | |
| 13 | Butadiene-acrylonitrile (Perbunan) | 11.7 | 4.5 | 0.43 | 0.19 | |
| 14 | Polybutadiene | $\cdots$ | 9.6 | 1.5 | 1.05 | |
| 15 | Polyvinyl acetate (glassy) | 9.52 | 2.10 | 0.051 | $\cdots$ | 0.0019 |

Various authors [94, 10, 59] have reported the permeability of liquids to be much greater than that for the same penetrant as a vapor. This is hard to understand if the activities at the polymer surface are the same, since the conditions within a polymer membrane should depend only on the activity of the phase in contact with the surface. However, the difference is supported by comparable tests using an azeotrope of methanol and benzene at 59.5°C, first with the liquid and then with the equilibrium vapor [10]. These showed the permeation rate with the liquid to be nearly twice as great as when the upstream face was exposed to the vapor. The pressures $p_1$ and $p_2$ were 760 and 35 mm Hg in both tests. The authors do not describe the polymer.

Perhaps the aspect of diffusion in polymers of greatest interest to chemical engineers is the possibility that membranes might be employed for the separation of the components of a mixture of gases or of liquids. Membranes do show selective permeability, and partial separations by selective permeation have been demonstrated. The engineering analyses which have been made, however, have generally indicated that the process would not be economical because permeabilities are low and enormous membrane areas would be required for commercial purposes. Perhaps the use of hollow polymer fibers could change this. Such fibers have been developed in recent years for use in desalination of brines by reverse osmosis; small hollow fibers can provide upward of 10,000 ft$^2$ of membrane surface in a cubic foot of equipment volume.

Blaisdell and Kammermeyer [13] have recently reported data on oxygen permeability of small tubular membranes of elastic polymers. Tubes of silicone rubber 0.381 mm OD and 0.254 mm ID expanded considerably due to applied internal gas pressure. The increase in surface and the thinning of the wall effected a twentyfold increase in oxygen permeability.

Figure 2.6 illustrates the basic separation process. This shows the feed and raffinate to be a liquid, though a gas might be fed. The extent of the enrichment varies almost linearly with the fraction of the feed gas which permeates the membrane. Weller and Steiner [104] found this to be true for the separation of oxygen from air using ethyl cellulose at 30°C with $p_1 = 1.0$ atm and $p_2$ near zero. As the ratio of permeate to raffinate approached zero, the former was found to contain 39 to 42 mole percent oxygen. Reilly, Henley, and Staffin [85] report several percent propane enrichment of the permeate in attempts to separate ethane and propane using polyethylene. Using an unspecified polymer at 100°C, workers at American Oil Company obtained a permeate containing 75 volume percent heptane from an equal-volume mixture of heptane and isooctane [10]. Brubaker and Kammermeyer [16] found considerable enrichment of the permeate in ammonia, using a feed containing hydrogen, ammonia, and nitrogen, with either polyethylene or Trithene. Using the data as a basis, these authors calculate that it should be possible to obtain 1 mole of permeate containing 64.2 percent $NH_3$ and 25.6 percent $H_2$ from 2 moles of feed containing 45.6 percent $NH_3$ and 26.2 percent $H_2$, with a single stage operating at $p_1 = 6.8$ atm and $p_2 = 1.0$ atm. It is estimated that 36,500 ft$^2$ of polyethylene would be needed to obtain a product rate of 360 ft$^3$ min. The process would appear

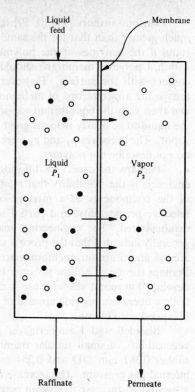

**FIGURE 2.6**
Separation process based on difference in
transport rates through a membrane.

to be more applicable to vapor separations, since the permeabilities may be so much
greater.

Apart from the large membrane area required, it is evident that a large
number of stages would be needed to obtain a relatively pure ammonia product.
The design of such a cascade would be similar to that for isotope separation as
described by Benedict [9]. Equations have been developed for use in predicting
the performance of a single stage with the gases either well mixed or in laminar
flow in the feed and product compartments [17, 104]. Blaisdell and Kammermeyer
[13] treat cocurrent and countercurrent operation of membrane processes for gas
separation. See also Ref. 74a.

Hot palladium is permeable to hydrogen and processes for the purification
of hydrogen by the use of palladium membranes have been developed and used
commercially [61, 24]. Evidently the palladium dissociates the hydrogen to form
hydrogen atoms, and it is the latter which diffuse in the metal.

# REFERENCES

1 AKGERMAN, A., and J. L. GAINER: *Ind. Eng. Chem. Fundam.*, **11:** 373 (1972).
2 ANDERSON, D. K., J. R. HALL, and A. L. BABB: *J. Phys. Chem.*, **62:** 404 (1958).
3 ANDRUSSOW, L.: *Z. Elektrochem.*, **54:** 566 (1950).
4 ARNOLD, J. H.: *Ind. Eng. Chem.*, **22:** 1091 (1930).
5 ASHWORTH, J. C., and R. B. KEEY: *Chem. Eng. Sci.*, **27:** 1383 (1972).
6 ASTARITA, G.: *Ind. Eng. Chem. Fundam.*, **4:** 236 (1965).
7 BARRER, R. M.: In "Advances in Separation Techniques," p. 93, Institute of Chemical Engineers, London, 1965.
8 BEARMAN, R. J.: *J. Chem. Phys.*, **32:** 1308 (1960); *J. Phys. Chem.*, **65:** 1961 (1961).
9 BENEDICT, MANSON: In R. E. Kirk and D. F. Othmer (eds.), "Encyclopedia of Chemical Technology," Interscience Encyclopedia, New York, 1950. [See also *Chem. Eng. Progr.*, **43:** 41 (1947).]
10 BINNING, R. C., R. J. LEE, J. F. JENNINGS, and E. C. MARTIN: *Ind. Eng. Chem.*, **53:** 45 (1961).
11 BIRD, R. B.: In J. W. Hoopes and T. B. Drew (eds.), "Advances in Chemical Engineering," vol. 1, pp. 155–239, Academic, New York, 1956.
12 BIRD, R. B., W. E. STEWART, and E. N. LIGHTFOOT: "Transport Phenomena," Wiley, New York, 1960.
13 BLAISDELL, C. T., and K. KAMMERMEYER: *AIChE J.*, **18:** 1015 (1972); *Chem. Eng. Sci.*, **28:** 1249 (1973).
14 BLANDER, M.: "Molten Salt Chemistry," pp. 590–593, Interscience, New York, 1964.
15 BROMLEY, L. A., and C. R. WILKE: *Ind. Eng. Chem.*, **43:** 1641 (1951).
16 BRUBAKER, D. W., and K. KAMMERMEYER: *Ind. Eng. Chem.*, **46:** 733 (1954).
17 BRUBAKER, D. W., and K. KAMMERMEYER: *Ind. Eng. Chem.*, **45:** 1148 (1953).
18 BYERS, C. H., and C. J. KING: *J. Phys. Chem.*, **70:** 2499 (1966).
19 CALDWELL, C. S., and A. L. BABB: *J. Phys. Chem.*, **60:** 51 (1956).
20 CHANG, S. Y.: S.M. thesis in chemical engineering, M.I.T., 1949.
21 CHEN, N. H., and D. F. OTHMER: *J. Chem. Eng. Data*, **7:** 37 (1962).
22 CRANK, J., and G. S. PARK (eds.): "Diffusion in Polymers," Academic, New York, 1968.
23 CULLINAN, H. T., JR.: *Ind. Eng. Chem. Fundam.*, **5:** 281 (1966).
24 DARLING, A. S.: *Inst. Chem. Eng., Symp. on the Less Common Means of Separation*, 163 (1963).
25 DAVIES, G. A., A. B. PONTER, and K. CRAINS: *Can. J. Chem. Eng.*, **45:** 372 (1967).
26 DENBIGH, K. G.: "The Thermodynamics of the Steady State," Methuen, London, 1951.
27 "Diffusion Data" (Inorganic Solids and Their Melts), Diffusion Information Center, Cleveland, Ohio.
28 DIM, A., and A. B. PONTER: *Chem. Eng. Sci.*, **26:** 1301 (1971).
29 DUDA, J. L., and J. S. VENTRAS: *AIChE J.*, **14:** 286 (1968).
30 DULLIEN, F. A. L.: *Ind. Eng. Chem. Fundam.*, **10:** 41 (1971).
31 DULLIEN, F. A. L.: *AIChE J.*, **18:** 62 (1972).
31a ERTL, H., R. K. GHAI, and F. A. L. DULLIEN, *AIChE J.*, **19:** 881 (1973).
32 FALLAT, R. J.: *Univ. Calif. Lawrence Radiat. Lab., Rep.* UCRL-8527, March 1959.
33 FENG, C., and W. E. STEWART: *Ind. Eng. Chem. Fundam.*, **12:** 143 (1973).
33a FENG, C. F., V. V. KOSTROV, and W. E. STEWART: *Ind. Eng. Chem. Fundam.*, **13:** 5 (1974).
34 FERRELL, R. T., and D. M. HIMMELBLAU: *AIChE J.*, **13:** 702 (1967).
35 FULLER, E. N., P. D. SCHETTLER, and J. C. GIDDINGS: *Ind. Eng. Chem.*, **58:** No. 5, p. 19 (1966).
36 GAINER, J. L., and A. B. METZNER: *AIChE–Inst. Chem. Eng. Joint Meet., London*, "Transport Phenomena," June 13–17, 1965.
37 GAINER, J. L.: *Ind. Eng. Chem. Fundam.*, **9:** 381 (1970).

38   GILLILAND, E. R.: *Ind. Eng. Chem.*, **26**: 681 (1934).

38a  GILLILAND, E. R., R. F. BADDOUR, G. P. PERKINSON, and K. J. SLADEK: *Ind. Eng. Chem. Fundam.*, **13**: 95, 100 (1974).

39   GUBBINS, K. E., K. K. BHATIA, and R. D. WALKER, JR.: *AIChE J.*, **12**: 548 (1966).

40   HABGOOD, H. W., and J. F. HANLAN: *Can. J. Chem.*, **37**: 843 (1959).

41   HALUSKA, J. L., and C. P. COLVER: *Ind. Eng. Chem. Fundam.*, **10**: 610 (1971).

42   HARNED, H. S., and OWEN, B. B.: "The Physical Chemistry of Electrolytic Solutions," ACS Monogr. 95, Reinhold, New York, 1950.

43   HARTLEY, G. S.: *Trans. Faraday Soc.*, **45**: 801 (1949).

44   HAYDUK, W., and W. D. BUCKLEY: *Chem. Eng. Sci.*, **27**: 1997 (1972).

45   HAYDUK, W., and S. C. CHENG: *Chem. Eng. Sci.*, **26**: 635 (1971).

45a  HAYDUK, W., and H. LAUDIE: *AIChE J.*, **20**: 611 (1974).

46   HENLEY, E. J., and J. M. PRAUSNITZ: *AIChE J.*, **8**: 133 (1962).

46a  HILDEBRAND, J. H., and R. H. LAMOREAUX: *Proc. Nat. Acad. Sci.*, **71**: 3321 (1974).

46b  HIMMELBLAU, D. M.: *Chem. Rev.*, **64**: 527 (1964).

47   HIRSCHFELDER, J. O., C. F. CURTISS, and R. B. BIRD: "Molecular Theory of Gases and Liquids," Wiley, New York, 1964.

47a  HISS, T. G., and E. L. CUSSLER: *AIChE J.*, **19**: 698 (1973).

48   HOSHINO, S.: *Inst. Chem. Eng.*, **11**: 353 (1971).

49   JOLLEY, J. E., and J. H. HILDEBRAND: *J. Amer. Chem. Soc.*, **80**: 1050 (1968).

50   KAMMERMEYER, K., and S. HWANG: *Can. J. Chem. Eng.*, **44**: 82 (1966).

51   KAMMERMEYER, K., and L. O. RUTZ: *Chem. Eng. Progr. Symp. Ser.* **55**(24): 163 (1959).

52   KENNARD, E. H.: "Kinetic Theory of Gases," McGraw-Hill, New York, 1938.

53   KING, C. J., L. HSUEH, and C.-W. MAO: *J. Chem. Eng. Data*, **10**: 348 (1965).

54   LEFFLER, R. A. J.: *J. Catal.*, **5**: 22 (1966).

55   LEFFLER, J., and H. T. CULLINAN, JR.: *Ind. Eng. Chem. Fundam.*, **9**: 84 (1970).

56   LI, N. N., and R. B. LONG: *AIChE J.*, **15**: 73 (1969).

57   LI, N. N., and E. J. HENLEY: *AIChE J.*, **10**: 666 (1964).

58   LI, S. U., and J. L. GAINER: *Ind. Eng. Chem. Fundam.*, **7**: 433 (1968).

59   LONG, R. B., *AIChE J.*, **4**: 445 (1965).

60   LUSIS, M. A., and G. A. RATCLIFF: *AIChE J.*, **17**: 1492 (1971).

61   MCBRIDE, R. B., and D. L. MCKINLEY: *Chem. Eng. Progr.*, **61**: No. 3, 81 (1965).

62   MARANGOZIS, J.: Ph.D. thesis in chemical engineering., Univ. of Toronto, 1961.

63   MARRERO, T. R., and E. A. MASON: *J. Phys. Chem. Ref. Data*, **1**: 3–118 (1972).

64   MICHAELS, A. S., and H. J. BIXLER: *J. Polymer. Sci.*, **50**: 393, 413 (1961).

65   MILLER, L., and P. C. CARMAN: *Trans. Faraday Soc.*, **57**: 2143 (1961).

66   MOELWYN-HUGHES, E. A.: "Physical Chemistry," Pergamon, New York, 1957.

67   NAIN, V. P. S., and J. R. FERRON: *Ind. Eng. Chem. Fundam.*, **10**: 610 (1971).

68   NAVARI, R. M., J. L. GAINER, and K. R. HALL: *AIChE J.*, **17**: 1028 (1971).

69   NEWMAN, J. S., D. BENNION, and C. W. TOBIAS: *Ber. Phys. Chem.*, **69**: 608 (1965).

70   NEWMAN, J. S.: In C. W. Tobias (ed.), "Advances in Electrochemistry and Electrochemical Engineering," vol. 5, Interscience, Wiley, New York, 1967.

71   O'CONNELL, J. P., and J. M. PRAUSNITZ: In "Advances in Thermo-Physical Properties at Extreme Temperatures and Pressures," p. 19, ASME, New York, 1969.

72   OSMERS, H. R., and A. B. METZNER: *Ind. Eng. Chem. Fundam.*, **11**: 161 (1972).

73   OTHMER, D. F., and M. S. THAKAR: *Ind. Eng. Chem.*, **45**: 589 (1953).

74   OTHMER, D. F., and H. T. CHEN: *Ind. Eng. Chem. Process Des. Dev.*, **1**: 249 (1962).

74a  PAN, C. Y., and H. W. HABGOOD: *Ind. Eng. Chem. Fundam.*, **13**: 323 (1974).

75   PAUL, D. R.: *Ind. Eng. Chem. Fundam.*, **6**: 217 (1967).

76   PEACEMAN, D. W.: Sc.D. thesis in chemical engineering, M.I.T., 1951.
77   POLLARD, W. G., and R. D. PRESENT: *Phys. Rev.*, **73**: 762 (1948).
78   POLSON, A.: *J. Phys. Colloid Chem.*, **54**: 649 (1950).
79   POWERS, J. E., and C. R. WILKE: *AIChE J.*, **3**: 213 (1957).
80   RAO, S. S., and C. O. BENNETT: *AIChE J.*, **17**: 75 (1971).
81   RATCLIFF, G. A., and M. A. LUSIS: *Ind. Eng. Chem. Fundam.*, **10**: 474 (1971).
82   RATCLIFF, G. A., and J. G. HOLDCROFT: *Trans. Inst. Chem. Eng.*, **41**: 315 (1963).
83   REDDY, K. A., and L. K. DORAISWAMY: *Ind. Eng. Chem. Fundam.*, **6**: 77 (1967).
84   REID, R. C., and T. K. SHERWOOD: "Properties of Gases and Liquids," 2d ed., McGraw-Hill, New York, 1966.
85   REILLY, G. W., E. J. HENLEY, and H. K. STAFFIN: *AIChE J.*, **16**: 353 (1970).
86   REMICK, R. R., and C. J. GEANKOPLIS: *Ind. Eng. Chem. Fundam.*, **12**: 214 (1973).
87   RICKLES, R. N.: *Ind. Eng. Chem.*, **58**(6): 18 (1966).
88   ROBINSON, R. A., and R. H. STOKES: "Electrolyte Solutions," 2d ed., Academic, New York, 1959.
89   ROGERS, C. E., V. STANNETT, and M. SZWARC: *J. Polymer Sci.*, **45**: 61 (1960).
90   ST. DENIS, C. E., and C. J. D. FELL: *Can. J. Chem. Eng.*, **49**: 885 (1971).
91   SATTERFIELD, C. N.: "Mass Transfer in Heterogeneous Catalysis," M.I.T. Press, Cambridge, Mass., 1970.
92   SCOTT, D. S.: *Ind. Eng. Chem. Fundam.*, **3**: 278 (1964).
93   SLATTERY, J. C., and R. B. BIRD: *AIChE J.*, **4**: 137 (1958).
94   STERN, S. A., J. T. MULLHAUPT, and P. J. GAREIS: *AIChE J.*, **15**: 64 (1969).
95   SVEHLA, R. A.: *NASA Tech. Rep.* R-132, Lewis Research Center, Cleveland, Ohio, 1962.
96   TOLLERT, H.: *Naturewiss.*, **41**: 277 (1954).
97   TUWINER, S. B.: "Diffusion and Membrane Technology," ACS Monogr. 156, Reinhold, New York, 1962.
97a  TYN, M. T.: *AIChE J.*, **20**: 1038 (1974).
98   VAN GEET, A. L., and A. W. ADAMSON: *J. Phys. Chem.*, **68**: 238 (1964).
99   VIGNES, A.: *Ind. Eng. Chem. Fundam.*, **5**: 189 (1966).
100  VINOGRAD, J. R., and J. W. MCBAIN: *J. Amer. Chem. Soc.*, **63**: 2008 (1941).
101  VIVIAN, J. E., and C. J. KING: *AIChE J.*, **10**: 220 (1964).
102  WAY, P. F.: Sc.D. thesis in chemical engineering, M.I.T., 1971.
103  WEISZ, P. B.: *Z. Phys. Chem.*, **11**: 1 (1957). (In English.)
104  WELLER, S., and W. A. STEINER: *Chem. Eng. Progr.*, **46**: 585 (1950).
105  WICKE, E., and R. KALLANBACH: *Kolloid Z.*, **97**: 135 (1941).
106  WILKE, C. R.: *Chem. Eng. Progr.*, **45**: 218 (1949).
107  WILKE, C. R., and P. C. CHANG: *AIChE J.*, **1**: 264 (1955).
108  WILKE, C. R., and C. Y. LEE: *Ind. Eng. Chem.*, **47**: 1253 (1955).
109  WISE, D. L., and G. HOUGHTON: *Chem. Eng. Sci.*, **23**: 1211 (1968).
110  WISE, D. L., and G. HOUGHTON: *Chem. Eng. Sci.*, **21**: 999 (1966).
111  WITHERSPOON, P. A., and L. BONOLI: *Ind. Eng. Chem. Fundam.*, **8**: 589 (1969).
112  ZANDI, I., and C. D. TURNER: *Chem. Eng. Sci.*, **25**: 517 (1970).

## PROBLEMS

*2.1*   Diffusion coefficients in many binary gas systems are conveniently measured by the use of a Stefan tube (see Fig. 3.12). This is a small glass tube of uniform cross section, open at one end. The vertical tube is partially filled with constituent $A$, a liquid at the test temperature. Constituent $B$, a gas, passes slowly over the open end of the tube. The vapor of $A$ diffuses out of the tube and the drop in liquid level is measured as a function of time. The tube is held at constant temperature.

A student plans to use this method to check the reported values of $D_{AB}$ for benzene in nitrogen at 26.1°C, at which temperature the vapor pressure of benzene is 100 mm Hg. The error in reading the liquid level with a cathetometer is such that the level should fall at least 1.0 cm during the experiment. The student would like to make his measurements of initial and final levels in a 24-h time period. To what initial levels should he fill the tube with liquid benzene? If the student measures the liquid level at a number of different times, how should he analyze his data?

Neglect accumulation of benzene vapor in the gas in the tube. The pressure is 1.0 atm, the nitrogen is pure, and the specific gravity of the liquid benzene is 0.8272 g/cm³.

2.2   A "diaphragm cell" is used to measure the diffusion coefficient of benzoic acid in water at 25°C. The cell consists of a cylindrical glass vessel 3 cm in diameter, divided into the two compartments by a porous glass disk 2 mm thick. The aqueous solution fills the lower compartment $A$ (76 cm³) and an equal volume of water is placed in the upper compartment $B$. Solute diffuses through the porous disk, the transport tending to equalize the concentrations in the two compartments. Measurements of the solute concentrations (g moles/l) at the start and after a period of some hours are given below. Similar tests were made with KCl and with benzoic acid in the same apparatus.

| | $t = 0$ | | Final | | |
|---|---|---|---|---|---|
| | $c_A$ | $c_B$ | $c_A$ | $c_B$ | $t$, hours |
| KCl | 0.1000 | 0 | 0.0912 | 0.0092 | 37 |
| Benzoic acid | 0.01250 | 0 | 0.01210 | 0.00068 | 31 |

The test with KCl was made to standardize the cell, using the well-established data on $D_{AB}$ for KCl–water. Concentrations changed slightly during each test, and since $D_{AB}$ varies somewhat with concentration, the derived values are "integral" or averages of the mutual diffusion coefficients. Ignoring this complication, determine $D_{AB}$ for benzoic acid in water under the experimental conditions. At 25°C, $D_{AB}$ for KCl in 0.05N aqueous solution is $1.864 \times 10^{-5}$ cm²/s.

2.3   Vignes' equation [Eq. (2.35)] relates the mutual diffusion coefficient to the composition of a binary liquid mixture.

(a)   Show that where this holds, the logarithm of the activity-corrected $D_{AB}$ must be linear in mole fraction $A$.

(b)   The following data are reported by Caldwell and Babb [19] for $D_{AB}$ at 25.3°C in the system benzene $(A)$–carbon tetrachloride $(B)$:

| $X_A$ | 0.02154 | 0.2502 | 0.5051 | 0.7498 | 0.9815 |
|---|---|---|---|---|---|
| $10^5 D_{AB}$ | 1.419 | 1.519 | 1.651 | 1.759 | 1.912 |

The system is nearly ideal: $\partial \ln a_A / \partial \ln X_A$ can be taken to be unity.

Prepare a graph using these data to test Vignes' equation. For this ideal mixture, does the correlation have any advantage over a graph of $D_{AB}$ vs. $X_A$, for example, for accurate interpolation and extrapolation?

2.4 The possibility of separating $N_2$ and He by selective permeation through a polymer membrane is being considered. The membrane will be cellulose acetate–butyrate 1 mill thick, operated at 21°C, with upstream and downstream pressure of 10 and 1.0 atm, respectively. The permeabilities to the two gases are reported to be $0.45 \times 10^{-9}$ for He and $0.07 \times 10^{-9}$ for $N_2$, the units being $(cm^3$ at 0°C, 1.0 atm$)$ $(cm)/(s)(cm^2)(cm$ Hg$)$. In order to get some idea of the separation possible, it is proposed to analyze the steady-state operation of a single stage in which the gases in both upstream and downstream compartments are well mixed, and in which half the feed is withdrawn as permeate (refer to Fig. 2.6). The feed gas will be 30 percent He, 70 percent $N_2$.

(a) Assuming permeabilities independent of pressure and the diffusion of each gas to be unaffected by the presence of the other, calculate the composition of the permeate.

(b) How many square feet of membrane will be required to obtain 1 lb mole/h of the helium-enriched permeate?

2.5 The diffusion cell described in Prob. 2.2 is used to measure $D_{AB}$ for HCl in water at 25°C; with 0.1N HCl on one side of the diaphragm and pure water on the other $D_{AB}$ is found to be $3.1 \times 10^{-5}$ cm²/s. From the test data it is calculated that the diffusion resistance of the porous disk is the same as that of a stagnant water layer 0.1 cm thick.

The experiment is to be repeated with 0.1N HCl on one side but with 0.1N NaCl in the solutions on both sides of the diaphragm. Flux will be determined by measuring the change of $H^+$ concentration with time. Estimate the value of $D_{AB}$ for HCl which might be expected under these conditions.

Only an approximate answer is desired: the solutions can be considered ideal and the ion gradients taken as linear with distance through the porous disk.

Is there a diffusion flux of sodium ion through the diaphragm?

2.6 Component $A$ diffuses unidirectionally under steady-state conditions in a binary mixture of ideal gases.

(a) By what fraction or percent is the flux $J$ (relative to the plane of no net molal flux) changed if the total pressure is doubled and (i) the local *concentration gradient* is held constant, (ii) the local gradient of *mole fraction A* is held constant, and (iii) the local gradient of the *partial pressure A* is held constant? The temperature is not changed.

(b) By what percent will the flux $J$ change if the local gradient of mole fraction $A$ and the total pressure are held constant but the temperature is increased from 25 to 250°C? The force constant $\varepsilon_{AB}/k$ for the binary is 250°K.

# 3
# RATE EQUATIONS FOR MOLECULAR DIFFUSION

## 3.0  SCOPE

The first part of the chapter presents equations expressing the rate of steady-state molecular diffusion in binary gas and liquid systems. The more general relations involving transient diffusion are then discussed, and two of the most important cases of transient diffusion in stagnant media presented. Finally, methods of analysis of several technically important examples of diffusion in laminar flow are described.

## 3.1  PRINCIPLE SYMBOLS USED IN CHAPTER 3

| | |
|---|---|
| $c$ | Concentration, g moles/cm$^3$ |
| $c_{av}$ | Average concentration, g moles/cm$^3$ |
| $c_i$ | Concentration after step change; concentration at interface, g moles/cm$^3$ |
| $c_0$ | Initial concentration, g moles/cm$^3$ |
| $c_T$ | Total concentration or molal density of mixture, g moles/cm$^3$ |
| $D_{AB}$ | Binary diffusion coefficient, cm$^2$/s |
| $D_{Am}$ | Effective diffusion coefficient, $A$ in a mixture, cm$^2$/s |

| | |
|---|---|
| $g$ | Local acceleration due to gravity, cm/s$^2$ |
| $h$ | Half-width of two-dimensional channel, cm |
| $J_A$ | Flux density of $A$ relative to plane of no net volume transport, g moles/(s) (cm$^2$) |
| $L$ | Length, cm |
| $N_A$ | Flux density of $A$ relative to apparatus, g moles/(s) (cm$^2$) |
| Re | Reynolds number, dimensionless |
| Sc | Schmidt number, $v/D_{AB}$ |
| $p$ | Partial pressure, atm |
| $p_{BM}$ | Logarithmic mean of $p_{B1}$ and $p_{B2}$, atm |
| $P$ | Total pressure, atm |
| $Q_A, Q_B$ | Rates of formation of species, $A$, $B$, g moles/(s) (cm$^3$) |
| $Q$ | Point source flow, g moles/s |
| $r$ | Radial distance, cm |
| $r_0$ | Radius of sphere, cm |
| $R$ | Gas constant, (cm$^3$) (atm)/(g mole) ($^\circ$K); also $= y/h$ |
| $S$ | Surface area, cm$^2$ |
| $t$ | Time, s |
| $T$ | Temperature, $^\circ$K |
| $U$ | Velocity, cm/s |
| $U_M$ | Velocity of plane of no net molal transport, cm/s |
| $U_y$ | Velocity in $y$ direction, cm/s |
| $V_{yw}$ | Rate of withdrawal through channel wall, cm/s |
| $V_A$ | Molal volume of $A$, cm$^3$/g mole |
| $\bar{V}$ | Partial molal volume, cm$^3$/g mole |
| $y$ | Distance in direction of diffusion, cm |
| $y_0$ | Layer or "film" thickness, $y_2 - y_1$, cm |
| $X, Y$ | Mole fraction, in liquid and gas, respectively |
| $\beta$ | $\pi^2 D_{AB} t/4 y_0^2$ |
| $\beta'$ | $D_{AB} t/y_0^2$ |
| $\beta''$ | $D_{AB} x/4 r_0^2 U_{av}$ |
| $\Gamma$ | Flow rate, g/(s) (cm) |
| $\Delta$ | $(c_i - c)/(c_i - c_0)$ = ratio of unaccomplished to total possible concentration change |
| $\Delta'$ | $(c_i - c_{av})/(c_i - c_0)$ |
| $\theta$ | Angle, rad |
| $\mu$ | Viscosity, poises |
| $v$ | Kinematic viscosity, $\mu/\rho$, cm$^2$/s |
| $\omega$ | Rotational speed, rad/s |
| $\rho$ | Density, g/cm$^3$ |
| $\rho_L, \rho_M$ | Molal density of liquid, g moles/cm$^3$ |
| erf($z$) | Error function, $= \dfrac{2}{\sqrt{\pi}} \displaystyle\int_0^z e^{-z^2}\, dz$ |

## Subscripts

| | |
|---|---|
| $A, B, C, n, i, j$ | Chemical species |
| 1, 2 | Distances or positions in direction of diffusion |
| $m$ | Mixture |

## 3.2 INTRODUCTION

Where diffusion occurs within a single phase, the quantities of principal interest are the diffusion fluxes and the concentrations of diffusing species at various points. These are functions of time, the geometry of the system, its velocity field, temperature, pressure, various initial and boundary conditions, and the nature of the diffusing species. To the extent that it is possible to do so, it is desirable to develop equations relating these variables in order to facilitate solutions to many practical problems.

Fick's law is expressed by the simple differential equation

$$J_A = -D_{Am}\frac{\partial c_A}{\partial x} \tag{3.1}$$

where the mutual diffusion coefficient $D_{Am}$ for species $A$ in a mixture is defined as the ratio of the unidirectional molal flux density $J_A$ to the negative of the gradient of the molal concentration in the direction of diffusion. To complete the definition of $D_{Am}$, it is specified that the flux density $J_A$ must be referred to a plane across which there is no net volume transport. This plane moves with respect to the fixed apparatus, even though the medium in which diffusion occurs is completely stagnant.

For practical purposes, it is clearly more desirable to obtain expressions for the flux relative to the apparatus than that relative to the plane of no net volume transport. The former will be represented by the symbol $N$ in order to avoid confusion with the latter, for which $J$ has been used. It is first necessary to relate the two.

Only binary mixtures of the two species $A$ and $B$ will be considered in this and the next section; diffusion in multicomponent mixtures will be taken up in Sec. 3.4. Furthermore, most of the equations to be developed will deal primarily with cases of diffusion in the $y$-direction only. These are simpler to understand, and extensions to three coordinates are not difficult in principle, though the mathematics may become complicated.

The velocity, in the $y$-direction, of the plane of no net volume transport, $\overline{U}_y$, is given by

$$\overline{U}_y = \sum_i N_i \overline{V}_i \tag{3.2}$$

For a binary mixture of $A$ and $B$,

$$\overline{U}_y = N_A \overline{V}_A + N_B \overline{V}_B \tag{3.3}$$

where $\overline{U}_y$, $N_A$, and $N_B$ are defined with reference to the fixed apparatus; $\overline{V}_A$ and $\overline{V}_B$ are the partial molal volumes of $A$ and $B$. Then from the definitions of $N$ and $J$,

$$N_A = \overline{U}_y c_A + J_A \tag{3.4}$$

$$N_A - c_A(N_A \overline{V}_A + N_B \overline{V}_B) = J_A = -D_{AB}\frac{dc_A}{dy} \tag{3.5}$$

The fluxes $N_A$ and $N_B$ are frequently of opposite sign, and the second term on the left is often negligible. In this case,

$$N_A = J_A = -D_{AB} \frac{dc_A}{dy} \tag{3.6}$$

This holds rigorously when the plane of no net volume transport is stationary with respect to the apparatus.

In many situations, as in isothermal gases at constant low pressures, the molal volumes are equal and the left side of Eq. (3.5) reduces to $N_A[1 - c_A V_A(1 + N_B/N_A)]$. For ideal gases the volume fraction $c_A V_A$ is equal to the mole fraction $Y_A$. Still further simplification is noted where $N_A$ and $N_B$ are equal but opposite in sign, in which case Eq. (3.5) reduces exactly to Eq. (3.6).

## 3.3  STEADY-STATE MOLECULAR DIFFUSION

**Slab**  Equations for steady-state diffusion of only one of the two species in binary mixtures of ideal gases are readily obtained by integration of Eq. (3.5), which reduces to

$$N_A(1 - Y_A) = -D_{AB} \frac{dc_A}{dy} = -\frac{D_{AB} P}{RT} \frac{dY_A}{dy} \tag{3.7}$$

with $V_A = V_B = RT/P$, $c_A = PY_A/RT$, and $N_B = 0$. Integration for the case of a slab of thickness $y_0$ gives

$$N_A = \frac{D_{AB}P}{RTy_0} \ln \frac{Y_{B2}}{Y_{B1}} = \frac{D_{AB}P}{RTy_0} \frac{(Y_{A1} - Y_{A2})}{Y_{BM}} = \frac{D_{AB}P}{RTy_0} \frac{(p_{A1} - p_{A2})}{p_{BM}} \tag{3.8}$$

where  $N_A$ = flux density from face 1 to face 2
  $Y_{A1}$ and $Y_{A2}$ = mole fractions
  $p_{A1}, p_{A2}$ = partial pressures at the two faces
  $Y_{BM}$ and $p_{BM}$ = logarithmic means of $Y_{B1}, Y_{B2}$ and $p_{B1}, p_{B2}$, respectively

Figure 3.1 illustrates the nature of the variation of the mole fraction $Y_A$ across the slab. Note that the curves are not linear, since $dY_A/dx$ is proportional to $1 - Y_A$. The gas $B$ is not diffusing, yet there is a large gradient of $Y_B$ maintained by the diffusion of $A$. Stevenson [57] gives the derivation of this case for three-dimensional geometry.

If both $A$ and $B$ are diffusing at constant fluxes $N_A$ and $N_B$ and at constant pressure and temperature, integration of Eq. (3.5) yields

$$N_A + N_B = \frac{D_{AB}P}{RTy_0} \ln \frac{(1 - Y_{A2})N_A - Y_{A2} N_B}{(1 - Y_{A1})N_A - Y_{A1} N_B} \tag{3.9}$$

In the case of equal-molal counterdiffusion ($N_A = -N_B$), Eq. (3.9) is indeterminate, but integration of Eq. (3.5) gives

$$N_A = -N_B = \frac{D_{AB}P}{RTy_0}(Y_{A1} - Y_{A2}) \tag{3.10}$$

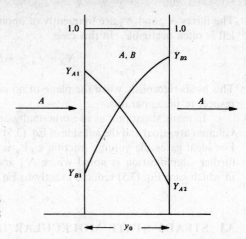

**FIGURE 3.1**
Steady-state diffusion in a binary gas slab
of thickness $y_0$; only gas $A$ diffusing
$(N_B = 0)$.

Integration of Eq. (3.5) to obtain Eqs. (3.8), (3.9), and (3.10) involves the assumption that $D_{AB}$ is independent of concentration. This is acceptable for dilute-gas systems, but in liquid mixtures the diffusion coefficient may vary substantially over the concentration range of interest.

In the special case of a binary mixture and slab geometry, integration of Eq. (3.5) leads to the form

$$y_0 = -\int_1^2 \frac{D_{AB}\, dc_A}{N_A(1 - c_A \overline{V}_A) - N_B c_A \overline{V}_B} \qquad (3.11)$$

Integration is possible if $D_{AB}$, $\overline{V}_A$, and $\overline{V}_B$ can be expressed as a function of $c_A$. An approximate result is obtained by using constant (average) values of $\overline{V}_A$ and $\overline{V}_B$ and expressing the variation of $D_{AB}$ with $c_A$ by the methods of Sec. 2.6. In most liquid systems the convection term can be neglected and Eq. (3.6) employed.

**Annulus**   Equations (3.8), (3.9), and (3.10) apply to steady-state diffusion in a binary gas mixture in the space between infinitely long concentric cylinders of radii $r_1$ and $r_2$ provided $y_0$ is replaced by $r_2 - r_1$; the flux densities $N_A$ and $N_B$ are then per unit area of a cylinder having a radius $r_M$ equal to the logarithmic mean of $r_1$ and $r_2$. The products of the flux densities and the radii are constant, so the fluxes at $r_1$ and $r_2$ are easily obtained.

**Sphere**   An equation for steady-state diffusion from the surface of a sphere into an infinite binary gas mixture is readily obtained for the case of equal-molal counter-diffusion. The flux at the sphere surface is equated to that at radius $r$. Integration between the limits $r = r_0$, $Y_A = Y_{A1}$, and $r = \infty$, $Y_A = Y_{A\infty}$, gives

$$(N_A)_{r=r_0} = \frac{D_{AB} P}{RT r_0}(Y_{A1} - Y_{A\infty}) \qquad (3.12)$$

This is the same as Eq. (3.10) for the slab, with $y_0$ replaced by $r_0$. If only $A$ diffuses $(N_B = 0)$ the result is the same as Eq. (3.8), again with $y_0$ replaced by $r_0$.

There is no steady-state solution for diffusion into an infinite medium from a cylinder (radially) or from a flat surface (normal to the surface).

**Other shapes** Equations for steady-state heat conduction in solids of various geometries are widely available in the literature. These may be employed for equal-molal counterdiffusion in constant-pressure binary gas systems if the thermal diffusivity $k/\rho C_p$ is replaced by $D_{AB}$. They serve as approximate solutions of Eq. (3.5) when the second term on the left is small compared with $N_A$, as, for example, the case of diffusion at low solute concentrations $(Y_A \ll 1)$.

Where diffusion and flow are to be related to cartesian coordinates, Eq. (3.6) takes the form

$$\vec{U}_m \nabla Y_A = D_{AB} \nabla^2 Y_A \tag{3.13}$$

This is for the steady-state diffusion of species $A$ in a binary mixture of ideal gases at constant temperature and pressure. If $A$ is supplied to a space of arbitrary shape and removed at the same rate, and if the boundary of the space is impermeable to $B$, then the flux of $B$ is everywhere zero. In this case a combination of the three-dimensional equivalents of Eqs. (3.3) and (3.5) gives

$$\vec{U}_m = \frac{-D_{AB} \nabla Y_A}{(1 - Y_A)} \tag{3.14}$$

This may be substituted in the preceding expression to obtain

$$\nabla Y_A \cdot \nabla Y_A + (1 - Y_A) \nabla^2 Y_A = 0 \tag{3.15}$$

Solution of this nonlinear second-order equation to obtain $Y_A$ as a function of the space coordinates is difficult. Stevenson [57], however, has shown how Eq. (3.15) can be reduced to the more readily solved Laplace equation by a simple substitution. Since $Y_A = 1 - Y_B$, Eq. (3.15) may be written in the form

$$\left(\frac{\partial Y_B}{\partial x}\right)^2 + \left(\frac{\partial Y_B}{\partial y}\right)^2 + \left(\frac{\partial Y_B}{\partial z}\right)^2 = Y_B\left(\frac{\partial^2 Y_B}{\partial x^2} + \frac{\partial^2 Y_B}{\partial y^2} + \frac{\partial^2 Y_B}{\partial z^2}\right) \tag{3.16}$$

Letting $Y_B = e^{-\alpha}$, this reduces to

$$\frac{\partial^2 \alpha}{\partial x^2} + \frac{\partial^2 \alpha}{\partial y^2} + \frac{\partial^2 \alpha}{\partial z^2} = \nabla^2 \alpha = 0 \tag{3.17}$$

Boundary conditions in terms of $\alpha$ are obtained from the known boundary values of $Y_B(= 1 - Y_A)$ through the relation $\alpha = -\ln Y_B$.

**EXAMPLE 3.1  DIFFUSION AND REACTION IN A GAS FILM**  A stagnant gas layer ("film") of thickness $y_0$ contains the three gases $A$, $B$, and $C$, the inert gas $C$ being present in large excess. Substance $A$ reacts very rapidly to produce 2 moles of $B$, the equilibrium given by $Y_A = k_e Y_B^2$ being maintained everywhere. At one face of the gas layer the composition is held constant at $Y_{A1} + Y_{B1} = Y_1$ and at the other face constant at $Y_2 = Y_{A2} + Y_{B2}$. Derive the equations describing the flux of $A$ and $B$ and the concentrations in the gas layer. (Note that this case corresponds to the application of the "film model" to the absorption of gaseous nitrogen oxides by a liquid absorbent.)

SOLUTION   Since $Y_A$ and $Y_B$ are small in comparison with $Y_C$, it will be assumed that the molal fluxes are adequately expressed by

$$N_A = -D_{AC}\frac{dc_A}{dy} = -k_A\frac{dY_A}{dy} \qquad (a)$$

and

$$N_B = -D_{BC}\frac{dc_B}{dy} = -k_B\frac{dY_B}{dy} \qquad (b)$$

where $k_A = D_{AC}P/RT$ and $k_B = D_{BC}P/RT$.

By a material balance on a differential gas layer of thickness $dy$,

$$dN_B = -2dN_A$$

whence

$$-k_B\frac{d^2Y_B}{dy^2} = 2k_A\frac{d^2Y_A}{dy^2}$$

Since $A$ and $B$ are in equilibrium at every $y$,

$$Y_A = k_e Y_B^2$$

$$\frac{d^2Y_A}{dy^2} = 2k_e\left[Y_B\frac{d^2Y_B}{dy^2} + \left(\frac{dY_B}{dy}\right)^2\right] \qquad (c)$$

and

$$\left(Y_B + \frac{k_B}{4k_e k_A}\right)\frac{d^2Y_B}{dy^2} + \left(\frac{dY_B}{dy}\right)^2 = 0 \qquad (d)$$

The solution is

$$\frac{k_B}{4k_e k_A}Y_B + \frac{1}{2}Y_B^2 = ay + b \qquad (e)$$

and

$$\frac{dY_B}{dy} = a\left(Y_B + \frac{k_B}{4k_e k_A}\right)^{-1} \qquad (f)$$

Since $Y_1 = Y_{A1} + Y_{B1}$ and $Y_2 = Y_{A2} + Y_{B2}$ are stipulated constants, the values of $Y_{B1}$ and $Y_{B2}$ may be found from the equilibrium expression. Substituting these limits,

$$b = \frac{k_B}{4k_e k_A}Y_{B1} + \frac{1}{2}Y_{B1}^2$$

and

$$a = \frac{1}{y_0}\left[\frac{k_B}{4k_e k_A}(Y_{B2} - Y_{B1}) + \frac{1}{2}(Y_{B2}^2 - Y_{B1}^2)\right]$$

Eq. (e) expresses $Y_B$ as a function of the distance $y$ through the gas layer. The value of $Y_A$ corresponding to any $Y_B$ is obtained from the equilibrium relation, and the fluxes from Eqs. (a), (b), and (f). Note that the fluxes are related by

$$N_B = \frac{k_B}{2k_e k_A Y_B}N_A = \frac{k_B}{2k_A\sqrt{k_e Y_A}}N_A \qquad (g)$$

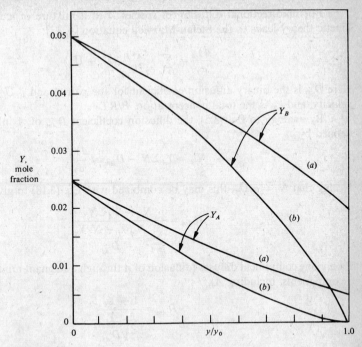

**FIGURE 3.2**
Diffusion of two reacting gases. Case $a$: $Y_B = 0.02$ at $y/y_0 = 1$; case $b$: $Y_B = 0$ at $y/y_0 = 1$. The curves are for $Y_1 = Y_{A1} + Y_{B1} = 0.075$, $k_A = k_B$, and $k_e = 10$.

Figure 3.2 illustrates the nature of the variation of gas concentrations $Y_A$ and $Y_B$ through the gas layer. Cases $a$ and $b$ represent conditions for two different values of $Y_2$: 0.024, and zero. Note that case $b$, in which both $Y_{A2}$ and $Y_{B2}$ are held at zero, indicates $dY_A/dy$ to be zero at $y = y_0$. Does this mean that no $A$ could be absorbed by a reacting liquid in contact with the gas layer?                                                                       ////

## 3.4  DIFFUSION IN MULTICOMPONENT MIXTURES

**Gases**  Diffusion coefficients in ideal-gas binary systems have been shown to be *properties* of the gas pair. By contrast, the effective diffusion coefficient $D_{Am}$ of the species $A$ in a multicomponent mixture depends not only on the nature and concentrations but on the fluxes of the other species present. It is possible, therefore, to have a flux of $A$ in the $y$ direction when the gradient $dc_A/dy$ is zero or even positive. Conversely, the flux may be zero or negative when $dc_A/dy$ is negative. The coefficient $D_{Am}$ can, however, be related to the fluxes of the several components and the binary coefficients $D_{ij}$, where the subscripts $i$ and $j$ refer to the different species in the mixture.

For unidirectional diffusion of species $A$ in a mixture of $n$ ideal gases, the kinetic theory leads to the Stefan-Maxwell equation [4, 17]:

$$\frac{dY_A}{dy} = \sum_{j \neq A}^{n} \frac{c_A c_j}{c_T^2 D_{Aj}} (\overline{U}_j - \overline{U}_A) \qquad (3.18)$$

Here $D_{Aj}$ is the binary diffusion coefficient for the pair $A$ and $j$, $\overline{U}$ is the diffusion velocity, and $c_T$ is the total concentration, $P/RT$.

By analogy to Eq. (3.5), the diffusion coefficient $D_{Am}$ of $A$ in the mixture is defined by

$$N_A = Y_A \Sigma N - D_{Am} c_T \frac{dY_A}{dy} \qquad (3.19)$$

Noting that $N_j = c_j \overline{U}_j$, this may be combined with Eq. (3.18) to give

$$D_{Am} = \frac{N_A - Y_A \Sigma N}{\displaystyle\sum_{j=1}^{n} \frac{N_A Y_j - N_j Y_A}{D_{Aj}}} \qquad (3.20)$$

If only one component diffuses (diffusion of $A$ through a stagnant mixture containing $n$ components, including $A$),

$$D_{Am} = \frac{1 - Y_A}{\displaystyle\sum_{j=B}^{n} \frac{Y_j}{D_{Aj}}} \qquad (3.21)$$

In the general case, with all of the $n$ components diffusing, there are $n - 1$ independent equations similar to Eq. (3.18). The solution of this set, which is difficult, has been the subject of much study, especially by H. L. Toor and his associates [61, 62, 55, 9, 63, 64, 22, 16]. Several approximate solutions are available which simplify the application of the rigorous equations with modest loss in accuracy.

The principal situations of practical interest are those in which (1) there is equal-molal counterdiffusion across a film ($\Sigma N = 0$), and (2) one of the gases does not diffuse. These are treated by Toor [61] for ternary mixtures; it is sufficient here to illustrate the problem by discussing only item (2) for the ternary system.

The ternary mixture of $A$, $B$, and $C$ is contained at $P$ and $T$ in a "slab" of thickness $y_0$. Gases $A$ and $B$ enter and leave at the two faces, but $C$ does not ($C$ is the "inert gas"). The mole fractions are held constant at the two faces: $Y_{A1}$, $Y_{B1}$, and $Y_{C1}$ at one face, and $Y_{A2}$, $Y_{B2}$, and $Y_{C2}$ at the other. It is desired to relate the fluxes $N_A$ and $N_B$ ($N_C = 0$) to the boundary concentrations and the gas properties.

Equation (3.18) may be written for each of the three components, but only two of these are independent, since $Y_A + Y_B + Y_C = 1$. Integration of the equation for $C$ and that for either $A$ or $B$, first performed by Gilliland [26], leads to

$$\frac{N_A}{D_{AC}} + \frac{N_B}{D_{BC}} = \frac{P}{RT y_0} \ln \frac{Y_{C2}}{Y_{C1}} \qquad (3.22)$$

and

$$N_A + N_B = \frac{D_{AB}P}{RTy_0} \ln \frac{[(N_A + N_B)/N_A]Y_{A2} - [(N_A + N_B)/N_B]QY_{B2} + Q - 1}{[(N_A + N_B)/N_A]Y_{A1} - [(N_A + N_B)/N_B]QY_{B1} + Q - 1}$$

(3.23)

where $\qquad Q = \dfrac{1/D_{AB} - 1/D_{AC}}{1/D_{AB} - 1/D_{BC}}$

Wilke [67] and Toor [61] give the same equations; Grenier [29] has rearranged them to give $N_A$ as functions of $N_A/N_B$, $D_{BC}/D_{AB}$, $D_{BC}/D_{AC}$, $Y_{A2}$, $Y_{A1}$, $Y_{C2}$, and $Y_{C1}$.

Since explicit solutions for $N_A$ and $N_B$ are not obtained, it is usually necessary to solve Eqs. (3.22) and (3.23) by trial and error, or to use one of the approximate methods outlined below. This applies not only to the case of a ternary with one gas not diffusing but to other applications of Eq. (3.18).

In the special case of $N_C = 0$ and $D_{AC} = D_{BC}$, the two equations (3.22) and (3.23) may be readily combined to eliminate $N_B$ and obtain an explicit solution for $N_A$. The result is

$$N_A = \frac{D_{BC}P}{RTy_0} \frac{[Y_{A2} - Y_{A1}(Y_{C2}/Y_{C1})^\alpha] \ln (Y_{C2}/Y_{C1})}{[(1 - Y_{C2}) - (1 - Y_{C1})(Y_{C2}/Y_{C1})^\alpha]}$$

(3.24)

where $\alpha = D_{BC}/D_{AB} = D_{AC}/D_{AB}$. This would apply, for example, to the simultaneous steady-state diffusion of ammonia and water vapor through a stagnant layer of hydrogen at room temperature.

Approximate procedures to reduce or eliminate the trial and error usually encountered in using the Stefan-Maxwell equations for multicomponent diffusion have been suggested by Wilke [67], Toor [61, 62], and Shain [53]. These are described in the Appendix. Except in the special situations noted by the authors, these appear to give results which compare closely with the rigorous solutions; they have not, however, been widely tested. Toor [63, 64] has obtained solutions to the linearized equations by the use of matrices. Hellund [30] appears to be the only one to treat transient diffusion in a multicomponent gas system.

**EXAMPLE 3.2  DIFFUSION IN A TERNARY GAS MIXTURE**  A condenser operates with a feed vapor consisting of a mixture of ammonia, water vapor, and hydrogen at 3.36 atm. At one point in the condenser the mole fractions of the three components are 0.30 ammonia, 0.40 water vapor, and 0.30 hydrogen. The liquid on the condensing surface at this point is at 200°F and contains 10 mole percent ammonia and 90 mole percent water, with negligible hydrogen. Employing the stagnant film model, which assumes only molecular diffusion, estimate the rate of condensation of water relative to that of ammonia.

Assume the gas at the liquid surface to be in equilibrium with liquid of the stated composition. An aqueous solution containing 10 mole percent ammonia at 200°F has a total vapor pressure of 2.18 atm; the equilibrium vapor (hydrogen-free) contains 70 mole percent ammonia and 30 mole percent water. Approximate values of the binary diffusion coefficients at one atmosphere (see Chap. 2) are $D_{AB} = 0.294$, $D_{AC} = 1.14$, and $D_{BC} = 1.30$ cm²/s ($A = $ ammonia, $B = $ water, and $C = $ hydrogen).

SOLUTION Since the hydrogen does not condense, this is an example of case 2; the diffusion of two gases in a ternary in which one gas is stagnant ($N_C = 0$). The gas laws apply at 200°F and 3.36 atm, and $Y_A + Y_B + Y_C = 1$ at every point.

Substituting in Eq. (3.18), with $N = Uc$, $c_T = P/RT$, and $N_C = 0$,

$$\frac{P}{RT}\frac{dY_A}{dy} = \frac{Y_A N_B - Y_B N_A}{D_{AB}} + \frac{Y_A N_C - Y_C N_A}{D_{AC}}$$

$$= Y_A\left(\frac{N_B}{D_{AB}} + \frac{N_A}{D_{AC}}\right) + Y_B\left(\frac{N_A}{D_{AC}} - \frac{N_A}{D_{AB}}\right) - \frac{N_A}{D_{AC}} \qquad (a)$$

$$\frac{P}{RT}\frac{dY_C}{dy} = \frac{Y_C N_B - Y_B N_C}{D_{BC}} + \frac{Y_C N_A - Y_A N_C}{D_{AC}} = Y_C\left(\frac{N_B}{D_{BC}} + \frac{N_A}{D_{AC}}\right) \qquad (b)$$

These have been integrated to give Eqs. (3.22) and (3.23), which can be employed directly, with substitution of the following quantities:

$$Y_{A1} = 0.30 \qquad Y_{B1} = 0.40 \qquad Y_{C1} = 0.30 \qquad Y_{A2} = 0.7 \times \frac{2.18}{3.36} = 0.455$$

$$Y_{B2} = 0.3 \times \frac{2.18}{3.36} = 0.195 \qquad \text{and} \qquad Y_{C2} = 1 - 0.455 - 0.195 = 0.35$$

From the defining equation, since $DP$ is independent of pressure,

$$Q = \frac{1/0.294 - 1/1.14}{1/0.294 - 1/1.30} = 0.96$$

Making appropriate substitutions in Eqs. (3.22) and (3.23),

$$\frac{N_A}{1.14} + \frac{N_B}{1.30} = \frac{1}{RTy_0}\ln\frac{0.35}{0.30} \qquad (c)$$

$$N_A + N_B = \frac{0.294P}{RTy_0}\ln\left\{\frac{(N_A + N_B)[0.455/N_A - (0.96 \times 0.195/N_B)] + 0.96 - 1}{(N_A + N_B)[0.30/N_A - (0.96 \times 0.40/N_B)] + 0.96 - 1}\right\} \qquad (d)$$

These may be solved by trial and error (first letting $p = N_B/N_A$ and dividing one equation by the other) to give $N_B/N_A = 4.68$.

Since $D_{BC}$ and $D_{AC}$ do not differ greatly, and both differ considerably from $D_{AB}$, an approximate solution may be obtained directly from Eq. (3.24). Let $D_{BC} = D_{AC} = (1.14 + 1.30)/2P = 1.22/P$; write Eq. (3.24) for $N_A$ and for $N_B$; divide one equation by the other:

$$\frac{N_B}{N_A} = \frac{Y_{B2} - Y_{B1}(Y_{C2}/Y_{C1})^{D_{BC}/D_{AB}}}{Y_{A2} - Y_{A1}(Y_{C2}/Y_{C1})^{D_{BC}/D_{AB}}}$$

$$= \frac{0.195 - 0.4(0.35/0.30)^{4.15}}{0.455 - 0.3(0.35/0.30)^{4.15}} = 4.95$$

Note that the mixture condensing is richer in ammonia than the liquid, which contains a molal ratio of water to ammonia of 9: the liquid passing this point in the condenser is being enriched in ammonia. Note also that the gas at the liquid surface is richer in both ammonia and hydrogen than the bulk gas mixture. The magnitude of the negative concentration gradients depends on $y_0$ and on the diffusion fluxes. Those stipulated are possible only for a particular value of $N_A y_0$, which can be calculated from the two equations. The actual condensation rates will vary inversely as $y_0$, or directly with the mass-transfer coefficient.

The example presented may be solved by any of the approximate methods described in the appendix. The results obtained by the several methods are compared in the following table.

| Method | Flux ratio, $N_B/N_A$ |
|---|---|
| Rigorous [Eqs. (3.22), (3.23)] | 4.68 |
| Eq. (3.24); $D_{BC} \approx D_{AC} = (D_{BC} + D_{AC})/2$ | 4.95 |
| Wilke | 5.15 |
| Toor | 5.63 |
| Shain | 5.20 |

No generalization about the relative merits of the approximate calculation procedures can be made from the results of this single numerical example.　　　　////

**Multicomponent diffusion in liquids**　There can be little question about the validity of Eq. (3.18) as applied to molecular diffusion in gases at low or moderate pressures, with known values of the binary coefficients. This has been found to agree within experimental error in several ternary gaseous systems [23, 33, 36, 22, 25].

For liquid mixtures, however, the situation is somewhat more complicated. Here it is customary to use multicomponent diffusion equations that are more closely related to experimental measurements of diffusion phenomena than Eq. (3.18).

Consider, for example, the determination of the coefficients of diffusion of the three-component mixture $A$, $B$, and $C$ in a diffusion cell consisting of two well-mixed chambers separated by a porous membrane. One charges the chambers initially with mixtures of different composition and analyzes the contents after permitting diffusion to occur for various lengths of time. From the observed concentration changes one can find the fluxes of any two of the components and the concentration gradients across the membrane as functions of time. (The third component's flux and concentration gradient are dependent quantities because the mole fractions must add up to unity.) By repeating the experiment it is possible to change the initial concentration gradient of $B$, for example, while holding the initial concentration gradient of $A$ constant, or vice versa. Thus, from the measurements one can obtain values of two independent coefficients in the expression for the flux,

$$J_A = -D'_{AA} \frac{dc_A}{dy} - D'_{AB} \frac{dc_B}{dy} \qquad (3.25)$$

The first term represents the flux of $A$ owing to its own concentration gradient; the second represents the possible effect of $B$'s concentration gradient on the flux of $A$. Obviously one has not ruled out the possibility that $C$'s gradient may also contribute to the flux of $A$. The choice of terms in Eq. (3.25) was arbitrary; the gradient of $C$ is simply the negative of the sum of the gradients of $A$ and of

**B.** In fact, some writers, notably Bird, Stewart, and Lightfoot [4] and Hirschfelder, Curtiss, and Bird [32], prefer to write Eq. (3.25) in an alternative form in which the flux of $A$ is expressed as a function of the concentration gradients of $B$ and $C$:

$$J_A = -D'_{AB} \frac{dc_B}{dy} - D'_{AC} \frac{dc_C}{dy} \tag{3.26}$$

Diffusion coefficients used in equations of the form of Eq. (3.25) and (3.26) are referred to as "practical coefficients," apparently because they are so closely related to convenient experimental measurement. The equations are also practical in the sense that they give the flux explicitly rather than implicitly, as in Eq. (3.18). They are therefore much easier to use in treating diffusion problems with equations of change such as Eqs. (3.36) and (3.37).

The $D'$ values are not equal to the binary coefficients which appear in Eq. (3.6) except in the special case in which the mixture is a binary. For multi-component mixtures the relationship between the $D'$'s and the binary diffusion coefficients is more complicated, although it exists [32]. Hirschfelder, Curtiss, and Bird [32, 17] and Bird, Stewart, and Lightfoot [4] have shown that for ternary mixtures of thermodynamically ideal components, the practical diffusion coefficients in the equation for the molal flux relative to the *mass* average velocity,

$$J_A^* = \frac{c_T^2}{\rho_M} \left( M_B D'_{AB} \frac{dX_B}{dy} + M_B D'_{AC} \frac{dX_C}{dy} \right) \tag{3.27}$$

are related to the binary diffusion coefficients $D_{ij}$ by

$$D'_{AB} = D_{AB} \left\{ 1 + \frac{X_C[(M_C/M_B)D_{AC} - D_{AB}]}{X_A D_{BC} + X_B D_{AC} + X_C D_{AB}} \right\} \tag{3.28}$$

$$D'_{AC} = D_{AC} \left\{ 1 + \frac{X_B[(M_B/M_C)D_{AB} - D_{AC}]}{X_A D_{BC} + X_B D_{AC} + X_C D_{AB}} \right\} \tag{3.29}$$

It is clear that among the practical coefficients the values depend on the order of the subscripts. For example, using the equations above, with $D_{AB} = D_{BA}$,

$$D'_{BA} - D'_{AB} = D_{AB} X_C \frac{(M_C/M_A)D_{BC} - (M_C/M_B)D_{AC}}{X_A D_{BC} + X_B D_{AC} + X_C D_{AB}} \tag{3.30}$$

indicating that the difference is greater the larger the mole fraction of the third component.

The derivation of Eq. (3.30) [4, 32] results from proving that the Stefan-Maxwell equation, Eq. (3.18), and the $n$-component equations like Eq. (3.27) are equivalent, at least for ideal mixtures. Thus, it would be possible to interpret measurements of diffusion for systems of $n$ components by means of Eq. (3.18), determining the values of the $n(n-1)/2$ binary independent values of $D$ to fit the observed fluxes and gradients; alternatively, one can determine the $n(n-1)/2$ independent values of the practical coefficient $D'$ to fit the same data.

For gases, $D_{ij}$ is nearly independent of composition and is to be preferred, since the values of $D'$ vary with composition, as indicated by the equations above.

For liquids, on the other hand, both sets of coefficients vary with composition; and the equations which give the fluxes explicitly may be preferred to the Stefan-Maxwell equation because in many applications to diffusion problems, especially when transient changes occur, it is much easier to employ equations such as (3.24). For example, both Toor [63] and Stewart and Prober [57a] have shown how to solve many such problems using the binary $D$ values. When the composition variations are not so great that average values can be employed for the evaluations of the practical diffusivities, he was able to reduce multicomponent diffusion calculations to equivalent binary computations.

## 3.5  TRANSIENT DIFFUSION—INTRODUCTION

Steady-state diffusion is a special case of the more general situation in which concentrations and fluxes vary with time. A common experimental procedure for the measurement of $D_{AB}$ in binary gas systems employs a cylindrical cavity divided by a partition which can be removed. Gas $A$ initially fills the cavity on one side of the partition and gas $B$ the cavity on the other side. The partition is removed and diffusion allowed to occur at constant temperature and pressure, without convection. Concentration at every point in the system changes with time, approaching limiting values equal to those which would be obtained by mixing the two original gas quantities. Unlike the cases considered in Secs. 3.3 and 3.4, the concentration at any point varies with time as well as position.

Since there are now three variables—concentration, time, and position—it is evident that the diffusion process must be described by partial rather than ordinary differential equations.

Consider, for example, the relatively simple case of the experiment described for the measurement of $D_{AB}$ in a binary system of ideal gases at constant temperature and pressure. The molal volumes $V_A$ and $V_B$ are constant and equal, and since the total volume is fixed, $N_A = -N_B$. The velocity $U_y$ is zero, and diffusion takes place only in the $y$ direction. Equation (3.5) reduces to Eq. (3.6). The flux $N_A$, however, is not constant, but varies continuously with time and with position.

Consider a thin slice of gas bounded by two planes normal to the direction of diffusion and separated by the distance $dy$. The rate of increase of the number of moles of $A$ contained in differential volume may be equated to the decrease in flux of $A$ across the two planes, to obtain the simple equation of continuity:

$$\frac{\partial c_A}{\partial t} + \frac{\partial N_A}{\partial y} = 0 \tag{3.31}$$

where $t$ represents time. Combining this with Eq. (3.6), the following is obtained:

$$\frac{\partial c_A}{\partial t} = D_{AB} \frac{\partial^2 c_A}{\partial y^2} \tag{3.32}$$

or
$$\frac{\partial Y_A}{\partial t} = D_{AB} \frac{\partial^2 Y_A}{\partial y^2} \tag{3.33}$$

which is a form of Fick's second law. Applying appropriate boundary conditions, this may be solved to relate the gas compositions $c_A$ or $Y_A$ to $t$ and $y$.

In the somewhat more general case in which the whole system is moving in the $y$ direction (relative to the apparatus), with the velocity $\overline{U}_M$ being the velocity of the plane of no net molal transport, the corresponding relation is

$$\frac{\partial c_A}{\partial t} + \overline{U}_M \frac{\partial c_A}{\partial y} = D_{AB} \frac{\partial^2 c_A}{\partial y^2} \tag{3.34}$$

or
$$\frac{\partial Y_A}{\partial t} + \overline{U}_M \frac{\partial Y_A}{\partial y} = D_{AB} \frac{\partial^2 Y_A}{\partial y^2} \tag{3.35}$$

These relations are valid only if $\overline{U}_M$ is constant, as when $\Sigma c = c_T$ is constant, i.e., where there is no volume change on mixing. The velocity $\overline{U}_M$ is frequently zero, as in the method described for measuring $D_{AB}$ in an ideal-gas system, and Eq. (3.35) reduces to Eq. (3.33).

Equations (3.32) and (3.34) are of the same mathematical form as the equations for heat conduction in solids, for which numerous solutions are available for various geometric shapes and boundary conditions; these solutions may be employed to treat problems of transient diffusion. The most comprehensive collections of solutions to Eq. (3.32) are to be found in the treatises of Crank [14] and of Carslaw and Jaeger [11]. Solutions to the equation for transient heat conduction in solids involve temperature in place of concentration and thermal diffusivity in place of $D_{AB}$ (thermal diffusivity, square centimeters per second, is the ratio of the thermal conductivity to the product of the specific heat and the density of the medium).

Equations (3.33) and (3.35) may be employed where diffusion and motion are in the $y$ direction only. The more general relations involving the three coordinates $x$, $y$, and $z$ are

$$\frac{\partial c_A}{\partial t} + (\mathbf{U}_m \cdot \nabla c_A) = D_{AB} \nabla^2 c_A \tag{3.36}$$

and
$$\frac{\partial Y_A}{\partial t} + (\mathbf{U}_m \cdot \nabla Y_A) = D_{AB} \nabla^2 Y_A \tag{3.37}$$

In the solution of various practical problems it is often helpful to transform these into cylindrical or polar coordinates.

Simultaneous diffusion and chemical reaction is the subject of Chap. 8. It is sufficient here to see how the equations for transient diffusion must be modified to allow for chemical reaction in the case of one-directional diffusion in a binary gas system at constant pressure and temperature.

Let $Q_A$ and $Q_B$ represent the instantaneous rate of *formation* of species $A$ and $B$ at position $y$, each expressed in gram moles per second per cubic centimeter.

The velocity $\bar{U}_M (= \bar{U}_y)$ of the plane of no net molal flux is not constant, and $(N_A + N_B)$ is not zero. From Eq. (3.4),

$$\frac{\partial N_A}{\partial y} = \frac{\partial}{\partial y}\left(\bar{U}_M c_A - D_{AB}\frac{\partial c_A}{\partial y}\right) \tag{3.38}$$

The velocity $\bar{U}_M$ is $(N_A + N_B)/c_T$ and $c_T$ is constant. A molal balance on a differential element of width $dy$ gives

$$\frac{\partial c_A}{\partial t} + \frac{\partial N_A}{\partial y} = Q_A \tag{3.39}$$

and

$$\frac{\partial c_T}{\partial t} = Q_A + Q_B - \frac{\partial}{\partial y}(N_A + N_B) = 0 \tag{3.40}$$

Combining Eqs. (3.38) to (3.40) and (3.34) leads to

$$\frac{\partial c_A}{\partial t} + \bar{U}_M\frac{\partial c_A}{\partial y} = Q_A - \frac{c_A}{c_T}(Q_A + Q_B) + D_{AB}\frac{\partial^2 c_A}{\partial y^2} \tag{3.41}$$

The more general result, in terms of the coordinates $x$, $y$, and $z$, is

$$\frac{\partial c_A}{\partial t} + \mathbf{U}_m \cdot \nabla c_A = Q_A - \frac{c_A}{c_T}(Q_A + Q_B) + D_{AB}\nabla^2 c_A \tag{3.42}$$

EXAMPLE 3.3 DISSOLUTION OF A SPHERE OF BENZOIC ACID    A solid sphere 1.0 cm in diameter is immersed in completely stagnant water at 25°C. The surface of the sphere is supplied with benzoic acid at such a rate that the concentration of the aqueous solution immediately adjacent to the surface is indefinitely maintained constant at $c = c_s = 0.0278$ g moles/l. (The problem is that of a sphere of benzoic acid dissolving by molecular diffusion into stagnant water of infinite volume, but somehow being maintained at constant diameter.) How long will it be for the flux of benzoic acid at the surface of the sphere to reach 99 percent of its steady-state value?

SOLUTION    The concentration of benzoic acid $(c)$ is very small compared with that of water, so Eqs. (3.6) and (3.32) may be employed, with $D_{AB} = 1.1 \times 10^{-5}$ cm²/s. (See Sec. 2.6) It is necessary to solve Eq. (3.32) (first converted to spherical coordinates) with the boundary conditions

$$c = c_s = 0.0000278 \text{ g moles/cm}^3 \qquad \text{at } r = r_0, t \geqslant 0$$
$$c = 0 \qquad\qquad\qquad\qquad\qquad \text{at } r = \infty$$

The solution to the corresponding heat-conduction problem is given by Carslaw and Jaeger [11], which may be translated for application to molecular diffusion to give

$$\frac{c}{c_s} = \frac{r_0}{r}\left(1 - \text{erf}\frac{r - r_0}{2\sqrt{Dt}}\right) \tag{a}$$

where $c$ is the acid concentration at $r = r$, and erf $x$, tabulated in many handbooks, is defined by

$$\text{erf } x = \frac{2}{\sqrt{\pi}} \int_0^x e^{-x^2} \, dx \tag{b}$$

The quantity of interest is not $c$ but the acid flux in the solution immediately adjacent to the surface. This is given by

$$(N_A)_{r=r_0} = -D_{AB}\left(\frac{dc}{dr}\right)_{r=r_0} \tag{c}$$

The concentration gradient is found by differentiation of Eq. (a).

$$\frac{\partial c}{\partial r} = \frac{c_s r_0}{r^2}\left(\text{erf }\frac{r - r_0}{2\sqrt{D_{AB}t}} - 1\right) - \frac{r_0 c_s}{r\sqrt{\pi D_{AB}t}} \exp \frac{-(r - r_0)^2}{4D_{AB}t}$$

whence

$$(N_A)_{r=r_0} = \frac{D_{AB}c_s}{r_0} + \frac{D_{AB}c_s}{\sqrt{\pi D_{AB}t}} \tag{d}$$

The first term on the left will be recognized as the solution for the steady-state flux, as given for gases by Eq. (3.12). The transient flux given by Eq. (d) will evidently differ from the steady-state flux by 1 percent when $\sqrt{\pi D_{AB}t}$ is 100 $r_0$, that is, at time

$$t = \frac{10000 \times 0.5^2}{\pi \times 1.1 \times 10^{-5}} = 72,200,000 \text{ s}$$

*Note*: This solution neglects the effect of "transpiration" on the diffusion flux. See Sec. 6.5. The density of the saturated solution of benzoic acid is so nearly that of water that the correction for the transpiration effect is significant only during the first 8 to 10 s, when the solution rate is large. The stipulation that the diameter remains constant is unrealistic: integration of Eq. (d) shows the total amount of benzoic acid dissolving in $7.22 \times 10^7$ s to be some 26 times that contained in a 1-cm sphere. (See Sec. 3.9.) According to the approximate analysis of Rosner and Epstein [50] the sphere would have dissolved completely in roughly $10^7$ s. ////

## 3.6 TRANSIENT DIFFUSION IN STAGNANT MEDIA

Solutions to Eq. (3.32) ($U_M = 0$) are given in the following paragraphs for several of the cases encountered most frequently.

**Semi-infinite solid (flat surface)** The stagnant medium has a face of infinite area and extends indefinitely in the direction $x$ and $z$. The concentration everywhere is $c_0$ at time zero. The surface concentration at $y = 0$ is suddenly changed from $c_0$ to $c_i$, at which it is held. Equation (3.32) describes the diffusion of solute in the $y$ direction into the medium. The well-known solution for this "penetration" case is

$$\frac{c_i - c}{c_i - c_0} \equiv \Delta = \frac{\text{unaccomplished change}}{\text{total possible change}} = \text{erf}\left(\frac{y}{2\sqrt{D_{AB}t}}\right) \tag{3.43}$$

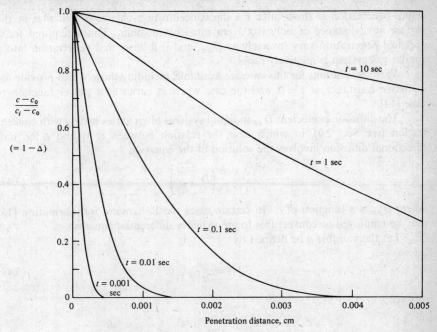

**FIGURE 3.3**
Solute penetration by diffusion into an infinite medium. Times indicated are for
diffusion into a liquid with $D_{AB} = 1.0 \times 10^{-5}$ cm²/s. For a binary gas system with
$D_{AB} = 0.1$ cm²/s, the times shown should be divided by 10,000.

At any instant the flux into the medium ($y = 0$) is

$$N_A = - D_{AB}\left(\frac{\partial c}{\partial y}\right)_{y=0} = (c_i - c_0)\sqrt{\frac{D_{AB}}{\pi t}} \tag{3.44}$$

and the total moles $A$ entering the medium from time $t = 0$ to $t = t$ is

$$\int_0^t N_A \, dt = 2(c_i - c_0)\sqrt{\frac{D_{AB}t}{\pi}} \tag{3.45}$$

The average flux over the same time period is

$$(N_A)_{av} = \frac{1}{t}\int_0^t N_A \, dt = 2(c_i - c_0)\sqrt{\frac{D_{AB}}{\pi t}} \tag{3.46}$$

The extent to which the penetration of solute into a semi-infinite plane medium
changes with time is illustrated by Fig. 3.3. This is a graph of Eq. (3.43) for
$D_{AB} = 10^{-5}$ cm²/s, which is a representative value for liquid systems. The rate of

solute penetration is slow: after 1 s the concentration change is half that at the surface at a distance of only 0.003 cm into the medium. Times required for a specified penetration vary inversely as $D_{AB}$, and it follows that penetration into a similar gas system is relatively rapid.

Various solutions for this case are available, including those which provide for a surface resistance at $y = 0$, and the case where $c_i$ varies as a power function of time [14].

The diffusion coefficient $D_{AB}$ in liquid systems often varies widely with concentration (see Sec. 2.6), in which case the relation between $c$, $y$, and $t$ for unidirectional diffusion involves the solution of the equation

$$\frac{\partial c}{\partial t} = \frac{\partial}{\partial y}\left(D_{AB}\frac{\partial c}{\partial y}\right) \tag{3.47}$$

where $D_{AB}$ is a function of $c$. In certain cases the Boltzmann transformation [14] may be employed to convert this to an ordinary differential equation.

Let the variable $\eta$ be defined by

$$\eta \equiv \frac{y}{2t^{1/2}} \tag{3.48}$$

whence

$$\frac{\partial c}{\partial y} = \frac{1}{2t^{1/2}}\frac{dc}{d\eta} \quad \text{and} \quad \frac{\partial c}{\partial t} = -\frac{y}{4t^{3/2}}\frac{dc}{d\eta}$$

Substituting in Eq. (3.47), with $D_{AB}$ stipulated to be a function of $c$ only,

$$-2\eta\frac{dc}{d\eta} = \frac{d}{d\eta}\left(D_{AB}\frac{dc}{d\eta}\right) \tag{3.49}$$

If the variation of $D_{AB}$ with $c$ is known, this ordinary differential equation may be solved to obtain $c$ as a function of $y/2t^{1/2}$, as for example, to obtain Eq. (3.43) for constant $D_{AB}$.

The method is applicable only when $D_{AB}$ depends solely on $c$ and when the initial and boundary conditions can be expressed in terms of $\eta$ alone—not $y$ and $t$ separately. It cannot be used, for example, in the case of the semi-infinite slab, since the boundary condition $c = c_i$ at $y = 0$ (any $t$) involves $y$ separately from $\eta$.

Secor [52] employed a microinterferometric method of measuring $D_{AB}$ in the liquid system dimethylformamide-polyacrylonitrile. The technique provided values of the concentration gradient as a function of distance, from which the variation of $c$ with $y$ was obtained. The data were analyzed to obtain $D_{AB}$ as a function of $c$ by the use of Eq. (3.49), from which $\eta$ may be eliminated to give

$$D_{AB} = \frac{-\int_0^c y\, dc}{2t(dc/dy)} \tag{3.50}$$

$D_{AB}$ in this system at 25°C was found to vary sevenfold over the range of polymer concentrations from 1.5 to 16 g/100 cm$^3$.

**Liquid jets**    Gas absorption by liquid jets has been studied extensively in recent years. In a typical experiment the liquid passes vertically through a vessel containing the pure solute gas. If the orifice through which the liquid enters is properly designed, the jet is cylindrical and free of ripples, with no velocity gradient across its diameter. A very small amount of the ambient gas is absorbed in the short contact time of a second or less, so the solute penetration depth is exceedingly small and the equations for diffusion into a body having a flat surface apply.

Using $CO_2$ and water, Cullen and Davidson [15] obtained excellent agreement between theory and experiment; the agreement was so good, in fact, that the experimental technique has been considered as a way of measuring the molecular diffusion coefficient of the solute gas in the liquid. Accurate gas-solubility data are necessary, there must be no surface resistance due to contamination, as by surface-active agents, and the gas must be pure.

Mass transfer into turbulent gas jets has been studied by Davies and Ting [19], whose results agreed well with Levich's [39] theory of eddy behavior at an interface.

EXAMPLE 3.4    ABSORPTION FROM A BUBBLE GROWING IN A LIQUID    A gas bubble grows at the tip of a nozzle held in a stagnant liquid, being supplied with pure gas at a constant volumetric rate. Absorption into the liquid proceeds simultaneously with the growth of the bubble. Derive an approximate expression for the rate of diffusion into the liquid, as a function of the time of bubble growth, for the special case of a sparingly soluble gas. In this case the amount of gas dissolved may be assumed to have a negligible effect on the volume of the growing bubble. It may also be assumed that the resistance to diffusion is confined to a thin film of liquid surrounding the growing bubble; the thickness of this layer is very small in comparison with the bubble radius. *Note*: This case is similar to that of the polarograph, where the current is diffusion-limited. In the case of the polarograph, the volume of the growing mercury drop is not a function of the cumulative mass transfer.

SOLUTION    (The analysis which follows is that given by Levich [39], in his description of the theory of the polarograph. The case of a gas bubble dissolving in a liquid with no gas being fed to it is analyzed by Duda and Vrentas [21]. Mass transfer from bubbles is discussed more generally in Chap. 6).

If the volume of the bubble, presumed spherical, grows at the constant rate $q$, the radius $R$ is given by

$$R = bt^{1/3}$$

where $b = (3q/4\pi)^{1/3}$. The outward radial velocity of the interface is $\bar{U}_R = dR/dt = (b/3)t^{-2/3}$, and the radial velocity at any distance $r$ from the center is

$$\bar{U}_r = \bar{U}_r \left(\frac{R}{r}\right)^2$$

The unsteady-state diffusion equation to be solved is

$$\frac{\partial c}{\partial t} + U_r \frac{\partial c}{\partial r} = D\left[\left(\frac{\partial^2 c}{\partial r^2}\right) + \left(\frac{2}{r}\frac{\partial c}{\partial r}\right)\right] \qquad (a)$$

with the boundary conditions $c(r, 0) = c(\infty, t) = c_0$, a constant, and $c(R, t) = c_i$, a constant. In this case the diffusion boundary layer near the surface is always very thin and the flow representing outward radial expansion makes it thinner. Then the influence of curvature, represented by the second term of the right of Eq. $(a)$ is negligibly small. Furthermore, the velocity near $r = R$ can be represented rather accurately by

$$U_r \approx U_R\left(1 - 2\frac{y}{r}\right) = \frac{b}{3}t^{-2/3} - \frac{2}{3}yt^{-1} \qquad (b)$$

where $y = r - R$ is the distance from the interface rather than from the center of the sphere. In terms of the new independent variables $y$ and $t$, the transport equation becomes

$$\frac{\partial c}{\partial t} - \frac{2}{3}\frac{y}{t}\frac{\partial c}{\partial y} = D\frac{\partial^2 c}{\partial y^2} \qquad (c)$$

Now if we introduce the similarity transformation to new variables, $z$ and $\tau$, according to the definitions

$$z = \left(\frac{D^3}{b^7}\right)t^{2/3}y \qquad \tau = \left(\frac{3}{7}\right)\left(\frac{D^7}{b^{14}}\right)t^{7/3}$$

the differential equation becomes

$$\frac{\partial c}{\partial \tau} = \frac{\partial^2 c}{\partial z^2}$$

with the boundary conditions $c(z, 0) = c(\infty, \tau) = c_0$ and $c(0, \tau) = c_i$. The well-known solution of this problem is

$$\frac{c - c_0}{c_i - c_0} = \text{erfc}\left(\frac{z}{2\tau^{1/2}}\right) \qquad (d)$$

and the instantaneous mass-transfer flux at the surface is

$$N_A = -D\left(\frac{\partial c}{\partial y}\right)_0 = \sqrt{\frac{7}{3}}(c_i - c_0)\sqrt{\frac{D}{\pi t}} \qquad (e)$$

which corresponds closely to the familiar result of the penetration theory, as in Eq. (3.44).

The factor $(7/3)^{1/2} = 1.53$ is due to the fluid motion which is induced by the sphere's expansion. The increase in mass-transfer flux is due to a velocity near the interface greater than that farther out in the fluid, which causes the diffusion boundary layer to be compressed as it is stretched around the spherical surface. Concentration gradients are therefore sharpened and mass transfer is increased. The total rate of mass transfer, obtained by multiplying $N$ from Eq. $(e)$ by the surface area, $4\pi R^2$, increases as $t^{1/6}$. The appearance of $D^{1/2}$ in the final result should be noted especially. The presence of fluid motion does not in itself rule out the familiar result of the penetration theory, in which $N$ is proportional to $D^{1/2}$, although certain flow patterns such as the laminar boundary layer on a flat plate do change the exponent on $D$ slightly, as in Eq. (3.68). ////

**Semi-infinite slab**   This is the case of a medium of thickness $2y_0$ having two parallel faces of infinite area. Initially the concentration is everywhere $c_0$. The surface concentrations $(y = 0, 2y_0)$ are suddenly changed to $c = c_i$ and maintained constant. Diffusion into the slab takes place, the concentration everywhere approaching $c = c_i$ at infinite time. Equation (3.32) is solved with the boundary conditions $c = c_0$ at $t = 0, y > 0$; $c = c_i$ at $t = \infty, y > 0$; $c = c_i$ at $y = 0, 2y_0, t > 0$; $\partial c/\partial y = 0$ at $y = y_0$, to give

$$\Delta = \frac{c_i - c}{c_i - c_0} = \frac{4}{\pi}\left[e^{-\beta}\sin\frac{\pi y}{2y_0} + \frac{1}{3}e^{-9\beta}\sin\frac{3\pi y}{2y_0}\right.$$
$$\left. + \frac{1}{5}e^{-25\beta}\sin\frac{5\pi y}{2y_0} + \cdots\right] \tag{3.51}$$

where

$$\beta = \frac{\pi^2 D_{AB}t}{4y_0{}^2}$$

The total solute entering the slab between $t = 0$ and $t = t$ may be obtained from the expression for the *average* concentration at time $t$, which is

$$c_{av} - c_0 = \frac{c_i - c_0}{2y_0}\int_0^{2y_0}(1 - \Delta)\,dy$$
$$= (c_i - c_0)\left\{1 - \frac{8}{\pi^2}\left[e^{-\beta} + \frac{1}{9}e^{-9\beta} + \frac{1}{25}e^{-25\beta} + \cdots\right]\right\} \tag{3.52}$$

If $\beta$ is greater than 1.2, this is given approximately by

$$1 - \Delta' \equiv \frac{c_{av} - c_0}{c_i - c_0} = 1 - \frac{8}{\pi^2}e^{-\beta} \tag{3.53}$$

Equation (3.52) is evidently useful in the interpretation of data obtained by the gas-diffusion experiment described in Sec. 3.5. If the cavities on both sides of the partition are of equal volume and filled with pure $A$ and pure $B$, respectively, then each half of the cavity corresponds to half of an infinite slab (there is no flux at the ends, where $y = \pm y_0$). If the partition is replaced at time $t$ and the gases in each half well mixed, the measured average compositions $c_{av}$ can be employed with Eq. (3.52) to obtain $\beta$ and $D_{AB}$.

If the diffusion occurs in a cylinder 120 cm tall with helium and methane at 5 atm at room temperature (without convection), nearly 2.5 h are required for the average helium concentration to fall to $0.7c_0$ in the upper half and to increase to $0.3c_0$ in the lower half [56]. This illustrates the slow rate of mixing by molecular diffusion, even in gases, which is usually more rapid than in liquids with similar geometry.

**Other shapes**   The semi-infinite solid and the infinite slab are but two of the many cases of heat conduction in solids and transport by molecular diffusion for which solutions to the Laplace equation have been obtained. In many of the simpler cases the expression obtained relates the dimensionless groups $\Delta$, $\beta$, and $y/y_0$ (or other position variable). Table 3.1 lists values of $\Delta$ as a function of $D_{AB}t/y_0^2$ for $y/y_0 = 1$ (i.e., the central plane, axis, or point) for several common shapes. In the case of the cylinder and sphere, $y_0$ represents the radius. Table 3.2 gives values of $\Delta'$ as a function of $D_{AB}t/y_0^2$ for the infinite slab, infinitely long cylinder, and sphere.

Newman [43] has shown that various other shapes may be handled by simple multiplication of the values of $\Delta$ or $\Delta'$ for the cases listed in these tables. Thus, for a rectangular parallelepiped of sides $2y_{01}$, $2y_{02}$, and $2y_{03}$, the value of $\Delta$ is simply the product of the three values of $\Delta$ for infinite slabs of thickness $2y_{01}$, $2y_{02}$, and $2y_{03}$ at the same $\beta$. The same rule may be followed for $\Delta'$. The value of $\Delta$ for a cylinder of finite length is the product of $\Delta$ for the infinite cylinder and $\Delta$ for an infinite slab of thickness equal to the cylinder length. (See also Ref. 12.)

**Surface resistance**　When the fluid around a solid object is suddenly changed to a higher temperature $T_i$, the surface temperature $T_s$ rises as heat is transferred. The boundary condition $T = T_i$ at $y = 0$ is replaced by the condition

$$h(T_i - T_s)\, dS = -k\left(\frac{dT}{dy}\right)_{y=0} dS \tag{3.54}$$

where $S$ = surface area
　　　$h$ = coefficient of heat transfer from ambient fluid to surface
　　　$k$ = thermal conductivity of the solid
An analogous process occurs when diffusion takes place from an ambient fluid of concentration $c_i$ into a stagnant medium initially at $c_0$. Solutions for such cases are available in engineering handbooks [44] involving a fourth dimensionless group, $k/hy_0$. These solutions may be employed for the corresponding diffusion problems,

**Table 3.1　TRANSIENT DIFFUSION IN SIMPLE SHAPES; DIMENSIONLESS CONCENTRATION AT THE CENTER**

Values of $\Delta = \dfrac{c_i - c}{c_i - c_0} = \dfrac{\text{unaccomplished concentration change}}{\text{limit of change at infinite time}}$

| $D_{AB}t/y_0^2$ | Infinite slab | Infinite square bar | Cube | Infinite cylinder | Cylinder, length = diameter | Sphere |
|---|---|---|---|---|---|---|
| 0 | 1 | 1 | 1 | 1 | 1 | 1 |
| 0.032 | 0.9998 | 0.9997 | 0.9995 | 0.9990 | 0.9988 | 0.9975 |
| 0.080 | 0.9752 | 0.9510 | 0.9274 | 0.9175 | 0.8947 | 0.8276 |
| 0.100 | 0.9493 | 0.9012 | 0.8555 | 0.8484 | 0.8054 | 0.7071 |
| 0.160 | 0.8458 | 0.7154 | 0.6051 | 0.6268 | 0.5301 | 0.4087 |
| 0.240 | 0.7022 | 0.4931 | 0.3462 | 0.3991 | 0.2802 | 0.1871 |
| 0.320 | 0.5779 | 0.3340 | 0.1930 | 0.2515 | 0.1453 | 0.0850 |
| 0.800 | 0.1768 | 0.0313 | 0.0055 | 0.0157 | 0.00277 | 0.0007 |
| 1.600 | 0.0246 | 0.0006 | | 0.00015 | | |
| 3.200 | 0.00047 | | | | | |

$k/hy_0$ being replaced by $D_{AB}/k_c\,y_0$, where $k_c$ is the mass-transfer coefficient, ambient fluid to surface [the coefficient is defined in the text following Eq. (3.65)].

Transient diffusion into several geometries from a finite amount of well-stirred fluid is discussed by Ma and Evans [42].

## 3.7   MASS TRANSFER IN LAMINAR FLOW

Mass transfer into or within gases and liquids in laminar flow is encountered in many situations of engineering importance. Since there is no mixing, transport of components of the flowing mixture is accomplished only by molecular diffusion superposed on the bulk flow at any point in the stream. Combination of expressions for the velocity field with the appropriate forms of the diffusion equation provides useful equations for concentrations and fluxes. If the flow is laminar, the velocity field is fixed by the geometry of the system and the physical properties of the fluid. In principle, at least, all problems of the type described are amenable to analytic treatment, though the mathematics is sometimes difficult.

In cases of diffusion into a fluid which is moving without velocity gradients within it, the solutions available for transient heat conduction, such as those described in Sec. 3.6, may be employed. A cylindrical liquid jet issuing from a sharp-edged orifice is the equivalent of a solid rod; solute absorption from an ambient gas is described by the equations for heat conduction or diffusion into an infinite cylinder for an exposure time equal to the exposed length divided by the jet velocity.

**Table 3.2   TRANSIENT DIFFUSION IN SIMPLE SHAPES; DIMENSIONLESS AVERAGE CONCENTRATION**

Values of $\Delta' = \dfrac{c_i - c_{av}}{c_i - c_0}$, where $c_{av} = \dfrac{1}{V_S}\displaystyle\int_0^{V_S} c\,dV$, $V_S$ being the volume of the shape

| $D_{AB}t/y_0^2$ | Infinite slab | Infinite cylinder | Sphere |
|---|---|---|---|
| 0 | 1 | 1 | 1 |
| 0.005 | 0.922 | 0.843 | 0.774 |
| 0.01 | 0.890 | 0.784 | 0.690 |
| 0.02 | 0.839 | 0.698 | 0.579 |
| 0.04 | 0.773 | 0.558 | 0.440 |
| 0.06 | 0.725 | 0.512 | 0.352 |
| 0.10 | 0.643 | 0.394 | 0.226 |
| 0.20 | 0.497 | 0.219 | 0.084 |
| 0.30 | 0.388 | 0.122 | 0.031 |
| 0.50 | 0.236 | 0.0384 | 0.0043 |
| 0.80 | 0.113 | 0.0068 | 0.0002 |
| 1.00 | 0.069 | 0.0021 | |
| 2.00 | 0.0058 | | |

Several examples of diffusion in fluids in steady laminar flow with known velocity fields are described below. The derivations are outlined but not given in detail, since all are available in the literature. In all cases the approximation is introduced that Eq. (3.3) applies with reference to a plane moving at the local stream velocity, with $D_{AB}$ independent of concentration.

**The falling film**  This well-known case is encountered in wetted-wall gas absorbers, condensers, and other process equipment. A liquid is fed at the top of a vertical tube or plate at the rate of $\Gamma$ grams per second ($= q$ cubic centimeters per second) along each centimeter of perimeter or width of the top edge. At low liquid rates the flow down the wall is laminar and a parabolic velocity profile is quickly established. The situation is illustrated by Fig. 3.4.

Expressions for the film thickness $y_0$, the velocity $U_x$ at the distance $y$ from the wall, and the average velocity are easily derived:

$$y_0 = \left(\frac{3\mu\Gamma}{\rho^2 g}\right)^{1/3} = \left(\frac{3vq}{g}\right)^{1/3} \tag{3.55}$$

$$U_x = \frac{\rho g}{\mu}\left(y_0 y - \frac{1}{2}y^2\right) \tag{3.56}$$

$$U_{av} = \frac{1}{y_0}\int_0^{y_0} U\,dy = \frac{\Gamma}{\rho y_0} = \frac{q}{y_0} \tag{3.57}$$

$$U_x = \frac{gy_0^2}{2v} = \frac{3}{2}U_{av} \qquad (\text{at } y = y_0) \tag{3.58}$$

where $\mu$, $v$, $\rho$, and $g$ represent liquid viscosity, kinematic viscosity ($\mu/\rho$), liquid density, and local gravity. Note that the surface velocity is $\frac{3}{2}U_{av}$.

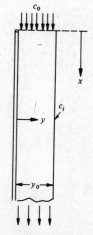

**FIGURE 3.4**
Free-falling laminar liquid film.

An important practical case is that where the liquid fed has the solute concentration $c_0$ and the surface of the falling film is maintained at the constant concentration $c_i$. Concentration at any point $x$, $y$ in the film is found by solving the equation

$$\frac{\partial c_A}{\partial t} + \overline{U}_x \frac{\partial c_A}{\partial y} = D_{AB} \frac{\partial^2 c_A}{\partial y^2} \tag{3.59}$$

with the substitution of $\overline{U}_x$ from Eq. (3.56). Equation (3.59) is similar to Eq. (3.34), but the convection term $\overline{U}_M(\partial c_A/\partial y)$ is replaced by $\overline{U}_x(\partial c_A/\partial x)$, which represents convection parallel to the wall. In this example, the diffusion flux is positive in the direction of decreasing $y$. The boundary conditions are $c = c_0$ at $t = 0$, $c = c_i$ at $y = y_0$. Integration across the film leads to a solution in terms of the average concentration $c_{av}$, where

$$c_{av} = \frac{1}{y_0} \int_0^{y_0} c \, dy \tag{3.60}$$

The result is [45, 43a]:

$$\Delta' \equiv \frac{c_i - c_{av}}{c_i - c_0} = 0.7857 \, e^{-5.121 \, \beta'} + 0.1001 \, e^{-39.21 \, \beta'}$$
$$+ 0.0360 \, e^{-105.6 \, \beta'} + 0.0181 \, e^{-204.7 \, \beta'}$$
$$+ \cdots \tag{3.61}$$

Here $\beta' = D_{AB}t/y_0^2$, in which $t$ is the time of exposure of the surface between the top and some point a distance $x$ below. The time $t$ is obtained by dividing the distance $x$ by the surface velocity, $\frac{3}{2}\Gamma/\rho y_0 [= \frac{3}{2}(q/y_0) = \frac{3}{2}\overline{U}_{av}]$.

For short exposure times the liquid concentration at the wall changes little and (with $c_i > c_0$) the penetration is confined to a thin layer near the surface. The local and average concentrations may be approximated by the use of Eqs. (3.51) and (3.52) with exposure time again obtained from the surface velocity.

Experimental data on mass transfer in falling liquid films are presented in Chap. 6. These include results obtained in studies of evaporation from pure liquids, gas absorption by the film, and mass transfer between the liquid and the wall.

**Laminar boundary layer on a flat plate (see Fig. 3.5)**    An incompressible fluid flowing at a uniform velocity passes over a flat plate of length $x_T$. The flow is laminar, and the plate is placed parallel to the direction of flow. The fluid approaching the plate has a solute concentration $c_0$ and velocity $\overline{U}_x = \overline{U}_0$; the

FIGURE 3.5
Laminar boundary layer on a flat plate.

solute concentration at the surface of the plate is maintained constant at $c_i$. The problem is to find the mass-transfer flux as a function of $\overline{U}_0$, the physical properties of the fluid, and the distance $x$ from the leading edge of the plate.

Again the solution is obtained by combining the velocity field with the diffusion equation, the basic equations being

$$\frac{\partial \overline{U}_x}{\partial x} + \frac{\partial \overline{U}_y}{\partial y} = 0 \tag{3.62}$$

$$\overline{U}_x \frac{\partial \overline{U}_x}{\partial x} + \overline{U}_y \frac{\partial \overline{U}_y}{\partial y} = v \frac{\partial^2 \overline{U}_x}{\partial y^2} \tag{3.63}$$

$$\overline{U}_x \frac{\partial c}{\partial x} + \overline{U}_y \frac{\partial c}{\partial y} = D \frac{\partial^2 c}{\partial y^2} \tag{3.64}$$

(The variation of fluid density with $x$, $y$, and $c$ is neglected.) The boundary conditions are $\overline{U}_x = \overline{U}_y = 0$ at $y = 0$; $c = c_0$ and $\overline{U}_x = \overline{U}_0$ at $x < 0$ or $y = \infty$; $c = c_i$ at $y = 0$. Diffusion in the $x$ direction is neglected.

The Pohlhausen solution is discussed at length by Schlichting [51] and is to be found in various places in the literature [49, 4, 39, 35]. The expressions obtained for the heat-transfer coefficient are readily transformed for use in mass transfer to give

$$\frac{k_c x}{D} = 0.332 \, (\mathrm{Re}_x)^{1/2} (\mathrm{Sc})^{1/3} \tag{3.65}$$

where $k_c$ is the *local* mass-transfer coefficient $N_A/(c_i - c_0)$ at the distance $x$ from the leading edge, and $\mathrm{Re}_x$ is defined as $x\overline{U}_0/v$. The average value of the mass transfer over the entire plate of length $x_T$ in the downstream direction is obtained as

$$(k_c)_{\mathrm{av}} = \frac{1}{x_T} \int_0^{x_T} k_c \, dx \tag{3.66}$$

whence

$$\frac{(k_c)_{\mathrm{av}} \, x_T}{D} = 0.664 \, (\mathrm{Re}_{xT})^{1/2} (\mathrm{Sc})^{1/3} \tag{3.67}$$

with $\mathrm{Re}_{xT}$ defined as $\overline{U}_0 x_T/v$. The result applies to mass transfer from plate to fluid, or fluid to plate. This is equivalent to $j_D = f/2$ (see Chap. 5).

**The rotating disk**   A plane circular disk is immersed in a fluid and caused to rotate at constant speed, with mass transfer occurring between the fluid and the surface of the disk (see Fig. 3.6). This case would appear to have little application in engineering, but it is the basis for a very useful experimental technique for measurements of mass and heat transfer and is of use in electrochemistry. It is one of the relatively few cases in which the surface is "uniformly accessible," i.e., where the mass-transfer coefficient is the same at all points on the surface. This can be shown to be true if the fluid motion over the surface is laminar.

As in other such cases, the situation is analyzed by combining an expression

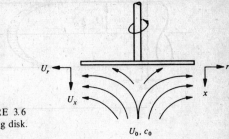

**FIGURE** 3.6
Rotating disk.

for the velocity field with the differential diffusion equation [39, 65, 28, 13]. The asymptotic solution for large Sc is

$$k_c = 0.62 \, D_{AB}^{2/3} \, v^{-1/6} \, \omega^{1/2} \tag{3.68}$$

where $\omega$ is the rotational speed in radians per second. This has been confirmed experimentally in various ways, as by measuring the dissolution of zinc in aqueous iodine solutions and determining the diffusion-limited current density when the disk is one electrode of a cell.

Mass transfer to rotating disks is discussed further in Sec. 6.8c.

**Taylor diffusion**  Several remarkable conclusions follow from Taylor's elegant analysis of solute diffusion in a fluid moving in laminar flow at a steady flow rate through a round tube. Two cases are discussed: (1) the sudden introduction of solute sufficient to provide a concentration $c_0$ over a very short length $\Delta x$ at the tube inlet, and (2) a sudden change of solute concentration from $c = 0$ to $c = c_0$ at the inlet, at time $t = 0$, after which the flow continues with solute concentration $c_0$ in the feed.

Case (1) is illustrated by Fig. 3.7. The velocity profile is parabolic, and the centerline velocity $\overline{U}_c$ is twice the average velocity $\overline{U}_{av}$. The source of width $\Delta x$ is essentially a line across the diameter, but this tends to be distorted into a parabola of width $\Delta x$ in the $x$ direction as flow continues. If no diffusion occurred, this pattern would persist, with a lengthening of the parabola. The concentration gradients lead to diffusion, however, and the equalization of concentration radially is very rapid if the tube diameter is very small in proportion to the length of travel of the injected solute. Longitudinal molecular diffusion is then negligible as compared with radial diffusion.

Near the apex of the parabola the diffusion is primarily toward the wall; near the wall it is primarily toward the center. Taylor explains the result as follows: "this means that the central part of the pipe fluid which is free of the dissolved substance passes into the zone where the concentration is rising. The dissolved substance is then absorbed till $c$ reaches its maximum value at $x = \overline{U}_{av} t$. The fluid then passes through the region where $c$ decreases with $x$ and finally leaves this zone, having yielded up the whole of the dissolved substance which it had acquired."

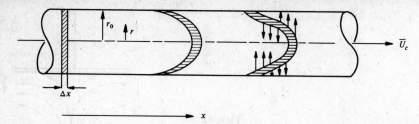

**FIGURE 3.7**
Taylor diffusion; laminar flow in a round table.

Radial molecular diffusion effectively *hinders* the dispersion in the axial direction, which would have otherwise been large because of the velocity profile. The axial distribution of the average concentration is found to be gaussian about a plane moving at the velocity $\bar{U}_{av}$:

$$c_{av} = \frac{c_0 \Delta x}{2\pi^{1/2}(k_1 t)^{1/2}} e^{-x_1^2/4k_1 t} \tag{3.69}$$

where $x_1 = x - \bar{U}_{av} t$, and $k_1$ is the "virtual" or effective coefficient of transport in the $x$ direction, given by

$$k_1 = \frac{r_0^2 \bar{U}_{av}^2}{48D} \tag{3.70}$$

This may be employed with the form of the diffusion equation using average concentration:

$$\frac{\partial c_{av}}{\partial t} = k_1 \frac{\partial^2 c_{av}}{\partial x^2} \tag{3.71}$$

where $c_{av}$ is the average concentration over the cross section of the tube at the distance $x$ from the inlet. As a result of radial diffusion, the peak concentration passing any point is quite sharp, whereas in the absence of molecular diffusion, $c_{av}$ is uniform over the length $\bar{U}_c t$ and decays inversely as $\bar{U}_c t$. Furthermore, *the mean concentration moves at the mean speed $\bar{U}_{av}$*, though parts of the fluid are moving more slowly and others more rapidly than the mean speed $\bar{U}_{av}$. This surprising conclusion is confirmed by Taylor's simple but elegant experiments.

As an exercise the student is urged to study and understand Taylor's classic papers in which these relations are developed [58, 59].

**Laminar flow in round tubes**  Fluid enters a round tube with a concentration $c_0$ and the parabolic velocity profile characteristic of laminar flow. The concentration at the wall is maintained at the constant value $c_i$ over a distance $x$, and solute enters or leaves the stream across the cylindrical surface. Concentration is obtained as a function of $r$ and $x$ by transposing Eq. (3.36) to cylindrical coordinates and solving with the boundary conditions $c = c_0$ at $t = 0$ and at $x = 0$, and $c = c_i$ at $r = r_0$ for $x > 0$. In terms of $c_{av}$ the solution is

$$\frac{c_i - c_{av}}{c_i - c_0} = 0.819\, e^{-14.6272\beta''} + 0.0976\, e^{-89.22\beta''} + 0.0135\, e^{-212.2\beta''} + \cdots \qquad (3.72)$$

where $\beta'' = D_{AB} x / 4 r_0^2 \bar{U}_{av}$. Jakob [34] gives slightly different values of the constants. This solution is adapted from that given by Drew [20] for the Graetz problem of heat conduction from a tube wall into a fluid flowing in laminar flow with a parabolic velocity distribution. It is the equivalent of the imaginary case of transient heat conduction into a cylindrical rod in which the heat capacity is a parabolic function of the distance $r$ from the axis.

The corresponding solution for laminar flow between parallel flat plates is given by Brown [8] and Butler and Plewes [10].

The Graetz analysis has been modified by Pigford [46] to allow for variations of fluid viscosity and density with temperature when the fluid is heated or cooled in laminar flow through a vertical tube. Pigford's analysis has been adapted by Goldman and Barrett [27] for the case of mass transfer to allow for variation of fluid properties and diffusion coefficient with concentration of the diffusing substance. Goldman and Barrett report experimental studies of the dissolution of a salt tube with laminar flow ($0.03 < \mathrm{Re} < 140$) of aqueous solutions of glycerol.

If the radial velocity gradient is ignored [$\bar{U}(r) = \bar{U}_{av} = $ constant], the problem is simply that of unsteady-state diffusion into a stagnant or solid cylinder. Levêque's solution [20, 37] for the analogous heat-transfer case involves rodlike flow in the core of the fluid but a linear velocity gradient very near the wall. Data [40] on dissolution of a slightly soluble tube into water in laminar flow have been found to agree better with the Levêque solution than with the more elegant Graetz analysis.

**Point source in a uniform velocity field (see Fig. 3.8)**  This case has many applications for both molecular and eddy diffusion, as in studies of the spread of gases from tall stacks and smoke generators. It applies to the data obtained by Walker and

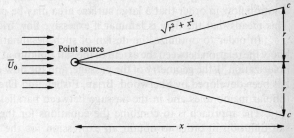

FIGURE 3.8
Diffusion from a point source in a flowing stream.

Westenberg [66], who have measured $D_{AB}$ in high-temperature gases by observing the radial diffusion from a point source of one gas spreading into a slowly rising stream of a second.

The concentration is given as a function of position downstream from the source by [68]

$$c = \frac{g_F}{4\pi D_{AB}\sqrt{x^2 + r^2}} \exp\left[-\frac{U_0}{2D_{AB}}\left(\sqrt{x^2 + r^2} - x\right)\right] \tag{3.73}$$

The velocity $U_0$ is constant everywhere. The material $A$ is supplied at a steady rate $g_F$ moles per second at a point (the origin); $c$ is the concentration in moles per cubic centimeter at the radius $r$ in the plane normal to the axis a distance $x$ from the origin in the direction of flow.

Data obtained by measuring the spread of concentration from a point source in a uniform flow field are conveniently analyzed by plotting $\log (c\sqrt{x^2 + r^2})$ vs. $(\sqrt{x^2 + r^2} - x)$; if $D_{AB}$, $g_F$, and $U_0$ are constant the points should fall on a straight line and an average value of $D_{AB}$ obtained from either the intercept or slope.

The solution for the case of an axially located point source in a stream flowing at uniform velocity through a finite round conduit is given by Bernard and Wilhelm [3].

**Desalination by reverse osmosis** One of the methods of recovering fresh water from sea water or inland brackish waters is based on the discovery of plastic membranes which allow the transport of water but little or no salt. Pressure in excess of the osmotic pressure is applied to the saline solution on one side of the membrane, causing water to pass through it to an outlet passage where the pressure is low. Water moving to and through the membrane surface transports salt, which tends to accumulate at the membrane surface until its concentration is such that the salt concentration gradient is sufficiently great to cause salt to diffuse back to the bulk liquid as fast as it is carried toward the surface by the water flux. This salt buildup at the surface is a serious disadvantage, since the effective osmotic pressure which must be exceeded is increased. The result is that a greater pressure must be applied to cause water transport, and the operating cost is increased.

The available membranes operate at low-water fluxes and the membrane area must be quite large. It is logical, therefore, to employ narrow passages for the solution flow in order that a large surface area may be packaged in a small volume. This means that the flow is laminar if excessive flow friction is to be avoided.

In order to optimize the design of such desalination units, it is desirable to know the relation between the extent of the salt buildup at the surface ("concentration polarization"), the geometry of the passage, and the flow rate. The applicable theory has been developed by Sherwood, Brian, Fisher, and Dresner [54] for flow in round tubular membranes and in the passage between parallel flat membranes.

The approach is to combine the equations for the velocity field and that for salt diffusion in order to obtain an expression for the local salt concentration at all points in the passage. Figure 3.9 illustrates the geometry in the case of parallel flat membranes.

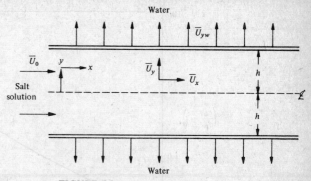

FIGURE 3.9

Reverse osmosis with membranes at walls of a flat channel.

For the case of uniform water flux through the membranes ($\overline{U}_{yw}$ = constant, cm/s) over the dimensionless length $L = x/h$ in the flow direction $x$, Berman [2] gives the velocity field in the forms

$$\frac{\overline{U}_x}{\overline{U}_0} = \frac{3}{2}\left(1 - \frac{\overline{U}_{yw}L}{\overline{U}_0}\right)(1 - R^2)\left[1 - \frac{\overline{U}_{yw}h}{420v}(2 - 7R^2 - 7R^4)\right] \tag{3.74}$$

and

$$\frac{\overline{U}_y}{\overline{U}_{yw}} = \frac{R}{2}(3 - R^2) - \frac{\overline{U}_{yw}y}{280v}(2 - 3R^3 + R^6) \tag{3.75}$$

where the principal symbols are explained by reference to Fig. 3.9; $\overline{U}_x$ and $\overline{U}_y$ are the velocity components in the $x$ and $y$ directions at any point, $R = y/h$; $\overline{U}_0$ is the mean velocity of the laminar flow at the inlet; and $v$ is the kinematic viscosity.

The steady-state diffusion equation, which follows from Eq. (3.36), is

$$\frac{\partial(\overline{U}_x c)}{\partial x} + \frac{\partial(\overline{U}_y c)}{\partial y} - D_{AB}\frac{\partial^2 c}{\partial y^2} = 0 \tag{3.76}$$

where $c$ is the salt concentration at any point, and $D_{AB}$ is the diffusion coefficient for salt in water. This neglects axial diffusion as compared with transport across the narrow channel. The last three equations are combined and solved with the initial and boundary conditions:

$$c = c_0 \quad \text{at } x = 0 \qquad c = \frac{D_{AB}}{\overline{U}_{yw}h}\left(\frac{\partial c}{\partial R}\right)_{R=1}$$

$$\text{at } y = h \qquad \text{and} \qquad \frac{\partial c}{\partial R} = 0 \quad \text{at } R = 0$$

The mathematics is difficult but the solution can be represented graphically by Fig. 3.10. The extent of the concentration polarization is measured by the ratio

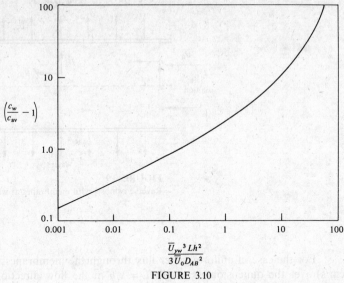

$$\left(\frac{c_w}{c_{av}} - 1\right)$$

$$\frac{\overline{U}_{yw}{}^3 L h^2}{3 \overline{U}_0 D_{AB}{}^2}$$

**FIGURE 3.10**
Concentration polarization in reverse osmosis.

$c_w/c_{av}$, where $c_w$ is the salt concentration at the membrane surface, and $c_{av}$ is the average salt concentration of the stream at the same distance $x$ from the inlet. The resulting relation provides the basis for design calculations necessary to balance the cost of the larger surface required when $c_w/c_{av}$ is large and the cost of the additional power to increase $\overline{U}_0$ and reduce the concentration polarization.

The corresponding problem for laminar flow through a tubular membrane is also solved in Ref. 54, and the solution for variable wall flux and incomplete rejection of salt at flat membranes is given by Brian [6, 7]. These analyses have been checked experimentally by several investigators, e.g., Thomas [60].

Though this problem is mathematically complex, it should be evident that it is basically similar to the others discussed in this section. In each, the analysis involves the combination of the velocity field with the appropriate form of the equation for molecular diffusion. The result is a relation describing concentration as a function of position, from which gradients and fluxes may be obtained.

**Laminar flow around a sphere** Mass transfer between a sphere and an ambient fluid is involved in many situations of importance. Transfer between catalyst pellets and a reacting gas, or between liquid and suspended crystals which are dissolving, approximates that for single spherical particles of similar size. Gas absorption from bubbles, and extraction from a liquid drop by a second liquid, are examples of mass transfer to spheres of fluid.

Because of its importance, the subject of mass transfer to both solid and fluid spheres has received much study, both theoretical and experimental. Friedlander [24], for example, shows how the theoretical velocity field may be combined with diffusion theory to provide a useful expression for mass-transfer coefficients in the case of solid spheres and flow at very low Reynolds numbers. This and related studies will be described in Chap. 6, where correlations of experimental data are presented for a wide range of flow conditions, including both laminar and turbulent regimes.

Example 3.5 illustrates the general approach to the development of a theory for the rather special case of the flow of liquid over a sphere as a thin laminar film.

EXAMPLE 3.5 It is required to calculate the rate of mass transfer into the laminar film of liquid flowing over a solid sphere. As in the several other cases noted, this involves diffusion into a moving stream in which the velocity field is known. Develop an expression for the rate of absorption of a gas into such a film, under conditions where it may be assumed that the liquid surface is saturated with the ambient gas being absorbed. The rate of absorption is to be expressed in terms of the geometry, flow rate, and relevant physical properties (see Fig. 3.11).

SOLUTION The steady-state flow of a newtonian liquid over a sphere is described by the following equations, easily obtained by equating the force of gravity to the resisting shear force on a differential element of fluid in the film:

$$(2\pi R \sin\theta)(Rd\theta)\rho yg \sin\theta = -(2\pi R \sin\theta)(Rd\theta)\mu \frac{d\overline{U}}{dy} \tag{a}$$

$$\rho g(\sin\theta)y\,dy = -\mu d\overline{U} \tag{b}$$

$$\overline{U}_i = \frac{y_0{}^2 g \sin\theta}{2v} = \frac{3g \sin^{-1/3}\theta}{4\pi R y_{OE}} \tag{c}$$

$$\overline{U} = \frac{g \sin\theta}{2v}(y_0{}^2 - y^2) = \overline{U}_i\left(1 - \frac{y^2}{y_0{}^2}\right) \tag{d}$$

$$\overline{U}_{av} = \frac{1}{y_0}\int_0^{y_0} \overline{U}\,dy = \frac{g(\sin\theta)y_0{}^2}{3v} = \frac{q}{2\pi R(\sin\theta)y_0} \tag{e}$$

$$y_0 = \left(\frac{3gv}{2\pi Rg \sin^2\theta}\right)^{1/3} = y_{OE} \sin^{-2/3}\theta \tag{f}$$

$$y_{OE} = \left(\frac{3gv}{2\pi Rg}\right)^{1/3} \tag{g}$$

The meaning of the symbols is suggested by Fig. 3.11: $y$ is the radial distance in from the liquid surface; $y_0$ is the film thickness and $y_{OE}$ the film thickness at the equator ($\theta = \pi/2$); $\overline{U}_i$ is the surface velocity and $\overline{U}$ the tangential velocity at $y$; $g$ is the acceleration due to gravity, $\mu$ the liquid viscosity, $v$ the kinematic viscosity, and $\rho$ the density; $q$ is the volumetric liquid feed rate at the top center.

Now write a solute balance on a differential ring of liquid having a perimeter $2\pi R \sin\theta$ (at constant height $\theta$) and a radial thickness $dy$. This element is indicated in cross section

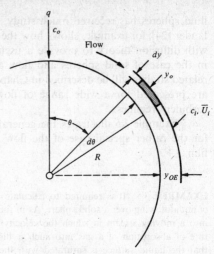

**FIGURE 3.11**
Laminar flow of liquid over a solid sphere.

by the solid rectangle on Fig. 3.11. Neglecting $y_0$ and $y$ in comparison with $R$, the solute input by circumferential flow is

$$\bar{U} c 2\pi R \sin \theta \, dy$$

and output is

$$\bar{U}\left(c + \frac{\partial c}{\partial \theta} \, d\theta\right) 2\pi R \sin \theta \, dy$$

At the same time there is radial diffusion into the element at the rate

$$-(2\pi R \sin \theta)(R d\theta) D \frac{\partial c}{\partial y}$$

where $D$ is the molecular diffusion coefficient. The radial diffusion out of the element is at the rate

$$-(2\pi R \sin \theta)(R d\theta) D \left[\frac{\partial c}{\partial y} + \frac{\partial}{\partial y}\left(\frac{\partial c}{\partial y}\right) dy\right]$$

Equating solute input and output for the element,

$$\bar{U}\left(\frac{\partial c}{\partial \theta}\right) = DR \frac{\partial^2 c}{\partial y^2} \tag{h}$$

This allows for radial diffusion but no radial flow into or out of the differential element, which means that its surface is in the curved plane of the stream lines of flow. The partial differential on the left, therefore, is to be taken at constant volumetric flow between any two values of $y$. The flow at any $\theta$ is found by integrating Eq. $(d)$ with the two values of $y$ as limits, and multiplying by the cross section of the element. Inspection of Eqs. $(d)$, $(c)$, and $(g)$ shows that the result must vary only with $y/y_0$, for which the symbol $p$ will be used $(q, \theta, R, v,$ and $g$ being constant).

Equation (h), therefore, becomes

$$\bar{U}_i(1 - p^2)\frac{y_0^2}{R}\left(\frac{\partial c}{\partial \theta}\right)_p = D\left(\frac{\partial^2 c}{\partial p^2}\right)_\theta \tag{i}$$

Let $\varphi$ be defined by the equation

$$\frac{d\varphi}{d\theta} = \frac{R}{\bar{U}_i y_0^2} \tag{j}$$

with $\varphi$ arbitrarily assigned the value zero at $\theta = 0$. Then Eq. (i) becomes

$$(1 - p^2)\left(\frac{\partial c}{\partial \varphi}\right)_p = D\left(\frac{\partial^2 c}{\partial p^2}\right)_\varphi \tag{k}$$

This is the final form of the desired differential expression.

Assuming equilibrium between the pure ambient gas and the liquid surface, the solute concentration at the surface may be assigned the constant value $c_i$ over the whole surface. The boundary conditions are then $c = c_i$ at $p = 0$, $\partial c/\partial y = 0$ at $p = 1$, and $c = c_0$ (the feed concentration) at $\theta = 0$. Equation (k) and the boundary conditions correspond exactly to the problem solved by Pigford [45] for mass transfer into a laminar liquid film on a vertical wall, and presented in Sec. 3.8. The result expresses $c$ as a function of $p$ and $\varphi$.

The rate of absorption in $g$ moles per second over the sphere from the top to a point of liquid withdrawal at $\theta = \theta_2$ is obtained from

$$\int N_A dS = -D\int_0^{\theta_2} 2\pi R^2 \sin\theta\left(\frac{\partial c}{\partial y}\right)_{y=0} d\theta$$

Substituting Eqs. (c), (f), and (j), this becomes

$$\int N_A dS = -\frac{3qD}{2}\int_0^{\varphi_2}\left(\frac{\partial c}{\partial p}\right)_{p=0} d\varphi \tag{l}$$

where the limit $\varphi_2$ is obtained from

$$\varphi_2 = \int_0^{\theta_2}\frac{R}{\bar{U}_i y_0^2} d\theta \tag{m}$$

This derivation is taken from Davidson and Cullen [18], who point out that the simple penetration equation [Eq. (3.43)] may be used in cases where the exposure time is short, in place of the more complicated series solution for flow with a parabolic velocity distribution in the film. To do this, $(\partial c/\partial p)_{p=0}$ is obtained by differentiating the equation.

$$\frac{c - c_0}{c_i - c_0} = \text{erfc}\left(\frac{p}{2\sqrt{D\phi}}\right) = 1 - \text{erf}\left(\frac{p}{2\sqrt{D\phi}}\right)$$

and the mass-transfer flux for the entire sphere is obtained by substitution in Eq. (l), with $\phi_2$ obtained from Eq. (m), the angle $\theta_2$ being $\pi$. The result is

$$\int_0^S N_A ds = 5.09(c_i - c_0)g^{1/6}v^{-1/6}R^{7/6}q^{1/3}D^{1/2} \tag{n}$$

The student may find it interesting to follow the suggested procedure and check this result.

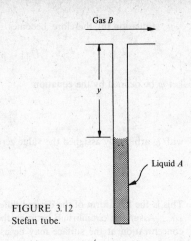

FIGURE 3.12
Stefan tube.

Davidson and Cullen carried out experiments in which pure gases were absorbed by water flowing in a laminar film over metal spheres. The data at low liquid rates, in the absence of ripples, agreed well with the theory. A similar study was carried out earlier by Lynn, Straatemeier, and Kramers [41]. Their results also agreed well with the theory, as did the measurements of Hikita and Nakanishi [31] of $CO_2$ absorption by water on single spheres.  ////

## 3.8  QUASI-STEADY-STATE DIFFUSION

The steady-state equations for molecular diffusion may be used with little error in many situations where conditions are changing with time. Unfortunately, it is often quite difficult to determine the error involved in doing this; the exact solution is needed to estimate the error and would be used if obtainable and simple.

**Stefan tube**  One experimental method of measuring $D_{AB}$ in binary gas systems is to place a liquid $A$ at the bottom of a tube of small diameter containing a gas $B$ (see Fig. 3.12). Pure $B$ is passed slowly over the top end of the tube, thus maintaining the partial pressure $p_{A2}$ zero at that point. The partial pressure $p_{A1}$ in the gas adjacent to the liquid surface is assumed equal to the vapor pressure of $A$ at the temperature of the experiment. Diffusion of $A$ through $B$ occurs in the gas-filled tube of the varying length $y$. The rate of diffusion is determined by the rate of fall of the level of the liquid, which has a known molal density $\rho_M$.

Though this is clearly a case of transient diffusion, the data obtained are usually interpreted by equating the steady-state flux, as given by Eq. (3.8), to the rate of vaporization:

$$N_A = \frac{D_{AB} P\,(p_{A1} - p_{A2})}{RTy \quad p_{BM}} = \rho_M \frac{dy}{dt} \tag{3.77}$$

whence
$$D_{AB} = \frac{RT p_{BM} \rho_M (y_t{}^2 - y_0{}^2)}{2P(p_{A1} - p_{A2})t} \tag{3.78}$$

where $y_0$ and $y_t$ are the heights of the gas space at $t = 0$ and $t = t$, respectively.

The use of the steady-state expression introduces two errors: (1) The distribution of $A$ through the gas space at the beginning of the test may be quite different from that prevailing under steady-state conditions, and (2) some of the observed drop in liquid level produces vapor which occupies the space originally filled with liquid and does not diffuse out of the tube.

These errors are very small. Lee and Wilke [38] have shown the first to be small by an analysis of the transient approach to the steady-state flux. The second is evidently small because the partial density of $A$ in the space $y_t - y_0$ is so small compared with that of the liquid. A third minor error, due to the radial velocity gradients in the vapor moving upward in the tube, is discussed by Rao and Bennett [48]. The effect of varying the gas-flow rate over the top of the tube is discussed in a recent critique of the method by Pommersheim and Ranck [47].

**Dissolution of a solid sphere**   Example 3.3 pertained to the radial diffusion of benzoic acid into water from the surface of a solid sphere 1.0 cm in diameter. For illustrative purposes it was stated that the diameter of the sphere did not change. If a solid sphere of benzoic acid were allowed to dissolve, there would be two transients involved, as in the Stefan tube: the changing concentration distribution radially, and the changing radius and dissolution surface of the sphere. The second leads to replacement of solid by nearly saturated solution. Furthermore, spherical surface of no net volume transport moves with respect to the sphere center.

The use of the steady-state equation [Eq. (d) of Example 3.3, with the last term omitted] could lead to very serious error in this case. As shown in Example 3.3, the transient concentration gradient approaches the steady-state condition extremely slowly. This is owing to the small value of $D_{AB}$ and to the fact that the molar concentration of acid in solution is relatively large (very much greater than that of gas $A$ in the gas space in the Stefan tube). The amount of solute contained in the shell originally occupied by solid is evidently significant, and the use of the steady-state equation would appear to introduce considerable error, even if the initial radial concentration distribution corresponded exactly to that for steady-state dissolution at the original radius.

Exact analysis of the situation described would present a difficult mathematical problem; it is not known that a solution has been obtained.

Brian and Hales [5] obtained numerical solutions of the combined flow and diffusion equations for Stokes flow past a sphere. The results show the manner in which mass transfer to or from the sphere varies, not only with bulk flow past the sphere but also with radial transpiration velocity and with rate of change of sphere diameter with time. This comprehensive analysis of transport to and from spheres will be described in Chap. 6, where it is compared with experimental data.

# APPENDIX

# METHODS OF OBTAINING APPROXIMATE SOLUTIONS TO THE STEFAN-MAXWELL EQUATIONS FOR MULTICOMPONENT DIFFUSION

The methods outlined apply to the solution of Eq. (3.18) for the steady-state diffusion in a multicomponent gas mixture contained in a flat slab or "film" of thickness $y_0$.

Wilke [67]. The flux of $A$ is given by

$$N_A = \frac{D''_A P}{RT y_0 p'_A} (p_{A1} - p_{A2}) \tag{A.1}$$

where $p_{A1}$ and $p_{A2}$ are the partial pressures of $A$ at the two faces. The quantity $p'_A$ is defined by

$$p'_A = \frac{P}{\phi_A} (\phi_A - Y_A)_M \tag{A.2}$$

where the subscript $M$ indicates the logarithmic mean of the values of $(\phi_A - Y_A)$ at the two faces. The symbol $\phi_A$ represents $N_A / \Sigma N$, all fluxes taken as positive in the direction of increasing $y$.

The diffusion coefficient $D''_A$ is given in terms of the binary coefficients by

$$D''_A = \left[ \frac{y''_B}{D_{AB}} = \frac{y''_C}{D_{AC}} + \frac{y''_D}{D_{AD}} \cdots \right]^{-1} \tag{A.3}$$

where

$$y''_B = \frac{p'_{AB}}{p'_{AB} + p'_{AC} + p'_{AD} + \cdots} \tag{A.4}$$

$$y''_C = \frac{p'_{AC}}{p'_{AB} + p'_{AC} + p'_{AD} + \cdots} \quad \text{etc.} \tag{A.5}$$

$$p'_{AB} = \frac{P}{\phi_{AB}} (y'_A + y'_B)(\phi_{AB} - y'_{AB}) \tag{A.6}$$

$$p'_{AC} = \frac{P}{\phi_{AC}} (y'_A + y'_C)(\phi_{AC} - y'_{AC}) \quad \text{etc.} \tag{A.7}$$

$$y'_{AB} = \frac{y'_A}{y'_A + y'_B} \qquad y'_{AC} = \frac{y'_A}{y'_A + y'_C} \quad \text{etc.} \tag{A.8}$$

$$\phi_{AB} = \frac{N_A}{N_A + N_B} \qquad \phi_{AC} = \frac{N_A}{N_A + N_C} \quad \text{etc.} \tag{A.9}$$

$$y'_A = \frac{y_{A1} + y_{A2}}{2} \qquad y'_B = \frac{y_{B1} + y_{B2}}{2} \quad \text{etc.} \tag{A.10}$$

This imposing set of relations is not difficult to apply in a numerical example, though some trial and error is usually involved. Wilke gives a sample calculation showing how the method is employed. (See Example 3.2.)

In extreme cases where the fluxes differ greatly, Eq. (A.7) may give a negative value of $p'_{AB}$, or $p'_{AC}$, etc. In such cases Wilke recommends that the negative value be replaced by zero for substitution in Eq. (A.5). Toor [61], however, thinks the negative value should be retained.

Toor [61]. (a) For the case of equal molal diffusion ($\Sigma N = 0$) in a ternary gas, Toor suggests the following:

$$N_A = \frac{D_{AC} P}{RT y_0} \phi_C (\delta_m y_{A1} - y_{A2})$$ (A.11)

$$N_B = \frac{D_{BC} P}{RT y_0} \phi_C (\delta_m Y_{B1} - Y_{B2})$$ (A.12)

where $\quad \delta_m = \exp\left[\left(1 - \frac{D_m}{D_{AB}}\right)(Y_{C1} - Y_{C2})\right]$ (A.13)

$$\phi_c = \frac{Y_{C1} - Y_{C2}}{(1 - Y_{C2}) - \delta_m(1 - Y_{C1})}$$ (A.14)

$$D_m = \frac{D_{AC} + D_{AB}}{2}$$ (A.15)

The approximation is introduced by the use of Eq. (A.14), so it is recommended that the choice of component $C$ be such that $D_{BC}$ and $D_{AC}$ are most nearly equal of the three possible choices. This is claimed to give good agreement with the exact solutions, though $N_C$, found by difference from $\Sigma N = 0$, may be quite in error if small.

(b) The corresponding procedure for the case of a ternary with one stagnant and two diffusing gases is based on the following:

$$N_A = \frac{D_{AC} P}{RT y_0} \phi'_C (\delta'_m Y_{A1} - Y_{A2})$$ (A.16)

$$N_B = \frac{D_{BC} P}{RT y_0} \phi'_C (\delta'_m Y_{B1} - Y_{B2})$$ (A.17)

where $\quad \delta'_m = \left(\frac{Y_{C2}}{Y_{C1}}\right)^{D_m/D_{AB}}$ (A.18)

and $\quad \phi'_C = \frac{\ln(Y_{C1}/Y_{C2})}{(1 - Y_{C2}) - \delta'_m(1 - Y_{C1})}$ (A.19)

Shain [53]. This is a modification of Wilke's procedure, involving an average of $D_{AM}/P(1 - Y_A \Sigma N/N_A)$ instead of separate averaging of $D_{AM}$ and $P(1 - Y_A \Sigma N/N_A)$:

$$N_A = \frac{P}{RT y_0} \frac{(Y_{A1} - Y_{A2})}{\left[\sum_{n \neq A}\left(Y_n - \frac{Y_A N_n}{N_A}\right)\frac{1}{D_{An}}\right]_M}$$ (A.20)

which applies to each of the $n$ constituents of the mixture. The subscript $M$ indicates the logarithmic mean of the quantity in square brackets, evaluated at the two faces of the gas space.

The three procedures described have not been tested by comparison with the rigorous equations over a wide enough range to provide an adequate judgment of their relative merits. Example 3.2 illustrates their application.

# REFERENCES

1 BARRER, R. M.: *Trans. Faraday Soc.*, **35**: 628 (1939).
2 BERMAN, A. S.: *J. Appl. Phys.*, **24**: 1232 (1953).
3 BERNARD, R. A., and R. H. WILHELM: *Chem. Eng. Progr.*, **46**: 233 (1950).
4 BIRD, R. B., W. E. STEWART, and E. N. LIGHTFOOT: "Transport Phenomena," Wiley, New York, 1960.
5 BRIAN, P. L. T., and H. B. HALES: *AIChE J.*, **15**: 419 (1969).
6 BRIAN, P. L. T.: *Proc. 1st Int. Symp. Water Desalination*, 349 (1965).
7 BRIAN, P. L. T.: M.I.T. Desalination Lab. *Rep.* 295-7, May 12, 1965.
8 BROWN, G. M.: *AIChE J.*, **6**: 179 (1960).
9 BURCHARD, J. K., and H. L. TOOR: *J. Phys. Chem.*, **66**: 2015 (1962).
10 BUTLER, R. M., and A. C. PLEWES: *Chem. Eng. Prog. Symp. Ser.* **50** (10): 121 (1954).
11 CARSLAW, H. S., and J. C. JAEGER: "Conduction of Heat in Solids," 2d ed., Oxford, New York, 1959.
12 CHORNY, R. C., and J. H. KRASUK: *Ind. Eng. Chem. Process Des. Dev.*, **5**: 206 (1966).
13 COCHRAN, W. G.: *Proc. Cambridge Phil. Soc.*, **30**: 365 (1934).
14 CRANK, J.: "The Mathematics of Diffusion," Clarendon, Oxford,
15 CULLEN, E. J., and J. F. DAVIDSON: *Trans. Faraday Soc.*, **53**: 113 (1957).
16 CULLINAN, H. T., and H. L. TOOR: *J. Phys. Chem.*, **69**: 3941 (1965).
17 CURTISS, C. F., and J. O. HIRSCHFELDER: *J. Chem. Phys.*, **17**: 552 (1949).
18 DAVIDSON, J. F., and E. J. CULLEN: *Trans. Inst. Chem. Eng. (London)*, **35**: 51 (1957).
19 DAVIES, J. T., and S. T. TING: *Chem. Eng. Sci.*, **22**: 1539 (1967).
20 DREW, T. B.: *Trans. AIChE*, **26**: 26 (1931).
21 DUDA, J. L. and J. S. VRENTAS: *AIChE J.*, **15**: 351 (1969).
22 DUNCAN, J. B., and H. L. TOOR: *AIChE J.*, **8**: 38 (1962).
23 FAIRBANKS, D. F., and C. R. WILKE: *Ind. Eng. Chem.*, **42**: 471 (1950).
24 FRIEDLANDER, S. K.: *AIChE J.*, **3**: 43 (1957); **7**: 347 (1961).
25 GETZINGER, R. W., and C. R. WILKE: *AIChE J.*, **13**: 577 (1967).
26 GILLILAND, E. R.: In T. K. Sherwood, "Absorption and Extraction," 1st ed., McGraw-Hill, New York, 1937.
27 GOLDMAN, M. R., and R. L. BARRETT: *Trans. Inst. Chem. Eng.*, **47**: T29 (1969).
28 GREGORY, D. P., and A. C. RIDDIFORD: *J. Chem. Soc.*, 3756 (1956).
29 GRENIER, PIERRE: *Can. J. Chem. Eng.*, **44**: 213 (1966).
29a HEINZELMANN, F. J., D. T. WASAN, and C. R. WILKE: *Ind. Eng. Chem. Fundam.*, **4**: 55 (1965).
30 HELLUND, E. J.: *Phys. Rev.*, **57**: 319, 328, 737 (1940).
31 HIKITA, H., and K. NAKANISHI: *Chem. Eng. (Japan)*, **23**: 513 (1959).
32 HIRSCHFELDER, J. O., CURTISS, C. F., and BIRD, R. B.: "Molecular Theory of Gases and Liquids," pp. 478–80, 485–88, Wiley, New York, 1954.
33 HOOPES, J. W.: Thesis in chemical engineering, Columbia University, 1951.
34 JAKOB, M.: "Heat Transfer," Wiley, New York, 1949.
35 KAYS, W. M.: "Convective Heat and Mass Transfer," McGraw-Hill, New York, 1966.
36 KEYES, J. J., and R. L. PIGFORD: *Chem. Eng. Sci.*, **6**: 215 (1957).

37 KNUDSEN, J. G., and D. L. KATZ: "Fluid Dynamics and Heat Transfer," McGraw-Hill, New York, 1958.

38 LEE, C. Y., and C. R. WILKE: *Ind. Eng. Chem.*, **46**: 2381 (1954).

39 LEVICH, V. G.: "Physicochemical Hydrodynamics," Prentice-Hall, Englewood Cliffs, N.J., 1962.

40 LINTON, W. H., and T. K. SHERWOOD: *Chem. Eng. Progr.*, **46**: 258 (1950).

41 LYNN, S., J. R. STRAATEMEIER, and H. KRAMERS: *Chem. Eng. Sci.*, **4**: 63 (1955).

42 MA, Y. H., and L. B. EVANS: *AIChE J.*, **14**: 956 (1968).

43 NEWMAN, A. B.: *Trans. AIChE*, **27**: 310 (1931).

43a OLBRICH, W. E., and J. D. WILD: *Chem. Eng. Sci.*, **24**: 25 (1969).

44 PERRY, J. H.: "Chemical Engineers' Handbook," 3d ed. and 5th ed., McGraw-Hill, New York, 1950 and 1974.

45 PIGFORD, R. L.: Thesis in chemical engineering, Univ. of Ill., 1941.

46 PIGFORD, R. L.: *Chem. Eng. Progr., Symp. Ser.*, **51** (17): 79 (1955).

47 POMMERSHEIM, J. M., and B. A. RANCK: *Ind. Eng. Chem. Fundam.*, **12**: 246 (1973).

48 RAO, S. S., and C. O. BENNETT: *Ind. Eng. Chem. Fundam.*, **5**: 573 (1966).

49 ROHSENOW, W. M., and H. Y. CHOI: "Heat, Mass, and Momentum Transfer," Prentice-Hall, Englewood Cliffs, N.J., 1961.

50 ROSNER, D. E., and M. EPSTEIN: *J. Phys. Chem.*, **74**: 4001 (1970).

51 SCHLICHTING, H.: "Boundary Layer Theory," McGraw-Hill, New York, 1960.

52 SECOR, R. M.: *AIChE J.*, **11**: 452 (1965).

53 SHAIN, S. A.: *AIChE J.*, **7**: 17 (1961).

54 SHERWOOD, T. K., P. L. T. BRIAN, R. E. FISHER, and L. DRESNER: *Ind. Eng. Chem. Fundam.*, **4**: 113 (1965).

55 SHUCK, F. L., and H. L. TOOR: *J. Phys. Chem.*, **67**: 540 (1963).

56 SMITH, A. S.: *Ind. Eng. Chem.*, **26**: 1167 (1934).

57 STEVENSON, W. H.: *AIChE J.*, **14**: 350 (1968).

57a STEWART, W. E., and R. PROBER: *Ind. Eng. Chem. Fundam.*, **3**: 224 (1964).

58 TAYLOR, G. I.: *Proc. Roy. Soc. (London)*, **A219**: 186 (1953); **A225**: 473 (1954).

59 TAYLOR, G. I.: *Proc. Phys. Soc. (London)*, **B67**: 857 (1954).

60 THOMAS, D. G.: *Ind. Eng. Chem. Fundam.*, **11**: 303 (1972).

61 TOOR, H. L.: *AIChE J.*, **3**: 198 (1957).

62 TOOR, H. L., and R. T. SEBULSKY: *AIChE J.*, **7**: 558 (1961).

63 TOOR, H. L.: *AIChE J.*, **10**: 448, 460 (1964).

64 TOOR, H. L., C. V. SESHADRI, and K. R. ARNOLD: *AIChE J.*, **11**: 746 (1965).

65 VON KÁRMÁN, TH.: *Z. Angew. Math. Mech.*, **1**: 244 (1921).

66 WALKER, R. E., and A. A. WESTENBERG: *J. Chem. Phys.*, **29**: 1139 (1958); **31**: 519 (1959); **32**: 436 (1960).

67 WILKE, C. R.: *Chem. Eng. Progr.*, **46**: 95 (1950).

68 WILSON, H. A.: *Proc. Cambridge Phil. Soc.*, **12**: 406 (1904).

## PROBLEMS

*3.1* Walker and Westenberg [66] describe a technique for measuring binary gas-diffusion coefficients which is especially applicable for measurements at high temperatures.

The apparatus consists of a vertical passage through which gas *B* flows very slowly upward. Screens are arranged to ensure plug or piston flow. Gas *A* is introduced through a tiny upward-pointing hypodermic tube fixed at the axis of the gas passage.

The steady flow of $A$ is set at a rate which does not disturb the flat velocity profile of the flowing $B$. Concentrations downstream from the point of injection of $A$ are measured using a fine probe and a conductivity cell. The cell reading in millivolts is proportional to the concentration of gas $A$. Very precise measurements are obtained.

A test with such a device is made with helium injected into flowing nitrogen at 25°C and 1.0 atm. The uniform-flow velocity is 40 cm/s. Concentrations are measured at several radial positions $y$ centimeters from the axis, at a plane normal to the duct axis 10.31 mm downstream from the helium injector.

The following data are obtained.

| $y$, cm | 0 | 0.08 | 0.161 | 0.296 | 0.414 | 0.521 |
|---|---|---|---|---|---|---|
| Cell reading, mV | 4.92 | 4.49 | 3.42 | 1.43 | 0.452 | 0.153 |

(a)  Using these data, determine the value of $D_{AB}$ for helium in nitrogen under the test conditions.

(b)  Compare the result with that predicted by statistical-mechanics theory (Chap. 2).

3.2  A gas mixture consists of hydrogen containing 10 mole percent pure olefin. In contact with a pelleted porous catalyst at 2.0 atm and 500°K, the olefin reacts to form the corresponding paraffin. Resistance to mass transfer from gas to catalyst pellet is negligible. The pores are small and diffusion is of the Knudsen type. The surface reaction is irreversible and first order in olefin concentration of the gas in the pore. The activation energy $(E)$ of the surface reaction is 20 kcal/g mole, and

$$k_v = \text{g moles/(s) (cm}^3 \text{ pellet) (g mole olefin/cm}^3 \text{ gas)}$$
$$= Ae^{-E/RT}$$

$A$ is independent of temperature and pressure.

It is desired to estimate the variation of the overall reaction rate (moles per second) with changes in temperature and pressure. Calculate the *percent* change in rate (a) per °K, and (b) per atmosphere increase in operating temperature and pressure (from 500°K, 2 atm, at 10 mole percent olefin).

*Note:* The catalyst is quite active, so the effectiveness factor $\eta$ is about 0.3; in this range $\eta$ is inversely proportional to the Thiele modulus $\phi$. The "effectiveness factor" is defined as the ratio of the actual reaction rate to the rate which would be attained if all the gas in the pores had the same concentration as that of the ambient gas in contact with the pellet. The Thiele modulus is defined by

$$\phi = R_p \left( \frac{k_v}{D_K} \right)^{1/2}$$

where $R_p$ is the radius of the pellet and $D_K$ is the Knudsen diffusion coefficient of olefin in the pores.

3.3  Cullen and Davidson [15] report measurements of the rate of absorption of $CO_2$ by a small cylindrical water jet. The water issued vertically downward from a small round orifice and was collected and removed through a tube of essentially the same diameter as the liquid jet. The gas surrounding the jet was pure $CO_2$ saturated with water.

The following data were obtained in one test: temperature, 25°C; partial pressure $CO_2$, 750 mm Hg; jet height, 5.0 cm; water rate, 12 cm$^3$/s; $CO_2$ absorption rate, $2.00 \times 10^{-4}$ g/s. The water fed contained no $CO_2$.

From these data calculate the molecular diffusion coefficient for $CO_2$ in water at 25°C. Assume that there was no mixing of liquid within the jet and that the cylindrical surface of the jet was in equilibrium with the ambient gas (the solubility of $CO_2$ in water at 25°C and 750 mm Hg pressure of $CO_2$ is 0.00150 g $CO_2$/cm$^3$ water). The diameter of the jet was not stated, but was evidently between 3 and 7 mm.

**3.4** A small temporary constant-humidity room consists of a space $10 \times 12 \times 8$ ft high enclosed by wallboard on the sides and top but with a sheet-metal floor. The room is surrounded by factory space with air at 97.5°F and 70 percent relative humidity. The air in the space is maintained at 97.5°F and 10 percent relative humidity by a silica-gel drying unit through which air from the space is continuously circulated and returned.

(a) With the room empty, what is the rate (pounds per hour) of water removal by the air-drying unit?

(b) The air-drying unit fails, and the humidity of the air in the space (initially 10 percent) rises. How long will it take for the humidity to rise to 45 percent?

*Data and assumptions:* The wallboard is nonhygroscopic, is 0.5 in thick, and has a porosity of 0.35. The total pressure, both inside and outside the room, is 760 mm Hg. The room is tight, except that the wallboard is porous. The tortuosity factor for diffusion through the porous wall is 4.0.

**3.5** Repeat Prob. 3.4 with the assumption that the wall is hygroscopic, the equilibrium moisture content being represented approximately by $c = 0.15h$, where $c$ is the moisture content as pounds of water per cubic foot of dry board, and $h$ represents percent relative humidity. Neglect any swelling and pore-size reduction due to moisture adsorption. Assume moisture transport to occur only by water vapor diffusion in air-filled pores.

*Note:* In this case a rigorous solution involving the stated assumptions presents certain mathematical difficulties. Having solved Prob. 3.4, however, it should be possible to obtain a numerical answer good to $\pm 25$ percent by mental arithmetic.

**3.6** A closed vessel is half full of water, which is agitated by a stirrer. The space over the liquid is filled with $CO_2$ saturated with water. A ball 1.0 cm in diameter floats on the surface exactly half submerged. Because of the motion of the liquid, the ball rotates on an axis which may be assumed to remain horizontal. As it rotates, fresh water is carried as a film on its surface and $CO_2$ is absorbed. The film is mixed with the bulk liquid when the surface submerges, and a new pure-water film is formed.

Estimate the rate of absorption by the rotating ball, expressed as gram moles per hour per square centimeter of flat liquid surface occupied by the ball. The solubility of $CO_2$ in water at the operating temperature is 0.037 g moles/l. The molecular diffusion coefficient for $CO_2$ in the water is $2.0 \times 10^{-5}$ cm/s. The rotational speed is 100 rpm. Neglect the movement of the water in the film on the ball surface exposed to the gas. The water film may be assumed to be sufficiently thick that essentially no $CO_2$ diffuses to the surface of the ball during one revolution.

**3.7** (a) A chemical engineer has a suggestion regarding the failure of the air-drying unit described in Prob. 3.4. He points out that a slow bulk flow of air outward through the porous wall would reduce or even eliminate the diffusion of water vapor into the room [Eq. (3.5)]. The only air available to pump into the room for this purpose is the factory air at 97.5°F and 70 percent relative humidity. The claim is that the moisture brought in with this air would be more than offset by the reduction in moisture diffusion through the wall. The moist-air addition would start after the dryer fails.

Do you think this proposal has merit? If so, determine the *constant* rate of air supply to the room in order that the maximum time (*after* the drying unit *fails*) be required for the humidity to rise from 10 to 45 percent in the mixed air in the room. Assume the pressure rise in the room to be negligibly small.

(b) Still another proposal involves the use of the air-drying unit to introduce outside air, completely dry, into the constant-humidity room. Instead of recycling air from the room through the dryer and back to the room, the drying unit would be supplied with factory air and deliver dry air to the room.

If this appears to be a useful idea, calculate the air rate required to maintain 10 percent relative humidity in the room, and the corresponding rate of water removal, as pounds per hour. Compare both results with the corresponding quantities for the original arrangement described in Prob. 3.4.

3.8 The permeability of vulcanized rubber to hydrogen is measured by a test on a homogenous hydrogen-free rubber sheet 0.10 cm thick free of pinholes. The rubber, 25 cm² in area, is clamped between two gas spaces in one of which hydrogen is maintained at a constant pressure of 1.0 atm. The gas space on the other (downstream) side of the rubber sheet has a fixed volume of 100 cm³ and is initially evacuated to below 1 μm Hg pressure.

The pressure in the second space is measured at intervals and is seen to rise as hydrogen dissolves, diffuses through the rubber, and appears in the downstream gas space. The entire apparatus is maintained at 25°C.

A graph of the pressure $p$ (in μm Hg) in the downstream gas space, as ordinate, vs. time in seconds shows a smooth curve starting at the origin and concave upward over a period of several minutes. This curve soon becomes asymptotic to a straight line which is well represented by the empirical relation

$$p \; (\mu m \; Hg) = 0.646 \, (t - 196)$$

where $t$ is the total elapsed time in seconds.

From these results, determine (a) the solubility of hydrogen in rubber (gram moles per cubic centimeter) under 1.0 atm hydrogen pressure, and (b) the diffusion coefficient (square centimeters per second) of hydrogen dissolved in the rubber.

*Suggestions:* Assume the concentration of hydrogen in rubber at the sheet surfaces to be in equilibrium with the gas at 1.0 atm and at $p$, respectively. Assume the solubility of hydrogen in rubber to obey Henry's law, and that the desired diffusion coefficient of hydrogen in rubber is independent of hydrogen concentration.

In the course of the test the downstream pressure never reached 2.0 mm Hg, and may be neglected in relation to the pressure of 1.0 atm on the other face. With these assumptions it is possible to obtain a series solution to the transient diffusion equation which is found to become linear when $t$ becomes very large. Comparison of this solution with the empirical result given above leads to the desired values of solubility and diffusion coefficient. If the mathematics proves to be difficult, consult Crank [14] or Barrer [1].

3.9 A petroleum hydrotreating unit consists of a fixed bed of cobalt-molybdate porous catalyst fed continuously with hydrogen and oil. The mixture fed contains 8,000 ft³ hydrogen per barrel of fresh liquid feed. Operation is at 1,000 psia and 700°F, with downflow of both gas and liquid. It is estimated that half of the oil is vaporized under the operating conditions. The liquid hourly space velocity (LHSV ≡ cubic feet of total oil fed, before any vaporization, per hour per cubic feet of bed) is 2.0, and the hydrogen consumption is 1,000 ft$^3$/bbl total oil, measured as liquid. The hydrogenation reaction is irreversible. The catalyst pellets are 1/4-in spheres.

Estimate the fraction of the hydrogen driving force lost in accomplishing diffusion of hydrogen across the film of liquid flowing over the catalyst pellets.

The oil not vaporized has a molecular weight of 206, a viscosity of 0.08 cP, and a specific gravity of 0.54. The surface of the flowing liquid may be assumed to be in equilibrium with the gas and to contain 0.16 mole fraction hydrogen. The diffusion coefficient for hydrogen in the hot oil is estimated to be 8.0 $\times$ 10$^{-4}$ cm$^2$/s. The bed has a void fraction of 0.41 and the ratio of bed volume to cross section is 1.0 ft.

*Suggestions:* The liquid flow in such a bed is complicated and the flow decreases as the reaction proceeds. However, only an engineering estimate is required, so it may be assumed that the oil flow and hydrogen consumption are typical of conditions somewhere near the middle of the bed. The hydrogen absorption is actually a transient process, but all that reacting reaches the pellet surface. Assume steady-state diffusion of hydrogen across the liquid film at the mean reaction rate. As a further approximation, assume each sphere to be fed with liquid at the top center.

3.10   The experiment described in Prob. 3.8 was devised by Barrer to measure the solubility and diffusion coefficient of a gas in a polymer, both values being obtained simultaneously from a single experiment.

A possible variation of this technique is the following: place the polymer in an evacuated chamber, admit gas quickly to a pressure of perhaps 1 to 5 atm, and then follow the change in gas pressure with time. It is proposed to test this idea with hydrogen and rubber at 25°C, in which system $D$ is 0.85 $\times$ 10$^{-5}$ and solubility is 0.0366 cm$^3$ (NTP)/(cm$^3$ rubber) (atm), as found in Prob. 3.8. The hydrogen-free rubber sample will be a flat sheet with impermeable edges, placed in a vessel which is to have a *gas space* of 1,000 cm$^3$. Both faces of the sheet will be exposed to the gas.

(a)   It is desired to design the test so the hydrogen pressure will drop from 780 to 740 in an hour, during which time the hydrogen dissolved in the rubber will reach 76.4 percent of its equilibrium value in hydrogen at 1 atm. What should be the thickness and face area of the rubber sample?

(b)   Suggest a modification of the experiment which would simplify a rigorous analysis.

3.11   Fine hollow fibers of an organic polymer are developed for the desalination of brines by "reverse osmosis." In one type of device a bundle of parallel fibers is confined in a cylindrical pipe several feet long. The fibers are sealed at one end and "potted" in plastic at the other end, where the water is withdrawn from the potting plastic "tube sheet." Saline water is fed to the shell side and half or more of the water in the feed passes through the fiber walls, flowing inside the fibers towards the plastic tube-sheet and out. Somewhat concentrated brine is removed from the shell side at the end opposite the feed. Operation is at steady conditions. The advantage of the use of fine hollow fibers is that several thousand square feet of membrane surface can be packed in one cubic foot.

In the design of this device the question arises about how long the bundle of fibers should be. In order to get some idea of the problem, it is proposed to analyze the operation with a feed of pure water. Water will be fed to the shell side at 600 psig and removed from the open ends of the fibers at atmospheric pressure.

(a)   Develop an expression for the linear water velocity at the fiber discharge as a function of the total length of the dead-end fibers.

(b) Using the data given below, calculate the water permeation rate for a 7-ft fiber bundle, as a fraction of the rate attainable with a bundle of infinite length.

*Data:* Water viscosity (20°C) 1.0 cP; inside diameter fibers, 25 $\mu$m; water permeability of fiber wall $= 0.87 \times 10^{-11}$ ft$^3$/(s) (lbf/ft$^2$) (ft$^2$ inside surface).

3.12 A reverse-osmosis cell for the desalination of salt water employs a thin plastic membrane which permits the passage of water but essentially no salt. An aqueous solution of NaCl passes over one surface of the membrane at 1,500 psia and product water is withdrawn from the other face at atmospheric pressure. The osmotic pressure which must be overcome is that of the solution in contact with the upstream surface of the membrane—not that of the bulk solution.

In one such device the bulk solution contains 0.00060 g moles NaCl per cm$^3$, and the water flux through the membrane is $104.4 \times 10^{-6}$ g moles/(s) (cm$^2$). What is the salt concentration immediately in contact with the surface on the high-pressure side (gram moles per cubic centimeter)?

*Data and approximations:* The dilute solutions all have a density of 1.0 g/cm$^3$; the total gram moles per cubic centimeters may be taken to be $\frac{1}{18}$; the molecular diffusion coefficient for NaCl in water at the operating temperature is $1.61 \times 10^{-5}$ cm$^2$/s; operation is isothermal at steady-state; the diffusional resistance for NaCl transport under conditions of no water flux into the membrane is equivalent to a stagnant solution film 0.010 cm thick.

3.13 A 10-in-ID vertical gas line has a $\frac{1}{2}$-in-ID horizontal sampling connection ending at a valve 2 ft out from the large line. Hydrogen flows through the large line at 100 psia and 15°C for several days, and the sampling line is purged and the valve closed. The gas flow is then switched from hydrogen to a reformer gas containing 70 mole percent $H_2$ and 30 mole percent $CH_4$, and the flow continues at 100 psia and 15°C.

An hour later a small gas sample is removed for analysis by opening the valve. Estimate the composition of the first small increment of gas withdrawn, assuming the gas in the short sampling line to have remained completely stagnant.

3.14 Referring to Example 3.3, what is the total amount of benzoic acid diffusing into the water in the calculated time period of $72.2 \times 10^6$ s?

3.15 Referring to Example 3.1, consider the case where the boundary at $y = y_0$ is permeable to A but not to B. Since B does not diffuse across the plane $y = y_0$, the diffusion equation requires that the concentration gradient $dY_B/dy$ be zero. Then by differentiation of the equilibrium expression, it follows that $dY_A/dy$ is also zero at $y = y_0$, since $Y_B$ is finite. From the diffusion equation for A, it would then appear that the flux of A is also zero at $y = y_0$, though this wall is permeable to A and the gas in contact with the wall has a finite concentration of A. Explain this apparent anomaly. [The article by J. O. Hirschfelder, *J. Chem. Phys.*, **26**: 274 (1957) may be helpful.]

# TURBULENT DIFFUSION

## 4.0 SCOPE

The fundamental nature of flow in the turbulent regime is reviewed, and the concept of an "eddy viscosity" is presented. This is followed by a discussion of transport by turbulent mixing, as encountered in process equipment. Taylor's classical development of the theory of turbulent diffusion is outlined and compared in a general way with experimental data on radial and axial diffusion in packed beds and open conduits such as long oil pipe lines. Methods of measuring turbulent diffusion are described very briefly, and representative data on both packed beds and open pipes are presented graphically.

## 4.1 PRINCIPAL SYMBOLS

$a$      Half-spacing between walls of flat duct, cm
$c$      Concentration, g moles/cm$^3$
$C_p$    Specific heat, cal/(g)(°C)
$d$      Diameter, cm
$d_p$    Effective particle diameter, cm

| | |
|---|---|
| $d_t$ | Tube or column diameter, cm |
| $D$ | Molecular diffusion coefficient, cm$^2$/s |
| $E$ | Eddy diffusion or dispersion coefficient, cm$^2$/s; $E_a$ in direction of flow (axial); $E_r$ normal to flow (radial) |
| $E_c$ | Coefficient of convective axial dispersion due to radial velocity gradient, cm$^2$/s |
| $E_D$ | Eddy-diffusion coefficient for mass, cm$^2$/s |
| $E_H$ | Eddy-diffusion coefficient for heat, cm$^2$/s |
| $E_t$ | Total or virtual diffusion coefficient, including both eddy diffusion and molecular diffusion, cm$^2$/s |
| $E_v$ | Eddy viscosity, cm$^2$/s |
| $f$ | Fanning friction factor |
| $g_c$ | Conversion factor = 32.2 (lb-m)(ft)/(lbf)(s$^2$) in English technical units; = 1.0 in cgs system |
| $k$ | Thermal conductivity, cal/(s)(cm$^2$)(°C/cm) |
| $L$ | Length, cm |
| $m$ | Ratio of void volume to wetted surface, cm |
| $n$ | Number of particles |
| $N_A$ | Molal flux, g moles/(s)(cm$^2$) |
| $p$ | Number of steps taken by particle |
| Pe | Peclet number, $d_p \bar{U}_i/E$; Pe$_a = d_p \bar{U}_i/E_a$; Pe$_r = d_p \bar{U}_i/E_r$ |
| Pr$_T$ | Turbulent Prandtl number, $E_v/E_H$ |
| $r$ | Radial distance, cm |
| $r_w$ | Radius of tube, cm |
| $R$ | Correlation coefficient for fluctuating velocities; $R_E$ (eulerian) given by Eq. (4.1); $R_L$ (lagrangian) |
| Re | Reynolds number, $d_p \bar{U}_{av}/v$, or $d_t \bar{U}_{av}/v$ |
| Sc$_T$ | Turbulent Schmidt number, $E_v/E_D$ |
| $t$ | Time, s |
| $T$ | Temperature, °C; total time, $p \, \Delta t$, s |
| $T_E$ | Eulerian scale of turbulence, cm |
| $T_L$ | $\int_0^\infty R_L \, d(\Delta t)$; Lagrangian integral time scale, s |
| $T_0$ | Time after which $R_L$ becomes negligible, s |
| $T_\varepsilon$ | Tortuosity in unconsolidated bed |
| $u$ | Fluctuating velocity in $x$ direction, cm/s |
| $u^+$ | Defined by Eq. (4.9) |
| $\bar{U}$ | Time-mean velocity (at a point) in $x$ direction, cm/s |
| $\bar{U}_{av}$ | Average flow velocity: volumetric flow rate divided by conduit cross section, cm/s |
| $\bar{U}_c$ | Velocity at axis, cm/s |
| $\bar{U}_{max}$ | Maximum velocity in conduit, cm/s |
| $v$ | Fluctuating velocity in $y$ direction, cm/s |
| $w$ | Fluctuating velocity in $z$ direction, cm/s |
| $x, y, z$ | Distance, cm |
| $y^+$ | Defined by Eq. (4.9) |
| $\alpha$ | $T - t$, s |
| $\gamma$ | Ratio of cell length to $d_p$ |
| $\varepsilon$ | Void fraction in packed bed |
| $v$ | Kinematic viscosity, cm$^2$/s |

$\rho$        Density, $g/cm^3$

$\tau$        Shear stress, $g/cm^2$

$\tau_w$        Shear stress at wall, $g/cm^2$

## 4.2 TURBULENCE

The flow of a gas or liquid past a surface is characterized as being either "laminar" or "turbulent." When laminar, as in the slow flow of a viscous liquid in a round tube, the layers of fluid slide over each other without mixing one with another. In principle, the velocity and direction of each element of fluid may be calculated at every position in the stream. Transport of solute from one layer to another may occur by molecular motion, but this tendency of concentrations to equalize is to be distinguished from mixing due to turbulence. The velocity at a point may vary with time yet the flow still be laminar: a regular vortex system behind an immersed solid object is not an example of turbulence.

Turbulence is characterized by rapid and *quite irregular* fluctuations of the velocity about the time-mean velocity at a point. These are not observable by ordinary instruments such as the Pitot tube but may be measured by hot-wire anemometry using wires which are so small and of such low heat capacity that the temperature of the wire or the heat flux from it can follow the rapid velocity fluctuations. Records of these fluctuations with time represent the "spectrum of turbulence." The mean speed at a point in an air stream may be 500 cm/s; if the flow is highly turbulent, the velocity measured at intervals of a few milliseconds may be 490, 527, 504, 518, 460, 510, 472, 548, ..., cm/s. The velocity fluctuations are irregular and random.

Fluid layers moving over each other produce a shearing stress in a plane parallel to the flow and perpendicular to the velocity gradient. In laminar flow the interaction of the moving layers and the resulting shear is due to molecular motion. Viscosity, which is a *property* of a newtonian fluid, is defined as the ratio of the shear stress to the velocity gradient in laminar flow. At higher speeds and shear rates the flow is turbulent, with poorly defined "eddies" of many sizes rolling and mixing with each other. In round pipes the critical Reynolds number is about 2,100. A fast-moving eddy may move into an adjacent slow-moving stream and mix with it, transporting momentum into the slow stream in a manner analogous to the molecular transport of momentum from one layer to another when the flow is laminar. The nature of the eddy motion is highly irregular and poorly understood; the intertwining and mixing of filaments of fluid moving at varying velocities must be included in the concept of eddies. A rough picture of the nature of eddy motion in turbulent shear flow may be obtained by observing smoke issuing from a chimney into a horizontal wind.

Eddies are pictured as continually forming, mixing, fragmenting, disappearing, and reforming. Furthermore, large eddies contain small eddies. Indeed, the eddy concept should be regarded as an heuristic aid rather than as an unequivocal

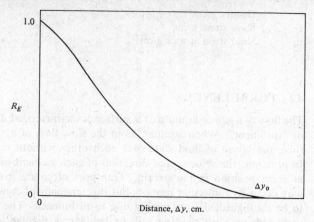

**FIGURE 4.1**
Coefficient of correlation of instantaneous velocities at two points separated by
distance $\Delta y$ in turbulent flow.

description of the mechanics of flow. Turbulent motion, therefore, is described in
terms of its *intensity*, which relates to the magnitude of the velocity fluctuations,
and its *scale*, which is a statistical measure of the size of the eddies. If $U$ is the
time-mean component of the flow in the $x$ direction, then the velocity at any
instant is $U \pm u$, where $u$ is the "fluctuating velocity" superposed on the mean.
Because $u$ has, by definition, a mean value of zero, it is most simply represented
by its root-mean-square value $\sqrt{\overline{u^2}}$, with the overbar indicating a time mean. The
ratio $\sqrt{\overline{u^2}}/U$ is termed the *intensity* of turbulence, often reported as a percentage.

Several *scales* of turbulence have been employed. The Prandtl "mixing length"
is a crude measure of the distance a large eddy travels before it breaks up and
loses its identity. Taylor defines a scale in terms of the correlation between the
velocities of two elements of the fluid. Let the stream be moving at a uniform mean
velocity $U$ in the $x$ direction, with two elements of fluid separated by a distance $\Delta y$
(along a line normal to the mean flow) having instantaneous velocities $U + u_y$ and
$U + u_{y+\Delta y}$. If the elements are very close together ($\Delta y$ small), they must be moving
at nearly the same velocities; and it may be said that there is a high degree
of correlation between the velocities $u_y$ and $u_{y+\Delta y}$. If they are far apart, their
velocities are unrelated, and the correlation is very poor. The correlation coefficient
$R_E$, defined by

$$R_E = \frac{\overline{u_y u_{y+\Delta y}}}{\sqrt{(\overline{u^2})_y} \sqrt{(\overline{u^2})_{y+\Delta y}}} = f(\Delta y) \tag{4.1}$$

$R_E$ varies from 1 to 0 as $\Delta y$ increases from 0 to $\infty$. The *eulerian scale* of turbulence $T_E$ is then defined by

$$T_E = \int_0^\infty R_E \, d(\Delta y) \tag{4.2}$$

The nature of the variation of $R_E$ with $\Delta y$, as measured by hot-wire anemometry, is illustrated diagrammatically by Fig. 4.1. $R_E$ goes to zero at some more or less definite value $\Delta y_0$ of $\Delta y$; this may evidently be substituted for infinity as the upper limit in Eq. (4.2). $R_E$ is known as the eulerian spatial correlation coefficient, and $T_E$ as the eulerian integral length scale. The latter is a measure of the size of the large eddies at a point in space. $R_E$ may be defined in terms of any direction in space, not necessarily on a line perpendicular to the mean flow.

Another scale of turbulence, particularly relevant to problems of turbulent diffusion, is based on the correlation between velocities of the same particle or fluid element at different times. This is defined in terms of the fluctuating velocities at times $t$ and $t + \Delta t$:

$$R_L = \frac{\overline{u_t \, u_{t+\Delta t}}}{\sqrt{(u^2)_t}\sqrt{(u^2)_{t+\Delta t}}} = f(\Delta t) \tag{4.3}$$

This is known as the lagrangian correlation coefficient; other correlation coefficients may be also expressed in terms of the fluctuating velocities $v$ and $w$. The meaning of the correlation coefficient $R_L$ is suggested by diagrams shown in Fig. 4.2 (after Van Driest [163]), in which $u_{t+\Delta t}$ at time $t + \Delta t$ is shown plotted vs. $u_t$ at time $t$.

A *lagrangian integral time scale* is defined by

$$T_L = \int_0^\infty R_L \, d(\Delta t) \tag{4.4}$$

A similar expression in terms of $y$ and $v$ defines a corresponding scale for the $y$ direction. A graph of $R_L$ vs. $\Delta t$ appears much like Fig. 4.1 of $R_E$ vs. $\Delta y$; $R_L$ goes to zero at some value $\Delta t_0$, which may replace infinity as the upper limit of the integral in Eq. (4.4). As will be shown later, $R_L$ may be obtained from measurements of heat and mass transport in a turbulent stream.

Turbulent flow is characterized as being *homogeneous* if the statistical properties do not vary in space, as *isotropic* if the statistical properties are invariant with direction, and as *stationary* if the statistical properties are time-independent. In turbulent pipe flow the shear stress increases with the distance from the axis of a pipe, and turbulence is damped in the region near the wall: the flow is neither homogeneous nor isotropic, but it may be stationary.

The two correlation coefficients $R_E$ and $R_L$ and the scales $T_E$ and $T_L$ are of particular interest in connection with turbulent diffusion. Many others have been defined, including those involving fluctuations of temperature or concentration

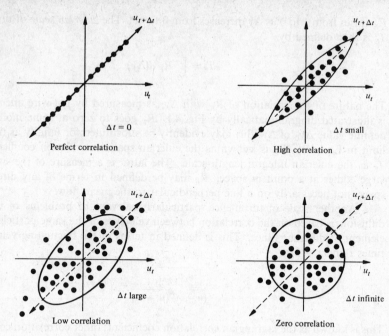

**FIGURE 4.2**
Representation of the meaning of a correlation coefficient—in this case $R_L$.

rather than velocity. Complete statistical representation requires far more information than is contained in the quantities defined above, even if the flow is homogeneous, isotropic, and stationary. Because of this, recourse to much simpler models is often necessary.

## 4.3 THE EDDY VISCOSITY

The molecular kinematic viscosity $v$ of a newtonian fluid is defined by

$$\tau g_c = -v \frac{\partial(\rho U)}{\partial y} \tag{4.5}$$

where $\tau$ is the shear stress on a plane normal to the direction $y$ in a fluid in plane laminar motion. The viscosity is not a function of the velocity, the velocity gradient, or the geometry of the conduit. $\tau$ has units of momentum flux density, and Eq. (4.5) may be looked upon as a rate equation, and $v$ as the coefficient of momentum transport, with momentum per unit volume as the potential.

In turbulent flow the effective shear stress also increases with increase in the velocity gradient but is augmented by eddy exchange; hence the proportionality constant is generally very much larger than $v$. Furthermore, it is *not a property of the fluid* but varies with the flow conditions. Recognizing these limitations, one may still define the ratio $E_t$ of the shear stress to the mass velocity gradient by the relation

$$\tau g_c = - E_t \frac{d(\rho \overline{U})}{dy} \tag{4.6}$$

where $E_t$ is the apparent, virtual, or total viscosity. The result of eddy interchange between parallel flowing elements in a velocity gradient is to increase rates of momentum transfer to much greater values than are observed in laminar flow with the same velocity gradient: $E_t$ is much larger than $v$. Molecular motion still prevails, and it is convenient to define an *eddy viscosity* $E_v$ by

$$\tau g_c = - (E_v + v) \frac{d(\rho \overline{U})}{dy} \tag{4.7}$$

For pipe flow, a simple force balance on a cylindrical fluid element of length $dx$ and radius $r$ leads to the relation

$$\tau = \frac{r}{r_w} \tau_w \tag{4.8}$$

where $\tau_w$ is the shear stress at the wall, $\frac{1}{2} f \rho \overline{U}_{av}^2 / g_c$, and $r_w$ is the pipe radius. Here $f$ is the Fanning friction factor. Evidently $E_v$ can be calculated from measured velocity gradients if $\overline{U}_{av}$ and $f$ are known. If the velocity deficiency $\overline{U}_c - \overline{U}$ is proportional to $r^2$, $E_t$ is independent of $r$. Here $\overline{U}_c$ is the velocity at the axis and $\overline{U}$ is the velocity at radius $r$. More realistic velocity profiles lead to the conclusion that $E_v$ increases from a finite value at the axis to a maximum at $r/r_w = 0.5 - 0.7$, and falls to zero at the wall. Both $\tau$ and $d(\rho \overline{U})/dy$ are zero at the axis in steady turbulent flow in pipes, but $E_v$ is not zero, as noted by Lynn [105] and Seagrave [133].

Figures 4.3 and 4.4 illustrate the manner in which $E_v$ (or more properly $E_t$, though $v$ is small relative to $E_v$) varies with position in the conduit, and with the Reynolds number. Figure 4.3 represents the data of Page, Schlinger, Breaux, and Sage [117] for air flow in a duct with flat parallel walls. These results were obtained with fully developed turbulent flow in a stream having a turbulent intensity of 0.03 at the axis. Experimental points are not shown, since the curves plotted are derived from measured values of $\tau_w$ (or $f$) and calculated slopes of measured velocity profiles. The kinematic viscosity of air at 1 atm and 38°C is 0.168 cm$^2$/s; even at the lowest Reynolds number, $E_v$ is much larger than $v$, except very near the walls of the duct.

Figure 4.4 shows the dimensionless ratio $E_v/v$ plotted against radial location in a round pipe, with curves for several values of Re [34, 128]. These results are calculated from general equations for velocity profiles in round pipes but agree reasonably well with the limited data available. Abbrecht and Churchill [1] obtained similar results from measurements of velocity profiles with air flowing in a tube at locations varying from 0.45 to 10 tube diameters downstream from a step increase in tube-wall temperature. Sleicher [142] using air in a 3.8-cm tube found less decrease from the maximum to the centerline values. Sesonske, Schrock, and

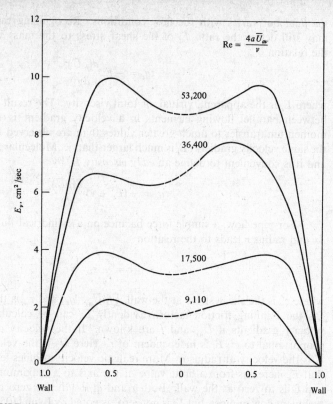

**FIGURE 4.3**
Variation of eddy viscosity for air flow at 38°C in a flat horizontal duct formed by plates 1.78 cm apart ($a = 0.89$ cm). Data of Page, Schlinger, Breaux, and Sage [117].

Buyco [135] show similar curves with approximately the same variation of maximum $E_v/\nu$ with Re. Beckwith and Fahien [15] show the same trends for water in a long 7.6-cm tube. Strunk and Tao's [148] results for air in a 5-cm tube are quite similar to Fig. 4.4 but show even greater decrease from the maximum values of $E_v/\nu$ as the axis is approached.

It is difficult to measure velocity profiles with the precision needed to yield reliable values of $E_v/\nu$, since evaluation of the derivative $d\overline{U}/dr$ is necessary before $E_v$ can be calculated from $\tau_w$ or $f$ by combining Eqs. (4.7) and (4.8). Figure 4.4 appears to represent a reasonable description of the manner in which $E_v$ varies with radial position, for either a gas or a liquid. The values near the maximum are the most reliable, since velocities very near the wall are difficult to measure, and the profile at the centerline is flat. Data on velocities near a wall are discussed in the next section.

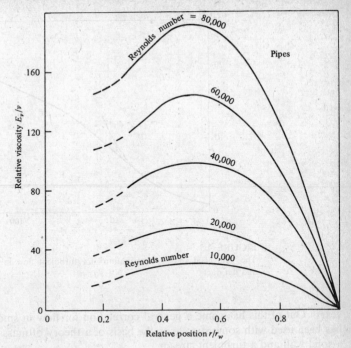

**FIGURE 4.4**
Ratio of eddy viscosity to molecular kinematic viscosity as a function of position in a round pipe [34, 128].

## 4.4 THE UNIVERSAL VELOCITY DISTRIBUTION IN SMOOTH TUBES

The interest of chemical engineers in mass transfer relates primarily to transfer between phases: between a fluid and a solid, or between two fluids. Conditions affecting mass transfer very near a phase boundary are of particular concern, since it is here that much of the resistance to mass transfer is found. Since the transport of mass and of momentum in turbulent flow occurs by similar mechanisms, it is useful to consider what is known about momentum transport near a phase boundary, especially since momentum transport has been studied so much more intensively than has mass transfer. The preceding section gives some guide on what to expect in the region of flow away from the wall of a duct or tube. In this section we shall look briefly at the data on flow patterns very near a wall, to provide a basis for a discussion of the "analogies" between heat, mass, and momentum transfer in Chap. 5.

The nature of velocity profiles for isothermal turbulent flow of gases and liquids in conduits has been the subject of a great amount of study over many

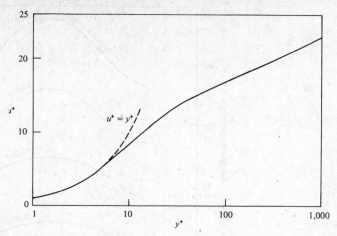

FIGURE 4.5

The "universal velocity distribution" for turbulent flow in smooth round pipes and flat ducts.

years. Out of this has come a general correlation for flow in smooth tubes which has been used with some success as the basis of a theory of mass transfer between a solid wall and a turbulent stream.

It has been found that velocity profiles for both liquids and gases in turbulent flow in glass or reasonably smooth metal tubes, for a very wide range of Reynolds number, can be represented by a single function relating the two dimensionless groups $u^+$ and $y^+$, where

$$u^+ \equiv \overline{U}\sqrt{\frac{\rho}{\tau_w g_c}} \quad \text{and} \quad y^+ \equiv \frac{y}{v}\sqrt{\frac{\tau_w g_c}{\rho}} \tag{4.9}$$

where $\tau_w$ = shear stress at the wall
  $y$ = distance from the wall
  $\overline{U}$ = time-mean velocity at $y$
$\rho$ and $v$ = the fluid density and kinematic velocity, respectively

$\tau_w g_c$ can evidently be replaced by $\frac{1}{2}f\rho \overline{U}_{av}^2$, where $\overline{U}_{av}$ is the volumetric flow rate divided by the cross section of the conduit, and $f$ is the Fanning friction factor. When this is done,

$$u^+ = \frac{\overline{U}}{\overline{U}_{av}}\sqrt{\frac{2}{f}} \quad \text{and} \quad y^+ = \frac{y\overline{U}_{av}}{v}\sqrt{\frac{f}{2}} \tag{4.10}$$

The nature of the relation between $u^+$ and $y^+$ for round pipes and flat ducts is shown in Fig. 4.5. This is based on experimental data, though data points are not shown, since there are literally thousands of them. The line is a simple representation of the results of an enormous amount of experimental work.

The general validity of the correlation represented by the solid line in Fig. 4.5 is well established by experiment. It should be noted, however, that there are relatively few reliable data for values of $y^+$ less than 5, and that the upper-right-hand branch does not apply near the center of a round pipe, since $du^+/dy^+$ is zero at the axis. Furthermore, there is substantial evidence of an additional influence of Re at high values of $y^+$, not allowed for by the single function shown [34, 24]. Ryan [137] suggested that a better correlation might be obtained by employing $\sqrt{f/2}\,y^+$ instead of $y^+$, but this has not been adequately tested. Rothfus and Monrad [125] found the correlation to be improved by plotting $u^+(\overline{U}_{av}/\overline{U}_{max})$ vs. $y^+(\overline{U}_{max}/\overline{U}_{av})$, and they provide graphs of $\overline{U}_{av}/\overline{U}_{max}$ vs. Re. The term $\overline{U}_{av}/\overline{U}_{max}$, which is the ratio of the bulk average velocity to the maximum velocity in the conduit, increases slowly with increase in Re.

To relate $y^+$ to pipe radius, note that at Re $= 20{,}000$, $f$ is 0.007, whence, from the definition of $y^+$,

$$\frac{y}{r_w} = 2\sqrt{\frac{2}{f}}\frac{y^+}{\text{Re}} = 0.00169y^+$$

Similarly, at Re $= 200{,}000$, $y/r_w = 0.00020y^+$. Thus the values of $y^+$ corresponding to the pipe axis are 591 at Re $= 20{,}000$, and 5,000 at Re $= 200{,}000$. (A convenient graph of $f$ as a function of Re is to be found in Perry's "Chemical Engineers' Handbook" [119].)

Combination of Eqs. (4.7) and (4.8) with the definitions of $u^+$ and $y^+$ leads to the relation

$$\frac{E_v}{v} = \frac{r}{r_w}\frac{dy^+}{du^+} - 1 \qquad (4.11)$$

Evidently $E_v/v$ can be obtained from the slopes of the curve shown in Fig. 4.5 and expressed as a single function of $y^+$. Though the correlation represented by Fig. 4.5 is excellent, the precision in estimating slopes is poor, and only an approximate relation between $E_v/v$ and $y^+$ can be obtained in this way. For convenience in a subsequent analysis, Von Kármán [166] noted that the line shown in Fig. 4.5 might be represented by three separate functions:

$$\begin{aligned}
u^+ &= y^+ & 0 < y^+ < 5 \\
u^+ &= -3.05 + 5.0 \ln y^+ & 5 < y^+ < 30 \\
u^+ &= 5.5 + 2.5 \ln y^+ & y^+ > 30
\end{aligned} \qquad (4.12)$$

These are easily differentiated to give simple but quite approximate expressions for $E_v/v$ as a function of $r/r_w$.

Von Kármán's division of the range of $y^+$ into three regions has led to the use of the terms "laminar sublayer," "buffer layer," and "turbulent core" in referring to the ranges of $y^+$ from 0 to 5, 5 to 30, and $>30$, respectively. It is evident from Fig. 4.5, however, that there are no sharply defined regions, and that both $u^+$ and its derivative are continuous functions of $y^+$.

## 4.5   THE EDDY-DIFFUSION COEFFICIENT

Mass transfer in a turbulent stream is essentially a mixing process, whereby mass (or heat) is transported by the mixing and blending of the eddies. For brevity, it will be referred to as "eddy diffusion," though its similarity to molecular diffusion lies only in the fact that the molar flux in a binary mixture is proportional to a concentration gradient under many but not all conditions. As suggested by the discussion of momentum transfer, it is usually very rapid, though near a phase boundary where the eddy motion is damped it may be unimportant as compared with the parallel process of diffusion by molecular motion.

A complete understanding of turbulent diffusion will necessarily depend on a quantitative description of turbulence, including the sizes and motion of the eddies. This is not now possible: the limited theory which exists is based on the random character of turbulence and on its statistical properties. There have been numerous excellent investigations of turbulent diffusion, however, so that much empirical information exists which is relevant to practical engineering problems.

It is obviously tempting to follow the pattern of molecular diffusion and attempt to define a diffusion coefficient as the ratio of molar flux of a species to the concentration gradient of the same species. This has been done, with the eddy-diffusion coefficient $E_D$, defined by the following relations, which neglect the usually small contributions of molecular diffusion or conduction to the flux:

$$E_t \approx E_D = -\frac{N_A}{\partial c_A/\partial y} \qquad \text{mass transfer} \qquad (4.13)$$

$$E_t \approx E_H = -\frac{q}{\partial(\rho C_p T)/\partial y} \qquad \text{heat transfer} \qquad (4.14)$$

The difficulty in this approach lies not only in the complex dependence of $E$ on the properties of the turbulent stream but also in the fact that the flux is not always proportional to the concentration or temperature gradient. The advantage is that the turbulent exchange processes for mass, heat, and momentum are sufficiently similar that much may be learned about $E_D$ and $E_H$ from the considerable amount of information concerning $E_v$.

The existing theory of turbulent diffusion is based in part on the concept that it is essentially a matter of bulk dispersion in which molecular diffusion can be neglected, and that the early theories of the random walk of particles in brownian motion may be relevant. These theories lead to expressions for the mean-square displacement $\overline{y^2}$ of particles, after a specified time interval, from some initial position. It is desirable, therefore, to show the connection between $E$ and the variance $\overline{y^2}$.

Consider first the random walk of a single particle along a straight line, as illustrated in Fig. 4.6$a$. The particle starts at point $O$ at zero time, moving in steps $\Delta y$ at constant velocity $\pm v$, requiring the time $\Delta t = \pm \Delta y/v$ for each step. On completion of each step the particle may either continue in the same direction or

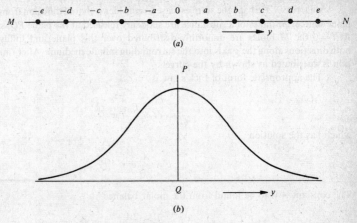

FIGURE 4.6
Diffusion from a point or line source ($y$ direction only).

reverse its direction with equal probability. After $p$ steps in time $p \, \Delta t$, its position is $\pm y$.

Now consider the situation in which a large number of particles behave in this same way, all starting at $O$ at the same instant and not interfering with each other. After time $p \, \Delta t$, these particles will then be distributed along the line $MN$ at different distances $\pm y$ on either side of $O$. Since the direction of the equal individual steps is random, the distribution of $y$ about $O$ is normal, or gaussian. The mean-square displacement of the particles from $O$ is $\overline{y^2}$; the standard deviation, or root-mean-square displacement, is $\sqrt{\overline{y^2}}$ where the overbar now denotes the mean for many particles.

Where Fick's law applies, $E_D$ and $\overline{y^2}$ are related by the simple equation, first derived by Einstein in 1905 [43]:

$$\overline{y^2} = 2E_D t \qquad (4.15)$$

Einstein's derivation is available in English in the book by Moelwyn-Hughes [113], who also gives derivations of the same equation by Smoluchowski, and by Langevin. Von Kármán [165] has presented a simple graphical derivation of the same relation. As will be shown in the next section, Fick's law is not always applicable to turbulent diffusion unless $E_D$ is defined as a function of $t$, so Eq. (4.15) has certain limitations in dealing with eddy diffusion.

EXAMPLE 4.1   Select a case of transient diffusion and show that the value of $\overline{y^2}$ after time $t$ is equal to $2E_D t$ if Fick's law holds.

SOLUTION   Perhaps the simplest case to consider is that of diffusion from an instantaneous plane source. $M$ moles of a substance are suddenly released at the plane $PQ$, shown in Fig. 4.6$b$. At $t = 0$ the $M$ moles are uniformly distributed over this plane and diffuse immediately in both directions along the $y$ axis into the surrounding infinite medium. After time $t$, the concentration is distributed as shown by the curve.

The appropriate form of Fick's law is

$$\frac{\partial c}{\partial t} = E_D \frac{\partial^2 c}{\partial y^2} \qquad (a)$$

which has the solution

$$c = \frac{A}{\sqrt{t}} e^{-y^2/4E_D t} \qquad (b)$$

The constant $A$ is to be found from the molar balance

$$M = \int_{-\infty}^{\infty} c \, dy = 2 \int_{0}^{\infty} c \, dy = \frac{2A}{\sqrt{t}} \int_{0}^{\infty} e^{-y^2/4E_D t} \, dy = 2A\sqrt{\pi E_D} \qquad (c)$$

From this it follows that

$$c = \frac{M}{2\sqrt{\pi t E_D}} e^{-y^2/4E_D t} \qquad (d)$$

Since $\frac{1}{2}M$ diffuses in each direction from $PQ$ ($c$ is an even function), the mean-square displacement is given by

$$\overline{y^2} = \frac{\int_{0}^{\infty} y^2 c \, dy}{\frac{1}{2}M} = \frac{1}{\sqrt{\pi E_D t}} \int_{0}^{\infty} y^2 e^{-y^2/4E_D t} \, dy = 2E_D t \qquad (e)$$

////

## 4.6  TURBULENT DIFFUSION

The rapid dispersion, or diffusion, of dust particles, smoke, or molecular species in a turbulent stream is one of the most characteristic features of turbulence. Since the dispersion phenomenon is so obviously related to the nature of turbulence, it has been widely studied over a period of many years, and the literature is quite voluminous. The review which follows will be limited to an outline of the existing theory of diffusion in a homogeneous turbulent stream in the absence of shear due to velocity gradients (as in the central region of a wind tunnel), and to certain experimentally observed characteristics of diffusion in fluids in shear flow in pipes and ducts.

The classical paper on the subject is that by G. I. Taylor, published in 1921 [151, 152]. The reviews by Frenkiel [49] and by Batchelor and Townsend [13] are excellent. The chemical engineering literature contains fine articles by Hanratty, Latinen, and Wilhelm [70]; Baldwin and Walsh [9]; and by Becker, Rosensweig, and Gwozdz [14]. The subject is treated at length in the book by Hinze [80].

Taylor first analyzes the random walk of particles along a line. Referring again to Fig. 4.6a, let us suppose that a cluster of $n$ particles starts at point $O$, each taking successive discrete steps of length $\pm\Delta y$ in time $\Delta t$, each particle being unaffected by the progress of the others. This process might simulate, for example, the dispersion, normal to the flow, of a small element of a gas species suddenly placed at a point in the center of a wind tunnel. If the line considered, lying normal to the flow in the $x$ direction, is allowed to move downstream at the velocity $\bar{U}$ of the free stream, the dispersion along the line with time corresponds to dispersion normal to the flow with downstream distance. If the dispersion of a cluster of particles can be described, it will be possible to analyze the case of dispersion of heat or mass from a continuous point or line source.

In time $p\,\Delta t$, each of the $n$ particles takes the same number of steps, after which each particle is at some distance $\Delta y_1 + \Delta y_2 + \cdots + \Delta y_p$ from 0. The absolute values of the distances $\Delta y$ are all equal, but some are negative and some positive, so the sum is different for each particle. The mean-square displacement of all the particles is then given by

$$\overline{y^2} = \frac{1}{n} \sum_{n=1}^{n} (\Delta y_1 + \Delta y_2 + \Delta y_3 + \cdots + \Delta y_p)_n^{\,2} \tag{4.16}$$

Let $R$ represent the correlation coefficient between two values $\Delta y_i$ and $\Delta y_j$ of $\Delta y$, where $R$ is defined by

$$R = \frac{\overline{\Delta y_i \Delta y_j}}{\overline{\Delta y^2}} \tag{4.17}$$

(If the particles tended to reverse direction only after several steps, the correlation between successive values of $\Delta y$, that is, $\Delta y_i$ and $\Delta y_j$, might be near unity; $R$ might then decrease to zero after many steps, indicating no correlation between the directions of the first and $p$th step.) Conditional correlations are neglected and the correlation between $\Delta y_i$ and $\Delta y_{i+2}$ is then $R^2$, that between $\Delta y_i$ and $\Delta y_{i+3}$ is $R^3$, etc. Introducing these values and simplifying the series, Eq. (4.16) becomes

$$\overline{y^2} = p\,\Delta y^2 + \frac{2}{n} \sum_{n=1}^{n} (\Delta y_1\,\Delta y_2 + \Delta y_1\,\Delta y_3 \cdots + \Delta y_1\,\Delta y_p)_n$$

$$+ \frac{2}{n} \sum_{n=1}^{n} (\Delta y_2\,\Delta y_3 + \Delta y_2\,\Delta y_4 \cdots + \Delta y_2\,\Delta y_p)_n$$

$$+ \frac{2}{n} \sum_{n=1}^{n} (\Delta y_3\,\Delta y_4 + \Delta y_3\,\Delta y_5 \cdots + \Delta y_3\,\Delta y_p)_n$$

$$+ \cdots$$

$$= p\,\Delta y^2 + 2\Delta y^2[(p-1)R + (p-2)R^2 + \cdots + R^{p-1}] \tag{4.18}$$

As might be expected, $n$ disappears since $\overline{y^2}$ is the mean for all $n$ particles. Summing the geometric series (4.18) by the standard procedure, and letting $\Delta y = v\,\Delta t$, where $v$ is the velocity in the $y$ direction,

$$\overline{y^2} = v^2\left\{\left(\frac{1+R}{1-R}\Delta t\right)T - \frac{2R(1-R^p)}{(1-R)^2}\Delta t^2\right\} \qquad (4.19)$$

where $T$ is the total time, $p\,\Delta t$.

Quoting Taylor [151], "By reducing $\Delta t$ indefinitely we can evidently make the case approximate to some sort of continuous migration, but in order that $\overline{y^2}$, $v$, and $T$ may be finite and tend to a definite limit as $\Delta t$ is decreased, it is necessary that $\dfrac{1+R}{1-R}\Delta t$ and $\dfrac{2R(1-R^p)\Delta t^2}{(1-R)^2}$ must also tend to a definite limit. That is to say, $1-R$ must be proportional to $\Delta t$." If in this limit $\Delta t/1-R$ approaches the value $A'$, then it may be shown by the use of L'Hôpital's rule that

$$\lim_{\Delta t \to 0} \overline{y^2} = v^2[2A'T - 2(A')^2(1 - e^{-T/A'})] \qquad (4.20)$$

The quantity $y$ is the displacement of a single particle from the origin $O$ after the time $p\,\Delta t = T$, and $\overline{y^2}$ is the mean-square displacement of a large number of particles.

Two limiting situations are suggested by Eq. (4.20). If the time $T$ considered is very short

$$\overline{y^2} \approx v^2 T^2 \qquad (4.21)$$

as may readily be shown by expanding $e^{-T/A'}$ in a power series in $T/A'$. The root-mean-square displacement is then equal to $vT$, which means that if $v$ and $\overline{U}$ are constant, $\sqrt{\overline{y^2}}$ spreads linearly with distance downstream from a point source.

When $T$ is large, corresponding to distances far downstream from a point source,

$$\overline{y^2} = 2A'Tv^2 \qquad (4.22)$$

Taylor notes "the constant $A'$ evidently measures the rate at which the correlation coefficient between the direction of an infinitesimal path in the migration and that of an infinitesimal path at a time $T$, say, later, falls off with increasing values of $T$." Equations (4.21) and (4.22) describe two limiting laws of dispersion: the first is applicable very near the source and the second, after long dispersion times. Experimental data which confirm these predictions will be cited in the next section.

Now consider the motion of fluid particles along the line $MN$ of Fig. 4.6$a$, and recognize that the motion is continuous with a continually fluctuating velocity $v$. (In the case of dispersion from a point source in a wind tunnel the velocity $\overline{v}$ is zero; $v$ is the fluctuating velocity in the $y$ direction, across the stream.)

The Lagrangian correlation coefficient of velocity fluctuations in the $y$ direction is defined by

$$R_L = \frac{\overline{v_t\,v_{t+\Delta t}}}{\sqrt{\overline{v_t^2}}\sqrt{\overline{v_{t+\Delta t}^2}}} = \frac{\overline{v_t\,v_{t+\Delta t}}}{\overline{v^2}} \qquad (4.23)$$

where $\overline{v^2}$ does not vary with time if the turbulence is stationary, so $R_L$ is a function of $\Delta t$ but not of $t$. Furthermore, since homogeneous turbulence has been assumed, $R_L$ is independent of position.

$R_L$ is an even function, so it is the same function of $-\Delta t$ as of $+\Delta t$. Then

$$\overline{v_T v_t} = \overline{v^2} R_L \tag{4.24}$$

where $R_L$ is a function of $\alpha \equiv T - t$. Let $y$ represent the displacement along the line $MN$ at time $T$. Then since $y = \int_0^L v \, dt$,

$$\frac{1}{2} \frac{d\overline{y^2}}{dT} = \overline{y \frac{dy}{dT}} = \overline{yv_T} = \overline{v_T \int_0^T v_t \, dt} = -\overline{v^2} \int_0^T R_L \, d\alpha \tag{4.25}$$

Consequently,

$$\overline{y^2} = 2\overline{v^2} \int_0^T \int_{\alpha=0}^{\alpha=t} R_L \, d\alpha \, dt = 2\overline{v^2} \int_0^T (T - \alpha) R_L(\alpha) \, d\alpha \tag{4.26}$$

The overbars in these equations indicate averages over a large number of eddies or particles, except that $\overline{v^2}$ and the bars in Eq. (4.23) indicate time means.

As noted earlier, the correlation coefficient $R_L$ tends to zero at some finite value of $T$, which we shall represent by $T_0$. In this case,

$$\int_0^{T_0} R_L \, dt \approx \int_0^\infty R_L \, d\alpha = T_L \tag{4.27}$$

where $T_L$ is the Lagrangian time scale in the $y$ direction. For very large $T$,

$$\lim_{T \to \infty} \overline{y^2} = 2\overline{v^2} T \int_0^{T_0} R_L \, dt + k_3 = 2\overline{v^2} T T_L + k_3 \tag{4.28}$$

where

$$k_3 = \lim_{t \to \infty} 2\overline{v^2} \int_0^\infty \alpha R_L(\alpha) \, d\alpha \tag{4.29}$$

*After large dispersion times* $\overline{y^2}$ is, therefore, linear in time $T$, and a graph of $\overline{y^2}$ vs. $T$ should have a slope $2\overline{v^2} T_L$ and intercepts on the $\overline{y^2}$ and $T$ axes which depend on the variation of $R_L$ with $\alpha$. In the limit, at long times, the constant $k_3$ becomes negligible and $\overline{y^2} \approx 2\overline{v^2} T T_L$. As shown in Sec. 4.5, $\overline{y^2} = 2E_D T$ when Fick's law holds; it follows that Fick's law may be employed for eddy diffusion if the dispersion time is large, and that when this is true,

$$E_D = \overline{v^2} T_L \tag{4.30}$$

On the other hand, *when $T$ is very small*, the correlation of velocities throughout the short dispersion time period is essentially perfect, and $R_L$ may be taken as unity. Then by direct integration of Eq. (4.25),

$$\overline{y^2} = \overline{v^2} T^2 \tag{4.31}$$

This also follows from the fact that in this limit $y = vT$.

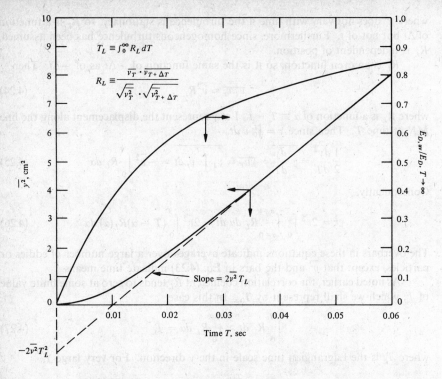

**FIGURE 4.7**
Displacement variance and average $E_D$ as functions of time, based on the assumed relation $R_L = e^{-T/T_L}$; $T_L = 0.1$ s; $\overline{v^2} = 8{,}000$ cm$^2$/s$^2$.

As shown by Fig. 4.7, a graph of $\overline{y^2}$ vs. $T$ may be expected to start parabolically at the origin (the point source of the diffusing material) and to approach the limit of a straight line with slope $2\overline{v^2}T_L$.

The detailed nature of the curve, as well as the location of the two intercepts, depends on the form of the correlation function. As an approximation, $R_L$ may be expressed by the empirical expression $R_L = e^{-T/T_L}$. Hanratty, Latinen, and Wilhelm [70] show that when this is done,

$$\overline{y^2} = 2\overline{v^2}T_L{}^2\left[\frac{T}{T_L} - (1 - e^{-T/T_L})\right] \tag{4.32}$$

The curve shown in Fig. 4.7 represents this equation for the particular case of $T_L = 0.010$ s and $\overline{v^2} = 8{,}000$ cm$^2$/s$^2$. With the assumed exponential decay of $R_L$, the intercept $k_3$ of the limiting asymptote is found from Eq. (4.29) to be $-2\overline{v^2}T_L{}^2$.

One may arbitrarily define an average diffusion coefficient $E_{D,\,av}$ to be used in Fick's law for the time period $O$ to $T$, but it is of little value, since it varies with $T$. Eliminating $\overline{y^2}$ by the use of Eq. (4.15), $E_{D,\,av}$ may be obtained from Eq. (4.32) and compared with the asymptotic value for long times, which is $\overline{v^2}T_L$. Figure 4.7 shows the manner in which this ratio increases with displacement time in the case of a particular numerical example. It is a function of the scale of turbulence, represented by $T_L$, but not of the intensity, which is proportional to $\sqrt{\overline{v^2}}$.

Summarizing this section, it should be noted that random dispersion of fluid particles follows different laws in the limiting cases of very short and very long dispersion times. The dispersion to be expected at intermediate times depends on the manner in which the correlation coefficient $R_L$ decays with time, and there is little information on this point. Equation (4.32) gives $\overline{y^2}$ for both the limits and the intermediate times, but it is based on a guess about the decay of $R_L$. Only when the dispersion times are large can Fick's law be employed.

The development of the theory and equations for turbulent diffusion presented in the foregoing is perhaps too greatly condensed; the reader may wish to refer to one or more of the many published papers in which this same analysis is given in more detail [70, 14, 163, 69, 151, 39, 12, 49, 103, 47, 90, 109, 59].

## 4.7 EXPERIMENTAL STUDIES OF TURBULENT DIFFUSION

The theory of turbulence provides no useful predictions about the scale or intensity of turbulence to be expected in a given flow system, and values of these are needed to use the theory of turbulent diffusion. Experimental measurements, however, have confirmed the general form of the Taylor theory in a remarkable way.

Most of the published investigations have employed a continuous line source or continuous point source of either heat or tracer material, with measurements of temperature or tracer concentration at points across the stream some distance downstream from the source. Figure 4.8 illustrates the nature of a typical experiment with a continuous point source in a turbulent stream flowing in a round pipe.

Precise comparison of theory with the data obtained in this way is difficult because the measurement of the lagrangian scale and the intensity of turbulence, employing hot-wire anemometry, involves relatively sophisticated techniques. Homogeneous and isotropic turbulence is difficult to attain; and when generated with screens in a wind tunnel, the turbulence decays with distance downstream. Furthermore, the tracer injector or hot-wire source tends to disturb the flow, and the velocity is not constant in the $y$ direction if there is shear; nevertheless, good results have been obtained using the procedure illustrated by Fig. 4.8.

If the time of travel $x/\overline{U}$ is sufficiently long, Fick's law applies and the spread of the tracer can be analyzed by comparing the data with available integrations of the heat conduction equations [86]. (Reference 61 provides a collection of such equations for instantaneous line, point, surface, and volume sources; the solution for the continuous point source with allowance for longitudinal

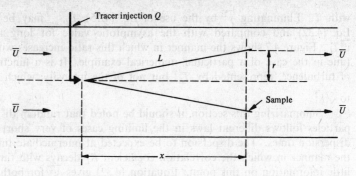

**FIGURE 4.8**
Experimental method of measuring turbulent diffusion from a continuous point source.

diffusion and for finite conduit radius is given by Hanratty, Latinen, and Wilhelm [70].) For the case of the continuous point source in a radially infinite stream,

$$c = \frac{Q}{4\pi E_D L}e^{-(U/2E_D)(L-x)} \approx \frac{Q}{4\pi E_D x}e^{-Ur^2/4E_D x} \tag{4.33}$$

where $c$ = tracer concentration (volume fraction) at $x$, $y$, if the approach stream contains none

$U$ = stream velocity

$Q$ = volumetric tracer flow, expressed in terms of the pure tracer

Diffusion in the $x$ direction is neglected in deriving this relation, which also assumes $E_D$ to be constant. The distance $r$ is the radial distance from the axis on which the injector is located.

In the more general case, but again neglecting longitudinal diffusion,

Continuous point source:

$$c = \frac{Q}{2\pi \overline{y^2} U}e^{-r^2/2\overline{y^2}} \tag{4.34}$$

Continuous line source:

$$c = \frac{Q}{\sqrt{2\pi \overline{y^2}}\ U^2 T_L}e^{-r^2/2\overline{y^2}} \tag{4.35}$$

where $T_L$ is defined by Eq. (4.27), and the mean-square displacement $\overline{y^2}$ is defined by

$$\overline{y^2} = \frac{1}{Q}\int_0^\infty 2\pi \overline{U} r^3 c\ dr \tag{4.36}$$

in which the concentration is symmetrical about the injector axis. Equations (4.34) and (4.35) may be applied to both limits, very long dispersion times and very short dispersion times, providing $\overline{y^2}$ is replaced by

Very short times:

$$\overline{y^2} = \overline{v^2}T^2 = \frac{\overline{v^2}x^2}{\overline{U}^2} \tag{4.37}$$

Very long times:

$$\overline{y^2} = 2\overline{v^2}TT_L = 2E_D T = \frac{2E_D x}{U} \tag{4.38}$$

Both Eqs. (4.34) and (4.35) neglect longitudinal diffusion and assume $r \ll x$. For long dispersion times, $E_D$ may be obtained from measurements of concentrations at the ejector axis, since from Eq. (4.33), $E_D = Q/4\pi L c_{r=0}$. It follows that Eq. (4.33) can be written in the form

$$\frac{c}{c_{r=0}} = e^{(-U/2E_D)(L-x)} \approx e^{-Ur^2/4E_D x} \tag{4.39}$$

Figure 4.9 represents data obtained by Baldwin and Walsh [9] showing the manner in which the variance of the lateral dispersion distance spreads with time (or distance) in pipe flow. These are data on the diffusion of heat from a line source in the form of a hot wire stretched across a diameter of a 20-cm pipe through which air was passed at a centerline velocity of 2,215 cm/s (72.6 ft/s). Temperature increase was measured over the central 6.4 cm of the stream ($r/r_w$ up to 0.32) at various distances downstream from the wire. After the first few centimeters the observed radial temperature dispersion was found to fit Eq. (4.35).

Various investigators studying diffusion of heat, tracer gases, dyes, and particulate solids from both line and point sources have obtained results similar in form to Fig. 4.9, thus providing quite substantial confirmation of the Taylor theory and the analysis summarized by Fig. 4.7.

Figure 4.10 shows data of quite another kind, reported by Flint, Kada, and Hanratty [47]. Tracer gas consisting of either pure hydrogen or pure carbon dioxide was supplied at a steady rate through a 1.9-mm injector at the axis of an 8-cm-ID pipe carrying air at about 1,410 cm/s (46 ft/s; Re = 72,000). The concentration distribution across the pipe is seen to be gaussian—the solid line represents Eq. (4.39) with $E_D = 6.65$ cm$^2$/s. Nearly identical distributions are obtained for both gases, though the molecular diffusion coefficient for hydrogen in air (0.69 cm$^2$/s) is much greater than for carbon dioxide (0.14 cm$^2$/s). Evidently molecular diffusion contributes little to the total transport under these conditions (see Sec. 4.8). Flint et al. also obtained results similar to Fig. 4.9 for diffusion of both hydrogen and $CO_2$ in air and of sodium chloride in a turbulent water stream.

In analyzing data of the type illustrated, or in predicting turbulent diffusion, it is necessary to have some guide about what is meant by "the long times" where the use of the eddy diffusion coefficient offers a simplification. If the exponential

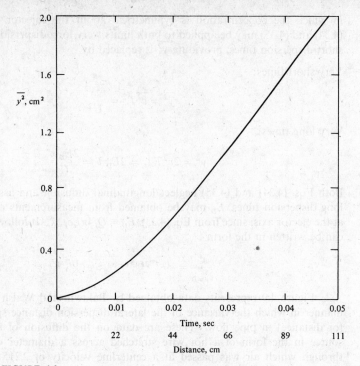

**FIGURE 4.9**
Data of Baldwin and Walsh [9] on the radial diffusion of heat in air from a continuous line source—turbulent flow in a round pipe.

decay of $R_L$ with $T$ is assumed, $[E_{D, \text{av}}/(E_D)_{T \to \infty}]$, which is then a function of $T/T_L$, is found to reach 0.9 when $T/T_L$ is 10. If this is an acceptable criterion of long time, then the minimum $T$ can be found from an estimate of $T_L$. For flow of air at 22 to 49 m/s near the centerline of their 20-cm pipe, Baldwin and Walsh found the turbulent intensity $\sqrt{\overline{u^2}}/\overline{U}$ and the lagrangian integral space scale $T_L\sqrt{\overline{v^2}}$ to be essentially constants equal to 0.035 and 0.74 cm, respectively. The ratio $\sqrt{\overline{v^2}}/\sqrt{\overline{u^2}}$ is typically about 0.8. Consequently for $\overline{U} = 2,215$ cm/s,

$$T_{\text{min}} = 10T_L = \frac{10 \times 0.74}{2215 \times 0.035 \times 0.8} = 0.12 \text{ s}$$

Reference to Fig. 4.9 suggests that this may be overly conservative. Perhaps $T_{\text{min}} \approx 5T_L$ is a better criterion of the time beyond which the long-time equations may be applied in engineering calculations.

Similar data from numerous other sources might have been used in place of those shown in Figs. 4.9 and 4.10 to illustrate the nature of turbulent diffusion. In

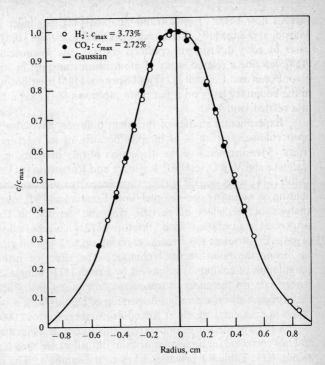

FIGURE 4.10
Tracer gas concentrations at various radial distances from the axis of an 8-cm tube carrying air at 46 ft/s (continuous point source) [47].

addition to that of Baldwin and Walsh, other studies of the diffusion of heat in air from a continuous line source have been reported by Schubauer [131], Uberoi and Corrsin [160], and Townsend [158]. Studies of diffusion from a point source in a turbulent stream include those of Towle, Sherwood, and Seder [156, 157], using carbon dioxide and hydrogen in air; Hanratty, Latinen, and Wilhelm [70], who used both a colored dye and small glass beads in water; Flint, Kada, and Hanratty [47], who used potassium chloride in water as well as carbon dioxide and hydrogen in air; Seagrave and Fahien [134], and Lee and Brodky [97], both employing dye in water; and Becker, Rosensweig, and Gwozdz [14], who introduced a fine-particle oil smoke into air.

Alexander, Baron, and Comings [2] report an extensive investigation of momentum, mass, and heat in turbulent-free jets, and their analysis is extended by Baron and Alexander [11].

Rosensweig, Hottel, and Williams [123] describe an ingenious technique for measuring turbulent concentration fluctuations in a free jet, which suggests studies of the relation between temperature, concentration, and velocity fluctuations in various flow fields. Forstall and Shapiro [48], Schlinger and Sage [129], Berry, Mason, and Sage [18], and Lynn, Corcoran, and Sage [106] studied mixing of two gases by eddy diffusion in confined coaxial jets. Weinstein, Osterle, and Forstall [168] describe a related study of momentum transport in coaxial flat gas jets.

Poreh and Cermak [121] and Spengos [145] used hot-wire heat sources placed in the boundary layer on a flat plate. Johnson [89] used a heated plate as a source in a related study.

Experimental studies of turbulent diffusion have also been made in water in open channels, which is of interest because of its relevance to sedimentation in rivers. Measurements of the dispersion of tiny droplets in water are reported by Kalinske and Van Driest [90], Kalinske and Robertson [91], and Van Driest; and by Orlob [116], who studied surface turbulence by observing the dispersion of particles floating on water in open-channel flow. Friedlander [50] has presented an interesting analysis of the effect of particle size and density on the motion of particles suspended in a turbulent fluid. Boothroyd [21] has reported studies of eddy diffusion of gases in turbulent gas streams carrying up to 8-g solid zinc particles per gram of gas, using the point-source technique. The effect of turbulence on the rate of coagulation of colloids is discussed by Frisch [51]. Suneja, Shea, and Wasan [150] report data on the effect of transverse flow of gas introduced or withdrawn from the porous wall on the radial dispersion of helium from a point source at the axis of a 6-in duct carrying air at Reynolds numbers from 12,000 to 66,000.

Values of the eddy conductivity (measured in the central section of a conduit and not corrected for molecular or axial diffusion) are shown plotted as $E_D$ vs. $\overline{U}_{av} d_t$ in Fig. 4.11; Table 4.1 provides a key to the symbols. The indicated correlation is remarkably good considering the wide range of experimental conditions and

Table 4.1  KEY TO FIGURE 4.11

| Key | Fluid | Diffusing material | Technique | $d_t$, cm | Ref. |
|---|---|---|---|---|---|
| ▦ | Air | Heat | Line source | 20 | 9 |
| ▲ | Air | Smoke | Point source | 20 | 14 |
| ● | Air | $CO_2$, $H_2$ | Point source | 15, 30 | 156, 157 |
| A | Air | $H_2$ | Point source | 8 | 47 |
| ◆ | Water | KCl | Point source | 8 | 47 |
| ▼ | Water | Dye | Point source | 7.6 | 97 |
| △ | Water | Dye | Point source | 10 | 134 |
| □ | Water | Heat | Traverse | 7.6 | 15 |
| ○ | Air, He, $CO_2$ | Water vapor | Traverse | 10.6* | 136 |
| B | Air | Heat | Traverse | 3.8 | 143 |
| ⊡ | Air | Heat | Traverse | 5.1 | 148 |
| ⊙ | Air | Water vapor | Traverse | 8.5 | 132 |
| △ | Air | Heat | Traverse | 3.6* | 117 |
| ▽ | Water | Salt | Point source | 5.0 | 63 |

* Duct with parallel flat walls; value of $d_t$ is twice the spacing between walls. The other conduits were round. $E_r$ is $E_D$ for diffusion normal to the flow direction.

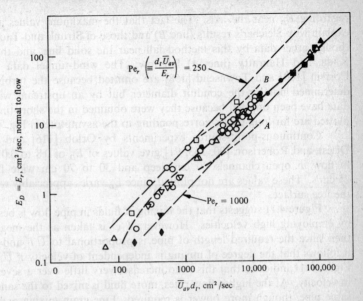

**FIGURE 4.11**
Eddy diffusion coefficients for transport normal to the flow direction in pipes and flat ducts.

techniques, though the several points for water in the lower left corner spatter badly. The choice of coordinates is based on the idea that $E_D$ is proportional to the product of the intensity and scale of turbulence, and that these might be represented very approximately by the velocity $\overline{U}_{av}$ and the conduit diameter $d_t$. Part of the scatter of points may be due to variation of turbulence intensity for a given $\overline{U}_{av}$. The Peclet number $d_t \overline{U}_{av}/E_D$ evidently falls between 250 and 1,000.†

Woertz' [136] data show no difference between results for water vapor transport in streams of He, $CO_2$, and air, though the density varied elevenfold. Towle's [156, 157] data show no difference between $CO_2$ and hydrogen as the tracer gas in air. The "traverse" technique involves measurement of concentration or temperature profiles as well as radial fluxes; such data provide values of $E_D$ at varying distances from the axis. The results obtained by this method provide curves similar to those shown in Fig. 4.4 for eddy viscosity. In these instances the values of $E_D$ plotted are the maximum values, since there is considerable uncertainty

† K. A. Smith [144] has suggested that since $\sqrt{\overline{v'^2}}$ is approximately proportional to $\overline{U}_{av}\sqrt{f/2}$, a better correlating variable would be $d\overline{U}_{av}\sqrt{f/2}$. This suggestion is supported by the data of Groenhof [63] on axial dispersion of a salt tracer in water in turbulent flow. Groenhof found the ratio $E_D/d\overline{U}_{av}\sqrt{f/2}$ to be essentially constant at the value 0.04 over a range of Re from 25,800 to 74,900.

regarding $E_D$ near the axis. The fact that the maximum values are plotted may explain why Sleicher's results (line $B$) and those of Strunk and Tao [148] are high, though other data by this method fall near the solid line, and the data of Flint, Kada, and Hanratty (line $A$) fall low. The wind-tunnel data of Uberoi and Corrsin [160] and Townsend [158] are omitted because the turbulence level was determined not by the conduit diameter but by an upstream wire grid. Other data have been omitted because they were obtained in the short-time region; those plotted are for long times, corresponding to the asymptote in Fig. 4.9.

Continuous-point-source experiments by Orlob [116] and Kalinske, Van Driest, and Robertson [90, 163, 91] give values of $E_D$ of 0.8 to 3.0 cm²/s in water in flow in open channels 30 cm deep and 30 to 70 cm wide flowing at 8 to 25 cm/s. These values are not plotted, since $E_D$ varies appreciably with depth below the free surface.

Figure 4.11 suggests that the mixing of fluids in pipe flow is best accomplished by employing high velocities. However, if $\overline{y^2}$ is taken as the measure of mixing, then since the required length of pipe is proportional to $\overline{U}T$ and so to $\overline{y^2}\,\overline{U}/E_D$, it follows that the degree of mixing is independent of velocity if $\overline{U}/E_D$ is constant. Figure 4.11 indicates that this ratio increases very little over a several fold change in velocity. At the higher flow rates, more fluid is mixed to the same degree in the same pipe, though more power is required. Fine-grain mixing over distances less than the eddy size is accomplished only by molecular diffusion, and concentration differences over short distances will persist much longer in liquids than in gases.

## 4.8   INTERACTION OF MOLECULAR AND TURBULENT DIFFUSION

Molecular diffusion takes place within and between eddies in turbulent flow, and the total transport is the result of both molecular motion and turbulent mixing. As in the foregoing analysis, it is commonly assumed that $E_D$ is independent of the molecular diffusion process, and that the coefficients are additive:

$$E_t = E_D + D \tag{4.40}$$

If there is no correlation between the turbulent fluctuations and the molecular agitations, the total variance of the dispersion is given [49] by

$$\overline{y_{\text{tot}}^2} = \overline{y_{\text{turb}}^2} + \overline{y_{\text{mol}}^2} \tag{4.41}$$

whence for very short times,

$$\overline{y_{\text{tot}}^2} = \overline{v^2}T^2 + 2DT \tag{4.42}$$

and for long times,

$$\overline{y_{\text{tot}}^2} = 2\overline{v^2}TT_L + 2DT \tag{4.43}$$

and

$$E_t = \overline{v^2}T_L + D \tag{4.44}$$

As $T \to 0$, $\overline{y_{\text{turb}}^2} / \overline{y_{\text{mol}}^2} \to 0$, and molecular diffusion predominates. This would be expected immediately downstream from a continuous point source. Further downstream, as the dispersion time becomes large, the molecular diffusion coefficient $D$ becomes negligible compared with $E_D$. This is true for the data shown in Fig. 4.11, where $E_D$ in gases is seen to range from 1 to 100 in mixtures in which $D$ is of the order of 0.1 cm$^2$/s. The values of $D/E_D$ are still smaller in the water tests, where $D$ is of the order of $10^{-6}$ to $10^{-5}$ and $E_D$ is $0.1 - 3$ cm$^2$/s.

The validity of Eq. (4.40) for long dispersion times has been questioned on the basis that there is, in fact, an interaction between the two types of dispersion. Batchelor and Townsend [13] discuss this point and conclude that the total dispersion must be greater than would be found if molecular and turbulent diffusion acted independently. Saffman [126], however, argues that Batchelor and Townsend's analysis is incorrect, and that the opposite conclusion should be accepted—that the interaction of the two processes *reduces* the total dispersion.

## 4.9  TURBULENT SCHMIDT AND PRANDTL NUMBERS

The dimensionless Prandtl and Schmidt numbers, $C_p \mu / k$ and $\nu / D$, have found wide applications in correlating data on heat and mass transfer between phases. Two analogous ratios are relevant to turbulent diffusion:

$$\text{Turbulent Schmidt number} = \text{Sc}_T = \frac{E_v}{E_D} \tag{4.45}$$

$$\text{Turbulent Prandtl number} = \text{Pr}_T = \frac{E_v}{E_H} \tag{4.46}$$

where $E_v$ is the eddy viscosity and $E_D$ and $E_H$ are the eddy diffusion coefficients for mass and heat, respectively.

These ratios are important, since much data on $E_v$ are available (see Sec. 4.3) from which $E_D$ and $E_H$ might be obtained if the effect of velocity, shear, and other variables on $\text{Sc}_T$ and $\text{Pr}_T$ could be established. The early "analogies" between mass and heat transfer (see Chap. 5) were developed with the assumption that both ratios were unity. This is evidently not so.

However, it is commonly believed that $E_D$ and $E_H$ must be equal. This is suggested by Fig. 4.11, which includes data on turbulent diffusion of both mass and heat. It is also to be expected that $\text{Sc}_T$ and $\text{Pr}_T$ would vary in the same way with position in the conduit, flow conditions, and molecular properties. However, since momentum transport is not governed by equations of precisely the same form as those for mass and heat transport (momentum is a vector whereas concentration is a scalar quantity), $E_v$ need not be the same as $E_H$ and $E_D$. The relatively few data available suggest that $\text{Sc}_T$ and $\text{Pr}_T$ lie between 0.6 and 1.0 in the region near the pipe axis, or at the radius where $E_v$ is a maximum. Values in this range have been found for gases [143, 128, 117, 48, 132, 148, 31a] and water [15, 22a]. By contrast, the corresponding ratios for liquid metals are found to be greater than unity

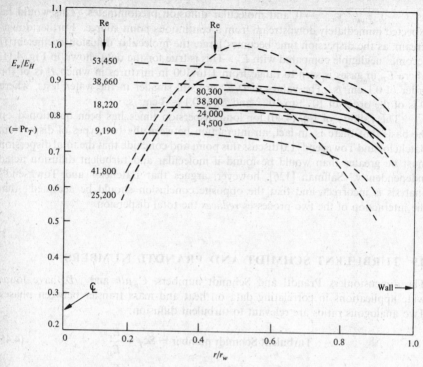

FIGURE 4.12
Variation of turbulent Prandtl number across pipe radius: solid, Sleicher [143]; dashed, Page et al. [117]; dotted, Strunk and Tao [148].

[25, 135]. $Pr_T$ in the central region of flow increases with Re in air [143, 24, 117] but decreases with increasing Re in the case of liquid metals [30]. Both $Pr_T$ and $Sc_T$ appear to be smaller for flow in flat-wall channels than in pipes [63].

Both $Sc_T$ and $Pr_T$ in air decrease with distance from the axis but evidently increase to large values in the region very near the wall [138]. Figure 4.12, after Sleicher, indicates the trend of $Pr_T$ in air with both $r/r_w$ and Re. See also Ref. 22a. The region very near the wall is most important in the development of the analogies but has not been adequately explored. The recent paper by Cebeci [31a] shows $Pr_T$ increasing as the wall is approached, reaching values greater than unity.

The simple theory of Prandtl pictures an eddy moving intact a distance corresponding to the "mixing length," thereupon being fragmented, losing its identity, and delivering mass, heat, or momentum. It is realized, however, that there can be dissipation of these quantities by molecular motion during the time of passage (the theory of the "leaky eddy"). This possibility has been analyzed in various ways by several writers [88, 104, 108, 40, 155, 26] to determine the variation of $Sc_T$

and $Pr_T$ with Sc and Pr, Re, and $E_v/v$. The observed increase in $Pr_T$ from less than unity in air to greater than unity in liquid metals is predicted, but in general these tentative theories appear to be inadequately developed for purposes of quantitative predictions.

## 4.10 MIXING AND DISPERSION IN PACKED BEDS

Diffusion and mixing of fluids flowing in packings and other porous media usually have adverse effects on the performance of such contacting devices as fixed-bed catalytic reactors, regenerators, gas absorbers, ion exchangers, chromatograph columns, and solvent-extraction equipment. The normal axial concentration or temperature gradient tends to be flattened if different elements of the fluid stream have different velocities; even with macroscopic plug flow, axial diffusion and mixing reduce concentration gradients in the direction of flow. Mean driving forces for mass or heat transfer are reduced and performance is adversely affected. Axial dispersion is undesirable in many other situations, as, for example, in secondary oil recovery and in the switch from one oil to another in pipe-line transport.

Dispersion normal to the direction of flow (radial) is usually helpful, in that concentration differences between small fluid elements tend to be equalized and so reduce the variations which enhance the unwanted axial (or "longitudinal") dispersion. Rapid radial transport is desirable when heat or mass is to be transferred to or from the confining wall of the bed, as in the case of an externally cooled exothermic tubular reactor containing solid catalyst particles.

The effect of axial dispersion and the spread of residence times on the performance of process equipment is the subject of a considerable literature. It is the purpose here to describe the phenomena causing dispersion and to indicate the nature of the available data on both axial and radial dispersion in single-phase flow of a fluid through a fixed bed of particles. The effect of axial dispersion on the performance of mass-transfer equipment, as in gas absorbers employing commercial packings, is discussed in Chap. 11. Section 4.11 discusses axial dispersion in flow through unpacked conduits.

Departures from piston or plug flow are the result of axial dispersion, due to one or more of the following: (1) a radial velocity gradient in the conduit, (2) eddy diffusion, or mixing, and (3) molecular diffusion. Taylor diffusion, discussed in Sec. 3.8, is the result of both a velocity gradient and radial molecular diffusion and mixing. Even in the absence of both molecular diffusion and mixing, a solute is distributed axially if there is a velocity gradient. The extent of this axial dispersion may be calculated if the velocity gradient is known (as in the case of laminar flow in a round tube, where the velocity is a parabolic function of the radius). Axial dispersion in liquids flowing in unpacked conduits is almost entirely due to velocity gradients. By contrast, piston flow is approached closely in the flow of a single phase through a bed of small particles of uniform size if the bed diameter is large in comparison with that of the particles. In this case the velocity profile is quite flat and both axial and radial dispersion are due to the combined effects of molecular diffusion and mixing.

Several "models" have been used to interpret and correlate experimental data on mixing in packed beds. The one most commonly employed assumes that the transport can be described by Fick's law, and that radial and axial "diffusion" coefficients (perhaps better termed "dispersion" coefficients), $E_r$ and $E_a$, independent of solute concentration, can be related to fluid properties, flow parameters, and the geometry of bed and packing. The cell model conceives the flow through the small voids between particles to be analogous to sequential flow through a large number of well-stirred vessels. Still a third model focuses on the ratio of the fluid physically back-mixed to the net flow forward.

These approaches provide empirical correlations of experimental data, useful in estimating the effects of mixing on the performance of chemical reactors and mass-transfer equipment. They do not provide or require knowledge of the actual physics of the dispersion process. Theoretical models of the small-scale mixing in beds have been developed by several writers (e.g., Refs. 115, 19, 38, 65, 10, 130, 99, 159, and 126), but none have provided a reliable quantitative basis for the prediction of mixing effects. This is not surprising, since the structure of beds and the geometry of flow passages are exceedingly complex, even in beds of uniform spheres. The difficulty of dealing with this complexity is clear from the excellent summary of bed structure by Haughey and Beveridge [75].

The problem is further complicated by any capacitance effect, as when the flow passages have side dead-end pores which take up and release solute by slow diffusion processes. This effect has been analyzed by Aris [7] and by Gunn and his colleagues [65, 66, 67].

As applied to flow in a cylindrical bed, the diffusion, or dispersion, model suggests the use of the following equation for the spread of a conserved solute in plug flow of a fluid through a bed of particles:

$$\frac{\partial c}{\partial t} = \frac{E_r}{r} \frac{\partial}{\partial r}\left[r\left(\frac{\partial c}{\partial r}\right)\right] + E_a \frac{\partial^2 c}{\partial x^2} - \overline{U}_i \frac{\partial c}{\partial x} \tag{4.47}$$

where $\overline{U}_i$ is the interstitial velocity, and $c$ the solute concentration. $\overline{U}_i$ is equal to the superficial velocity divided by the void cross section of the bed, the latter being the same as the void volume fraction $\varepsilon$. $E_a$ and $E_r$ are the axial and radial dispersion coefficients, respectively. Note that $E_a$ and $E_r$ are based on the void cross section. This is similar to the differential equation leading to Eq. (4.33) for the diffusion from a point source, where $E_r = E_D$ was assumed constant and $E_a$ taken to be zero.

Although Eq. (4.47) employs the interstitial velocity $\overline{U}_i$ and dispersion coefficients based on the void (or flowing stream) cross section, it is evident that the ratio $\overline{U}_i/E$ is the same as $\overline{U}_{av}/\varepsilon E$, where $\varepsilon$ is the void fraction. This follows since $\overline{U}_{av} = \varepsilon \overline{U}_i$, providing the stream fills the voids. The product $\varepsilon E$ is a dispersion coefficient based on the total cross section of the bed, as is $\overline{U}_A$.

The methods used to measure $E_r$ and $E_a$ involve comparisons of solute distribution in feed and effluent streams from a test bed, with steady flow. Solutions of Eq. (4.47), with boundary conditions appropriate to the test procedure, are compared with the experimental data to infer values of the dispersion coefficients.

Axial dispersion is usually measured by one of four methods: (1) introduce a step change in the solute concentration in the feed, and follow the breakthrough of solute in the effluent, (2) introduce a sudden pulse of high-solute concentration in the feed and measure the spread of solute concentration in the effluent as the pulse leaves the bed, (3) maintain a solute concentration in the feed which varies sinusoidally with time and measure the frequency response, and (4) introduce a steady flow of constant solute concentration at a point near the bed outlet and measure the concentration at one or more points upstream. When using method 2, it is not necessary to apply a perfect pulse distribution across the flow in essentially zero time, since Aris [6] and Levenspiel and Bischoff [99] have developed analyses applicable to the use of a finite pulse, requiring concentration-time measurements of the solution entering the bed.

Radial dispersion is most easily measured by feeding a continuous stream of tracer solute from a point source at the axis of the bed and measuring the radial spread of tracer concentration at some downstream location. This is analogous to the measurement of eddy diffusion in an open conduit, as described in Sec. 4.7.

The needed solutions of Eq. (4.47) are available in many places in the literature including Levenspiel and Bischoff's excellent 105-page review of the whole subject, with applications to the performance of chemical reactors [99]. Another very good review is that of Gunn [64]. The derivation of the equation applicable to the sudden step change in feed concentration has involved a debate about appropriate boundary conditions [64, 96, 35, 8, 100, 99], but Danckwerts' [35] simple form is usually employed. Danckwerts, Levenspiel, and Smith [100] and many later writers give the solution for the instantaneous pulse case. The student should not have difficulty in deriving the relation between concentration and distance applicable to case (4) for axial dispersion (see Refs. 4, 56, 169, 68, 84). Equation (4.33) can be applied to radial dispersion from a continuous point source if the solute reaches the wall in negligible amount; allowance for the wall "reflection" is included in the analysis by Bernard and Wilhelm [17]. Other solutions have been published for $E_r \neq E_a$ [19, 94, 66, 99]; continuous axial tracer stream of finite diameter [66, 115, 99]; and for tracer input from the walls confining the bed [115]. Kramers and Alberda [96] and various later writers describe the use of the frequency response technique.

*The cell model*, in its simplest form, considers the several small fluid streams entering a void in the packing to be thoroughly mixed before the combined flow passes to the next cavities downstream. Consider the following experiment: water is passed at a steady flow rate through a bed of particles of uniform size. The feed is changed suddenly to a dilute salt solution and the flow continues at the same steady rate. If there were no axial dispersion, the effluent concentration would evidently show a sudden step increase to that of the salt solution fed. Because of axial diffusion and mixing, however, the salt breakthrough is not sudden but exhibits a sigmoid curve of effluent concentration vs. time.

The salt concentration $c_n$ in the effluent from the $n$th cell from the bed inlet is related to that $(c_{n-1})$ entering the $n$th cell by

$$\bar{U}_i c_{n-1} - \bar{U}_i c_n = V_c \frac{dc_n}{dt} \tag{4.48}$$

where $V_c$ is the cell volume per unit void cross section of the bed. This is easily integrated for the first cell and the result used to obtain $c$ in the effluent from the second cell. Repetition of this procedure leads to an expression for the concentration in the bed effluent, providing that the number of cells corresponding to the bed length is known.

An alternative equation relating concentration, time, and bed length can also be obtained by solving Eq. (4.47), with appropriate boundary conditions. From the result (radial gradients being ignored) one obtains a relation between $c$, distance $x$, and time $t$, based on the dispersion model.

The two models give different forms for the relation $f(c, x, t) = 0$, applicable to the conditions of the experiment described. Both are readily transformed to give the probability density function, which expresses the probability that a molecule of salt or of water (salt and water are associated in constant proportion in the feed) will appear in the effluent at a time $t$ after the step change. If the pulse technique is used, this probability is proportional to the effluent concentration observed at time $t$ following the introduction of the pulse of tracer.

Aris and Amundson (8; see also Ref. 100) have compared the probability density functions found in this way for the two models. The dispersion model leads to a gaussian distribution and the cell model to a Poisson distribution, but the two are essentially indistinguishable if the number of cells is large. The two probability distributions superpose at large $n$ if $V_c$ and $E_a$ are related by

$$E_a = \frac{\overline{U}_i \varepsilon \gamma^2 d_p^{\ 2}}{2V_c} \tag{4.49}$$

or

$$\text{Pe}_a = \frac{\overline{U}_i d_p}{E_a} = \frac{2}{\gamma} \tag{4.50}$$

Here $\text{Pe}_a$ is the axial Peclet number, $\overline{U}_i$ ($= \overline{U}_{av}/\varepsilon$) is the interstitial and $\overline{U}_{av}$ the average velocity (assumed macroscopically uniform across a plane normal to the flow), $d_p$ is the diameter of the packing particle, and the bed length corresponding to one cell is $\gamma d_p$. The cell volume $V_c$ is $\varepsilon \gamma d_p$. Assuming one cell per void in a bed of spheres, $\gamma$ is 0.7 to 1.0, depending on the packing geometry, so $\text{Pe}_a$ is 2.0 to 2.8. $\text{Pe}_a$ is defined in terms of $\overline{U}_i$ (not $\overline{U}_{av}$), and $E_a$ is the dispersion coefficient based on unit area of the voids normal to the flow. However, as noted above, $\overline{U}_{av} = \varepsilon \overline{U}_i$ and $\varepsilon E$ is the dispersion coefficient based on the total bed cross section, if the stream fills the voids. As shown below, experimental data support the conclusion that $\text{Pe}_a$ is approximately 2.0 at particle Reynolds numbers greater than about 500.

Experimental data on dispersion in packed beds have been published by a very large number of investigators. Data available through 1962 are summarized by Levenspiel and Bischoff in their review. Data on radial dispersion of gases and liquids in fixed beds are reported in Refs. 17, 46, 171, 139, 16, 64, 115, 122, and 66. Radial dispersion of heat has been measured by Plautz and Johnstone [120] in packed beds and by Gabor [54, 55] in fluidized beds. Articles reporting data on axial dispersion include Refs. 107, 96, 29, 41, 102, 146, 27, 147, 164, 79, 45, 72, 20,

37, 110, 82, 114, 64, 161, 139, 115, 71, 60, 42, 73, 66 and 32. Axial dispersion of heat has also been measured [173, 62, 140]. Published data have been collected and summarized in graphical form (usually as graphs of $Pe_a$ or $Pe_r$ vs. Re ($= d_p \bar{U}_{av}/\nu$) by various authors [171, 27, 52, 41, 79, 28, 98, 95, 172, 118, 64, 164, 82, 122, 38, 32, 99, 98].

Studies of axial dispersion of one or both phases in *two-phase flow* in packings (including "trickle beds") are reported in Refs. 127, 57, 87, 101, 164, 111, 83, 23, 78, 3, 53, 57. References 64, 127, and 53 deal with dispersion in columns packed with Raschig rings and Berl saddles in sizes $\frac{1}{4}$ to 1.0 in. Reference 167 reports data on axial dispersion of water in several small packings as affected by the counterflow of mercury. The results agree well with the correlation proposed in Ref. 164.

Data on axial dispersion (treated as back-mixing) in agitated contactors (Mixco, RDC) are to be found in Refs. 68, 112, 164, 170, 74, 149 and 169. References 44 and 83 deal with axial dispersion in bubble columns.

In spite of the large amount of data now available, there is presently no general correlation which may be used with confidence to predict either $E_a$ or $E_r$. At very low flow rates, both coefficients approach the values appropriate to molecular diffusion ($E_a = E_r = D_{AB}/T_\varepsilon$, where $T_\varepsilon$ is the "tortuosity" of the bed structure; $T_\varepsilon$ is about 1.4 in unconsolidated porous media). In the other limit, as Re exceeds 100 to 500, the mixing in the voids is rapid, and $Pe_a$ and $Pe_r$ become constant, independent of further increase in Re, being then equal to about 2 and 11, respectively ($E_r$ is about $\frac{2}{11}$ of $E_a$ at fixed $d_p$ and $\bar{U}_{av}$). Many commercial reactors and contacting devices operate in this range.

The situation in the intermediate range is confused, partly because of the complexities of the phenomena and partly because of the inadequacies of the data. The large majority of the published tests were made with beds of small diameter and relatively large ratios $d_p/d_t$ (where $d_t$ is the column diameter); as emphasized by Hiby [79], the dispersion coefficients increase rapidly with $d_p/d_t$ because of the greater flow velocities along the walls (piston flow does not prevail). A semiquantitative description of the trends in the intermediate region, however, has been presented in the excellent articles by Wilhelm [172], who is responsible for much of the best work on dispersion in packings, Hiby [79], and Miller and King [110].

In attempting to understand the confusion of data, it is helpful to consider the changes which occur as the flow velocity is reduced from a value in the turbulent region. The excellent photographs of similar flow phenomena presented by Hiby provide a basis for such an understanding. Two of these photographs are shown as Figs. 4.13 and 4.14. The streams entering a void are well mixed at Reynolds numbers greater than about 500, and $Pe_a$ is approximately 2. There is little diffusion, only convective transport, between successive voids or "cells," and no back-mixing [79]. As Re is reduced, the mixing within each cell becomes incomplete and multiple filaments of fluid of differing concentration pass downstream. This tends to increase $E_a$ (for example, in the step-change tracer experiment, mixing across the front tends to occur over a longer axial distance than the length of one cell). The effect of the convective transport of the unmixed filaments tends to be offset, however,

FIGURE 4.13
Photograph showing the paths followed by illuminated aluminum particles in water
flowing down through an array of 2-cm cylinders simulating a packed bed.
Re = 20. (*Courtesy Prof. J. W. Hiby, Aachen.*)

by molecular diffusion between filaments passing through each cell. The net result
is greatly dependent on the magnitude of the molecular diffusion coefficient (or on
$Sc \equiv \nu/D_{AB}$). In the case of gases $D_{AB}$ is usually large enough to accomplish mixing
within the cell to a similar extent as turbulence did at large Re. $Pe_a$ remains
approximately constant, decreasing little from its initial value of 2 until molecular
diffusion from cell to cell becomes important at Re ≈ 1.

Molecular diffusion in liquids is so slow, however, that the filaments of
different concentrations persist and $E_a$ increases as Re is reduced below 500. However,
as Re is decreased from 300 to 10, $Pe_a$ remains approximately proportional to Re,
indicating that $E_a$ is roughly constant in this region. $Pe_a$ drops to 0.3 to 0.6 at
Re = 0.1 to 10 but rises somewhat as Re is further reduced. This increase in $Pe_a$
(decrease in $E_a$ with decreasing velocity) is the result of molecular diffusion within
the cells, though molecular diffusion from cell to cell (with consequent decrease
in $Pe_a$) does not become dominant until very low values of Re are reached (about
Re = 0.0001 for Sc = 730).

*Radial* dispersion behaves as predicted by a random-walk analysis. The particle
divides the oncoming stream element, diverting it radially, portions being deflected
in or out with equal probability. The radial flow and mixing will clearly be proportional to the axial velocity; if the scale of the deflection is presumed to be

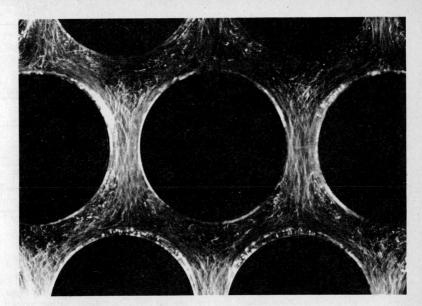

FIGURE 4.14
Flow through a simulated packed bed, as in Fig. 4.13, but at Re = 1,500. (*Courtesy Prof. J. W. Hiby, Aachen.*)

proportional to $d_p$, the radial Peclet number $Pe_r$ is then constant. As Re is reduced, the stream impinging on the particle is not well mixed; the unmixed filaments tend to flow around the particle and not move radially to the next particle. The result is a decrease in $E_r$ and an increase in $Pe_r$, for liquids, to a relatively large value. Gases, however, tend to be well mixed at much larger values of Re, so the molecular diffusion limit is encountered first and the increase in $Pe_r$ in the intermediate range of Re is not observed.

The general picture of the trends described is illustrated by Fig. 4.15 (after Wilhelm [172]). This represents an attempt at a general correlation of existing data for both axial and radial dispersion. It is obviously unsatisfactory as a basis for reliable predictions of $E_a$ or $E_r$, especially in the intermediate range between the condition of unmixed cells and dispersion by molecular diffusion at low values of Re. More careful and extensive studies of the intermediate range are needed.

It seems doubtful that the complicated phenomena involved in fluid flow in packings will ever be satisfactorily correlated in terms of the three parameters Pe, Re, and Sc, even for single-phase flow. Ebach and White [41] found no change in $E_a$ at constant liquid velocity when the viscosity was varied twenty-six fold. Gunn and Pryce [66] report wide discrepancies between the usual correlations for random packings and their data on gas flow in beds of spheres packed in regular cubic or rhombohedral arrangements. The discrepancies reported by Edwards and

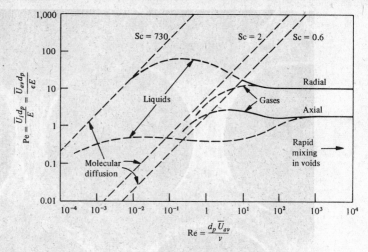

**FIGURE 4.15**
Approximate representation of a large amount of published data on radial and axial dispersion in randomly packed beds of uniform spheres; flow of a single phase. The dashed straight lines represent the molecular-diffusion asymptotes, for which $Pe = (Re)(Sc)T_\varepsilon/\varepsilon$. The lines shown are for $T_\varepsilon = \sqrt{2}$ and $\varepsilon = 0.4$.

Richardson [42] for axial gas dispersion in beds of sand and broken fragments of plastic are perhaps reminders that the effect of particle shape has been little studied, and that essentially no work has been done on beds of particles of mixed sizes. The single dimension $d_p$ cannot be expected to describe the flow channel adequately. Even with spheres, repacking the bed leads to appreciable differences in the results of duplicate tests [66]. The subject is complex and its clarification will require a great deal of work; in the meantime, the empirical representation of data giving $Pe_a$ and $Pe_r$ as functions of Re and Sc will serve many engineering needs.

The practical reason for interest in dispersion in packed beds is its effect on the performance of many types of equipment, especially chemical reactors and mass-transfer equipment. These effects can generally be calculated if the dispersion coefficients are known. Danckwerts [35] shows, for example, how the conversion with a first-order chemical reaction in a tubular flow reactor can be expected to decrease because of axial dispersion of the reactant. The same effect is encountered in mass-transfer equipment, where solute is transferred out of a stream by mass transfer to another phase—a first-order process.

Discussion of this important subject will be deferred to Chap. 11, where a section is devoted to methods of allowing for axial dispersion in columns employing such commercial packings as Raschig rings and Berl saddles. The discussion and example are specific to packed gas-liquid contacting devices but should serve to show how the problem can be handled in other types of mass-transfer operations, such as liquid extraction, where the effects are particularly serious [164].

## 4.11  AXIAL DISPERSION IN PIPE LINES

The velocity profile in single-phase flow is exceedingly flat in a bed of uniform particles, uniformly packed, especially in commercial equipment where the ratio $d_t/d_p$ of tube to particle diameter is large. Axial dispersion increases rapidly as a radial velocity gradient develops, and axial transport in unpacked tubes is predominantly by convection. Even if molecular and eddy diffusion in the axial direction are absent, the convective transport caused by the presence of a velocity gradient can be represented by Fick's law. If the velocity profile is known, the effective axial dispersion coefficient can be calculated.

In laminar flow this phenomenon is Taylor diffusion, described in Sec. 3.8. For laminar flow in round tubes, the parabolic profile leads to Eq. (3.65) for the virtual diffusion coefficient. Appreciable velocity gradients exist in turbulent flow, and the resulting convective transport predominates; both molecular and eddy diffusion are normally negligible. The situations in packed beds and open tubes, therefore, are quite different: in open tubes with turbulent flow, only convective transport is important; in packed beds, axial molecular and eddy diffusion are the controlling mechanisms, and convective transport due to velocity gradients across the bed can usually be neglected (though some of the scatter of the data on packed beds is doubtless due to neglect of convective transport, especially in tests where $d_t/d_p$ is small).

Adopting the so-called universal velocity profile, Taylor [152] has solved Eq. (4.47) with the substitution of $\overline{U}$ as a function of $r$ and $\overline{U}_{av}$, and the omission of the term containing $E_a$. The radial eddy diffusion coefficient was taken to be equal to the eddy viscosity, obtained from the assumed velocity profile. This leads to an expression for the axial convective transport, which is of the form of Fick's law, and provides an expression for the effective or virtual axial-dispersion coefficient:

$$\frac{E_c}{\overline{U}_{av} d_t} = 3.57 \sqrt{f} = \phi(Re) \tag{4.51}$$

where $f$ is the Fanning friction factor. The symbol $E_c$ is used to represent the coefficient of axial dispersion in cases where transport is due primarily to convective flow stemming from velocity gradients.

Tichacek, Barkelew, and Baron [154] have modified Taylor's analysis by employing critically selected and smoothed velocity profiles from the literature and assuming radial eddy and molecular diffusion to be additive. This also leads to a relation between $E_c/\overline{U}_{av} d_t$ and $f$ which they present in graphical form. Neither this nor Taylor's analysis is expected to hold for Re less than about 10,000, since at low velocities the appreciable thickness of the laminar sublayer at the wall holds

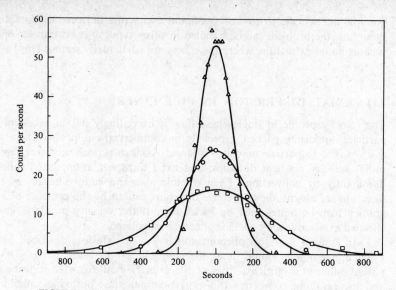

FIGURE 4.16
Tracer concentration in a 10-in oil pipeline measured at three distances downstream from the tracer pulse injection: △, 13.8 mi; ○, 43.1 mi; □, 108.5 mi [85].

solute which is released slowly by molecular diffusion—an effect not allowed for in the derivations. Aris' analysis [5] allows for molecular diffusion in the axial direction.

Numerous investigators have employed tracer techniques to measure $E_c$ for turbulent flow of gases and liquids in pipes of various diameters and lengths. One of the most interesting of the published studies is that of Hull and Kent [85] who introduced radioactive tracer pulses into commercial oil pipe lines and used Geiger counters at stations along the line to measure the passing wave of tracer concentration. Figure 4.16, from their report, shows the nature of the data obtained with oil flowing at 2.68 ft/s in a 10-in line at stations 13.8, 43.1, and 108.5 mi from the point of tracer addition.

As Taylor [152] has shown, the concentration spread should follow gaussian curves after 100 pipe diameters, the concentration being given by

$$c = A'' t^{-1/2} \exp\left(\frac{-x_1{}^2}{4E_c t}\right) \tag{4.52}$$

where $A''$ = constant which is proportional to amount of tracer in pulse

$t$ = time for midpoint or peak of curve to reach measuring station

$x_1$ = distance upstream or downstream from peak concentration [in practice, concentration is measured as a function of time at a fixed point; $x_1 = \overline{U}_{av} t$

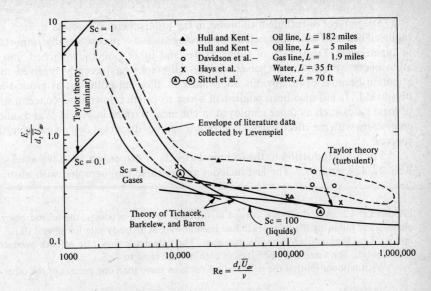

**FIGURE 4.17**
Axial diffusion data for turbulent flow in open pipes expressed in terms of the Peclet number for axial convection.

Data of the type shown in Fig. 4.16, even when reported only as relative concentration (counts per second), are evidently sufficient to calculate $E_c$.

Figure 4.17 illustrates the status of the existing correlations of experimental data on axial dispersion in turbulent flow. The three solid lines represent the theories of Taylor, and of Tichacek et al. The dotted envelope includes the large number of data points collected and plotted in this way by Levenspiel [98a]. A few points and a line [85, 36, 77, 141, 58] representing more recent data have been added. The solid triangle at Re = 24,000 represents a tracer test by Hull and Kent on a 10-in oil line 182 mi long. The agreement with the theory is remarkable, especially since the line went through mountainous country and included not only pipe elbows and bends but pumping stations.

Axial dispersion in the turbulent flow of water through a series of $1\frac{1}{4}$-in 90° elbows has been studied by Cassell and Perona [31]. The axial dispersion coefficients were found to be 8 to 61 percent greater than in a straight pipe, the experimental points falling within the Levenspiel envelope in Fig. 4.17.

Kenney and Thwaites [92] measured $E_a$ for water containing 100 ppm of a drag-reducing polymer in flow through a 1.9-cm pipe 728 cm long. Their data are in fairly good agreement with Taylor for 10,000 < Re < 30,000 but check better with the Tichacek, Barkelew, and Baron curve in the range of Re from 3,000 to 10,000. Axial dispersion in the flow of non-newtonian "power-flow" fluids with turbulent flow in straight tubes is treated by Krantz and Wasan [96a].

The subject of turbulent diffusion in the atmosphere will not be treated, even though it has been widely studied by meteorologists and is of considerable importance in engineering. Dispersion of sulfur oxides and other pollutants carried into the atmosphere by combustion gases leaving tall stacks has recently received much attention because of its obvious importance in affecting pollution at ground-level downwind. It has also been studied in order to predict downwind concentrations of toxic gases such as those employed in chemical warfare in World War I, and in connection with the effectiveness of screening smokes dispersed downwind from their sources.

For an introduction to the technical aspects of dispersion from tall stacks, see Refs. 33, 124, and 162. The last includes an extensive bibliography, with abstracts.

EXAMPLE 4.2   An oil pipeline is used to transport two oil products, the second being fed immediately following the first, which had been flowing at a steady rate for several days. The line is 24 mi long and 12 in inside diameter. The volumetric flow rate of both products is 48,300 bbl/d. The kinematic viscosities of both oils are close to 4 cS.

What amount (bbl) of the delivered oil contains more than one percent of the other oil?

SOLUTION

$$U_{av} = \frac{48,300 \times 42}{24 \times 3600 \times 7.48 \times 0.785} = 3.99 \text{ ft/s}$$

$$Re = \frac{1 \times 3.99 \times 62.3}{4 \times 0.000672} = 92,700$$

$E_c/d_t\,\overline{U}_{av}$ is estimated from Fig. 4.17 to be 0.41;

$$E_c = 0.41 \times 1 \times 3.99 = 1.64 \text{ ft}^2/\text{s}$$

The oils are dispersed both ways across the initially sharp front separating them. This dispersion follows Fick's law:

$$\frac{\partial c}{\partial t} = E_c \frac{\partial^2 c}{\partial x_1^2}$$

where $\partial c/\partial t$ is the time rate of change of concentration in an element moving at the velocity $U_{av}$, which is the velocity of the plane of 50 percent contamination, and $x_1$ is the axial distance from this plane. The boundary conditions are

$$c = 1 \quad t = 0, x_1 > 0$$
$$c = 0 \quad t = 0, x_1 < 0$$

and the solution is

$$c = \tfrac{1}{2} + \tfrac{1}{2} \operatorname{erf}\left(\frac{x_1}{2\sqrt{E_c t}}\right)$$

The concentration (volume fraction first oil) is symmetrical about the plane $x_1 = 0$, so the contaminated length $2x_1$ between $c = 0.01$ and $c = 0.99$ is given by

$$0.99 - 0.01 = 0.98 = \operatorname{erf}\left(\frac{x_1}{2\sqrt{E_c t}}\right)$$

whence from tables of the error function,

$$\frac{x_1}{2\sqrt{E_c\,t}} = 1.645 = \frac{x_1}{2}\sqrt{\frac{U_{av}}{E_c L}}$$

where $L$ is the length of the line.

$$2x_1 = 4 \times 1.645 \sqrt{\frac{1.64 \times 24 \times 5280}{3.99}} = 1500 \text{ ft}$$

The contaminated oil is that contained in this length of pipe. It amounts to

$$\frac{1,500 \times \pi \times 1 \times 7.48}{4 \times 42} = 209 \text{ bbl}$$

////

# REFERENCES

1 ABBRECHT, P. H., and S. W. CHURCHILL: *AIChE J.*, **6:** 268 (1960).
2 ALEXANDER, L. G., T. BARON, and E. W. COMINGS: *Univ. Ill. Eng. Exp. Sta. Bull.* 413 (1953).
3 ANDERSON, K. L., and R. E. GILBERT: *Ind. Eng. Chem. Fundam.*, **5:** 430 (1966).
4 ARGO, W. B., and D. R. COVA: *Ind. Eng. Chem. Process Des. Dev.*, **4:** 352 (1965).
5 ARIS, R.: *Proc. Roy. Soc. (London)*, **A234:** 67 (1956).
6 ARIS, R.: *Chem. Eng. Sci.*, **9:** 266 (1959).
7 ARIS, R.: *Chem. Eng. Sci.*, **10:** 80 (1959).
8 ARIS, R., and N. R. AMUNDSON: *AIChE J.*, **3:** 280 (1957).
9 BALDWIN, L. V., and T. J. WALSH: *AIChE J.*, **7:** 53 (1961).
10 BARON, T.: *Chem. Eng. Progr.*, **48:** 118 (1952).
11 BARON, T., and L. G. ALEXANDER: *Chem. Eng. Progr.*, **47:** 181 (1951).
12 BATCHELOR, G. K., and R. M. DAVIES: "Surveys in Mechanics," Cambridge, New York, 1956.
13 BATCHELOR, G. K., and A. A. TOWNSEND: "Surveys in Mechanics," p. 352, Cambridge, New York, 1956.
14 BECKER, H. A., R. E. ROSENSWEIG, and J. R. GWOZDZ: *57th Annu. Meet. AIChE, Boston, Mass.*, preprint 32c, Dec. 6-10, 1964.
15 BECKWITH, W. F., and R. W. FAHIEN: *55th Nat. Meet., AIChE, Houston*, preprint 33a, Feb. 7-11, 1965.
16 BEEK, J.: "Advances in Chemical Engineering," vol. III, Academic, New York, 1962.
17 BERNARD, R. A., and R. H. WILHELM: *Chem. Eng. Progr.*, **46:** 233 (1950).
18 BERRY, V. J., D. M. MASON, and B. H. SAGE: *Ind. Eng. Chem.*, **45:** 1596 (1953).
19 BISCHOFF, K. B., and O. LEVENSPIEL: *Chem. Eng. Sci.*, **17:** 245, 257 (1962).
20 BLACKWELL, R. J., J. R. RAYNE, and W. M. TERRY: *Trans. AIME*, **216:** 1 (1959).
21 BOOTHROYD, R. G.: *Trans. Inst. Chem. Eng. (London)*, **45:** T297 (1967).
22 BOSWORTH, R. C. L.: *Phil. Mag. (VII)*, **40:** 314 (1949).
22a BRINKWORTH, B. J., and P. C. SMITH: *Chem. Eng. Sci.*, **28:** 1847 (1973).
23 BRITTAN, M. I.: *Chem. Eng. Sci.*, **22:** 1019 (1967).
24 BRODKY, R. S.: *AIChE J.*, **9:** 448 (1963).
25 BROWN, H. E., B. H. ARMSTEAD, and B. E. SHORT: *Trans. ASME*, **79:** 279 (1957).
26 BURGERS, J. M.: In J. O. Hinze (ed.), "Turbulence," p. 304, McGraw-Hill, New York, 1959.
27 CAIRNS, E. J., and J. M. PRAUSNITZ: *Chem. Eng. Sci.*, **12:** 20 (1960).

28  CARBERRY, J. J.: *Chem. Proc. Eng.*, **44:** 306 (1963).

29  CARBERRY, J. J., and R. H. BRETTON: *AIChE J.*, **4:** 367 (1958).

30  CARR, A. D., and R. E. BALZHISER: *Brit. Chem. Eng.*, **12:** (1), 13 (1967).

31  CASSELL, R. E., and J. J. PERONA: *AIChE J.*, **15:** 81 (1969).

31a  CEBECI, T.: *ASME Paper* 72WA/HT 13 (1972).

32  CHUNG, S. F., and C. Y. WEN: *AIChE J.*, **14:** 857 (1968).

33  CONNER, W. D., and J. R. HODKINSON: "Optical Properties and Visual Effects of Smoke-Stack Plumes," U.S. Dept. of Health, Education, and Welfare, Cincinnati, Ohio, 1967.

34  CORCORAN, W. H., J. B. OPFELL, and B. H. SAGE: "Momentum Transfer in Fluids," Academic, New York, 1956.

35  DANCKWERTS, P. V.: *Chem. Eng. Sci.*, **2:** 1 (1953).

36  DAVIDSON, F. F., D. C. FARQUHARSON, J. Q. PICKEN, and D. C. TAYLOR: *Chem. Eng. Sci.*, **4:** 201 (1955).

37  DEISLER, P. F., JR., and R. H. WILHELM: *Ind. Eng. Chem.*, **45:** 1219 (1953).

38  DE LIGNY, C. L.: *Chem. Eng. Sci.*, **25:** 1177 (1970).

39  DRYDEN, H. L.: *Ind. Eng. Chem.*, **31:** 416 (1939).

40  DWYER, O. E.: *AIChE J.*, **9:** 261 (1963).

41  EBACH, E. A., and R. R. WHITE: *AIChE J.*, **4:** 161 (1958).

42  EDWARDS, M. F., and J. F. RICHARDSON: *Chem. Eng. Sci.*, **23:** 109 (1968).

43  EINSTEIN, A.: *Ann. Phys.*, **17:** 549 (1905); **19:** 371 (1906).

44  EISSA, S. H., M. M. EL-HALWAGI, and M. A. SALEH: *Ind. Eng. Chem. Proc. Des. Dev.*, **10:** 31 (1971).

45  EVANS, E. V., and C. N. KENNEY: *Trans. Inst. Chem. Eng. (London)*, **44:** T189 (1966).

46  FAHIEN, R. W., and J. M. SMITH: *AIChE J.*, **1:** 28 (1955).

47  FLINT, D. I., H. KADA, and T. J. HANRATTY: *AIChE J.*, **6:** 325 (1960).

48  FORSTALL, W., and A. H. SHAPIRO: *J. Appl. Mech.*, **17:** 399 (1950).

49  FRENKIEL, F. N.: "Advances in Applied Mechanics," vol. 3, Academic, New York, 1953.

50  FRIEDLANDER, S. K.: *AIChE J.*, **3:** 381 (1957).

51  FRISCH, H. L.: *J. Phys. Chem.*, **60:** 463 (1956).

52  FROMENT, G. F.: *Ind. Eng. Chem.*, **59:** 18 (1967).

53  FURZER, I. A., and R. W. MICHELL: *AIChE J.*, **16:** 380 (1970).

54  GABOR, J. D.: *AIChE J.*, **11:** 127 (1965).

55  GABOR, J. D., B. E. STRANGELAND, and W. J. MECHAM: *AIChE J.*, **11:** 130 (1965).

56  GILLILAND, E. R., and E. A. MASON: *Ind. Eng. Chem.*, **41:** 1191 (1949).

57  GLASER, M. B., and M. LITT: *AIChE J.*, **9:** 103 (1963).

58  GOLDSCHMIDT, V. W., and M. K. HOUSEHOLDER: *Ind. Eng. Chem. Fundam.*, **8:** 172 (1969).

59  GOLDSTEIN, S.: "Modern Developments in Fluid Mechanics," vol. 2., Oxford, New York, 1939.

60  GOTTSCHLICH, C. F.: *AIChE J.*, **9:** 88 (1963).

61  GRAVES, C. C., and D. W. BAHR: *Lewis Lab. NACA Rep.* 1300, 1957.

62  GREEN, D. W., R. H. PERRY, and R. E. BABCOCK: *AIChE J.*, **10:** 645 (1964).

63  GROENHOF, H. C.: *Chem. Eng. Sci.*, **25:** 1005 (1970).

64  GUNN, D. J.: *Chem. Eng. (London)*, p. CE153, June 1968.

65  GUNN, D. J.: *Trans. Inst. Chem. Eng. (London)*, **47:** T351 (1969).

66  GUNN, D. J., and C. PRYCE: *Trans. Inst. Chem. Eng. (London)*, **47:** T341 (1969).

67  GUNN, D. J., and R. ENGLAND: *Chem. Eng. Sci.*, **26:** 1413 (1971).

68  GUTTOFF, E. B.: *AIChE J.*, **11:** 712 (1965).

69  HANRATTY, T. J.: *AIChE J.*, **2:** 42 (1956).

70  HANRATTY, T. J., G. LATINEN, and R. H. WILHELM: *AIChE J.*, **2:** 372 (1956).

71   HARLEMAN, D. R. F., and R. R. RUMER: *J. Fluid Mech.*, **16:** 385 (1963).
72   HARLEMAN, D. R. F., P. F. MEHLHORN, and R. R. RUMER: *Proc. Amer. Soc. Civil Eng.*, *J. Hydraul. Div.*, Paper HY-2, p. 67, March 1963.
73   HASSEL, H. L., and A. BONDI: *AIChE J.*, **11:** 217 (1965).
74   HAUG, H. F.: *AIChE J.*, **17:** 585 (1971).
75   HAUGHEY, D. P., and G. S. G. BEVERIDGE: *Can. J. Chem. Eng.*, **47:** 130 (1969).
76   HAWTHORN, R. D.: *AIChE J.*, **6:** 443 (1960).
77   HAYS, J. R., K. B. SCHNELLE, JR., and P. A. KRENKEL: Paper presented at the 55th Nat. Meet. AIChE, Houston, Texas, February, 1965.
78   HAZLEBECK, D. E., and C. J. GEANKOPLIS: *Ind. Eng. Chem. Fundam.*, **2:** 310 (1963).
79   HIBY, W.: *Inst. Chem. Eng. (London) Symp. Interaction between Fluids and Particles*, 1962.
80   HINZE, J. O.: "Turbulence," McGraw-Hill, New York, 1959.
81   HOCHMAN, J. M., and E. EFFRON: *Ind. Eng. Chem. Fundam.*, **8:** 63 (1969).
82   HOFFMANN, H.: *Chem. Eng. Sci.*, **14:** 193 (1961).
83   HOOGENDOORN, C. J., and J. LIPS: *Can. J. Chem. Eng.*, **43:** 125 (1965).
84   HUANG, J-H, and J. M. SMITH: *Ind. Eng. Chem. Fundam.*, **2:** 189 (1963).
85   HULL, D. E., and J. W. KENT: *Ind. Eng. Chem.*, **44:** 2745 (1952).
86   JAKOB: MAX: "Heat Transfer," vol. I, Wiley, New York, 1949.
87   JAMESON, G. J.: *Trans. Inst. Chem. Eng. (London)*, **44:** T198 (1966).
88   JENKINS, R.: Thesis, Calif. Inst. Tech., 1949; "Heat Transfer and Fluid Mechanics Institute," Stanford University Press, Palo Alto, Calif., 1951.
89   JOHNSON, D. S.: *J. Appl. Mech.*, **E24:** 2 (1957); **E26:** 325 (1959).
90   KALINSKE, A. A., and E. R. VAN DRIEST: *Proc. 5th Int. Congr. Appl. Mech.*, 1938.
91   KALINSKE, A. A., and J. M. ROBERTSON: *Eng. News-Record*, April 10, 1941.
92   KENNEY, C. N., and G. R. THWAITES: *Chem. Eng. Sci.*, **26:** 503 (1971).
93   KEYES, J. J., JR.: *AIChE J.*, **1:** 305 (1955).
94   KLINKENBERG, A., H. J. KRAJENBRINK, and H. A. LAUWERIER: *Ind. Eng. Chem.*, **45:** 1202 (1953).
95   KRAMERS, H., and K. R. WESTERTERP: "Elements of Chemical Reactor Design and Operation," Academic, New York, 1963.
96   KRAMERS, H., and G. ALBERDA: *Chem. Eng. Sci.*, **2:** 173 (1953).
96a  KRANTZ, W. B., and D. T. WASAN: *Ind. Eng. Chem. Fundam.*, **13:** 56 (1974).
97   LEE, J., and R. S. BRODKY: *AIChE 55th Annu. Meeting, Chicago*, Preprint 14, Dec. 2–6, 1962.
98   LEVENSPIEL, O.: "Chemical Reaction Engineering," Wiley, New York, 1962.
98a  LEVENSPIEL, O.: *Ind. Eng. Chem.*, **50:** 343 (1958).
99   LEVENSPIEL, O., and K. G. BISCHOFF: In T. B. Drew, J. W. Hoopes, Jr., and T. Vermeulen (eds.), "Advances in Chemical Engineering," vol. 4, p. 95, Academic, New York, 1963.
100  LEVENSPIEL, O., and W. K. SMITH, *Chem. Eng. Sci.*, **6:** 227 (1957).
101  LI, N. N., and E. N. ZIEGLER: *Ind. Eng. Chem.*, **59:** 30 (1967).
102  LILES, A. W., and C. J. GEANKOPLIS: *AIChE J.*, **6:** 591 (1960).
103  LIN, C. C.: in "High Speed Aerodynamics and Jet Propulsion," vol. V, p. 240, Princeton, N.J., 1955.
104  LYKOUDIS, P. S., and Y. S. TOULOUKIAN: *Trans. ASME*, **80:** 653 (1958).
105  LYNN, S.: *AIChE J.*, **5:** 566 (1959).
106  LYNN, S., W. H. CORCORAN, and B. H. SAGE: *AIChE J.*, **3:** 11 (1957).
107  MCHENRY, K. W., JR., and R. H. WILHELM: *AIChE J.*, **3:** 83 (1957).
108  MARCHELLO, J. M., and H. L. TOOR: *Ind. Eng. Chem. Fundam.*, **2:** 8 (1963).
109  MICKELSON, W. R.: *J. Fluid Mech.*, **7:** 397 (1960); *NACA Tech. Note* 3570, October 1955.
110  MILLER, S. F., and C. J. KING: *AIChE J.*, **12:** 767 (1966).

111   MIYAUCHI, T., and T. VERMEULEN: *Ind. Eng. Chem. Fundam.*, **2:** 113, 304 (1963).

112   MIYAUCHI, T., H. MITSUTAKE, and I. HASASE: *AIChE J.*, **12:** 508 (1966).

113   MOELWYN-HUGHES, E. A.: " Physical Chemistry," 2d ed., Pergamon, New York, 1961.

114   MOON, J. S., A. HENNICO, and T. VERMEULEN: *Univ. Calif. Lawrence Radiat. Lab.*, *Rep.* UCRL-10928, October 23, 1963.

115   OLBRICH, W. E., J. B. AGNEW, and O. E. POTTER: *Trans. Inst. Chem. Eng. (London)*, **44:** T207 (1966).

116   ORLOB, G. T.: *Trans. ASCE*, **126:** 397 (1961).

117   PAGE, F., JR., W. G. SCHLINGER, D. K. BREAUX, and B. H. SAGE: *Ind. Eng. Chem.*, **44:** 424 (1952).

118   PERKINS, T. K., and O. C. JOHNSTON: *Soc. Petrol. Eng. J.*, **3:** 70 (1963).

119   PERRY, J. H. (ed.): "Chemical Engineers' Handbook," 4th ed., p. 5–20, McGraw-Hill, New York, 1963.

120   PLAUTZ, D. A., and H. F. JOHNSTONE: *AIChE J.*, **1:** 193 (1955).

121   POREH, M., and J. E. CERMAK: *Int. J. Heat Mass Transfer*, **7:** 1083 (1964).

122   ROEMER, G., J. S. DRANOFF, and J. M. SMITH: *Ind. Eng. Chem. Fundam.*, **1:** 284 (1962).

123   ROSENSWEIG, R. E., H. C. HOTTEL, and G. C. WILLIAMS: *Chem. Eng. Sci.*, **15:** 111 (1961).

124   ROSS, L. W.: *Hydrocarbon Proc.*, **47**(8): 143 (1968).

125   ROTHFUS, R. R., and C. C. MONRAD: *Ind. Eng. Chem.*, **47:** 1144 (1955).

126   SAFFMAN, P. G.: *J. Fluid Mech.*, **7:** 194 (1960); **8:** 273 (1960).

127   SATER, J. E., and O. LEVENSPIEL: *Ind. Eng. Chem. Fundam.*, **5:** 86 (1966).

128   SCHLINGER, W. G., V. J. BERRY, J. L. MASON, and B. H. SAGE: *Ind. Eng. Chem.*, **45:** 662 (1953).

129   SCHLINGER, W. G., and B. H. SAGE: *Ind. Eng. Chem.*, **45:** 657 (1953).

130   SCHMALZER, D. K., and H. E. HOELSCHER: *AIChE J.*, **17:** 104 (1971).

131   SCHUBAUER, G. B.: *NACA Tech. Rep.* 524 (1935).

132   SCHWARZ, W. H., and H. E. HOELSCHER: *AIChE J.*, **2:** 101 (1956).

133   SEAGRAVE, R. C.: *AIChE J.*, **11:** 748 (1965).

134   SEAGRAVE, R. C., and R. W. FAHIEN: *U.S. Atom. Energy Comm.* IS-419 (1961).

135   SESONSKE, A., S. L. SCHROCK, and E. H. BUYCO: *AIChE-ASME Heat Transfer Conf.*, Boston, Preprint 25, August 11–14, 1963.

136   SHERWOOD, T. K., and B. B. WOERTZ: *Ind. Eng. Chem.*, **31:** 1034 (1939).

137   SHERWOOD, T. K., and J. M. RYAN: *Chem. Eng. Sci.*, **11:** 81 (1959).

138   SIMPSON, R. L., D. G. WHITTEN, and R. J. MOFFAT: *Int. J. Heat Mass Transfer*, **13:** 125 (1970).

139   SINCLAIR, R. J., and O. E. POTTER: *Trans. Inst. Chem. Eng. (London)*, **43:** T3 (1965).

140   SINGER, E., and R. H. WILHELM: *Chem. Eng. Progr.*, **46:** 343 (1950).

141   SITTEL, C. N., JR., W. D. THREADGILL, and K. B. SCHNELLE, JR.: *Ind. Eng. Chem. Fundam.*, **7:** 39 (1968).

142   SLEICHER, C. A.: "Modern Chemical Engineering," vol. 1, Reinhold, New York, 1963.

143   SLEICHER, C. A.: *Trans. ASME*, **79:** 789 (1957); **80:** 693 (1958).

144   SMITH, K. A.: Personal communication, 1967.

145   SPENGOS, A. C.: *Colo. Agr. Mech. Coll., Ft. Collins, Colo. Sci. Rep.* 1 (CER No. 50ACS4) (1956).

146   STAHEL, E. P., and C. J. GEANKOPLIS: *AIChE J.*, **10:** 174 (1964).

147   STRANG, D. A., and C. J. GEANKOPLIS: *Ind. Eng. Chem.*, **50:** 1305 (1958).

148   STRUNK, M. R., and F. F. TAO: *AIChE J.*, **10:** 269 (1964).

149   SULLIVAN, G. A., and R. E. TREYBAL: *Chem. Eng. J.*, **1:** 302 (1970).

150   SUNEJA, S. K., R. H. SHEA, and D. T. WASAN: *AIChE J.*, **18:** 194 (1972).

151   TAYLOR, G. I.: *Proc. London Math. Soc.*, **20:** 196 (1921–22).

152 TAYLOR, G. I.: In G. K. Batchelor (ed.), "Scientific Papers," vol. II, Cambridge, New York, 1959.

153 TAYLOR, H. M., and E. F. LEONARD: *AIChE J.*, **11**: 686 (1965).

154 TICHACEK, L. J., C. H. BARKELEW, and T. BARON: *AIChE J.*, **3**: 439 (1957).

155 TIEN, C. L.: *J. Heat Transfer*, **83C**: 389 (1961).

156 TOWLE, W. L., and T. K. SHERWOOD: *Ind. Eng. Chem.*, **31**: 457 (1939).

157 TOWLE, W. L., T. K. SHERWOOD, and L. A. SEDER: *Ind. Eng. Chem.*, **31**: 462 (1939).

158 TOWNSEND, A. A.: *Proc. Roy. Soc. (London)*, **224A**: 487 (1954).

159 TURNER, G. A.: *Chem. Eng. Sci.*, **7**: 156 (1958); **10**: 14 (1959).

160 UBEROI, M. S., and S. CORRSIN: *NACA Tech. Note* 2710 (1952).

161 URBAN, J. C., and A. GOMEZPLATA: *Can. J. Chem. Eng.*, **47**: 353 (1969).

162 U.S. Dept. of Health, Education, and Welfare, NAPCA, "Tall Stacks," *Publ.* APTD 69-12 (1969).

163 VAN DRIEST, E. R.: *J. Appl. Mech.*, **67**: A-91 (1945).

164 VERMEULEN, T., J. S. MOON, A. HENNICO, and T. MIYAUCHI: *Chem. Eng. Progr.*, **62**: 95 (1966).

165 VON KÁRMÁN, TH.: *J. Roy. Aeronaut. Soc.*, **41**: 1109 (1937).

166 VON KÁRMÁN, TH.: *Trans. ASME*, **61**: 705 (1939).

167 WATSON, J. S., and L. E. MCNEESE: *Ind. Eng. Chem. Process Des. and Dev.*, **11**: 120 (1972).

168 WEINSTEIN, A. S., J. F. OSTERLE, and W. FORSTALL: *ASME Paper* 55 A-60 (1955).

169 WESTERTERP, K. R.: *Chem. Eng. Sci.*, **17**: 373 (1962).

170 WESTERTERP, K. R., and P. LANDSMAN: *Chem. Eng. Sci.*, **17**: 363 (1962).

171 WILHELM, R. H.: *Chem. Eng. Progr.*, **49**: 150 (1953).

172 WILHELM, R. H.: *Pure Appl. Chem.*, **5**: 403 (1962).

173 YAGI, S., D. KUNII, and N. WAKAO: *AIChE J.*, **6**: 543 (1960).

# PROBLEMS

**4.1** A gas-liquid reaction is to be carried out in an experimental vertical, cylindrical, cocurrent bubble column. In a preliminary test to determine the extent of back-mixing in such a reactor, a salt tracer is employed as follows: pure water is fed at the bottom of the column at a steady rate, along with air introduced through a sintered disk of stainless steel. Salt is employed as a tracer, being introduced near the top of the column as a 30 weight percent solution in water fed at a constant rate of 5 cm³/s. The liquid near the bottom of the column is sampled and analyzed.

The superficial gas and liquid rates are 6.95 and 0.82 cm/s, respectively. The column is 48 in tall and 4 in in diameter. Under these flow conditions the gas bubbles occupy 39 percent of the reactor volume. The entire operation is at steady state.

From published data it is predicted that the axial-dispersion coefficient will be found to be 238 cm²/s. If so, what salt concentration will be found in a sample withdrawn near the bottom of the column (after the water feed is mixed)? Assume plug flow of both gas and liquid, and no volume change with change in salt concentration.

**4.2** A clear liquid having a specific gravity of 1.0 g/cm³ and a viscosity of 0.04 poise flows at a steady rate of 3.14 cm³/s through a round tube 200 cm long and 1.0 cm ID. The feed is suddenly switched to the same liquid dyed red, and the flow is continued at the same rate.

Prepare a graph of effluent concentration $c$ vs. time $t$, where $c$ = fraction dyed feed in the effluent, and $t$ = time after the change of feeds.

4.3   Air at 20°C and 1.0 atm flows at an average velocity of 50 ft/s through a round conduit 1.0 ft in diameter. The velocity across nearly the entire diameter is represented approximately by the relation

$$U_c - U = ar^2$$

where $U$ = velocity at radius $r$
   $U_c$ = velocity at the axis
   $a$ = constant

Furthermore, the mean velocity $U_{av}$ is 82 percent of the maximum velocity, that is, $U_c = 50/0.82 = 61$ ft/s.

   Estimate the eddy-diffusion coefficient $E_D$ for a tracer in this air stream, near the pipe axis. Compare with the value given by Fig. 4.11. Neglect shear and transport due to molecular motion, and assume the turbulent Schmidt number to be unity.

4.4   Pure carbon dioxide gas is fed at the rate of 160 cm³/s from a continuous point source into the center of the air stream flowing at an average velocity of 50 ft/s in the 1-ft pipe described in Prob. 4.3. Prepare a graph of volume percent $CO_2$ vs. $r$, representing a concentration traverse across a diameter 6 ft downstream from the point source.

   Use the value of $E_D$ obtained in Prob. 4.3. Neglect molecular diffusion and any effect of the $CO_2$ injection on the air-flow pattern. Assume the velocity to be essentially constant at 50 ft/s across the diameter traversed.

4.5   Repeat Prob. 4.4 for a traverse 6 in downstream from the point source of $CO_2$. Assume the turbulence to be isotropic with an intensity of 4 percent. As an approximation, assume $R_L$ to decay linearly from 1.0 to zero in 0.02 s.

4.6   Flint, Kada, and Hanratty [47] report the following measurements of the dispersion of $CO_2$ from a point source in the center of a 3.15-in-ID pipe carrying air at 80°F at a Reynolds number of 72,000:

| $r$, in | 0 | 0.05 | 0.10 | 0.15 | 0.20 | 0.25 | 0.30 |
|---|---|---|---|---|---|---|---|
| Volume, % $CO_2$ | 3.69 | 3.65 | 3.08 | 2.31 | 1.49 | 0.805 | 0.40 |

The samples were withdrawn at an axial distance of 5 in downstream from the point source. The velocity near the axis (to $r = 0.3$ in) was perhaps 1.2 times the average velocity.

   From these data obtain an approximate mean value of the radial eddy-diffusion coefficient and compare with Fig. 4.11. Why is the discrepancy substantial in this case?

4.7   It is a common observation that jets of steam and other vapors issuing into room air, as well as stack gases leaving a tall chimney, show sharp lines of demarcation between the opaque plume and the transparent air around them. How can this fact be reconciled with Sec. 4.7, in which the concentration of effluent gas is shown to have a gaussian distribution about the axis of flow? (If the concentration falls off along a gaussian curve, the opacity of the plume should show a diffuse edge, or no edge at all.)

   Perhaps as a term paper rather than as a problem, this question might be developed quantitatively.

4.8   A column of large diameter is packed to a depth of 5 ft with nonporous ceramic spheres 5 mm in diameter. Water containing 1.0 percent dissolved sodium chloride at

20°C flows through the bed at a steady superficial mass velocity of 500 lb/(h)(ft²). The void fraction in the bed is 0.40.

The feed is instantaneously switched from salt solution to pure water and the flow continued at the same steady rate. Using the "longitudinal dispersion model" and estimated values of $E_a$, prepare a graph of bed effluent salt concentration vs. time.

Repeat for the case of air flow at 20°C, 1.0 atm, and the same mass velocity: The original gas contains 1.0 volume percent of a tracer gas of similar density; the feed is suddenly changed to pure air.

4.9 An aqueous dye solution is introduced into water flowing in steady flow in a round pipe through a thin-walled injector tube of finite radius $r_0$, placed at the pipe axis. The velocity of the dye solution fed and that of the water stream are both $\overline{U}$ feet per second, assumed to be essentially constant, independent of radial position.

Set up the appropriate partial differential equation and obtain a solution expressing the dye concentration $c$ as a function of radius $r$ and the downstream distance $x$. Allow for the finite radii of both injector tube and pipe.

(The corresponding problem with allowance for an appropriate turbulent-flow velocity profile across the pipe is both more realistic and more difficult. It is not known that this has or can be solved.)

4.10 Space limitations have precluded the inclusion of more than a brief mention of turbulence and diffusion in the atmosphere. Yet this is a subject of increasing importance to chemical engineers, since the dispersion of atmospheric pollutants has become a problem of major concern to industry. An excellent subject for a term paper would be the collection and summary of the quantitative information required for the design of a stack to handle a specified quantity of pollutant-containing gas in order to meet prescribed ground-level concentration limits.

4.11 Smoke leaving a power-plant stack spreads rapidly as it moves downwind, with a total cone angle of roughly 15°. If the gas issued from a point source and its spread were due only to molecular diffusion (say $D = 0.2$ cm²/s), what fraction of it would pass through a circle 2 ft in diameter a distance of 1 mi downwind if the wind speed is 10 mi/h?

# 5

# MASS TRANSFER AT A PHASE BOUNDARY

## 5.0 SCOPE

The preceding chapters have dealt primarily with mass transfer within a single phase, out of contact with a phase boundary. In most separation processes, and in many other practical situations where mass transfer is important, it is transfer from one phase to another that is involved. Only one side of the phase boundary need be considered in the case of the evaporation of a pure liquid into a gas; mass transfer within the pure liquid is not involved. In liquid extraction and in many other processes, mass transfer between bulk fluid and the interface occurs in both phases. The simpler case of transfer within a single phase, to or from a phase boundary, will be treated first.

The basic tools of the analyst or the design engineer are the following:

*1* Stoichiometry and the law of conservation of mass, basic to material balances
*2* The law of conservation of energy, and the use of thermodynamic data to construct energy balances
*3* Equilibrium concepts and data
*4* Rate equations and correlations of data on mass-transfer rates

It is with the last of these that this and the next chapter are concerned.

Several examples of transfer from a phase boundary to a moving fluid were discussed in Chap. 3. These were limited to cases where the transport is by molecular diffusion and the flow field known, as when the flow is laminar. In most practical situations the flow is turbulent and the flow field inadequately specified. Both molecular and eddy diffusion are involved and rigorous treatment is not usually possible.

The present chapter deals primarily with mass transfer at a differential or unit area of the phase boundary, with flux into or out of a single phase. Chapter 9 will consider the design problem: analysis of the performance of mass-transfer devices in which conditions vary over a large area of the phase boundary.

## 5.1 PRINCIPAL SYMBOLS

| | |
|---|---|
| $a$ | Activity of component of mixture |
| $c$ | Concentration, g moles/cm$^3$ |
| $C_p$ | Heat capacity, cal/(g)(°C) |
| $d$ | Diameter, or other characteristic dimension, cm |
| $D$ | Diffusion coefficient; $D_{AB}$ = diffusion coefficient for solute $A$ diffusing in mixture of $A$ and $B$, cm$^2$/s |
| $E_D$ | Eddy-diffusion coefficient, cm$^2$/s |
| $E_v$ | Eddy viscosity, cm$^2$/s |
| $f$ | Fanning friction factor |
| $g_c$ | Conversion factor; $g_c = 1$ in the cgs system; $g_c = 32.2$ (lbm)(ft)/(lbf)(s$^2$) in technical engineering units |
| $G_M$ | Molal mass velocity, g moles/(s)(cm$^2$) |
| $h$ | Heat-transfer coefficient, g cal/(s)(cm$^2$)(°C) |
| $H$ | Henry's-law constant, $Y/c$ |
| $J$ | Molecular diffusion flux, g moles/(s)(cm$^2$) |
| $k$ | Thermal conductivity, g cal/(s) (cm$^3$) (°C/cm) |
| $k^*$ | Mass-transfer coefficient defined by Eq. (5.1) |
| $k_c$ | Mass-transfer coefficient defined by Eq. (5.1), cm/s |
| $k_D$ | Rate coefficient defined by Eq. (5.37) |
| $k_G$ | Mass-transfer coefficient defined by Eq. (5.1) |
| $k_i$ | Coefficient of mass transfer across resistance at interface between phases, cm/s |
| $k_Y$ | Mass-transfer coefficient defined by Eq. (5.1) |
| $k_a$ | Overall coefficient using activity as potential |
| $k_y$ | Overall coefficient using mole fraction as potential |
| $M$ | Molecular weight |
| $N_A$ | Molal flux of solute $A$, g moles/(s) (cm$^2$) |
| $N_T$ | Total flux of all components, g mole/(s) (cm$^2$) |
| Nu | Nusselt number, $hd/k$ |
| $p$ | Partial pressure, atm |
| $p_{BM}$ | Logarithmic mean partial pressure of species not diffusing in a binary gas, atm |
| $P$ | Total pressure, atm |
| $q$ | Heat flux, g cal/(s) (cm$^2$) |
| $r$ | Radial distance, cm |

| | |
|---|---|
| $r_w$ | Tube radius, cm |
| $R$ | Gas constant = $82.07 \, (\text{cm}^3) \, (\text{atm})/(\text{g mole}) \, (^\circ\text{K})$ |
| Re | Reynolds number, $d\overline{U}_{av} \rho/\mu = d\overline{U}_{av}/\nu$ |
| $s$ | Fractional surface renewal rate, $s^{-1}$ |
| Sc | Schmidt number, $\mu/\rho D = \nu/D$ |
| St | Stanton number, $k_c/\overline{U}_{av}$, or $h/C_p \rho \overline{U}_{av}$ |
| $t$ | Time, s |
| $T$ | Temperature, $^\circ$C or $^\circ$K |
| $u^+$ | Dimensionless velocity, defined by Eq. (4.9) |
| $\overline{U}_{av}$ | Time-mean velocity averaged over flow, cm/s |
| $\overline{U}_x, \overline{U}_y$ | Time mean velocities in $x$ or $y$ directions, cm/s |
| $x$ | Distance in direction parallel to surface, cm |
| $y$ | Distance in direction of diffusion, cm |
| $y_0$ | Film thickness, cm |
| $y^+$ | Dimensionless distance from wall, defined by Eq. (4.9) |
| $Y$ | Mole fraction |
| $Y_{BM}$ | Logarithmic mean mole fraction of nondiffusing component in a binary gas, $= p_{BM}/P$ |
| $\alpha$ | Evaporation, or "sticking," coefficient |
| $\mu$ | Viscosity, g/(s) (cm); poises |
| $\nu$ | Kinematic viscosity, $\text{cm}^2/\text{s}$ |
| $\rho$ | Density, $\text{g/cm}^3$ |
| $\tau$ | Shear stress, $\text{dyn/cm}^2$ |

## 5.2 MODELS

Conditions in the immediate region of an interface between phases are difficult to observe or to explore experimentally. In such a situation it is helpful to develop a mathematical model of the process, starting with the known basic facts. The result of the analysis is then compared with those experimental measurements which it is possible to make. Good agreement suggests that the model may have been realistic.

The most important relevant fact pertaining to mass transfer to or from a turbulent stream is that the resistance to transfer is confined largely to a thin region adjacent to the interface. Here resistance refers to the ratio of the gradient of an appropriate potential to the mass flux, where "flux" means rate of transport per unit area. The flux of interest is the net flux normal to the phase boundary, and the gradient is defined in terms of distance from it. This gradient is steep near the interface.

The mass-transfer flux stops when equilibrium is established between the bulk fluid and the interface, so it is reasonable to employ a potential which is proportional to the distance from equilibrium. Activity or chemical potential might be adopted as the potential to be used in a rate equation, but the common choice is concentration, with units of moles per unit volume. There are three reasons for this choice: in almost all situations the flux goes to zero as the concentrations equalize; it is a valid potential for both eddy diffusion and for molecular diffusion in gases at low pressures; and it ties in conveniently with the stoichiometry involved in the design problem.

The three simplest and best known of the models will be described briefly. These are (1) the "stagnant-film" model, (2) the "penetration" model, and (3) the turbulent boundary-layer model.

**The stagnant-film model**   When fluid flows over a phase boundary, the local velocity approaches zero at the surface. Furthermore, in this limit the only flow normal to the surface is usually that corresponding to the mass-transfer flux. The fluid in *immediate* contact with the fixed surface can properly be said to be stagnant. The fact that resistances to mass transfer (and heat transfer) are confined largely to a region quite close to the phase boundary suggests the idea of a thin stagnant film adjacent to a surface over which the fluid flow is turbulent. If this thin layer is stagnant, the transport must be by molecular diffusion. The picture then is that of a stagnant film through which transport is solely by molecular diffusion, and which is of such a thickness $y_0$ as to explain the experimentally observed magnitude of the mass-transfer resistance.

With this picture it is logical to treat mass transfer at a phase boundary by the use of the equations for molecular diffusion presented in Chap. 3. Consider, for example, the evaporation of water into turbulent nitrogen stream flowing over a fixed wet surface. If steady-state conditions prevail, and if the possible effect of the simultaneous heat transfer is ignored, then Eq. (3.8) should apply:

$$N_A = \frac{D_{AB}P}{RTy_0} \frac{(Y_{A1} - Y_{A2})}{Y_{BM}} = \frac{D_{AB}P}{RTy_0} \frac{(c_{A1} - c_{A2})}{c_{BM}} \tag{3.8}$$

Here $A$ refers to water and $B$ to nitrogen.

The situation is illustrated by Fig. 5.1. The concentrations $c_{A1}$ and $c_{A2}$ pertain to water vapor in nitrogen at the wet surface and in the bulk of the flowing gas stream; $c_{BM}$ is the logarithmic mean of the corresponding nitrogen concentrations. All the potential drop from surface to flowing gas occurs in the stagnant film of thickness $y_0$. It is evident that $y_0$ is the *effective* thickness of a stagnant gas layer which offers a resistance to molecular diffusion equal to the observed resistance to mass transfer.

The film model is some 70 yr old, having been proposed by Nernst [117] in 1904. It has been applied to both heat and mass transfer, and it formed the basis of Whitman's "two-film theory of gas absorption" in 1923 [184, 96]. It was recognized very early that the concept was a gross oversimplification of the actual conditions near a phase boundary; Whitman noted that a sharp boundary at $y_0$ was implied, "although actually no such sharp demarcation exists." Though the model suggests no basis for the prediction of $y_0$, needed to calculate transport rates, it has been remarkably useful in several applications. It gives quite reliable predictions of the rate of mass transfer where there is a simultaneous chemical reaction, compared with the rate under the same conditions without reaction. It does equally well in predicting the effect of large mass-transfer rates on heat transfer.

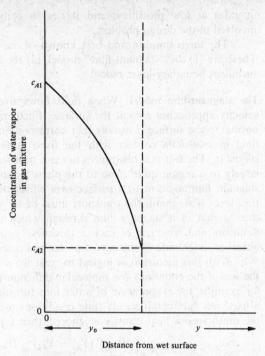

**FIGURE 5.1**
The film model for mass transfer at a phase boundary.

A major fault of the film model is that it predicts a first-power dependence of the mass flux on the molecular diffusion coefficient, that is, $N_A \alpha D$. Actually, it appears that turbulence dies out gradually as the surface is approached, and that eddy diffusion, dominant in the bulk flow, likewise fades away, so that in the mathematical limit, at the fixed surface $(y \to 0)$, the transport is solely by molecular diffusion. Thus $N_A \alpha D^0$ at the outer boundary of the film, and $N_A \alpha D^{1.0}$ at the fixed surface, or "wall." If a power function $D^n$ is employed to represent the overall process, it may be expected that $n$ will fall between zero and unity. This is found to be the case.

The correlations of mass-transfer data which employ dimensionless groups stem from the film theory. If $y_0$ has any physical meaning, it must depend on the flow conditions, and hence on the Reynolds number. For a given profile of eddy diffusivities near the wall, determined by the flow, the relative importance of $D$ will depend on its magnitude.

It is convenient to employ a mass-transfer coefficient, defined as the ratio of the molal flux $N_A$ to the potential or driving force. The coefficient may be expressed in various ways; the four most common forms are defined by the following:

$$N_A = k_c(c_{A1} - c_{A2}) = k_Y(Y_{A1} - Y_{A2}) = k^* \frac{Y_{A1} - Y_{A2}}{Y_{BM}} = k_G(p_{A1} - p_{A2}) \quad (5.1)$$

For the special case of a binary gas system with only one species diffusing, Eq. (3.8) gives

$$N_A = \frac{k_Y}{P}(p_{A1} - p_{A2}) = \frac{D_{AB}P}{RTy_0 Y_{BM}}(Y_{A1} - Y_{A2})$$

$$= k^* \ln \frac{P - p_{A2}}{P - p_{A1}} = k^* \ln \frac{1 - Y_{A2}}{1 - Y_{A1}}$$

$$= \frac{k_c}{RT}(p_{A1} - p_{A2}) = \frac{k_c P}{RT}(Y_{A1} - Y_{A2}) \quad (5.2)$$

For this case, the four coefficients are related by

$$k^* = \frac{k_c P Y_{BM}}{RT} = k_Y Y_{BM} = k_G P Y_{BM} = \frac{D_{AB}P}{RTy_0} \quad (5.3)$$

In the cgs system $k_y$ and $k^*$ have the units g moles per second per square centimeter; $k_c$ is in centimeters per second; $k_G$ has the units gram moles per second per square centimeter per atmosphere. Equation (5.1) defines the coefficients and is not restricted to binary gas systems.

For binary gas systems with one component diffusing, the best choice among the rate equations is

$$N_A = k^* \ln \frac{1 - Y_{A2}}{1 - Y_{A1}} \quad (5.4)$$

The coefficient $k_c$ is more commonly used for transport in liquid or dilute gas systems when $Y_{BM}$ is near unity.

For the film model, $y_0$ may be replaced by one of the coefficients to develop a convenient dimensionless group useful in correlating experimental data. Like $y_0$, this group may be expected to depend on the Reynolds number and $D$. Incorporating $D$ in the dimensionless Schmidt number, $v/D$, the following is suggested by dimensional analysis:

$$\frac{k_c Y_{BM} d}{D}\left( = \frac{k^* RTd}{PD} = \frac{k_Y RT Y_{BM} d}{PD}\right) = f(\text{Re, Sc}) \quad (5.5)$$

Here $d$ is a length appropriate to the geometry of the system. Numerous correlations for different geometries and flow conditions have employed this form; the dimensionless group on the left is sometimes referred to as the Sherwood number.

**The penetration model**  The film model neglects accumulation of the diffusing species in the film: the local flux across each small area is constant. Higbie in 1935 [71] pointed out that much industrial gas-liquid contacting equipment operates with repeated brief contacts of the two phases, and that the contact times are too short to permit the attainment of a steady state. As a gas bubble rises through a liquid, the liquid in immediate contact with the gas is replaced in a time approximately equal to that required for the bubble to rise one bubble diameter. Higbie developed his penetration theory to allow for the transient nature of solute diffusion from bubble to liquid. He suggested that a similar situation exists in packed towers, where liquid flowing over a single packing element is exposed briefly to the gas before the liquid is mixed and the process repeated.

The simplest version of the penetration theory pictures a small fluid element of uniform solute concentration $c_0$ being brought into contact with a phase boundary for a fixed time $t$. The concentration at the phase boundary is maintained constant at $c_{A1}$, as by desorption from a second phase of pure solute. Diffusion from interface to bulk fluid proceeds as a transient process, the rate decreasing with time. The analysis is the same as that of diffusion or conduction into a semi-infinite solid, treated in Sec. 3.6.

Neglecting convective flux, the expression for the average flux over the time $t$ is the following:

$$N_A = 2(c_{A1} - c_0)\sqrt{\frac{D}{\pi t}} \tag{3.49}$$

from which the time-average mass-transfer coefficient is obtained as

$$k_c = \frac{N_A}{(c_{A1} - c_0)} = 2\sqrt{\frac{D}{\pi t}} \tag{5.6}$$

In practice the time $t$ is seldom known, so the model cannot be used to predict mass-transfer rates except in special cases. The same difficulty is encountered in the film theory, which involves the unknown film thickness $y_0$.

The penetration theory, like the film theory, is sometimes successful in predicting the change in flux or $k_c$ when certain conditions are changed in mass-transfer equipment. For example, the calculated ratio of $k_c$ with simultaneous chemical reaction to $k_c$ for physical mass transfer is in good agreement with experimental results (see Chap. 8).

The penetration model predicts that the flux and $k_c$ should vary as the square root of the molecular diffusivity, whereas the film model indicates the first power. The square root is nearer the truth in many instances. Gilliland [54], for example, found proportionality to $D^{0.56}$ in the evaporation of several liquids into turbulent air streams in a wetted-wall column. Vivian and King [173] found $D^{0.50}$ for the desorption of slightly soluble gases from water in a tower packed with 1.5-in Raschig rings. Other investigators have reported powers of $D$ ranging generally from 0.5 to 0.75, and the Chilton-Colburn analogy, discussed below, suggests $D^{2/3}$.

A number of reports, especially in the Russian literature, suggest that the rate of mass transfer to a highly agitated liquid surface may be completely independent of the molecular diffusion coefficient of the transported solute. Thus Kishinevsky and

Serebryanski [80] found no effect of varying $D$ in the absorption of hydrogen, nitrogen, and oxygen by water stirred at 1,700 rpm. Lewis [94] found the same for transport between two stirred immiscible liquids. However, McManamey, Davis, Woollen, and Coe [106], using pairs of immiscible liquids in an apparatus similar to that employed by Lewis, concluded from their own data, together with those of Lewis, that the individual coefficients were proportional to $D^{0.5}$. It is possible that at the high agitation rates employed by Kishinevsky and Serebryanski small drops of one phase were dispersed in the other, where they equilibrated before coalescing and returning to the first phase.

The derivation of Eq. (5.6) assumes the fluid contacting the interface to be quite stagnant during its brief exposure. If the flow is laminar, the parabolic velocity gradient can be taken into account. This was done in analyzing the case of penetration into a falling liquid film in Sec. 3.7. With short contact times the depth of penetration by molecular diffusion into liquids is so small that the simpler analysis based on stagnant liquid is adequate.

Arngelo, Lightfoot, and Howard [2] have extended the penetration theory for application to cases where the surface varies with time, as in forming drops.

**The surface-renewal theory** An important extension of the penetration theory was published by Danckwerts in 1951 [34]. Whereas Higbie had taken the exposure time to be the same for all the repeated contacts of the fluid with the interface, Danckwerts employed a wide spectrum of times and averaged the varying degrees of penetration.

Fresh fluid elements are assumed to remain in contact with the surface for variable times $t$, which may be anything from zero to infinity. The fractional rate of renewal, $s$, of the area exposed to penetration is assumed to remain constant, in which case the surface–age distribution function is shown to be

$$\phi = se^{-st} \tag{5.7}$$

Here $\phi$ represents the probability that any element of area will be exposed the time $t$ before being replaced by fresh mixed fluid from the bulk. The mean steady-state flux normal to the phase boundary is then

$$N_A = (c_{A1} - c_0)\sqrt{D}\int_0^\infty \frac{se^{-st}}{\sqrt{\pi t}}\,dt = (c_{Ai} - c_0)\sqrt{Ds} \tag{5.8}$$

and

$$k_c = \sqrt{Ds} \tag{5.9}$$

The fractional surface renewal rate $s$ may be estimated from mass-transfer data and Eq. (5.9). For example, Hutchinson and Sherwood [75] obtained a value of $k_c$ of 0.00147 cm/s for the absorption of pure hydrogen in water in a small vessel stirred at 300 rpm at 25°C. Since $D$ is $6.3 \times 10^{-5}$ cm$^2$/s, the value of $s$ derived from Eq. (5.9) is 0.034 s$^{-1}$. At 1,000 rpm $k_c$ was 0.00303, corresponding to $s = 0.145$ s$^{-1}$. Since values of $s$ are not generally available, its appearance in the analysis presents the same problems as $y_0$ and $t$ of the film and Higbie models.

It is not generally believed that fluid eddies reach a fixed interface, such as the wall of a tube through which a fluid is passing in turbulent flow, but there is increasing evidence that this may happen. The surface-renewal concept appears more applicable, however, to the interface between a gas and a stirred liquid. Indeed, this case was pictured by Danckwerts in developing the analysis. Watching the surface of a swiftly moving river, or of a well-stirred liquid, it is not hard to discern fluid elements which come up from below and then appear to move back down after brief periods of contact with the air at the surface.

The only experimental measurements of the surface renewal rate as gas absorption is occurring would appear to be those of Lamb, Springer, and Pigford [88, 159] using an "interface impedance bridge." The absorption of pure $SO_2$ was measured in two identical vessels each containing 10 l of water. The gas volumes in each were varied sinusoidally and simultaneously, causing rapid continuous absorption and desorption of $SO_2$ from each of the two liquid surfaces. The water in one vessel was stagnant, while the other was stirred. Strain-gauge tranducers were employed to follow the varying small difference in pressure between the two gas spaces; a frequency-response analysis of the data provided values of the surface-age distributions as a function of frequency for the stirred vessel.

The results obtained with a clean surface and a stirring rate of 230 rpm are shown in Fig. 5.2. The solid curve is based on Eq. (5.7) with $s = 2.81$ s$^{-1}$. Similar results were obtained at 150 rpm giving $s = 1.09$ s$^{-1}$. This study provides direct support of the form of surface-age distribution proposed by Danckwerts.

The phenomenon of surface renewal can be observed by watching the surface of a stirred liquid which has been sprinkled with a fine powder. Surface renewal is shown by the momentary clearance of small surface areas. Davies and Khan [38] have measured the rate of clearance (frequency per square centimeter) as a function of stirrer speed by this technique.

**The turbulent boundary-layer model**  This model pertains to transport between a fixed surface (wall) and a turbulent stream of fluid. There is no slip at the wall and turbulence is damped out in the fluid immediately in contact with it. In this limit transport is by molecular diffusion only, since there is no mixing or eddy diffusion normal to the wall. Eddy diffusion is rapid in the turbulent stream at large distances from the wall, where the contribution of molecular diffusion is relatively insignificant.

It is proposed that both molecular and eddy diffusion play a role in the intermediate region, and that at any distance $y$ from the wall the rate of mass transfer can be expressed by

$$N_A = -(D + E_D)\frac{dc}{dy} \qquad (5.10)$$

This expression assumes simply that the flux is proportional to the concentration gradient normal to the wall. It neglects the convection term in the equation for molecular diffusion and assumes that molecular and eddy diffusion take place in parallel. The last point has been questioned (see Sec. 4.8), though it seems not unreasonable.

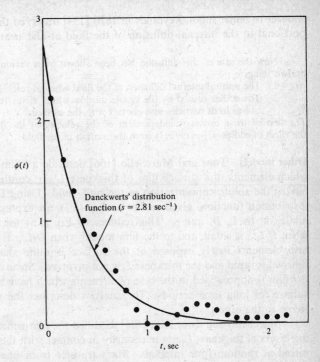

FIGURE 5.2
The surface-renewal model; experimental data on surface-age distribution function vs. surface-element age for water stirred at 230 rpm.

The region of importance is so thin that even in the case of turbulent flow in a round pipe the flux $N_A$ can be taken to be essentially independent of $y$. If $E_D$ can be expressed as a function of $y$, the equation can be integrated to relate $c$ to $y$ and the steady-state flux. In the case of the round pipe, this would give $c$ as a function of distance from the axis, from which the bulk average concentration is obtained if the velocity profile is known. The mass-transfer coefficient $k_c$ is then obtained from $N_A$, the average concentration, and the concentration at the wall.

The utility of this model evidently depends on a knowledge of $E_D$ as a function of $y$. There appear to have been no direct measurements of this relation, though it can be derived from a few studies where both the flux and the concentration profile were measured simultaneously. The usual approach is to obtain $E_D$ indirectly through its relation to the eddy viscosity, which can be derived from velocity profiles. This is the basis of the several quantitative expressions for the analogies between mass, heat, and momentum transfer. These are discussed in Sec. 5.3.

On a historical note, it is interesting that the three models which have been described perhaps owe their origin to Osborne Reynolds. Writing about heat transfer to boiler tubes, Reynolds in 1870 [134] observed that the heat flux "is proportional to the internal diffusion of the fluid at and near the surface" and adds:

> Now the rate of this diffusion has been shown from various considerations to depend on two things:
> 1.  The natural internal diffusion of the fluid when at rest,
> 2.  The eddies caused by the visible motion which mixes the fluid up and continually brings fresh particles into contact with the surface.
>
> The first of these causes is independent of the velocity of the fluid.... The second cause, the effect of eddies, arises entirely from the motion of the fluid....

**Other models**  Toor and Marchello [166] describe a "film-penetration" model, in which elements of a surface film of thickness $L$ are continually replaced by fluid having the solute concentration of the bulk fluid. Using Danckwert's surface–age distribution function, given earlier as Eq. (5.7), an expression is obtained which relates $k_c$ to $L$, $D$, and $s$. This reduces to Eq. (5.6) for the penetration model when $D/L^2 s$ is small, and to the film model when $D/L^2 s$ is large. The penetration into elements newly exposed at the surface is quite shallow when the surface renewal is rapid and the thickness $L$ is not involved. Steady-state diffusion through the film is approached in the case of elements which have been in contact with the surface for long time periods. The analysis describes the transition from one extreme to the other.

Marchello and Toor later [108] modified the film-penetration model, postulating layers of thickness $L$, not necessarily in contact with the surface, which are well mixed at random time intervals. Mass transfer from one layer to another then takes place by molecular diffusion. The rate equation, which is of the form of Eq. (5.10), leads to an expression relating $k_c$ to $\sqrt{Ds}$ and $D/L^2 s$.

Hanratty [64] applies Danckwerts' surface-renewal model to mass transfer at a fixed surface and compares calculated concentration profiles near the wall with the experimental data of Lin, Moulton, and Putnam [97]. Good agreement is obtained by using a value of $s$ derived from experimental values of $k_c$.

Harriott [65] pictures eddies from the main turbulent stream replacing fluid at varying distances $H$ from the wall. Mass transfer to and away from the wall takes place during the time periods between eddy renewals. Assuming various frequency distributions of the distance $H$, and the holding times at $H$, Harriott compares calculated relations between $k_c H/D$ and $H/\sqrt{Dt}$ with those derived from the film and penetration models. The indicated variation of $k_c$ with $D$ is in better agreement with experimental data than that predicted by most other models.

Ruckenstein [137] visualizes penetration into a laminar layer moving along the wall for a short distance before being displaced. The analysis is developed for heat transfer but is readily translated for application to mass transfer. Thomas [163] reports that it is not applicable for Pr or Sc greater than about 6.

Perlmutter [123] modifies the surface-renewal model in several ways. Two methods of allowing for a surface resistance, or "nonequilibrium interface condition," are described and analyzed, one of which has been treated by Danckwerts [34]. Perlmutter points out the analogy between the frequency distribution of surface contact times and the residence-time distribution in a continuous stirred tank and modifies the distribution function to correspond to the residence-time distribution for two continuous stirred tanks in series.

King [79] introduces a variable eddy diffusivity into the penetration model as applied to mass transfer into a liquid from a free liquid surface. $E_D$ is taken to be a power function of the distance from the surface, with the results of the analysis indicating that the exponent is 4. Contact time $t$ is left as a variable, though the introduction of an age-time distribution function is discussed. The analysis, which is concerned largely with the variation of $k_c$ with $D$, shows a smooth transition from the penetration to the turbulent boundary layer model as the contact time is increased.

Lamont and Scott [89] have proposed an interesting model for mass transfer into the surface of a turbulent liquid which suggests power input to the liquid as a correlating parameter.

Still another variation of the combined film and penetration model is described by Wasan and Ahluwalia [178]. This was shown to compare favorably with data on heat transfer between solid surfaces in fluidized beds of solid particles.

The most recent of these models, due to Pinczewski and Sideman [124a], is important because it requires only constants obtainable from data on fluid mechanics, without reliance on mass- or heat-transfer measurements. The wall region is considered to be a mosaic of patches of periodically replaced developing boundary layers which are partially steady and partially transient. Complete fluid renewal is considered to occur only occasionally, resulting in a thin fluid layer adjacent to the tube wall (to $y^+ \approx 1$) which is infrequently replenished. The result of the analysis at Sc > 500 agrees with the data of Harriott and Hamilton and with the latter's correlating equation.

Scriven [144] has written an excellent critical review of many of the earlier models. He suggests a strategy for the development of better models, emphasizing the role of convective diffusion in distorting concentration profiles. In particular, he points out that strong surface dilation and contraction at fluid interfaces can strongly influence transport and suggests that the effect on the motion of the micro-flow elements might be modeled by a single population of steady irrotational stagnation flows.

## 5.3 ANALOGIES BETWEEN MASS, HEAT, AND MOMENTUM TRANSFER [185, 147]

The close similarity of the processes of mass, heat, and momentum transfer in fluids is suggested by the fact that the basic equations expressing the fluxes have the same form:

$$J_A = -D_{Am} \frac{\partial c_A}{\partial y} \tag{3.1}$$

$$q = -\frac{k}{C_p \rho} \frac{\partial (C_p \rho T)}{\partial y} \tag{5.11}$$

$$\tau g_c = -v \frac{\partial (U_x \rho)}{\partial y} \tag{5.12}$$

These relations pertain to fluxes in the $y$ direction when the transport is effected by the motion or vibration of the molecules. The three processes are quite different on the molecular level, but the basic equations have the same form. It is perhaps not easy to picture the shear stress $\tau$ as a flux of momentum, but it is the result of momentum transport, and the coefficient $v$, the kinematic viscosity, has the same dimensions (square centimeters per second) as $D_{Am}$; so also does the "thermal diffusivity" $k/C_p\rho$.

The total diffusion flux relative to a fixed plane was shown in Chap. 3 to be

$$N_A = \overline{U}_y c_A + J_A \tag{3.4}$$

The corresponding equations for heat and momentum, where there is a convective diffusional flux, are:

$$q = \overline{U}_y C_p \rho T - \frac{k}{C_p \rho} \frac{\partial (C_p \rho T)}{\partial y} \tag{5.13}$$

and

$$\tau g_c = \overline{U}_y{}^2 \rho - v \frac{\partial (\overline{U}_x \rho)}{\partial y} \tag{5.14}$$

The velocity of the convective flux in the direction of transport is $\overline{U}_y$; $\overline{U}_x$ is the velocity of the fluid normal to the direction of the flux. Note that the potentials in each case are volume concentrations: $c_A$ is expressed in gram moles per cubic centimeter, $C_p\rho T$ is in gram calories/cubic centimeter, and $\overline{U}_x\rho$ is in gram centimeter per second per cubic centimeter.

The similarity of the forms of the three rate equations makes it possible to apply an analysis of one of the processes to either of the other two. Thus the comprehensive collection of solutions of the Laplace equation for unsteady-state heat conduction by Carslaw and Jaeger [23] can be applied to equivalent geometries and boundary conditions in problems of molecular diffusion.

Most industrial processes in which mass transfer is important involve fluids in turbulent flow, and the rigorous analyses which are often possible for fluids which are stagnant or in laminar flow, are not applicable. However, the extensive studies of fluid mechanics have provided a wealth of information regarding momentum transport in turbulent flow, and many attempts have been made to extend the analogies described above so that what is known about fluid mechanics might be employed to correlate and predict mass- and heat-transfer coefficients when the flow is turbulent.

The analogy between mass and heat transfer is generally valid and often very helpful. Empirical correlations of heat-transfer data are frequently of the form

$$\text{Nu} = f(\text{Re, Pr}) \tag{5.15}$$

These may be employed to estimate mass-transfer coefficients by substituting $k_c Y_{BM} d/D$ for Nu, and the Schmidt number Sc for Pr. Mass-transfer coefficients are sometimes easier to measure than heat-transfer coefficients. Perhaps the principal difficulty with this analogy is the fact that the temperature difference across a boundary layer where heat transfer is occurring leads to a variation in temperature-dependent physical properties, not encountered in isothermal mass transfer.

Extension of the analogy from heat and mass transfer to momentum transfer presents a number of difficulties which have never been resolved on a sound theoretical basis. Temperature and concentration are scalar quantities and momentum is a vector. The Reynolds modification of the Navier-Stokes equation for turbulent flow in the $x$ direction is [86]

$$\bar{U}_x \frac{\partial \bar{U}_x}{\partial x} + \bar{U}_y \frac{\partial \bar{U}_x}{\partial y} + \bar{U}_z \frac{\partial \bar{U}_x}{\partial z} = \frac{1}{\rho} \left[ \frac{\partial}{\partial x} \left( \mu \frac{\partial \bar{U}_x}{\partial x} - \overline{\rho u_x u_x} \right) \right.$$

$$\left. + \frac{\partial}{\partial y} \left( \mu \frac{\partial \bar{U}_x}{\partial y} - \overline{\rho u_x u_y} \right) + \frac{\partial}{\partial z} \left( \mu \frac{\partial \bar{U}_x}{\partial z} - \overline{\rho u_x u_z} \right) \right] - \frac{g_c}{\rho} \frac{\partial \bar{P}}{\partial x} \quad (5.16)$$

This is derived from a force balance on an element of fluid, and similar equations apply to $\bar{U}_y$ and $\bar{U}_z$. Here $\bar{U}_x$, $\bar{U}_y$, and $\bar{U}_z$ are the time mean velocities in the three directions and $u_x$, $u_y$, and $u_z$ the corresponding fluctuating velocities. The corresponding equation for heat transfer with no dissipation of kinetic energy to heat is

$$\bar{U}_x \frac{\partial \bar{T}}{\partial x} + \bar{U}_y \frac{\partial \bar{T}}{\partial y} + \bar{U}_z \frac{\partial \bar{T}}{\partial z} = \frac{1}{\rho C_p} \left[ \frac{\partial}{\partial x} \left( k \frac{\partial \bar{T}}{\partial x} - \rho C_p \overline{u_x t} \right) \right.$$

$$\left. + \frac{\partial}{\partial y} \left( k \frac{\partial \bar{T}}{\partial y} - \rho C_p \overline{u_y t} \right) + \frac{\partial}{\partial z} \left( k \frac{\partial \bar{T}}{\partial z} - \rho C_p \overline{u_z t} \right) \right] \quad (5.17)$$

in which $t$ is the temperature fluctuation about the time mean $\bar{T}$.

In the derivation of a similar expression for mass transfer, the difficulty is encountered that diffusion velocities must be referred to a plane of no net molal transport (rather than no net mass transport) if diffusion coefficients are to be essentially independent of concentration. Bedingfield and Drew [11] have shown that the equation based on molal linear velocities can be transposed and written in terms of mass linear velocities so as to compare with Eq. (5.17). This is done by letting the diffusion potential be expressed as $\ln M$ instead of concentration. For a binary mixture $M = Y_A M_A + Y_B M_B$. For low concentrations of $A$, $Y_A$ may be employed as the potential in place of $\ln M$. The relation is then

$$\bar{U}_x \frac{\partial \bar{Y}_A}{\partial x} + \bar{U}_y \frac{\partial \bar{Y}_A}{\partial y} + \bar{U}_z \frac{\partial \bar{Y}_A}{\partial z} = \frac{\partial}{\partial x} \left( D \frac{\partial \bar{Y}_A}{\partial x} - \overline{u_x Y_A'} \right)$$

$$+ \frac{\partial}{\partial y} \left( D \frac{\partial \bar{Y}_A}{\partial y} - \overline{u_y Y_A'} \right) + \frac{\partial}{\partial z} \left( D \frac{\partial \bar{Y}_A}{\partial z} - \overline{u_z Y_A'} \right) \quad (5.18)$$

For a binary mixture at low concentrations of $A$, comparison of Eqs. (5.17) and (5.18) shows them to be identical if $k/C_p \rho$ and $D$ are constant and equal, i.e., if Pr = Sc. Equation (5.16), however, includes an additional term for the gradient of the mean pressure. If this last is negligible, all three equations are of the same form if $\mu/\rho = k/C_p \rho = D$, that is, if Pr = Sc = 1.0.

**The analogy to momentum transport** The general approach to the problem of obtaining a relation between rates of momentum transport and the other two processes is to obtain $E_v$ from correlations of velocity profiles, assume or develop a relation between $E_v$ and $E_D$, and integrate Eq. (5.10) or its heat-transfer equivalent. The procedure will be developed for the special case of fully developed turbulent flow in a smooth round pipe. Transport is from the wall to the main stream. Various simplifying assumptions will be made to keep the derivation from becoming too involved.

The tube has a radius $r_w$ and the mean velocity is $\overline{U}_{av}$. The fluid density is $\rho$ and the kinematic viscosity is $v$. Turbulence is isotropic, so $u_x{}^2 = u_y{}^2 = u_z{}^2$. The coordinate in the axial or flow direction is $x$ and that in the radial direction $y$ (from the wall) or $r$ (from the axis). The shear at a plane a distance $y$ from the wall is given by Eq. (4.7), and the wall shear $\tau g_c$ replaced when convenient by $\frac{1}{2}f\rho\overline{U}_{av}{}^2$, where $f$ is the Fanning friction factor.

The analysis which follows is developed for mass transfer between the tube wall and the turbulent stream, the flux being given by Eq. (5.10). Since there is good reason to believe that the eddy diffusivities for mass and heat are essentially equal, the same development applies to heat transfer. The heat flux $q$ replaces $N_A$, the total diffusivity $(k/\rho C_p + E_H)$ replaces $(D + E_D)$, and the potential is $\rho C_p T$ instead of $c$.

In order to integrate Eq. (5.10), it is necessary to obtain $E_D$ as a function of $y$. This will be done by using the universal velocity distribution in smooth tubes, which was described in Sec. 4.4 and shown as Fig. 4.5. Derivatives of this function are employed in Eq. (4.7) to obtain $E_v(y)$, and $E_v$ is assumed equal to $E_D$.

The coordinates $u^+$ and $y^+$ are defined by

$$u^+ \equiv \overline{U}\sqrt{\frac{\rho}{\tau_w g_c}} = \frac{\overline{U}}{\overline{U}_{av}}\sqrt{\frac{2}{f}} \quad \text{and} \quad y^+ \equiv \frac{y}{v}\sqrt{\frac{\tau_w g_c}{\rho}} = \frac{(r_w - r)\overline{U}_{av}}{v}\sqrt{\frac{f}{2}} \quad (4.9)$$

For steady flow in a round tube the pressure is uniform across the diameter, and a simple force balance gives

$$\tau(r) = \frac{r}{r_w}\tau_w \qquad (4.8)$$

From Eqs. (4.9) and (4.7) it follows that

$$\frac{E_v}{v} = \frac{r}{r_w}\frac{dy^+}{du^+} - 1 \qquad (4.11)$$

Two additional approximations can now be used. Since the molecular transport is important very near the wall but not at large $y$, it may be assumed that some value of $y^+$, denoted by $y_1^+$, can be selected such that for $y^+ > y_1^+$, $E_v/v$ is very large compared with unity. Integration of Eq. (4.11) to $y^+ > y_1^+$ then gives

$$u^+ = \int_0^{y_1^+} \frac{dy^+}{(E_v/v) + 1} + \int_{y_1^+}^{y^+} \frac{dy^+}{E_v/v} \qquad (5.19)$$

The idea of breaking the integral in this way is due to Rannie [128]. Furthermore, since the region of principal interest is so close to the wall, the approximation is made that $r/r_w$ in Eq. (4.11) is unity, and that the flux $N_A$ is constant through the boundary layer, that is, $N_A$ will be used to designate the molal flux at the wall.

The flux equation, Eq. (5.10), is now integrated, assuming $E_D$ to become large in comparison with $D$ at the same limit that $E_v$ becomes large compared with $v$. The result is

$$\frac{c_w - c}{N_A} \sqrt{\frac{\tau_w g_c}{\rho}} = \int_0^{y^+} \frac{dy^+}{E_v/v + D/v} + \int_{y_1^+}^{y^+} \frac{dy^+}{E_D/v} \qquad (5.20)$$

Then making the final assumption that $E_D$ and $E_v$ are equal and that $\overline{U} = \overline{U}_{av}$ and $c = c_{av}$ at $y_1^+$, the last term of Eq. (5.20) is replaced by the last term of Eq. (5.19), giving

$$\frac{(c_w - c_{av})\overline{U}_{av}}{N_A} = \frac{2}{f} + \sqrt{\frac{2}{f}} \int_0^{y_1^+} \left( \frac{1}{E_D/v + D/v} - \frac{1}{E_D/v + 1} \right) dy^+ \qquad (5.21)$$

The left side will be recognized as the reciprocal of the Stanton number St, which is $k_c/\overline{U}_{av}$ for mass transfer and $h/C_p \rho \overline{U}_{av}$ for heat transfer.

Equation (5.21) is a general form of the analogy. To complete the derivation and obtain explicit expressions for $k_c$ and St, it is necessary to obtain $E_D$ $(= E_v)$ as a function of $y^+$, as from the universal velocity profile and Eq. (4.11). When this is done the result is of the form

$$\frac{1}{\text{St}} = \frac{2}{f} + \sqrt{\frac{2}{f}} \, g(\text{Sc}) \qquad (5.22)$$

The effect of Re on St is due to variations of the friction factor.

The choice of the limit $y_1^+$ presents no problem since the integrand goes nearly to zero well before the value of $y^+$ corresponding to $c_{av}$ is reached.

The final limit $y^+$ in the integrations to obtain Eqs. (5.19) and (5.20) was taken to be that at which $c$ is $c_{av}$, the average concentration of diffusing species in the flowing stream. A better procedure would be to use Eq. (5.20) to obtain a concentration profile across the tube diameter and integrate $c\overline{U}$ over the tube cross section to obtain $c_{av}$ involved in the definition of $k_c$. This was done by Deissler (see below). The loose identification of the limit with $c_{av}$ is acceptable, however, since the integrand of Eq. (5.21) becomes relatively insignificant at values of $y^+$ greater than about 200 (at Re = 50,000, $y^+ = 30$ corresponds to 2.4 percent of the tube radius for any fluid).

The right-hand side of Eq. (5.21) represents the sum of two resistances: The first term on the right is the resistance to mass transfer in the turbulent stream beyond $y_1^+$, while the second is the resistance of the region near the wall where $E_v$ and $E_D$ are important. It is seen that the second vanishes when Sc is unity, in which case

$$\text{St} \equiv \frac{k_c}{\overline{U}_{av}} = \frac{f}{2} \qquad (5.23)$$

or, for heat transfer,

$$\text{St} \equiv \frac{h}{C_p \rho \overline{U}_{av}} = \frac{f}{2} \qquad (5.24)$$

This last is equivalent to the assumption that the momentum loss equivalent to the wall friction, divided by the momentum of the stream, is equal to the ratio of the heat transferred in the section to the heat which would be transferred were the fluid to come to thermal equilibrium with the wall of the pipe.

Equations (5.23) and (5.24) represent the well-known *Reynolds analogy*. This simple form agrees fairly well with experimental data on heat transfer to air (Pr = 0.74) and mass transfer in gases (Sc $\approx$ 0.60 to 3). It does not apply where there is form drag in addition to "skin friction" and the coefficient of total drag is substituted for $f$. Nor does it hold if Sc or Pr are much greater or less than unity. Toor [165a] has modified the Reynolds analogy for turbulent transport in multicomponent systems.

Numerous authors have developed expressions for $E_v/v$ as a function of $y^+$ for use in the integration of Eq. (5.21) to obtain $g(\text{Sc})$ and to put the analogy in the form of a quantative relation between St, $f$, and Sc (or Pr). The results of several of these will be reviewed briefly.

Von Kármán [174] was evidently the first to employ the universal velocity profile and the general procedure outlined above. As noted in Sec. 4.4, he represented the "universal" $u^+ \approx y^+$ function by three simple equations, which could be used with Eq. (4.11) to obtain $E_v$ as a function of $y^+$. The limit $y_1^+$ was taken to be 30; very near the wall ($0 < y^+ < 5$) $u^+$ was taken equal to $y^+$ (whence $E_v = 0$ in this region). The result of the analysis was Eq. (5.22) with

$$g(\text{Sc}) \equiv 5\sqrt{\frac{2}{f}}(\text{Sc} - 1) + \ln\left[1 + \frac{5}{6}(\text{Sc} - 1)\right] \tag{5.25}$$

This result was shown to agree well with data on heat transfer with turbulent flow in smooth tubes in the range $0.73 < \text{Pr} < 35$.

Differentiation of Von Kármán's empirical representation [Eq. (4.12)] of the $u^+ \approx y^+$ function leads to unrealistic discontinuities in the variation of $E_v$ with $y^+$. More serious, however, is the fact that it predicts values of St much lower than observed for mass transfer at large Schmidt numbers. In this region the resistance is concentrated in a layer very close to the wall, where it now seems evident that a very slight amount of mixing can greatly enhance the transport. Since there are few data on the velocity profile at $y^+ < 5$, most of the attempts to improve on Von Kármán have employed empirical or semiempirical expressions for $E_v$ in this region.

Deissler [42] has proposed the following semitheoretical expression to represent conditions at small $y^+$:

$$\frac{E_v}{v} = n^2 u^+ y^+ \left(1 - e^{-n^2 u^+ y^+}\right) \qquad 0 < y^+ < 26 \tag{5.26}$$

with 0.124 as the value of the empirical constant $n$. The Von Kármán logarithmic velocity profile [Eq. (4.12)] was employed for $y^+ > 26$. Taking $E_D = E_v$, $N_A/(N_A)_w = \tau/\tau_w$, and $D$ negligible for $y^+ > 26$, Deissler first integrated Eq. (5.10) to obtain $c$ as a function of $y^+$. Using this result together with the universal velocity profile, he obtained integrated (velocity-weighted) average stream potentials, and so related St, Re, and Sc. The final result of this procedure showed good agreement with a large collection of experimental data on mass and heat transfer at Reynolds numbers of 10,000, 25,000, and 50,000 over the range of Pr or Sc from 0.7 to 3,000. In the limit as $Sc \to \infty$, Deissler's analysis simplifies to

$$St = \frac{2n\sqrt{f}}{\pi\, Sc^{3/4}} = 0.111 \sqrt{\frac{f}{2}}\, Sc^{-3/4} \tag{5.27}$$

This corresponds to Eq. (5.22) with negligible resistance in the main stream, and $g(Sc) = 9(Sc)^{3/4}$. The matter of the limiting value of the power on Sc as $Sc \to \infty$ has interested many writers [157, 73, 74, 81].

Wasan and Wilke [180, 185] have pointed out that many of the proposed eddy-viscosity functions do not satisfy the criterion [167] which states that the turbulent contribution to the Reynolds stress $\overline{u_x u_y}$ near the wall is proportional to $y^n$, where $n$ is not less than 3. By using the equation of mean motion and the well-established logarithmic velocity distribution for large $y^+$, Wasan, Tien, and Wilke [179] derived semitheoretical expressions for the continuous variation of velocity and eddy viscosity in the wall region.

The results of these studies may be summarized by two equations:

$$u^+ = y^+ - 1.04 \times 10^{-4}(y^+)^4 + 3.03 \times 10^{-6}(y^+)^5 \tag{5.28}$$

and
$$\frac{E_v}{v} = \frac{4.16 \times 10^{-4}(y^+)^3 - 15.15 \times 10^{-6}(y^+)^4}{1 - 4.16 \times 10^{-4}(y^+)^3 + 15.15 \times 10^{-6}(y^+)^4} \tag{5.29}$$

Both relations are restricted to the range $0 < y^+ < 20$; both are smooth continuous functions without discontinuities. Equation (5.29) gives a value of $E_v/v$ at $y^+ = 20$ agreeing with that calculated from the logarithmic distribution, given as Eq. (4.12), for $y^+ > 30$.

Employing Eq. (5.29) for $E_v/v$ in Eq. (5.21), and assuming $E_v = E_D$, the result is

$$\frac{1}{St} = \frac{2}{f} + \sqrt{\frac{2}{f}} \left[ \int_0^{20} \frac{dy^+}{1/Sc + E_v/v} - 13.0 \right] \tag{5.30}$$

whence $g(Sc)$ is the quantity enclosed in the square brackets. Wasan and Wilke tabulate values of the integral for values of Sc from 0.1 to 10,000.

Wasan and Wilke extended their analysis to the entry region of a round tube, where $k_c$ is decreasing with length of passage. Using a computer, they obtained solutions of the transient diffusion flux equation for $Sc = 1$ and 2.5. These showed St to fall sharply near the tube inlet, approaching closely the value obtained from Eq. (5.30) for fully developed turbulence at about eight tube diameters from the inlet. Meyerink and Friedlander [111], in their experimental study of the dissolution of benzoic acid, found $k_c$ to level off to an essentially constant value at about four tube diameters from the tube inlet. In their system Sc was 8,800.

**Other studies of the transport analogies** In addition to the three forms of the analogy described above, there are numerous other treatments of the relation between mass or heat transfer and fluid friction in turbulent flow [91, 61, 128, 131,

158, 107, 170, 136, 135, 100, 53, 81, 139, 183, 118, 63, 101, 16]. Reference 101 extends the analogy concept to nonnewtonian power-law fluids.

Because of the lack of a sound theory and understanding of the nature of the eddy transport very near the wall of a tube through which the flow is turbulent, most authors start with a relation between $u^+$ and $y^+$ which is more or less empirical at low $y^+$. Differentiation of this function then yields $E_v/v$ by the use of Eq. (4.11). Various other writers have simplified the analysis by starting with an empirical expression for $E_D/v$ as a function of $y$ or $y^+$. When this is done the result cannot be termed a complete analogy, since it does not relate, except indirectly, to momentum transport. Analyses following this procedure stem from the turbulent boundary layer model.

Murphree in 1932 [115] assumed $E_D$ to vary as the cube of the distance from the wall, and the integration of the flux equation yielded heat-transfer coefficients in good agreement with data on air, water, and oils over a limited range of Pr.

Lin, Moulton, and Putnam [97] also assumed $E_D$ proportional to $y^3 (0 < y^+ < 5)$. A linear expression was employed for $5 < y^+ < 33$, and the Reynolds analogy beyond $y^+ = 33$. The integrated result, of the form of Eq. (5.22), agreed well with a variety of heat- and mass-transfer data over a very wide range of values of Pr and Sc. Calculated concentration profiles were in agreement with experimental interferometric measurements near a phase boundary in the diffusion-controlled electrolytic deposition of cadmium on a mercury surface.

Sleicher [156] obtained $E_H$ from temperature profiles across the tube diameter in an experimental study of heat transfer to turbulent air streams. With an allowance for a variation of $E_H/E_v$ with $y$, he obtained computer solutions showing the variation of Nu with Re and Pr. The results agreed well with Deissler and with experimental data in the range $0.7 < \mathrm{Pr} < 7.5$. In a more recent paper, Notter and Sleicher [118] obtained the following empirical correlation from a critical study of experimental data on mass and heat transfer in turbulent flow in pipes:

$$\frac{E_D}{v} = \frac{0.00090(y^+)^3}{[1 + 0.0067(y^+)^2]^{1/2}} \qquad 0 < y^+ < 45 \qquad (5.31)$$

For Pr or Sc greater than 100, this corresponds approximately to

$$\mathrm{St} = 0.0149 \mathrm{Re}^{-0.12} \mathrm{Sc}^{-2/3} \qquad (5.32)$$

This gives values of St 11 to 50 percent greater than the Chilton-Colburn analogy (see below) in the range $2,000 < \mathrm{Re} < 100,000$.

The analogy published by Chilton and Colburn in 1934 [26] is perhaps the most useful, and certainly the simplest, of the many expressions relating mass, heat, and momentum transfer. The equation, written separately for mass and heat transfer, is

For mass transfer:

$$\mathrm{St} = \frac{k_c}{U_{av}} = \frac{f}{2} \mathrm{Sc}^{-2/3} \qquad (5.33)$$

For heat transfer:

$$\frac{h}{C_p \rho \overline{U}_{av}} = \frac{f}{2} \, Pr^{-2/3} \tag{5.34}$$

This was perhaps based on (1) the observation that the simple Reynolds analogy held for heat transfer when Pr was near unity, (2) the apparent validity of the simple empirical function $1.0 \, Pr^{-2/3}$ to represent heat transfer data over a limited range of Pr, (3) the assumption of a close analogy between heat and mass transfer, which suggested that $k_c$ should vary with Sc in the same way that $h$ varies with Pr, and (4) the fact that $Pr^{-2/3}$ had been shown theoretically to apply to transport through a laminar boundary layer. In more recent years Eq. (5.33) has been shown to agree surprisingly well with newer data on mass transfer over a range of values of Sc from 0.6 to more than 3,000. This is illustrated by Fig. 5.3, discussed below.

Equation (5.33) can be derived [171] by the use of Eq. (5.10) and the empirical assumption that $E_D/\nu = 1.77(f/2)^{3/2}(y^+)^3$.

Levich [91] has developed a theoretical expression for mass transfer in turbulent flow based on four zones or layers: a diffusion layer, a viscous sublayer, a turbulent boundary layer, and the main turbulent stream. This approach is discussed and the derivation presented in some detail in the book by Davies [37].

The analysis by Pinczewski and Sideman [124a], described briefly in Sec. 5.2, is not an analogy stemming from Eqs. (5.16) to (5.18) but is based on a model of the transport near the wall. For the lower of two ranges of Sc the result is:

$$St = \frac{0.0097(1.10 + 0.44 \, Sc^{-1/3} - 0.70 \, Sc^{-1/6})}{Re^{0.1} \, Sc^{1/2}[1 + 0.064(1.10 + 0.44 \, Sc^{-1/3} - 0.70 \, Sc^{-1/6})]} \qquad 1 < Sc < 500 \tag{5.34a}$$

This agrees well with the data, giving values of St falling only a few percent below the Friend and Metzner line on Fig. 5.3. As in the case of the true analogies, it is based only on fluid mechanics data.

**Comparison with experimental data**    Figure 5.3 compares several of the analogies with representative data on both heat and mass transfer. It is impractical to attempt a comparison of all the published analogies and all the relevant data, but Fig. 5.3 provides a reasonably fair picture.

The open circles represent data on heat transfer in tubes at $Re = 10,000$ for gases, water, oils, molten salt, organic liquids, and aqueous solutions of sugars. They are taken from Friend and Metzner [53], who made a critical literature search. The six circles at $0.46 < Pr < 0.74$ are for gases; all the open circles at $Pr > 1$ are for heat transfer to the various liquids. Viscosity is the main variable affecting Pr for liquids, and the range is limited by the difficulty of obtaining turbulent flow if the viscosity is high.

The solid points at large Sc represent the excellent data of Meyerink and Friedlander [111] and of Harriott and Hamilton [66] on the dissolution of tubes formed from slightly soluble organic solids. The first used benzoic acid, cinnamic

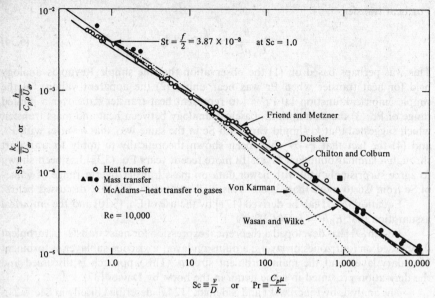

FIGURE 5.3
Variation of the Stanton number with Sc or Pr for flow of gases and liquids in tubes at Re = 10,000.

acid, and aspirin with water. Harriott and Hamilton used benzoic acid with aqueous solutions of Methocel and glycerine in order to vary the viscosity. The solutes employed gave Sc = 850 to 930 in water at 25°C. (Cermak and Beckman [25] report similar data for flow in an annulus.) The three solid points in the vicinity of Sc ≈ 1 are from Gilliland's [54] study of the vaporization of several liquids into a turbulent air stream in a wetted-wall column. Not shown are the data of Hubbard and Lightfoot [72], which agreed well with the Chilton-Colburn line over a range of Re from 7,000 to 60,000 and Sc from 1,700 to 30,000.

The data cover an enormous range, yet the mass- and heat-transfer data fall in a single fairly narrow band. As noted by Notter and Sleicher [118], the data are particularly sensitive to wall roughness at high Pr or Sc. The analogies of Diessler and of Wasan and Wilke do well in the lower range of Sc but deviate by 20 to 50 percent from the data points at the highest Sc. The simple Chilton-Colburn equation represents the data about as well as the more elegant forms. Von Kármán's analysis fails above Sc ≈ 10, presumably because he made no allowance for eddy diffusion at $y^+ < 5$, which is the region of importance when Sc is large. At Re ≈ 10,000, Eq. (5.32) gives values of St 23 percent greater than the Chilton and Colburn relation, in excellent agreement with Friend and Metzner and with the data points. The mass-transfer data for gases fall considerably above the heat-transfer data, although McAdams' equation [104] for heat transfer to gases falls somewhat above the data points used by Friend and Metzner.

The top line represents the general correlation for heat transfer proposed by Friend and Metzner [53]. These authors used the general form of Eq. (5.22) modified empirically so as to fit the heat-transfer data. Their equation is

$$St = \frac{f/2}{1.20 + 11.8\sqrt{f/2}\,(Pr - 1)\,Pr^{-1/3}} \tag{5.35}$$

As is to be expected, the solid line representing Eq. (5.35) fits the open circles well, since it is based on these data. Extended, it provides an excellent representation of the mass-transfer data at high Sc, which Friend and Metzner did not consider. Equation (5.32) does as well at Re = 10,000.

In preparing Fig. 5.3, the lines representing the principal analogies were based on values of $g(Sc)$ given by the authors. These are summarized briefly in Table 5.1. Only in the case of Deissler does $g(Sc)$ vary with Re; at Re = 100,000 Deissler's curves indicate $g(Sc)$ to be about 1.4 at Sc = 1 and some 10 percent greater than the tabulated values at larger Schmidt numbers.

The effect of variations in Re are not shown in Fig. 5.3, and may be greater than suggested by the Deissler analogy. A recent study [1a] of mass transfer in narrow channels provided data which for laminar flow agreed exactly with the Graetz-Leveque theory as extended to parallel-wall channels by Newman [117a]. For turbulent flow over a smooth surface, the ratio $j_D/(f/2)$ varied irregularly from about 0.6 to 0.95 in the range 2,500 < Re < 250,000. However, with surface roughness increased up to 63 $\mu$m, the same ratio peaked at 2.4 at Re = 3300. The data, which are excellent, provided *local* mass-transfer and friction coefficients, the mass transfer results being obtained by the electrochemical technique with aqueous ferricyanide reduction, in channels with equivalent diameters of 0.096 and 0.038 cm.

**Summary**   The development of the analogies has proved to be a subject of great interest to many investigators. The evident inadequacies of the resulting expressions

Table 5.1   VALUES OF THE FUNCTION $g(Sc)$ APPEARING IN Eq. (5.22), AS THE RESULT OF DERIVATIONS BASED ON THE ANALOGY TO MOMENTUM TRANSFER; THE ANALYSIS IN EACH CASE IS FOR FULLY DEVELOPED TURBULENT FLOW IN SMOOTH TUBES

| Sc or Pr | Von Kármán | Deissler* | Wasan and Wilke |
|---|---|---|---|
| 1 | 0 | 0 | 0 |
| 10 | 55.7 | 43 | 60.6 |
| 100 | 517 | 283 | 340 |
| 1,000 | 5,025 | 1,560 | 1,643 |
| 10,000 | 50,040 | (8,860) | 7,550 |

* Approximate values calculated from Deissler's Fig. 3 at Re = 10,000.

stem primarily from the lack of a sound theory of turbulence in flow near a phase boundary.

Most of the derivations introduce the assumption that $E_D = E_v$. In all likelihood this is not so, particularly near a tube wall. Furthermore, $E_v/E_D$ (the ratio of the turbulent Schmidt number to the turbulent Prandtl number) apparently varies with Sc and Pr. The variation of $E_v/E_D$ with various factors has been studied by several workers [156, 76, 165, 102, 132, 169], but there appears to be no complete and useful theory.

The various expressions of the form of Eq. (5.22) indicate that St should vary as $\sqrt{f}$ at large values of Sc. Most of the experimental data on mass transfer at tube walls was obtained at Reynolds numbers from 10,000 to 50,000 over which range $f$ decreases only some 30 percent. The Chilton-Colburn relation suggests a power of 1.0 instead of 0.5. The experimental data of Eisenberg, Tobias, and Wilke [48] on mass transfer from a rotating cylinder at Sc = 870, which covered more than a thousandfold variation in Re, showed excellent agreement with the Reynolds-Chilton-Colburn use of $f$ with a power unity.

Sherwood and Ryan [149] have suggested that the general method of developing the analogy to momentum transfer would lead to the appearance of $f/2$ instead of $\sqrt{f/2}$ in Eq. (5.22) if the universal velocity profile expressed $u^+$ as a function of $(y\overline{U}_{av}/v)(f/2)$ rather than $(y\overline{U}_{av}/v)(\sqrt{f/2})$. The usual definitions of $u^+$ and $y^+$ are based largely on dimensional considerations, and there seems to be no theoretical reason why this modification of $y^+$ should not be made. It would, in fact, provide a better correlation of data on velocity profiles at large $y^+$ for widely different Reynolds numbers.

As noted earlier, the failure to develop a sound and complete analogy is due primarily to lack of understanding of flow in the immediate region of the wall. The use of Pitot tubes and hot-wire anemometers has failed to provide reliable data at $y^+$ less than 2 or 3. Optical techniques, pioneered by Fage and Townend [50], are beginning to provide the missing picture, which appears to be highly complicated [150, 126, 30, 52, 141, 84, 143, 82, 116, 29, 97]. The techniques include dye injection, photolysis of a dye, interferometry, photography of tracer particles or bubbles, electroluminescence, and the use of laser beams.

The data of Nedderman [116] and Fowles [150], based on optical measurements of tracer particles well within what has been called the "laminar sublayer" at the wall, showed wide fluctuations in velocity and direction. These appeared to be random, and the data at a fixed value of $y^+$ had to be analyzed statistically to be interpreted.

S. J. Kline and his colleagues [82, 84] have photographed the motion of swarms of tiny gas bubbles. Figure 5.4 [83] is one of their pictures, looking vertically at the flow of water along the bottom of an open water channel. The water depth is 10 in, the mean flow about 0.5 ft/s, and the bubbles are at $y^+ = 2.7$. The evidence from this and many other photographs shows the structure to be streaky in the region very near the wall.

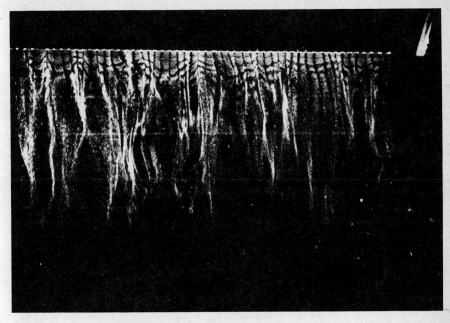

FIGURE 5.4
Swarms of tiny hydrogen bubbles in water flowing along the bottom of an open
water channel. Water depth, 10 in, mean velocity 0.5 ft/s; bubbles near bottom at
$y^+ = 2.7$. (*Photograph by courtesy of Prof. S. J. Kline.*)

In an early paper describing studies with this technique, Kline and Runstadler
[84] concluded that the evidence supports "a picture of the turbulent shear layer
generated by repeated instability of the wall layers occurring over the entire wall
in a turbulent boundary layer. The instability takes the form of a longitudinal
vortex-island of hesitation pattern that breaks up near the edge of the viscous
sublayer." And later [82],

What is most important about this structure is that it "bursts" intermittently. This bursting
causes a strong interaction between the inner and outer layers.... The data now available
make it clear that the region $0 < y^+ < 7$ is decidedly unsteady and three-dimensional right to
the wall. Measurements now go to $y^+ = 0.1$ and there is no evidence of a steady, two-
dimensional layer even there.

Further studies of these bursts and streaks in the wall region are reported by
Offen and Kline [118a].

It seems evident that the concept of a laminar sublayer should be discarded
and a much better description of flow near the wall be developed. Until this is done,
efforts to develop new forms of the traditional analogies may not prove profitable.

## 5.4 MASS TRANSFER AT LARGE FLUXES AND AT HIGH CONCENTRATION LEVELS

The convective flux becomes important relative to the diffusion flux if (1) diffusion occurs in a stream moving rapidly in the direction of diffusion, as in Example 5.1, or (2) if the convective flux is due only to diffusion itself, but the concentration is high. Referring to Eq. (3.4), $\overline{U}_y c_A$ may be large relative to $J_A$ if $c_A$ is small but $\overline{U}_y$ made large by forced flow, or if $\overline{U}_y$ is small but $c_A$ large. These two effects will be discussed by considering an isothermal binary gas system and employing the film model as the basis for analysis. Experimental data indicate that the film model provides an excellent first approximation for this purpose.

First consider the case of an applied convective flux superposed on that due to diffusion. As an example, assume that air is introduced through the porous wall of a tube through which air flows. The inner surface of the porous tube is wet with water and maintained at a constant temperature. The air supplied through the porous wall is assumed to become saturated with water immediately as it enters (at $y = 0$). Superposed on this air flux, $N_B$, carrying water vapor, is the diffusion flux of water vapor, $J_A$. At $y = 0$, the mole fraction water is $Y_{A1}$; at $y = y_0$ it is $Y_{A2}$. The total flux of water vapor from the wet surface to the airstream is then given by Eq. (3.4):

$$N_A = N_T Y_{A1} + J_A \tag{3.4}$$

where $N_T$ is the total flux of all components (in this case $N_T = N_A + N_B$). The question of concern is: does the term $Y_{BM}$ in Eqs. (3.8) and (5.2) provide properly for the effects of high fluxes and high-solute concentration levels in real situations, including turbulent flow in industrial equipment?

The case of high fluxes, as in the example described, may be analyzed by employing the film model. From a solute (water-vapor) balance on a differential slice of the film,

$$\frac{d^2 Y_A}{dy^2} - \frac{N_T RT}{D_{AB} P} \frac{dY_A}{dy} = 0 \tag{5.36}$$

Solving this and differentiating to obtain $(dY_A/dy)$, one obtains

$$J_A = k_D(Y_{A1} - Y_{A2}) = -\frac{D_{AB} P}{RT}\left(\frac{dY_A}{dy}\right) = \frac{N_T(Y_{A1} - Y_{A2})}{[\exp(N_T RT y_0/D_{AB} P)] - 1} \tag{5.37}$$

The coefficient $k_D$ is not a mass-transfer coefficient in the usual sense; Eq. (5.37) gives only $J_A$ from which $N_A$ is obtained by Eq. (3.4).

In the limit, as $N_T \to 0$, $(k_D)_{N_T \to 0} = D_{AB} P/RT y_0 = k^*$. Furthermore, if there is no applied convective flux, that is, if $N_T = N_A$, Eq. (5.37) reduces to Eq. (3.8). It is evident that the flux of $A$ is increased by a convective flux away from the wall if $N_T > N_A$ ("blowing"). Conversely, the solute flux away from the wall may be greatly reduced by an applied convective flux toward the wall ("sucking").

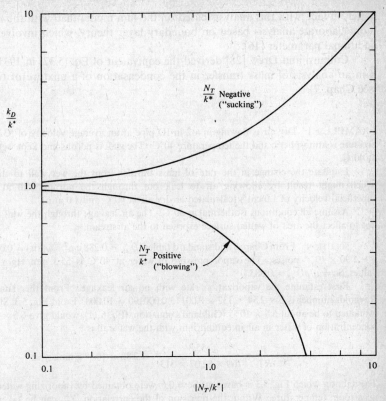

**FIGURE 5.5**
Effect of transverse flow (sucking or blowing) on mass transfer between a wall and a flowing fluid. Mass transfer away from the wall; $N_T$ positive for blowing and negative for sucking.

It follows from Eq. (5.37) that the ratio $k_D/k^*$ depends only on the group

$$\frac{N_T RT y_0}{D_{AB} P} = \frac{N_T}{k^*}$$

Figure 5.5 shows the relation, as presented by Mickley, Ross, Squyres, and Stewart [112]. The upper curve is an extension of the lower curve to negative values of the abscissa ($N_T$ and $N_A$ in opposite directions).

Similar analyses using the film model for heat and momentum transfer lead to relations of the same form. In the case of heat transfer, the ordinate is $h/h^*$, where $h^*$ is the heat-transfer coefficient with no mass transfer, and the abscissa is $\bar{U}_1 \rho C_p$. Data on heat transfer from hot air to a porous wall through which air is either blown or sucked are in good agreement with Fig. 5.5 [112]. The data agree better, in fact, with this analysis based on the film model than with the results of a more elaborate analysis based on boundary layer theory, which involves Sc as an additional parameter [16].

Colburn and Drew [28] derived the equivalent of Eq. (5.37) in 1937 and used it in an analysis of mass transfer in the condensation of a mixture of two vapors (see Chap. 7).

EXAMPLE 5.1 Dry air is flowing in a 2-in-ID pipe at an average velocity of 332 cm/s. The pressure is atmospheric and the temperature 40°C. The wall is porous and kept wet with water at 40°C.

Estimate the change in the rate of mass transfer from the wet wall to the airstream which might result by allowing air to leak out through the porous wall at a uniform superficial velocity of 1.0 cm/s (calculated as dry air at 40°C and 1.0 atm).

Assume all conditions isothermal at 40°C. The air leakage through the wall is presumed not to affect the area of wetted surface exposed to the airstream.

SOLUTION From Chap. 2 and standard tables: $D_{AB} = 0.288$ cm$^2$/s, $\rho$(air) = 0.00113 g/cm$^3$, $\mu = 1.90 \times 10^{-4}$ poises. The vapor pressure of water at 40°C is 55.3 mm Hg. From these values, Sc $= \mu/\rho D_{AB} = 0.583$.

First estimate the vaporization rate with no air leakage. From the data given, the Reynolds number is $2 \times 2.54 \times 332 \times 0.00113/0.000190 = 10,000$. From Fig. 5.3, St $= k_c/\overline{U}_{av}$ is estimated to be about $5.5 \times 10^{-3}$ (Gilliland's equation, (Eq. 6.11), would give $6.5 \times 10^{-3}$.) The concentration of water in air in equilibrium with the wet wall is $c_{A1}$:

$$c_{A1} = \frac{p}{RT} = \frac{55.3}{760 \times 82.07 \times 313} = 2.83 \times 10^{-6} \text{ g moles/cm}^3$$

The data on which Fig. 5.3 is based, at Sc $\approx 0.6$, were obtained by vaporizing water into air at near room temperature. Within the precision of the correlation, $Y_{BM}$ can be taken to be the same as in the present case ($Y_{BM} = p_{BM}/P = 0.96$) and no $Y_{BM}$ correction is needed. With no air leakage through the wall,

$$N_A = k_c(c_{A1} - c_{A2}) = 5.5 \times 332(2.83 \times 10^{-6} - 0) \times 10^{-3}$$
$$= 5.17 \times 10^{-8} \text{ g moles/(s)(cm}^2)$$

The dry air flow through the wall is 1.0 $P/RT = 38.9 \times 10^{-6}$ g moles/(s)(cm$^2$); this is $-(N_T - N_A)$, since both $N_T$ and $N_A$ are defined as positive for a flux from the wall toward the flowing stream.

Substituting in Eq. (3.4), with $Y_{A2} = 0$, $N_A = N_T Y_{A1} + k_D(Y_{A1} - 0) = (55.3/760)(N_T + k_D)$ $= 0.0728(N_T + k_D) = 38.9 \times 10^{-6} + N_T$, g moles/(s)(cm$^2$). The coefficient $k^*$ is $k_c P Y_{BM}/RT$ [Eqs. (5.1) and (5.2)]:

$$k^* = 5.5 \times 10^{-3} \times 332 \times 1.0 \times \frac{0.96}{82.07} \times 313 = 6.82 \times 10^{-5}$$

whence
$$\frac{N_T}{k^*} = 0.0785 \frac{k_D}{k^*} - 0.615$$

This last is solved by trial, using Fig. 5.5, to obtain $k_D/k^* = 1.28$ (at $N_T/k^* = -0.51$). Finally $N_A = 0.0728 (N_T + k_D) = 0.0728 \, k^*(-0.51 + 1.28) = 3.82 \times 10^{-6}$ g moles/(s)(cm$^2$). The rate of transport of water vapor from wall to airstream is reduced from $5.17 \times 10^{-6}$ to $3.82 \times 10^{-6}$, i.e., by 26 percent. The air leaking through the wall is saturated with water, so the total evaporation amounts to $3.82 \times 10^{-6} + 38.9 \times 10^{-6} \times 0.0728/(1 - 0.0728) = 6.87 \times 10^{-6}$ g moles/(s)(cm$^2$).

*Additional questions:* 1 What rate of air flowing through the porous wall would be required to prevent *any* water vapor from reaching the airstream flowing in the pipe? 2 Show that the problem can be solved by the use of Eq. (3.9), as an alternative to the procedure followed above.  ////

**Mass transfer at high concentrations of the diffusing species**  The analysis based on the film model leads to Eqs. (3.8) and (5.2), which are for molecular transport of one component of an ideal gas binary when the second does not diffuse. The term $p_{BM}$ is sometimes referred to as the "film-pressure factor," and the incorporation of $p_{BM}$ (or $Y_{BM}$) in the denominator of the rate equation stems from the assumption that $J_A$ is proportional to the negative of the concentration gradient. If this last is true, the validity of using $p_{BM}$ to allow for the convective flux does not depend on the transport mechanism, and it belongs in the equation for eddy diffusion [181]. Its justification for cases of high solute-concentration levels with turbulent flow must be based on experimental data, and these are not plentiful.

The film model suggests that the film-pressure factor should apply to mass transfer from a gas to a liquid or solid interface, and that $k^*$ should be independent of $Y_{BM}$. Its use has become common practice, and many authors have employed the dimensionless group $k_c d\, Y_{BM}/D_{AB}$ in the correlation of data on mass transfer in turbulent gas systems. Whether or not it applies in liquid systems is not known; solute concentrations are often low, and where high the picture is confused by the considerable variation of $D_{AB}$ with concentration.

Only a few investigators have carried out experimental studies of the effect of $p_{BM}/P$ (or $Y_{BM}$) in gas systems. Gilliland [54] operated a wetted-wall column over a wide range of total pressures, but $Y_{BM}$ was varied over too narrow a range to determine its effect. The same is true of Shulman and Margolis [153] results for the sublimation of naphthalene. Cairns and Roper [22] used an adiabatic wetted-wall column with air and water, and obtained data for turbulent air flow over a range of $Y_{BM}$ from 0.15 to 0.97. They concluded that $k_c$ was inversely proportional to $(Y_{BM})^{0.83}$. Others [182, 154, 152] have reported exponents on $Y_{BM}$ in the range 0.67 to 1.0, though the data do not justify a precise value.

Vivian and Behrmann [172] absorbed ammonia from a turbulent stream of nitrogen into water, aqueous sulfuric acid, and aqueous ammonia solutions, using a very short wetted-wall column. By varying the ammonia concentration in both gas and liquid streams at a nearly constant Reynolds number of 3,200, they were able to measure $k_Y$ at values of $Y_{BM}$ ranging from 0.068 to 0.934. The Schmidt number varied only a few percent over this concentration range in the $N_2$–$NH_3$ system. The data obtained are shown in Fig. 5.6, plotted as $k_Y/G_M$ vs. $Y_{BM}$. Although the points scatter somewhat, these results show $k_Y$ to vary inversely as $Y_{BM}$ to the power 1.0. The transport rate varied as $Y_{BM}$ was varied, but there appeared to be no correlation between the flux and the effect of $Y_{BM}$ on $k_Y$.

The importance of the film-pressure factor is also confirmed by the recent

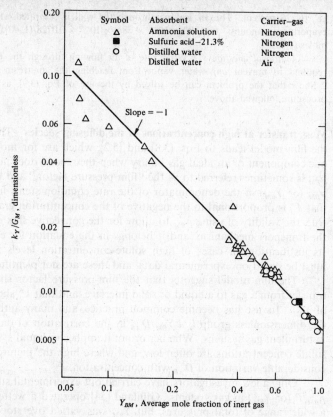

**FIGURE** 5.6
Effect of inert-gas concentration on the rate of gas absorption in a short wetted-wall column. (*Data of Vivian and Behrmann.*)

work of Yoshida and Hyōdō [186]. Water was allowed to evaporate into air flowing in a 2.9-cm-ID wetted-wall column 100 cm long, the air being fed at temperatures as high as 417°C. Figure 5.7 shows their data plotted as $(St)(Sc)^{2/3}$ vs. $Y_{BM}$. The Stanton number St is $k_c/\overline{U}_{av}$ and does not include the $Y_{BM}$ term. Sc varied only a few percent over the concentration range studied. The indicated slope of $-0.92$ is not far from that suggested by the film model for diffusion in an ideal gas binary.

Yoshida and Hyōdō also used pure superheated steam, in which case the water temperature was 100°C and heat transfer from the vapor determined the evaporation rate. This was also the case with very hot dry air having a wet-bulb temperature of almost 100°C. Since the heat-transfer coefficient for water vapor is greater than for air, the rate of vaporization of water into superheated steam was

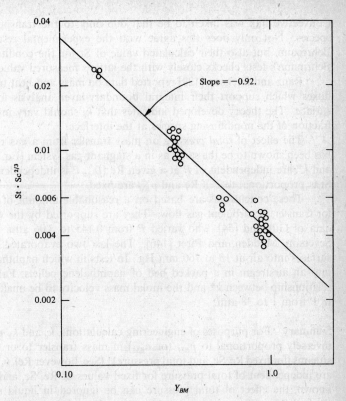

**FIGURE 5.7**
Effect of inert-gas concentration on the rate of evaporation of water into air at 66 to 417°C; Sc = 0.54 to 0.62; Re = 3,200 to 13,000. (*Data of Yoshida and Hyōdō.*)

greater than into dry air. At low temperatures the rate decreased with increase in air humidity; at high temperatures the reverse was found. At 170°C and Re = 8,000 to 10,000, evaporation rate was not affected by changes in humidity of the air supplied.

Shulman and Delaney [152] evaporated carbon tetrachloride into hot air in a very short packed column under conditions such that $Y_{BM}$ was varied from about 0.5 to 0.9. Though their data scatter appreciably, the results were correlated somewhat better by the use of $Y_{BM}^{2/3}$ than with $Y_{BM}^{1.0}$.

The analogy approach has been used by Wasan and Wilke [181] in a theoretical study of the role of $Y_{BM}$ on mass transfer in turbulent gas flow. Equation (5.29), which is their expression for the variation of the eddy-diffusion coefficient with distance from the wall, was employed to follow the decrease of St from a pipe inlet to fully developed turbulent flow. They found St to vary inversely with $Y_{BM}$ in several cases of binary gas mixtures with Sc = 0.2, 1.0, and 2.5.

Convective flux was taken to be that due only to the transport of the diffusing species. Not only does this agree with the experimental results of Vivian and Behrmann, but also their calculated value of St for the conditions of Vivian and Behrmann's tests checks closely with the latter's measured values.

Ranz and Dickson [129] reported data on mass and heat transfer at very high fluxes which support their integral boundary-layer analysis for flow over a flat surface. The theory developed indicates that $k_c$ should vary inversely as the mass fraction of the nondiffusing species at the interface.

The effect of *total pressure* on mass transfer from a gas to a liquid or solid has been shown to be the same as in a stagnant-gas system [Eq. (3.8)]. Thus $k_c p_{BM}$ and $k^*$ are independent of $P$ at a given Re ($D_{AB}P$ is independent of pressure), and St is proportional to $P$ if Re and $p_{BM}$ are fixed.

These conclusions are based on a reasonable amount of mass-transfer data for transfer in turbulent-gas flow. They are supported by the wetted-wall column data of Gilliland [54], who varied $P$ from 0.145 to 3.06 atm, Phillips [124], and Severson, Madden, and Piret [146]. The last two evaporated liquids from small surfaces into air at 15 to 760 mm Hg. In tests in which naphthalene was sublimed into an airstream in a packed bed of naphthalene pellets, Fallat [51] found the relationship between $k^*$ and the molal mass velocity to be unaffected by variations in $P$ from 1 to 38 atm.

**Summary**  For purposes of engineering calculations, $k_c$ and $k_Y$ may be taken to be inversely proportional to $p_{BM}$ (or $c_{BM}$) in mass transfer to or from turbulent-gas streams (for fixed Re, Sc, and total pressure). (See, however Ref. 6.) These coefficients are independent of total pressure for fixed values of Re, Sc, and $p_{BM}$. As far as is known, the effect of total pressure can be ignored in liquid systems. Although the influence of $c_{BM}$ does not appear to have been studied experimentally in liquids, theory suggests that its effect should be similar to that for gases.

## 5.5  THE TWO-FILM THEORY

Most industrial processes in which mass transfer is important involve transport of solutes from one phase to another. This is the case in liquid-extraction, distillation, gas-absorption, and other separation processes. The solute is transported from the bulk of one phase to the phase boundary, or interface, and then from the interface to the bulk of the second phase. In many cases the foregoing treatment of transport within a single phase can be applied to each of the boundary layers, or films, adjacent to the interface.

The flux equations are analogous to Ohm's law, and the ratio of the potential, or driving force, to the flux represents a resistance. The additivity of the transport resistances on the two sides of an interface was first proposed by Whitman and Lewis [96, 184]. Whitman's original paper carried the title The Two-Film Theory of Gas Absorption, but the main point of the article was the additivity of the resistances; the analysis presented did not depend on the validity of the film theory.

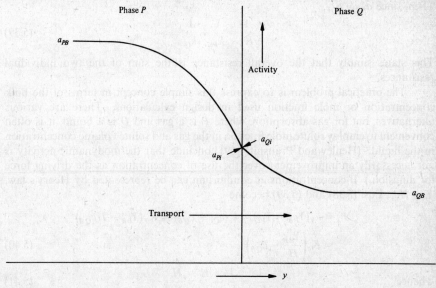

FIGURE 5.8
Activity gradients near a phase boundary.

In its simplest form, the additivity concept assumes that there is no resistance to transport at the actual interface, i.e., within distances corresponding to molecular mean free paths in the two phases on either side of the phase boundary. This is equivalent to assuming that the phases are in equilibrium at the actual points of contact at the interface. The validity of this assumption is discussed below.

For practical purposes of process design, it is convenient, if not essential, to express transport rates in terms of the bulk phase concentrations employed in the stoichiometry of the process. It is helpful, therefore, to define an overall mass-transfer coefficient, which is the reciprocal of the sum of the resistances in series. To do this it is necessary to employ potentials in the same units for both phases.

Figure 5.8 illustrates the situation at a point in a mass-transfer device where a solute is being transferred at constant rate from phase $P$ to phase $Q$, the bulk of each being well mixed. The activity of the solute in phase $P$ is $a_{PB}$ and in $Q$ it is $a_{QB}$. Assuming equilibrium at the interface, the activities $a_{Pi}$ and $a_{Qi}$ are equal at the plane of contact.

If the two solutions are ideal, activity can be employed as the potential instead of concentration, and two "individual" mass-transfer coefficients and an "overall" coefficient defined by the steady-state flux equations

$$N_A = k_P(a_{PB} - a_{Pi}) = k_Q(a_{Qi} - a_{QB}) = K_a(a_{PB} - a_{QB}) \qquad (5.38)$$

Then, since $a_{Pi} = a_{Qi}$,

$$\frac{1}{K_a} = \frac{1}{k_P} + \frac{1}{k_Q} \tag{5.39}$$

This states simply that the overall resistance is the sum of the two individual resistances.

The practical problem is to express this simple concept in terms of the bulk concentration or mole fraction used in design calculations. There are various alternatives, but for gas absorption, where $P$ is a gas and $Q$ is a liquid, it is often convenient to employ solute mole fraction in the gas and solute volume concentration in the liquid. (Henley and Prausnitz [69] conclude that thermodynamic activity is not necessarily an improvement over the use of concentration as the driving force for diffusion.) If concentrations at equilibrium can be represented by Henry's law, $Y_i = Hc_i$, Eqs. (5.38) and (5.39) become

$$N_A = k_Y(Y_{PB} - Y_{Pi}) = k_c(c_{Qi} - c_{QB}) = K_Y(Y_{PB} - Hc_{QB})$$

$$= K_c\left(\frac{Y_{PB}}{H} - c_{QB}\right) \tag{5.40}$$

whence
$$\frac{1}{K_Y} = \frac{1}{k_Y} + \frac{H}{k_c} = \frac{H}{K_c} \tag{5.41}$$

It follows from this last equation that if Henry's law does not apply, then $H$ varies with the concentration at the interface, and both $K_Y$ and $K_c$ will vary with concentration, even if the individual coefficients $k_Y$ and $k_c$ do not. If the gas is highly soluble in the liquid, $H$ will be small and $K_y \approx k_Y$. In this case the liquid-side resistance is negligible and the flux is $k_Y(Y_{PB} - Hc_{QB}) = k_Y(Y_{PB} - Y_{QB}^*)$, where $Y_{QB}^* = Hc_{QB}$ is the gas mole fraction corresponding to equilibrium with the bulk liquid. If the gas is relatively insoluble ($H$ large), the gas-side resistance $1/k_Y$ becomes negligible in comparison with the liquid-side resistance $H/k_c$, so $K_c \approx k_c \approx HK_Y$. The relative magnitudes of the individual resistances evidently depend on gas solubility, as represented by the Henry's-law constant. This explains the common statements that "the liquid-side resistance is controlling" in the absorption of a relatively insoluble gas, and the "gas-side resistance is controlling" when a relatively soluble gas is absorbed (or desorbed). The ratio $k_c/k_Y$ is evidently involved in any specification of the limiting value of $H$ at which one or the other individual resistance can be ignored.

The application of the two-film model to mass transfer between two immiscible liquids, as in solvent extraction, follows easily from the above treatment for gas-liquid systems; $H$ is replaced by a distribution coefficient, and $Y$ with concentration in the second liquid phase.

The additivity of the resistances in the two phases has been questioned by various writers, notably by Abramzon and Ostrovskii [1]. In a sense, the concept is correct by definition. However, the individual coefficients $k_c$ and $k_Y$, as determined from experiments in which all the resistance is in one of the two phases, may be substantially different when there are appreciable resistances in both phases.

Furthermore, it is obvious that Eq. (5.41) cannot be used to obtain $k_Y$ from $k_c$ and $k_Y$ if there is (1) an additional resistance introduced, as may be caused by the concentration of a surfactant at the interface, (2) the development of ripples or interfacial turbulence which increase $k_c$ or $k_Y$ over the single-phase values, or (3) the occurrence of a chemical reaction near the phase boundary, not encountered when the individual coefficients are measured.

Goodgame [55] checked Eq. (5.41) experimentally for a gas-liquid system, as did Gordon [58] and McManamay [105] for liquid-liquid systems in stirred vessels. However, King [78] and Szekely [161] point out that it may not apply if $k_c$ or $k_Y$ vary with contact time at renewable surfaces, or if these coefficients vary over a finite surface. The sum of the average resistances is not the same as the average of the sums. In the application of the penetration theory, both King and Szekely note that the addition of a gas-side resistance increases $k_c$ because the penetration depth is reduced for a fixed contact time. King emphasizes the point that Eq. (5.41) will give low values of the average $k_Y$ if $Hk_Y/k_c$ varies from point to point over the surface. These two effects tend to offset each other, but the latter may introduce large errors in cases where there is a wide distribution of surface lifetimes, giving large variations of $k_c$ over the surface, as in irrigated packings [78]. If $k_Y$ is constant over the surface, the Danckwerts surface-renewal model (see Sec. 5.2) shows that the two effects cancel, and that Eq. (5.41) is correct.

The principle of additivity of resistances cannot be used unless all resistances are properly accounted for. If a surfactant is present at the interface, its diffusional resistance must, of course, be included. Furthermore, the presence of a surfactant may change $k_c$ or $k_Y$, or both. Even with a clean interface, the effect of mass transfer may cause interfacial turbulence which greatly enhances the individual coefficients in one or both of the phases in contact. Olander [119] found that interfacial turbulence caused a fourfold increase in the rate of mass transfer of nitric acid from isobutanol to water in a stirred vessel. The nature of interfacial turbulence is discussed in Sec. 5.6.

## 5.6 MASS TRANSFER AT THE INTERFACE BETWEEN PHASES

The simple two-film theory described in the preceding section employs the assumption that the phases are in equilibrium at the interface, i.e., that there is no diffusional resistance at the phase boundary [Eqs. (5.39) and (5.41)]. Significant interfacial resistances are encountered in some systems, however, as in the case of liquids containing surfactants which tend to concentrate at the surface. Furthermore, the diffusion of solute sometimes causes an interfacial turbulence which is not related to the turbulence in the bulk of the flowing fluid. This last tends to increase the rate of transfer and is equivalent to a negative interfacial resistance. These effects have received much study in recent years, but their quantitative treatment is not yet possible.

### The Limiting Transport Rate to a Gas

Consider a system in which liquid water is in contact with pure water vapor, the two phases being in equilibrium. There is no net transport from one phase to the other, and the gas pressure is equal to the vapor pressure of water at the temperature of the liquid surface. The gas molecules are moving at high speed and some of those which strike the liquid "stick" and are incorporated in the liquid. Since equilibrium prevails, evaporation must occur at the same rate. The rate of collision of the gas molecules with the surface is readily calculated from kinetic theory, and some fraction $\alpha$ of these remains in the liquid while the fraction $1 - \alpha$ rebounds to the gas. Assuming that the rate of evaporation is not influenced by the presence of the gas, one may conclude that the rate of vaporization into an absolute vacuum must be $\alpha$ times the rate of collision with the liquid from the gas saturated at the surface temperature. Evidently evaporation into an absolute vacuum must proceed at a finite rate.

This reasoning led Hertz [70], Knudsen [85], and Langmuir [90] to the following expression for the maximum possible rate of transport from a surface to a gas:

$$N_A = 1,006\alpha(2\pi MRT_S)^{-1/2}(p_S - p_G)$$
$$= 44.3\alpha(T_S M)^{-1/2}(p_S - p_G) \qquad \text{g moles/(s)(cm}^2) \qquad (5.42)$$

where $p$ is in atm and $T$ is °K.

This maximum rate is very large in comparison with mass-transfer rates encountered in most industrial equipment. If $\alpha$ is unity, a free-water surface at 20°C evaporating into an absolute vacuum would retreat at the rate of 2.6 mm/s. Only in high-vacuum and space technology is such a flux approached.

Schrage [142] was evidently the first to point out the possible importance of this phenomena in chemical engineering. Even if the flux is small, some of the overall potential must be used in accomplishing the phase change, i.e., the gas immediately in contact with the surface is not in equilibrium with it. The interfacial resistance, $1/k_i$ is

$$\frac{1}{k_i} = \frac{(2\pi MRT_S)^{1/2}}{1,006\alpha} = 0.0225\frac{(MT_S)^{1/2}}{\alpha} \qquad (5.43)$$

where $k_i$ is the mass-transfer coefficient for transport across the interface, in units of gram moles per second per square centimeter per atmosphere. In the case of the water surface at 20°C, with $\alpha = 1$, $k_i$ is 0.612 g mole/(s)(cm$^2$)(atm), or 14,700 cm/s. Schrage has shown, as have Delaney and Eagleton [43], that even such small resistances may amount to several percent of the total resistance to transport in gas-liquid contacting equipment if the transport rate is large, as in vacuo. The same type of resistance is encountered at any phase boundary, but in liquid-liquid contact it is negligible in comparison with the much larger resistances usually found in such systems.

The possible importance of the interfacial barrier described by Eq. (5.43) clearly depends on the magnitude of the other resistances in series. It is generally quite negligible if the mass-transfer rate is small. In an elegant study of mass transfer into quiescent liquids, Ward and Brooks [176] used an interferometer technique to measure the change of solute concentration as a function of distance from a phase boundary. With benzoic acid dissolving in water, a short extrapolation of the concentration curves gave an interfacial concentration of $0.02796N$; the separately measured static saturation value was $0.0280N$. The liquid at the surface was evidently in equilibrium with the solid, within the precision of these excellent data. The mass-transfer rate was low, however, and the diffusional resistance of the stagnant water was relatively large.

Numerous workers have studied gas absorption into small cylindrical jets of water (e.g., Cullen and Davidson [32]). In the large majority of the most careful studies, the results checked closely with the penetration theory with no allowance for a surface resistance, though the mass-transfer rates were large. Using a short water jet, Cullen and Davidson absorbed $CO_2$ at rates of more than $0.0017$ g/(s) (cm$^2$), or $0.28$ lb moles/(h)(ft$^2$), which are greater than found in most industrial equipment, yet observed no evidence of an interfacial barrier. The limiting rate given by the Hertz-Knudsen equation becomes important in practice only when the transfer rates are exceptionally high.

The practical application of Eq. (5.42) requires values of the evaporation coefficient, $\alpha$ (also called the "sticking," or "accommodation," coefficient). Not only is there no useful theory to employ in predicting $\alpha$, there is also no easy way to experimentally measure it. Experimental determination of $\alpha$ requires the measurement of the surface temperature, usually by means of thermocouples or thermistors. This leads to errors, since the temperature gradient at the surface can be very steep; Littlewood and Rideal [99] question the validity of most of the values of $\alpha$ reported because of questionable surface temperatures. Published values of $\alpha$ for liquids range from $1.0$ to $0.02$, or even lower, and values as low as $10^{-9}$ have been reported for solids. Delaney, Houston, and Eagleton [44] report values of $\alpha$ for water of $0.042$ at $0°C$ and $0.027$ at $43°C$. Maa [109], using a laminar jet and an ingenious method not requiring a probe to measure the surface temperature, obtained $\alpha \approx 1.0$ for water. It is conceivable that most of the published data are in error, and that $\alpha$ is essentially unity for all simple liquids.

In the case of solids, the molecules must disengage from the crystalline or other molecular structure, and this process introduces the equivalent of an additional resistance. This may explain the very low values of $\alpha$ reported for some solids, when data are interpreted by the use of Eq. (5.42). Bennett and Lewis [12] measured the rate of dissolution in mercury of rotating cylinders of different metals. The rate of dissolution of lead agreed with a correlation of data on several solids, including cast benzoic acid dissolving in water, but the rate for zinc was some 40 percent lower. This result may possibly be explained on the basis of a finite resistance to the dissociation of the zinc crystals to form mobile molecules. The rate of dissolution of zinc did, however, increase substantially with rotational speed of the cylinder, so that molecular disengagement resistance could not have been dominant. Konak [86a] notes that the rate of dissolution of inorganic crystals is usually increased by stirring, that crystals dissolve with a rounding of corners and edges,

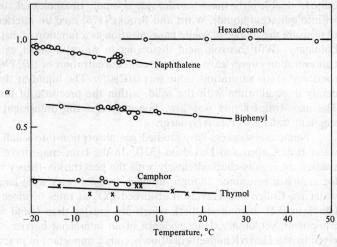

FIGURE 5.9
Evaporation coefficients for subliming solids.

and that inpurities which retard crystal growth do not necessarily affect the rate of dissolution. He concludes that surface reaction-controlled dissolution of crystals is very unlikely.

It is well known that crystals grow from solutions at different rates on different crystal faces, though the ambient solution is the same. Furthermore, crystal growth rates are sometimes very low even when the mass-transfer coefficient is large. This is illustrated, for example, by the data of Cartier, Pindzola, and Bruins [24] on the growth of single crystals of citric and itaconic acids with such high rates of flow of solution that the resistance to mass transfer from solution to crystal face had been essentially eliminated. Clontz, Johnson, McCabe, and Rousseau [27] report that in the crystallization of magnesium sulfate heptahydrate the "incorporation resistance" was large at the growing 110 face but small at the 111 face.

Johannes [148] reports values of $\alpha$ for several solids, obtained by measurements of the rates of sublimation in vacuo. His study is unique in that he measured the temperature of the subliming surface by the user of a radiometer, so avoiding the errors which Littlewood and Rideal believe to be so serious in most of the published studies. Johannes' data are shown in Fig. 5.9. A large collection of published values of $\alpha$ for liquids and solids is tabulated by Paul [121].

**Interfacial turbulence**   The spontaneous emulsification of petroleum oils in water, which occurs without agitation of the two phases, has been known for many years. The role of this and related phenomena in mass transfer near a liquid phase boundary, however, was not brought to the attention of chemical engineers until about 1950. Lewis [93] and Lewis and Pratt [95] noted ripples, erratic pulsations, and

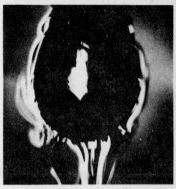

FIGURE 5.10
Schlieren photograph of activity at the
surface of a drop of water in contact with
unstirred ethyl formate. (*Austin, Ying,
and Sawistowski; photograph courtesy of
Dr. H. Sawistowski.*)

Ethyl acetoacetate - Water

surface motion of drops in the course of measurements of interfacial tension by the
pendant-drop method. Lewis observed abnormally high mass-transfer rates where
marked "interfacial turbulence" was seen.

Surface ripples caused by mass transfer are easily observed if a drop of acetone
is added to the edge of a small pool of water on a watch glass. The irregular
rapid rippling of the surface dies out as the acetone diffuses into the water and
concentrations become equalized. Photographs of drops formed within a second
immiscible liquid and horizontal views of liquid-liquid interfaces often show regions
of violent activity and even tiny drops in the region very close to the phase boundary. Figures 5.10 and 5.11 are examples; other similar pictures have been published
[87, 151, 162, 59, 13]. Within seconds after contact of the phases "the interface
starts teeming with activity as though it were a living organism" [120]. It is clear
that the turbulence noted is not that induced by motion of the bulk fluid, as are
the eddies near the interface described in Sec. 5.2. It is also evident that these
phenomena can be expected to increase the mass-transfer flux over that found when
the region near the interface is stagnant or in laminar flow; this has been demonstrated experimentally by numerous investigators.

Surface ripples and interfacial turbulence have been observed at liquid interfaces in contact with gases as well as at points of contact of two liquids. It
appears that these phenomena are always associated with simultaneous mass transfer,
and the effects are more pronounced when the mass transfer is rapid. They are most
common in ternary or multicomponent systems but also noted in some partially
miscible binary systems [7]. In some cases the appearance of tiny drops can be
explained as due to the "salting-out" of the solution of one liquid in a solute-
rich binary, the solubility being exceeded as the dissolved liquid diffuses into a layer
of lower solute concentration [40]. In other cases small drops have been seen to
be ejected violently from one liquid into the other. In one spectacular example,
Wei [151] observed a drop of benzene containing acetic acid, which was rising
slowly in a column of ammonia in water, to jump suddenly sidewise, while at the

**FIGURE 5.11**
Interferometer photographs of the development of interfacial turbulence after contact of $CO_2$ with the surface of a solution of 0.046 g moles/1 monoethanolamine in water. (*Courtesy of Prof. J. C. Berg.*)

same time releasing a tiny satellite drop. The surface eruptions causing drops to "kick" have been studied by Haydon [68]. Interfacial turbulence has been reported where an amalgam is in contact with an aqueous electrolyte and a solute transferred from metal to aqueous phase electrochemically [20].

In some cases the surface activity is strong with mass transfer in one direction but completely absent when the solute diffuses in the opposite direction. The most pronounced interfacial turbulence is observed when a chemical reaction occurs simultaneously with mass transfer, as in the extraction of acetic acid from *i*-butyl alcohol by water containing ammonia [151]. Surface-active, or "wetting," agents which tend to concentrate at the phase boundary greatly reduce or even prevent the development of interfacial turbulence. A number of cases have been reported where interfacial turbulence increases the mass-transfer rate severalfold.

The phenomena which have been described are by no means well understood. They evidently stem from the random variations of interfacial tension at the interface which result from local concentration variations as mass transfer occurs, and so depend in part on the rate of change of the interfacial tension, or surface pressure, with solute concentration. The instability which develops causes ripples and sometimes regularly shaped roll cells which create circulation between the surface and the bulk liquid. This is known as the Marangoni effect [15]. The significant feature of the phenomenon is the hydrodynamic instability of the surface motion, which continues as long as mass transfer takes place [160]. Evidence that an electric field may induce interfacial turbulence is reported by Austin, Banczyk, and Sawistowski [8]. In studies of mass transfer between water and isobutanol, they found applied voltages to greatly enhance the mass-transfer coefficient in the organic

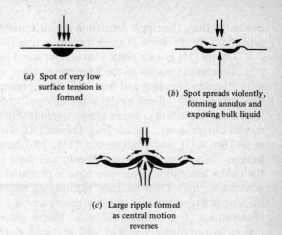

(a) Spot of very low
surface tension is
formed

(b) Spot spreads violently,
forming annulus and
exposing bulk liquid

(c) Large ripple formed
as central motion
reverses

FIGURE 5.12
Interfacial turbulence at driving forces greater than the critical driving force. (*After Ellis and Biddulph.*)

phase: and they speculate that local variations in charge density lead to surface tension variations and so to Marangoni instability. A review of the effects of electrical fields on fluid motion (electrohydrodynamics), including instabilities at interfaces, is to be found in Ref. 110.

The manner in which the Marangoni effect operates to develop ripples is well described by Ellis and Biddulph [49]. These authors measured the amplitude of the ripples formed on the surface of water as acetone was being absorbed from air. Rapid absorption at a small spot on the surface causes a reduction in $\sigma$, the interfacial tension, and the surface spreads radially. At low mass-transfer rates the change in $\sigma$ and $d\sigma/dr$ is small; the spot spreads slowly as the surface is depleted of acetone by diffusion, with a slight rippling which dies away. In the words of Ellis and Biddulph:

As the driving force and the amount of preferential mass transfer increases, so also the surface tension difference increases, and greater rippling results. At some critical driving force, the rate of mass transfer is sufficient to instantaneously create points of very steep surface tension gradient. The resulting spreading, as shown in Fig. 5.12, is so rapid that the momentum of the spreading liquid is sufficient to break the center of the point source, and expose subjacent liquid drawn from below the surface. There now exists an expanding annulus of low surface tension liquid, with a center of high surface tension. This increased surface tension at the center tends to reverse the spreading motion and so liquid flows from the bulk, and from the spreading film, to the center of the annulus. Before the momentum of these streams is destroyed they cause a build-up of liquid into a ripple.

If the surface tension of the binary mixture is reversed, that is, mass transfer causes an increase in surface tension, then a localized point of preferential mass transfer has a higher surface tension than its surrounding interface liquid. This point shows no tendency to spread and may even contract slightly. Therefore, since no motion has been initiated there is no ripple formation and the interface remains steady.

Evidently, then, the ripple formation which enhances the rate of mass transfer will depend on the direction of mass transfer across the interface.

Davies [35] gives a somewhat similar word picture of the phenomenon.

Theoretical studies of the Marangoni effect were presented by Pearson [122] in 1958 and by Sternling and Scriven [160] in pioneering contributions. A simplified two-dimensional roll-cell model was employed to develop a quantitative theory for the onset of instability, which agrees qualitatively with various subsequent experimental observations. Levich [91], Davies [40], Gross and Hixson [62], Lightfoot et al. [101, 175], and Ruckenstein [138, 140] have also made theoretical contributions. More recently, Brian, Smith, and Ross [17, 18, 19] have suggested that the Gibbs adsorption layer may have a profound stabilizing influence on Marangoni convection. Their analyses incorporate the effect of the Gibbs adsorption in the earlier hydrodynamic stability theory and are more in line with experimental observations. It is evident, however, that a great deal needs to be done before interfacial turbulence becomes well understood and the theory developed to the point where it is useful in engineering design.

A regular pattern of roll cells would be expected to develop on a liquid surface as the spots of low interfacial tension spread and meet. Such cells are analogous to the Bénard cells seen when a thin layer of water is heated from below. They have been photographed in various mass-transfer studies but seldom found to be regular. Figure 5.13, one of the best of such pictures, is from the excellent investigation by Orell and Westwater [120]. The polygons have three to seven sides, and their edges are fringed with tiny droplets. The cells disappear in the later stages of the transient mass-transfer process, being replaced by rippled areas of irregular geometry.

Davies and Rideal [40] state that "no system has yet been found where, under close observation, spontaneous emulsification did not form during the transfer of a third component." No effect of either enhancement or depression of the mass-transfer rate is observed, however, in gas-liquid or liquid-liquid contacting for very short times with clean surfaces. Numerous workers employing laminar jets have checked the penetration theory almost exactly without making any allowance for a surface effect, for both gas absorption and for transfer between the jet and a surrounding liquid [32, 177, 46, 145, 77, 130, 127]. Similar results have been obtained with longer contact times in a wetted-wall column [103]. The agreement with theory is so good, in fact, that the short laminar jet has been used to measure molecular diffusion coefficients in water [46].

The general subject of surface-tension-driven phenomena, including the role of the Marangoni effect on mass transfer at a phase boundary, is the subject of an excellent review by Levich and Krylov [92].

**Surfactants** Many substances in solution tend to concentrate at the liquid surface and change the interfacial or surface tension. These surface-active agents ("surfactants," or "wetting agents") cause marked changes in $\sigma$ when present in extremely small amounts. Even a monolayer on the surface develops a structure which tends to immobilize the surface, reducing or eliminating fine-scale surface

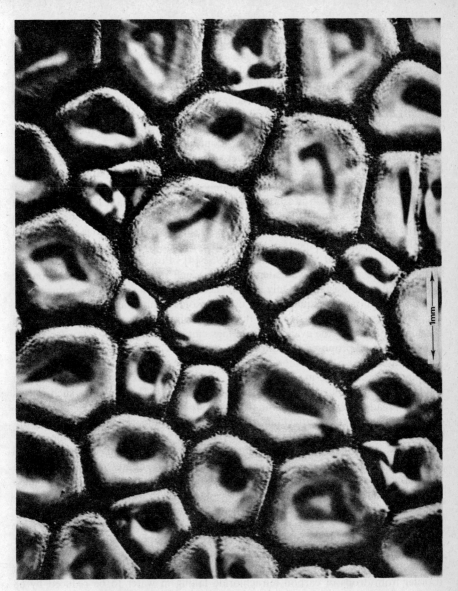

FIGURE 5.13
Roll cells at an interface 12 min after contacting ethyl acetate with ethylene glycol containing 10 weight percent acetic acid. (*Orell and Westwater; photograph courtesy of Prof. J. W. Westwater.*)

motion. The presence of such a surface layer has two important effects on the rate of mass transfer through the surface: it reduces and often eliminates the Marangoni effect while at the same time introducing a surface resistance to diffusion across the interface. The reduction in mass-transfer rate can be large.

A very simple experiment will indicate the magnitude of the effect. Fill two identical litre breakers with distilled water. Add a tiny pinch of hexadecanol (cetyl alcohol) to one and set the two aside for a week or more. By the time evaporation has reduced the level of the pure water by 4 in, the level in the beaker with the hexadecanol will have dropped only about an inch.

The use of hexadecanol (many other surfactants behave similarly) to conserve water supplies by adding it to reservoirs is an obvious technical application. This has been studied extensively but is not widely practiced. Wind and waves destroy the surface film and wash it ashore, so that additional surfactant must be added at frequent intervals. Some of the problems encountered in practice are described in a report to the American Water Works Association [3]; see also Refs. 133, 21, and 45. The more scientific aspects of evaporation control by monolayers are discussed in papers presented at a symposium in 1962 [4]. Possibly one of the difficulties is the increase in surface-water temperature which partially compensates for the greater transport resistance. Only limited success has been achieved in reducing the evaporation of organic liquids by the use of insoluble monolayers [14].

Surfactants greatly reduce the rate of vaporization of very small drops, and their use in reducing the evaporation of fogs employed to protect crops from frost damage has been proposed [5]. Surfactants reduce the size of dispersed gas bubbles, the increased area possibly more than offsetting the added transport barrier [188]. They might be effective in improving the absorption of oxygen or other gases in bubble columns, as in the activated sludge process [187] (providing such agents are not already present in substantial amounts).

The role played by surfactants in laboratory and industrial mass-transfer equipment has received much study in recent years and is described well in the review by Davies and Rideal [40]. The principal effects of surfactants are those noted earlier: the formation of a relatively rigid interface, with suppression or elimination of interfacial turbulence [36], and the introduction of a surface barrier to mass transfer. Surfactants are known to inhibit crystal growth from supersaturated solutions. The reduction of surface tension reduces liquid holdup in packings [155]. In bubbles the effect is to reduce the liquid flow over the surface of a rising bubble, with the result that the velocity of free rise is appreciably reduced. In the case of drops the internal circulation is reduced or eliminated, and small drops behave as rigid spheres. The early work of Lindland and Terjesen [98] showed mass transfer from drops of carbon tetrachloride falling in water to be reduced by 68 percent with the addition of only $6 \times 10^{-5}$ g surfactant per 100 ml water. Similar reductions of the rate of mass transfer by the addition of surfactants are reported by Thompson [164] who studied the absorption of $NH_3$, $SO_2$, and $CO_2$ into water in an unstirred container. Others have found the rate in various systems to reach or pass through a minimum as the surfactant concentration was increased [40, 39].

At short contact times the surfactant does not diffuse to the surface and build an absorbed layer. Consequently, agitated systems with rapid surface renewal at the phase boundary often show little or no effect of added surfactant (though Goodridge and Bricknell [56] found $k_c$ to be substantially reduced by the addition of long-chain alcohols with gas absorption in a stirred vessel at high rpm). Very small concentrations of surfactants improve wetting and eliminate rippling in wetted-wall columns and in film flow over spheres, without introducing a significant surface-diffusion barrier. Liquid jets and very short wetted-wall columns sometimes show a buildup of immobilized surfactant near the liquid outlet at the bottom. It seems unlikely that added surfactants influence the rates of mass transfer in packed gas absorbers, though packed towers used as liquid-liquid extractors may have problems due to emulsification.

At the present time it seems futile to attempt further generalizations. The complex interactions of hydrodynamics, surface chemistry, and diffusion result in phenomena which are extremely difficult to analyze. The student interested in the challenge will find Davies' review helpful. Some of the more interesting of the relevant articles appearing in recent years include Refs. 67, 56, 60, 125, 41, 113, 57, 114, 10, and 92, though there are dozens of others.

If the value of the interfacial resistance is a known constant, the concept of additivity of resistances would suggest that the quantity $1/k_i$ be added to the right-hand side of Eq. (5.41) for application to cases of steady-state mass transfer between two phases.

For transient diffusion from a region of constant composition into a stagnant fluid of infinite depth, the allowance for a constant resistance $1/k_i$ at $y = 0$ modifies Eq. (3.46), leading to [33, 67]

$$\frac{c(y, t) - c_0}{c_i - c_0} = \operatorname{erfc} \frac{y}{2\sqrt{D_{AB}t}} - \exp\left(\frac{k_i y}{D_{AB}} + \frac{k_i^2}{D_{AB}}t\right) \times \operatorname{erfc}\left(\frac{y}{2\sqrt{D_{AB}t}} + k_i\sqrt{\frac{t}{D_{AB}}}\right) \quad (5.44)$$

The instantaneous molar flux $N_A$ is obtained from this result after differentiating to obtain $\partial c/\partial y$ at $y = 0$. The total moles of $A$ transferred in time $t$ is then found as $\int_0^t N_A \, dt$. If the initial concentration $c_0$ of $A$ is zero, the result is

$$\int_0^t N_A \, dt = 2c_i\sqrt{\frac{D_{AB}t}{\pi}} + \frac{c_i D_{AB}}{k_i}\left[\operatorname{erfc}\left(k_i\sqrt{\frac{t}{D_{AB}}}\right) \times \exp\left(\frac{k_i^2 t}{D_{AB}}\right) - 1\right] \quad (5.45)$$

In these equations $c_i$ is the concentration in the second phase corresponding to equilibrium with the first phase of constant composition from which the solute $A$ is diffusing.

Equations (5.44) and (5.45) may be compared with the corresponding Eqs. (3.42) and (3.45) for the same situation with no interfacial resistance ($k_i$ infinite). Duda and Vrentas [47] solve the same case with $k_i$ stipulated to be time-variant. The solution for transient diffusion between two cells of infinite extent with no interfacial barrier is given by Crank [31]. The solution for the same case with allowance for an interfacial resistance is given in Ref. 41. Tung and Drickamer [168] solve this case for cells of finite depth normal to the diffusion path.

# REFERENCES

1 ABRAMZON, A. A., and M. V. OSTROVSKII: *Zh. Prikl. Khim.*, **36** (4): 789, 793 (1963).

1a ACOSTA, R. E.: Ph.D. thesis in chemical engineering, Univ. Calif., Berkeley, November 1973.

2 ANGELO, J. B., E. N. LIGHTFOOT, and D. W. HOWARD: *AIChE J.*, **12**: 751 (1966).

3 ANONYMOUS: Survey of Methods of Evaporation Control, *J. Amer. Water Works Assoc. Task Group* Rep. **53**, no. 2, p. 157 (1963).

4 ANONYMOUS: In V. K. Lamer (ed.), "Retardation of Evaporation by Monolayers," Academic, New York, 1962.

5 ANONYMOUS: *Chem. Eng. News*, p. 20, September 12, 1966.

6 ASANO, K., and S. FUJITA: *Chem. Eng. Sci.*, **26**: 1187 (1971).

7 AUSTIN, L. J., W. E. YING, and H. SAWISTOWSKI: *Chem. Eng. Sci.*, **21**: 1109 (1966).

8 AUSTIN, L. J., L. BANCZYK, and H. SAWISTOWSKI: *Chem. Eng. Sci.*, **26**: 2120 (1971).

9 BAKKER, C. A. P., P. M. VAN BUYTENEN, and W. J. BEEK: *Chem. Eng. Sci.*, **21**: 1039 (1966).

10 BAKKER, C. A. P., and F. H. FENTENER VAN VLISSINGEN: *Chem. Eng. Sci.*, **22**: 1349 (1967).

11 BEDINGFIELD, C. H., and T. B. DREW: *Ind. Eng. Chem.*, **42**: 1164 (1950).

12 BENNETT, J. A. R., and J. B. LEWIS: *U. K. Atomic Energy Authority* Rep. A.E.R.E. CE/R 1998, 1957.

13 BERG, J. C., and C. R. MORIG: *Chem. Eng. Sci.*, **24**: 937 (1969).

14 BERNETT, M. K., L. A. HALPER, N. L. JARVIS, and T. M. THOMAS: *Ind. Eng. Chem. Fundam.*, **8**: 150 (1970).

15 BIKERMAN, J. J.: "Surface Chemistry," p. 81, Academic, New York, 1948.

16 BIRD, R. B., W. E. STEWART, and E. N. LIGHTFOOT: "Transport Phenomena," p. 674, Wiley, New York, 1960.

17 BRIAN, P. L. T.: *AIChE J.*, **17**: 765 (1971).

18 BRIAN, P. L. T., and J. R. ROSS: *AIChE J.*, **18**: 582 (1972).

19 BRIAN, P. L. T., and K. A. SMITH: *AIChE J.*, **18**: 231 (1972).

20 BRIMACOMBE, J. K., A. D. GRAVES, and D. INMAN: *Chem. Eng. Sci.*, **25**: 1817 (1970).

21 CADENHEAD, D. A.: *Ind. Eng. Chem.*, **61** (4): 22 (1969).

22 CAIRNS, R. C., and G. H. ROPER: *Chem. Eng. Sci.*, **3**: 97 (1954).

23 CARSLAW, H. S., and J. C. JAEGER: "Conduction of Heat in Solids," 2d ed., Oxford, New York, 1959.

24 CARTIER, R., D. PINDZOLA, and P. E. BRUINS: *Ind. Eng. Chem.*, **51**: 1409 (1959).

25 CERMAK, J. O., and R. B. BECKMANN: *AIChE J.*, **15**: 250 (1969).

26 CHILTON, T. H., and COLBURN, A. P.: *Ind. Eng. Chem.*, **26**: 1183 (1934).

27 CLONTZ, N. A., R. T. JOHNSON, W. L. MCCABE, and R. W. ROUSSEAU: *Ind. Eng. Chem. Fundam.*, **11**: 368 (1972).

28 COLBURN, A. P., and T. B. DREW: *Trans. AIChE* **33**: 197 (1937).

29 COLLELO, R. G., and G. S. SPRINGER: *Int. J. Heat Mass Transfer*, **9**: 1391 (1966).

30 CORINO, E. R., and R. S. BRODKEY: *J. Fluid Mech.*, **37** (1): 1 (1969).

31 CRANK, J.: "The Mathematics of Diffusion," Oxford, New York, 1956.

32 CULLEN, E. J., and J. F. DAVIDSON: *Trans. Faraday Soc.*, **53**: 113 (1957).

33 DANCKWERTS, P. V.: *Research*, **2**: 494 (1949).

34 DANCKWERTS, P. V.: *Ind. Eng. Chem.*, **43**: 1460 (1951).

35 DAVIES, J. T.: *Proc. Roy. Soc. (London)*, **290A**: 515 (1966).

36 DAVIES, J. T.: *AIChE J.*, **18**: 169 (1972).

37 DAVIES, J. T.: "Turbulence Phenomena," Academic, New York, 1972.

38 DAVIES, J. T., and W. KHAN: *Chem. Eng. Sci.*, **20**: 713 (1965).

39 DAVIES, J. T., and G. R. A. MAYERS: *Chem. Eng. Sci.*, **16**: 55 (1961).

40 DAVIES, J. T., and E. K. RIDEAL: "Interfacial Phenomena," chap. 7, Academic, New York, 1961. The same material appeared in "Advances in Chemical Engineering," vol. 4, p. 1, Academic, New York, 1963.

41 DAVIES, J. T., and J. B. WIGGILL: *Proc. Roy. Soc. (London)*, **255A**: 277 (1960).

42 DEISSLER, R. G.: Nat. Adv. Comm. Aeronaut. Rep. 1210 (1955).

43 DELANEY, L. V., and L. C. EAGLETON: *AIChE J.*, **8**: 418 (1962).

44 DELANEY, L. V., R. W. HOUSTON, and L. C. EAGLETON: *Chem. Eng. Sci.*, **19**: 105 (1964).

45 DRESSLER, R. G.: *Ind. Eng. Chem.*, **56** (7): 36 (1964).

46 DUDA, J. L., and J. S. VRENTAS.: *AIChE J.*, **14**: 286 (1968).

47 DUDA, J. L., and J. S. VRENTAS: *Chem. Eng. Sci.*, **22**: 27 (1967).

48 EISENBERG, M., C. W. TOBIAS, and C. R. WILKE: *Chem. Eng. Progr.* Symp. Ser. **51** (16): 1 (1955).

49 ELLIS, S. R. M., and M. BIDDULPH: *Chem. Eng. Sci.*, **21**: 1107 (1966).

50 FAGE, A., and H. C. H. TOWNEND: *Proc. Roy. Soc. (London)*, **A135**: 656 (1932).

51 FALLAT, R. J.: *Univ. Calif. Lawrence Radiat Lab. Rep.* UCRL-8527, March 1959.

52 FRANKTISAK, F. A. PALADE DE IRIBARNE, J. W. SMITH, and R. L. HUMMEL: *Ind. Eng. Chem. Fundam.*, **8**: 160 (1969).

53 FRIEND, W. L., and A. B. METZNER: *AIChE J.*, **4**: 393 (1958).

54 GILLILAND E. R., and T. K. SHERWOOD: *Ind. Eng. Chem.*, **26**: 516 (1934).

55 GOODGAME, T. H., and T. K. SHERWOOD: *Chem. Eng. Sci.*, **3**: 37 (1954).

56 GOODRIDGE, F., and D. J. BRICKNELL: *Trans. Inst. Chem. Eng. (London)*, **40**: 54 (1962).

57 GOODRIDGE, F., and I. D. ROBB: *Ind. Eng. Chem. Fundam.*, **4**: 49 (1965).

58 GORDON, K. F., and T. K. SHERWOOD: *Chem. Eng. Progr. Symp. Ser.*, **50** (10): 15 (1954).

59 GORE, R. W.: Paper presented at the 54th Annual Meeting of the AIChE, New York, December 1961.

60 GHOSH, D. N., and D. D. PERLMUTTER: *AIChE J.*, **9**: 474 (1963).

61 GOWARIKER, V. R., and F. H. GARNER: *U. K. Atomic Energy Auth. Rep.* AERE-R4197, 1962.

62 GROSS, B., and A. N. HIXON: *Ind. Eng. Chem. Fundam.*, **8**: 288 (1969).

63 HANN, O. T., and O. C. SANDALL: *AIChE J.*, **18**: 527 (1972).

64 HANRATTY, T. J.: *AIChE J.*, **2**: 359 (1956).

65 HARRIOTT, P.: *Chem. Eng. Sci.*, **17**: 149 (1962).

66 HARRIOTT, P., and R. M. HAMILTON: *Chem. Eng. Sci.*, **20**: 1073 (1965).

67 HARVEY, E. A., and W. SMITH: *Chem. Eng. Sci.*, **10**: 274 (1959).

68 HAYDON, D. A.: *Proc. Roy. Soc. (London)*, **A243**: 483 (1958).

69 HENLEY, E. J., and J. M. PRAUSNITZ: *AIChE J.*, **8**: 133 (1962).

70 HERTZ, H.: *Ann. Phys.* **17**: 177 (1882).

71 HIGBIE, R.: *Trans. AIChE*, **31**: 365 (1935).

72 HUBBARD, D. W., and E. N. LIGHTFOOT: *Ind. Eng. Chem. Fundam.*, **5**: 370 (1966).

73 HUBBARD, D. W.: *AIChE J.*, **14**: 354 (1968).

74 HUGHMARK, G. A.: *AIChE J.*, **14**: 352 (1968).

75 HUTCHINSON, M. H., and T. K. SHERWOOD: *Ind. Eng. Chem.*, **29**: 836 (1937).

76 JENKINS, R.: "Heat Transfer and Fluid Mechanics Institute," Stanford, Stanford, Calif. 1951.

77 KIMURA, S., and T. MIYAUCHI: *Chem. Eng. Sci.*, **21**: 1057 (1966).

78 KING, C. J.: *AIChE J.*, **10**: 671 (1964).

79 KING, C. J.: *Ind. Eng. Chem. Fundam.*, **5**: 1 (1966).

80 KISHINEVSKY, M. KH., and V. T. SEREBRYANSKI: *J. Appl. Chem. (USSR)*, **29**: 43 (1956).

81 KISHINEVSKY, M. KH.: *Int. J. Heat Mass Transfer*, **8**: 1181 (1965).

82 KLINE, S. J.: In L. S. Dzung (ed.), "Flow Research on Blading," p. 372, Elsevier, Amsterdam, 1969.

83  KLINE, S. J.: Personal communication, April 1970.

84  KLINE, S. J., and P. W. RUNSTADLER: *Trans. ASME, J. Appl. Mech.,* **26E:** 166 (1959).

85  KNUDSEN, M.: *Ann. Phys.,* **47:** 697 (1915).

86  KNUDSEN, J. G., and D. L. KATZ: "Fluid Dynamics and Heat Transfer," McGraw, New York, 1958.

86a  KONAK, A. R.: *Chem. Eng. Sci.,* **29:** 1785 (1974).

87  KROEPELIN, H., H. J. NEUMANN, and E. PRÖTT: *5th World Petrol. Congr. Sec. III Paper 24* (1959).

88  LAMB, W. B., T. G. SPRINGER, and R. L. PIGFORD: *Ind. Eng. Chem. Fundam.,* **8:** 823 (1969).

89  LAMONT, J. C., and D. S. SCOTT: *AIChE J.,* **16:** 513 (1970).

90  LANGMUIR, I.: *Phys. Rev.* **2:** 329 (1913).

91  LEVICH, V. G.: "Physicochemical Hydrodynamics," Prentice-Hall, Englewood Cliffs, N.J., 1962.

92  LEVICH, V. G., and V. S. KRYLOV: In W. R. Sears and M. Van Dyke (eds.), "Annual Review of Fluid Mechanics," vol. 1, p. 293, Annual Reviews, Inc., Palo Alto, Calif., 1969.

93  LEWIS, J. B.: *Trans. Inst. Chem. Eng. (London),* **31:** 323, 325, (1953).

94  LEWIS, J. B.: *Chem. Eng. Sci.,* **3:** 248, 260 (1954).

95  LEWIS, J. B., and H. C. R. PRATT: *Nature,* **171:** 1155 (1953).

96  LEWIS, W. K., and W. G. WHITMAN: *Ind. Eng. Chem.,* **16:** 1215 (1924).

97  LIN, C. S., R. W. MOULTON, and G. L. PUTNAM: *Ind. Eng. Chem.,* **45:** 636, 640 (1953).

98  LINDLAND, K. P., and S. G. TERJESEN: *Chem. Eng. Sci.,* **5:** 1 (1956).

99  LITTLEWOOD, R., and E. RIDEAL: *Trans. Faraday Soc.,* **52:** 1598 (1956).

100  LOITSIANSKY, L. H.: *Appl. Math. Mech.,* **24:** 950 (1960).

101  LUDVIKSSON, V., and E. A. LIGHTFOOT: *AIChE J.,* **14:** 620 (1968).

102  LYKOUDIS, P. S., and Y. S. TOULOUKIAN: *Trans. ASME,* **80:** 653 (1958).

103  LYNN, S., J. R. STRAATEMEIER, and H. KRAMERS: *Chem. Eng. Sci.,* **4:** 49, 58, 63 (1955).

104  MCADAMS, W. H.: "Heat Transmission," 2d ed., McGraw-Hill, New York, 1942.

105  MCMANAMAY, W. J.: *Chem. Eng. Sci.,* **15:** 251 (1961).

106  MCMANAMEY, W. J., J. T. DAVIES, J. M. WOOLLEN, and J. R. COE: *Chem. Eng. Sci.,* **28:** 1061 (1973).

107  MARTINELLI, R. C.: *Trans. ASME,* **69:** 947 (1947).

108  MARCHELLO, J. M., and H. L. TOOR: *Ind. Eng. Chem. Fundam.,* **2:** 8 (1963).

109  MAA, J. R.: *Ind. Eng. Chem. Fundam.,* **6:** 504 (1967); **9:** 283 (1970).

110  MELCHER, J. R., and G. I. TAYLOR: In W. R. Sears and M. Van Dyke "Annual Review of Fluid Mechanics," vol. 1, p. 111, Annual Reviews, Inc., Palo Alto, Calif., 1969.

111  MEYERINK, E. S. C., and S. K. FRIEDLANDER: *Chem. Eng.,* **17:** 121 (1962).

112  MICKLEY, H. S., R. C. ROSS, A. L. SQUYRES, and W. E. STEWART: *NACA Tech. Note 3208* (1954).

113  MIREV, D., D. ELENKOV, and C. BALAREV: *Compt. Rend. Acad. Bulgare Sci.,* **14** (4): p. 349 (1961).

114  MUDGE, L. K., and W. J. HEIDEGER: *AIChE J.,* **16:** 602 (1970).

115  MURPHREE, E. V.: *Ind. Eng. Chem.,* **24:** 726 (1932).

116  NEDDERMAN, R. M.: *Chem. Eng. Sci.,* **16:** 120 (1961).

117  NERNST, W.: *Z. Phys. Chem.,* **47:** 52 (1904).

117a  NEWMAN, J. S.: "Electrochemical Systems," Prentice-Hall, Englewood Cliffs, N.J., 1973.

118  NOTTER, R. H., and C. A. SLEICHER: *Chem. Eng. Sci.,* **26:** 161 (1971).

118a  OFFEN, G. R., and KLINE, S. J.: Thermosciences Div., Dept. Mech. Eng., Stanford Univ. *Rep. MD-31.*

119  OLANDER, D. R.: *Chem. Eng. Sci.,* **19:** 67 (1964).

120  ORELL, A., and J. W. WESTWATER: *AIChE J.*, **8**: 350 (1962).

121  PAUL, B.: *J. Amer. Rocket Soc.*, p. 1321, September 1962.

122  PEARSON, J. R. A.: *J. Fluid Mech.*, **4**: 489 (1958).

123  PERLMUTTER, D. D.: *Chem. Eng. Sci.*, **16**: 287 (1961).

124  PHILLIPS, O.: Sc.D. thesis in chemical engineering, M.I.T., 1957.

124a  PINCZEWSKI, W. V., and S. SIDEMAN: *Chem. Eng. Sci.*, **29**: 1969 (1974).

125  PLEVAN, R. E., and J. A. QUINN: *AIChE J.*, **12**: 894 (1966).

126  POPOVICH, A. T., and R. L. HUMMEL: *AIChE J.*, **13**: 854 (1967).

127  QUINN, J. A., and P. G. JEANNIN: *Chem. Eng. Sci.*, **15**: 243 (1961).

128  RANNIE, W. D.: *J. Aeronaut. Sci.*, **23**: 485 (1956).

129  RANZ, W. E., and P. E. DICKSON: *Ind. Eng. Chem. Fundam.*, **4**: 345 (1965).

130  REHM, T. R., A. J. MOLL, and A. L. BABB: *AIChE J.*, **9**: 760 (1963).

131  REICHARDT, H.: *Z. Angew. Math. Mech.*, **31**: 208 (1951).

132  REICHARDT, H.: *Nat. Adv. Comm. Aeronaut., Tech. Memo.*, 1408 (1957).

133  REISER, C. O.: *Ind. Eng. Chem. Process Des. Dev.*, **8**: 63 (1969).

134  REYNOLDS, O.: *Proc. Manchester Lit. Phil. Soc.*, **14**: 7 (1874); reprinted in "Papers on Mechanical and Physical Subjects," vol. 1, p. 81, Cambridge, New York, 1900.

135  RIBAUD, G.: *Compt. Rend.*, **242**: 959 (1956).

136  RUBESIN, M. W.: *Nat. Adv. Comm. Aeronaut. Tech. Note* 2917 (1953).

137  RUCKENSTEIN, E.: *Chem. Eng. Sci.*, **7**: 265 (1958).

138  RUCKENSTEIN, E.: *Chem. Eng. Sci.*, **19**: 505 (1964).

139  RUCKENSTEIN, E.: *Chem. Eng. Sci.*, **22**: 474 (1967).

140  RUCKENSTEIN, E.: *Int. J. Heat Mass Trans.*, **11**: 1753 (1968).

141  RUNSTADLER, P. W., S. J. KLINE, and W. C. REYNOLDS: *Thermosciences Div., Dept. Mech. Eng., Stanford Univ. Rep.* MD-8, 1963.

142  SCHRAGE, R. W.: "A Theoretical Study of Interphase Mass Transfer," Columbia Univ., New York, 1953.

143  SCHRAUB, F. A., and S. J. KLINE: *Thermosciences Div., Dept. Mech. Eng., Stanford Univ. Rep.* MD-12, 1965.

144  SCRIVEN, L. E.: *Chem. Eng. Educa.*, **2**: 150 (Fall 1968),; 26 (Winter 1969),; 94 (Spring 1969).

145  SCRIVEN, L. E., and R. L. PIGFORD: *AIChE J.*, **4**: 439 (1958).

146  SEVERSON, D. E., A. J. MADDEN, and E. L. PIRET: *AIChE J.*, **5**: 413 (1959).

147  SHERWOOD, T. K.: *Chem. Eng. Progr. Symp. Ser.* **55** (25): 71 (1959).

148  SHERWOOD, T. K., and C. JOHANNES: *AIChE J.*, **8**: 590 (1962).

149  SHERWOOD, T. K., and J. M. RYAN: *Chem. Eng. Sci.*, **11**: 81 (1959).

150  SHERWOOD, T. K., K. A. SMITH, and P. E. FOWLES: *Chem. Eng. Sci.*, **23**: 1225 (1968).

151  SHERWOOD, T. K., and J. C. WEI: *Ind. Eng. Chem.*, **49**: 1030 (1957).

152  SHULMAN, H. L., and L. J. DELANEY: *AIChE J.*, **5**: 290 (1959).

153  SHULMAN, H. L., and J. E. MARGOLIS: *AIChE J.*, **3**: 157 (1957).

154  SHULMAN, H. L., and R. G. ROBINSON: *AIChE J.*, **6**: 469 (1960).

155  SHULMAN, H. L., C. F. ULLRICH, N. WELLS, and A. Z. PROULX: *AIChE J.*, **1**: 259 (1955).

156  SLEICHER, C. A.: *ASME, Paper* 57-HT9 (1957).

157  SON, J. S., and T. J. HANRATTY: *AIChE J.*, **13**: 689 (1967).

158  SPALDING, D. B.: *Proc. Inst. Mech. Eng.* (*London*), **168**: 545 (1954); *J. Appl. Mech.* **28**: 455 (1961).

159  SPRINGER, T. G., and R. L. PIGFORD: *Univ. Calif., Berkeley*, UCRL Rep. 18995, November 1969.

160  STERNLING, C. V., and L. E. SCRIVEN: *AIChE J.*, **5**: 514 (1959).

161  SZEKELEY, J.: *Chem. Eng. Sci.*, **20**: 141 (1965).

162  THOMAS, W. J., and E. MCK. NICHOLL: *Chem. Eng. Sci.*, **22:** 1877 (1967).

163  THOMAS, L.: *Chem. Eng. Sci.*, **26:** 1271 (1971).

164  THOMPSON, D. W.: *Ind. Eng. Chem. Fundam.*, **9:** 243 (1970).

165  TIEN, C. L.: *J. Heat Transfer*, **83c:** 389 (1961).

165a  TOOR, H. L.: *AIChE J.*, **6:** 525 (1960).

166  TOOR, H. L., and J. M. MARCHELLO: *AIChE J.*, **4:** 97 (1958).

167  TOWNSEND, A. A.: "The Structure of Turbulent Shear Flow," Cambridge, New York, 1956.

168  TUNG, L. H., and H. G. DRICKAMER: *J. Chem. Phys.*, **19:** 1075 (1951).

169  TYLDESLEY, J. R., and R. S. SILVER: *Int. J. Heat Mass Trans.*, **11:** 1325 (1968).

170  VAN DRIEST, E. R.: "1955 Heat Transfer and Fluid Mechanics Institute," Univ. of Calif. at Los Angeles, 1955.

171  VIETH, W. R., J. H. PORTER, and T. K. SHERWOOD: *Ind. Eng. Chem. Fundam.*, **2:** 1 (1963).

172  VIVIAN, J. E., and W. C. BEHRMANN: *AIChE J.*, **11:** 656 (1965).

173  VIVIAN, J. E., and C. J. KING: *AIChE J.*, **10:** 221 (1964).

174  VON KÁRMÁN, TH.: *Trans. ASME*, **61:** 705 (1939).

175  WANG, K. H., V. LUDVIKSSON, and E. A. LIGHTFOOT: *AIChE J.*, **17:** 1402 (1971).

176  WARD, A. F. H., and L. H. BROOKS: *Trans. Faraday Soc.*, **48:** 1124 (1952).

177  WARD, W. J., and J. A. QUINN: *AIChE J.*, **11:** 1005 (1965).

178  WASAN, D. T., and M. S. AHLUWALIA: *Chem. Eng. Sci.*, **24:** 1535 (1969).

179  WASAN, D. T., C. L. TIEN, and C. R. WILKE: *AIChE J.*, **9:** 567 (1963).

180  WASAN, D. T., and C. R. WILKE: *Int. J. Heat Mass Trans.*, **7:** 87 (1964).

181  WASAN, D. T., and C. R. WILKE: *AIChE J.*, **14:** 577 (1968).

182  WESTKAEMPER, L. E., and R. R. WHITE: *AIChE J.*, **3:** 69 (1957).

183  WHITE, D. A.: *Chem. Eng. Sci.*, **24:** 911 (1969).

184  WHITMAN, W. G.: *Chem. Met. Eng.*, **29:** 146 (1923).

185  WILKE, C. R.: *Proc. Ann. Mtg. Soc. Chem. Eng. (Japan)*, p. 513, April 1965.

186  YOSHIDA, T., and T. HYŌDŌ: *Ind. Eng. Chem. Process Des. Dev.*, **9:** 207 (1970).

187  ZIEMINSKI, S. A., and R. R. LESSARD: *Ind. Eng. Chem. Process Des. Dev.*, **8:** 69 (1969).

188  ZIEMINSKI, S. A., M. M. CARON, and R. B. BLACKMORE: *Ind. Eng. Chem. Fundam.*, **6:** 233 (1967).

# PROBLEMS

*5.1*  Derive Eq. (5.37).

*5.2*  A student proposes a "new theory" of mass transfer from the inner wall of a pipe to a fluid in turbulent flow. The student suggests that the total resistance to mass transfer ($1/k_c$) be represented as the sum of two resistances $r_1$ and $r_2$ and makes three assumptions:

(1)  $r_1$, depending only on molecular diffusion, is equal to $y_1/D$, and the ratio of the fictitious film thickness $y_1$ to the pipe diameter $d$ is inversely proportional to Re.

(2)  $r_2$ involves only eddy diffusion.

(3)  the ratio of $r_1$ to the total resistance $r_1 + r_2$ is proportional the the one-third power of Sc.

Assuming further that the Reynolds analogy holds for Sc = 1, show that this model leads to the Chilton-Colburn analogy.

*5.3*  Following a careful study of the published data on velocity profiles for turbulent flow in round pipes, L. F. Flint [*Chem. Eng. Sci.*, **22**: 1127 (1967)] suggests the following equations to represent $u^+ = f(y^+)$:

$$u^+ = 2.5 \ln (1 + 0.4y^+) + 7.3 \left[ 1 - \exp \left( \frac{-y^+}{11} \right) - \frac{y^+}{11} \exp (-0.33y^+) \right] \qquad 0 < y^+ < 150$$

$$u^+ = 2.5 \ln \left[ 1.5 \left( \frac{1 + z^2}{1 + 2z^2} \right) y^+ \right] + 5.0 \qquad y^+ > 150$$

Where $z = (a^+ - y^+)/a^+$, where $a^+ \equiv y^+$ at the pipe axis.

Using these results, calculate the value of the Stanton number at Re = 50,000 and Sc = 100. From Fig. 5.3 it appears that the experimental value is about 0.00024 at Re = 10,000.

The following approximations may be employed:
(1) Resistance is localized near the wall, so $r/r_w$ may be taken to be 1.0; (2) $E_v = E_H = E_D$; (3) all of the resistance is in the wall region with $y^+ < 150$; and (4) the Fanning friction factor is 0.0052 at Re = 50,000, and 0.008 at Re = 10,000.

5.4 Develop a new "analogy" between mass and momentum transfer for turbulent flow in a pipe by assuming the velocity profile near the wall to be represented by the simple relation

$$u^+ = 50 \ln (1.0 + 0.020y^+) \qquad 0 < y^+ < 40$$

and that the transport resistance in the turbulent core is negligible for $y^+ > 40$.

Using this approach, obtain values of the Stanton number for Sc = 10, 100, 1,000, at Re = 10,000. Compare these results with corresponding values given by the Chilton-Colburn analogy. At Re = 10,000, $f = 0.0080$.

5.5 In a study of flow and heat transfer to air flowing in a pipe, C. A. Sleicher [*Trans. ASME*, **80**: 693 (1958)] measured both velocity and temperature gradients across the pipe diameter. He found that the velocity gradients could be well represented by the simple equation

$$u^+ = \frac{1}{0.091} \tan^{-1} (0.091y^+)$$

which was limited, however, to $0 < y^+ < 45$.

He obtained eddy conductivities for heat from the temperature profiles and concluded that $E_H/E_V$ was about 1.4, though it varied across the pipe diameter.

Using these results, and assuming $E_H = E_D$, derive an analogy relating the Stanton number $(k_c/\overline{U}_{av})$ to the Schmidt number $(\mu/\rho D$, or $v/D)$ for Re = 10,000, at which $f = 0.0078$ in smooth tubes with fully developed turbulent flow.

Employ your result to calculate values of the Stanton number for Schmidt numbers of 1, 10, 100, and 1,000. Compare the results with Stanton numbers calculated from the Chilton-Colburn analogy and with the following approximate experimental results (at Re = 10,000):

| Sc | 1 | 10 | 100 | 1,000 |
|---|---|---|---|---|
| $k_c/\overline{U}_{av}$ | 0.0036 | 0.0010 | 0.00029 | 0.000042 |

5.6   Experimentally, it is found that $k_c$ is 2.1 cm/s for evaporation of water at 25°C into a turbulent airstream flowing counterflow to the water film in a wetted-wall column (Re $\approx$ 6,000) at 1.0 atm.

   If the accommodation, or sticking, coefficient is 1.0, what percent of the observed mass-transfer resistance is represented by the surface resistance of the water molecules to leave the liquid?

5.7   A power plant proposes to scrub its stack gases to remove $SO_2$ for pollution control. A countercurrent scrubber will be employed, using an aqueous solution of soda ash, operated at 120°F and 1.0 atm. A plastic "egg-crate" packing is being considered but has not been tested for $SO_2$ absorption. This packing has been used, however, in commercial water-cooling towers, and data are available.

   It is proposed to use gas and solution flow rates of 3,000 and 4,000 lb/(h)(ft²), respectively, at which the mass-transfer coefficient for cooling-tower operation is 900 lb water/(h)(ft³)(unit $\Delta H$). This coefficient, on a packed volume basis, involves the use of absolute humidity (lb water/lb dry air) as the potential causing mass transfer.

   Employ this cooling tower coefficient to estimate $k_G a$ in units of pound moles per hour per cubic foot per atmosphere for $SO_2$ from stack gases in the same equipment. The stack gases contain 1,800 ppm $SO_2$ and have an average molecular weight of 30. All the resistance to mass transfer is on the gas side of the gas-liquid interface.

5.8   Moist air is to be dehumidified by passing it over a bank of staggered 1.0-in tubes which are wet with a strongly hygroscopic aqueous solution of lithium bromide. Cooling water inside the tubes maintains the solution at 30°C. The tube spacing is such that the maximum air velocity between tubes in a row is 15 ft/s. It is desired to remove 90 percent of the water-vapor content of the air fed saturated at 30°C.

   Employ available data on heat transfer for air flow over tube banks to estimate a value of $k_c$ (centimeters per second) for use in the design of this equipment.

5.9   A stirred autoclave is used for the hydrogenation of an organic compound dissolved in water. Hydrogen is fed as needed to keep the pressure constant at 200 psia, the gas being dispersed as bubbles by the impeller. A solid catalyst is provided in the form of fine particles maintained in suspension in the liquid. The reaction on the surface of the catalyst is first order in hydrogen and irreversible.

   In batch tests it is found that the conversion to the hydrogenated form is 90 percent complete in 4.0 h when the catalyst loading is 4 g/l; under the same operating conditions 90 percent conversion is attained in 3.0 h when 6.5 g/l of the catalyst is used.

   What time would be required to attain 90 percent conversion under the same conditions if 10 g/l catalyst were used?

   It may be assumed that the catalyst is sufficiently active that the hydrogen concentration in the liquid in immediate contact with the catalyst surface is essentially zero.

# RATES OF MASS TRANSFER AT SURFACES WITH SIMPLE GEOMETRY

## 6.0 SCOPE

Most industrial mass-transfer equipment involves a flowing stream in which the concentration of the material being transferred changes appreciably along the flow path. In principle, the design of such equipment requires an integration based on known local transfer rates. The present chapter presents brief summaries of what is known about these local rates at surfaces of simple geometry, as for single spheres, cylinders, falling films, and bubbles. The flow is turbulent in most cases of practical interest and empirical correlations must be employed in the absence of adequate theory.

The literature is now so voluminous that it is quite impractical to attempt new general correlations in the course of writing a text of this kind. It has been necessary to select typical summaries of experimental data, or to quote from the very limited number of good reviews. The choice of what to present is necessarily arbitrary, but it is believed that the data and correlations which follow are typical of the published information which may be useful for engineering purposes. In many cases it is possible to infer mass-transfer coefficients from correlations of heat-transfer data as, for example, by taking $j_D$ to be equal to $j_H$.

## 6.1  PRINCIPAL SYMBOLS

*Note:* Units in the cgs system are suggested, but essentially all the relations in this chapter are dimensionless, so any consistent set of units may be employed.

$c$       Concentration, g moles/cm$^3$; $c_i$, at phase boundary; $c_0$, in fluid fed

$d$       Diameter, cm; $d_p$, diameter of spherical particle

$D$      Diffusion coefficient, cm$^2$/s; $D_{AB}$, coefficient for diffusion of $A$ in binary gas mixture containing $A$ and $B$

$f$       Fanning friction factor

Fr     Froude number, $\overline{U}_{av}(gy_0)^{-1/2}$

$g$      Local acceleration due to gravity, cm/s$^2$

$h$      Heat-transfer coefficient, g cal/(s)(cm$^2$)($°$C)

$H$     Henry's-law coefficient, defined where used in text

$j_D$     $(k_c/\overline{U}_{av})Sc^{2/3}$; in binary gas system with $A$ diffusing, $j_D = (k_c/\overline{U}_{av})(p_{BM}/P)Sc^{2/3}$

$j_H$     $(h/C_p\rho\overline{U}_{av})Pr^{2/3}$

$k$      Thermal conductivity, g cal/(s)(cm$^2$) ($°$C/cm)

$k_c$     Mass-transfer coefficient, cm/s; $k_{cT}$, at terminal velocity of drop; $k_c^*$, with no convective flux or transpiration in direction of diffusion.

$n$      Number of gram moles

$N$     Molal flux, g moles/(s)(cm$^2$)

Nu    Nusselt number, $hd/k$

Nu$_F$   Nusselt film-thickness group $= y_0[(g \sin \theta/v^2]^{1/3} [(\rho - \rho_c)/\rho]^{1/3}$

$p_{BM}$   Logarithmic mean partial pressure of nondiffusing $B$ in binary of $A$ and $B$, atm

$P$      Total pressure, atm

Pe    Peclet number, $d\overline{U}/D$; Pe$_{r_0}$, radial Peclet number; Pe$_T$, transpiring Peclet number

Pr     Prandtl number, $C_p\mu/k$

$r$       Radial distance, cm; $r_0$, radius of tube or sphere

Re    Reynolds number, $d\overline{U}_{av}\rho/\mu$; Re$'$, based on velocity relative to that of surface of liquid film; Re$_T$ based on length of plate; Re$_x$ based on downstream distance from leading edge of plate; Re $\equiv 4\Gamma/\mu$ for falling films.

Sc    Schmidt number, $\mu/\rho D = v/D$

St     Stanton number for mass transfer, $k_c/\overline{U}_{av}$

$t$       Time, s; $t_f$, time to form drop

$T$      Temperature, $°$K

$\overline{U}$      Time mean velocity in $x$ direction, cm/s; $\overline{U}_{av}$, average over flow cross section; $\overline{U}_T$, terminal velocity in free fall; $\overline{U}_{TS}$, terminal velocity calculated by Stokes' law; $\overline{U}_S$, surface velocity

$V$      Volume, cm$^3$

We   Weber number, $\overline{U}(\sigma/\rho y_0)^{-1/2}$

$x$      Distance in flow direction, cm; $x_t$, distance from leading edge of plate to point of transition to turbulent boundary layer; $x_T$, plate length; $x_0$, distance from leading edge to beginning of mass-transfer surface

$y$      Distance in direction of diffusion, cm; $y_0$, film thickness

$Y$     Mole fraction

$\beta'$     $Dt/y_0^2$

$\Gamma$     Liquid feed, g/(s)(cm width)

$\Delta'$    Fraction unaccomplished approach to equilibrium

$\varepsilon$     Void fraction; work input to fluid, erg/(s)(g)

$\theta$     Angle of inclination from the horizontal

$\mu$     Viscosity, poises [g/(s)(cm)]; $\mu_c$, of continuous phase; $\mu_D$, of dispersed phase

| $v$ | Kinematic viscosity, $\mu/\rho$, cm$^2$/s |
|---|---|
| $\rho$ | Density, g/cm$^3$; $\rho_c$, of continuous phase; $\rho_d$, of dispersed phase |
| $\Delta\rho$ | Difference in density, $|\rho_d - \rho_c|$, g/cm$^3$ |
| $\sigma$ | Surface tension, dyn/cm; $\sigma_i$, interfacial tension |
| $\omega$ | Rotational speed, rad/s |

## 6.2 FLAT SURFACES

The case to be considered is that of mass transfer to or from a flat plate as fluid flows over it in a direction parallel to the plate, with negligible pressure gradient. A laminar boundary layer builds up near the leading edge; but if the main flow is turbulent, the flow near the plate becomes turbulent at some downstream distance $x_t$. If the leading edge is sharp and there is no artificially induced turbulence, this transition occurs at $Re_x = 200,000$ to $500,000$ [116]. Here $Re_x$ is a Reynolds number based on the downstream distance $x$ from the leading edge. This transition from laminar to turbulent boundary layer is to be expected in most practical situations.

The laminar boundary layer has been discussed in Chap. 3, where the local coefficient $k_c$ and the average $(k_c)_{av}$ are given as Eqs. (3.65) and (3.67). If the transition Reynolds number is 350,000, then for air at 20°C and 1.0 atm flowing over a flat plate at 10 ft/s, transition may be expected at a downstream distance $x_t$ of about 170 cm. Transition often occurs much earlier, however, due to blunt edges or roughness.

If the total length of the plate $x_T$ is greater than the transition distance $x_t$, then the average mass-transfer coefficient for the whole plate will be some weighted mean of the values for the laminar and turbulent boundary layer regions. Rohsenow and Choi [179] used Eq. (3.65) to $Re_x = 320,000$ and the Chilton-Colburn analogy $(j_D = j_H = f/2)$ for the turbulent region. The result is

$$\frac{(k_c)_{av}\, x_T}{D} = 0.037 Sc^{1/3}[Re_T^{0.8} - 15,500] \tag{6.1}$$

An empirical equation was used to represent the friction factor, with $f$ varying as $x^{-0.2}$. Similar equations are given by Eckert and Drake [44]. Equation (6.1) is derived for the case of mass transfer over the entire plate surface, with $Sc > 0.5$.

Most of the data on mass transfer from flat surfaces has been obtained in gases by measuring the rates of evaporation of liquids or the sublimation of solids. A graphical correlation [187] of the older data showed $j_H$, $j_D$, and $f/2$ to be essentially equal, and this result is supported by more recent data of the same type. A best line through the data points for gases is represented approximately by

$$j_D = (k_c)_{av} \frac{p_{BM}}{U_{av} P} Sc^{2/3} = j_H = \frac{f}{2} = 0.037 Re_T^{-0.2} \qquad 8,000 < Re_T < 300,000 \tag{6.2}$$

where $j_D$, $j_H$, and $f$ are average values for the whole plate. This is supported by the more recent data of Davies and Walters [37] for the evaporation of aniline, and of Träss [189], who measured the rate of sublimation from a plate coated with naphthalene. Träss measured *local* values of $k_c$ along the plate ($2 \times 10^4 < \text{Re}_x < 10^7$) and found close agreement of $j_D$ with $f/2$, where both $f$ and $j_D$ are *local* coefficients. Träss based $\text{Re}_x$ on the distance downstream from the origin of turbulence. The heat transfer data of Seban and Doughty (Fig. 8.16 of Ref. 179) show good agreement between $j_H$ and $f/2$ for points beyond the transition, and also quite a good check with the laminar boundary layer equation at smaller values of $\text{Re}_x$.

It may be noted that accurate values of $k_c$ are difficult to measure by any evaporation technique, since the evaporating surface tends to cool below that of the gas stream, and the surface temperatures necessary to the calculation of $k_c$ are not easily measured.

In some situations the mass- or heat-transfer surface is some distance downstream from the leading edge of the plate. Maisel [147], for example, evaporated water into air from a porous surface flush with a long plate in a wind tunnel, the wet areas being preceded by a dry plate surface extending a distance $x_0$ downstream from the leading edge. This situation is obviously complicated by the fact that the wet area may be exposed to a laminar boundary layer, a turbulent boundary layer, or partially to both.

Maisel found that his data ($13 < x_0 < 103$ cm; $18.1 < x_T < 117$ cm) could be represented closely by the empirical form

$$j_D = 0.0415 \text{Re}_T^{-0.2} \left[ 1 - \left( \frac{x_0}{x_T} \right)^{0.8} \right]^{-0.11} \tag{6.3}$$

For $x_0 = 0$ this is of the same form as Eq. (6.2) but the constant is 11 percent greater. A similar relation is proposed by Jakob and Dow [107] and Tessin and Jakob [201]. Equation (6.2) is of the same general form as a theoretical equation for the *local* coefficient in a laminar boundary layer. The derivation of this for heat transfer is given by Levich (pp. 102–106 of Ref. 133).

In cases where the mass-transfer surface is downstream from a dry or inert surface, it would appear that both $x_0$ and $x_T$ should be separate independent variables and that no function of the ratio $x_0/x_T$ alone could serve to account for their influence. If the flow and the transition point are fixed, small values of $x_0$ and $x_T$ might correspond to a laminar boundary layer over the entire plate. Large values of $x_0$ and $x_T$, giving the same ratio $x_0/x_T$, could correspond to a situation where, with the same free-stream conditions, the mass-transfer surface is entirely in the region of the turbulent boundary layer.

The work of Träss, referred to above, included measurements of sublimation rates from a plate in a wind tunnel with the air flow at supersonic speeds. The lower curve on Fig. 6.1, based on tests at Mach numbers of 0.43, 2.0, and 3.5, shows the Stanton number $\text{St}_c = k_c/\overline{U}_{av}$ in compressible flow to be smaller than the Stanton number $\text{St}_i$ for incompressible flow. The upper curve shows the trend of the corresponding ratios for heat and momentum transfer [43]. The curves would evidently coincide for $\text{Sc} = \text{Pr} = 1$.

There are numerous reports of data on vaporization of liquids placed in pans on the floor of a small tunnel. Plewes et al. [19, 169] measured rates of

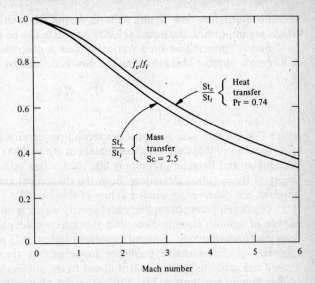

Mach number

**FIGURE 6.1**
Ratio of friction, heat-transfer, and mass-transfer coefficients in compressible flow to those for incompressible flow at the same Reynolds number. Comparison at $(Re_T = Re_t) = 10^6$. From Träss [189].

sublimation of several organic solids from the bottom surface of a square duct through which the flow of air was *laminar*. They derive the theoretical equation for this case, which is a variation of the Graetz solution for heat or mass transfer with laminar flow in round tubes. Dawson and Träss [38] report data on mass transfer from the bottom of a square duct with water in turbulent flow.

## 6.3  FALLING LIQUID FILMS

Liquid films, falling over an inclined or vertical surface under the influence of gravity, are sometimes employed in mass-transfer equipment. The pressure drop for gas flow through a wetted-wall column is low because there is little form drag, and equipment incorporating wetted-wall tubes is particularly advantageous for vacuum distillation, evaporation, condensation from a mixed vapor, and perhaps also for degassing liquids. However, the difficulty of distributing liquid feed uniformly to the perimeters of a number of tubes in parallel presents a design problem.

Liquid flow down a vertical plate (see Fig. 3.4) is essentially the same as for liquid flow down either the inside or outside surface of a vertical tube, providing the tube radius is large compared with the thickness of the flowing liquid film. The principal difference is due to drag and surface tension effects at the edge of a plate of finite width; the flow is not uniform across the width. When surface tension effects are important, the liquid velocity may actually be greatest near the edges.

Steady laminar flow on a vertical surface is described by Eqs. (3.55) to (3.58). A Reynolds number characterizing the flow is defined by

$$\text{Re} = \frac{4\Gamma}{\mu} = \frac{4\overline{U}_{av}\, y_0}{v} \tag{6.4}$$

where $\Gamma$ is the flow rate in grams per second per centimeter perimeter.

Jackson's [105] data on three liquids at Re < 4,000 checks Eq. (3.55) for $y_0$, but Dukler and Bergelin [42] report film thicknesses up to 50 percent greater than theory at Re = 4,000. Deviations from the theoretical velocity profile [Eq. (3.56)], however, are observed at smaller values of Re [79, 95].

The development of capillary and gravity waves greatly complicates the simple picture of smooth laminar flow, and the interrelated phenomena have not been described quantitatively. Fulford's excellent review [57] provides a comprehensive summary of the extensive published investigations through 1964. Levich, who himself has made important contributions to the subject, discusses various aspects of film flow in his book [133]. Following the pioneering study by Grimley [79], many papers on the nature and behavior of the waves which develop on falling films have appeared [1, 45, 27, 172, 194, 200].

Gravity waves appear as Fr reaches 0.6 to 2; capillary waves, at a Weber number of about unity. At high liquid rates the flow becomes turbulent. The transition is not abrupt, however, since the thin film continues in laminar flow in the region immediately adjacent to the fixed wall. Many observers have found the transition to occur in the range 250 < Re < 500. At Re < 250 the flow is laminar, the velocity profile is parabolic, and the ratio $\overline{U}_s/\overline{U}_{av}$ of surface to average velocity is 1.5. Values of this ratio as low as 1.1 have been measured when the flow is turbulent. The appearance of waves at low flow rates does not mean that the flow is turbulent.

The gradual transition from laminar to turbulent flow is illustrated by Fig. 6.2, where the Nusselt film-thickness group vs. Re is plotted. The right-hand scale gives the corresponding film thickness for water on a vertical surface, in contact with air.

The thin films corresponding to very low flow rates are often observed to be smooth and the flow to be steady. Small additions of surfactants produce the same type of laminar flow at much greater flow rates. It is suspected that film flow on a vertical surface may be inherently unstable at any flow rate, though waves may not develop in the time of passage over a surface several feet tall.

Mass transfer into a nonwavy laminar film is subject to rigorous analysis by the penetration theory (see Sec. 3.7). Waves on a laminar film cause a certain amount of mixing of liquid near the surface, and this considerably increases the rate of mass transfer. The waves do not appreciably increase the surface or interfacial area [177, 198].

Gas or liquid flow over a falling film produces a surface drag which changes

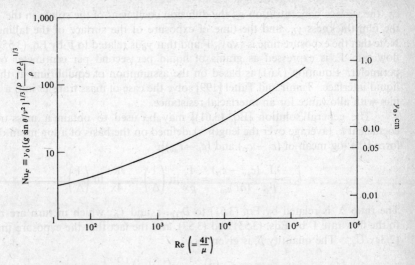

**FIGURE 6.2**
Thickness of liquid films flowing by gravity over inclined surfaces. The scale at the right is for water at 20°C on a vertical surface.

the flow pattern, but the effect is not great at typical gas rates. In countercurrent flow of air and water in a wetted-wall column, the ratio $\overline{U}_s/\overline{U}_{av}$ remains close to 1.5 up to gas Reynolds numbers of 24,000 or more. It is also reported that the film thickness in cocurrent flow is reduced less than 10 percent by gas velocities up to 8 m/s.

Many investigators have used surface-active agents in water in order to eliminate waves and so obtain reproducible data on mass transfer to films. Unfortunately, relatively little work has been done to determine the effect of these additives on the film flow pattern, the onset of waves, and the transition to turbulent flow.

### (a) The Absorption or Desorption of a Sparingly Soluble Gas

The case to be considered is typified by gas absorption in a wetted-wall column, where the liquid is in steady *laminar flow*, without waves. The gas-side resistance is negligible and the gas concentration constant (the case of decreasing concentration due to absorption from the flowing gas appears not to have been analyzed). The solute concentration at the liquid surface is everywhere $c_i$, corresponding to static equilibrium with the gas. Solute is diffusing from the liquid surface into the flowing liquid film.

Figure 3.4 illustrates the situation, and Eqs. (3.55), (3.56), and (3.57) describe the liquid flow. Equation (3.61) is a general solution, as $\Delta' = f(\beta')$, giving the average concentration of the effluent liquid as a function of the feed concentration

$c_0$, the surface concentration $c_i$, the diffusion coefficient of the solute in the liquid, the film thickness $y_0$, and the time of exposure of the surface of the falling film. Note that the exposure time is $\frac{2}{3}x\rho y_0/\Gamma$ and that $y_0$ is related to $\Gamma$ by Eq. (3.55). The flow rate, $\Gamma$, is expressed as grams of liquid per second per centimeter of tube perimeter. Equation (3.61) is based on the assumption of equilibrium at the gas-liquid interface. Tamir and Taitel [199] solve the case of mass transfer into a falling film with allowance for an interfacial resistance.

The general solution [Eq. (3.61)] may be used to obtain a mass-transfer coefficient $k_c$ (average over the length $x$), defined on the basis of a log mean driving force [the log mean of $(c_i - c_0)$ and $(c_i - c_{av})$]:

$$k_c = \frac{\Gamma}{\rho x} \frac{(c_{av} - c_0)}{(\Delta c)_{lm}} = \frac{\Gamma}{\rho x} \ln\left(\frac{1}{\Delta'}\right) = \frac{\nu Re}{4x} \ln\left(\frac{1}{\Delta'}\right) \qquad (6.5)$$

The ratio $\Delta'$ is related by Eq. (3.61) to $D_{AB}$, $t$, and $y_0$, which in turn are related to the flow rate $\Gamma$ by Eqs. (3.55) and (3.57), and the fact that the exposure time $t$ is $(2/3)x/\overline{U}_{av}$. The quantity $\beta'$ is given by

$$\beta' = 2\left(\frac{4}{3}\right)^{4/3} Dx\left(\frac{g}{\nu^5 Re^4}\right)^{1/3} \qquad (6.6)$$

For water at 20°C and $D = 2 \times 10^{-5}$ cm$^2$/s, this reduces to

$$\beta' = 1.25x\, Re^{-4/3}$$

where $x$ is the height of the vertical surface, in centimeters.

As an approximation for *short contact times* and small values of $D$, one may employ the simple penetration theory [Eq. (3.46)] to obtain

$$k_c = \frac{2\sqrt{D/\pi t}}{1 - \Delta'} \ln\frac{1}{\Delta'} \qquad (6.7)$$

which reduces to Eq. (3.46) as $\Delta' \to 1$, in which case

$$\frac{k_c \rho x}{\Gamma} = 1.695\sqrt{\beta'} \qquad (6.8)$$

Similarly, an approximation based on Eq. (3.53) for penetration into a semi-infinite slab with *long exposure times* ($\beta' > 1.2$) is

$$k_c = \frac{\Gamma}{\rho x}\left[\ln\frac{\pi^2}{8} + \beta'\right] \approx \frac{y_0}{t}\beta' \approx 2.41\frac{D}{y_0} \qquad (6.9)$$

The corresponding approximation for *long contact times*, taking the velocity gradient to be parabolic (with $\beta'$ large) is

$$k_c = \frac{2}{3}\frac{y_0}{t}\left[\ln\frac{4}{\pi} + 5.121\frac{Dt}{y_0^2}\right] \approx 3.41\frac{D}{y_0}$$

or

$$\frac{k_c x\rho}{\Gamma} \approx 0.24 + 5.1\beta' \qquad (6.10)$$

Selected experimental data on mass transfer between a gas and a falling film are compared with theory in Fig. 6.3. The considerable spread of the data appears to be due primarily to the development of waves or ripples at the surface of the falling liquid film. Hodson [98] and Hurlburt [104] added no surfactant to the water, and their data fall well above the theoretical curve. Lynn, Straatemeier, and Kramers' data [143], which check the theory closely, are for $SO_2$ absorption in water containing 0.05 weight percent soluble surfactant. The data of Emmert and Pigford [47] on the absorption and desorption of oxygen and carbon dioxide by water containing a surfactant agree well with theory at the higher water-flow rates. The data of Hikita [90], not shown, on $CO_2$ absorption by water containing a surfactant, check the theory closely in the range $20 < 1/\beta' < 1,000$ but fall some 50 percent high at $1/\beta' = 5$.

The role of the surfactant is to depress or eliminate wave formation; the interfacial resistance introduced by its presence appears to be negligible, or, in any case, of much less effect than that due to elimination of the ripples. Emmert and Pigford [47] and Hikita found the addition of small amounts of surfactants to reduce $k_c$ by as much as 30 to 50 percent. The effect tended to disappear as the film flow became turbulent at high water rates. The nature of the surface roll waves which cause the enhanced rates of mass transfer found when rippling of the film occurs has been studied [165, 221, 5, 171].

Equation (3.61) suggests that $k_c$ should vary as a power of $D$ which ranges from 0.5 at high liquid rates ($\beta' \ll 1$) to nearly the first power at $\beta' = 1.0$. Hikita, Nakanishi, and Kataoka [92], who absorbed six different gases in water films, report the power 0.5 at Re = 100 (laminar) and 0.38 at Re = 1,000. This trend, in the opposite direction from that indicated by the laminar theory, points up the fact that Eq. (3.61) cannot be used at large $\beta'$ if the flow is turbulent (Re greater than about 500).

Prasher and Fricke [174a], following King [121a], introduce an allowance for eddy diffusion in the falling film, expressing $E_D$ as a power function of $1 - y$. Both studies led to expressions for $k_c$ in terms of dimensionless groups involving energy dissipation per unit volume of fluid, which is $g\Gamma/\rho y_0$ in the case of a film falling vertically. Both obtained fairly good correlations of published data.

Many authors have used very short wetted-wall columns in experimental studies of mass transfer between gases and liquids. Lynn, Straatemeier, and Kramers [142] report a thickening of the film and an apparent stagnation of the surface at the bottom 0.2 to 1.7 cm near the liquid outlet. With suitable corrections for this stagnant region, their data checked Eq. (6.8). Vivian and Peaceman [209] using columns 2.87 cm ID and 1.88 to 4.25 cm tall desorbed $CO_2$ from water and chlorine from dilute aqueous hydrochloric acid. Their data fell 10 to 30 percent lower than predicted by Eq. (6.8), over a range of $\Gamma$ from 0.1 to 1.5 g/(s)(cm).

Lamourelle and Sandall [127] measured the rate of absorption of several gases ($CO_2$, $O_2$, $H_2$, He) by water films in *turbulent* flow, finding no effect of the height of the liquid fall on $k_c$. Wasan et al. [213] report that additions of less than 1 percent of water-soluble polymers markedly reduced $k_c$ as compared with pure water at the same $\Gamma$.

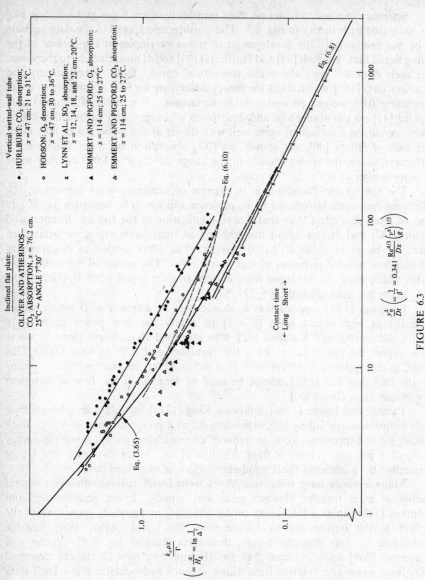

FIGURE 6.3
Mass transfer to the surface of a falling film (diffusion resistance in the liquid).

## (b)  Mass Transfer from a Solid Surface to a Falling Film

Mass transfer between a vertical tube wall or an inclined flat surface to a falling liquid film is a case having perhaps little practical application, but it has relevance to trickle-flow operation of beds of catalyst particles. The corresponding case of heat transfer from a wall to a falling film, however, is one of considerable industrial interest.

The applicable theory for the heat-transfer case was derived by Nusselt [162] some 50 yr ago. The derivation starts with Eq. (3.59), and is similar to that leading to Eq. (3.61). The velocity parallel to the surface is taken to be parabolic in $y$ and zero at $y = 0$. The boundary conditions, however, are different: $c = c_i$ at $y = 0$ rather than at $y = y_0$. The series solution obtained by Nusselt provides a relation between two dimensionless groups of heat-transfer variables, the Nusselt and Graetz numbers. These are tabulated by Norris and Streid [161] for the case of heat transfer from the walls of a flat duct to a fluid in laminar flow, which is the same mathematical case as the falling film if the film thickness $y_0$ is taken equal to half the spacing between the duct walls. Brown [16] provides a computer solution to the flat-duct case, with precise calculation of six eigenfunctions and eigenvalues. The results can be used to calculate heat-transfer coefficients at the walls.

The extensive table provided by Norris and Streid gives the Nusselt number as a function of the Graetz number. These have been converted to the dimensionless groups $k_c \rho x / \Gamma$ and $\beta'$, and a limited number of values are given in Table 6.1. The indicated theoretical relation is represented by the line on Fig. 6.4.

Figure 6.4 compares the laminar theory with two sets of data on quite different systems. Hikita, Nakanishi, and Asai [91] measured the rate of dissolution of iron by a falling film of $0.01N$ $H_2SO_4$, and of zinc by $0.01N$ $KI-I_2$, in vertical tubes of the metals 30 to 120 cm tall. The envelope enclosing their 70 to 75 data points is shown in Fig. 6.4. Kramers and Kreyger [124] dissolved cast benzoic acid sections 0.5 to 8 cm long set in the surface of an inclined metal plane. The data plotted were obtained with a plate angle of $45°$. These results would appear to have been obtained with turbulent flow, since the Reynolds numbers were in the range of 1,000 to 7,600.

Table 6.1  **COORDINATES OF THEORETICAL LINE REPRESENTING MASS TRANSFER FROM A SOLID SURFACE TO A LAMINAR FALLING FILM**

| $\dfrac{1}{\beta'}$ | 0.0936 | 0.187 | 0.655 | 0.938 | 1.87 | 5.63 |
|---|---|---|---|---|---|---|
| $\dfrac{k_c \rho x}{\Gamma}$ | 30.4 | 7.6 | 4.4 | 3.14 | 1.61 | 0.595 |
| $\dfrac{1}{\beta'}$ | 12.2 | 37.5 | 197 | 562 | 1,870 | 3,750 |
| $\dfrac{k_c \rho x}{\Gamma}$ | 0.323 | 0.144 | 0.0469 | 0.0223 | 0.0104 | 0.00628 |

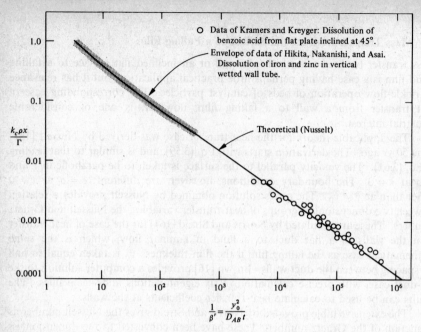

FIGURE 6.4

Mass transfer from solid surfaces to falling liquid films.

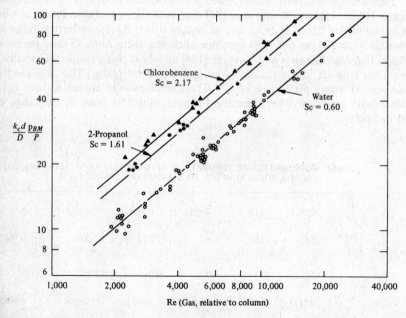

FIGURE 6.5

Data of Gilliland [75] on vaporization of liquids into air in a wetted-wall column.

Figure 6.4 shows a remarkably good check of the theory with data obtained over a wide range of solute properties, film flow rates, and geometries. The correlation is very much better than that shown in Fig. 6.3 for transfer to or from the surface of the falling film. The reason for this is evident; the surface waves and ripples which lead to surface mixing do not affect the laminar character of the flow in the immediate vicinity of the wall. The molecular diffusivities are very small in liquid systems, and the solute penetrates but a short distance into the liquid film in the few seconds time of contact. The agreement of the laminar theory with the data for turbulent films is particularly interesting.

Kramers and Kreyger suggest an alternative to the Nusselt theory based on the assumption that the liquid velocity near the wall is proportional to the distance from the wall surface. This is intended to apply to films in turbulent flow. Figure 6.4, however, indicates that their data at high Re agree well with the Nusselt equation.

The analogous problem of heat transfer from a vertical surface to falling oil and water films has been studied experimentally by Bays and McAdams [8] and McAdams, Drew, and Bays [144].

## (c) Transfer between the Surface of a Liquid Film and a Gas Stream in a Wetted-Wall Column

The gas-side resistance in wetted-wall columns has been studied by numerous investigators, using several techniques. Some have absorbed a constituent of a flowing gas mixture under conditions where the liquid-side resistance could be neglected; several have studied binary rectification. Perhaps the most convenient method involves the measurement of the rate of evaporation of the liquid.

Most of the published data agree approximately with the early work of Gilliland [75], who evaporated water and eight different organic liquids into air flowing over a wetted surface 117 cm long in a column having an inside diameter of 2.54 cm. Figure 6.5 shows a few of his data, which fell on parallel lines for each of the nine liquids when plotted in the manner shown. An excellent correlation of the data from nearly 400 tests was obtained in the form of the empirical equation

$$\frac{k_c d}{D_{AB}} \frac{p_{BM}}{P} = 0.023 \, \text{Re}^{0.83} \, \text{Sc}^{0.44} \qquad (6.11)$$

where $k_c$ = gas-side mass-transfer coefficient based on logarithmic mean of driving forces at two ends of column, cm/s

$d$ = tube diameter, cm

$D_{AB}$ = molecular diffusion coefficient for the vapor in gas, cm$^2$/s

Sc = Schmidt number for vapor in air

Re = air-stream Reynolds number based on the velocity of the air *relative to metal tube*

Both cocurrent and countercurrent flow were employed, at pressures ranging from 0.1 to 3 atm.

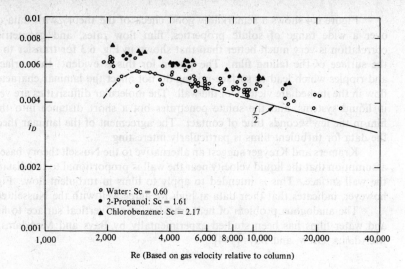

FIGURE 6.6
Data of Gilliland [75] replotted for comparison with the Chilton-Colburn analogy.

Gilliland did not vary the tube diameter, which is introduced in Eq. (6.11) only to form a dimensionless group. The results support the inclusion of the group $p_{BM}/DP$, though the product $DP$ is independent of pressure and $p_{BM}$ was not greatly different from $P$ except for some of the tests with water. Gas viscosity was varied little, but gas density varied some thirtyfold with $P$.

The Schmidt group varied over a fairly narrow range: from 0.60 for water-air to 2.17 for chlorobenzene-air. Equation (6.11) indicates that $k_c$ varies as $D^{0.56}$, though the power 0.50 is found in more recent studies in similar systems. The variation of Sc from 0.6 to 2.17 is hardly great enough to determine the exponent with precision, though when the data are replotted as $j_D$ vs. Re in Fig. 6.6 there is seen to be a definite trend of $j_D$ with Sc. This graph also shows Gilliland's values of $j_D$ to fall some 0 to 30 percent above the line for $f/2$, whereas the Chilton-Colburn analogy would suggest $j_D = f/2$. Since the values of Sc were in the vicinity of unity, the forcing of the data to fit the power 0.5 on Sc would hardly change the constant 0.023 in Eq. (6.11).

The liquid Reynolds numbers in Gilliland's tests were in the range of 150 to 1,560, and the liquid-film-surface velocities were appreciable in comparison with the gas velocities, especially since both cocurrent and countercurrent flow were employed. Yet the correlation obtained was much better when Re was based on gas velocity relative to the pipe than when gas velocities relative to the liquid surface were employed. Later workers have questioned this result and have obtained better correlations using gas velocity relative to the calculated liquid-surface velocity.

Kafesjian et al. [114] distinguish between situations where the liquid film is smooth and those where ripples are noted. Their conclusion, based on their own

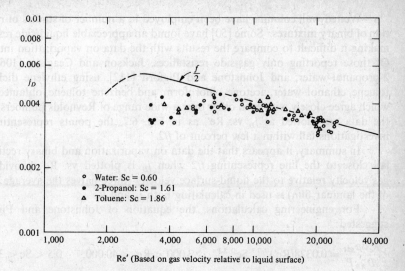

Re′ (Based on gas velocity relative to liquid surface)

**FIGURE 6.7**
Data of Jackson and Ceaglske [106] on vaporization of liquids into air in a wetted-wall column. (The velocity of the gas relative to that of the surface of the falling film is employed in both $j_D$ and Re′.)

and other data for the evaporation of water with countercurrent flow, would suggest the empirical equation

$$\frac{k_c \, dp_{BM}}{D_{AB}P} = 0.00814 \, Re^{0.83} \, Sc^{0.44} \left(\frac{4\Gamma}{\mu}\right)^{0.15} \tag{6.12}$$

for conditions where the film is rippling. This checks Eq. (6.11) for $4\Gamma/\mu = 1,000$, which is in the range of liquid Reynolds numbers used by Gilliland. For smooth flow with no ripples (as when the water contains a surfactant), their correlation can be expressed by

$$\frac{k_c \, dp_{BM}}{D_{AB}P} = 0.0163(Re')^{0.83} \, Sc^{0.44} \tag{6.13}$$

with no effect of liquid Reynolds number. It should be noted that Re appearing in Eq. (6.12) is based on gas velocity relative to the pipe, whereas Re′ in Eq. (6.13) employs the velocity relative to the surface of the laminar water film. Both equations are based on data in the range $30 < 4\Gamma/\mu < 800$ to 1,200.

The extensive data of Jackson and Ceaglske [106] on the vaporization of three liquids are shown in Fig. 6.7. Their data points fall well below Gilliland's results shown in Fig. 6.6, and $j_D$ is generally less than $f/2$. The gas velocity appearing in $j_D$ and in Re′ is that relative to the surface of the falling film.

Wetted-wall columns have been employed in a number of studies of rectification of binary mixtures. Some [50] have found an appreciable liquid-side resistance, making it difficult to compare the results with the data on vaporization into a gas. Of those reporting only gas-side resistances, Jackson and Ceaglske [106], using 2-propanol–water, and Johnstone and Pigford [112], using ethylene dichloride–toluene, ethanol–water, acetone–chloroform, and benzene–toluene, obtained results which agree closely with each other over a wide range of Reynolds numbers. When the data are plotted as $j_D$ vs. Re′, as in Fig. 6.7, the points representing both investigations fall within a few percent of $f/2$.

In summary, it appears that the data on vaporization and binary rectification fall close to the line representing $f/2$ when $j_D$ is plotted vs. Re′, providing the gas velocity relative to the liquid-surface velocity (three-halves the average velocity of the laminar film) is used in calculating both $j_D$ and Re′.

For engineering calculations, the equation of Johnstone and Pigford is suggested:

$$\frac{k_c \, dp_{BM}}{D_{AB} P} = 0.0328(\text{Re}')^{0.77} \, \text{Sc}^{0.33} \qquad 3{,}000 < \text{Re}' < 40{,}000; \qquad 0.5 < \text{Sc} < 3 \quad (6.14)$$

or,
$$j_D = 0.0328(\text{Re}')^{-0.23} \tag{6.15}$$

Alternatively, $j_D$ may be taken as equal to $f/2$, where $f$ is the Fanning friction factor for flow in smooth tubes. Note that the velocity of the gas relative to the calculated surface velocity of the laminar film is to be used in both $j_D$ and Re′ in Eqs. (6.14) and (6.15). If the surface velocity cannot be calculated, a good approximation can be obtained by the use of Eq. (6.11), in which Re is based on the gas velocity relative to the pipe.

The data presented in the foregoing are for vaporization and rectification in wetted-wall columns where the gas flow is turbulent. The Graetz equation for transport into a laminar stream with a parabolic velocity profile should apply in the laminar regime, with Re < 2,000. Gilliland found, however, that his data obtained at low gas Reynolds numbers compare more closely with an equation for rodlike flow, corresponding to flow with uniform velocity across the diameter. Conditions were nearly isothermal and flow both cocurrent and countercurrent, so free convection should not have distorted the presumed parabolic velocity profile. Predictions of mass transfer at Re < 2,000 can be made by the use of the values of $\Delta'$ given in Table 3.2 as a function of $Dt/y_0^2$. In this procedure $y_0$ is to be taken as the tube radius, and $t$ as the time of passage through the tube at the mean or average velocity of the gas.

## 6.4  MASS TRANSFER BETWEEN A FLUID AND A SOLID SPHERE

The sphere in a flowing fluid is found in many types of engineering equipment. It may be a solid immersed in a stream of gas or liquid, in which case the mass-transfer resistance is usually confined to the external boundary layer. Suspended drops or bubbles, however, present diffusional resistances in both phases.

The flow field around a fixed sphere is exceedingly complicated in most practical situations, and theoretical developments have been confined primarily to flow at very low velocities. Boundary-layer separation develops as the fluid velocity increases, the ring of separation moving back from the upstream stagnation point, with the development of a wake behind the sphere as the flow becomes more and more turbulent.

If the fluid is stagnant, mass transfer occurs only by molecular diffusion. Example 3.3 gives an analysis of transient diffusion in this case. If the fluid medium is infinite in extent, it is theoretically possible for a steady state to be established. Let $r$ be the radial distance from the center of the sphere and $r_0$ the sphere radius. The molal flux $N$ through the spherical shell at radius $r$ $(r > r_0)$ is given by

$$N(4\pi r_0{}^2) = -D(4\pi r^2)\frac{dc}{dr}$$

Integration between the limits $c = c_i$ at $r = r_0$ and $c = c_\infty$ at $r = \infty$ leads to

$$\frac{2Nr_0}{D(c_i - c_\infty)} = \frac{2k_c r_0}{D} = \frac{k_c d_p}{D} = 2 \tag{6.16}$$

(The corresponding case of diffusion into infinite space from one side of a round disk [157] leads to the constant $8/\pi$ in place of 2.) This relation, due to Langmuir, provides a limiting value of $k_c$; any fluid motion will cause $k_c$ to be greater.

The flow field in laminar "creeping flow" around a sphere has been successfully treated analytically, and the results combined with diffusion theory to obtain mass-transfer coefficients [133, 126, 53, 15, 222, 54]. (In this and similar situations analyses of heat transfer by thermal conduction are directly applicable to mass transfer by molecular diffusion, and vice versa.) The results of these studies are summarized in Fig. 6.8, in which the group $k_c d_p/D$ vs. the Peclet number, Pe $[= d_p \bar{U}/D = (\text{Re})(\text{Sc})]$ is plotted. The asymptote at zero flow is 2, as given by Eq. (6.16). The curve in the middle region represents the results of the theoretical analysis of mass transfer in creeping flow by Brian and Hales [15], which is perhaps the best of the several theoretical studies of the problem. (Brian and Hales' curve in this region is nearly identical with the earlier one by Friedlander [53, 54]. For values of Pe less than 10,000 it is well represented by

$$\frac{k_c d_p}{D} = (4.0 + 1.21\,\text{Pe}^{2/3})^{1/2} \tag{6.17}$$

The dotted upper branch represents the equation

$$\frac{k_c d_p}{D} = 1.0\,\text{Pe}^{1/3} \tag{6.18}$$

which was obtained by Levich [133] and Friedlander; Levich gives the constant as 1.01; Friedlander 0.991. Pfeffer and Happel [168] discuss a number of other theoretical treatments of creeping flow around spheres.

The theoretical results presented in the foregoing are for mass transfer for

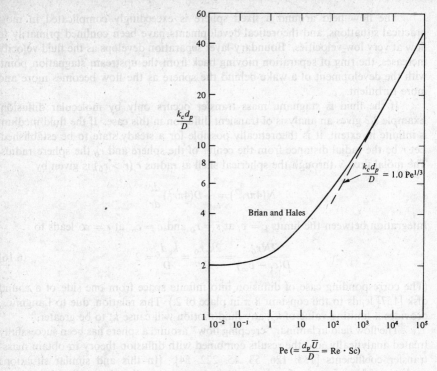

**FIGURE 6.8**
Mass transfer from spheres to fluids at low relative velocities.

spheres of constant diameter and negligible convective flux due to diffusion. The case of varying sphere diameter and transpiration flux is treated by Brian and Hales (see below).

EXAMPLE 6.1   A solid sphere is falling in a liquid at its terminal velocity in Stokes' flow. The sphere diameter is 100 $\mu$m and its density is 2.0 g/cm³. The liquid density is 1.0 g/cm³ and the liquid viscosity is 1.0 cP. What is the value of the mass-transfer coefficient $k_c$ (cm/s) for the transfer of solute from sphere surface to liquid if the molecular diffusion coefficient is $10^{-5}$ cm²/s?

SOLUTION   The terminal velocity is obtained from Stokes' law as

$$U_{TS} = \frac{g d_p^2 \, \Delta \rho}{18 \mu} = \frac{981 \times 0.01^2 \times (2-1)}{18 \times 0.01} = 0.545 \text{ cm/s}$$

$$Pe = \frac{d_p \bar{U}}{D} = \frac{0.01 \times 0.545}{10^{-5}} = 545$$

From Fig. 6.8 it appears that this is in the range where Eq. (6.17) applies. From either the equation or the graph, $k_c d_p/D = 9.2$;

$$k_c = \frac{9.2 \times 10^{-5}}{0.01} = 0.0092 \text{ cm/s}$$

*Note:* This calculation may be generalized to obtain an equation giving $k_c$ for spheres in free fall in the range of values of Pe over which Eq. (6.18) applies [23]:

$$\text{Pe} = \frac{d_p U}{D} = \frac{g d_p^3 \Delta \rho}{18 \mu D} = \left(\frac{k_c d_p}{D}\right)^3$$

$$k_c = 0.38 \left(\frac{g \mu \Delta \rho}{\rho^2}\right)^{1/3} \text{Sc}^{-2/3}$$

More generally, Fig. 6.8 would be used in place of Eq. (6.18) to relate $k_c$ to Pe. Harriott [85] follows the same approach using a correlation such as represented by Fig. 6.9 (see below). The curved section of the line on Fig. 6.8 (Pe < 1,000) applies for Re < 1 if Sc > 1,000 (as in many liquid systems). In gases where Sc $\approx$ 1, the break occurs at Re $\approx$ 1,000.                    ////

A very large number of published studies have reported measurements of the rate of mass transfer between a single sphere and a fluid stream. The techniques employed include sublimation of a solid, evaporation of a liquid into a gas, and dissolution of a solid or liquid into a liquid. Apparently no studies of gas absorption by single spheres under conditions where the gas-side-resistance controls have been reported. Most of the data are for the evaporation of drops of pure liquids, since the experimental technique is simple and small drops (or larger drops containing surfactant) behave as rigid spheres. In addition, the considerable body of information on heat transfer to spheres can be applied generally to mass transfer by substituting $k_c d_p/D$ for Nu, and Sc for Pr.

Since the pioneering studies of Frössling [56] and Vyrubov [211] there have been several attempts to develop correlations of the data for general estimation purposes. Perhaps the best of these are the works of Ranz and Marshall [176], Steinberger and Treybal [195], and Rowe, Claxton, and Lewis [181]. In each case the result was a graph or equation relating $k_c d_p/D$ (or $k_c d_p p_{BM}/DP$ for gases) to Re and Sc. That of Ranz and Marshall, in particular, has been widely quoted and used.

Figure 6.9 illustrates the three correlations referred to above. The Ranz and Marshall line well represents the large mass of data, though Rowe, Claxton, and Lewis would predict somewhat higher values of $k_c$ in the lower range of $(\text{Re})(\text{Sc})^{2/3}$. The general correlation is considered to be very good in light of the wide range of Re and Sc for both gases and liquids. Also shown is a dashed line representing Williams' [219] correlation of published data on heat transfer to spheres, which is seen to be in fair agreement with the mass-transfer data.

The only data points shown are those of Rowe, Claxton, and Lewis for dissolution of 1.27- and 3.81-cm spheres of benzoic acid in water (Sc = 1,200 to 2,590). These fall considerably above Steinberger and Treybal's line for Sc = 1,500.

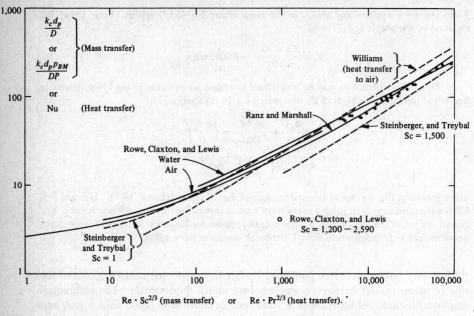

$$\left.\begin{array}{c} \frac{k_c d_p}{D} \\ \text{or} \\ \frac{k_c d_p p_{BM}}{DP} \end{array}\right\} \text{(Mass transfer)}$$

or

Nu    (Heat transfer)

Williams (heat transfer to air)

Ranz and Marshall

Steinberger, and Treybal
Sc = 1,500

Rowe, Claxton, and Lewis
Water
Air

Steinberger
and Treybal
Sc = 1

o  Rowe, Claxton, and Lewis
Sc = 1,200 − 2,590

Re · Sc$^{2/3}$ (mass transfer)    or    Re · Pr$^{2/3}$ (heat transfer).

**FIGURE 6.9**
Mass and heat transfer to spheres.

The latter was based in part on data obtained using viscous solutions in which the Schmidt numbers were as large as 70,000. The recent data of Gibert et al. [73], who used the electrochemical technique with liquids, would extend the curves shown up and to the right, in reasonable agreement with an extrapolation of the data of Rowe, Claxton, and Lewis. Though Fig. 6.9 shows a remarkably general correlation, it probably indicates values of $k_c$ which are too large in the case of viscous liquids.

The mass-transfer coefficient $k_c$ appearing in the ordinate of Fig. 6.9 is the average for the entire surface of the sphere. Local values vary severalfold, being greatest at the forward stagnation point and smallest near the waist. Lee and Barrow [128] show the local coefficient to be approximately one-quarter as large at the rear as at the upstream stagnation point, and still smaller at the waist. See also Garner and Grafton [63] and Sage et al. [17, 60].

Various other aspects of mass transfer to spheres have been studied. The effect of free convection can be very large at low fluid velocities; the rate of dissolution of a solid sphere in a nearly stagnant liquid is many times that predicted by Eq. (6.16) [65, 66, 185]. The effect of the deviation of the shape from that of a true sphere ("sphericity") has been investigated [191, 101]. Rates of evaporation of drops at quite high temperatures is the subject of several papers [155, 40, 167, 129]. The enhancement of $k_c$ by vibrating or rotating the sphere has been reported [159, 160]. Mass transfer to a single sphere in an array has received attention because of the importance of packed absorbers and catalytic reactors [59, 74, 113, 182]. The rate of solution of uranium spheres in molten cadmium at 500 to 600°C has been measured [205]. Increase in the intensity of turbulence (see Sec. 4.2) of the flowing fluid has been shown to have an appreciable, if not enormous, effect on $k_c$ [17, 18, 60], and some of the spread of the data on Fig. 6.9 may be due to differences in the turbulence levels in streams employed by the several investigators.

### Spheres: Effects of Transpiration and Changing Diameter

The preceding section has dealt with mass transfer to spheres of constant diameter with negligible convective flux (transpiration). In most practical applications, as in the evaporation of a drop, growth of crystals, or dissolution of solid particles, the particle diameter changes as mass transfer occurs, and the transient nature of the process must be taken into account. Furthermore, the convective flux can be important at high transport rates.

If the flow is turbulent, it is perhaps reasonable to develop a differential equation expressing the instantaneous rate as a function of the sphere diameter, using the Ranz and Marshall correlation, with allowance for transpiration as described in Sec. 5.4. This can then be integrated between the limits of the initial and final diameters.

In the case of spheres in free fall, or suspended in an agitated tank, the low relative velocity may correspond to Stokes' flow. Figure 6.8 shows the theoretical line for this situation for spheres of constant diameter with no allowance for transpiration, as derived by Brian and Hales. These authors have extended their analysis to include the effects of transpiration and of changing sphere diameter.

For a sphere of fixed diameter, the effect of steady transpiration on diffusion into an infinite stagnant medium is given by

$$\frac{k_c}{k_c^*} = \frac{\mathrm{Pe}_T}{(k_c\, d_p/D)^* \{\exp\left[\mathrm{Pe}_T(D/k_c\, d_p)^*\right] - 1\}} \tag{6.19}$$

where $\mathrm{Pe}_T$ is the transpiring Peclet number, $d_p\, \overline{U}_r/D$, based on the radial transpiration velocity $\overline{U}_r$. The asterisks refer to conditions of no transpiration and fixed $d_p$. This relation is of the same form as Eq. (5.37). For Stokes' flow the basic differential equations for diffusion and flow were solved by finite-difference techniques. Over the range of bulk flow Peclet numbers (Pe) from 30 to 1,000, the numerical solutions gave results nearly identical to those obtained by the use of Eq. (6.19). Evidently, in this range, Eq. (6.19) can be used, with $(k_c\, d_p/D)^*$ obtained from Fig. 6.8.

It is important to note that $k_c$ obtained in this way [as in the case of $k_D$ in Eq. (5.37)] represents only the diffusive transport: The convective flux must be added (see Sec. 5.4).

For spheres in Stokes' flow with no transpiration, the effect of changing diameter is expressed in terms of the radial Peclet number, $\mathrm{Pe}_{r_0} \equiv (2r_0/D)(dr_0/dt)$, where $r_0$ represents the sphere radius, which is changing with time at the rate

$dr_0/dt$. An analytical solution has been obtained for a growing sphere with $Pe = 0$ by Frank [52], and computer solutions of the basic differential equations have been obtained by Brian and Hales for spheres in Stokes' flow which are either growing or shrinking. The results are given in the form of graphs of $k_c/k_c^*$ vs. $Pe_{r_0}(D/k_c d_p)^*$, which show that $k_c$ can be much greater than $k_c^*$ if the sphere is growing, and much less than $k_c^*$ if shrinking.

To allow for the combined effects of transpiration and changing sphere diameter, Brian and Hales suggested the product of the separate correction ratios: $k_c/k_c^*$ with allowance for both effects will be the product of $k_c/k_c^*$ for no transpiration but changing $r_0$, and $k_c/k_c^*$ for fixed $r_0$ but significant transpiration.

Additional treatments of mass transfer from advancing or receding surfaces are to be found in Refs. 9, 20, and 206.

## 6.5  SOLID PARTICLES SUSPENDED IN AGITATED VESSELS

Mass transfer between a fluid and small suspended particles is important in many industrial situations. Liquid-phase hydrogenation in a slurry of catalyst particles involves a mass-transfer resistance at the particle surface which may control the reaction rate if the catalyst is highly active. The dissolution of crystals, crystallization in well-mixed evaporators, and the combustion of powdered coal all require transport between fluid and suspended particles.

Space limitations prohibit treatment of the many cases where this general situation is encountered. As one example, however, mass transfer from solid particles suspended in a liquid in an agitated vessel will be discussed briefly. The vessel is usually a cylinder with several vertical baffles fixed to the walls, the agitation being provided by an impeller of the turbine or propeller type.

There are many variables involved: geometry of the vessel, nature of the baffles, type of impeller, speed of rotation (or power input), and slurry density all vary with the vessel design and the type of operation. Physical variables include liquid density, viscosity, and molecular diffusivity of the diffusing solute. The size distribution, shape, and density of the suspended particles may also be important. The resistance to diffusion into a porous particle and the intrinsic rate of crystal growth or chemical reaction obviously complicate the situation. It is not surprising that there is no reliable general correlation of mass-transfer coefficients for such systems.

Since the pioneering study by Hixson and Baum [94] there have been numerous studies of both mass and heat transfer to solid particles suspended in liquids in agitated vessels [21, 151, 7, 85, 197, 158, 103a, 15]. Some used spheres and some crystals treated as spheres of equivalent diameter. Attempts to correlate the resulting data have usually involved (1) inspection of calculated slip velocities, (2) relating of $k_c$ to power input per unit volume, or (3) empirical correlations involving dimensionless groups. In a recent article, Miller [152] measured mass transfer from suspended particles in 1-, 10-, and 100-gal baffled vessels, with power inputs of 0.05 to 19 hp/1,000 gal. He used his and other data to suggest procedures for scaleup of similar mass-transfer equipment.

The slip velocity is the velocity of the fluid relative to the particle. Harriott [85] introduced the idea of calculating a minimum $k_c (\equiv k_{cT})$ by taking the slip velocity as the terminal velocity of the particle falling under the influence of gravity and adopting this as the velocity to be used in the Reynolds number to get $k_c$ from established correlations of data on mass transfer to fixed spheres. For the latter purpose he used a modified Frössling equation:

$$\frac{k_c d_p}{D} = 2 + 0.6 \, \text{Re}^{1/2} \, \text{Sc}^{1/3} \tag{6.20}$$

The terminal velocity was calculated from the established drag coefficients for spheres. This is most easily done by using Stokes' law and applying a correction factor to the calculated velocity, as described by Lapple on pp. 1018–1020 of the third edition of Perry's "Chemical Engineers' Handbook." The method of calculating the terminal velocity is as follows:

1   Calculate the terminal velocity from Stokes' law,

$$U_{TS} = \frac{d_p^2 |\rho_d - \rho_c| g}{18 \mu_c}$$

2   Calculate the Reynolds number using $U_{TS}$,

$$\text{Re}_{TS} = \frac{d_p \bar{U}_{TS} \rho_c}{\mu_c} = \frac{g d_p^3 \rho_c |\rho_d - \rho_c|}{18 \mu_c^2}$$

3   Obtain $\bar{U}_T/\bar{U}_{TS}$ from the following table, or from Lapple:

| $\text{Re}_{TS}$ | 1 | 10 | 100 | 1,000 | 10,000 | 100,000 |
|---|---|---|---|---|---|---|
| $\bar{U}_T/\bar{U}_{TS}$ | 0.9 | 0.65 | 0.37 | 0.17 | 0.07 | 0.023 |

4   Obtain the terminal velocity as $U_T = U_{TS} \times (U_T/U_{TS})$. For a 5-mm sphere ($d_p = 0.5$) of density 2, falling in water at 20°C ($\rho_c = 1.0$, $\mu_c = 0.01$ poises),

$$U_{TS} = \frac{981 \times 0.5^2 (2-1)}{18 \times 0.01} = 1,362 \text{ cm/s}$$

and $\text{Re}_{TS} = 68,100$. So $\bar{U}_T/\bar{U}_{TS}$ is 0.028 and $\bar{U}_T = 38$ cm/s.

Figure 6.10 shows values of $k_{cT}$ calculated in this way for a typical system (water at 20°C and $D = 10^{-5}$ cm$^2$/s). Terminal velocity increases with increase of $d_p$, but $k_c$ decreases with increasing $d_p$ at constant velocity; this explains the fact, noted by several investigators, that $k_c$ varies little with changes in $d_p$ over much of the range of practical interest.

The evidence about how $k_c$ varies with the rotational speed of the impeller is confused by the fact that the particles tend to rest on the bottom of the vessel at low speed, and this evidently occurred in some of the experimental studies which have been published (the speed required to suspend particles can be estimated by an equation due to Zwietering [223]). Though Harriott's extensive data covered a wide range of physical systems and impeller speeds in baffled vessels of three sizes,

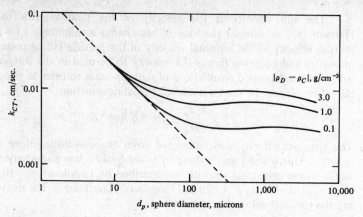

**FIGURE 6.10**
Mass transfer to spheres in free fall at their terminal velocities. Liquid density 1.0 g/cm³; viscosity 0.01 poise; $D = 1.0 \times 10^{-5}$ cm²/s; Sc = 1,000. *(After Harriott [85].)*

the ratio $k_c/k_{cT}$ fell within the relatively narrow range 1.5 to 8 in all tests when the particles were fully suspended.

Harriott's results suggest a method of obtaining a rough estimate of $k_c$ in baffled agitated vessels containing solid particles suspended in a liquid: calculate $k_{cT}$ and multiply by 2. Since the ratio $k_c/k_{cT}$ is empirical, it is necessary to use Harriott's procedure in calculating $k_{cT}$: use the method presented in Perry to obtain the terminal velocity, and Eq. (6.20) for $k_{cT}$.

Once the particles are suspended, $k_c$ increases slowly as impeller speed is further increased. It is probably correct, as several authors have suggested, that the most economical use of power for mass-transfer enhancement involves operation at a speed just sufficient to ensure particle suspension. This would not be true in gas-liquid systems, where power increase results in greater mass-transfer area.

Since $k_c$ depends primarily on the velocity of the liquid relative to the mass-transfer surface, and on the turbulence level in the liquid, it seems logical to attempt the development of a correlation involving the power input per unit volume as a parameter. This is an important objective, since $k_c$ and power input are of direct interest to the design engineer.

Kolmogoroff's theory of isotropic turbulence suggests that kinetic energy supplied by the impeller is dissipated primarily by viscous interaction of the smallest eddies, whose turbulent motion is isotropic. This range of the turbulent energy spectrum is independent of the geometry of the vessel and impeller and, for any one fluid, dependent only on the power supplied. This is the range of eddy sizes which affect $k_c$.

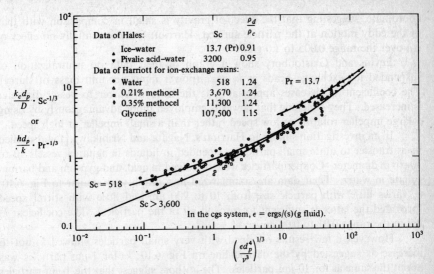

FIGURE 6.11
Heat- and mass-transfer data for particles suspended in agitated vessels.

Several authors have suggested the applicability of Kolmogoroff's concepts to the situation in stirred vessels with suspended particles. Comparison with published data has proved difficult, since few investigators have measured power input simultaneously with mass-transfer tests. (Power input can be estimated, however, by the methods of Rushton, Costich, and Everett [183].)

Hales [15] has developed a correlation between $k_c$ and power input, using dimensionless groups based in part on the Kolmogoroff theory. Levins and Glastonbury [134] and Van den Berg [208] report similar studies.

Figure 6.11 shows Hales' correlation, based on his own data and those of Harriott for mass transfer to suspended ion-exchange resin spheres. The dimensionless group in the abscissa involves particle diameter $d_p$, the kinematic viscosity $v$, and the agitation power per unit mass of fluid $\varepsilon$. Hales' data covered heat transfer to suspended spheres of ice and mass transfer in the dissolution of pivalic acid. The data of Barker and Treybal on dissolution of boric and benzoic acids (Sc = 735 to 1328) fall some 25 percent low, but the results of Wilhelm, Conklin, and Sauer [217] and Nagata et al. [154] on the dissolution of sodium chloride (Sc = 560 to 950) agree (when corrected for transpiration and shrinkage) with the curve at the upper right, or an extension of it. Harriott's data on dissolution of boric, benzoic, and butyl benzoic acid particles fall 15 to 50 percent low.

Figure 6.11 indicates (as does Fig. 6.10) that $d_p$ is important in the range of small particles sizes but that its relevance almost disappears with large particles. It is interesting to note that the density difference $\Delta\rho$ does not appear in either coordinate, suggesting that the effect of gravity is small in comparison with that of the eddy motion at the particle surface. Harriott found essentially no effect of $\Delta\rho$ over the range 0.005 to 1.0 g/cm$^3$.

Levins and Glastonbury show a graph of their data on neutralization of suspended ion-exchange beads as $k_c$ vs. $\varepsilon$ (power input per unit mass of slurry). The coefficient $k_c$ increases appreciably as the ratio of stirrer to vessel diameter is increased. They conclude that power is employed more advantageously by using a large impeller rotating at low speed rather than a small impeller at high speed.

In an investigation similar to Harriott's, Nagata and Nishikawa [154a] studied mass transfer to quite small particles suspended in liquids in agitated vessels 10 to 30 cm in diameter. Copper particles were dissolved in acid, and gypsum and barium sulfate in water. Their data are summarized in a graph very similar to Fig. 6.10; $k_c$ varies little with particle size from 10 to 200 $\mu$m and little with stirrer speed (provided the latter is sufficient to suspend all of the particles). The coefficient $k_c$ fell in the range 0.01 to 0.04 cm/s.

However, a few results obtained with very small particles showed $k_c$ not to increase as suggested by the dashed line on Fig. 6.10: $k_c$ for 1-$\mu$m particles was about the same as for 10-$\mu$m particles. The authors suggest that the 1-$\mu$m particles may be smaller than the dimensions of the microeddies, and affected only by viscous shear in the tiny eddies in which they are trapped—not by the turbulence of the bulk liquid.

Various published reports have described studies of mass transfer to spheres or surfaces fixed in place in vessels containing agitated liquids. Johnson and Huang [110], Colton and Smith [30], and Marangozis and Johnson [149] treat mass transfer from the bottom of the vessel, and Askew and Beckmann [2] treat mass transfer from the vertical wall. Keey and Glen [120] measured mass transfer from spheres fixed with the vessel rotating, and from fixed spheres with agitation by an impeller. In another study, LeLan et al. [130] compared rates of mass transfer to a small sphere held at various locations in a stirred vessel.

## 6.6 DROPS

Equipment designed to promote mass transfer between two liquids, as in extraction processes, can be made quite compact if one liquid is dispersed and suspended in the second as small drops. The advantage lies in the establishment of a very large surface of contact between the two phases. Liquid sprays in gases are effective for the same reason, and are commonly employed for gas absorption, humidification, and spray drying. It is usually difficult, however, to design such equipment to ensure effective countercurrent flow of the two phases.

One liquid may be introduced into a second by flow through nozzles or porous plates, forming drops generally 1 to 6 mm in diameter, which rise or fall through the second (continuous) phase. Much smaller drops are formed if mechanical means of dispersion are employed.

Mass transfer from single drops rising or falling in a second liquid has been

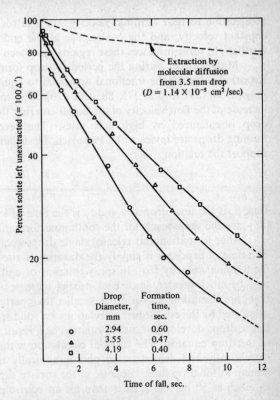

FIGURE 6.12
Data of Licht and Pansing [137] on extraction of acetic acid from water drops falling through methyl isobutyl ketone in columns of various heights.

extensively studied. Much less has been published on spray columns, involving swarms or clouds of droplets, and on liquid-liquid extraction in agitated vessels. The drop mechanics and the mass-transfer phenomena have been found to be quite complex, and the voluminous literature provides only a semiquantitative basis for design. For a summary of the research accomplishments pertaining to both drop mechanics and mass transfer, see the excellent review by Heertjes and De Nie [89].

In the typical experimental study, the continuous phase is contained in a vertical transparent tube. The second liquid is introduced through a small capillary at top or bottom, forming a succession of drops which rise or fall through the column under the influence of gravity. At the end opposite the feed they coalesce and are removed as a continuous phase. Figure 6.12 illustrates the type of data obtained in studies of this kind. The experiment itself is complicated because there are four regions to consider: the drop formation with simultaneous mass transfer (small satellite drops may be formed as the larger drop is released); a period of acceleration to near terminal velocity; the rise (or fall) of the drop at essentially constant velocity; and the coalescence at the end, again with simultaneous mass transfer. In some instances these regions have been isolated and studied separately.

Mass transfer during the period of drop formation can be quite substantial. Reported values of the fractional approach to equilibrium (the $1 - \Delta'$ of Chap. 3) range from 0.08 to 0.29 [76, 88, 137, 170]. The mass-transfer rate increases with increase of the flow velocity of the liquid entering the drop, since mixing within the drop is enhanced by high flow rates. The several theories of diffusion into forming drops are reviewed by Popovich, Jervis, and Träss [170], whose own data support the relation

$$1 - \Delta' = \frac{8}{d_p} \sqrt{\frac{Dt_f}{\pi}} \tag{6.21}$$

Here $d_p$ is the final drop size and $t_f$ is the time of formation. Heertjes and De Nie [88] give the same result for the continuous formation of successive drops.

After formation and release, single drops soon reach their terminal velocity. The time of exposure is simply the distance of rise or fall of the drop divided by the terminal velocity $U_T$. In spray columns or agitated vessels, however, the time may be much greater, since back-mixing increases dispersed-phase holdup. To a first approximation, $U_T$ for drops smaller than perhaps 1 to 2 mm diameter can be calculated by the established correlations for solid spheres (see Sec. 6.5). Somewhat larger drops develop internal circulation as a result of frictional drag in the region of the drop equator. The toroidal circulation within the drop has been predicted theoretically and demonstrated photographically using suspended colloidal aluminum in the drop fluid. Because of the surface slip, the terminal velocity may be as much as 50 percent greater than for an equivalent solid sphere. The presence of surface-active agents or small amounts of impurities greatly reduce the tendency to develop circulation and, consequently, the effect of circulation on the terminal velocity.

As drop size increases, the terminal velocity $U_T$ reaches a maximum and then slowly decreases. As the size increases beyond that for which $U_T$ is a maximum, the drop becomes distorted and oscillates between oblate and prolate spheroids, or irregular shapes. The larger drops often show indentations of the trailing surface. Still larger drops tend to break up with the formation of two or more smaller drops.

Hu and Kintner [102] report extensive drop-velocity data and suggest a correlation involving dimensionless groups: the drag coefficient, Weber and Reynolds numbers, and a "physical properties" group. The complicated phenomena of drop mechanics are described in Kintner's excellent review [122], which relates particularly to liquid extraction.

Diffusion rates within the drop depend on transport both by molecular motion and fluid mixing. Very small drops are essentially stagnant, and transport is by molecular diffusion. Drops of intermediate size develop laminar toroidal internal circulation, which reduces the path length for molecular diffusion. In very large drops the laminar circulation is replaced by what appears to be quite violent internal mixing, resulting from the kneading effect of drop oscillation. Much of what is known about circulation within drops comes from a series of studies by Garner and his colleagues at Birmingham (e.g., Refs. 62, 64, 67, 68, 69, and 70).

The motion of single drops under gravity is exceedingly complicated, involving many variables, even in "clean" systems. As a further complication, however, it is well known that extremely small amounts of surface-active impurities can greatly affect drop behavior [184]. Surfactants tend to make drops more rigid, so that larger drops behave more as solid spheres than is the case in clean systems. The result is to modify the diffusion mechanism, both within the drop and in the continuous-phase boundary layer. In light of these complications, it is perhaps not surprising that flow patterns and diffusion mechanisms have not yet been assembled to provide a quantitative description of mass transfer to and from single drops, let alone the operation of spray columns of industrial interest. The developing theory does, however, provide some useful guides and limits.

Mass transfer inside a completely stagnant drop can be expected to follow the equation for transient diffusion, which for spheres is

$$1 - \Delta' = 1 - \frac{6}{\pi^2} \sum_{n=1}^{\infty} \frac{1}{n^2} \exp\left( -\frac{4n^2\pi^2 Dt}{d_p^2} \right) \tag{6.22}$$

Values of $\Delta'$ as a function of $4Dt/d_p^2$ are given in Table 3.2. This would be expected to apply to the overall mass transfer in situations where there is no surface resistance ("barrier") and negligible resistance in the continuous phase. It sets a limit, in the sense that drop oscillations and internal circulation will always increase the diffusion rate as compared with that in a stagnant drop. The upper dotted curve in Fig. 6.12 represents Eq. (6.22) for the conditions of the tests with acetic acid in water reported by Licht and Pansing. It is evident that the rate of extraction was several times greater than calculated in this way, which is typical of all but very small drops.

In passing it may be noted that if $c_i$ is constant and a logarithmic mean of initial and final values of $c_i - c_{av}$ is used, the average mass-transfer coefficient $k_c$ for a spherical drop rising or falling for time $t$ is related to $\Delta'$ by

$$k_c = \frac{d_p}{6t} \ln\left( \frac{1}{\Delta'} \right) \qquad \text{average over time } t \tag{6.23}$$

In this transient process, $k_c$ varies with time. When transfer has gone 63 percent of the way to equilibrium ($\Delta' = 1/e$; $4Dt/d_p^2 = 0.055$), then $k_c d_p/D = 12.1$. This gives some guidance as to the relative resistances of the dispersed and continuous phases, since correlations of the latter involve $k_c$ (continuous phase) $\times d_p/D$.

The effect of circulation within the drop on the rate of diffusion has been the subject of many analytical studies. Hadamard [83] described the flow field, and Kronig and Brink [125] applied the diffusion equation to obtain an expression for $\Delta'$ as a function $Dt/d_p^2$.

Levich [133] has an excellent treatment of the theory of circulating drops, to which he has contributed very considerably himself. Boyadzhiev, Elenkov, and Kyuchukov [13] employ a simplified form of Eq. (6.22) with the introduction of an effective diffusion coefficient which is $\overline{R}$ times the molecular diffusion coefficient $D$.

Two different sets of drop-extraction data were correlated in this way, $\overline{R}$ ranging from 2 to 10. $\overline{R}$ was found to be a function of a modified Reynolds number based on the surface velocity of the circulating drop at its equator, as developed by Levich.

*Drop oscillation* is to be expected in most systems if the drop Reynolds number is greater than 500 to 1,000, but no good criterion is known. High drop velocity, large diameter, low interfacial tension, and low drop viscosity promote oscillation in clean systems. The ratio $\overline{R}$ of effective to molecular diffusion coefficients appears to lie in the range 1 to 4 for circulating drops but to be 10 or more in oscillating drops. An oscillating drop has a low resistance to internal diffusion but is considerably distorted from the spherical shape.

A theoretical analysis of diffusion in oscillating drops has been described by Handlos and Baron [84] and developed further by Olander [164], Patel and Wellek [166], and others. This leads to the expression

$$k_c = \frac{0.00375\overline{U}}{1 + \mu_d/\mu_c} \tag{6.24}$$

where $\overline{U}$ is the drop velocity relative to the continuous phase. Data on mass transfer with oscillating drops are reported in Refs. 84, 87, 135, 180, and 166.

If the *continuous-phase resistance* is dominant, the extraction of solute from a drop is still a transient process. If the drop is well mixed, or if the solute equilibrium strongly favors the dispersed phase and the solute concentration in the continuous phase is constant, then

$$1 - \Delta' = 1 - \exp\left(-\frac{6k_c t}{d_p}\right) \tag{6.25}$$

where $1 - \Delta' = $ fractional approach to equilibrium
$\quad\quad t = $ contact time
$\quad\quad k_c = $ external or continuous-phase mass-transfer coefficient

It is not yet possible to predict $k_c$ with any assurance, though certain limits can be set. If the drop is essentially rigid, as when the ratio $\mu_d/\mu_c$ of dispersed to continuous-phase viscosity is high, if it is quite small, or if drop circulation is eliminated by the presence of a surfactant, then $k_c$ may be estimated by Eq. (6.18) (Re < 10) or by the Ranz and Marshall correlation (Fig. 6.9) for solid spheres. The terminal velocity must first be obtained, in either case.

Applying the penetration theory, one may assume the contact time to be the time required for the drop to rise or fall one drop diameter. Then from Eq. (3.45) it is not difficult to derive the relation

$$\frac{k_c d_p}{D} = 1.13\sqrt{\frac{d_p \overline{U}}{D}} = 1.13\,\text{Pe}^{1/2} = 1.13\,\text{Re}^{1/2}\,\text{Sc}^{1/2} \tag{6.26}$$

This equation was obtained by Boussinesq from potential flow theory. Garner and Tayeban point out that flow separation occurs near the drop waist and that the wake behind the drop carries solute, so that the coefficient 1.13 should be reduced

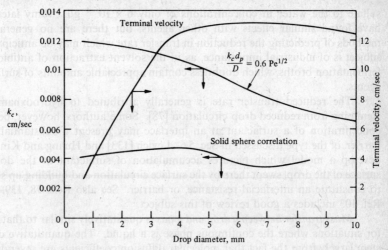

FIGURE 6.13

Continuous-phase mass-transfer coefficients for drops, calculated from Eq. (6.26) (taking the constant to be 0.6 instead of 1.13) and compared with values for solid spheres. The example is nitrobenzene falling in water at terminal velocities reported by Hu and Kintner [102]. $\mu_d = 0.0174$, $\mu_c = 0.00884$ poises; $\rho_d = 1.195$, $\rho_c = 0.997$ g/cm$^3$; $D = 10^{-5}$ cm$^2$/s; Sc $= 886$.

to perhaps 0.6. Figure 6.13 compares values of $k_c$ calculated in this way with those obtained from Fig. 6.9 for the particular example of nitrobenzene drops falling in water. Experimental terminal velocities $\overline{U}_T$, as reported by Hu and Kintner [102] can be used only for quite small drops. Where experimental values of $\overline{U}_T$ are lacking, the Hu and Kintner correlation of drag coefficients for drops may be used.

The drop circulation which develops in clean systems with drops of intermediate sizes increases the continuous phase, $k_c$, for two reasons: the terminal velocity is larger than for rigid drops, and the mass transfer is enhanced by the slip of the drop surface (the boundary layer is thinned, quite apart from the increased terminal velocity). (See, however, Ref. 204.) The oscillation of the larger drops leads to lower terminal velocities than for spheres but increases the ratio of surface to volume. That the continuous-phase coefficient should be several times as great for medium-sized drops as for solid spheres, as indicated by Fig. 6.13, has been confirmed by several investigators [12, 70, 78, 212].

The considerable literature on both theoretical and experimental studies of the continuous-phase resistance is covered in the excellent review by Sideman and Shabtai [190].

In an early study of the role of surfactants, Lindland and Terjesen [138] found the rate of extraction of iodine by water from 2 to 7 mm drops of carbon tetrachloride to be reduced some 70 percent by adding the surfactant oleyl-p-anisidine sulfate to the water in concentrations of only $6 \times 10^{-4}$ g/l. Many later workers have found similar effects with other agents, but there are no generally useful methods of predicting the reduction in transfer rate which may be anticipated. The subject is of industrial importance, as in the solvent extraction of antibiotics from fermentation broths, which doubtless contain appreciable amounts of surface-active agents.

The reduced transfer rate is generally attributed to hydrodynamic effects stemming from reduced drop circulation [78]. Some authors, however, believe that concentration of a surfactant at an interface may present a substantial diffusion barrier, of the type discussed in Sec. 5.6. Levich [133] and Huang and Kintner [103] develop a model which pictures accumulation of surfactant at the downstream surface of the drop, swept there by the surface circulation and building up sufficiently to constitute an interfacial resistance, or barrier. See also Refs. 78, 139, and 204; Ref. 103 includes a good review of this subject.

*Mass transfer between a drop and a gas* is qualitatively similar to that described for situations where the continuous phase is a liquid. The quantitative differences stem largely from the fact that molecular diffusion coefficients are several orders of magnitude greater in gases than in liquids, resulting in much greater continuous-phase coefficients, and that the viscosity ratio $\mu_d/\mu_c$ is enormous, reducing but not eliminating the internal circulation within the drop.

Much of the data on which the correlation shown by Fig. 6.9 is based were for evaporation of drops in flowing gases, and this graph is useful for the estimation of $k_c$ in the continuous phase. The principal variables in most systems are drop size and relative velocity, with Sc generally falling in the range 0.6 to 2.6. In contrast with liquid-liquid systems, there is an important effect of total pressure on the diffusion potential for a fixed gas composition, and hence on transfer rate. In free fall in air, water drops larger than 8 to 10 mm tend to distort enough to break up. Measurements of the rates of evaporation of small drops less than 1 $\mu$m in diameter are reported by Davis and Chorbajian [37a].

Mass transfer within the drops is enhanced by the internal circulation which is set up by the gas-liquid surface shear forces. This has little effect on the evaporation of a pure liquid but is of importance in the absorption or desorption of a sparing-soluble gas. The early studies of Whitman, Long, and Wang [216] and Hatta and Baba [86] showed the absorption of $CO_2$ into 5-mm water drops falling freely to be 50 to 70 times greater than could be explained by transient molecular diffusion in a sphere. Similar results were obtained by Garner and Lane [67] for the absorption of $CO_2$ by drops of various liquids supported in a wind tunnel, and by Constan and Calvert [32] for absorption of $SO_2$ by supported drops of glycerine and glycols. Little or no circulation was observed in the case of the viscous liquids.

The terminal velocities of drops in gases are very high, and the time of exposure very short in most spray-type mass-transfer equipment. Equilibrium is not usually approached closely if the solubility is low, in spite of the high ratio of surface to volume. Water drops 5.5 mm in diameter falling freely through 52 cm of $CO_2$ (in 0.326 s) were only 22 percent saturated [216]. With the more viscous glycols, only 6 to 8 percent saturation is attained under similar conditions [32].

There is considerable evidence that internal circulation and mass-transfer rates are quite large during the periods of drop formation, release, and acceleration. Spray-type mass-transfer equipment should, as suggested by Hatta and Baba, be designed to collect and respray the liquid a number of times. This is particularly true if the solute is relatively insoluble in the liquid; actual values of $k_c$ for the continuous (gas) phase are very large, so a single spray serves if the solute is highly soluble, or if a fast reaction occurs in the liquid phase. Incidently, it should be noted that drop surface temperatures may be raised or lowered appreciably by heat effects associated with mass transfer with drops in a gas—very much less so in liquid-liquid systems.

Drops are of technical interest primarily because dispersion provides such a large surface in equipment of small volume. Studies of single drops are for the purpose of understanding the mechanism of the mass-transfer phenomena and are of limited use for design except for the important matter of selection of equipment type.

As noted above, liquids dispersed in a gas provide a very large surface of contact between the two phases, and sprays are employed for gas absorption, deaeration, and desorption of small amounts of gases which are sparingly soluble in the liquid. However, it has been shown recently by Simpson [190a] that most of the mass transfer takes place at the surface of the liquid sheet near the spray nozzle. Before it beaks up to form drops this sheet becomes exceedingly thin as it expands radially.

Simpson desorbed $CO_2$ from water, using a single commercial centrifugal spray nozzle in a 45-cm-ID cylindrical column at room temperature and 20 to 60 mm Hg total pressure. With the spray nozzle fixed at heights of only 2, 4, and 6 cm the approach to equilibrium as 85 to 93 percent; with a nozzle height of 26 cm the approach was 97 to 99 percent. The large role played by the liquid sheet at the nozzle is both surprising and of obvious importance.

## 6.7  BUBBLES

Dispersions of gases in liquids are widely employed for gas absorption in chemical reactors involving the reaction of a gas with a liquid. As in the case of liquid-liquid dispersions, the interfacial area can be very large: 3-mm spherical bubbles with a gas holdup of 25 percent provide 152 $ft^2/ft^3$—more than most tower packings. The performance of agitated vessels, flooded bubble columns, and perforated plates will be discussed briefly in Chap. 11. This section will treat only single bubbles.

Bubbles behave very much like drops, but their buoyancy and velocity of rise are greater. Mass transfer within the bubble is rapid because molecular diffusivities in gases are large, and the resistance to mass transfer on the liquid side of the interface is controlling.

In studies of single bubbles, the gas is introduced at the bottom of a liquid column through an orifice or small capillary. If the gas flow is very slow, the bubble is released when the buoyant force just overcomes the surface tension, at which time the bubble diameter is given by

$$d_p = \left(\frac{6d_0\sigma}{\Delta\rho g}\right)^{1/3} \tag{6.27}$$

where $d_0$ = orifice diameter

$\sigma$ = surface tension

$\Delta\rho$ = difference in density between liquid and gas

This applies when the liquid wets the orifice; it indicates a modest increase of $d_p$ with increase in $d_0$. Inertial effects become dominant at the larger flow rates employed with industrial multiorifice spargers, however, and the initial bubble size depends almost not at all on the orifice size. The role of the orifice in determining bubble size is treated at some length in Valentin's book "Absorption in Gas-Liquid Dispersions" [207]. As the bubbles rise through the liquid, they develop a size distribution resulting from the breakup of the larger bubbles and the tendency of bubbles of all sizes to coalesce.

Very small bubbles ($d_p <$ about 0.1 cm in purified water) behave as rigid spheres and rise steadily with a terminal velocity $U_T$ which is proportional to $d_p^2$. As in drops, there is internal circulation, and the velocity is somewhat greater than for solid spheres. (There is some evidence that mass transfer to or from a bubble influences the rate of free rise [36, 132].) Traces of surface-active impurities, however, tend to stop this internal circulation. Bubbles of intermediate sizes (roughly 0.2 to 1.5 cm, in water) become flattened and distorted, with oblate spheroidal or ellipsoidal shapes. These oscillate and "wabble" as they rise. Large bubbles ($d_p > 1.5$ cm) form spherical caps with shapes similar to the silhouette of an umbrella and rise steadily. The terminal velocity increases only some 35 percent as $d_p$ is increased from 1.5 to 4.8 cm. In the case of a nonspherical bubble, $d_p$ represents the diameter of a sphere of the same volume.

The terminal velocity depends on so many variables, including some which are unknown or uncontrollable (trace surface-active agents), that no generalization is possible. Figure 6.14 shows typical experimental results obtained with several systems [22, 24, 82, 108]. It appears that the terminal velocity is roughly constant at 22 to 26 cm/s over the relatively wide range of bubble sizes from 0.2 to 1.5 cm, which is the range of greatest industrial interest. These data are for water; $U_T$ in viscous liquids is very much less for $d_p < 0.8$ cm, but viscosity appears to have little effect with larger bubbles [22, 207]. The various theoretical attempts to express $U_T$ as a function of $d_p$ and the liquid properties are summarized in Ref. 22.

Mass transfer to or from single bubbles in free rise in a liquid has been studied by a great number of investigators. Potential flow theory has been applied to develop theoretical expressions for $k_c$, with allowances for the circulation within the bubble, the eccentricity of distorted bubbles, the spherical caps characterizing the very large bubbles, and the tendency of traces of surface-active agents to collect at the bottom surface of a rising bubble. These developments are summarized briefly in Refs. 22, 24, 108, and 21. See also Levich [133] and the theoretical study of local and average coefficients by Oellrich et al. [163].

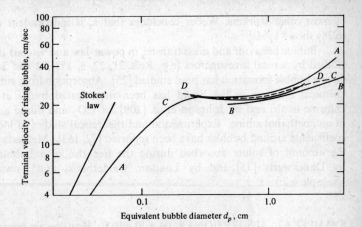

FIGURE 6.14
Terminal velocities of gas bubbles in pure water. $A$. Haberman and Morton (air) [82]; $B$. Calderbank and Lochiel ($CO_2$) [24]; $C$. Calderbank, Johnson, and Loudon ($CO_2$) [22]; $D$. Johnson, Besic, and Hamielec ($CO_2$) [108].

In brief, it appears that $k_c$ may be obtained from Eq. (6.18) if $d_p$ is less than about 0.1 cm. If Stokes' law applies, $k_c$ is independent of $d_p$ (see Example 6.1). Various authors have found that Eq. (6.26) fits the data fairly well for $d_p > 0.5$ cm [21]. With $U_T$ essentially constant, this means that $k_c$ is proportional to $d_p^{-1/2}$, and $k_c a$ is proportional to $d_p^{3/2}$. Calderbank and Lochiel [24] and Johnson, Besic, and Hamielec [108] have shown $k_c a$ nearly proportional to $d_p^2$ for $d_p > 0.4$ cm.

Although Eq. (6.18) gives a fair estimate of $k_c$, it appears that an empirical modification due to Johnson, Besic, and Hamielec gives a more reliable prediction:

$$\frac{k_c d_p}{D} = 1.13 \, Pe^{1/2} \left( \frac{d_p}{0.45 + 0.2d_p} \right) \tag{6.28}$$

The data of several investigators for absorption of $CO_2$, $C_2H_4$, and $C_4H_8$ show the constant 1.13 to be quite independent of Re over the range 500 to 8,000. Leonard and Houghton [132] report no effect of surface-active agents for $d_p > 0.6$ cm.

The difficulty in obtaining a good correlation of $k_c$ for bubbles is enormous, since there are so many variables. In view of this, it is interesting that for non-reacting gases in pure water with $D$ in the vicinity of $1 \times 10^{-5}$ to $2 \times 10^{-5}$ cm$^2$/s, much of the data show $k_c$ to fall in the relatively narrow range of 0.0015 to 0.010 cm/s, for $d_p$ 0.2 to 5.0 cm, though Weller [215] reports values as large as 0.037 cm/s for bubbles having equivalent diameters of 6 to 9 mm. In contradiction with the results of most other workers, Weller concludes that $k_c$ is independent of $D$. Equation (6.28) shows $k_c \alpha D^{1/2}$.

Bubble behavior and mass transfer in power-law and viscous liquids have been studied by several investigators (e.g., Refs. 21, 22, 3, 153, 82 and 214). Absorption during bubble formation has been studied [25]. Absorption from single bubbles with chemical reaction in the liquid has been investigated by Li et al. [136] using chlorine in air, and by Johnson et al. [109] for $CO_2$ and dilute aqueous solutions of monoethanol amine. Experimental and theoretical studies of local mass-transfer coefficients around bubbles have been reported [72, 163]. Methods of calculation of the amount of solute absorbed during the free rise of a bubble are developed by Danckwerts [35] and by Loudon, Calderbank, and Coward [141]. (See Example 6.2.)

EXAMPLE 6.2   ABSORPTION FROM A BUBBLE   How does the presence of an insoluble gas affect the amount of absorption of a soluble gas from a bubble containing a mixture of the two? Derive an expression relating fractional absorption of the soluble constituent and the time of contact of the bubble with a liquid absorbent. Compare the *fractional absorption* of the bubble containing no insoluble gas with that for a pure gas, the ratio of the two being expressed in terms of suitable variables.

In order to simplify the analysis, assume the bubbles always saturated with solvent, zero solute concentration in the liquid, constant total pressure, and constant bubble shape. The diffusional resistance is entirely in the liquid, and the variation of $k_c$ with bubble size may be neglected.

SOLUTION   (The analysis which follows is due to Danckwerts [35].)

First consider the stoichiometry. Let $V$ represent bubble volume and $n$ the number of moles, with $V_0$ being the initial volume and $n_A$, $n_B$, $n_C$, and $n_T$ being the moles soluble gas, insoluble gas, solvent vapor, and total moles, respectively. Then

$$n_A + n_B + n_C = n_T$$

and $n_B = n_{T0} Y_{B0} = (PV_0/RT)Y_{B0}$, $n_C = n_T Y_C$, where $Y$ represents mole fraction. No $B$ dissolves, so $n_B$ is constant, and $Y_C$, the ratio of solvent vapor pressure to total pressure, is also constant. The concentration $c_A$ of solute gas is then given by

$$c_A = \frac{n_A}{V} = \frac{n_T}{V}\left(1 - \frac{V_0}{V} Y_{B0} - Y_C\right)$$

Now let $Y'_{B0}$ represent the initial mole fraction insoluble gas, on a solvent vapor-free basis: $Y'_{B0} = Y_{B0}/(1 - Y_C)$. Then

$$c_A = \frac{n_T}{V}(1 - Y_C)\left(1 - \frac{V_0}{V} Y'_{B0}\right) \qquad (a)$$

The surface area of the constant-shape bubble is $\alpha V^{2/3}$, so the mass-transfer-rate equation is

$$-\frac{dn_A}{dt} = \alpha k_c V^{2/3}\left(\frac{c_A}{H} - 0\right) \qquad (b)$$

where $H$ is the Henry's-law constant $c_A(\text{gas})/c_A(\text{liquid})$ at equilibrium. Since $n_A = n_T - n_{T0}\,Y_{B0} - n_T\,Y_C$,

$$-(1 - Y_C)\frac{dn_T}{dt} = \frac{\alpha k_c}{H}\,V^{2/3}(1 - Y_C)\left(1 - \frac{V_0}{V}\,Y'_{B0}\right)\frac{n_T}{V}$$

or

$$-\frac{dV}{dt} = \frac{\alpha k_c}{H}\,V^{2/3}\left(1 - \frac{V_0}{V}\,Y'_{B0}\right) \qquad (c)$$

If the bubble contains no insoluble gas, $Y'_{B0}$ is zero, and integration gives

$$\alpha k_c\,\frac{t}{H} = 3(V_0^{1/3} - V^{1/3}) \qquad (d)$$

from which the time $\tau$ for the bubble of pure gas to disappear completely is seen to be $3HV_0^{1/3}/\alpha k_c$.

Equation $(c)$ can be integrated and the limits $V = V_0$ at $t = t_0$, $V = V$ at $t = t$, substituted to obtain the relation between $V$, $t$, and the constant $Y'_{B0}$ for a bubble containing insoluble gas. Since $n_A = n_T(1 - Y_C)[1 - (V_0/V)Y'_{B0}]$, and $n_T = PV/RT$, this result provides the function $n_A/n_{A0} = f_1(t/\tau, Y'_{B0})$. In the case of the pure gas, Eq. $(d)$ gives $n_A/n_{A0} = f_2(t/\tau)$.

At the outset, the fractional absorption $(1 - n_A/n_{A0})$ approaches zero at $t \to 0$ in both cases. The fractional absorption approaches unity in both cases as $t \to \infty$, so in both limits the ratio of the fractional absorption in one case to that in the other case is again unity.

Danckwerts has calculated numerical results for intermediate values of $t$, following the analysis outlined above, and shows a graph of the ratio of fractional absorption with insoluble gas present to fractional absorption from a bubble of pure gas vs. $t/\tau$ for various values of $Y'_{B0}$. Over the range of $Y'_{B0}$ from 0.2 to near unity, the ratio reaches minima of 0.88 to 0.90, at $t/\tau = 0.35$ to 0.55. Within 12 percent, therefore, the fractional absorption of the soluble gas is the same (for the same $\alpha k_c$), whether or not the initial bubble contains insoluble gas. Within the limitations of the assumptions, the rate of absorption in both cases is essentially proportional to the initial partial pressure of the soluble gas.

Loudon, Calderbank, and Coward [141] have extended this analysis allowing for the change in $k_c$ with bubble size, and for the varying total pressure as the bubble rises in a deep pool. Their conclusion is similar to Danckwert's. Hirose and Moo-Young [93] analyze the related problem of absorption from bubbles of a binary mixture of soluble gases. $\quad$ ////

## EXAMPLE 6.3 EFFECT OF TRANSPIRATION AND SHRINKAGE ON MASS TRANSFER FROM A BUBBLE

Section 6.4 described Hales' analysis of the effects of transpiration and shrinkage on mass transfer from a sphere in Stokes' flow. It is of interest to estimate the possible importance of these effects on absorption from a gas bubble.

Consider the case of a 0.1-cm-diameter spherical bubble of pure $CO_2$ in free rise at 12 cm/s in pure water at 20°C and 1.0 atm; $D$ is $2 \times 10^{-5}$ cm$^2$/s and $k_c^*$ is 0.003 cm/s. Are the corrections for transpiration and shrinkage important?

SOLUTION $\quad$ The Henry's-law constant is 1,420 atm/mole fraction (Perry), whence $H = c_A/c_L$ at equilibrium is 1.06.

$$-\frac{dV}{dt} = -4\pi r_0^2\,\frac{dr_0}{dt} = \frac{\alpha k_c\,V^{2/3}}{H} = \frac{4\pi r_0^2 k_c\,V^{2/3}}{V^{2/3}H}$$

which gives

$$-\frac{dr_0}{dt} = \frac{k_c}{H}$$

The radial Peclet number is $(2r_0/D)(dr_0/dt)$, so the parameter $Pe_{r_0}(D/k_c^* d_p)$ used by Hales is $(-k_c/k_c^* H)$. Brian and Hales' Fig. 6 [15] is a graph of $k_c/k_c^*$ vs. $Pe_{r_0}(D/k_c^* d_p)$, from which, for $H = 1.06$, $k_c/k_c^*$ is about 0.7. $(Pe = 0.1 \times 12/2 \times 10^{-5} = 6 \times 10^4.)$

The transpiration velocity at $r_0$ is

$$U_r = -\frac{1}{4\pi r_0^2}\frac{dV}{dt}\frac{c_A}{c_L} = -\frac{\alpha k_c V^{2/3}}{4\pi r_0^2} = -k_c$$

whence $\qquad Pe_T = \frac{d_p U_r}{D} = -\frac{k_c d_p}{D}\quad$ and $\quad Pe_T\left(\dfrac{D}{k_c^* d_p}\right)$ is $\dfrac{k_c}{k_c^*}$

Figure 4 of Brian and Hales is a graph of $k_c/k_c^*$ vs. $Pe_T(D/k_c^* d_p)$, from which it appears that $k_c/k_c^*$ is about 0.75. The correction for both transpiration and shrinkage gives $k_c/k_c^* = 0.7 \times 0.75 = 0.53$. The indicated correction would be enormous for a bubble of a highly soluble gas such as ammonia.

In the case of the 0.1-cm bubble in free rise, $Re \approx 120$. This is well out of the Stokes flow region and Brian and Hales' analysis does not apply. Ward, Träss, and Johnson [212] found a Stokes-flow analysis to check data on transport to spheres at Reynolds numbers up to 10, so it seems likely that a substantial correction for transpiration and shrinkage may be required for typical bubbles of diameters 0.2 to 1.0 cm in free rise, especially if the gas is highly soluble.

Note that the bubble diameter and velocity do not enter into an analysis based on Brian and Hales—these come in only in the consideration of whether or not the procedure is applicable. ////

## 6.8  CYLINDERS AND DISKS

### (a)  Turbulent Flow Normal to Single Cylinders

For flow normal to a fixed cylinder, both mass- and heat-transfer data are conveniently represented by a graph of $j_D$ or $j_H$ vs. Re. The 1952 version of this book showed the early data of Lohrisch [140], on water vapor absorption, and of Powell [173, 174], on vaporization of water, to agree well with McAdams' correlation for heat transfer. The latter was based on extensive data obtained by many investigators, using small wires and cylinders up to 11.4 cm in diameter.

Figure 6.15 shows the McAdams correlation as curve $B$, with the Lohrisch and Powell points omitted. A few representative newer data are shown, which support the McAdams curve and tend to confirm the equivalence of $j_H$ and $j_D$. Curve $A$ represents the data of Vogtlander and Bakker [210] on mass transfer to wires in flowing liquids with high values of Sc. Curve $C$ represents the data of Johnson and Joubert [111] on heat transfer from a 15.2-cm cylinder placed normal to an air stream. Also shown are three points from Sogin and Subramanian [192], who sublimed naphthalene into air at high Reynolds numbers. The other three points are the recent data of Kestin and Wood [121] for the sublimation of p-dichlorobenzene into air flowing with a turbulent intensity of 0.28 percent. The correlation indicated by Fig. 6.15 covers a remarkably wide range of values of Sc, cylinder diameter, and fluid velocity for both mass and heat transfer.

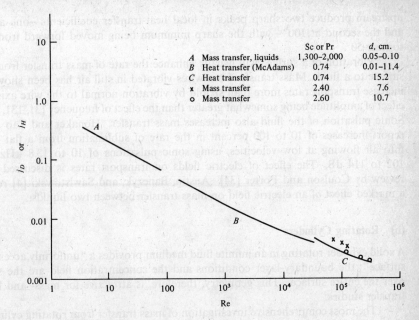

|   | | Sc or Pr | d, cm. |
|---|---|---|---|
| A | Mass transfer, liquids | 1,300–2,000 | 0.05–0.10 |
| B | Heat transfer (McAdams) | 0.74 | 0.01–11.4 |
| C | Heat transfer | 0.74 | 15.2 |
| x | Mass transfer | 2.40 | 7.6 |
| o | Mass transfer | 2.60 | 10.7 |

**FIGURE 6.15**
Mass and heat transfer for cylinders placed normal to the flow of gases and liquids.
Coefficients are average values for the whole surface.

As in other flow systems, heat- and mass-transfer coefficients averaged over the cylindrical surface are affected by changes in the intensity and scale of the turbulence of the flowing stream. The early work of Comings, Clapp, and Taylor [31] showed $j_H$ to increase some 25 percent when grid-type turbulence generators were placed upstream of the cylinder. Maisel [147] found $j_D$ to increase 50 percent as the turbulent intensity of the stream was increased from 3.5 to 24 percent. Many other investigators have reported similar results (e.g., Ref. 156 on heat transfer to spheres). Some of the scatter of published data plotted as in Fig. 6.15 is doubtless due to differences in stream turbulence.

*Local* heat- and mass-transfer coefficients vary considerably over the perimeter of the cylinder, usually being greatest at the forward stagnation point but falling to a low minimum at the point of separation (in the vicinity of 80°). The coefficients then increase to maxima at the rear of the cylinder (180°) and are nearly as large as at the front stagnation point. Similar variation is noted for both mass- and heat-transfer coefficients. Kestin and Wood [121], in studies of local coefficients for the sublimation of *p*-dichlorobenzene, found increased turbulence intensity to introduce new maxima in the vicinity of 110 to 120° where the local coefficients were greater than either at the front stagnation point or at the rear. Sogin and Subramanian [192] noted similar peaks in the local heat-transfer coefficient in the same region at high Reynolds numbers. Johnson and Joubert [111] show that vortex generators upstream produce two sharp peaks in local heat-transfer coefficients—one at 60° and the second at 100°—with the sharp minimum being moved forward from 80° to near 50°.

Vibration of a surface will usually enhance the rate of mass transfer from the surface to a fluid. Mass transfer from wires vibrated in still air has been shown to increase transport rates more than sixfold by vibration normal to the wire axis, the effect of amplitude being somewhat greater than the effect of frequency [31, 131, 123]. Sonic pulsation of the fluid also increases mass transfer. Honaker and Tao [100] report increases of 10 to 100 percent in the rate of sublimation from a flat plate into air flowing at low velocities, using sonic pulsations of 10 to 13.8 kHz and 102 to 114 dB. The effect of electric fields on transport rates is discussed in a review by Coulson and Porter [34]. Austin, Banczyk, and Sawistowski [4] report a marked effect of an electric field on mass transfer between two liquids.

### (b)  Rotating Cylinders

A solid cylinder rotating in an infinite fluid medium provides a "uniformly accessible surface"; the boundary-layer conditions and the concentration field are the same over the entire surface. This geometry, therefore, is attractive for mass- and heat-transfer studies.

The most comprehensive investigation of mass transfer from rotating cylinders is that of Eisenberg, Tobias, and Wilke [46]. Cylinders of diameters 1.98 to 5.98 cm were rotated in fixed cylindrical vessels having inside diameters of 6.3, 10.1, and 13.9 cm. Data were obtained for the dissolution of benzoic and of cinnamic acid and for two electrolytic redox reactions. Reynolds numbers, based on the diameter and peripheral speed of the inner rotating cylinder, varied from 112 to 241,000; Sc ranged from 835 to 11,490.

Though Taylor vortices in the annulus might be expected to complicate such a study, an excellent correlation of the data was obtained using a Reynolds number as defined above; the variation of the dimensions of the gap between the two cylinders from 2.07 to 5.96 cm showed no effect. The observed mass-transfer coefficient varied as $Sc^{-0.644}$.

Curve $BB$ on Fig. 6.16 represents the Eisenberg data. The ordinate $j'_D$ is a modified $j_D$ defined as $j'_D = (k_c/\overline{U})Sc^{0.644}$; the Reynolds number is based on the diameter and the peripheral velocity of the rotating cylinder. Subsequent measurements of benzoic acid dissolution by Ryan [188] agree with Eisenberg's results for $Re > 600$. Also shown on Fig. 6.16 are curves representing data on heat transfer from rotating cylinders to air and to liquids. These fall above the mass-transfer line, perhaps because of the relatively greater effect of free convection.

One of the interesting features of Fig. 6.16 is the rather close agreement of the mass- and heat-transfer data with curve $FF$, representing $f/2$, where $f$ is the friction data obtained by Theodorsen and Regier [202] from torque measurements. In this case of skin friction without form drag, the Chilton-Colburn analogy holds well: $j_D \approx j_H \approx f/2$. The correlation is remarkable in that it brings together data on heat transfer, mass transfer, and friction for cylinders rotating in gases and liquids over a range of Sc and Re of more than a thousandfold.

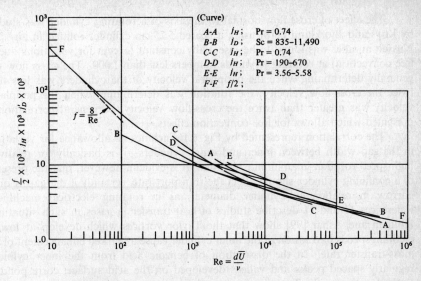

**FIGURE 6.16**
Heat-, mass-, and momentum-transfer correlations for cylinders rotating in gases and liquids. Data: $AA$, Dropkin and Carmi [41]; $BB$, Eisenberg, Tobias, and Wilke [46]; $CC$, Kays and Bjorkland [117, 118]; $DD$, and $EE$, Seban and Johnson [186]; $FF$, Theodorsen and Regier [202].

Bennett and Lewis [10] report data on dissolution of benzoic acid into water and also for dissolution of tin, lead, and zinc into mercury; the rotating cylinders were 1.0 cm in diameter. The results agree quite well with curve $BB$ of Fig. 6.16 except in the case of zinc. The zinc surface became etched and rough, indicating a surface resistance to breakdown of the zinc crystallites. Similar results were obtained when the vessel rather than the cylindrical specimen was rotated. Data on the dissolution of copper, nickel, and copper-nickel alloys into molten lead are reported by Stevenson and Wulf [196]. In the case of the alloys, the rate decreased as dissolution progressed, evidently because the composition of the surface changed as the more soluble component was preferentially dissolved.

Kappessar [115] provides data on oxygen transport in water to a rotating 6.35-cm polished metal cylinder, using an electrolytic technique. The results check well with the Eisenberg correlation (Fig. 6.16) over the range $1,000 < \text{Re} < 300,000$. Russian workers [80] have recently reported data on the dissolution of 3.5- and 6.0-cm cylinders of benzoic acid rotating in water and in 35 to 80 weight percent aqueous glycerol. When plotted as $j_D$ vs. Re, as on Fig. 6.16, the water data agree quite well with curve $BB$. However, the points for the viscous glycerol solutions fall 20 to 50 percent lower, $j_D$ evidently decreasing with increase in viscosity at a fixed Reynolds number. The authors concluded the exponent on the Schmidt group should be 0.73 rather than 0.644.

The effect of cross flow normal to the axis of a rotating cylinder was studied by Kays and Bjorkland [118] using a heated 5.75-cm cylinder rotated in air. The Nusselt number was found to be essentially constant (except for variations due to free convection) at rotating Reynolds numbers less than 1,000. The cross flow was generally determining when the peripheral velocity of the cylinders was less than twice the cross-flow velocity; the rotation was determining when the peripheral velocity was greater than twice the cross-flow velocity. A general correlation is presented which allows for free-convection effects.

The correlation represented by Fig. 6.16 includes no allowance for variations in the gap width between inner and outer cylinders—it is basically for relatively high-speed rotation in an infinite medium. It is evident, however, that the presence of a confining cylinder should affect the transport rate, certainly if the gap is quite narrow in relation to cylinder diameter, as in rotating electrical machinery. References 11 and 71 describe studies of heat transfer in gases in such situations. Holman and Ashar [99] show that the Taylor vortices which develop at low to medium speeds of rotation of the inner cylinder cause a marked enhancement of the mass-transfer rate. In the dissolution of benzoic acid from the inner cylinder, regularly spaced peaks and valleys developed on the acid surface, corresponding to the location of the Taylor vortices.

## (c) Rotating Disks

Rotating disks are frequently employed in electrochemical research, since with laminar flow the surface is "uniformly accessible," i.e., the thickness of the diffusion boundary layer (and the local mass-transfer coefficient) is the same at all points on the surface. The theory of mass transfer between the rotating surface and a fluid, for both laminar and turbulent regimes, is presented in considerable detail by Levich [133], who also describes the use of rotating disks in electrochemistry.

As noted in Sec. 6.4, the steady-state diffusion from one face of a round disk, into an infinite stagnant medium, is given by

$$\frac{k_c d}{D} = \frac{8}{\pi} \tag{6.29}$$

where $d$ is the disk diameter.

As a disk is rotated, the ambient fluid is drawn toward the disk face and then forced away radially over the surface by the centrifugal force developed by the surface drag (see Fig. 3.6). The flow field is complicated, but exact solutions to the Navier-Stokes equations can be obtained for *laminar motion*, as shown by Levich (see also Refs. 29, 99a, and 77). The asymptotic solution for $Sc \to \infty$ is

$$\frac{k_c d}{D} = 0.879 Re^{1/2} Sc^{1/3} \tag{6.30}$$

This is identical with Eq. (3.68). The mass-transfer coefficient $k_c$ is everywhere the same over the surface of the disk of diameter $d$. The Reynolds number is here defined in terms of the disk diameter and the linear speed at the periphery ($U = \omega d/2$, where $\omega$ is the rotational speed in radians per second). Re is defined in various other ways in the literature, often as $\omega r_0^2/v$, where $r_0 = d/2$. Re as defined above is twice this.

Sparrow and Gregg [193] also obtained Eq. (6.30) for laminar flow with large Sc. These authors furthermore obtained solutions for values of Sc from 0.01 to 100 [Eq. (6.30) is valid for Sc greater than about 100].

Cornet and Kaloo [33], in a report of oxygen transport to a rotating disk in salt water, as measured electrochemically, show *turbulent flow* to develop as Re is increased above $6 \times 10^5$. Levich indicates that the critical Re lies in the range $2 \times 10^4$ to $10^5$. Cornet and Kaloo's data for turbulent flow can be expressed approximately by

$$\frac{k_c d}{D} = 5.6 \text{Re}^{1.1} \text{Sc}^{1/3} \qquad 6 \times 10^5 < \text{Re} < 2 \times 10^6; \quad 120 < \text{Sc} < 1,200 \quad (6.31)$$

Levich gives a semitheoretical equation for the *turbulent range* which suggests that $k_c d/D$ should vary at $(\text{Re})^{0.4}(\text{Sc})^{0.25}$. Levich also describes electrochemical studies by I. A. Bagotskaya, who found $k_c$ proportional to the first power of rotational speed in the case of turbulent flow on a rotating disk.

## 6.9 SINGLE-PHASE FLOW IN PACKED BEDS

Data on the rate of mass transfer between beds of particles and a flowing gas or liquid are needed in the design of the many industrial devices used for adsorption, leaching, ion exchange, chromatography, and, of particular importance, hetero-geneous catalysis and catalyst regeneration. Numerous studies have been carried out with the object of measuring mass-transfer coefficients in packed beds and correlating the results. Most of these studies have involved sublimation of solids or evaporation of liquids from porous pellets, with gas flow through the bed. With the use of liquids, the technique of partial dissolution of the solid particles forming the bed is common.

Heat- and mass-transfer resistances between catalyst pellets and the flowing stream may not be significant in cases where the catalyst is only moderately active, but the external resistances become of importance if the catalyst is highly active (low effectiveness factor). In the latter case the temperature and concentration differences between pellet surface and the fluid may be quite large.

Evaporation or sublimation into a gas is rapid if the particles are small, since the transfer coefficients are large and the bed has a high ratio of pellet surface to bed volume. The result is that the gas normally approaches equilibrium (saturation) in flow through only a few inches of bed. This is why so many of the reported investigations with gases have been carried out using beds only 1 to 5 in long; it is necessary to measure the driving force at the outlet. This is much less of a problem with liquids.

As in many other systems, the Chilton-Colburn analogy holds to the extent that $j_H$ and $j_D$ are essentially identical for the same bed geometry and flow conditions. Data on heat transfer in beds of particles, therefore, supplement the mass-transfer measurements in establishing a base for the prediction of $k_c$. The relation between the fluid friction and the transfer coefficients is not well developed, though theoretical studies have appeared [26, 48, 175]. A theoretical analysis of mass transfer at particle surfaces within a bed is described by Pfeffer and Happel [168].

Representative data on mass transfer in packed beds are shown in Figs. 6.17 and 6.18 as graphs of $j_D$ vs. Re. Table 6.2 provides a key to experimental methods and conditions. Each line shown represents a large number of data points, which in some cases spatter appreciably on both sides. With one exception, each set of data is for spheres or cylinders of uniform size. Re is defined in terms of the *superficial* fluid velocity $\overline{U}_{av}$ and a dimension $d_p$, which is the diameter of a sphere having the same surface or volume as the particle. The agreement between the several investigators is poor, but the experimental difficulties are considerable and the spread of the lines is perhaps not much greater than the experimental error.

Superposition of Figs. 6.17 and 6.18 reveals generally good agreement of the results for gases and for liquids. A single straight line (shown dashed) does a fair job of representing all the data. This is represented by the following equation, which may be employed for engineering estimates:

$$j_D = 1.17 \left( \frac{d_p \overline{U}_{av} \rho}{\mu} \right)^{-0.415} \qquad 10 < \text{Re} < 2{,}500 \qquad (6.32)$$

**Table 6.2** **REPRESENTATIVE DATA ON MASS TRANSFER IN PACKED BEDS OF SPHERES AND CYLINDRICAL PELLETS WITH SINGLE-PHASE FLUID FLOW\* (KEY TO FIGS. 6.17 AND 6.18)**

| Line | System | $d_p$ range, cm | Sc | Investigator |
|---|---|---|---|---|
| | Gases: | | | |
| 1 | Water–air | 0.23–1.9 | 0.61 | Gamson et al. [61] |
| | | | | Wilke and Hougen [218] |
| 2 | $C_{10}H_{12}$–air, He | 0.64 | 2.7, 4.1 | Fallat [51] |
| 3 | $C_{10}H_{12}$–air | 0.4–0.8 | 2.6 | Bar-Ilan and Resnick [6] |
| 4 | $C_{10}H_{12}$–air | 0.64–1.27 | 2.6 | Bradshaw and Bennett [14] |
| 5 | Water–air | 1.6 | 0.61 | de Acetis and Thodos [39] |
| 6 | Water–air | 1.6 | 0.61 | McConnachie and Thodos [145] |
| 7 | $C_{10}H_{12}$–air | 0.069–1.4 | 2.6 | Chu et al. [28] |
| 8 | Water–air | 1.6 | 0.61 | Gupta and Thodos [81] |
| 9 | Various† | 0.78 | 0.61–5.1 | Hobson and Thodos [97] |
| | Liquids: | | | |
| 1 | Water–isobutanol | 0.91 | 865 | |
| | Water–methyl ethyl ketone | 0.91, 1.61 | 776 | Hobson and Thodos [96] |
| 2 | Water–benzoic acid | 0.63 | (1850?) | Williamson et al. [220] |
| 3 | Water–benzoic acid | 0.056–0.21‡ | (1850?) | Evans and Gerald [49] |
| 4 | Water–2-naphthol | 0.32–0.64 | 1189–1456 | McCune and Wilhelm [146] |
| 5 | Benzene–salicylic acid | | 340–434 | |
| | n-Butanol–succinic acid | 0.063–1.27 | 10,100–12,300 | Gaffney and Drew [58] |
| | Acetone–succinic acid | | 159–185 | |

\* A similar graph, with data from additional sources, is given by Marcussen [150].

† Water–air; n-butanol–air; n-butanol–$N_2$; toluene–air; n-octane–air; n-octane–$N_2$; n-butanol–$CO_2$; toluene–$CO_2$; n-butanol–$H_2$; n-dodecane–$H_2$.

‡ Screened granules of mesh sizes 24–28, 20–24, 12–14, and 8–10.

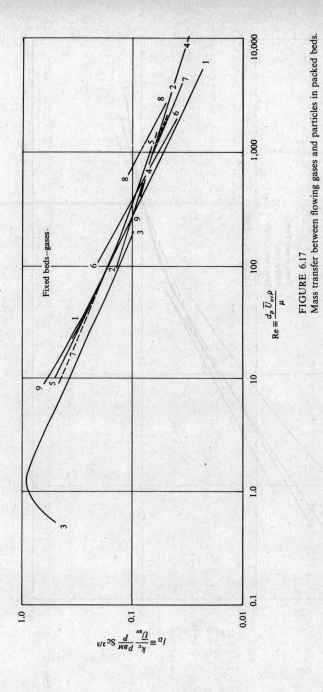

FIGURE 6.17
Mass transfer between flowing gases and particles in packed beds.

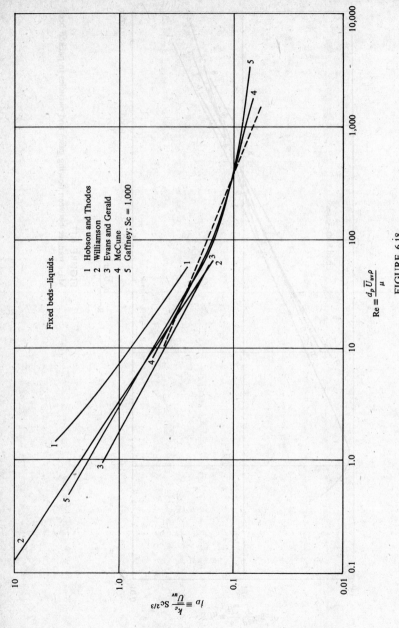

Fixed beds—liquids.

1 Hobson and Thodos
2 Williamson
3 Evans and Gerald
4 McCune
5 Gaffney; Sc = 1,000

$Re \equiv \dfrac{d_p \overline{U}_{av} \rho}{\mu}$

$j_D \equiv \dfrac{k_c}{\overline{U}_{av}} Sc^{2/3}$

FIGURE 6.18
Mass transfer between flowing liquids and particles in packed beds.

This also agrees well with Froment's collection of published data [55]. However, a recent study of published heat-transfer data [80a], using several hundred data points, suggests that a better correlation is obtained by plotting $\varepsilon j_H$ vs. $d_p \bar{U}_{av} \rho / \mu$.

Correlations of the type shown by Figs. 6.17 and 6.18 fail to allow for variations in the void fraction $\varepsilon$, which in fixed beds of spheres and pellets is typically 0.4 to 0.44 but may range from 0.3 to 0.5. McConnachie and Thodos [145] used "distended" beds of spheres strung on wires to vary $\varepsilon$ from 0.416 to 0.778; the data were brought together by introducing $1 - \varepsilon$ in the denominator of Re. Chu et al. [28] and Riccetti and Thodos [178] used the same method to obtain a single line for both fixed and fluidized beds. Gupta and Thodos [81] suggest graphs of $\varepsilon j_D$ vs. Re. Data · on mass transfer in liquids at very low values of Re are reported by Mandelbaum and Böhm [148], who found it necessary to introduce the Graetz group to allow for free-convection effects due to density differences at the low flow rates.

The shape or "shape factor" of nonspherical particles is a variable about which little is known. Bar-Ilan and Resnick [6] found much lower values of $j_D$ than suggested by Fig. 6.17 in tests on fixed beds of very small granular particles $(d_p < 0.05 \text{ cm})$.

Most of the investigators have analyzed their data on the assumption that plug flow prevailed, with no allowances for radial variation in fluid velocity or axial dispersion. See, however, Ref. 14. Good surface-temperature measurements are difficult. Gamson et al. and Wilke calculated $k_c$ from their data on evaporation of liquids on the assumption that the surface was at the wet-bulb temperature of the air; the probability that it was somewhat higher [39] suggests that the line representing their data may be high at the low Reynolds numbers. If so, the ratio $j_H/j_D$ is perhaps greater than the value 1.076 which they report.

Mass transfer from a single sphere in an array of similar spheres has been studied by several investigators. Thoenes and Kramers [203] used both gases and liquids with eight packing geometries. An approximate equation representing some 438 tests is given as

$$\varepsilon j_D = 1.0 \left( \frac{\text{Re}}{1 - \varepsilon} \right)^{-1/2} \qquad 0.25 < \varepsilon < 0.5; \quad 40 < \text{Re}/(1 - \varepsilon) < 4,000 \qquad (6.33)$$

The indicated negative slope is greater than shown by Eq. (6.32), but for $\varepsilon = 0.4$ the two equations agree at Re $= 550$.

Another study of mass transfer to a single sphere in a randomly packed array of 2.54-cm spheres is reported by Jolls and Hanratty [113] who used electrochemical techniques in a liquid system with Sc $= 1,700$. The data are well represented by the equation

$$j_D = 1.59 \text{Re}^{-0.44} \qquad 140 < \text{Re} < 900 \qquad (6.34)$$

This gives values of $j_D$ 13 to 20 percent greater than calculated by Eq. (6.32) in the applicable range of Re.

## EXAMPLE 6.4

(a) Water at 20°C flows through a packed bed of 5-mm-diameter spheres of benzoic acid, at a superficial velocity of 1.0 ft/s. Assuming piston flow, estimate the packed height required for the entering pure water to become 90 percent saturated with benzoic acid.

(b) Repeat for the case of dry air at 1.0 atm entering a bed of water-wet porous 5-mm spheres, the superficial velocity again being 1.0 ft/s. Assume evaporation to occur at the surface of the wet spheres at a uniform and constant temperature which is essentially the wet-bulb temperature. Percent saturation refers to saturated air at this temperature.

*Data:* At 20° the viscosity and density of the dilute solution of benzoic acid in water are 0.01 poises and 1.0 g/cm$^3$, respectively. The molecular diffusion coefficient is $0.77 \times 10^5$ cm$^2$/s. In the case of air, the density and viscosity of dry air may be used, with $D_{AB} = 0.233$ cm$^2$/s. The void fraction in the bed of spheres is 0.41.

SOLUTION

(a) Using the data given for the aqueous system,

$$\text{Re} = \frac{d_p \, \overline{U}_{av} \rho}{\mu} = \frac{0.5 \times 30.48 \times 1.0}{0.01} = 1{,}524 \qquad \text{Sc} = \frac{0.01}{1 \times 0.77 \times 10^{-5}} = 1{,}300$$

From either Eq. (6.32) or Fig. 6.18, $j_D = 0.056$;

$$k_c = j_D \, \overline{U}_{av} \, \text{Sc}^{-2/3} = \frac{0.056 \times 30.48}{1{,}300^{2/3}} = 0.0143 \text{ cm/s}$$

The ratio of surface to volume of a single sphere is $6/d_p$, and since the bed void fraction is 0.41,

$$a = \frac{\text{surface}}{\text{unit bed volume}} = \frac{6}{0.5}(1 - 0.41) = 7.08 \text{ cm}^2/\text{cm}^3$$

For piston flow through the bed with a constant-surface concentration $c_i$ at the surface,

$$\overline{U}_{av} \, dc = k_c a (c_i - c) dx$$

where $\overline{U}_{av}$ = superficial fluid velocity
$c$ = solute concentration in the flowing fluid
$x$ = distance from the inlet

Integration gives

$$x = \frac{\overline{U}_{av}}{k_c a} \ln \frac{c_i - c_1}{c_i - c_2}$$

The liquid in contact with surface is saturated and the inlet concentration $c_1$ is zero, so the effluent will be 90 percent saturated when $c_2 = 0.9c_i$:

$$x = \frac{30.48}{0.0143 \times 7.08} \ln \frac{1}{0.1} = 693 \text{ cm (22.7 ft)}$$

(b) Following the same calculation for the air-water system, with $\rho = 0.00121$, $\mu = 0.000172$ poises, Sc = 0.61, Re = 107, $j_D = 0.17$, and

$$k_c = \frac{0.17 \times 30.48}{0.61^{2/3}} = 7.2 \text{ cm/s}$$

whence 

$$x = \frac{30.48}{7.2 \times 7.08} \ln 10 = 1.38 \text{ cm (0.54 in)}$$

The purpose of this example is to bring out the fact that first-order processes (evaporation, dissolution, chemical reaction at the surface with diffusion controlling) in beds of small particles approach equilibrium in a few inches in gas systems but are slow in liquid systems. The principal reason is the enormous difference in $D$.  ////

# REFERENCES

1   ANSHUS, B. E., and S. L. GOREN: *AIChE J.*, **12:** 1004 (1966).
2   ASKEW, W. S., and R. B. BECKMANN: *Ind. Eng. Chem. Process Des. Dev.*, **4:** 311 (1965).
3   ASTARITA, G., and G. APAZZO: *AIChE J.*, **11:** 815 (1965).
4   AUSTIN, L. J., L. BANCZYK, and H. SAWISTOWSKI: *Chem. Eng. Sci.*, **26:** 2120 (1971).
5   BANERJEE, S., E. RHODES, and D. S. SCOTT: *Chem. Eng. Sci.*, **22:** 43 (1967).
6   BAR-ILAN, M., and W. RESNICK: *Ind. Eng. Chem.*, **49:** 313 (1957).
7   BARKER, J. J., and R. E. TREYBAL: *AIChE J.*, **6:** 289 (1960).
8   BAYS, G. S., JR., and W. H. MCADAMS: *Ind. Eng. Chem.*, **29:** 1240 (1937).
9   BEEK, W. J., and H. KRAMERS: *Chem. Eng. Sci.*, **17:** 909 (1962).
10  BENNETT, J. A. R., and J. B. LEWIS: *AIChE J.*, **4:** 418 (1958); *J. Chimie Phys.*, **55:** 83 (1958).
11  BJORKLAND, I. S., and W. M. KAYS: *ASME Paper 58-A-99* (1958).
12  BOWMAN, C. W., D. M. WARD, A. I. JOHNSON, and O. TRÄSS: *Can. J. Chem. Eng.*, **39:** 9 (1961).
13  BOYADZHIEV, L., D. ELENKOV, and G. KYUCHUKOV: *Can. J. Chem. Eng.*, **47:** 42 (1969).
14  BRADSHAW, R. D., and C. O. BENNETT: *AIChE J.*, **7:** 48 (1961).
15  BRIAN, P. L. T., and H. B. HALES: *AIChE J.*, **15:** 419 (1969).
16  BROWN, G. M.: *AIChE J.*, **6:** 179 (1960).
17  BROWN, R. A. S., and B. H. SAGE: *J. Chem. Eng. Data*, **6:** 355 (1961).
18  BROWN, R. A. S., K. SATO, and B. H. SAGE: *J. Chem. Eng. Data*, **3:** 263 (1958).
19  BUTLER, R. M., and A. C. PLEWES: *AIChE Symp. Ser.*, (10): 121 (1955).
20  CABLE, M.: *Chem. Eng. Sci.*, **22:** 1393 (1967).
21  CALDERBANK, P. H.: In N. Blakebrough (ed.), "Biochemical and Biological Eng. Science," chap. 5, Academic, New York, 1967; in V. W. Uhl and J. B. Gray (eds.), "Mixing: Theory and Practice" vol. II, chap 1, Academic, New York, 1967; also in *Chem. Eng. (London)*, no. 212, p. CE209, October 1967.
22  CALDERBANK, P. H., S. L. JOHNSON, and J. LOUDON: *Chem. Eng. Sci.*, **25:** 235 (1970).
23  CALDERBANK, P. H., and S. J. R. JONES: *Trans. Inst. Chem. Eng. (London)*, **39:** 363 (1961).
24  CALDERBANK, P. H., and A. C. LOCHIEL: *Chem. Eng. Sci.*, **19:** 485 (1964).
25  CALDERBANK, P. H., and R. P. PATRA: *Chem. Eng. Sci.*, **21:** 719 (1966).
26  CARBERRY, J. J.: *AIChE J.*, **6:** 460 (1960).
27  CERRO, R. L., and S. WHITAKER: *Chem. Eng. Sci.*, **26:** 785 (1971).
28  CHU, J. C., J. KALIL, and W. A. WETTEROTH: *Chem. Eng. Progr.*, **49:** 141 (1953).
29  COCHRAN, W. G.: *Proc. Cambridge Phil. Soc.*, **30:** 365 (1934).
30  COLTON, C. K., and K. A. SMITH: *AIChE J.*, **18:** 958 (1972).
31  COMINGS, E. W., J. T. CLAPP, and J. F. TAYLOR: *Ind. Eng. Chem.*, **40:** 1076 (1948).
32  CONSTAN, G. L., and S. CALVERT: *AIChE J.*, **9:** 109 (1963).
33  CORNET, I., and U. KALOO: *Proc. 3d Int Congr. Metallic Corrosion, Moscow*, **3:** 83 (1966).
34  COULSON, J. M., and J. E. PORTER: *Trans. Inst. Chem. Eng. (London)*, **44:** T388 (1966).
35  DANCKWERTS, P. V.: *Chem. Eng. Sci.*, **20:** 785 (1965).
36  DATTA, R. L., D. H. NAPIER, and D. M. NEWITT: *Trans. Inst. Chem. Eng. (London)*, **28:** 14 (1950).
37  DAVIES, D. R., and T. S. WALTERS: *Proc. Phys. Soc. (London)*, **B65:** 640 (1952).

37a  DAVIS, E. J., and E. CHORBAJIAN: *Ind. Eng. Chem. Fundam.*, **13:** 272 (1974).

38  DAWSON, D. A., and O. TRÄSS: *Int. J. Heat Mass Transfer*, **15:** 1317 (1972).

39  DE ACETIS, J., and G. THODOS: *Ind. Eng. Chem.*, **52:** 1003 (1960).

40  DOWNING, C. G.: *AIChE J.*, **12:** 760 (1966).

41  DROPKIN, D., and A. CARMI: *Trans. ASME*, **79:** 474 (1957).

42  DUKLER, A. E., and O. P. BERGELIN: *Chem. Eng. Progr.*, **48:** 557 (1952).

43  ECKERT, E. R. G.: *ASME Paper* 55-A31 (1955).

44  ECKERT, E. R. G., and R. M. DRAKE, JR.: "Heat and Mass Transfer," McGraw-Hill, New York, 1959.

45  EINARSSON, A., and A. A. WRAGG: *Chem. Eng. Sci.*, **26:** 1289 (1970).

46  EISENBERG, M., C. W. TOBIAS, and C. R. WILKE: *Chem. Eng. Progr. Symp. Ser.*, **51**(16): 1 (1955).

47  EMMERT, R. E., and R. L. PIGFORD: *Chem. Eng. Progr.*, **50:** 87 (1954).

48  ERGUN, S.: *Chem. Eng. Progr.*, **48:** 227 (1952).

49  EVANS, G. C., and C. F. GERALD: *Chem. Eng. Progr.*, **49:** 135 (1953).

50  EVERITT, C. T., and H. P. HUTCHINSON: *Trans. Inst. Chem. Eng.* (*London*), **45:** 9 (1967).

51  FALLAT, R. J.: *Univ. Calif., Berkeley* Rep. UCRL-8527, March, 1959.

52  FRANK, F. C.: *Proc. Roy. Soc.* (*London*), **A201:** 586 (1950).

53  FRIEDLANDER, S. K.: *AIChE J.*, **3:** 43 (1957).

54  FRIEDLANDER, S. K.: *AIChE J.*, **7:** 347 (1961).

55  FROMENT, G. F.: In "Chemical Reaction Engineering," p. 19, Advances in Chemistry Series 109, American Chemical Society, 1972.

56  FRÖSSLING, N.: *Beitr. Geophys.*, **52:** 170 (1938).

57  FULFORD, G. D.: In T. B. Drew, J. W. Hoopes, Jr., T. Vermeulen, and G. R. Cokelet (eds.), "Advances in Chemical Engineering," vol. 5, p. 151, Academic, New York, 1964.

58  GAFFNEY, B. J., and T. B. DREW: *Ind. Eng. Chem.*, **42:** 1126 (1950).

59  GALLOWAY, T. R., and B. H. SAGE: *Chem. Eng. Sci.*, **25:** 495 (1970).

60  GALLOWAY, T. R., and B. H. SAGE: *Int. J. Heat Mass Transfer*, **11:** 539 (1968).

61  GAMSON, B. W., G. THODOS, and O. A. HOUGEN: *Trans. AIChE*, **39:** 1 (1943).

62  GARNER, F. H.: *Trans. Inst. Chem. Eng.* (*London*), **28:** 88 (1950).

63  GARNER, F. H., and R. W. GRAFTON: *Proc. Roy. Soc.* (*London*), **224A:** 64 (1954).

64  GARNER, F. H., and P. J. HAYCOCK: *Proc. Roy. Soc.* (*London*), **252A:** 457 (1959).

65  GARNER, F. H., and J. M. HOFFMAN: *AIChE J.*, **7:** 148 (1961).

66  GARNER, F. H., and R. B. KEEY: *Chem. Eng. Sci.*, **9:** 218 (1959).

67  GARNER, F. H., and J. J. LANE: *Trans. Inst. Chem. Eng.* (*London*), **37:** 162 (1959).

68  GARNER, F. H., and A. H. P. SKELLAND: *Trans. Inst. Chem. Eng.* (*London*), **29:** 315 (1951).

69  GARNER, F. H., and M. TAYEBAN: *An. Real Soc. Espan. Fis. Quim.*, **B56:** 479 (1960). (In English.)

70  GARNER, F. H., and M. TAYEBAN: *An. Real Soc. Espan. Fis. Quim.*, **B56:** 491 (1960). (In English.)

71  GAZLEY, C.: *ASME Paper* 56-A-128 (1956).

72  GIBERT, H., and H. ANGELINO: *Chem. Eng. Sci.*, **28:** 855 (1973).

73  GIBERT, H., J. P. COUDERC, and H. ANGELINO: *Chem. Eng. Sci.*, **27:** 45 (1972).

74  GILLESPIE, B. M., E. D. CRANDALL, and J. J. CARBERRY: *AIChE J.*, **14:** 483 (1968).

75  GILLILAND, E. R., and T. K. SHERWOOD: *Ind. Eng. Chem.*, **26:** 516 (1934).

76  GREGORY, C. L.: Sc.D. thesis in chemical engineering, M.I.T., 1957.

77  GREGORY, D. P., and A. C. RIDDIFORD: *J. Chem. Soc.*, 3756 (1956).

78  GRIFFITH, R. M.: *Chem. Eng. Sci.*, **12:** 198 (1960).

79  GRIMLEY, S. S.: *Trans. Inst. Chem. Eng.* (*London*), **23:** 228 (1945).

80   GRUSINTSEV, G. I., T. S. KORNIENKO, and M. KH. KISHINEVSKI: *Theor. Found. Chem. Eng.*, **4**(6): 820 (1970). Trans. by the Consultants Bureau, New York.

80a  GUPTA, S. N., R. B. CHAUBE, and S. N. UPHADHYAY: *Chem. Eng. Sci.*, **29**: 839 (1974).

81   GUPTA, A. S., and G. THODOS: Paper presented at the 56th Annual Meeting of the AIChE, Houston, Texas, Dec. 1–5, 1963.

82   HABERMAN, W. L., and R. K. MORTON: *U.S. Navy Dept., David W. Taylor Model Basin Rep.* 802 (1953).

83   HADAMARD, J.: *Compt. Rend.*, **152**: 1735 (1911).

84   HANDLOS, A. E., and T. BARON: *AIChE J.*, **3**: 127 (1957).

85   HARRIOTT, P.: *AIChE J.*, **8**: 93 (1962).

86   HATTA, S., and A. BABA: *J. Soc. Chem. Ind. (Japan)*, **38**: (10), 546B (1935).

87   HAYDON, D. A.: *Proc. Roy. Soc. (London)*, **A243**: 483, 492 (1958).

88   HEERTJES, P. M., and L. H. DE NIE: *Chem. Eng. Sci.*, **21**: 755 (1966).

89   HEERTJES, P. M., and L. H. DE NIE: In C. Hanson (ed.), "Recent Advances in Liquid-Liquid Extraction," chap. 10, Pergamon, New York, 1971.

90   HIKITA, H.: *Chem. Eng. (Japan)*, **23**: 23 (1959).

91   HIKITA, H., K. NAKANISHI, and S. ASAI, *Chem. Eng. (Japan)*, **23**: 28 (1959).

92   HIKITA, H., K. NAKANISHI, and T. KATAOKA: *Chem. Eng. (Japan)*, **23**: 459 (1959).

93   HIROSE, T., and M. MOO-YOUNG: *Chem. Eng. Sci.*, **25**: 729 (1970).

94   HIXSON, A. W., and S. J. BAUM: *Ind. Eng. Chem.*, **33**: 478, 1433 (1941).

95   HO, F. C. K., and R. L. HUMMEL: *Chem. Eng. Sci.*, **25**: 1225 (1970).

96   HOBSON, M., and G. THODOS: *Chem. Eng. Progr.*, **45**: 517 (1949).

97   HOBSON, M., and G. THODOS: *Chem. Eng. Progr.*, **47**: 370 (1951).

98   HODSON, J. R.: S.M. thesis in chemical engineering, M.I.T., 1949.

99   HOLMAN, K. K., and S. T. ASHAR: *Chem. Eng. Sci.*, **26**: 1817 (1971).

99a  HOMSY, R. V., and J. S. NEWMAN: *AIChE J.*, **19**: 929 (1973).

100  HONAKER, D. E., and L. C. TAO: *Ind. Eng. Chem. Fundam.*, **9**: 325 (1970).

101  HSU, N. T., K. SATO, and B. H. SAGE: *Ind. Eng. Chem.*, **46**: 870 (1954).

102  HU, S., and R. C. KINTNER: *AIChE J.*, **1**: 42 (1955).

103  HUANG, W. S., and R. C. KINTNER: *AIChE J.*, **15**: 735 (1969).

103a HUGHMARK, G. A.: *AIChE J.*, **20**: 202 (1974).

104  HURLBURT, H. Z.: Sc.D. thesis in chemical engineering, M.I.T., 1949.

105  JACKSON, M. L.: *AIChE J.*, **1**: 231 (1955).

106  JACKSON, M. L., and N. H. CEAGLSKE: *Ind. Eng. Chem.*, **42**: 1188 (1950).

107  JAKOB, M., and W. DOW: *Trans. ASME*, **68**: 123 (1946).

108  JOHNSON, A. I., F. BESIC, and A. E. HAMIELEC: *Can. J. Chem. Eng.*, **47**: 559 (1969).

109  JOHNSON, A. I., A. E. HAMIELEC, and W. T. HOUGHTON: *Can. J. Chem. Eng.*, **45**: 140 (1967).

110  JOHNSON, A. I., and C. J. HUANG: *AIChE J.*, **2**: 412 (1956).

111  JOHNSON, T. R., and P. N. JOUBERT: *Trans. ASME, J. of Heat Transfer*, p. 91, February 1969.

112  JOHNSTONE, H. F., and R. L. PIGFORD: *Trans. AIChE*, **38**: 25 (1942).

113  JOLLS, K. R., and T. J. HANRATTY: *AIChE J.*, **15**: 199 (1969).

114  KAFESJIAN, R., C. A. PLANK, and E. R. GERHARD: *AIChE J.*, **7**: 463 (1961).

115  KAPPESSAR, R. R.: D.Eng. thesis in mechanical engineering, Univ. of Calif., Berkeley, 1970.

116  KAYS, W. M.: "Convective Heat and Mass Transfer," McGraw-Hill, New York, 1966.

117  KAYS, W. M., and I. S. BJORKLAND: *Dept. Mech. Eng., Stanford Univ., Tech. Rep.* 27, 1955.

118  KAYS, W. M., and I. S. BJORKLAND: *ASME Paper* 56-A-71, 1956.

119  KAYS, W. M., and I. S. BJORKLAND: *Trans. ASME*, **80**: 70 (1958).

120  KEEY, R. B., and J. B. GLEN: *AIChE J.*, **12**: 401 (1966).

121 KESTIN, J., and R. T. WOOD: *ASME Paper* 70-WA/HT-3 (1970).

121a KING, C. J.: *Ind. Eng. Chem. Fundam.*, **5:** 1 (1966).

122 KINTNER, R. C.: "Drop Phenomena Affecting Liquid Extraction," Advances in Chemical Engineering, pp. 52–92, vol. 4, Academic, New York, 1963.

123 KNIGHT, I. C., and D. A. RATKOWSKY: *AIChE J.*, **11:** 370 (1965).

124 KRAMERS, H., and P. J. KREYGER: *Chem. Eng. Sci.*, **6:** 42 (1956).

125 KRONIG, R., and J. C. BRINK: *Appl. Sci. Res.*, **A2:** 142 (1950).

126 KRONIG, R., and J. BRUIJSTEN: *Appl. Sci. Res.*, **A2:** 439 (1951).

127 LAMOURELLE, A. P., and O. C. SANDALL: *Chem. Eng. Sci.*, **27:** 1035 (1972).

128 LEE, K., and H. BARROW: *Int. J. Heat Mass Transfer*, **8:** 403 (1965).

129 LEE, K., and D. J. RYLEY: *Trans. ASME, J. Heat Transfer*, p. 445, November 1968.

130 LELAN, A., H. GIBERT, and H. ANGELINO: *Chem. Eng. Sci.*, **27:** 1979 (1972).

131 LEMLICH, R., and M. R. LEVY: *AIChE J.*, **7:** 240 (1961).

132 LEONARD, J. H., and G. HOUGHTON: *Chem. Eng. Sci.*, **18:** 133 (1963).

133 LEVICH, V. G.: "Physicochemical Hydrodynamics," Prentice-Hall, Englewood Cliffs, N.J., 1962. (In English.)

134 LEVINS, D. M., and J. R. GLASTONBURY: *Chem. Eng. Sci.*, **27:** 537 (1972).

135 LEWIS, J. B., and H. C. R. PRATT: *Nature*, **171:** 1155 (1953).

136 LI, P-S., F. B. WEST, W. H. VANCE, and R. W. MOULTON: *AIChE J.*, **11:** 581 (1965).

137 LICHT, W., JR., and W. F. PANSING: *Ing. Eng. Chem.*, **45:** 1885 (1953).

138 LINDLAND, K. P., and S. G. TERJESEN: *Chem. Eng. Sci.*, **5:** 1 (1956).

139 LOCHIEL, A. C.: *Can. J. Chem. Eng.*, **43:** 40 (1965).

140 LOHRISCH, W.: *Mitt. Forsch.*, **322:** 46 (1929).

141 LOUDON, J. R., P. H. CALDERBANK, and I. COWARD: *Chem. Eng. Sci.*, **21:** 614 (1966).

142 LYNN, S., J. R. STRAATEMEIER, and H. KRAMERS: *Chem. Eng. Sci.*, **4:** 58 (1955).

143 LYNN, S., J. R. STRAATEMEIER, and H. KRAMERS: *Chem. Eng. Sci.*, **4:** 49 (1955).

144 MCADAMS, W. H., T. B. DREW, and G. S. BAYS, JR.: *Trans. ASME*, **62:** 627 (1940).

145 MCCONNACHIE, J. T. L., and G. THODOS: *AIChE J.*, **9:** 60 (1963).

146 MCCUNE, L. K., and R. H. WILHELM: *Ind. Eng. Chem.*, **41:** 1124 (1949).

147 MAISEL, D. S., and T. K. SHERWOOD: *Chem. Eng. Progr.*, **46:** 131, 172 (1950).

148 MANDELBAUM, J. A., and V. BÖHM: *Chem. Eng. Sci.*, **28:** 569 (1973).

149 MARANGOZIS, J., and A. I. JOHNSON: *Can. J. Chem. Eng.* **40:** 231 (1962); **41:** 133 (1963).

150 MARCUSSEN, L.: *Chem. Eng. Sci.*, **25:** 1487 (1970).

151 MILLER, D. N.: *Ind. Eng. Chem.*, **56**(10): 18 (1964).

152 MILLER, D. N.: *Ind. Eng. Chem. Process Des. Dev.*, **10:** 365 (1971).

153 MOO-YOUNG, M., and T. HIROSE: *Can. J. Chem. Eng.*, **50:** 128 (1972); *Ind. Eng. Chem. Fundam.*, **11:** 281 (1972).

154 NAGATA, S., I. KAWAGUCHI, S. YABUTA, and M. HARADA: *Soc. Chem. Eng. (Japan)*, **24:** 618 (1960).

154a NAGATA, S., and M. NISHIKAWA: *Proc. First Pacific Chem. Eng. Congr.*, Soc. Chem. Eng. (*Japan*) and *AIChE*, Kyoto, October 10–14, 1972, pt 3, sess. 18, p. 301.

155 NARASIMHAN, C., and W. H. GAUVIN: *Can. J. Chem. Eng.*, **45:** 181 (1967).

156 NEWMAN, L. B., E. M. SPARROW, and E. R. G. ECKERT: *ASME Paper* 71-Ht-8 (1971).

157 NEWMAN, J. S.: *J. Electrochem. Soc.*, **113:** 501 (1966).

158 NIENOW, A. W.: *Can. J. Chem. Eng.*, **47:** 248 (1969).

159 NOORDSIJ, M. P., and J. W. ROTTE: *Chem. Eng. Sci.*, **22:** 1475 (1967).

160 NOORDSIJ, M. P., and J. W. ROTTE: *Chem. Eng. Sci.*, **23:** 657 (1968).

161 NORRIS, R. H., and D. D. STREID: *Trans. ASME*, **36:** 525 (1940).

162 NUSSELT, W.: *Z. Ver. Deut. Ing.*, **67:** 206 (1923).

163  OELLRICH, L., H. SCHMIDT-TRAUB, and H. BRAUER: *Chem. Eng. Sci.*, **28:** 711 (1973).

164  OLANDER, D. R.: *AIChE J.*, **12:** 1018 (1966).

165  OLIVER, D. R., and T. E. ATHERINOS: *Chem. Eng. Sci.*, **23:** 525 (1968).

166  PATEL, J. M., and R. M. WELLEK: *AIChE J.*, **13:** 384 (1967).

167  PEI, D. C. T., and W. H. GAUVIN: *AIChE J.*, **9:** 375 (1963).

168  PFEFFER, R., and J. HAPPEL: *AIChE J.*, **10:** 605 (1964).

169  PLEWES, A. C., R. M. BUTLER, and H. E. MARSHALL: *Chem. Eng. Progr.*, **50:** 77 (1954).

170  POPOVICH, A. T., R. E. JERVIS, and O. TRÄSS: *Chem. Eng. Sci.*, **19:** 357 (1964).

171  PORTALSKI, S.: *Ind. Eng. Chem. Fundam.*, **3:** 49 (1964).

172  PORTALSKI, S., and A. J. CLEGG: *Chem. Eng. Sci.*, **26:** 773 (1971); **27:** 1257 (1972).

173  POWELL, R. W.: *Trans. Inst. Chem. Eng.* (*London*), **18:** 36 (1940).

174  POWELL, R. W., and E. GRIFFITHS: *Trans. Inst. Chem. Eng.* (*London*), **13:** 175 (1935).

174a  PRASHER, B. D., and A. L. FRICKE: *Ind. Eng. Chem. Process Des. Dev.*, **13:** 336 (1974).

175  RANZ, W. E.: *Chem. Eng. Progr.*, **48:** 247 (1952).

176  RANZ, W. E., and W. R. MARSHALL, JR.: *Chem. Eng. Progr.*, **48:** 141, 173 (1952).

177  REKER, J. R., C. A. PLANK, and E. R. GERHARD: *AIChE J.*, **12:** 1008 (1960).

178  RICCETTI, R. E., and G. THODOS: *AIChE J.*, **7:** 442 (1961).

179  ROHSENOW, W. M., and H. Y. CHOI: "Heat, Mass and Momentum Transfer," Prentice-Hall, New York, 1961.

180  ROSE, P. M., and R. C. KINTNER: *AIChE J.*, **12:** 530 (1966).

181  ROWE, P. N., K. T. CLAXTON, and J. B. LEWIS: *Trans. Inst. Chem. Eng.* (*London*), **43:** 14 (1965).

182  ROWE, P. N., and K. T. CLAXTON: *Trans. Inst. Chem. Eng.* (*London*), **43:** 321 (1965).

183  RUSHTON, J. H., E. W. COSTICH, and H. J. EVERETT: *Chem. Eng. Progr.*, **46:** 467 (1950).

184  SAWISTOWSKI, H.: In C. Hanson (ed.), "Recent Advances in Liquid-Liquid Extraction," chap. 9, Pergamon, New York, 1971.

185  SCHENKELS, F. A. M., and J. SCHENK: *Chem. Eng. Sci.*, **24:** 585 (1969).

186  SEBAN, R. A., and H. A. JOHNSON: *NASA Memo* 4-22-59W (1958).

187  SHERWOOD, T. K., and R. L. PIGFORD: "Absorption and Extraction," 2d ed., McGraw-Hill, New York, 1952.

188  SHERWOOD, T. K., and J. M. RYAN: *Chem. Eng. Sci.*, **11:** 81 (1959).

189  SHERWOOD, T. K., and O. TRÄSS: *Trans. ASME, J. of Heat Transfer*, **82C:** 313 (November 1960).

190  SIDEMAN, S., and H. SHABTAI: *Can. J. Chem. Eng.*, **42:** 107 (1964).

190a  SIMPSON, S. G.: Ph.D. thesis in chemical engineering, Univ. Calif., Berkeley, 1975.

191  SKELLAND, A. H. P., and A. R. H. CORNISH: *AIChE J.*, **9:** 73 (1963).

192  SOGIN, H. H., and V. S. SUBRAMANIAN: *ASME Paper* 60-WA-193 (1960).

193  SPARROW, E. M., and J. L. GREGG: *Trans. ASME, J. of Heat Transfer*, **81:** 249 (1959).

194  STAINTHORP, F. P., and R. S. W. BATT: *Trans. Inst. Chem. Eng.* (*London*), **45:** T372 (1967).

195  STEINBERGER, R. L., and R. E. TREYBAL: *AIChE J.*, **6:** 227 (1960).

196  STEVENSON, D. A., and J. WULF: *Trans. Met. Soc. AIMME*, **221:** 279 (1961).

197  SYKES, P., and A. GOMEZPLATA: *Can. J. Chem. Eng.*, **45:** 189 (1967).

198  TAILBY, S. R., and S. PORTALSKI: *Trans. Inst. Chem. Eng.* (*London*), **38:** 324 (1960).

199  TAMIR, A., and Y. TAITEL: *Chem. Eng. Sci.*, **26:** 799 (1971).

200  TELLES, A. S., and A. E. DUKLER: *Ind. Eng. Chem. Fundam.*, **9:** 412 (1970).

201  TESSIN, W., and M. JAKOB: *Trans. ASME*, **75:** 473 (May 1953).

202  THEODORSEN, T., and A. REGIER: *Nat. Adv. Comm. Aeronaut. Rep.* 793 (1945).

203  THOENES, D., JR., and H. KRAMERS: *Chem. Eng. Sci.*, **8:** 271 (1958).

204  THORSON, G., and S. G. TERJESEN: *Chem. Eng. Sci.*, **17:** 137 (1962).

205 TRAYLOR, E. D., L. BURRIS, and C. J. GEANKOPLIS: *Ind. Eng. Chem. Fundam.*, **4:** 119 (1965).

206 UNAHABHOKHA, R., A. W. NIENOW, and J. W. MULLIN: *Chem. Eng. Sci.*, **26:** 357 (1971).

207 VALENTIN, F. H. H.: "Absorption in Gas-Liquid Dispersions," E. and F. N. Spon, Ltd., London, 1967.

208 VAN DEN BERG, H. J.: in "Chemical Reaction Engineering," Advances in Chemistry Series 109, American Chemical Society, 1972.

209 VIVIAN, J. E., and D. W. PEACEMAN: *AIChE J.*, **2:** 437 (1956).

210 VOGTLANDER, P. H., and C. A. P. BAKKER: *Chem. Eng. Sci.*, **18:** 583 (1963).

211 VYRUBOV, D. N.: *J. Tech. Phys. (Moscow)*, **9:** 1923 (1939).

212 WARD, D. M., O. TRÄSS, and A. I. JOHNSON: *Can. J. Chem. Eng.* **40:** 164 (1962).

213 WASAN, D. T., M. A. LYNCH, K. J. CHAD, and N. SRINIVASAN: *AIChE J.*, **18:** 928 (1972).

214 WELLEK, R. M., and C-C HUANG: *Ind. Eng. Chem. Fundam.*, **9:** 480 (1970).

215 WELLER, K. R.: *Can. J. Chem. Eng.*, **50:** 49 (1972).

216 WHITMAN, W. G., L. LONG, JR., and H. Y. WANG: *Ind. Eng. Chem.*, **18:** 363 (1926).

217 WILHELM, R. H., L. H. CONKLIN, and T. C. SAUER: *Ind. Eng. Chem.*, **33:** 453 (1941).

218 WILKE, C. R., and O. A. HOUGEN: *Trans. AIChE*, **41:** 445 (1945).

219 WILLIAMS, G. C.: In W. H. McAdams, "Heat Transmission," 2d ed., p. 236, McGraw-Hill, New York, 1942.

220 WILLIAMSON, J. E., K. E. BAZAIRE, and C. J. GEANKOPLIS: *Ind. Eng. Chem. Fundam.*, **2:** 126 (1963).

221 WRAGG, A. A., and A. EINARSSON: *Chem. Eng. Sci.*, **25:** 67 (1970).

222 YUGE, T.: *Trans. ASME J. Heat Transfer*, **82C:** 214 (1960).

223 ZWIETERING, TH. N.: *Chem. Eng. Sci.*, **8:** 244 (1958).

## PROBLEMS

6.1 A vertical 2-in-ID tube 6 ft tall is employed as a wetted-wall absorber. It is fed 28.5 cm³/s pure water at 25°C, supplied uniformly around the top perimeter. Falling as a uniform laminar film, the water absorbs $CO_2$, which is supplied pure at 10 atm.

(a) Estimate the absorption rate as pounds $CO_2$ per hour.

(b) Repeat on the assumption that a wire device is used to mix the water film thoroughly at a point 3 ft down from the top. Assume the film to be in fully developed laminar flow in both top and bottom sections.

(c) Repeat for the case of the same total water flow divided equally between two tubes 3 ft tall.

6.2 In Example 6.4 it was calculated that dry air would become 90 percent saturated with water after flowing through a bed packed to a depth of only 1.4 cm with 5 mm water-wet spheres.

How tall would a 5.0-in-ID wetted-wall column have to be to attain the same 90 percent of saturation at the water temperature? Assume the temperature of the surface of the water film to be constant and the average temperature of the air to be 20°C. The mean air velocity will be 1.0 ft/s and the pressure 1.0 atm, as in Example 6.4. The surface velocity of the falling water film is calculated to be 50 cm/s.

6.3 A hollow porous ceramic sphere 1.0 cm in diameter is placed in a turbulent stream of dry air flowing at 15 ft/s and 1.0 atm. The sphere is kept wet with water and thermocouples are used to measure the temperature of the surface. The objective is a precise measurement of the wet-bulb temperature of the air. Since $k_c$ varies with distance from the front stagnation point, the sphere is slowly rotated.

It is found that the measured temperature rises as the water evaporates, so the hollow center is fed continuously with water through a small tube. If the actual wet-bulb temperature is 15.0°C, at what rate (grams per hour) should water be supplied to maintain the surface of the sphere thoroughly wet but not dripping?

6.4   A cylindrical stirred vessel contains 5 ft$^3$ of 0.005$N$ sodium hydroxide in water. This is neutralized in a batch test by the addition of 2 lb of strong cationic exchange resin in the acid form. The resin is in the form of swollen resin spheres 300 $\mu$m in diameter having a specific gravity of 1.2 g/cm$^3$. Power input by the stirrer is 1.0 hp/1,000 gal.

How long after the resin addition will the NaOH be 96 percent neutralized? Assume that resin addition and dispersion is instantaneous and all the mass-transfer resistance is external to the particles. The specific gravity and density of the solution are 1.0 g/cm$^3$ and 0.01 poise; $D$ for NaOH in water is $1.9 \times 10^{-5}$ cm$^2$/s.

6.5   As a laboratory exercise, a student plans to carry out a simple mass-transfer experiment by subliming naphthalene from a rotating cylinder. A cast cylinder of solid naphthalene 3.0 cm in diameter will be carefully machined and fixed to a drill press in the shop. The rpm will be 1,800 and the vapor buildup in the ventilated room will be neglected. It is estimated that the diameter will have to decrease to at least 2.9 cm in order that the mass-transfer rate can be obtained either by weighing the cylinder or by measuring the change in the cylinder diameter with a micrometer. The air temperature is 20°C and the cooling of the surface due to sublimation can be assumed negligible.

For how long should the test last? *Data:* Specific gravity of solid naphthalene is 1.15 g/cm$^3$; vapor pressure at 20° is approximately 0.037 mm Hg; $D$ for naphthalene–air is 0.071 cm$^2$/s.

6.6   A deep fermenter is supplied with pure oxygen sparged in at the bottom. It is desired to obtain an estimate of the percent oxygen absorption.

In order to make such an estimate, assume that oxygen is supplied at the bottom of a vessel containing oxygen-free water at 25°C, the bubbles as released being 4 mm in diameter. The total depth of the water is 50 ft and the top is at atmospheric pressure. For purposes of calculation, assume that the bubbles behave as single bubbles and use Eq. (6.28) to estimate $k_c$.

6.7   A crystal supported in a flowing stream of supersaturated solution is found to grow at the linear rate of 0.010 mm/min. The solution contains 4.00 g moles/l, and the velocity past the 1.0 mm crystal is 1.0 ft/s. The liquid density is 1.1 g/cm$^3$, its viscosity 5 cP, and the saturation concentration 3.93 moles/l at the conditions of the test. The solid has a specific gravity of 1.5 g/cm$^3$ and a molecular weight of 150. The molecular diffusion coefficient for the solute at 3.9 to 4.0 g moles/l is estimated to be $0.5 \times 10^{-5}$ cm$^2$/s.

Does the process of incorporating the solute molecules into the crystal habit constitute a significant hindrance to crystal growth rate?

6.8   A catalytic reactor for the conversion of para- to orthohydrogen consists of a straight round tube 10 ft long, the inside walls of which are coated with a 2.0 mm layer of porous reduced nickel oxide-on-alumina catalyst. The inside diameter of the catalyst-coated tube is 1.0 in.

Hydrogen containing 50 percent $p$-H$_2$ and 50 percent $o$-H$_2$ is passed into the tube at 25°C and 1.0 atm at a mean velocity of 42.3 ft/s.

Estimate the effluent composition as mole percent para-H$_2$.

*Data:* The enthalpy change for the para-ortho hydrogen conversion is negligible. The equilibrium mixture at 25°C is 25 percent $p$-H$_2$. The rate of conversion on the catalyst surface is known to be proportional to $(c - c_e)$, where $c_e$ is the equilibrium $p$-H$_2$ concentration, and $c$ is the concentration of $p$-H$_2$ in contact with the catalyst surface. (Concentrations as moles per unit volume.) The intrinsic rate constant for the catalyst used is 16,600 g moles/(s)(cm$^3$ porous catalyst) [unit $(c - c_e)$ g moles/cm$^3$ gas].

The porous catalyst has a BET surface of 131 m$^2$/g, a porosity of 0.39 cm$^3$/cm$^3$ and a bulk density of 1.47 g/cm$^3$. Values of gas viscosity and the molecular diffusion coefficient $D$ are to be calculated from the theoretical equations based on the Lennard-Jones potential function. The constants $\varepsilon/k$ and $\sigma$ for H$_2$ are 33.3°K and 2.968Å, respectively. The "tortuosity" factor for diffusion in the porous catalyst is approximately 4.0.

6.9 A petroleum naphtha is being hydrodesulfurized in a steady-flow adiabatic fixed-bed catalytic reactor packed with 1/8-in beads of porous cobalt molybdate-on-alumina catalyst. The performance is as follows:

| | |
|---|---|
| Feed | 49.4° API naphtha (specific gravity, 0.7803); boiling range 315 to 490°F; average molecular weight = 166 |
| Pressure | 420 psia |
| Average bed temperature | 700°F |
| Feed rate | 4310 bbl/day (1 bbl = 42 gal) |
| Bed dimensions | 5 ft ID; 20.6 ft high |
| Hydrogen feed | 3,000 ft$^3$ (60°, 1 atm)/bbl |
| Sulfur in naphtha feed | 1.42 weight percent |
| Sulfur in hydrocarbon product | 0.20 weight percent |

It is now proposed to build a new unit for the same purpose with the same feed, feed rate, and operating conditions. However, since the pressure drop in the first unit was considered to be excessive, it is proposed to build the new unit with a shorter bed of larger diameter. Specifically it is proposed that the new unit have a bed 7 ft in diameter with the same catalyst volume.

(a)  Does mass transfer from gas to pellet present a significant resistance in either bed?

(b)  What weight percent sulfur can be expected in the effluent from the new 7-ft bed?

*Assumptions:*  Radial velocity gradients are negligible: assume "piston" flow. The naphtha is completely vaporized; there is no liquid in the feed.

The reaction is well represented as the desulfurization of thiophene and is first order and irreversible, producing H$_2$S.

The viscosity of the hydrogen–hydrocarbon mixture at the operating temperature and pressure is approximately 0.036 cP.

The void fraction in the bed (not including pores in pellets) is 0.40. Thiophene has a critical temperature of 580°K and a critical volume of about 233 cm$^3$/g mole.

6.10 A flat horizontal plate is placed in a wind tunnel with dry air flowing parallel to the surface at 50 ft/s, 15.4°C, and 1.0 atm. The plate has a sharp leading edge, and a shallow cavity holds a 2.5-mm-thick layer of an oil–benzene solution, the surface of the liquid being flush with the floor of the tunnel. The liquid initially contains 6 mole percent benzene, and the oil is essentially nonvolatile.

What fraction of the benzene will have evaporated at points 0, 5, 10, 15, and 20 cm downstream from the leading edge after 1.93 h exposure to the air flow?

*Data and assumptions:* The oil–benzene solution does not mix vertically or longitudinally. The wet-bulb effect can be neglected and the surface considered to be at 15.4°C, at which the vapor pressure of benzene is 60 mm Hg. Raoult's law holds for the solution. The liquid density and molecular weight remain nearly constant at 0.9 g/cm$^3$ and 240, respectively. The molecular diffusion coefficient of benzene in the liquid is $0.9 \times 10^{-5}$ cm$^3$/s.

# SIMULTANEOUS HEAT AND MASS TRANSFER

## 7.0  SCOPE

The expressions representing the rate of heat transfer between a fluid and a solid surface in the presence of simultaneous mass transfer to the surface are derived in this chapter. Use is made of the results for the design of partial condensers and dehumidifiers, including conditions for which the rates of mass transfer are slow and fog forms as the temperature of the vapor mixture falls. The theory of the wet-bulb thermometer is reviewed briefly; and the heat- and mass-transfer relationships are applied, in an approximate form valid for small driving forces, to the design of water cooling towers both for countercurrent and crossflow conditions.

## 7.1  PRINCIPAL SYMBOLS

$A$      Interfacial area, $ft^2$

$B$      Width of packed space in crossflow cooling tower, ft

$C_i$      Slope of enthalpy-temperature curve, Btu/(lb)(°F)

$C_0$      Ackermann factor, defined after Eq. (7.2)

$C_p$      Molar heat capacity of a component, cal/(g mole)(°K); $C_{pA}$ for diffusing species; $C_{p,av}$ for mixture average; $C_{pL}$ for liquid

| | |
|---|---|
| $c_p$ | Average specific heat of fluid mixture, cal/(g)(°C) |
| $c_s$ | Humid heat, as in Eq. (7.16) |
| $D$ | Diffusion coefficient in fluid, cm$^2$/s |
| $F$ | Mass velocity of fog, lb mole/(h)(ft$^2$) |
| $G$ | Mass velocity of inert gas in condenser, lb mole/(h)(ft$^2$) |
| $G_T$ | Total gas-flow rate in cross-flow tower, lb/h |
| $H$ | Humidity, g vapor/g inert gas |
| $H_G$ | Height of a transfer unit, ft |
| $h$ | Heat-transfer coefficient cal/(cm$^2$)(s)(°C), or Btu/(h)(ft$^2$)(°F) |
| $i$ | Enthalpy of air–water-vapor mixture, Btu/lb air |
| $j_{mass}, j_{heat}$ | Colburn factors for mass and heat transfer, respectively |
| $k$ | Thermal conductivity of fluid phase, cal/(cm)(s)(°C) |
| $k'$ | Mass-transfer coefficient based on humidity driving force, lb/(h)(ft$^2$)(lb/lb) |
| $k_G^*$ | Mass-transfer coefficient defined by Eq. (7.8), independent of driving force |
| $L$ | Mass velocity of condensate phase, lb mole/(h)(ft$^2$) |
| Le | Lewis number of fluid = Pr/Sc |
| $L_T$ | Total liquid-flow rate, lb/h |
| $M$ | Mass velocity of gas mixture in condenser, lb moles/(h)(ft$^2$), equal to $G + V$ |
| $\overline{M}$ | Average molecular weight of fluid mixture; $\overline{M}_1$ is value at interface, $\overline{M}_2$ in turbulent gas |
| $N$ | Number of transfer units in crossflow cooling tower |
| $N_A, N_B$ | Molar fluxes of components $A$ and $B$ from interface, g moles/(s)(cm$^2$), or lb mole/(h)(ft$^2$) |
| P | Pressure of system, atm |
| Pr | Prandtl number of fluid |
| $p$ | Partial pressure of condensible component in gas mixture, atm |
| $q_H$ | Enthalpy flux toward interface, cal/(cm$^2$)(s), or Btu/(h)(ft$^2$) |
| R | Parameter used in equations for crossflow cooling tower; also gas constant |
| Sc | Schmidt number of fluid |
| $T$ | Absolute temperature, °K |
| $t$ | Fluid temperature at a point; $t_0$ is the standard-state value; $t_1$ and $t_2$ apply at interface and in bulk of fluid, respectively, °F |
| $t_W$ | Temperature of coolant in condenser, °F |
| $t_{WB}$ | Wet-bulb temperature, °F |
| $U$ | Heat-transfer coefficient for liquid phase in cooling tower, Btu/(h)(ft$^2$)(°F) |
| $V$ | Mass velocity of condensible vapor in condenser, lb mole/(h)(ft$^2$); for cooling towers, volume of packing, ft$^3$ |
| $W$ | Flow rate of coolant in condenser, lb/h |
| $X$ | Total height of cooling tower, ft |
| $X_A$ | Mole fraction $A$ in liquid |
| $y$ | Distance from bulk of gas toward interface; $y_0$ is value at interface |
| $Y$ | Total width of packing in crossflow cooling tower, ft |
| $Y_A$ | Mole fraction of component $A$ in fluid |
| $Z$ | Ratio of molar fluxes in condenser, equal to $N_A/(N_A + N_B)$ or composition of incremental condensate |
| $\beta$ | Correction factor from Bedingfield and Drew [cf. Eq. (7.10)] |
| $\Delta$ | Enthalpy driving force, $i^* - i$, Btu/lb dry air |
| $\lambda$ | Latent heat of vaporization, cal/g mole, or Btu/lb mole |
| $\rho$ | Fluid density, g mole/cm$^3$, or lb mole/ft$^3$ |

## 7.2 THE EFFECT OF MASS TRANSFER ON THE RATE OF HEAT TRANSFER AT THE SAME SURFACE

In Sec. 5.4 the theory was developed for the molecular diffusion of two gaseous components through a film. The theory of enthalpy transport through such a film is very similar.

Referring to Fig. 7.1, we see that the total flux of enthalpy into a differential element of thickness $dy$ is made up of two parts. One is the conduction heat flux, $-k(dt/dy)$; the other is the flux of enthalpy due to diffusion, $N_A C_{pA}(t - t_0) + N_B C_{pB}(t - t_0)$, where $t_0$ is a standard-state temperature to be selected later. When these quantities are evaluated for the flux vectors entering and leaving the differential element and their difference is set equal to zero, we find that the temperature distribution must satisfy

$$k \frac{d^2t}{dy^2} - (N_A C_{pA} + N_B C_{pB}) \frac{dt}{dy} = 0 \tag{7.1}$$

The solution, satisfying the conditions that $t = t_1$ at the interface where $y = 0$ and $t = t_2$ at the bulk-gas boundary of the film, is

$$t(y) = t_1 + (t_2 - t_1) \frac{\exp (C_0 y/y_0) - 1}{\exp (C_0) - 1} \tag{7.2}$$

where $C_0 = (N_A C_{pA} + N_B C_{pB})/h$ and $h = k/y_0$. The conduction flux of heat at the interface is found from this result as

$$q_c = -kt'(0) = h(t_2 - t_1) \frac{C_0}{\exp (C_0) - 1} \tag{7.3}$$

Note that $q_c$ is not equal to the total enthalpy flux in the gas at the interface unless the standard-state temperature is $t_1$; in general, the total, $q_H$, is equal to $q_c + (N_A C_{pA} + N_B C_{pB})(t_1 - t_0)$. If enthalpies are computed relative to a standard state at the bulk-gas temperature $t_2$, the total enthalpy flux from the surface is

$$q_H = h(t_1 - t_2) \frac{C_0}{1 - \exp (-C_0)} \tag{7.4}$$

The function of $C_0$ appearing in Eq. (7.3) is often called the "Ackermann correction" for mass transfer [1]. It was also used by Colburn [10].

Equation (7.3) is of the same form as that used by Mickley et al. [35] and plotted as Fig. 5.5. Thus the figure shows the effect of the ratio $(N_A C_{pA} + N_B C_{pB})/h$ on the ratio of the conduction heat flux to the heat flux without mass transfer. Evidently the presence of mass transfer can either raise or lower the rate of heat conduction depending on the direction of mass transfer. Thus, a surface exposed to a hot gas can be partially protected from rapid heating from the bulk gas if the surface is kept wet with a volatile liquid, which evaporates and creates a positive, opposing mass transfer ("film" or "sweat" cooling). For many of the applications of the theory in this chapter, the mass transfer is toward the surface, as in condensers, and $C_0$ is negative. The table below compares the coordinates of Fig. 5.5 for mass and heat transfer.

| Transfer process | Definition of $C_0$ | Ordinate of Fig. 5.5 |
|---|---|---|
| Mass transfer | $\dfrac{(N_A + N_B)RTy_0}{D_{AB}P} = \dfrac{N_A + N_B}{\rho k_D^*}$ | $\dfrac{J_A}{k_D^*(Y_{A1} - Y_{A2})} = \dfrac{-y_0 \, Y_A'(0)}{Y_{A1} - Y_{A2}}$ |
| Heat transfer | $\dfrac{(N_A C_{pA} + N_B C_{pB})y_0}{k} = \dfrac{N_A C_{pA} + N_B C_{pB}}{h}$ | $\dfrac{q_c}{h(t_1 - t_2)} = \dfrac{-y_0 \, t'(0)}{t_1 - t_2}$ |

The value of $J_A$, calculated from Fig. 5.5 for mass transfer, must be increased (or decreased) by the rate of convective transfer, as in Example 5.1; similarly, the conduction rate found from Fig. 5.5 or Eq. (7.3) in a heat-transfer problem must be increased (or decreased) by the rate of convective flow of enthalpy in the gas at the surface. For processes in which a phase change occurs at the interface, as in evaporation or condensation, there is an additional latent enthalpy effect at the interface, as in Eq. (7.7) below. Equations (7.3) and (7.4) apply only on the gas side of the interface.

If the total mass-transfer rate is small, Eq. (7.3) assumes a simple form in the limit:

$$\lim_{N_T \to 0} q_c = h(t_1 - t_2) - \tfrac{1}{2}(N_A C_{pA} + N_B C_{pB})(t_1 - t_2) \tag{7.5}$$

Equation (7.5) shows that in the limit of small mass-transfer driving forces, the amount of heat conduction near the interface is slightly changed. The similar expression for $q_H$

$$\lim_{N_T \to 0} q_H = h(t_1 - t_2) + \tfrac{1}{2}(N_A C_{pA} + N_B C_{pB})(t_1 - t_2) \tag{7.6}$$

shows that the total enthalpy flux is slightly greater, as if the molecules leaving the surface at $t$ undergo only half the maximum temperature change before they reach the bulk temperature at $y = y_0$. The effect of mass transfer can be either positive or negative depending on the sign of the specific heat-flux vector, $N_A C_{pA} + N_B C_{pB}$.

The total flux of enthalpy into the surface of a condenser which receives mass-transfer fluxes $N_A$ and $N_B$ is the sum of $q_H$ and the latent enthalpy change at the surface due to condensation:

$$Q = -q_H - N_A \lambda_A - N_B \lambda_{B1} \tag{7.7}$$

If only component $A$ condenses,

$$Q = h(t_2 - t_1)\frac{C_0}{e^{C_0} - 1} + k_g^* P \lambda_{A1} \ln \frac{1 - Y_{A1}}{1 - Y_{A2}} \tag{7.8}$$

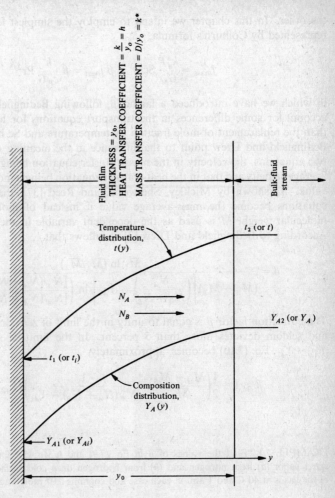

FIGURE 7.1
Diagram of a vapor film experiencing heat conduction and simultaneous diffusion toward an interface.

where $\lambda_{A1}$ is the latent heat at the surface temperature $t_1$. $Q$ represents the enthalpy flux into the condensate film, nearly equal to that into the coolant fluid inside the condenser surface.

Since Eq. (7.8) contains both mass-transfer and heat-transfer coefficients and there is a relationship between the two through the analogies, as discussed in Sec. 5.3, it is obvious that $Q$ can be expressed either in terms of one coefficient or the other. In this chapter we intend to employ the simplest form of the analogy, represented by Colburn's formula,

$$j_{\text{mass}} = \frac{k_g^* P}{G_M} \, \text{Sc}^{2/3} = \beta \, j_{\text{heat}} = \beta \, \frac{h}{c_p G} \, \text{Pr}^{2/3} \tag{7.9}$$

in which we have introduced a factor, $\beta$, following Bedingfield and Drew [2] to account for some differences in the transport equations for heat and mass other than the replacement of mole fraction by temperature and Sc by Pr. In particular, Bedingfield and Drew point to the difference in the meaning of "velocity" in the two equations, the velocity in the mass-transfer equation being a molecule-number average velocity and that in the heat-transfer equation being a molecule-mass average value. As shown by Mickley, Sherwood, and Reed [37], the velocities in both equations become the mass-average value if instead of mole fraction, average molecular weight $\overline{M}$ is used as the dependent variable in the diffusion equation. According to Bedingfield and Drew [2], it follows that

$$\beta = \frac{\overline{M}_1 \ln (\overline{M}_1/\overline{M}_2)}{(M_A - M_B)\left[\left(\dfrac{N_A}{N_A + N_B}\right) - Y_{A1}\right] \ln \left(\dfrac{[N_A/(N_A + N_B) - Y_{A2}]}{[N_A/(N_A + N_B) - Y_{A1}]}\right)} \tag{7.10}$$

The correction factor $\beta$ is equal to unity in the limit of zero driving force for mass and seldom deviates more than 5 percent. In the limit of small driving force, $Y_{A2} - Y_{A1}$, Eq. (7.10) becomes, approximately,

$$\beta \approx 1 - \frac{1}{2}\left(\frac{M_A - M_B}{\overline{M}_1} - \frac{1}{N_A/(N_A + N_B) - Y_{A1}}\right)(Y_{A2} - Y_{A1}) \tag{7.11}$$

EXAMPLE 7.1   Find the values of $\beta$ in Eq. (7.9) and $h/\overline{M}c_p k_g^* P$ for the condensation of water vapor (a) from nitrogen and (b) from hydrogen on a cold surface at 20°C. The bulk of the gas is at 40°C and 1 atm in each case and contains 0.07 mole fraction water vapor.

SOLUTION   (a) At 20°C the vapor pressure of water is 17.53 mm, corresponding to $Y_{A1} = 0.0230$. Thus, the average molecular weights are $\overline{M}_1 = 27.77$ and $\overline{M}_2 = 27.30$. These are so close together that the approximate form, Eq. (7.11), can be used. Then

$$\beta = 1 - 0.5\left(\frac{-10}{27.77} - \frac{1}{1 - 0.023}\right)(0.07 - 0.023) = 1.0325$$

From Table 2.4, $D_{AB} = 0.256$ cm²/s at 24.4°C. Using the density and viscosity of nitrogen at the same temperature, Sc = 0.620. From Eckert and Drake [15], Pr = 0.713 at the same temperature, giving

$$\frac{h}{\overline{M}c_p k_g^* P} = \frac{(0.620/0.713)^{2/3}}{1.0325} = 0.882$$

(b) When nitrogen is replaced by hydrogen, the two average molecular weights are $\overline{M}_1 = 2.384$ and $M_2 = 3.136$. From Eq. (7.10),

$$\beta = \frac{2.384 \ln (2.384/3.136)}{(16.000)(1 - 0.023) \ln [(1 - 0.07)/(1 - 0.023)]} = 0.895$$

(The approximate expression gives $\beta = 0.866$.)

Using physical properties for hydrogen and a diffusion coefficient from Table 2.4, $Sc = 1.10$ and $Pr = 0.706$. Thus,

$$\frac{h}{\overline{M} c_p k_g^* P} = \frac{(1.10/0.706)^{2/3}}{0.895} = 1.47 \qquad ////$$

## 7.3 THE THEORY OF THE WET-BULB THERMOMETER

One of the simplest applications of the theory of simultaneous transport of enthalpy and mass through the fluid near an interface is for the calculation of the steady-state temperature of a wet-bulb thermometer. This is an ordinary mercury-in-glass thermometer or thermocouple, the temperature-sensing element of which is covered with a porous wick that is kept wet with a pure liquid; the liquid is the same as the condensible component in an otherwise inert gas which flows over the thermometer. The gas is warmer than the wet bulb and heat is transferred to the cool surface. Simultaneously, some of the liquid evaporates, diffusing into the gas. Such instruments are widely used for the measurement of the humidity of air.

Use of the wet-bulb thermometer for measurement of humidity can be traced as far back as 1792. The phenomenon has been treated theoretically by many well-known scientists and engineers, leading sometimes to argument and disagreement about the significance of the data. Sherwood [52] has given an account of The Curious History of the Wet-Bulb Hygrometer, which recalls some of the important steps in the development. The following is a development of the theory which shows how the final simple and practically important relationship between humidity and wet-bulb temperature can be found from a very general starting point which should be applicable to mixtures other than air and water vapor.

At steady state the inward flux of heat is just equal to the enthalpy requirement at the surface:

$$h(t - t_{WB}) \frac{C_0}{1 - e^{-C_0}} = \lambda_{WB} k_g^* P \ln \left( \frac{1 - Y}{1 - Y_{WB}} \right) \qquad (7.12)$$

Introducing the analogy relationship between the heat- and mass-transfer coefficients according to Eq. (7.9),

$$(t - t_{WB}) \frac{C_0}{1 - e^{-C_0}} = \lambda_{WB} \frac{\beta}{C_p} Le^{2/3} \ln \left( \frac{1 - Y}{1 - Y_{WB}} \right) \qquad (7.13)$$

in which the Lewis number Le is equal to Pr/Sc or to $DC_p \rho / k$ [25]. Equation (7.13) is seen to be an equation determining $t_{WB}$ for fixed $Y$ and $t$ because the interface mole fraction $Y_{WB}$ is determined by the vapor-pressure curve of the evaporating liquid and the total pressure.

Ordinarily in the use of the wet-bulb thermometer, the temperature difference and the partial pressure difference are small, permitting considerable simplification in Eq. (7.13). Introducing series approximations for the Ackermann factor on the left, Eq. (7.13) becomes

$$t - t_{WB} = \frac{\lambda_{WB}}{C_{pA}} \frac{\beta Le^{2/3} (C_{pA}/C_{p,av}) \ln \left[ (1 - Y)/(1 - Y_{WB}) \right]}{1 - \beta Le^{2/3} (C_{pA}/C_{p,av}) \ln \left[ (1 - Y)/(1 - Y_{WB}) \right]} \tag{7.14}$$

Introducing series approximations for the right side,

$$t - t_{WB} = \frac{\lambda_{WB}}{C_{p,av}} Le^{2/3} (Y_{WB} - Y) \tag{7.15}$$

The errors involved in the approximations are smaller than two or three percent for the air-water system at typical wet-bulb test conditions. If we further introduce the humidity $H = M_A Y / M_B (1 - Y)$ and the mass humid heat capacity $c_s = c_B + H c_A$, Eq. (7.15) becomes

$$t - t_{WB} = \frac{(\lambda_{WB}/M_A)}{c_s} Le^{2/3} (H_{WB} - H) \tag{7.16}$$

and the series approximation error is even smaller.

Equation (7.16) is often obtained by writing the heat- and mass-transfer rates as $h(t - t_{WB})$ and $k'(H_{WB} - H)$, respectively. The energy balance at the interface then gives Eq. (7.16), with the coefficient of the humidity difference on the right being replaced by $(\lambda_{WB}/M_A)(k'/h)$ or $(\lambda_{WB}/M_A c_s)(c_s k'/h)$. The group $h/c_s k'$ is often called the "psychrometric ratio." From the analogy between heat and mass transfer it should be equal to $Le^{2/3}$, as in Eq. (7.16). Measurements for water and air in wetted-wall columns give values of $(\lambda_{WB}/M_A)(H_{WB} - H)/(t - t_{WB}) = h/k'$ from $0.95c_s$ to $1.12c_s$. Several authors, including Wilke and Wasan [59], have computed values of the psychrometric ratio for various mixtures from assumed structures of the turbulent flow near surfaces in an effort to improve upon the simple function of the Lewis number used in Eq. (7.16). The results are more complicated but may be useful when greater precision is required. When the ratio of Schmidt to Prandtl groups is smaller than unity, the psychrometric ratio is smaller than $Le^{2/3}$.

When Eq. (7.16) is applied practically to the air-water system, for which $Le = 0.713/0.620 = 1.150$ and $Le^{2/3} = 1.10$, the Lewis-number term is omitted from the equation. This is justified by several sets of experimental data for this system [30]. Such apparent disagreement with the theory should not be thought to mean that the theory is wrong in principle, however. Recall that in the typical wet-bulb experiment, heat is transferred to the cool bulb by radiation from warmer surroundings or by conduction along the thermometer stem, tending to make the evaporation rate greater than that assumed in deriving the equation here. Note, also, that an experimental error in observing $t_{WB}$ has an effect on each side of the equation since $H_{WB}$ is affected, and the two effects are additive.

Note that if $Le^{2/3}$ is set equal to unity in Eq. (7.16), the equation can be rearranged as

$$c_s t + \left(\frac{\lambda_{WB}}{M_A}\right) H = c_s t_{WB} + \left(\frac{\lambda_{WB}}{M_A}\right) H_{WB} \qquad (7.17)$$

This indicates that if small changes in heat capacity with temperature are disregarded, the enthalpy of saturated air per unit mass of $B$ at the wet-bulb surface is equal to that of the bulk-air–$H_2O$ mixture. Since the psychrometric ratio is found experimentally to be near unity, the wet-bulb temperature is essentially the same as the "temperature of adiabatic saturation" in the case of air and water. This remarkable coincidence was pointed out first by Carrier [7] and is the basis for the construction of commonly used humidity charts and for widely used design procedures for water cooling towers, as will be seen below. For the air-water system, unsaturated air, brought into contact with liquid water at the temperature $t_{WB}$, will remain at the same enthalpy per unit mass of air even though its temperature falls and its humidity increases. As this occurs, $t_{WB}$ will not change, although $t$ and $H$ will. This is not the case for other systems, such as organic vapors in air.

## 7.4 THE DESIGN OF PARTIAL CONDENSERS

The removal of a condensible component from an inert gas in a partial condenser is often needed in chemical processing. The rates of formation of condensate on the cold heat-transfer surface are determined both by the rate of sensible heat transfer to the surface and the rate of diffusion. The latter rate may be relatively fast when the percentage of inert components in the vapor is small but often becomes the determining rate when removal of the condensible component $A$ is nearly complete. Especially when $A$ has a small diffusion coefficient, removal of heat may be so rapid that diffusion can not keep up and fog droplets may form in the gas phase. Such suspended droplets may be very small and hard to remove from the gas stream.

Figure 7.2 shows a differential section of a condenser, useful for writing an energy balance. Let $G$, $V$, and $F$ be the molar mass velocities of inert gas ($B$), condensible vapor ($A$), and fog, respectively. The total molar mass velocity is $M = G + V + F$. The enthalpy flow rate of the vapor mixture, relative to pure liquid $A$ and inert gas at temperature $t_0$, is

$$GC_{pB}(t - t_0) + V[\lambda_0 + C_{pA}(t - t_0)] + FC_{pL}(t - t_0)$$

and its differential is

$$MC_{p,av}\, dt + [\lambda_0 + C_{pA}(t - t_0)]\, dV + C_{pL}(t - t_0)\, dF$$

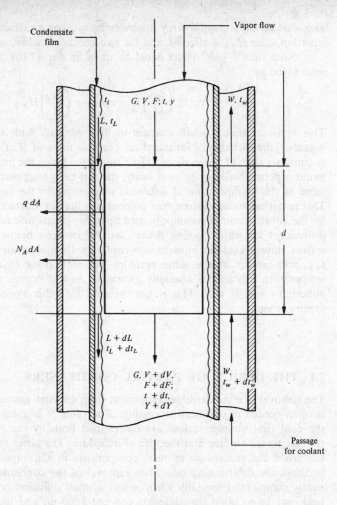

FIGURE 7.2

Differential section of a partial condenser.

where $C_{p,av}$ is the mole-weighted average of $C_{pB}$, $C_{pA}$, and $C_{pL}$. The steady-state energy balance for a system enclosing the vapor space of differential volume is

$$MC_{p,av}\, dt - \lambda\, dF + C_{pA}(t - t_i)(dV + dF) + q\, dA = 0 \qquad (7.18)$$

where $\lambda = \lambda_0 + (C_{pA} - C_{pL})(t - t_0)$ is the molar latent heat at the bulk-gas temperature, $dA$ is the differential condenser surface area, $q$ is the heat flux at the interface, and $t_i$ is the temperature of the gas-liquid interface. From a material balance,

$$-N_A\, dA = dV + dF \qquad (7.19)$$

$A$ represents the surface area per unit of tube cross section, or $4l/d$ for a cylindrical tube. Using Dalton's law to relate the partial pressure of $A$ in the gas to the quantity of vapor, $V/G = p/(P - p) = Y/(1 - Y)$ and

$$dV = G \frac{P\,dp}{(P-p)^2} = G \frac{dY}{(1-Y)^2} = M \frac{dY}{1-Y} \qquad (7.20)$$

Equations (7.3) and (5.4) are available for evaluating $q$ and $N_A$, respectively.

The interface temperature lies between $t_W$, the temperature of the coolant, and $t$, and is to be found from an energy balance at the interface. Using an overall heat-transfer coefficient $U$ to represent the thermal resistances of the condensate layer (neglecting the enthalpy change of the liquid film), the tube wall and a fluid film on the coolant side,

$$U(t_i - t_W) = q + \lambda_{Ai} N_A$$
$$= h(t - t_i) \frac{C_0}{1 - e^{-C_0}} + \lambda_{Ai} k_g^* P \ln \frac{1 - Y_i}{1 - Y} \qquad (7.21)$$

Since $Y_i$ is related to $t_i$ by the vapor pressure relationship for component $A$, Eq. (7.21) can be solved for $t_i$ by successive approximations.

An energy balance for a system enclosing vapor, condensate, and coolant streams shows that

$$MC_{p,\mathrm{av}}\,dt + [\lambda_0 + C_{pA}(t - t_0) - C_{pL}(t_i - t_0)]\,dV$$
$$+ C_{pL}(t - t_L)\,dF + C_{pL} L\,dt_L = WC_{pW}\,dt_W \qquad (7.22)$$

or, if the latent heat at the gas temperature is introduced,

$$MC_{p,\mathrm{av}}\,dt + [\lambda + C_{pL}(t - t_i)]\,dV + C_{pL}(t - t_L)\,dF + C_{pL} L\,dt_L = WC_{pW}\,dt_W \qquad (7.23)$$

Equation (7.23) can be used to find $dt_W$ after the other differentials have been determined from previous equations.

## 7.5  CONDENSATION FROM A SUPERHEATED GAS

When there is no fog carried by the gas stream, $dF = 0$ and $N_A\,dA = dV$. Then the material balance gives

$$\frac{1}{P-p}\frac{dp}{dA} = \frac{1}{1-Y}\frac{dY}{dA} = -\left(\frac{k_g^* P}{M}\right) \ln \frac{1 - Y_i}{1 - Y} \qquad (7.24)$$

The energy balance gives

$$\frac{dt}{dA} = -\left(\frac{h}{MC_{p,\mathrm{av}}}\right)(t - t_i) \frac{C_0}{(e^{C_0} - 1)} \qquad (7.25)$$

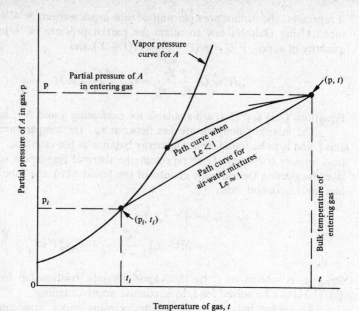

FIGURE 7.3
Behavior of path curves in partial condensers.

The slope of the path curve on $pt$ coordinates is found by dividing one equation by the other and using Eq. (7.9) to express $k_g^*/h$. The result is

$$\frac{dp}{dt} = (P - p)\, \text{Le}^{2/3} \frac{(e^{C_0} - 1)}{C_0} \frac{\ln\left[(P - p_i)/(P - p)\right]}{t - t_i}$$

$$= (P - p)\left(\frac{C_{p,\text{av}}}{C_{pA}}\right) \frac{\left[(P - p_i)/(P - p)\right] \exp\left[(C_{pA}/C_{p,\text{av}})\, \text{Le}^{2/3}\right] - 1}{t - t_i} \quad (7.26)$$

If $p - p_i$ is small, Eq. (7.26) is closely approximated by

$$\frac{dp}{dt} = \text{Le}^{2/3}\left(\frac{p - p_i}{t - t_i}\right)\left\{1 + 1/2\left(\frac{C_{pA}}{C_{p,\text{av}}}\, \text{Le}^{2/3} - 1\right)\frac{p - p_i}{P - p} + \cdots\right\} \quad (7.27)$$

Obviously the quantity in brackets on the right will be very nearly unity in many practical situations.

Figure 7.3 shows the location of path curves in partial condensers for the air-water system (Le $\approx$ 1.0) and for systems such as benzene-air for which, because of low diffusivities, Le < 1. For air and water the path curve points exactly toward the interface conditions, represented by a point on the equilibrium curve; if the interface conditions were constant, the whole path curve would be a straight line connecting the $p$ and $t$ values for the gas entering the condenser to the interface point. For benzene and air, however, each segment of the curved path line has a smaller slope than the chord connecting the points $p$, $t$ and $p_i$, $t_i$. For such systems it is very possible for the bulk of the gas to become saturated before $t$ reaches $t_i$. Then a fog may form in the gas if condensation nuclei are present. Even for air and water vapor, the straight path line can intersect the convex equilibrium curve if $t_i$ is small enough. Calculation methods for locating the path curves are also discussed by Brac [5], Olander [44], and Mizushina et al. [38–40]. Computations using analog computers are presented by Coughanowr and Stensholt [13] and O'Brien and Franks [43]. Experimental results are given by Cairns [6], Porter and Jeffreys [48], Smith and Robson [54], and Stern and Votta [56].

## 7.6 THE FORMATION OF FOG IN PARTIAL CONDENSERS

The theory of homogeneous nucleation [58, 29] applies to vapor mixtures that contain no dirt particles, droplets, or gas ions. When these are present they provide sites for growth by condensation of vapor molecules, even at small degrees of supersaturation; but if they are absent, both experiment and theory show that the partial pressure of $A$ can reach a value several times the vapor pressure before nucleation is observed even at a low rate. For process conditions, however, nuclei are often available as a result of operations conducted upstream in liquid-vapor equipment, combustion chambers, or reactions in which solids are produced or are present. According to Steinmeyer [55], formation of fog will begin in most process condensers when the bulk partial pressure falls to the saturation value. This point of view was also taken by Schuler and Abell [51], who reported data on the condensation of $TiCl_4$ from nitrogen in a single tube condenser. Ivanov [21] and Schrodt [49, 50], on the other hand, believe that some subcooling is possible. The following development is based on the assumptions (1) that fog formation begins as soon as the vapor mixture reaches saturation at its bulk temperature, and (2) fog particles, once formed, are carried by the gas stream and are not deposited on the condenser surface. For a discussion of the rate of deposition of fog particles from gas streams see Hutchinson, Hewitt, and Dukler [20].

When fog is present the changes in bulk partial pressure and temperature are related by the Clausius-Clapeyron equation,

$$\frac{dp}{dt} = \frac{\lambda p}{RT^2} \tag{7.28}$$

Now it is possible to find $dF$ from Eqs. (7.18), (7.19), and (7.20) such that Eq. (7.28) will be satisfied along the path curve. Introduction of the rate expressions for $q$ and $N_A$ yields

$$\frac{dF}{dA} = \frac{C_{pA}[(t - t_i)/(e^{C_0} - 1)] - C_{p,av}(RT^2/\lambda)[(P - p)/p]}{\lambda + C_{p,av}(RT^2/\lambda)[(P - p)/p]} N_A \tag{7.29}$$

$$\frac{dV}{dA} = -\frac{\lambda + C_{pV}[(t - t_i)/(e^{C_0} - 1)]}{\lambda + C_{pV}(RT^2/\lambda)[(P - p)/p]} N_A \tag{7.30}$$

$$\frac{dp}{dA} = \frac{P - p}{M}\frac{dV}{dA} = -\frac{\lambda + C_{pV}[(t - t_i)/(e^{C_0} - 1)]}{\lambda + C_{p,av}(RT^2/\lambda)[(P - p)/p]} N_A \tag{7.31}$$

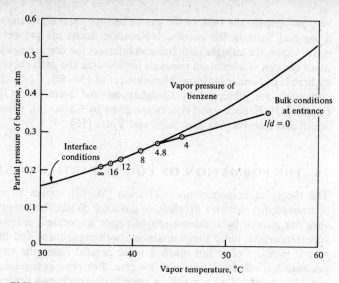

**FIGURE 7.4**
Calculated path curves during condensation of benzene vapor from nitrogen.

Obviously both the temperature difference $t - t_i$ and the partial pressure difference affect $dp/dA$ because of the constraint provided by Eq. (7.28)—because the partial pressure can fall even if there were no mass transfer if heat transfer causes the vapor temperature to fall. If $p - p_i$ is small, series approximation of Eq. (7.31) shows that

$$\frac{dp}{dA} = -\frac{k_g^*}{M}(P - p)\frac{p - p_i + (C_{p,av}/\lambda)\,\mathrm{Le}^{-2/3}(P - p)(t - t_i)}{1 + (C_{p,av}RT^2/\lambda^2)[(P - p)/p]} \qquad (7.32)$$

$$\frac{dV}{dA} = -k_g^*\frac{p - p_i + (C_{p,av}/\lambda)\,\mathrm{Le}^{-2/3}(P - p)(t - t_i)}{(C_{p,av}RT^2/\lambda^2)[(P - p)/p]} \qquad (7.33)$$

**EXAMPLE 7.2** Find the surface area of a condenser tube needed to remove benzene vapor from air [50a]. The mixture entering the condenser is at 54.5°C and is composed of 17.4 lb moles/(h)(ft²) of air and 9.37 lb moles/(h)(ft²) of benzene, corresponding to 0.35 mole fraction benzene. The temperature of the condensate film is constant and equal to 36.6°C. The mass-transfer coefficient is $k_g^* = 1.294$ lb mole/(h)(ft²)(atm) and the pressure is 1 atm. Physical properties include $\lambda = 8167$ cal/g mole, $C_{pA} = 23.45$ cal/(g mole)(°C), $C_{pL} = 32.7$, and $C_{pB} = 6.98$; Le = 0.707.*

> * Note added in proof: Based on the properties of benzene and air (Sc = 1.65, Pr = 0.713) the value of the Lewis number should be 0.713/1.65 = 0.432. If this value had been used, the results of the calculations would not have been changed qualitatively although the quantity of fog would have been greater.

SOLUTION   This problem is solved most easily on a digital computer, using a procedure for the simultaneous integration of the differential equations for $V$, $F$, $t$, and $p$ as functions of $A$. Initial values of the four dependent variables are known at $A = 0$. Equation (3.8) gives the mass-transfer rate and Eq. (7.4) the heat-transfer rate. As long as no fog is present, i.e., near the tube's entrance, Eqs. (7.25) and (7.27) give the rates of change of $t$ and $p$. These are valid until $p$ becomes equal to the vapor pressure of benzene at $t$. After this occurs Eqs. (7.25) and (7.27) are replaced by Eqs. (7.28) and (7.31), respectively.

Figures 7.4 and 7.5 show the results of the calculations, using the Runge-Kutta-Gill procedure for the solution of four equations. The path curve in Fig. 7.4 is very slightly curved for $l/d < 4.8$. According to Fig. 7.5 the bulk vapor stream becomes saturated at $l/d = 4.8$, after which fog builds up until, at $l/d \to \infty$, $F/V = 0.0091$ mole/mole.

Figure 7.5 also compares the results obtained using the rate equations in their most general form, as described, with those obtained with equations such as Eq. (7.27) in which approximations for small driving forces have been introduced. Note that for the conditions assumed, the value of $l/d$ required to remove 90 percent of the maximum amount of benzene is 31 percent greater according to the approximate equations. The path curve on Fig. 7.4 is hardly affected by the use of linearized rate expressions.        ////

## 7.7   EFFECT OF VARIATIONS IN INTERFACE TEMPERATURE

Although for the sake of simplicity it was assumed in previous sections that $t_i$ was known and was constant, $t_i$ in fact usually needs to be found by solving Eq. (7.21) at each point. Then the change in $t_W$ needs to be found from Eq. (7.23). Solution of Eq. (7.21) is accomplished readily with the iterative numerical procedure of Newton and Raphson. We define $F(t_i)$ by

$$F(t_i) = \frac{U}{h}(t_i - t_w) - (t - t_i)\frac{C_0}{1 - e^{-C_0}} - \frac{\lambda_{Ai}}{C_{p,av}}\beta(\text{Le})^{2/3}\ln\left(\frac{P - p_i}{P - p}\right) \quad (7.34)$$

$F(t_i)$ is zero when the correct solution for $t_i$ has been obtained. Successive approximations to $t_i$ are obtained from

$$t_i^{(k+1)} = t_i^{(k)} - \frac{F(t_i^{(k)})}{F'(t_i^{(k)})} \quad (7.35)$$

where $F'(t_i) = dF/dt_i$ is given approximately by

$$F'(t_i) = \frac{U}{h} + 1 + \frac{\lambda_{A_i}{}^2(\text{Le})^{2/3}p_i}{C_{p,av}RT^2(P - p_i)} \quad (7.36)$$

The integration of the differential equations for the five variables, $p$, $t$, $V$, $F$, and $t_W$, along the condenser surface is carried out readily with a digital computer. An illustration is included in the examples at the end of the chapter.

Although the design of a condenser based on Eq. (7.4), etc., is easy and efficient when a computer is available, an approximate graphical design originally given by Mickley [36] is also attractive because it gives an intuitively clear explanation of the principal variables that affect the design. The method employs the simplified

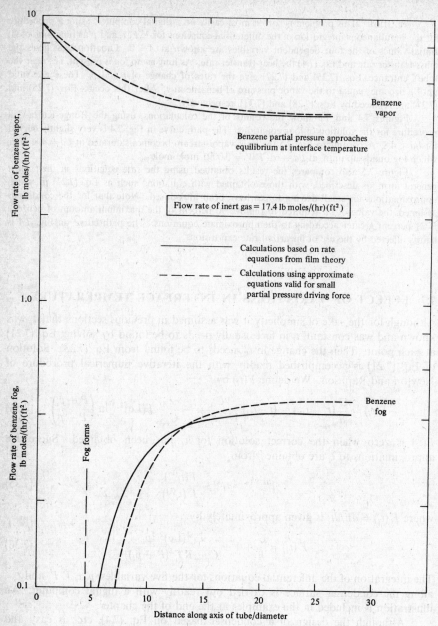

FIGURE 7.5
Calculated flow rates of vapor and fog in benzene partial condenser.

linear equations for the heat- and mass-transfer rates. It is assumed that no fog is formed.

The rates of change of gas temperature and humidity are represented by

$$Gc_s \, dt = h(t_i - t) \, dA \qquad (7.37a)$$

and
$$G \, dH = k'(H_i - H) \, dA \qquad (7.37b)$$

The total rate of heat transfer to the liquid condensate layer, per unit area, is

$$Wc_W \frac{dt_W}{dA} = h(t - t_i) + \lambda k'(H - H_i) = k' \left[ \frac{h}{k'c_s} c_s(t - t_i) + \lambda_i(H - H_i) \right] \qquad (7.38)$$

For the air-water system, since $h/k'c_s \approx 1$, this simplifies to

$$U(t_i - t_W) = Wc_W \frac{dt_W}{dA} = k'[(c_s t + \lambda_i H) - (c_s t_i + \lambda_i H_i)] = k'(i - i_i) \qquad (7.39)$$

Equation (7.39) is a remarkable result, for it shows that the rate of transport of enthalpy to the coolant is very nearly equal to the product of the *mass*-transfer coefficient $k'$ and an *enthalpy driving force*. [Note that gas enthalpy per unit mass of air is $i = c_s t + \lambda H$ Btu per pound of dry air, as in Eq. (7.17).] Enthalpies of saturated air-water vapor mixtures are tabulated by Perry [47]. From an energy balance around the gas stream,

$$- G \, di = k'(i - i_i) \, dA \qquad (7.40)$$

Merkel [34] or Hirsch [17] were apparently the first to show that use of the enthalpy driving force results in great simplification in the design of air-water cooling towers.

From Eqs. (7.40) and (7.36), the slope of the path curve on $it$ coordinates is

$$\frac{di}{dt} = \frac{i - i_i}{t - t_i} \qquad (7.41)$$

From Eqs. (7.39) and (7.40) the slope of a tie line connecting the points $i$, $t_W$ and $i_i$, $t_i$ is

$$\frac{i_i - i}{t_i - t_W} = - \frac{U}{k'} \qquad (7.42)$$

The slope of the operating line representing the bulk properties, $i$ vs. $t_W$, is

$$\frac{di}{dt_W} = \frac{Wc_W}{G} \qquad (7.43)$$

Equations (7.41), (7.42), and (7.43) permit the rapid construction of curves representing the changes in gas enthalpy and temperature as well as coolant and interface temperatures, as shown by Mickley [36].

Figure 7.6 is a typical plot of enthalpy vs. temperature for the air-water system. $PQ$ is the locus of saturated states and is assumed to represent interface conditions $(i_i, t_i)$. $AB$ is the operating line representing the points $i$, $t_W$. $EM$ is a

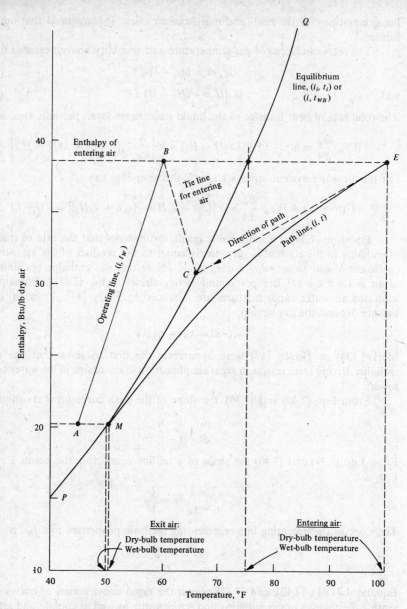

**FIGURE 7.6**
Graphical construction of dehumidification path line according to Mickley [36].

path curve $(i, t)$ which starts at the known state of the moist air entering the condenser. $BC$ is a typical tie line having a slope equal to $-(U/k')$.

The graphical construction is begun at point $E$, representing the entering air-water vapor mixture. Point $B$ has coordinates $i$, $t_W$ and the tie line is drawn from it toward the equilibrium line. The intersection at point $C$ has the coordinates $i_i$, $t_i$ and a straight line drawn from $E$ to $C$ gives the direction of the initial tangent to the path curve. The path curve is extended in this direction a short distance to a new point to the left of $E$, and the construction is repeated with a new tie line and a new direction of the path line.

The construction does not give the required condenser surface area, the humidity of the exit air, or the quantity of condensate, but these quantities can be found after the construction is complete. The humidity is found from a humidity chart when the wet- and dry-built temperatures are known; the quantity of water condensed from the air is found from a material balance. The surface area needed is found from the graphical integration of Eq. (7.40),

$$A = \frac{G}{k'} \int \frac{di}{i - i_i} \tag{7.44}$$

using the values of $i$ and $i_i$ found graphically.

In the graphical construction indicated in Fig. 7.6, the air mixture approaches saturation very closely before it leaves. Fog formation is not indicated for the particular conditions assumed, but it could occur if the coolant temperatures were lower or the entering air enthalpy were greater. Usually such conditions are to be avoided; if they occur, the assumptions $(F = 0)$ in Eqs. (7.36) to (7.44) will be invalid and the more general methods will be required, as illustrated earlier.

## 7.8  THE CONDENSATION OF MIXED VAPORS

When the vapor in a condenser is a mixture of two condensible components having different vapor pressures, the rate of condensation and the composition of the incremental condensate are determined partly by the resistance to diffusion in the vapor phase. Colburn and Drew [10] showed that solution of such problems can be based on the film theory and an assumption of interfacial equilibrium. They wrote Eq. (3.9) in the form

$$N_A + N_B = k_g^* P \ln \frac{Z_A - Y_{Ai}}{Z_A - Y_A} \tag{7.45}$$

where $Z_A = N_A/(N_A + N_B)$ is the ratio of moles of $A$ condensing to total moles condensing, or the mole fraction of $A$ in the small increment of condensate that is deposited on the condenser surface. $Y_A$ is the mole fraction of $A$ in the bulk vapor mixture; $Y_{Ai}$ is the mole fraction at the interface and, for a binary mixture, is a function of pressure and liquid composition only.

Colburn and Drew gave an interesting numerical example showing the application of Eq. (7.45) to the cooling of a mixture of methanol and water vapor. At low cooling rates $(N_A + N_B \to 0)$ the condensate is essentially in equilibrium

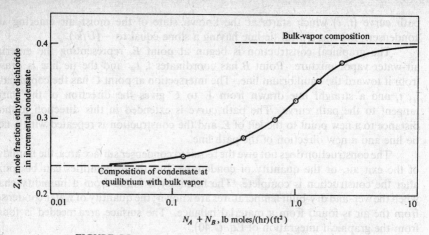

FIGURE 7.7
Calculated effect of condensation rate on composition of condensate: toluene–ethylene dichloride at atmospheric pressure.

with the bulk vapor; for a vapor containing 70 mole percent methanol this "static dew" contains 34 percent methanol. On the other hand, at large cooling rates $(N_A + N_B \to \infty)$, the condensate is much richer in methanol, approaching the bulk vapor composition. Under such conditions the condensation of both components produces such a large vapor velocity toward the surface, $(RT/P)(N_A + N_B)$, that differences in the velocity of the two molecules, $(RT/P)(N_A/Y_A)$ and $(RT/P)(N_B/Y_B)$, are very small in comparison. Thus, obtaining an enriched condensate in a partial condenser requires that the condensation proceed slowly.

EXAMPLE 7.3   A vapor mixture of 40 mole percent ethylene dichloride and 60 percent toluene is fed at 130°C and 1 atm to a condenser. Find the composition of the first drop of condensate that forms on the cold surface as a function of the total rate of condensation. The relative volatility is 2.14 and the mass-transfer coefficient for the gas phase is $k_g^* = 1.2$ lb mole/(h)(ft²)(atm).

SOLUTION   We write Eq. (7.45) for ethylene dichloride $(A)$ as

$$N_A + N_B = 1.2 \ln \frac{Z_A - Y_{Ai}}{Z_A - 0.4} \qquad (a)$$

and express the interfacial equilibrium by

$$Y_{Ai} = F(X_{Ai}) = F(Z_A) = \frac{2.14 Z_A}{1 + 1.14 Z_A} \qquad (b)$$

The calculations are carried out easily if values of $Z_A$ are assumed and $N_A + N_B$ computed.

Figure 7.7 shows the results. The interfacial temperature is fixed by the interfacial compositions. This and the bulk gas temperature determine the rate of sensible heat transfer to the surface. Assuming $Le = 0.7$ for the system and $C_p = 23$ Btu/(lb mole)(°F) for each compound, we calculate $h = (1.2)(23)/0.7^{2/3} = 35$ Btu/(h)(ft²)(°F). The latent heats are both about $\lambda = 13{,}500$ Btu/lb mole and the total rate of heat removal from the interface must be

$$1.8 \times 35(130 - t_i)\frac{C_0}{1 - e^{-C_0}} + 13{,}500(N_A + N_B) \qquad \text{Btu/(ft}^2\text{)(h)}$$

where $C_0 = (N_A + N_B)(23)/35$. For example, when $N_A + N_B = 1.0$ lb mole/(h)(ft²), $Z_A = 0.325$, $Y_{Ai} = 0.506$ and $t_i = 97.3$°C, giving $Q = 1.8 \times 35(130 - 97.3)$ $[0.657/(1 - e^{-0.657})] + 13{,}500 = 16{,}310$ Btu/(h)(ft²). $\qquad\qquad ////$

Equation (7.45) determines the condensate incremental composition at only one point in a condenser tube. The bulk liquid composition varies with distance in the direction of condensate flow according to a material balance, depending on the direction of vapor flow. If the vapor flows downward, cocurrent with the condensate, the allowable vapor velocity is greater but the difference between compositions of emergent vapor and liquid streams is small, corresponding to equilibrium between them. If the vapor flows upward, the vapor velocity is limited by flooding, but the countercurrent action permits the condensate leaving at the bottom of the condenser to be much richer in the high-boiling component than the vapor leaving at the top.

Figure 7.8 is an example of data obtained by Kent and Pigford [24], who measured compositions of the vapor at 1-ft intervals inside a vertical-tube condenser, using the countercurrent flow of vapor and condensate. The binary system was ethylene dichloride ($A$) and toluene. Total-reflux operation was used, vapor leaving the top of the tube being condensed externally and returned to the top of the condenser. In this way, liquid compositions were made equal to local vapor compositions as shown in the figure. Incremental condensate compositions could be computed from a material balance, $d(V Y_A) = Z_A \, dV$, or

$$Z_A = Y_A - V\left(\frac{dY_A}{dV}\right) = Y_A^2 \frac{d(V/Y_A)}{dV} \tag{7.46}$$

where $V =$ moles total vapor/(hr)(ft²). The figure shows that $Z_A$ values computed this way were nearly equal to values of $X_A^*$ (the mole fraction at equilibrium with the bulk vapor mixture) except at the top of the tube, where the condensation was most rapid. There, $Z_A$ lay between $X_A^*$ and $Y_A$, as expected when vapor-phase diffusion has a limiting effect. Values of $Y_{Ai}$ shown on the figure were computed from Eq. (7.45) using values of $k_g^*$ estimated from the vapor velocity in the tube and fluid properties. Values of $X_{Ai}$ follow directly from the equilibrium relationship and indicate that, especially near the bottom of the tube where the condensate layer was thickest, there was appreciable resistance to diffusive mixing in the liquid stream.

Kent and Pigford concluded that the liquid-phase mass-transfer coefficients found from their condensation tests were approximately in agreement with $k_L$ values reported separately for wetted-wall columns in which condensation was absent. Values of $H_L = L/k_L\rho_L$ were found to be in the range 2 to 3 ft at liquid Reynolds numbers ($4\Gamma/\mu$) of about 400.

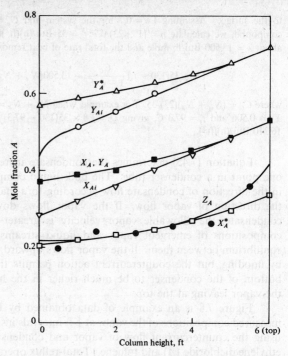

**FIGURE 7.8**
Compositions in a vertical, countercurrent condenser tube [24]: toluene and ethylene dichloride ($A$) at atmospheric pressure, total reflux.

Condensers of the type tested can clearly be used to produce changes in gas composition that are greater than the changes equivalent to a single theoretical stage, provided that the condensation surface is large enough both to remove the required amount of heat and also to permit easy diffusion in the vapor and liquid phases. Fractionation in such a condenser can be accomplished sometimes more economically than in a regular, adiabatic distillation column with a total condenser at the top, owing to the fact that the heat can be removed from the partial condenser partly at a higher temperature.

## 7.9 THE MECHANISM OF VAPOR-PHASE NUCLEATION IN PARTIAL CONDENSERS

In the previous section, changes in vapor-mixture conditions in the gas stream were computed on the assumption that the bulk of the vapor never becomes subcooled below its dew-point temperature. This can be true only if there are a sufficient number of condensation nuclei present. That such an assumption may not be far wrong, except in very clean process streams, is suggested by the work of Johnstone, Kelley, and McKinley [22]. Their data indicate that fog particles may form near the cold surface in a condenser even though the bulk gas is at a temperature *above* its saturation point, at least when condensation nuclei are added to the gas stream. This is possible when a part of the gas film near the interface is subcooled at some points along the temperature and concentration profiles. Figure 7.9 shows three sets of profiles, in each of which the gas mixture temperature $t$ and its dew-point temperature $t_D$ (a function of its local mole fraction of condensible vapor) can be compared. On the assumption of equilibrium at the gas-liquid interface, the two profiles must coincide there. If the $t$ curve begins with a slope greater than the initial slope of the $t_D$ curve, the whole film will be superheated and no fog can form; if the difference in slopes is reversed, however, a portion of the film may be supersaturated even though the bulk gas is not.

If fog particles form in the film, they may evaporate when they move into the bulk gas, causing no practical trouble. This formation may affect the rates of diffusion and heat conduction, however, causing the simple rate expressions to be inaccurate. In any event, it appears likely that there are regions in partial condensers which are more sensitive to growth of nuclei than the bulk gas stream itself and that the assumption made in Sec. 7.4 that fog occurs at the instant the bulk vapor becomes saturated may be not only safe but also realistic. As demonstrated by Colburn and Edison [11], fog formation can be eliminated from the gas leaving a partial condenser by placing a hot wire along the axis of the tube to keep the bulk gas hot.

The interfacial slopes of the $t$ and $t_D$ profiles in Fig. 7.9 can be found from the rate expressions previously developed:

$$-k\left.\frac{dt}{dy}\right)_i = h(t - t_i)\frac{C_0}{1 - e^{-C_0}} \tag{7.47}$$

$$-\left.\frac{dt_D}{dy}\right)_i = -\left.\frac{dt_D}{dp_A}\right)_i \left.\frac{dp_A}{dy}\right)_i = \frac{RT_i^2}{\lambda p_{Ai}} \cdot \frac{k_g^* P}{D_{AB}} \ln\left(\frac{P - p_i}{P - p}\right) \tag{7.48}$$

in which the Clausius-Clapeyron equation has been used to evaluate the first of the derivatives in Eq. (7.48). If we use Eq. (7.9) to relate $h$ to $k_g^*$ we find, after some algebra, that the condition for absence of supersaturation anywhere in the gas phase is

$$t - t_i > \frac{RT_i^2 C_{p,\text{av}}}{\lambda Y_{Ai} C_{pA} \text{Le}}\left[1 - \left(\frac{1 - Y_A}{1 - Y_{Ai}}\right)^{(C_{pA}/C_{p,\text{av}})\beta \text{Le}^{2/3}}\right] \tag{7.49}$$

where $\text{Le} = \text{Pr/Sc} = C_{p,\text{av}}(P/RT)D_{AB}/k$ as before. If the driving force, $Y_A - Y_{Ai}$, is small, a good approximation to Eq. (7.49) is

$$t - t_i > \frac{RT_i^2}{\lambda \text{Le}^{1/3}} \cdot \frac{Y_A - Y_{Ai}}{Y_{Ai}(1 - Y_{Ai})} \tag{7.50}$$

Equation (7.50) indicates that the boundary between the regions for possible and impossible fog formation on a $Y_A t$ diagram is a straight line passing through the interface point $Y_{Ai}, t_i$.

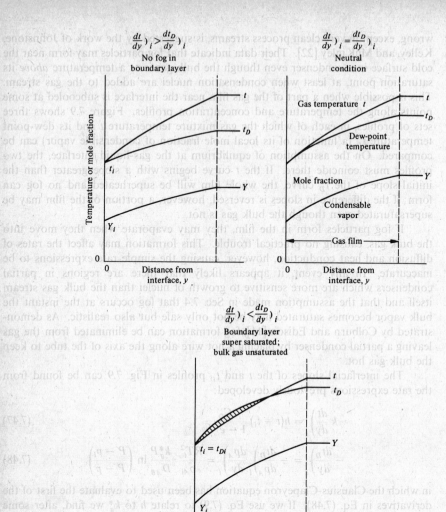

FIGURE 7.9
Temperature and composition profiles in gas film, showing possibility of fog formation in film with bulk gas unsaturated.

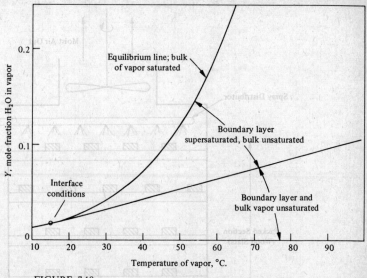

**FIGURE 7.10**
Calculated conditions in boundary layer for nitrogen and water vapor at 1 atm.

**EXAMPLE 7.4**  The temperature of the interface in a condenser removing water from nitrogen is at 15°C and the pressure is 1 atm absolute. Plot curves representing saturation of the bulk gas mixture and showing the bulk conditions permitting fog formation in the gas film. Physical properties include $\lambda = 10,600$ cal/g mole, $Pr = 0.713$, $Sc = 0.620$, $Y_{Ai} = 0.01682$ mole fraction $H_2O$ vapor, $C_{pA} = 8.97$ and $C_{pB} = 6.94$ cal/(mole)(°K).

SOLUTION   A series of values of $Y_A$ is assumed and $t$ is calculated from 7.50 used as an equation. For example, at $Y_A = 0.03$ we obtain

$$\beta = 1 - \frac{1}{2}\left(\frac{18 - 28}{28} - \frac{1}{1 - 0.01682}\right)(0.03 - 0.01682) = 1.0090$$

$$Le = \frac{0.713}{0.620} = 1.150$$

$$1 - \left(\frac{1 - Y_A}{1 - Y_{Ai}}\right)^{(C_{pA}/C_{p,av})\beta(Le)^{2/3}} = 0.01922$$

$$t > 15.0 + \frac{(1.987)(273 + 15)^2(6.94)(0.01922)}{(10,600)(0.01682)(8.47)(1.150)} = 27.7°C$$

Similar calculations can be made for other values of $Y_A$.

Figure 7.10 is a graph of the results. The figure shows that even for this system having a comparatively large value of the Lewis number, there is a wide range of gas temperatures above the bulk dew point within which at least some small part of the gas volume is subcooled.                                                                    ////

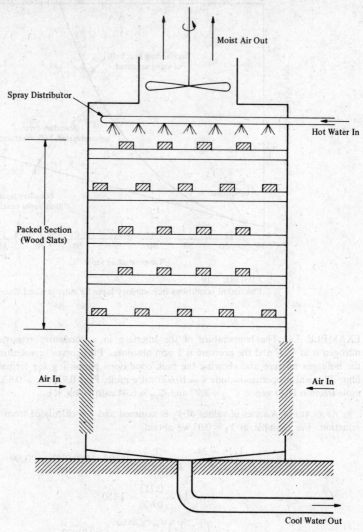

FIGURE 7.11
Induced-draft, countercurrent cooling tower.

## 7.10 WATER COOLING TOWERS

Evaporation of water into air in order to cool water, using a packed tower, is a very widely practiced process for treating recirculated water needed for condensers, heat exchangers, and other process equipment. Such cooling towers are among the largest mass-transfer devices in service. They may operate either countercurrently

FIGURE 7.12
Natural draft, crossflow cooling tower.

or with the air entering at the side of the column and flowing across the water stream as it falls through the packing. The motive force needed to overcome fluid friction opposing airflow is provided either by forced- or induced-draft fans, as in Fig. 7.11, or by natural convection owing to the light weight of humidified air, as in Fig. 7.12. See Berman [3], Chilton [8], Crankshaw [14], Parker and Krenkel [46], and McKelvey and Brooke [33].

The rate of transfer of enthalpy from the warm water to the air stream is governed by Eq. (7.39). For the design of a tower, however, it is convenient to introduce an overall mass-transfer coefficient, defined by equating the rate of enthalpy transfer to $K'(i^* - i)$, where $i^*$ is the enthalpy of air that would be saturated at the bulk water-stream temperature, as shown in Fig. 7.13. The overall driving force can be expressed as the sum of separate driving forces across the gas and liquid streams.

$$i^* - i = (i_i - i) + (i^* - i_i) \tag{7.51}$$

from which

$$\frac{1}{K'} = \frac{1}{k'} + \frac{1}{U} \frac{i^* - i_i}{t_w - t_i} \tag{7.52}$$

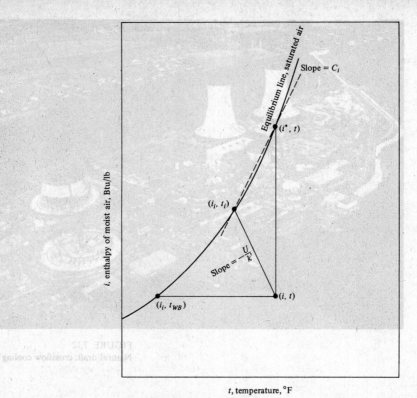

*i*, enthalpy of moist air, Btu/lb

Equilibrium line, saturated air

Slope = $C_i$

$(i^*, t)$

$(i_i, t_i)$

Slope = $-\dfrac{U}{k}$

$(i_i, t_{WB})$

$(i, t)$

*t*, temperature, °F

**FIGURE 7.13**
Driving forces for enthalpy exchange—the air-water system.

$U$ is the heat-transfer coefficient on the water side of the interface. The factor multiplying $1/U$ on the right is the slope of a chord connecting two nearby points on the equilibrium curve, as indicated in Fig. 7.13. Its value, hereafter called $C_i$, lies between 0.48 and 0.58 Btu/(lb air)(°F) for the important range of liquid temperatures from 45 to 55°F which are important in condensers. In cooling towers, however, the liquid temperatures are usually greater and $C_i$ is larger, as shown in Table 7.1. Unless $U$ is very small—a rare occurrence—it is quite satisfactory to use $C_i = 0.53$ for the design of condensers.

Introducing $K'$ into the energy balance around a differential section of height $dx$ in a countercurrent packed tower, we find

$$W \, dt_w = G \, di = K'a(i^* - i) \, dx \tag{7.53}$$

from which the required height of packing follows. It can be computed either from

$$X = \frac{G}{K'a} \int_{i_{in}}^{i_{out}} \frac{di}{i^* - i} \approx \frac{G}{K'a} \frac{i_{out} - i_{in}}{(i^* - i)_{lm}} \tag{7.54}$$

or from

$$X = \frac{L}{K'a} \int_{t_{w,\text{out}}}^{t_{w,\text{in}}} \frac{dt_w}{i^* - i} \tag{7.55}$$

A third alternative also exists, although it is not as useful as Eq. (7.54) or (7.55). Since the air enthalpy is a function only of the wet-bulb temperature, we may replace the enthalpy driving force, $i^* - i$, by $C_i(t_w - t_{wB})$, giving

$$X = \frac{L}{K'a} \int \frac{dt_w}{C_i(t_w - t_{wB})} \tag{7.56}$$

where $C_i$, a function of temperature, is evaluated at the arithmetic average of $t_w$ and $t_{wB}$. Although for accurate calculations, $C_i$ should be kept under the integral when it varies, it is sometimes useful to use a single value, taken at the average of the four values of $t_w$ and $t_{wB}$ at top and bottom of the tower. Thus Eq. (7.56) becomes

$$X = \frac{L}{K'aC_{i,\text{av}}} \int \frac{dt_w}{t_w - t_{wB}} \tag{7.57}$$

indicating that the "wet-bulb approach," $t_w - t_{wB}$, is a driving force governing the depth of packing. If the water temperature closely approaches the air wet-bulb temperature at some point in or at the end of the packing, the value of $t_w - t_{wB}$

Table 7.1 VALUES OF ENTHALPY OF SATURATED STATES AND SLOPES OF EQUILIBRIUM LINE FOR AIR-WATER SYSTEM AT 1 ATM [47]

| $t$, °F | $i$, Btu/lb* | $di/dt$, Btu/(lb)(°F) |
|---|---|---|
| 40 | 15.230 | 0.454 |
| 45 | 17.650 | 0.507 |
| 50 | 20.301 | 0.557 |
| 55 | 23.22 | 0.616 |
| 60 | 26.46 | 0.684 |
| 65 | 30.06 | 0.763 |
| 70 | 34.09 | 0.855 |
| 75 | 38.61 | 0.960 |
| 80 | 43.69 | 1.082 |
| 85 | 49.43 | 1.224 |
| 90 | 55.93 | 1.389 |
| 95 | 63.32 | 1.580 |
| 100 | 71.73 | 1.802 |
| 105 | 81.34 | 2.061 |
| 110 | 92.34 | 2.364 |
| 115 | 104.98 | 2.72 |
| 120 | 119.54 | 3.14 |
| 125 | 136.4 | 3.64 |
| 130 | 155.9 | 4.25 |
| 135 | 178.9 | |

* Standard states: Dry air at 0°F and liquid water at 32°F.

there will be critical. Thus it is understandable that commercial manufacturers of cooling towers often specify their performance by giving the maximum number of degrees Fahrenheit that the exit water stream should exceed the wet-bulb temperature of the air that is available in the particular locality.

EXAMPLE 7.5 Calculate the cross-sectional area and the depth of packing required in a slat-packed water-cooling tower to cool 10,000 gal/h of water initially at 130°F to 90°F by counter-current contact with air at atmospheric pressure, having a wet-bulb temperature of 70°F. The air rate will be 32 percent greater than the minimum rate and the superficial air mass velocity will be 1,710 lb/(h)(ft$^2$).

SOLUTION   Referring to Table 7.1, we find that the enthalpy of the entering air is 34.09 Btu/lb. To determine the minimum air rate, draw a straight line in Fig. 7.14 tangent to the equilibrium curve and passing through the point 34.09 Btu/lb, 90°F. Its slope is 2.8 Btu/(lb) (°F), from which the minimum air rate is (10,000)(8.33)/2.8 = 29,800 lb/h. The actual gas rate is (1.32)(29,800) = 39,300 lb/h. If the allowable gas superficial mass velocity is 1,710 lb/(h)(ft$^2$), the required tower cross section is 39,300/1,710 = 23.0 ft$^2$. The enthalpy of the exit air is found from an energy balance:

$$i = 34.09 + \frac{(10^4)(8.33)}{39,300}(130 - 90) = 118.77 \text{ Btu/lb}$$

The number of enthalpy transfer units is calculated according to Eq. (7.54). Since the equilibrium line is curved, the logarithmic mean enthalpy difference will overestimate the average driving force. A numerical evaluation of the integral based on Simpson's rule is shown in the following table.

| $t_w$, °F | $i$, Btu/lb | $i^*$, Btu/lb | $1/(i^* - i)$ |
|---|---|---|---|
| 90  | 34.09  | 55.93  | $0.0458 \times 1 = 0.0458$ |
| 100 | 55.26  | 71.73  | $0.0608 \times 4 = 0.2432$ |
| 110 | 76.43  | 92.34  | $0.0628 \times 2 = 0.1256$ |
| 120 | 97.60  | 119.54 | $0.0456 \times 4 = 0.1824$ |
| 130 | 118.77 | 155.9  | $0.0269 \times 1 = 0.0269$ |
|     |        |        | $\overline{\phantom{0000}0.6239}$ |

$$\int_{34.09}^{118.77} \frac{di}{i^* - i} = \frac{(118.77 - 34.09)(0.6239)}{(3)(4)} = 4.40$$

An alternative calculation using the log-mean driving force—not necessarily quicker—gives nearly the same result if the 40° temperature is divided into two intervals: 90 to 110, and 110 to 130. For the first interval, $\Delta i_{lm} = 18.75$ Btu/lb and the integral is (76.43 − 34.09)/18.75 = 2.26; for the second interval, $\Delta i_{lm} = 25.1$ Btu/lb and the integral is (118.77 − 76.43)/25.1 = 1.69, the total integral over the full range is 2.26 + 1.69 = 3.95. If only one log-mean driving force is used, the integral is found to be only 2.94.

Figure 7.15 shows at $G/L = 39,300/83,300 = 0.472$ that $H_{OG} = 8.6$ ft for slat packing. This corresponds to a packed height of (8.6)(4.4) = 38 ft. The calculation can also be based on Fig. 7.14

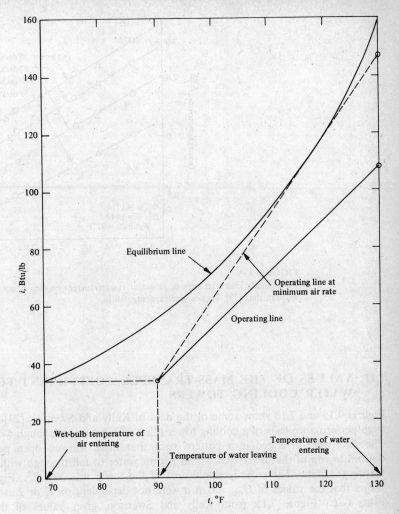

FIGURE 7.14
Graph for the design of a countercurrent water cooling tower.

and Eq. (7.55). From the table above, the integral in Eq. (7.55) is $(10.0)(0.6239)/3 = 2.08$, the required value of the "tower characteristic" $K'\alpha X/L$. From the figure at $L/G = 2.12$, the required number of decks is about 50. This corresponds to a packed height of $X = 50 \times 9/12 = 37.5$ ft.

Practical calculations based on the driving force, $t_w - t_{wB}$, and Eq. (7.56) are less accurate because of the variation in the slope $C_i$ of the equilibrium line. Using a single log-mean driving force of $16.6°F$ and estimating $C_i \approx 1.88$ from Table 7.1 at $t_{av} = 101.6°F$, we find that the integral is 2.41 and $X = (8.6)(2.12)(2.41)/1.88 = 23$ ft—much too low.    ////

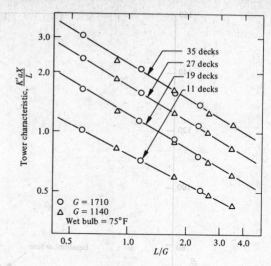

**FIGURE 7.15**

Performance characteristics of air-water countercurrent cooling tower filled with wooden slats [23]. Vertical deck spacing, 9.0 in.

## 7.11 VALUES OF THE MASS-TRANSFER COEFFICIENT FOR WATER COOLING TOWERS

Figures 7.15 and 7.18 show some of the data of Kelly and Swenson [23] resulting from performance tests of a cooling tower filled with horizontal, rough wood slats. The slats were arranged in a pattern such that there was no open path from bottom to top of the tower that would permit water to fall through without contacting the slats. The cross-sectional area of the tower was 32.6 ft². Figure 7.18 shows that the values of $H_{OG} = G/K'a$ were not dependent on $G$ or $L$ at constant value $G/L$. Figure 7.15, from Kelly and Swenson, gives values of the tower characteristic $K'aX/L$ appearing in Eq. (7.55). The values of this group are very nearly proportional to the number of horizontal decks of slats, although Kelly and Swenson found that at the limit of number of slats approaching zero there was a residual value of $K'aX/L = 0.07$, probably owing to the small end effect of drops of water raining from the packing. The value of $K'a$ fell only about 2 percent for each 10°F rise in water temperature. The values observed carefully on the test apparatus agreed with performance data for larger plant towers.

Apparently the most complete investigation of mass- and heat-transfer-rate coefficients for the simultaneous transfer of heat and mass in tower packings is that of McAdams, Pohlenz, and St. John [32], who used a small tower made from 4-in steel pipe filled with 1-in carbon Raschig rings to depths of 6, 9, and 12 in.

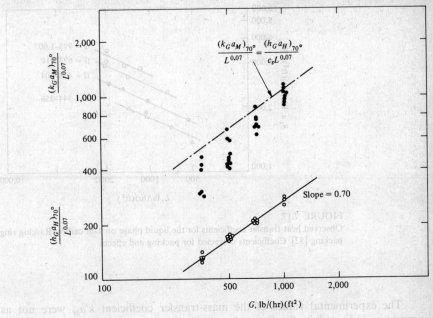

**FIGURE 7.16**
Observed mass- and heat-transfer coefficients for gas phase between 1-in carbon
Raschig-ring packing [32]. Coefficients corrected to a gas-film temperature of 70°F
and for packing end effects.

Adiabatic-saturation runs were conducted by adjusting the inlet water temperature
to obtain no change in $t_w$ as measured at the top and the bottom of the tower.
Such data yielded values of the gas-film coefficients, unaffected by the heat-transfer
resistance of the liquid stream, and permitted a determination of the tower's end
effects, equivalent to an additional packed depth of 0.60 ft.

The data shown in Fig. 7.16 were obtained after adjustment of the apparent
coefficients to obtain values for the packing alone. The empirical equation representing the heat-transfer data was found to be

$$ha_H = 1.78G^{0.70}L^{0.07}e^{0.0023t_f} \qquad (7.58)$$

where the film temperature $t_f$ is the arithmetic average of $t$ and $t_i$ in degrees
Fahrenheit. Equation (7.58) is equivalent to

$$H_G = \frac{Gc_s}{ha_H} = 0.129G^{0.3}L^{-0.07}\exp\left[0.0023(70 - t_f)\right] \qquad \text{ft} \qquad (7.59)$$

For the average conditions of the experiments [$G = 700$, $L = 1,500$ lb/(h)(ft$^2$),
$t_f = 70°F$], $H_G = 0.55$ ft.

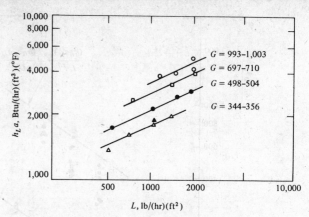

**FIGURE 7.17**
Observed heat-transfer coefficients for the liquid phase on 1-in carbon Raschig ring packing [32]. Coefficients corrected for packing end effects.

The experimental values of the mass-transfer coefficient $k'a_M$ were not as accurate as indicated in the figure, but lay below the values expected from the heat–mass-transfer analogy. [This lack of agreement implies that $H_G$ for mass or enthalpy transfer was somewhat greater than the values from Eq. (7.59).] The expected values are shown by the broken line. McAdams et al. [32] suggested that the difference could be accounted for by assuming that the surface area per unit volume for heat transfer, $a_H$, exceeded that for mass transfer, $a_M$, because the former included the dry area of each packing piece while the latter did not. The deviation of $k'a_M$ from $c_s ha_H$ became smaller as the gas rate increased and as more of the packing became wet. At low gas velocities the observed constant water temperature was 3°F higher than the temperature of adiabatic saturation, indicating the same effect of the greater area for heat transfer; the temperature excess fell to about 1°F at the highest values of $G$. The discrepancy between $c_s ha_H$ and $k'a_M$ is not likely to be significant in the design of commercial towers, where usually less than 20 percent of the total enthalpy exchanged is due to sensible heat changes in the gas stream.

By conducting water cooling runs in which $t_w$ changed by 30°F, McAdams, Pohlenz, and St. John [32] determined values of the waterside heat-transfer coefficient, $h_L a$. Their results are shown in Fig. 7.17. When these values are divided by $k'a_M$ to find the negative slope of the tie line on Fig. 7.13, the resulting values ranged from 1.4 to 2.7 Btu/(ib)(°F) for $500 < G < 2,000$ lb/(h)(ft$^2$), according to the equation

$$\frac{h_L a}{k'a_M} = 0.092 \, L^{0.43} \tag{7.60}$$

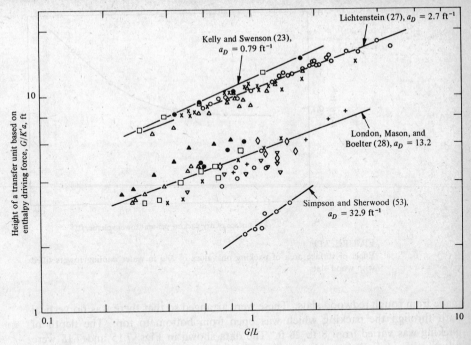

**FIGURE 7.18**
Experimental values of $G/K'a$ for slat-packed cooling towers from various sources [23, 27, 28, 53].

For the conditions of the experiments, the liquid phase offered 27 to 46 percent of the total resistance to enthalpy transfer between the phases.

Simpson and Sherwood [53] reviewed the data available on $K'a$ values for packed cooling towers before 1946 and reported results for small towers of their own. These were all of rectangular cross section, approximately 41 by 16 in, and were filled with 36 to 41.4 in of various rough wood slats, $\frac{1}{8}$-in-thick Masonite sheets, or wire mesh. Figure 7.18 compares their results with those of London, Mason, and Boelter [28], who used parallel $\frac{7}{8}$-in-wide by $2\frac{3}{4}$-in-high "ovate" red-wood slats in decks $\frac{1}{2}$ in apart in a tower having a cross-sectional area of 7.6 ft², and of Lichtenstein [27], whose tower was 6 by 6 ft in cross section with tiers of wood slats $\frac{3}{8}$ in wide by 2 in tall spaced parallel with 15-in clearance between tiers. The slats were displaced $\frac{3}{8}$ in sidewise in adjacent tiers to promote good water-flow distribution. See also Wrinkle [60].

Probably the best available data are those of Kelly and Swenson [23]. The wood tower used for the measurements had an inside plan area of 32.6 ft² and was insulated to reduce heat losses. Eight different packing decks were used, all

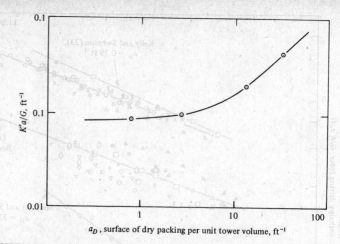

**FIGURE 7.19**

Effect of surface area of packing on values of $K'a$ in water cooling towers filled with wood slats.

made from rough redwood slats. These were arranged so that there was no vertical path through the packing which was open from bottom to top. The depth of packing was varied from 8 to 26 ft. The data shown in Figs. 7.15 and 7.18 were taken with slats 2' in wide horizontally by $\frac{3}{8}$ in thick, the vertical spacing between decks was 9 in and the horizontal spacing between slats was 8 in. Other designs included slats in vertical orientation, square slats oriented with their sides vertical and at 45°, and vertical spacings between decks to 24 in.

Kelly and Swenson's [23] results showed that the values of $K'aV/L_T$ fell only about 2 percent for each 10°F increase in water temperature, depending somewhat on the type of slat packing used. End effects were apparently small, $K'aV/L_T$ being very nearly proportional to the number of decks used. The data were correlated by the empirical equation

$$\frac{K'aX}{L_T} = 0.07 + bN\left(\frac{G}{L}\right)^n \qquad (7.61)$$

where $b$ and $n$ are constants varying with packing type and $N$ is the number of decks used. The largest value of $b'$ ($= 0.135$), corresponding to the most effective packing at $L/G = 1$, was obtained with decks made from $\frac{1}{2}$- by 1-in slats fastened on edge to the top and bottom of 2-in-deep stringers. The slats in each layer were spaced at intervals of $2\frac{1}{4}$ in and the slats in the bottom layer were centered in the spaces between slats in the top layer. Such decks were spaced at 24 in vertically. The value of $b$ for the packing used to obtain the data in Figs. 7.15 and 7.18 was 0.060, the smallest value observed.

Figure 7.19 shows values of the reciprocal of the HTU value read from Fig. 7.18 at $G/L = 1$, plotted against the surface area of the wood slat packing per unit volume, $a_D$. For closely spaced slats having a large value of $a_D$, the $K'a$ product is approximately proportional to $a_D$. For wide-spaced slats, however, there is only a small effect of $a_D$, probably because most of the effective interfacial area is in the form of water droplets and spray produced by impact of water streams falling onto the slats.

Very few mass-transfer data have been published for crossflow conditions in cooling towers, but fragmentary evidence indicates that values of $K'a$ are nearly the same for similar packings if evaluated at equal values of $L$ and $G$, each a mass velocity, and not a total mass-flow rate. The areas are the horizontal and vertical cross sections of the packed space, respectively perpendicular to the directions of gas and liquid flow. At equal values of $G$, the pressure drop appears to be smaller for horizontal flow than for vertical flow of the gas stream. See Coffey and Horne [9], Geibel [16], and Park and Vance [45]. A few data for spray towers are given in Niederman [42], and in Pigford and Pyle [62].

## 7.12 THE CROSSFLOW COOLING TOWER

Although countercurrent flow of water and air produces the greater temperature change of the water temperature, given the total flow rates of water and air and the air's inlet wet-bulb temperature, crossflow designs are sometimes used. In this arrangement, air enters opposite sides of the tower, flowing horizontally through the descending water stream. The two air streams join in an internal passage and leave through the top of the tower.

Figure 7.20 is a picture of a commercial crossflow tower and Fig. 7.21 is a schematic diagram of the crossflow situation. In Fig. 7.21, $B$ is the depth of packing perpendicular to the page. A steady-state energy balance for a differential volume $B\, dx\, dy$ shows that

$$LBX\left(\frac{dx}{X}\right)\left(\frac{\partial t_w}{\partial y}\right)dy = K'a(i - i^*)B\, dx\, dy \tag{7.62}$$

and

$$GBY\left(\frac{dy}{Y}\right)\left(\frac{\partial i}{\partial x}\right)dx = K'a(i^* - i)B\, dx\, dy \tag{7.63}$$

or

$$\frac{\partial t_w}{\partial y} = \left(\frac{K'a}{L}\right)(i^* - i) \tag{7.64}$$

and

$$\frac{\partial i}{\partial x} = \left(\frac{K'a}{G}\right)(i - i^*) \tag{7.65}$$

If we define $\Delta = i^* - i$ and assume that the equilibrium line is straight, so that $di^*/dt_w = C_i$ is constant, we can show that Eqs. (7.64) and (7.65) reduce to

$$\frac{\partial^2 \Delta}{\partial x\, \partial y} = RN\frac{\partial \Delta}{\partial x} + N\frac{\partial \Delta}{\partial y} = 0 \tag{7.66}$$

FIGURE 7.20
A commercial crossflow cooling tower. (*Courtesy of The Enjay Chemical Co., Linden, N.J.*)

where $x/X$ and $y/Y$ have been replaced temporarily by $x$ and $y$, respectively, and where $R = C_i G_T/L_T$ and $N = K'aV/G_T$. A solution of Eq. (7.66), consistent with boundary conditions on $\Delta(x, y)$ such that $\Delta(0, y) = \Delta(x, 0) = i_{in}^* - i_{in}$, is

$$\Delta = (i_{in}^* - i_{in}) \exp\left(-N\frac{x}{X} - NR\frac{y}{Y}\right) I_0\left(2N\sqrt{R\frac{x}{X}\frac{y}{Y}}\right) \qquad (7.67)$$

From this expression for the driving force as a function of $x$ and $y$, Eqs. (7.64) and (7.65) can be used to find the functions $t_w(x, y)$ and $i(x, y)$. The results can be integrated to find the mixed-average water temperature, $t_{w,\,out}$; the result shows that the group $P = C_i(t_{w,\,in} - t_{w,\,out})/(i_{in}^* - i_{in})$ is a function of the dimensionless quantities $R$ and $N$.

A convenient way to express the final result is that customarily used for crossflow heat-exchanger calculations. McAdams [31] and Bowman, Mueller, and Nagle [4] give a graph showing the correction factor $F$ to be applied to the logarithmic-mean driving force to account for the crossflow conditions, as a function

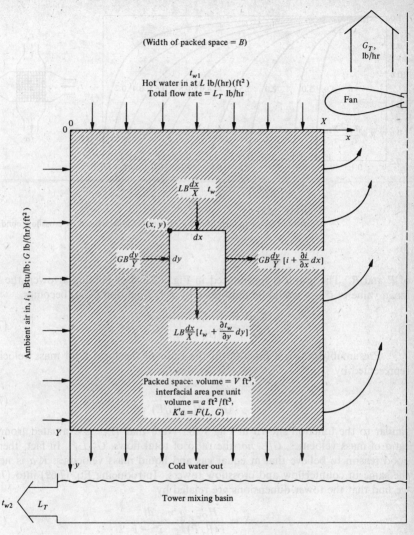

FIGURE 7.21
Sketch of a crossflow cooling tower.

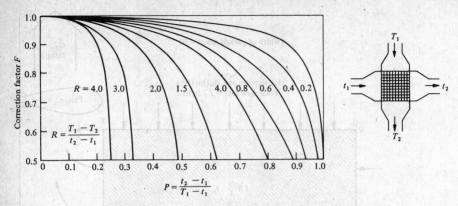

FIGURE 7.22
Correction factor to logarithmic-mean driving force caused by crossflow conditions [4].

of $P$ and $R$. The graph is reproduced in Fig. 7.22. For cooling a tower, the log-mean value of $i^* - i$ would be multiplied by $F$.† Equation (7.54) becomes

$$\frac{K'aV}{G_T} = \frac{1}{F(R, P)} \int_{i_{in}}^{i_{out}} \frac{di}{i^* - i} \tag{7.68}$$

Presumably $K'a$ is a function of the ratio of the local fluid mass velocities, represented by

$$\frac{K'a}{G} = \frac{1}{H_{OG}(G/L)} \tag{7.69}$$

similar to the function shown in Fig. 7.18. Note that $H_{OG}$ is evaluated from the ratio of mass velocities, $G/L$, *not* the ratio of total flows, $G_T/L_T$. In fact, there is good reason to believe that at equal gas and liquid mass velocities, $K'a$ is nearly the same in counterflow and crossflow towers. Introducing Eq. (7.69) into (7.68), we find that the tower dimensions are related by

$$X = \frac{H_{OG}}{F(R, P)} \int_{i_{in}}^{i_{out}} \frac{di}{i^* - i} \tag{7.70}$$

EXAMPLE 7.6  Find the dimensions of the packed space in a crossflow cooling tower designed to reduce the temperature of 5,000 gal/min of water from 110 to 95°F using air having a wet-bulb temperature of 80°F.

---

† The correction factor $F$ is closely related to the function $J$, used in the theory of the break-through curve for fixed-bed adsorption discussed in Chap. 10. The exact relationship between the functions and the independent variables which determine their values is left for the student to work out.

SOLUTION

(1) Calculate the number of transfer units for countercurrent flow.

$$i_{in} = 43.69 \text{ Btu/lb} \qquad \text{at } t_{wB, in} = 80°F$$
$$i_{in}^* = 92.34 \text{ Btu/lb} \qquad \text{at } t_{w, in} = 110°F$$
$$i_{out}^* = 63.32 \text{ Btu/lb} \qquad \text{at } t_{w, out} = 95°F$$

From a graph similar to Fig. 7.14 we find that the minimum value of $G_T/L_T$ is $(110 - 95)/(92.34 - 43.69) = 0.3083$. If we use twice the minimum quantity of air, we shall be operating with $G_T/L_T = 0.6166$, making

$$i_{out} = 43.69 + \frac{110 - 95}{0.6166} = 68.02 \text{ Btu/lb}$$

The average slope of the equilibrium line over the range of water temperature is $(2.364 + 1.580)/2 = 1.972$, making $R = C_i G_T/L_T = 1.216$. Thus, the operating line and the equilibrium line are nearly parallel. The enthalpy driving forces at the top and bottom of the countercurrent column are

$$\Delta i_{top} = 92.34 - 68.02 = 24.32 \text{ Btu/lb}$$
$$\Delta i_{bottom} = 63.32 - 43.69 = 19.63$$

and the log-mean driving force is 21.91 Btu/lb. The number of transfer units is $(68.02 - 43.69)/21.91 = 1.11$.

(2) Calculate the correction owing to crossflow conditions. Using Fig. 7.22 with $R = 1.216$, $P = (1.972)(110 - 95)/(92.34 - 43.69) = 0.6080$, we find that $F = 0.765$. Therefore the required number of transfer units for crossflow is $1.11/0.765 = 1.45$.

(3) Find tower dimensions. We assume that the gas and liquid mass velocities will be those used in typical test work on the countercurrent tower of Kelly and Swenson, as in Fig. 7.15.

$$G = 1,710 \text{ lb/(h)(ft}^2)$$
$$L = 1.7(1,710) = 2,907 \text{ lb/(h)(ft}^2)$$

where 1.7 is the slope of the operating line on Fig. 7.14. This makes $G/L = 1710/2907 = 0.588$ and Fig. 7.18 gives $H_{OG} = 9.4$ ft. The width of the packed space in the direction of gas flow is

$$X = (9.4)(1.45) = 13.6 \text{ ft}$$

The vertical height of the packing is found from the ratio of total flows:

$$0.6166 = \frac{G_T}{L_T} = \frac{GBY}{LBX} \qquad Y = \frac{(0.6166)(2,907)(13.6)}{1,710} = 14.3 \text{ ft}$$

The tower depth, perpendicular to the gas and liquid flows, is

$$B = \frac{L_T}{LX} = \frac{(5,000)(8.33)(60)}{(2907)(13.6)} = 63.2 \text{ ft}$$

If the tower is constructed with packing on each side of a central air plenum chamber, the total depth would be 31.6 ft. The volume of packing is 12,291 ft$^3$, and the rate of transfer of heat from the water stream is

$$(5,000)(8.33)(60)(110 - 95) = 3.75 \times 10^7 \text{ Btu/h}$$

# REFERENCES

1 ACKERMANN, G.: *Ver. Deutsch Ing., Forschungs.*, **382**: 1–16 (1937).
2 BEDINGFIELD, C. H., and T. B. DREW: *Ind. Eng. Chem.*, **42**: 1164 (1950).
3 BERMAN, L. D.: "Evaporative Cooling of Circulating Water," Pergamon, New York, 1961.
4 BOWMAN, R. A., A. C. MUELLER, and W. M. NAGLE: *Trans. ASME*, **62**: 284–94 (1940).
5 BRAC, G. H.: *Chem. Eng.* **60**: 223 (1963); **61**: 238 (1964).
6 CAIRNS, R. C.: *Chem. Eng. Sci.*, **2**: 127 (1953).
7 CARRIER, W. H.: *Trans. ASME*, **33**: 1005 (1911).
8 CHILTON, H.: *Inst. Elec. Eng. London*, **99**: 1952.
9 COFFEY, B. H., and G. A. HORNE: *J. Am. Soc. Refrig. Eng.*, **7**: 173 (1920).
10 COLBURN, A. P., and T. B. DREW: *Trans. AIChE*, **33**: 197 (1937).
11 COLBURN, A. P., and A. C. EDISON: *Ind. Eng. Chem.*, **33**: 475 (1941).
12 COLBURN, A. P., and O. A. HOUGEN: *Ind. Eng. Chem.*, **26**: 1178 (1934).
13 COUGHANOWR, D. R., and E. O. STENSHOLT: *Ind. Eng. Chem. Process Des. Dev.*, **3**: 369 (1964).
14 CRANKSHAW, C. J.: *Proc. Inst. Mech. Eng. (London)*, **178**(1): 927 (1963–1964).
15 ECKERT, E. R. G., and R. M. DRAKE: "Analysis of Heat and Mass Transfer," p. 732, McGraw-Hill, New York, 1972.
16 GEIBEL, C.: *Mitt. Forsch.*, **242**: 1–98 (1921).
17 HIRSCH, M.: "Die Trockentechnik," Springer, Berlin, 1927.
18 HIRSCHFELDER, J. O., C. F. CURTISS, and R. B. BIRD: "Molecular Theory of Gases and Liquids," pp. 708–720, Wiley, New York, 1954.
19 HULDEN, B.: *Chem. Eng. Sci.*, **7**: 60 (1959).
20 HUTCHINSON, P., G. F. HEWITT, and A. E. DUKLER: *Chem. Eng. Sci.*, **26**: 419 (1971).
21 IVANOV, M. E.: *Int. Chem. Eng.*, **2**: 282 (1962).
22 JOHNSTONE, H. F., M. KELLEY, and O. L. MCKINLEY: *Ind. Eng. Chem.*, **42**: 2298 (1950).
23 KELLY, N. W., and L. K. SWENSON: *Chem. Eng. Progr.*, **52**: 263 (1956).
24 KENT, E. R., and R. L. PIGFORD: *AIChE J.*, **2**: 363 (1956).
25 LEWIS, W. K.: *Trans. AIChE*, **20**: 9 (1927).
26 LEWIS, W. K.: *Trans. ASME*, **44**: 329 (1922).
27 LICHTENSTEIN, J.: *Trans. ASME*, **65**: 779 (1943).
28 LONDON, A. M., W. E. MASON, and L. M. K. BOELTER: *Trans. ASME*, **62**: 41 (1940).
29 LOTHE, J., and G. M. POUND: *J. Chem. Phys.*, **36**: 2080 (1962).
30 MARK, J. G.: data quoted by Sherwood, T. K., *Trans. AIChE*, **28**: 107 (1932).
31 MCADAMS, W. H.: "Heat Transmission," 3d ed., p. 195, pp. 355–365, McGraw-Hill, New York, 1954.
32 MCADAMS, W. H., J. B. POHLENZ, and R. C. ST. JOHN: *Chem. Eng. Progr.*, **45**: 241 (1949).
33 MCKELVEY, K. K., and M. BROOKE: "The Industrial Cooling Tower," Elsevier, Amsterdam, 1959.
34 MERKEL, F.: *Forschung.*, **275** (1926).
35 MICKLEY, H. S., R. C. ROSS, A. L. SQUYRES, and W. E. STEWART: *NACA Tech. Note* 3208 (1954).
36 MICKLEY, H. S.: *Chem. Eng. Progr.*, **45**: 739 (1949).
37 MICKLEY, H., T. K. SHERWOOD, and C. E. REED: "Applied Mathematics in Chemical Engineering," McGraw-Hill, New York, 1957.
38 MIZUSHINA, T., N. HASHIMOTO, and M. NAKAJIMA: *Chem. Eng. Sci.*, **9**: 195 (1959).
39 MIZUSHINA, T., K. ISHII, and H. UEDO: *Int. J. Heat Mass Transfer*, **1**: 95 (1954).

40  MIZUSHINA, T., M. NAKAJIMA, and T. OSHIMA: *Chem. Eng. Soc. (Japan)*, **13**: 7–17 (1960).

41  NESTER, D. M.: *Chem. Eng. Progr.*, **67**: 49 (July 1971).

42  NIEDERMAN, H. H.: *Heating, Piping, Air Conditioning*, **13**: 591 (1941).

43  O'BRIEN, N. G., and R. G. FRANKS: *Chem. Eng. Progr. Symp. Ser.*, **31**: 37 (1960).

44  OLANDER, D. R.: *Ind. Eng. Chem.*, **53**: 121 (1961).

45  PARK, J. E., and J. M. VANCE: Computer Model of Crossflow Towers, in "Cooling Towers," p. 122, American Institute of Chemical Engineers, 1972.

46  PARKER, F. L., and P. A. KRENKEL: "Thermal Pollution: Status of the Art," National Center for Research and Training in Hydrologic and Hydraulic Aspects of Water Pollution Control, Vanderbilt University, Nashville, Tenn., 1969.

47  PERRY, J. H.: "Chemical Engineers' Handbook," 3d ed., pp. 760–765, McGraw-Hill, New York, 1950.

48  PORTER, K. E., and G. V. JEFFREYS: *Trans. Instn. Chem. Eng. (London)*, **29**: 195 (1951).

49  SCHRODT, J. T., and E. R. GERHARD: *Ind. Eng. Chem. Fundam.*, **4**: 46 (1965).

50  SCHRODT, J. T., and E. R. GERHARD: *Ind. Eng. Chem. Fundam.*, **7**: 281 (1968).

50a  SCHRODT, J. T.: *2nd. Ind. Eng. Chem. Process Design Develop.*, **11**: 20 (1972).

51  SCHULER, R. W., and J. B. ABELL: *Chem. Eng. Progr. Symp. Ser.*, **52**: 51 (1956), no. 18.

52  SHERWOOD, T. K.: *Chem. in Canada*, 19–21 (June 1950).

53  SIMPSON, W. M., and T. K. SHERWOOD: *Refrig. Eng.*, **52**: 535 (1946).

54  SMITH, J. C., and H. T. ROBSON: *Proc. Gen. Disc. Heat Transfer, Instn. Mech. Engrs. (London)*, 38 (1951).

55  STEINMEYER, D. E.: *Chem. Eng. Progr.*, **68**: 64–68 (1972).

56  STERN, F., and F. VOTTA: *AIChE J.*, **14**: 928–933 (1968).

57  VIVIAN, J. E., and BEHRMANN: *AIChE J.*, **11**:656 (1965).

58  VOLMER, M.: "Kinetics of Phase Formation," Edwards, Ann Arbor, Mich., 1945.

59  WILKE, C. R., and D. T. WASAN: *AIChE-IChE Symp. Ser. 6, Instn. Chem. Engrs. (London)*, 21–26, 1965.

60  WRINKLE, R. B.: Performance of Counterflow Cooling Tower Cells, in "Cooling Towers," American Institute of Chemical Engineers, p. 118, 1972.

61  ZIVI, S. M., and B. B. BRAND: *Refrig. Eng.*, **64**: 31 (1956).

62  PIGFORD, R. L., and C. PYLE: *Ind. Eng. Chem.*, **43**: 1649 (1951).

## PROBLEMS

*7.1*  Show that the fractional error in the measured value of the psychrometric ratio, $\Gamma = c_s k'/h$, owing to an error $\Delta t_{WB}$ in observation of the wet-bulb temperature, is

$$\frac{\Delta \Gamma}{\Gamma} = \frac{-\Delta t_{WB}}{t - t_{WB}} \left( 1 + H_{WB} \frac{\lambda_{WB} \Gamma}{M_A c_s T_{WB}^{2}} \right)$$

*7.2*  Show that the temperature and humidity of a mixture of inert gas and a condensible vapor leaving a partial condenser are given approximately by

$$\ln \frac{t_{\text{in}} - t_i}{t_{\text{out}} - t_i} = \frac{hA}{Gc_s}$$

and

$$\ln \frac{H_i - H_{\text{in}}}{H_i - H_{\text{out}}} = \frac{K'A}{G}$$

when $t_i$ is constant. Also find an expression for the approach of the wet-bulb temperature of an air-water mixture to $t_i$.

**7.3** In the use of the Colburn-Hougen method [12] for the design of a partial condenser, it is assumed that as the gas mixture cools, owing to transfer of sensible heat to the cold surface, it stays saturated and no fog forms. Thus, it is possible to find the partial pressure driving force causing mass transfer from the gas temperature and the interface temperature. The object of this problem is to explore the conditions in which this assumption may be true.

Consider a binary gas mixture composed of a condensible component $A$ and a fixed gas $B$, which is insoluble in the liquid condensate layer. The bulk-gas partial pressure $p_A$ and the temperature $t$ lie on the equilibrium curve $p^* = p^*(t)$. The temperature of the interface between liquid condensate and vapor is a lower value, $t_i$. Obviously, $p_i = p^*(t_i)$, where $p^*$ is the same function as the one already introduced. As a result of the differences in partial pressure and temperature, substance $A$ tends to diffuse out of the gas and sensible heat flows out of the gas producing, respectively, changes in the partial pressure and the temperature.

Find the permissible value of the gas bulk temperature $t$, which for a fixed interface temperature $t_i$ must not be exceeded if the path curve $p_A(t)$ is not to pass into the supercooled region.

**7.4** A thin metal plate 1 ft long and infinitely broad is immersed in a stream of hot air at 1000°F which flows at a velocity of 50 ft/s tangentially across the plate in the direction of the 1-ft dimension. The upper surface of the plate is porous and is kept wet with liquid water, thus cooling both the top and the bottom surfaces. The plate is so thin and its thermal conductivity is so high that the top and bottom surfaces are at the same temperature.

(a) What is the steady-state temperature of the plate?

(b) How many pounds of water evaporates from the upper surface per foot of plate in the direction across the direction of gas flow?

You may assume that the water fed to the plate is at the plate's steady-state temperature, as in a wet-bulb thermometer.

**7.5** Tests on a small forced-draft countercurrent cooling tower gave the following results:

Inlet air dry-bulb temperature: 80°F
Inlet air wet-bulb temperature: 70°F
Outlet air wet-bulb temperature: 90°F
Outlet air dry-bulb temperature: impossible to measure
Inlet water temperature: 110°F
Outlet water temperature: 80°F

(a) Assuming air and water rates to be unchanged, what is the maximum temperature of the water delivered by this tower in the summer, when the worst condition will be an air supply having a dry-bulb temperature of 90°F and a wet-bulb temperature of 78°F? The cooling load is to be kept constant, i.e., the heat removed from the water stream is the same in the summer as in the winter.

(b) Do the results obtained in this computation correspond to a constant value of the wet-bulb approach as used by tower manufacturers?

(c) Construct a path curve for the gas as it flows through the tower and find the exit air humidity and dry-bulb temperature for the test conditions.

**7.6** Laboratory tests of a small countercurrent, packed water cooling tower by A. Shaines (M.Ch.E. thesis, University of Delaware) yielded the following values of the number of overall enthalpy transfer units,

$$N_{OG} = \int \frac{di}{i - i^*}$$

## HUMIDIFICATION OF AIR WITH WARM WATER*

| $C_{i,av} G/L$ | $H_{OG}$, ft | Temperature water in, °F |
|---|---|---|
| 0.375 | 0.275 | 91.3 |
| 0.46 | 0.305 | 99.2 |
| 0.51 | 0.335 | 103.9 |
| 0.56 | 0.34 | 107.8 |
| 0.61 | 0.34 | 114.2 |
| 0.65 | 0.32 | 118.0 |
| 0.80 | 0.355 | 124.6 |
| 0.855 | 0.355 | 127.7 |
| 0.955 | 0.365 | 132.3 |
| 0.97 | 0.365 | |
| 1.065 | 0.375 | |

Tower packing: Raschig Rings 8 in deep
Liquid rate: 1,540 lb/(h)(ft$^2$)
Gas rate: 442 lb/(h)(ft$^2$)
$C_{i,av} = \int C_i \, di/(i^* - i) / \int di/(i^* - i)$

* Data of A. Shaines (M.Ch.E. thesis, University of Delaware, 1952)

Note that the values vary slightly depending on the temperature of the inlet water or the inlet gas at the ends of the tower, indicating that the apparent rate of heat transfer was not precisely proportional to the overall enthalpy driving force.

Using these values, find values of the quantities $H_G$ and $H_L$. ($H_G$ represents the height of a transfer unit for enthalpy transfer across the gas phase, assumed constant, and $H_L$ represents the heat-transfer resistance of the liquid phase, also assumed constant.)

7.7 Using a digital computer, Zivi and Brand [61] have calculated the temperatures of the parallel streams of water leaving a crossflow cooling tower. For example, using $K'a = 500$ lb/(h)(ft$^2$)(lb/lb), $G = 1,500$ lb air/(h)(ft$^2$), $L = 1,500$ lb water/(h)(ft$^2$), $X = 6$ ft, $Y = 6$ ft, $t_{WB,in} = 78°F$, and $t_{W,in} = 95°F$, they obtained the following values of the exit water temperature as a function of position across the bottom of the packing.

| $x/X$ | $t_{W,out}$, °F |
|---|---|
| 0 | 79.2 |
| 0.24 | 80 |
| 0.51 | 81 |
| 0.78 | 82 |
| 1.00 | 82 |

Calculate the average exit water temperature produced by the tower using the theoretical method of this chapter and compare the result with the values reported by Zivi and Brand.

7.8 (a) Show that, for a binary system, the changes in vapor composition from top to bottom of a countercurrent partial condenser are represented by

$$\ln \frac{V_{bottom}}{V_{top}} = \int_{Y_{A,bottom}}^{Y_{A,top}} \frac{[e^{(N_A + N_B)/k_g^* P} - 1] \, dY_A}{Y_A - Y_{Ai}}$$

When the diffusional resistance of the condensate phase is negligible, $Y_{Ai} = F(X_A)$ and the integral can be evaluated if the condensation rate is known as a function of $Y_A$.

(b) When the diffusional resistance of the condensate is appreciable, show that the interface composition can be found by solving the equation

$$\frac{Z_A - X_A}{Z_A - X_{Ai}} = \left(\frac{Z_A - Y_{Ai}}{Z_A - Y_A}\right)^{(k_L \rho_L / k_g^* P)}$$

From a binomial expression in powers of $Z_A^{-1}$ and for small values of $k_g^* P / k_L \rho_L$ show that

$$\frac{Y_{Ai} - Y_A}{X_{Ai} - X_A} = b\left[1 - \frac{1}{2Z_A}\frac{(1 + b - 2m)Y_A + (1 + b - 2b/m)mX_A}{m - b}\right]$$

where $b = k_g^* P / k_L \rho_L$ and $m = dY_i/dX_i$ is assumed constant over the range $Y_A \leq Y_{Ai} \leq Y_A^*$. Note that in the limit for equimolar counter diffusion (no condensation), the equation reduces to the classical result.

7.9 In order to compensate for the variation in the driving force in a crossflow air-water cooling tower, it has been suggested that it may be beneficial to reduce the water-stream's mass velocity fed to the packing above the air exit, where the driving force is small, and to increase the flow above the air inlet. Will such measures increase the mass-transfer efficiency of a tower, assuming that maximum mass velocities are not exceeded at any point?

# MASS TRANSFER AND
# SIMULTANEOUS CHEMICAL REACTION

## 8.0  SCOPE

In many practical situations, mass transfer is accomplished in order to bring chemical reagents together so that a reaction can take place. Thus, for example, reactions can be used to provide more rapid and more extensive solution of a gas into a liquid than could be obtained by purely physical solution alone; reactions in polymers cause an increase in the molecular weight but sometimes depend upon the supply or removal of volatile compounds for the reaction to proceed; and synthesis of nitric acid requires a reaction in the liquid phase between water and dissolved gaseous nitrogen oxides. The reaction may affect the solubility of such gaseous compounds and in addition, the *rate* of solution can be increased if the reaction is fast enough to compete with diffusion in the critical region near an interface. It is the object of this chapter to provide some understanding of the ways in which diffusion and homogeneous reaction rates couple with each other during mass transfer across an interface, leading sometimes to an increase in the mass-transfer coefficient.

The important theoretical developments include the kinetically first-order chemical reaction of the dissolved molecules as they diffuse away from the interface into a liquid phase and the bimolecular, second-order reaction of dissolved gas molecules with a nonvolatile reagent which is present in the liquid phase and which diffuses toward the interface to meet the arriving gas molecules. It is shown that the reaction effect can be quite different in the two cases and that the rate of transfer may not be proportional to a driving force, especially when bimolecular reactions occur. Examples of the application of the theory are discussed, including the rates of absorption of nitrogen oxides into water and acid solutions, the absorption of carbon dioxide into alkaline buffer systems, and the oxidation of sulfite ion in aqueous solution.

## 8.1  PRINCIPAL SYMBOLS

| | |
|---|---|
| $a$ | Interfacial area per unit of packed volume, $ft^2/ft^3$ |
| $a$ | $(k_I/D)^{1/2}$ in Eq. (8.10) |
| $B_0$ | Bulk concentration of nonvolatile reagent $B$ in liquid, g mole/$cm^3$ |
| $C$ | Concentration in the liquid, g mole/$cm^3$; $C_i$ is the value at the interface |
| $D$ | Diffusion coefficient in liquid phase, $cm^2/s$ |
| $H$ | Henry's-law coefficient, g mole/$(cm^3)$(atm) |
| $H_L$ | Height of a transfer unit for liquid phase, ft |
| $K$ | Equilibrium constant for first-order reaction, $A \to B$ |
| $K_G$ | Overall mass-transfer coefficient based on partial pressure in the gas phase, lb mole/$(h)(ft^2)$(atm) |
| $K_p$ | Equilibrium constant for gas-phase reaction |
| $K_W$ | Ion product for water, (g ion/l)$^2$ |
| $K_2$ | Second ionization constant for carbonic acid, g ion/l |
| $k_L$ | Mass-transfer coefficient for liquid phase, including effect of the chemical reaction, g mole/$(s)(cm^2)$ (g mole/$cm^3$), or cm/s |
| $k_L^*$ | Steady-state value for $k_L$ for first-order reaction in liquid, equal to $(Dk_I)^{1/2}$ |
| $k_L^\circ$ | Mass-transfer coefficient without chemical reaction effect, cm/s |
| $k_I$ | First-order-reaction-rate constant, $sec^{-1}$ |
| $k_{II}$ | Second-order-reaction-rate constant, $cm^3$/(g mole)(s) |
| $L$ | Mass velocity of liquid, lb/$(ft^2)$(s) |
| $L_M$ | Molar mass velocity of liquid, lb mole/$(h)(ft^2)$ |
| $M^2$ | Quotient $k_{II} B_0 D_A/k_L^{\circ 2}$ in Eq. (8.50) |
| $N$ | Rate of transfer across interface, g mole/$(s)(cm^2)$; also total number of transfer units |
| $n$ | Number of transfer units in packed absorber |
| $p$ | Partial pressure in gas, atm or mm |
| $Q$ | Volumetric flow rate to absorber, $cm^3/s$ |
| $R$ | Rate of absorption of gas in stirred tank, g mole/s; gas constant |
| $R_i$ | Reaction rate of component $i$, g mole/$(s)(cm^3)$ in Eq. (8.7) |
| $r$ | Ratio of diffusion coefficients, $D_B/D_A$, in Eqs. (8.44) and (8.45) |
| $r$ | Relative rate of absorption, actual rate $\div$ rate for no reaction, effluent saturated with dissolved gas |
| $S$ | Stoichiometric ratio, $B_0/vC_{Ai}$ |
| $s$ | Fractional rate of surface replacement, $s^{-1}$ |
| $t$ | Time, s |
| $U_s$ | Velocity of liquid surface, cm/s |

| $u$ | Fluid velocity, cm/s |
|---|---|
| $V$ | Volume of liquid phase, cm$^3$ |
| $x$ | Distance along axis of packed absorber, cm or ft |
| $y_0$ | Film thickness, cm |
| $y$ | Distance into liquid film, cm |
| $Z$ | Total depth of packing, ft |

*Greek Letters*

| $\varepsilon$ | Fraction of tower·volume filled with liquid |
|---|---|
| $\theta$ | Nominal holding time in stirred-tank absorber, s |
| $\nu$ | Stoichiometric coefficient in chemical reaction; kinematic viscosity, cm$^2$/s |
| $\rho_L$ | Density of liquid, lb/ft$^3$ |
| $\Phi$ | Function used in Eq. (8.54), found from turbulent velocity distribution near an interface |
| $\phi$ | Reaction factor, equal to rate of mass-transfer rate in the presence of a reaction to rate without reaction |

## 8.2  INTRODUCTION[1]

Chemical processes which incorporate diffusion usually involve chemical reactions producing new substances. Reactants must overcome a diffusional resistance to arrive at the site of the reaction, and the products must diffuse away to permit the reaction to continue. When such steps occur consecutively, they may be treated separately and the series of resistances added together to express the resistance to the total driving force. Such a case occurs, for example, in the heterogeneous reaction by which sulfur dioxide reacts with oxygen on the surface of a solid catalyst to form sulfur trioxide. The gaseous oxygen and sulfur dioxide do not react chemically until they reach the solid surface, but their concentrations at the reaction site may be influenced by the resistance to their diffusion through the gas mixture and through the opposing current of reaction products moving away from the catalyst. The diffusion resistances are not affected at all by the reaction phenomena, however, and may be treated in the ways described in earlier chapters.

Often, though, diffusion and reaction occur in the same region, and the two rate phenomena are coupled so closely that they have to be treated simultaneously in the same differential rate equations. These phenomena are the principal subject of this chapter. They are of considerable importance not only in gas scrubbing operations, many of which will be discussed quantitatively, but also in chemical synthesis, in polymer processing, in homogeneous and heterogeneous catalysis, in ion exchange—in fact, in any such operation in which mass exchange from one phase to another must take place in order to bring the phases closer to chemical equilibrium.

Consider, for example, the production of a polyester by the homogeneous reaction of an organic dibasic acid and a difunctional alcohol. Unless water or similar small-product molecules are removed by diffusion from the reaction mass, the reaction may reach an equilibrium in which polymerization is incomplete. The speed of the reaction and the yield and molecular weight of the polymer will depend on the speed with which diffusion occurs. The reaction and diffusion occur together; each rate affects the concentration of the product at every point in the mixture.

Consider also the process by which ethylene oxide and water react in contact with an ion-exchange resin to form ethylene glycol. The catalytic acid sites are inside the bead of resin and the reactant must diffuse through the pores of the resin bead to reach them. If the bead is too large or the porosity too small, the concentrations of the reactants in the center of the bead will be much smaller than in the fluid outside the bead and part of the acid will be used at much reduced efficiency. Here, too, the diffusion effect and the reaction rate must be considered together rather than consecutively. Each rate affects the other.

In gas absorption, many of the most attractive industrial operations involve a rapid chemical reaction between the dissolved gas and the liquid phase. There are two reasons for this: (1) by permitting the dissolved gas to react, the capacity of a unit volume of liquid for dissolving the gas can be greatly increased, and (2) the reaction may also increase the mass-transfer coefficient if it is fast enough to occur appreciably near the interface as the gas dissolves. If the reaction is a reversible one, the liquid reagent can be recovered for reuse in the absorber by heating it and steam-stripping to carry the evolved gas away, as illustrated in Fig. 11.1. Owing to the increased gas-absorbing capacity of a reacting solution, less liquid has to be circulated through the absorber and stripper than if no reaction occurred.

In the manufacture of ammonia, hydrogen is often produced by the partial combustion of a hydrocarbon, yielding a gas mixture containing largely carbon dioxide, hydrogen, and nitrogen. The $CO_2$ has to be removed, and a common practice is to wash the gas with an alkaline liquid in a packed column or a plate absorber. Typical alkaline reagents used include water solutions of potassium or sodium carbonate–bicarbonate or of monoethanolamine. These react partially with the dissolved gas, most of which is held in chemical combination with the reagent until the solution reaches the stripping tower. There the reaction reverses itself owing to either a higher temperature or a lower partial pressure of $CO_2$ in the gas.

The following table shows a few selected data which illustrate the effect of increasing alkalinity of the solution on the mass-transfer coefficient for a packed-column absorber.

Note that the coefficient listed in the table is one based on the gas-phase composition even though the resistance to mass exchange is primarily in the liquid. Thus, the coefficient given is properly thought of as the product of the liquid-phase coefficient and the Henry's-law coefficient, viz., $K_L a H$. Values of $K_L a$ can not be found from the data unless uncertain estimates of H in the chemically reactive solution are employed. Notice that even very strongly alkaline reagents offer appreciable resistance to the solution of $CO_2$. Despite the very fast reactions by which free $CO_2$ is destroyed chemically, there is nevertheless a very appreciable resistance to the passage of $CO_2$ into the solution; most of the resistance is in the liquid phase. This implies that the reaction must take place at a finite rate in an appreciable volume of the liquid rather than on the interface itself. The half-life of a dissolved $CO_2$ molecule is extremely brief in the strongest of these solutions, and the reaction rate per unit of liquid volume is so large that the rate of consumption of $CO_2$ should be many times the absorption rate if all the liquid were reacting. Even so, the rate of absorption is not more than two orders of magnitude greater in 2-N KOH solution than it is in water. Consequently, the reaction must be taking place in a very thin layer of liquid rather than in the whole mass of the liquid phase. This is owing to the fact that unless diffusion of $CO_2$ away from the interface occurs, no volume of the liquid is available for the reaction. The faster the diffusion, the deeper the reaction layer at the interface; the faster the reaction, the thinner the zone near the interface which has to be supplied with $CO_2$ molecules by diffusion. The two phenomena have to go hand in hand to produce an effect of the reaction on the rate of absorption.

Note, however, that the mass-transfer coefficient shown in Table 8.1 for weakly alkaline solutions of sodium carbonate and sodium bicarbonate is smaller than the coefficient for water, despite the presence of hydroxyl ions in these solutions. The reaction rate is not great enough for the dissolved $CO_2$ molecules to be destroyed chemically in a thin region near the interface, so the whole volume of the liquid stream may be required for the reaction. The reduction in the coefficient below the value for water is doubtless owing to the somewhat greater viscosity of the salt solution, leading to a smaller diffusion coefficient for $CO_2$, and to reduced solubility of free $CO_2$ in the salt solution.

Table 8.1 SELECTED ABSORPTION COEFFICIENTS FOR $CO_2$ IN VARIOUS SOLVENTS (PACKED TOWERS FILLED WITH RASCHIG RINGS)*

| Solvent | $K_g a$, lb mole/(h)(ft³)(atm) | Reference |
|---|---|---|
| Water | 0.05 | 90 |
| 1-N sodium carbonate, 20% Na as bicarbonate | 0.03 | 41 |
| 3-N diethanolamine, 50% converted to carbonate | 0.4 | 23 |
| 2-N sodium hydroxide, 15% Na as carbonate | 2.3 | 96 |
| 2-N potassium hydroxide, 15% K as carbonate | 3.8 | 94 |
| Hypothetical perfect solvent having no liquid-phase resistance and having infinite chemical reactivity | 24.0 | 90 |

* Basis: $L = 2,500$ lb/(h)(ft²); $G = 300$ lb/(h)(ft²); $T = 77°F$; pressure, 1.0 atm.

## 8.3 EFFECT OF A CHEMICAL REACTION IN GAS ABSORPTION

When we consider the gradual increase in the rate of solution of a gas in a chemically reactive liquid as the reactivity gradually increases, we see that several factors contribute to the rate of solution, as Table 8.1 illustrates. First, at vanishingly small values of the reaction rate constant, the liquid becomes saturated with the physically dissolved and unreacted gas; it leaves the apparatus with the reaction far from complete. Second, for an intermediate range of reaction rates per unit volume of solution, the whole mass of liquid is available for reaction, little buildup of unreacted gas occurs anywhere in the solution, and the rate of absorption is proportional to the total liquid holdup in the apparatus. Finally, for very rapid chemical reactions, the mass-transfer coefficient is increased because the reaction occurs near the gas-liquid interface where diffusion is critical; the rate of solution is proportional to the total surface exposed in the apparatus rather than to the volume of liquid.

To bring out these factors more clearly, let us consider the simple example of a pure gas $A$ passing through a stirred-tank chemical reactor. The gas is dispersed into the liquid by an agitator, forming a steady stream of bubbles which present a large surface to the liquid. Within limits, an increase in the gas-flow rate produces a nearly proportional increase in the total surface. The liquid flows through continuously at the volumetric rate $Q$ cubic centimeters per second. The extent of mixing in the tank is good enough to make the concentration of dissolved $A$ in the liquid the same at all points, equal to $C$ moles per cubic centimeter. The volume of liquid in the tank is $V$ cubic centimeters. Assuming an irreversible, first-order reaction of $A$ in the liquid, a steady-state material balance shows that

$$k_L aV(C_i - C) = QC + Vk_1 C \tag{8.1}$$

Define the nominal liquid-holding time in the tank by $\theta = V/Q$ and let $\phi = k_L/k_L^\circ$, the reaction factor. (To allow for the fact that the mass-transfer coefficient may be affected by the reaction rate, we introduce

$$\phi = \frac{k_L}{k_L^\circ} = \phi(k_1) \tag{8.2}$$

where $k_L^\circ$ is the coefficient for purely physical absorption.)

Solving for the exit concentration,

$$C = \frac{k_L^\circ a\theta\phi}{1 + k_1\theta + k_L^\circ a\theta\phi} C_i \tag{8.3}$$

The total rate of solution of $A$ is found from Eq. (8.3), using the expression

$$R = k_L aV(C_i - C) = \frac{k_L^\circ a\theta\phi(1 + k_1\theta)}{1 + k_1\theta + k_L^\circ a\theta\phi} C_i Q \tag{8.4}$$

The factor $C_i Q$ on the right of Eq. (8.4) represents the rate of solution if the liquid leaves the tank physically saturated with $A$.

A convenient form of Eq. (6.4) is obtained by solving for the reciprocal of the relative rate of solution, $R/QC_i = R\rho/LC_i = r$, obtaining

$$r^{-1} = \frac{1}{k_L^\circ a\theta\phi} + \frac{1}{1 + k_i\theta} \tag{8.5}$$

The two terms on the right may be thought of as representing the resistances to interfacial mass transfer and to chemical reaction rate, respectively. $k_L^\circ a\theta$ corresponds to the number of transfer units for purely physical mass transfer and $\theta$ is the reciprocal space velocity. Note that for many transfer units and a negligible reaction rate, $r$ approaches unity; for $k_L^\circ a\theta$ large, $r$ may be determined by $k_i\theta$ until $1 + k_i\theta$ and $k_L^\circ a\theta$ are comparable; for larger values of the rate coefficient $k_I$, the rate of solution will be dominated by the first term.

Figure 8.1 is a plot of Eq. (8.5) showing the variations in the rate of solution of $A$ as the first-order reaction-rate constant $k_I$ varies over 14 orders of magnitude. Three curves are shown. Curve $A$ is for $k_L^\circ$ and $a$ approximately equal to values observed by Calderbank [16], Robinson and Wilke [81], and Westerterp et al. [109]; the space velocity $\theta^{-1}$ is chosen so that the liquid passes from the tank about 90 percent saturated with dissolved gas in the absence of reaction. Beginning at about $k_I = 10^{-3}$ s$^{-1}$, reaction in the bulk of the liquid occurs appreciably and the rate begins to rise accordingly, until at $r = 10$ the rate becomes nearly constant over about a tenfold range of $k_I$; afterward the rate goes up approximately as $k_I^{1/2}$ owing to the increase of $\phi$. [In order to carry out the needed numerical estimates of $\phi$ we have used Eq. (8.20), anticipating some of the results of the diffusion-reaction theory.] Below $k_I \approx 10^{-3}$ and $r \approx 1.0$, the rate is nearly independent of $\theta$ or of the total volume of liquid in the reactor; in these ranges, $r$ is proportional to the total surface exposed.

Comparison of curves $A$ and $B$ in Fig. 8.1 shows the effect of increasing the total reactor volume while keeping the total interface ($a\theta$) constant. For a tenfold increase in volume, there is only a fourfold increase in $r$ in the intermediate range of $k_I$; there is no effect at all for $k_I$ greater than unity. Curve $C$ shows the effect of a change in the physical mass-transfer coefficient, $k_L^\circ$, keeping the number of transfer units constant. Comparison of curves $A$ and $C$ shows that there is no effect until $k_I > 10^2$ s$^{-1}$, where the increase in $k_I$ begins to increase $\phi$ appreciably. For large values of $k_I$ the rate is one-tenth that for the larger $k_L^\circ$ value, showing that tests of absorber-reactors using methods which yield values of the product $k_L^\circ a$ are not adequate to determine the effectiveness of the same absorbers for use as rapid chemical reactors; values of $k_L^\circ$ and $a$ are needed separately.

These calculations bring out the fact that, as Astarita [7] has pointed out, there are two different time scales which affect the performance of all gas absorbers in which chemical reactions occur: the time scale of the reaction itself, determined by its rate constant, and the diffusion time scale, determined in the calculations above by $D/k_L^{\circ 2}$. In addition, there is also a third time scale set by the size of the

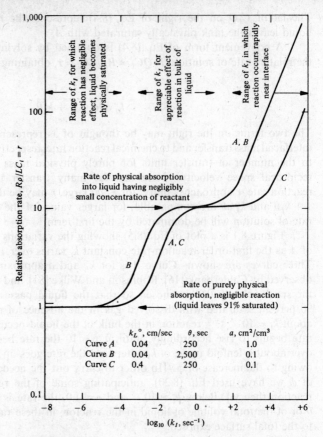

**FIGURE 8.1**
Effect of reactivity on rate of gas absorption in stirred-tank reactor.

equipment itself, here the nominal holding time or reciprocal space velocity, $V/Q$, which has an effect only within an intermediate range of reactivities.

**EXAMPLE 8.1**  A packed column filled with 1-in Raschig rings is used to carry out the chemical reaction of a pure gas with a continuous stream of liquid which flows through the packing at a rate of 5,000 lb/(h)(ft²). The reaction is first-order in the liquid phase. The liquid holdup is 7 percent of the total volumn of the column, the mass-transfer coefficient is $k_L^\circ = 0.01$ g mole/(cm²)(s)(mole/cm³), and the packing height is equivalent to 10 transfer units.

(*a*) Find an equation expressing the total rate of absorption of the gas into the liquid as a function of $k_1$, the first-order reaction-rate constant.

(*b*) Find the approximate value of $k_1$ at which the rate is limited by mass transfer without appreciable back pressure of dissolved gas.

(*c*) Find the range of $k_1$ over which the rate in (*b*) is nearly correct.

SOLUTION    (a) The differential material balance for a thin slice of the packing is

$$k_L a(C_i - C) = \varepsilon k_1 C + \frac{L_M}{\rho} \frac{dC}{dx}$$

where $\varepsilon$ is the fraction by volume of liquid in the column and $x$ is the vertical distance through the packing. Introducing the number of transfer units, $n = x\rho k_L^o a/L_M$, and the total holdup time of liquid passing through the column, $\theta = \varepsilon x\rho/L_M$, the differential equation becomes

$$\frac{dC}{dn} + \frac{k_1 \theta}{N} + \phi C = \phi C_i \qquad C(0) = 0$$

The solution is

$$C = C_i \left( \frac{N\phi}{N\phi + k_1 \theta} \right) \left\{ 1 - \exp\left[ -(N\phi + k_1 \theta) \frac{n}{N} \right] \right\}$$

where $N$ = total number of transfer units. The total rate of absorption is

$$R = \int_0^z k_L a(C_i - C)\, dx = \frac{L_M C_i}{\rho} \left( \frac{N\phi}{N\phi + k_1 \theta} \right) \left[ k_1 \theta + \left( \frac{N\phi}{N\phi + k_1 \theta} \right) (1 - e^{-(N\phi + k_1\theta)}) \right] \qquad (8.6)$$

(b) At the specified liquid rate, the superficial linear velocity of the liquid stream is $(5{,}000)(30.4)/(62.3)(3{,}600) = 0.68$ cm/s. The height of a transfer unit is 1.2 ft or 36.6 cm (cf. Chap. 11); the fraction by volume of the packing that is filled with liquid is 0.07 (Furnas and Bellinger [41]), making

$$\theta = \frac{(0.07)(36.6)(10)}{0.68} = 37.6 \text{ s}$$

Equation (8.7) for the mass-transfer enhancement factor becomes

$$\phi^2 = 1 + \frac{Dk_1}{k_L^{o2}} = 1 + \frac{D}{k_L^{o2}\theta}(k_1 \theta) = 1 + 0.0048 \, k_1 \theta$$

based on a diffusivity of $1.8 \times 10^{-5}$ cm$^2$/s.

The term on the right of Eq. (8.6) containing the exponential function is negligible. Thus, to answer question b we may solve the simpler equation

$$\frac{\phi k_1 \theta}{N\phi + k_1 \theta} = 1$$

by trial and error, obtaining $k_1 \theta = 73$ or $k_1 = 1.94$ s$^{-1}$ for the solution reactivity at which the absorption rate reaches the expected value for physical absorption into a liquid kept free of dissolved gas by a slow reaction in the bulk.

(c) Similar trial-and-error calculations for relative rates equal to 1.1 and 0.9 give $k_1 \theta = 104$ and 47, respectively, corresponding to $k_1 = 2.8$ and 1.2 s$^{-1}$, respectively.      ////

Although the conclusions from Fig. 8.1 are for a typical stirred-tank reactor and are based on a series of simplifying assumptions, they are qualitatively correct even when other conditions prevail. Plate columns and stirred-tank reactors are qualitatively similar, as brought out in a problem for the student at the end of the chapter, the important parameters which set the diffusion time scales not being greatly different. In order for the reaction to affect the mass-transfer coefficient in any of these kinds of equipment, it is usually necessary that the reaction half-life be smaller than about $D/k_L^{\circ 2}$ or about 0.01 to 0.001 s, the basis of which will appear later; for most practically realizable circumstances, mass transfer between phases has an appreciable effect on the total rate.

In most of what follows in this chapter, attention is directed primarily to the highest range of chemical reactivity in the liquid phase in gas-liquid contactors. This is the more interesting range for study of rate phenomena and, moreover, it is the range of practical importance for several of the gas-absorption processes that are of primary industrial interest.

## 8.4   THE THEORY OF SIMULTANEOUS DIFFUSION AND CHEMICAL REACTION NEAR AN INTERFACE

To account for the influence of a chemical reaction on the concentration of a diffusing species in the important region near an interface between phases, one must include a reaction-rate expression in the unsteady-state diffusion equation, such as Eq. (3.33), obtaining

$$D \frac{\partial^2 C_i}{\partial y^2} = \frac{\partial C_i}{\partial t} + u \frac{\partial C_i}{\partial x} + R_i(C) \tag{8.7}$$

where the new term, $R_i$, represents the rate of loss of component $i$ owing to chemical reaction per unit volume of the fluid. Obviously, development of the theory of diffusion-affected reactions can take on as many forms as there are different expressions for the $R_i$ and different fluid-flow situations. For the sake of simplicity, several assumptions are usually made about the surface geometry and the fluid motion, as outlined in Chap. 5 for the "penetration theory," the "surface replacement theory," or the "film theory." For the first two, the results are expressed by Eqs. (3.45), (5.3), and (5.6); for the film theory the expression for the purely physical mass-transfer coefficient is simply $k_L^{\circ} = D/y_0$. In each of these idealized pictures of interface behavior, simplifications are made in the first two terms on the right of Eq. (8.5); but in the earlier discussions we have assumed always that the reaction-rate term was absent. Now we need to see how the expressions for rate of interphase transfer are changed owing to the reaction term, using the same assumptions as before about the other terms in the equation. Often it will be convenient in expressing the results of our calculations to compute theoretical values of the "reaction factor" $\phi$, which we define as the ratio of the mass-transfer coefficient according to the theory with the reaction term included to the coefficient from the same theory omitting the reaction. It seems very probable that even though the assumptions about fluid motion embodied in the theories may not be precisely correct, the reaction factors estimated from the theory may be accurate. If this is true, it should be possible to apply the reaction factors which we have estimated theoretically to experimental values of the mass-transfer coefficients measured for purely physical mass transfer so as to correct such data for the reaction kinetic effect.

One may recall that in each type of interface diffusion theory, as outlined in Chap. 5, some quantities appear in the final expressions for mass-transfer coefficients that ordinarily cannot be specified from theory alone. For example, in the film theory, we need to know the film thickness, $y_0$; in the Higbie penetration theory, we must have the interphase exposure time, $t$; and for the surface-replacement theory, we need to know the surface element mean lifetime, $s$. Estimation of each of these quantities requires use of an experimental value of the mass-transfer coefficient without chemical reaction, $k_L^\circ$. Thus,

$$y_0 = \frac{D}{k_L^\circ} \tag{8.8a}$$

$$t = \left(\frac{4D}{\pi k_L^{\circ 2}}\right)^{1/2} \tag{8.8b}$$

$$s = \left(\frac{k_L^{\circ 2}}{D}\right)^{1/2} \tag{8.8c}$$

These same estimates can be used with the reaction present, since the existence of a reaction in the fluid phase should have only a minor effect, if any, on the flow situation. Obviously, in applying the three theories to purely physical mass transfer, it does not matter which one is best if we must use a measured $k_L^\circ$ to evaluate their constants (the prediction of the effect of changes in $D$ are a different matter); but when we come to the estimation of a reaction effect, we must choose between the theories and we must estimate the corresponding parameters using one of Eqs. (8.8). This is for the reason that either $y_0$, $t$, or $s$ will appear in the theories in combination with the reaction-rate constants, indicating that when there are separate and independent time scales both for diffusion and for reaction, the latter being given by the reaction rate, we must have information on the diffusion-flow time scale if we are to compare reaction and diffusion effects. Fortunately, it turns out that the differences between the theoretical estimates of $\phi$ are usually not significant, and it is not a vital matter, nor is it possible experimentally, to distinguish between the theories.

EXAMPLE 8.2   Estimate the apparent values of the time of exposure of an element of liquid surface to the gas (a) in a packed column filled with 1-in ceramic Raschig ring packing and (b) in a sieve tray column holding 1 in of clear water with a superficial gas velocity of 1 ft/s, using the following information: (1) Data for the liquid-phase resistance to the desorption of oxygen from water in the Raschig ring packing were given by Sherwood and Holloway [90], who obtained a value of $H_L = 1.05$ ft at 25° C and a liquid-flow rate of 4,440 lb(h) (ft$^2$), equal to 0.6 cm$^3$/(s) (cm$^2$); (2) For the same packing, temperature, and flow rate, Danckwerts and Sharma [28] give $k_L^\circ = 0.0113$ cm/s. They also report an apparent interfacial area of 1.12 cm$^2$/cm$^3$ of packed space, 0.048 pieces of packing per cubic centimeter of packed space, and

1.8 cm² of dry packing surface per cubic centimeter of packed space. (3) Andrew [112], quoted by Danckwerts and Sharma [28], gives the following equation for the purely physical mass-transfer coefficient in the liquid phase on a bubbling tray.

$$k_L^\circ = 11\left(\frac{U_G D^2}{Z_c^2}\right)^{1/4} \quad \text{cm/s}$$

where $U_G$ = superficial gas velocity, cm/s;

$D$ = diffusion coefficient, cm²/s;

$Z_C$ = depth of clear liquid on tray, cm

SOLUTION   (a) If we use the wetted area reported by Danckwerts and Sharma with $H_L = 1.05$ ft, we find that, for the packed column,

$$k_L^\circ = \frac{L}{\rho_L H_L a} = \frac{0.6}{(1.05)(30.5)(1.12)} = 0.0167 \text{ cm/s}$$

$$t = \frac{4D}{\pi k_L^{\circ 2}} = \frac{(4)(2.4 \times 10^{-5})}{(3.14)(0.0167)^2} = 0.110 \text{ s}$$

Using the value of $k_L^\circ$ from Danckwerts and Sharma we have

$$t = \frac{(4)(1.9 \times 10^{-5})}{(3.14)(0.0113)^2} = 0.195 \text{ s}$$

Based on the dimension of the packing, if there are 0.048 pieces per cubic centimeter, each piece occupies on the average $1/0.048 = 20.8$ cm³, corresponding to a horizontal cross-sectional area of 8.19 cm² and a vertical height of 2.54 cm. The total wetted perimeter inside and out is about 14.9 cm, giving for the water flow per unit of perimeter the value $q = (0.6)(8.19)/14.9 = 0.33$ cm³/(cm) (s). At this flow rate the surface velocity of the laminar liquid layer can be found from the formula

$$U_s = \left(\frac{9q^2 g_L}{8v}\right)^{1/3} = \left[\frac{(9)(0.33)^2(981)}{(8)(0.008937)}\right]^{1/3}$$

$$= 23.8 \text{ cm/s}$$

The time required for flow down a 1-in vertical surface is $2.54/23.8 = 0.107$ s.

(b) Substituting Andrew's equation into the expression for $t$, we have

$$t = \frac{4D}{\pi}\left(\frac{Z_c^{1/2}}{11 U_G^{1/4} D^{1/2}}\right)^2 = 0.01 \frac{Z_c}{U_G^{1/2}} \text{ s}$$

Using $U_G = 30.5$ cm/s and $Z_C = 2.54$ cm, we find $t = 0.0048$ s for the sieve tray.

////

## 8.5   THE FILM THEORY FOR A FIRST-ORDER IRREVERSIBLE REACTION

One of the earliest theoretical developments in this field was given by Hatta [47]. Equation (6.7) becomes

$$D\frac{\partial^2 C}{\partial y^2} = k_1 C \qquad C(0) = C_i \qquad C(y_0) = 0 \tag{8.9}$$

and the solution is

$$C = C_i \frac{\sinh [a(y_0 - y)]}{\sinh a y_0} \tag{8.10}$$

where $a = (k_I/D)^{1/2}$. The rate of mass transfer is found from the derivative $dC/dy$ at the interface, giving

$$N = -D\left(\frac{dC}{dy}\right)_{y=0} = \frac{DC_i}{y_0} \frac{ay_0 \cosh ay_0}{\sinh ay_0} \tag{8.11}$$

Since $D/y_0 = k_L^\circ$ and, defining $k_L = N/C_i$, we have

$$\phi = ay_0 \coth ay_0 \tag{8.12}$$

It can easily be shown that $\phi$ is unity at the limit $k_I = 0$, as expected; for large values of $ay_0$, that is, for very reactive solutions, the hyperbolic cotangent is nearly unity and $\phi \approx ay_0$. The approximation is within 1 percent for $ay_0 > 2.6$. The corresponding expression for the rate of mass transfer across the interface is

$$N_i = (k_I D)^{1/2} C_i \tag{8.13}$$

Note that $N_i$ according to Eq. (8.13) is independent of the rarely known film thickness, $y_0$; it depends only on the physical constants $k_I$ and $D$, which can be measured independently or estimated. The thickness of the film, $y_0$, does not matter if $ay_0$ is great enough. Then the reactive molecules, having dissolved at the interface, disappear chemically before they have diffused very far into the fluid. Nearly none are left as the diffusion current reaches the inner boundary of the diffusion layer. The corresponding $k_L$, defined as $k_L^*$, is given by

$$k_L^* = (Dk_I)^{1/2} \tag{8.14}$$

Equation (8.13) can be used for the experimental determination of $Dk_I$ if measurements of the transfer rate can be made in an apparatus which has a known surface area of the interface.

The analysis of Hatta is easily extended to reversible reactions, as shown by Huang and Kuo [51].

## 8.6 THE THEORY OF FIRST-ORDER REACTIONS ACCORDING TO THE SURFACE-REPLACEMENT MODEL

Following the suggestion of Danckwerts [119], as outlined in Chapter 5, the constant rate of mass transfer across an interface which is continuously being replaced with fresh elements of fluid from deep in the nearby fluid is found by averaging the rate of transfer into an element of age $t$ over all ages, using the probability distribution of surface ages given in Eq. (5.4). Thus we shall expect to compute

$$N_{av} = s\int_0^\infty e^{-st} N(t) dt = -Ds \int_0^\infty e^{-st} \frac{\partial C}{\partial y}(0, t) dt$$

$$= -Ds\left(\frac{d}{dy}\right)_{y=0} \left[ s \int_0^\infty e^{-st} C(y, t) dt \right] \tag{8.15}$$

Equation (8.15) shows that we need to find the Laplace transform of the concentration of the reacting material rather than the concentration itself.

Equation (8.7) takes the form

$$D \frac{\partial^2 C}{\partial y^2} = \frac{\partial C}{\partial t} + k_1 C \tag{8.16}$$

for an irreversible reaction. The corresponding equation for the $s$-multiplied Laplace transform of $C$ is

$$D \frac{d^2 \overline{C}}{dy^2} = sC + k_1 C \qquad \overline{C}(0) = C_i \qquad \overline{C}(\infty) = 0 \tag{8.17}$$

and the solution is

$$\overline{C} = C_i \exp\left( -\sqrt{\frac{k_1 + s}{D}} \, y \right) \tag{8.18}$$

for which, after evaluating the derivative at the interface,

$$\overline{N} = C_i \sqrt{D(k_1 + s)} \tag{8.19}$$

Equation (8.19) differs from Eqs. (5.5) and (5.6) only by the appearance of $k_1 + s$ instead of $s$ alone. The corresponding formula for the reaction factor is

$$\phi = \left( 1 + \frac{k_1}{s} \right)^{1/2} = \left( 1 + \frac{Dk_1}{k_L^{\circ 2}} \right)^{1/2} \tag{8.20}$$

in which the last expression is obtained by introducing the value of $s$ found from Eq. (8.8c).

The method outlined above can be applied to multiple reactions that are first-order. For the reversible pair $A \rightleftharpoons B$, with $B$ not volatile at the interface, the result is given by Eq. (8.30) below. The derivation of Eq. (8.30) is left for the student.

## 8.7  FIRST-ORDER REACTIONS DURING UNSTEADY-STATE DIFFUSION INTO SEMIINFINITE MEDIUM: THE PENETRATION THEORY

The third model of the fluid motion at an interface is that suggested by Higbie [49] when he was a graduate student. His proposal, advanced originally to account for the mass transfer into a liquid from a gas bubble rising inside a tube, was that during the time of contact of the bubble with the liquid phase around it, the liquid layer is stagnant and is so thick that no barrier is encountered by diffusing molecules. The liquid phase therefore behaves as if it were an infinitely deep, stationary medium for which Eq. (8.7) becomes

$$D \frac{\partial^2 C}{\partial y^2} = \frac{\partial C}{\partial t} + k_1 C \tag{8.21}$$

if the reaction is first-order and irreversible. The boundary and initial conditions on $C(y, t)$ are $C(0, t) = C_i$, and $C(\infty, t) = C(y, 0) = 0$. Finding the solution is equivalent to finding the inverse Laplace transform expressed by Eq. (8.17), and the instantaneous rate of transfer across the interface is the inverse transform of Eq. (8.19),

$$N(t) = C_i(Dk_1)^{1/2}\left[\text{erf } (k_1 t)^{1/2} + \frac{e^{-k_1 t_1}}{(\pi k_1 t)2}\right] \tag{8.22}$$

For small values of $k_1 t$,

$$N(t) = C_i\left(\frac{D}{\pi t}\right)^{1/2}(1 + k_1 t + \cdots) \tag{8.23}$$

The first two factors on the right of Eq. (8.23) represent the rate for purely physical mass transfer. For large values of $k_1 t$

$$N(t) \approx C_i(Dk_1)^{1/2}\left[1 + \frac{e^{-k_1 t}}{2\sqrt{\pi(k_1 t)^3}} + \cdots\right] \tag{8.24}$$

Note that the leading term makes $N(t)$ a constant having the value as in the Hatta film theory [Eq. (8.13)].

Usually we are more interested in the average rate of absorption over the total time of exposure, $t$, than in the instantaneous rate, $N(t)$. The total, cumulative amount absorbed during the time $t$ is given by

$$Q = \int_0^t N(t)dt = C_i(Dk_1)^{1/2}t\left[\left(1 + \frac{1}{2k_1 t}\right)\text{erf } (k_1 t)^{1/2} + \frac{e^{-k_1 t}}{\sqrt{\pi k_1 t}}\right] \tag{8.25}$$

a result originally obtained by Danckwerts [25]. The time-average rate of absorption is $Q(t)/t$; the corresponding time-average mass-transfer coefficient, $k_L$, is equal to $Q/tC_i$ and is readily calculated from the function in Eq. (8.25). For small values of $k_1 t$,

$$Q = C_i\left(\frac{4D}{\pi t}\right)^{1/2}t\left(1 + \frac{k_1 t}{3} - \frac{(k_1 t)^2}{30} + \frac{(k_1 t)^3}{210} + \cdots\right) \tag{8.26}$$

in which the second factor on the right is the coefficient for purely physical mass transfer as given by Eq. (5.3). For large values of $k_1 t$,

$$Q \approx C_i(Dk_1)^{1/2}t\left[1 + \frac{1}{2k_1 t} - \frac{e^{-k_1 t}}{2\sqrt{\pi(k_1 t)}}\left(1 - \frac{3}{k_1 t} + \cdots\right)\right] \tag{8.27}$$

When $k_1 t > 2$, the sum of the first two terms in the last factor on the right does not differ from the true result by more than 1.1 percent and the mass-transfer coefficient corresponding to Eq. (8.27) is nearly constant and equal to $k^* = (Dk_1)^{1/2}$.

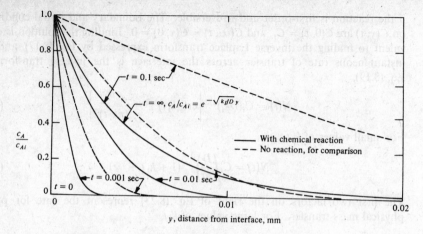

FIGURE 8.2
Calculated concentration profiles of diffusing and reacting molecules near an interface.

Figure 8.2 compares the concentration distribution in the liquid with and without an irreversible first-order reaction having a rate constant $k_1$ equal to $1.0\ \mathrm{s}^{-1}$ and $D = 2 \times 10^{-5}\ \mathrm{cm}^2/\mathrm{s}$. The dashed lines on the figure show the gradual penetration of the dissolved molecules into the liquid when there is no reaction, as in Fig. 3.3; comparison with the solid lines shows the change owing to the reaction. At each position in the liquid the concentration is lower with the reaction than without; as a result, the negative concentration gradient at the interface is greater and the rate of transfer is increased. As time elapses, the concentration profiles with reaction take on a different shape and approach a limiting exponential profile for which reaction and diffusion are exactly in step at each value of $y$. The profile remains steady because the rate of diffusion, $-D(\partial C/\partial y)$, at each $y$ is exactly equal to the total rate of reaction throughout the more distant parts of the liquid $\int_y^\infty k_1 C\, dy$. When this occurs there is no further decrease in the mass-transfer rate into the liquid surface and $k_L$ reaches its least value, $(Dk_1)^{1/2}$, as shown in Eq. (8.27).

EXAMPLE 8.3   Brian, Vivian, and Habib [13] measured the rate of absorption of gaseous chlorine into water using a wetted-wall absorber having a total wetted height of 4.68 cm. In one of their runs at 25°C and with a gas partial pressure of about 0.028 atm and a water-flow rate of 11.5 cm³/(cm) (min), they observed a liquid-phase mass-transfer coefficient equal to 0.016 cm/s. Assuming that the rate of the reverse reaction was negligible under the conditions of the experiment, find the first-order rate constant for the chemical reaction

$$\mathrm{Cl_2 + H_2O} \xrightarrow{k_1} \mathrm{Cl^- + H^+ + HOCl}$$

SOLUTION  The steady-state downward velocity of the liquid surface in the column can be computed from the kinematic viscosity of water, $v$, the acceleration of gravity $g_L$, and the volumetric flow rate per unit of wetted perimeter, $q$, using the formula

$$U_s^3 = \frac{9q^2 g_L}{8v} = \frac{(9)(11.5/60)^2(981)}{(8)(0.00894)} = 4{,}550 \text{ (cm/s)}^3$$

or $U_s = 16.6$ cm/s. This gives a time of exposure of the liquid surface equal to $4.68/16.6 = 0.282$ s. From Eq. (8.27), representing the total rate of absorption accompanied by an irreversible first-order reaction, we have

$$k_L = Q/t = (Dk_1)^{1/2}\left(1 + \frac{1}{2k_1 t}\right) = \left(\frac{D}{t}\right)^{1/2}\left[(k_1 t)^{1/2} + \frac{1}{2(k_1 t)^{1/2}}\right]$$

Using the diffusivity of dissolved $Cl_2$ equal to $1.48 \times 10^{-5}$ cm$^2$/s reported by Vivian and Peaceman [101], we have $(D/t)^{1/2} = 0.00724$ cm/s and the above equation becomes a quadratic equation for $(k_1 t)^{1/2}$. We find that $k_1 = 13.3$ s$^{-1}$.

Using all their data, Brian, Vivian, and Habib obtained $k_1 = 13.6$ s$^{-1}$, which they found was in good agreement with several values measured directly at lower temperatures.    ////

When the reaction is of the form $A \rightleftharpoons B$ and $B$ is a nonvolatile product, the penetration theory requires that the two equations

$$D_A \frac{\partial^2 C_A}{\partial y^2} = \frac{\partial C_A}{\partial t} + k_1(C_A - K^{-1}C_B) \tag{8.28}$$

$$D_B \frac{\partial^2 C_B}{\partial y^2} = \frac{\partial C_B}{\partial t} - v k_1(C_A - K^{-1}C_B) \tag{8.29}$$

be solved simultaneously with the boundary conditions

$$C_A(0, t) = C_{Ai} \qquad C(y, 0) = C(\infty, t) = \frac{C_T}{K + 1}$$

$$\frac{\partial C_B}{\partial y}(0, t) = 0 \qquad C(y, 0) = C(\infty, t) = \frac{KC_T}{K + 1}$$

where $C_T$ represents the sum of $A$ and $B$ initially in the bulk of the liquid, the separate concentrations being assumed at equilibrium with each other.

Introducing the Laplace transform, as before, the equation

$$-D\left(\frac{d\overline{C}_A}{dy}\right)_{y=0} = \left(C_{Ai} - \frac{C_T}{1 + K}\right)\sqrt{Ds} \frac{(1 + K^{-1})\sqrt{k_1(1 + K^{-1}) + s}}{\sqrt{s} + K^{-1}\sqrt{k_1(1 + K^{-1}) + s}} \tag{8.30}$$

is obtained for the transform of the surface absorption rate of $A$ when $D_A = D_B$ and $v = 1$. This is equal to the actual rate when the Danckwerts model is assumed. Inversion of the transform gives a result which can be expressed as

$$k_L = \sqrt{\frac{4D}{\pi t}}\left\{1 + \sqrt{\frac{\pi K^5}{4(K + 1)^2(K - 1)k_1 t}} \exp\left(\frac{k_1 t}{K(K - 1)}\right)\left[\operatorname{erf}\sqrt{\frac{K}{K - 1}k_1 t}\right.\right.$$
$$\left.\left. - \operatorname{erf}\sqrt{\frac{k_1 t}{K(K - 1)}}\right] - \sqrt{\frac{\pi K^3}{4(K + 1)k_1 t}}\operatorname{erf}\sqrt{\frac{(K + 1)}{K}k_1 t}\right\} \tag{8.31}$$

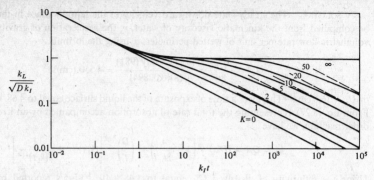

FIGURE 8.3
Effect of reversible chemical reaction on mass-transfer coefficients.

when the time-average rate is calculated. The mass-transfer coefficient $k_L$ is based on the driving force, $C_{Ai} - C_T/(K + 1)$, which contains concentrations of $A$ alone, not $B$. Equation (8.30) can be written in a similar form. Equation (8.31) gives limiting values of $k_L$ as follows:

$$\lim_{k_1 \to 0} k_L = k_L^\circ \qquad \lim_{\substack{k_i \to \infty \\ K \neq \infty}} k_L = (1 + K)k_L^\circ \qquad \lim_{\substack{k_i \to \infty \\ K = \infty}} k_L = \sqrt{Dk_1} \qquad (8.32)$$

The second of these agrees with a result obtained by Crank [22], who assumed that $C_B = KC_A$ everywhere in the diffusion space. The reaction rate term was eliminated from the pair of Eqs. (8.28) and (8.29) by subtraction; and then, with equal diffusivities and the equilibrium relationship between concentrations, a single second-order equation was obtained for the sum of the concentrations, $C_A + C_B$. The same result was obtained by Olander [71], who also extended the method to more complicated reactions.

Figure 8.3 shows the variation of the mass-transfer coefficient with the reaction-rate constant according to Eq. (8.31). When the reaction is irreversible, the topmost curve applies. The rate is very high initially because of the very sharp concentration gradients which occur near the interface just after it is first exposed. These disappear gradually, however, and the rate approaches a constant value. When there is a finite value of the equilibrium constant, however, there is a second region of falling rate at large values of $k_1 t$ within which the mass-transfer coefficient behaves like the purely physical coefficient but with the multiplying factor $1 + K$.

EXAMPLE 8.4  Based on the forward rate constant for the homogeneous reaction,

$$Cl_2 + H_2O \xrightarrow{\ k_1\ } H^+ + Cl^- + HOCl$$

found from absorption rate data in Example 8.2, and using the Henry's-law coefficients for molecular chlorine and the chemical equilibrium constants reported by Vivian and Whitney

[100], find the rates of desorption of chlorine from water at 50°C when the equilibrium partial pressure of chlorine over the solution is (a) 0.5 atm and (b) 0.01 atm. The liquid is in contact with an insoluble, inert gas which contains no chlorine. The physical characteristics of the desorption apparatus are the same as those in Example 8.2.

SOLUTION   The physical constants at 50°C, based on extrapolations from the values reported by Vivian and Whitney below 25°C, are $H = 0.0135$ g mole/(l)(atm)(K) $= 1.78 \times 10^{-4}$ (g mole/l)$^2$. Using the data of Brian, Vivian, and Habib [13], the forward rate constant in the reaction above is 49 s$^{-1}$ at 50°C and the liquid-phase diffusion coefficient for molecular chlorine in water is $2.61 \times 10^{-5}$ cm$^2$/s. The kinematic viscosity is about $0.568 \times 10^{-2}$ cm$^2$/s, making the time of exposure in the wetted-wall column 0.243 s.

According to the reaction-rate expression used by Brian, Vivian, and Habib, the rate of the reverse of the reaction given above is proportional to the third power of the concentration of reacted chlorine if the concentrations of the chloride ion and of hypochlorous acid are equal. The theory leading to Eq. (8.31) was based on the assumption that the rate was first-order in each direction and is not directly applicable. Nevertheless, an approximation can be made which will be valid more accurately as $k_1 t$ becomes greater and local equilibrium is more accurately maintained at all points in the liquid.

Consider the behavior of the concentration function in the rate expression when none of the concentrations deviates very much from its value at chemical equilibrium. Then we may expand the nonlinear function of concentrations of products as follows:

$$[Cl_2] - K^{-1}[H^+][Cl^-][HOCl]$$

$$= [Cl_2]\left(1 + \frac{[Cl_2] - [Cl_2]_0}{[Cl_2]_0}\right) - K^{-1}\left(1 + \frac{[HOCl] - [HOCl]_0}{[HOCl]_0}\right)^3 [HOCl]_0^{\,3}$$

$$\approx \left([Cl_2] - [Cl_2]_0 - \sqrt[3]{\frac{3[Cl_2]_0^{\,2}}{K}}\,([HOCl] - [HOCl]_0)\right)$$

where the subscript 0 refers to initial concentrations. We have used

$$K = \frac{[HOCl]_0^{\,3}}{[Cl_2]_0}$$

which follows from the assumption that initially the whole mass of liquid is at chemical equilibrium before exposure to the gas. The approximate expression is linear in the pseudo-concentrations, $[Cl_2]-[Cl_2]_0$ and $[HOCl]-[HOCl]_0$, and Eq. (8.31) can be applied if all concentrations of component $A$ are replaced by the excess of the local molecular chlorine concentration over its initial value and the equilibrium constant in Eq. (8.31) is replaced by $(K/3[Cl_2]_0^{\,2})^{1/3}$. The stoichiometry need not be changed because according to the chemical equation, a change in $[Cl_2]-[Cl_2]_0$ by the reaction produces an equal and opposite change in $[HOCl]-[HOCl]_0$.

At 0.5 atm $[Cl_2]$ equilibrium partial pressure, $[Cl_2]_0 = 0.5 \times 0.023 = 0.0137$ g mole/l; at 0.01 atm the value is 0.000273 g mole/l. This makes the equivalent value of $K$ equal to $7.62 \times 10^{-4}/(3)(0.0137)^2 = 1.105$ and 15.1 at equilibrium partial pressures of 0.5 and 0.01 atm, respectively.

At $k_1 t = 49.0 \times 0.243 = 11.9$, inspection of Fig. 8.3 shows that the two conditions lie in different ranges of the diagram. The diffusion-plus-reaction situation in the more concentrated solution is nearly that of purely physical diffusion; the less concentrated solution corresponds to a point nearly at the upper boundary of the curves, representing a steady-state concentration profile in the liquid and a time-independent mass-transfer rate which is nearly independent of the chemical equilibrium constant.

Since

$$(Dk_1)^{1/2} = 0.0358 \text{ cm/s}$$

The calculated rates of mass transfer are

$$-(0.0137 \times 10^{-3} \text{ g mole/cm}^3)(0.54)(0.0358) = 2.65 \times 10^{-7} \text{ g mole/(cm}^2)(\text{s})$$

and

$$-(0.000273 \times 10^{-3} \text{ g mole/cm}^3)(1.0)(0.0358) = 9.76 \times 10^{-9} \text{ g mole/(cm}^2)(\text{s})$$

at 0.5 and 0.01 atm, respectively. Note that when the liquid concentration is high, the value of $k_L$ is reduced to 54 percent of its maximum value, owing to the finite value of the pseudo-equilibrium constant; there is no reduction when the concentration is low. Such variations are characteristic of nonlinear rate expressions. Brian, Vivian, and Habib computed these effects accurately for the absorption process and found that their observed rates of absorption per unit of gas partial pressure were reduced when the gas concentrations were increased.   ////

The steady-state analysis, i.e., the film theory, may be extended easily to a somewhat more general class of reactions if the steady-state balance between diffusion and reaction is assumed. Then, even if the reaction rate is expressed by the formula

$$R = k_n C^n \tag{8.33}$$

with $n > 0$, it is relatively easy to obtain the asymptotic value of $k_L$. This is illustrated by a problem at the end of the chapter. The result is

$$-D\left(\frac{dC}{dy}\right)_{y=0} = \sqrt{2D \int_0^{C_{Ai}} R(C)dC}$$

$$= \sqrt{2Dk_n \int_0^{C_{Ai}} C^n \, dC} = \sqrt{\frac{2}{n+1} Dk_n C_{Ai}^{n+1}} \tag{8.34}$$

$$k_L = \sqrt{\frac{2}{n+1} Dk_n C_{Ai}^{n-1}} \tag{8.35}$$

Reactions of zero-order are a special case which has been treated by Astarita [4] and Astarita and Marrucci [5]. The absorption rate in the steady state is equal to $[2R_0 D(C_{Ai} - C_0)]^{1/2}$ where $R_0$ is the constant rate of the reaction. The mass-transfer coefficient is

$$k_L = \sqrt{\frac{2R_0 D}{C_{Ai} - C_0}} \tag{8.36}$$

Notice that in every case except the ones involving zero-order or $n$th-order reactions the mass-transfer rate is proportional to the driving force. This is because the combination of linear expressions for reaction rates with linear expressions for diffusion rates leads to linear differential equations. We shall see in a following section that different results are obtained when the reaction is a bimolecular.

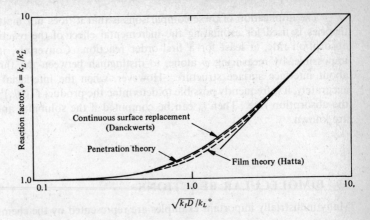

FIGURE 8.4
Comparison of three theories of interfacial mass transfer accompanied by first-order
irreversible chemical reaction.

## 8.8 COMPARISON OF THE FILM, SURFACE-REPLACEMENT, AND PENETRATION THEORIES FOR FIRST-ORDER REACTIONS

As noted in Eq. (8.8), a measured value for the physical mass-transfer coefficient $k_L^\circ$ is the common source of estimates of the quantities $y_0$, $s$, and $t$ appearing in the three theories. The question now arises of whether there are significant differences in the estimates of the reaction factor $\phi$.

First, comparing the penetration and surface-replacement models, one finds that they do not agree perfectly at all values of the reaction-rate constant. At small times of exposure $t$, the penetration theory indicates, according to Eq. (8.26), that the fractional increase in the rate owing to the reaction is equal to $k_1 t/3$; the surface-replacement theory indicates that the fractional increase is $k_1/2s$. Thus, to achieve the same initial slope of the curves of $k_L$ vs. $k_1$, we should choose $st = 3/2$. On the other hand, forcing the two theories to give the same value of the physical coefficient, as in Eq. (8.8), requires that we take $st = 4/\pi$. Thus, the two theories will not agree perfectly over all ranges of the reaction-rate constant $k_1$ for either choice. However, they agree exactly at large values of $k_1$, despite the choice of $st$, because the steady-state diffusion-reaction pattern has been reached. Using $st = 4/\pi$, the agreement at intermediate values is very good, as shown in Fig. 8.4. The maximum vertical difference between the two curves is about 2.6 percent.

The film theory also does not differ significantly. Figure 8.4 shows that after fitting the film and surface-replacement theories together at $k_L = 0$ as in Eq. (8.12), the calculated values of $\phi$ do not differ by more than 5.9 percent over the whole range of $k_1$. Values of $k_L$ computed from the penetration theory lie between those from the other two.

The implication of these comparisons is that it does not matter which of the theories is used for estimating the incremental effect of the reaction rate on the absorption rate, at least for a first-order reaction. Conversely, it is very nearly impossible, by measuring $\phi$ alone, to distinguish between the three assumptions about interface surface structure. However, when the interfacial area is known accurately, it is frequently possible to determine the product $C_i(Dk_{\mathrm{I}})^{1/2}$ by measuring the absorption rate. Then $k_{\mathrm{I}}$ can be computed if the solubility and the diffusivity are known.

## 8.9   BIMOLECULAR REACTIONS

Many industrially important examples are represented by the chemical equation

$$A + vB \longrightarrow C$$

where $B$ represents a nonvolatile reagent which is present initially in the liquid phase and which diffuses toward the gas-liquid interface to meet $A$ as it dissolves. The reaction rates in many such examples can be represented by the formula

$$R = k_{\mathrm{II}} C_A C_B \tag{8.37}$$

Equation (8.7) has to be used twice, once for $A$ molecules and once for $B$. Its solution has not yet been obtained analytically, except for some special cases which happen to be of great practical interest. In general, use must be made of digital computers to obtain numerical solutions [12, 72]. Inasmuch as the differential equations are not linear, the result for the Danckwerts model is not obtained directly and indeed has usually not been obtained at all. Approximate analytical solutions have been found by van Krevelen and Hoftijzer [98] using the film model.

Figure 8.5 shows the types of concentration distributions which exist near a gas-liquid interface in the region where $A$ and $B$ are reacting. It is seen that $B$'s concentration falls off somewhat near the interface owing to the consumption of it there by the reaction. The slope $\partial B/\partial y$ is zero at the interface, however, because $B$ is not diffusing out of the liquid. The drop in $B$'s concentration must be more or less equal to the fall in $A$'s concentration divided by the mole number $v$. Now suppose the initial concentration of $B(= B_0)$ is so large compared with the largest concentration of $A(= C_{Ai})$ that the difference in concentration of $B$ is only a small fraction of $B_0$. Then it will be a good approximation to replace the variable concentration of $B$ in the differential equation for $A$ by the constant value, $B_0$. Under these conditions the equation will be linear and will, in fact, be the same equation which was solved to obtain Eqs. (8.12) and (8.13). These equations may be used, therefore, if the first-order chemical-reaction-rate constant is replaced by its pseudo first-order value, $k_{\mathrm{II}} B_0$. In particular, if $k_{\mathrm{II}} B_0 t$ is sufficiently great, the mass-transfer coefficient for $A$ will be given by

$$k_{LA} = \sqrt{D_A k_{\mathrm{II}} B_0 t} \tag{8.38}$$

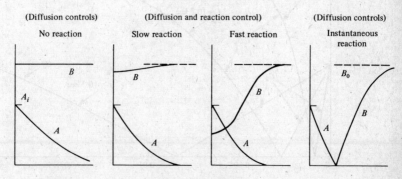

Effect of irreversible reaction

$$A + B \longrightarrow \text{products}$$

| (Diffusion controls) | (Diffusion and reaction control) | (Diffusion controls) |
|---|---|---|
| No reaction | Slow reaction     Fast reaction | Instantaneous reaction |

Distance from interface

**FIGURE 8.5**
Types of concentration distribution near an interface in a chemically reactive liquid, bimolecular reaction: $A + vB \rightarrow$ products.

One other extreme case leads to a comparatively simple solution. This occurs when the reaction-rate constant $k_{\mathrm{II}}$ is so large that $A$ and $B$ react immediately and completely upon contact in the solution. Then the product $C_A C_B$ must be zero everywhere in the liquid, $C_B$ being zero near the interface where $C_A$ is finite and $C_A$ being zero deep in the solution where $C_B \approx B_0$. Thus, a sharp reaction plane forms in the liquid, as illustrated in Figs. 8.5 and 8.6. Initially it is located at the interface, but owing to the very fast rate of absorption of $A$, the plane moves rapidly into the liquid to a position represented by the equation $y' = 2(\alpha t)^{1/2}$ where $y'$ is the distance from the interface to the reaction plane and $\alpha$ is a constant. Behind the moving zone, $A$'s concentration is represented by the transient diffusion equation without a reaction term; the same is true for $B$ in the region to the right of the plane. The solution of the problem is therefore represented by two error-function expressions,

$$C_A = A + B \operatorname{erf} \frac{y}{2D_A t} \qquad 0 \le y \le y' \tag{8.39}$$

$$C_B = E + F \operatorname{erf} \frac{y}{2D_B t} \qquad y' \le y \le \infty \tag{8.40}$$

The two equations must be fitted together at $y = y'$, where $C_A(y') = C_B(y') = 0$ and where

$$v D_A \frac{\partial C_A}{\partial y} + D_B \frac{\partial C_B}{\partial y} = 0 \tag{8.41}$$

$$\left(\frac{B_0}{v C_{Ai}}\right) \sqrt{\frac{D_B}{D_A}} \, e^{\alpha/D_A} \operatorname{erf} \left(\frac{\alpha}{D_A}\right)^{1/2} = e^{\alpha/D_B} \operatorname{erf} \left(\frac{\alpha}{D_B}\right)^{1/2} \tag{8.42}$$

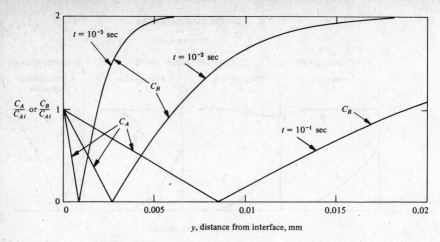

FIGURE 8.6
Concentration profiles of reactants near an interface—infinitely fast chemical reaction.

the mass-transfer coefficient of $A$ can be found from

$$k_{LA} = \frac{\sqrt{4D_A/\pi t}}{\text{erf }(\alpha/D_A)^{1/2}} \tag{8.43}$$

Although the solution appears complicated, it has a simple approximate form when $D_A \approx D_B$:

$$k_{LA} = r^{-1/2}\left(1 + r\frac{B_0}{vC_{Ai}}\right)\left(\frac{4D_A}{\pi t_D}\right)^{1/2} \tag{8.44}$$

Equation (8.44) is exact if $r = D_B/D_A = 1$, as is nearly true in most applications to gas absorption.

Equation (8.43) was obtained by Danckwerts [24] and is similar to a closely related equation occurring in the theory of heat condition to a moving boundary when ice melts [52]. See also Crank [22] for a more detailed discussion.

The strong influence of the concentration ratio $B_0/C_{ai}$ in increasing the mass-transfer coefficient above the value for purely physical absorption is owing to the fact that this ratio determines the position of the reaction plane at $y = y'$. The larger $B_0$ is, the closer this plane remains to the interface. As a result the rate of absorption depends on two concentrations and is given by

$$N_A = r^{-1/2}\left(C_{Ai} + \frac{r}{v}B_0\right)k_L^\circ \tag{8.45}$$

Equation (8.45) indicates that unlike the results obtained for purely linear processes such as purely physical absorption or absorption with first-order reaction, the rate

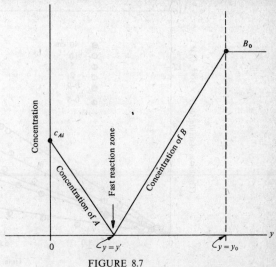

FIGURE 8.7
Concentration profiles at steady state in film.

is not now proportional to the concentration of the substance being absorbed. In fact, if the gas partial pressure of $A$ is very small and $C_{Ai}$ is nearly zero, the rate may be large owing to the effect of $B_0$. It should not be surprising therefore, in the analysis of mass-transfer data from absorber tests in which certain reactions occur, to find that the measured coefficients, calculated by dividing the observed rate by the gas-phase partial pressure of $A$, may vary with the gas and the liquid compositions.

The treatment of the bimolecular reaction near an interface described above is an improvement in some ways over the older treatment of Hatta [47], which was based on the film model. Figure 8.7 shows the profiles of concentration which would occur at steady state in a film of fixed thickness, $y_0$. The location of the reaction zone is determined by the stoichiometry requirement expressed by Eq. (8.41). Since the concentrations are represented by straight lines, the derivatives near the reaction zone can be expressed easily in terms of $C_{Ai}$ and $B_0$. The result is

$$\phi = 1 + rS \tag{8.46}$$

where $r = D_B/D_A$ and $S = B_0/\nu C_{Ai}$. It is very striking that the result is so close to Eq. (8.45). The principal difference lies in the incorrect dependence of $\phi$ on the diffusivities; the influence of the two concentrations is very nearly right.

It is not difficult to extend the film analysis to other reaction situations, such as consecutive or parallel reactions. Some results of this kind have been obtained by Astarita and Gioia [6] and by Weber and Nielson [105].

Still another useful modification of the film analysis can be made to allow for the effect of the electrostatic charges on ions, such as $OH^-$, on their diffusion rates

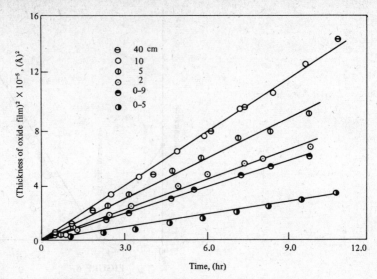

FIGURE 8.8
Growth of ZnO layer in metallic zinc exposed to gaseous oxygen at 400°C.
Oxygen partial pressures as per code on figure.

through solutions containing other ions. (See Chapter 3 in this connection.) Sherwood and Wei [91] and Sherwood and Ryan [92] developed an equation representing the rate of mass transfer of an undissociated acid substance HA into a water solution containing hydroxyl ions, as in NaOH (cf. Sec. 8.10).

EXAMPLE 8.5   Moore and Lee [66] have measured the growth of a layer of ZnO on a sheet of metallic zinc which was immersed in gaseous oxygen and was held in a furnace at 400°C. Figure 8.8 shows some of their results and indicates that oxidation under the conditions employed follows the "parabolic growth law," the square of the oxide layer thickness being proportional to the elapsed time of exposure.

Moore and Lee suggest that the following chemical reactions occur at the gas-solid surface:

$$O_{2(g)} \longrightarrow O_{2(ads)} \tag{A}$$

$$O_{2(ads)} + 4e^- \longrightarrow 2\,O^{--} \tag{B}$$

$$O^{--} + Zn^{++} \longrightarrow ZnO \tag{C}$$

The second step represents the formation of oxygen ions by reaction of adsorbed oxygen with electrons which came from the base metal. The third step is the fast reaction of these ions with zinc anions which have diffused through the oxide layer, having been formed at the metal-oxide interface.

On the assumption that the rate of the process is governed by the rate of diffusion of zinc ions, show that the parabolic rate law is to be expected theoretically and determine the value of the product anion diffusivity and concentration difference using the data at an oxygen partial pressure of 10 cm Hg, at which $dy'^2/dt = 1.77 \times 10^{-10}$ cm²/s.

SOLUTION   We need to find $y'$, the depth of penetration of zinc ions from the metal-oxide interface toward the reaction plane, i.e., toward the gas-oxide interface, using Eq. (8.42). Since $O^{--}$ is held at constant concentration at the reaction site as $y'$ increases, we use $D_B = 0$; $B_0$ becomes equal to the molar density of the reaction product, ZnO, since it governs the rate of increase of volume of the product layer per mole of $Zn^{++}$ reacting. $D_A$ is the diffusivity of zinc ions and $C_{Ai}$ is the concentration of $Zn^{++}$ in ZnO at the metal surface. As the ionic reaction is very fast, $C_A$ is very nearly zero at the reaction plane.

By using an asymptotic series to evaluate the complementary error function of $(\alpha/D_B)^{1/2}$ on the right of Eq. (8.42), the equation for the velocity coefficient $\alpha$ is obtained,

$$qe^{q^2} \operatorname{erf}(q) = \frac{1}{\sqrt{\pi}} \frac{vC_{Ai}}{B_0} = \frac{1}{\sqrt{\pi}} \frac{[Zn^{++}]_i}{[ZnO]} \tag{8.47}$$

where $q = (\alpha/D_A)^{1/2}$. From the data on Fig. (8.8), the growth rate is very small, indicating that $q$ is a small quantity. Then Eq. (8.47) has a simple approximate solution. The convergent series for the exponential and the error functions give

$$q(1 + q^2 + \cdots) \frac{2}{\sqrt{\pi}} \left( q - \frac{q^3}{3} + \cdots \right) = \frac{1}{\sqrt{\pi}} \frac{vC_{Ai}}{B_0}$$

or, approximately,

$$q^2 = \frac{1}{2} \frac{vC_{Ai}}{B_0}$$

Note that this simple result is easily obtained directly by assuming that the diffusion of $A$ through the growing product layer is always at steady state. Then we may equate the rate of diffusion of $A$ across the distance $y_0$ to the rate of consumption of $A$ by the reaction

$$D_A \frac{C_{Ai} - 0}{y_0} = \frac{B_0}{v} \frac{dy'}{dt}$$

from which

$$y'^2 = 2D_A \frac{vC_{Ai}}{B_0} t$$

which is the simplified form of Eq. (8.42) valid for $D_B = 0$ and $C_{Ai}/B_0 \ll 1$.

The slope of the line on the figure implies that

$$D_{Zn^{++}} \frac{[Zn^{++}]_i}{[ZnO]} = 0.88 \times 10^{-10} \text{ cm}^2/\text{s}$$

and, using the crystalline density of pure ZnO,

$$D_{Zn^{++}} [Zn^{++}]_i = 0.88 \times 10^{-10} \frac{5.5}{81.38} = 0.6 \times 10^{-11} \text{ g mole}/(\text{cm})(\text{s})$$

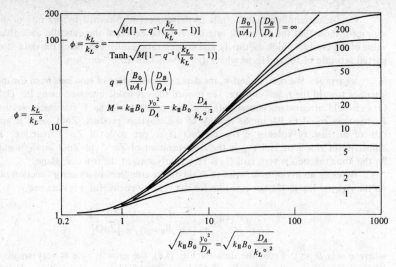

**FIGURE 8.9**
Reaction factor for a bimolecular reaction near an interface.

The diffusion theory indicates that with the assumptions used, the slope of the lines on Fig. 8.8 should be independent of the oxygen partial pressure as long as the rate of formation of ions by reaction $B$ is sufficiently rapid. Moore and Lee [66] suggest that this may not be true unless the surface concentration of adsorbed oxygen is sufficiently great, as it apparently was at the higher oxygen pressures. Wagner and Grunewald [103] offer a different explanation of the same phenomena in which diffusion (conduction) of electrons and ionized oxygen through the oxide layer is the controlling rate.                                    ////

Although Eqs. (8.38) and (8.44) may be expected to represent the solution of the bimolecular transient reaction problem in some special, extreme cases, there are obviously intermediate conditions ($k_{II} B_0 t$ finite and $B_0/vC_{Ai}$ not very large) where the nonlinear partial differential equations have to be solved as they stand. Some numerical solutions for a small range of the parameters were obtained by Perry and Pigford [72], but the most complete and useful results are those of Brian, Hurley, and Hasseltine [12]. Their results are not substantially different from Fig. 8.9. Note that the curves, corresponding to different constant values of the stoichiometry parameter $B_0/vC_{Ai}$, all lie beneath the one for the pseudo first-order condition. Note also that the lines approach horizontal asymptotic levels when the abscissa is large, corresponding to the conditions of sharp reaction zone. Figure 8.10 shows how a quantity proportional to the absorption rate of $A$ varies as $C_{Ai}$ changes while the solution composition $B_0$ is held constant. It is evident that there are conditions where the rate is not proportional to $A$'s concentration.

An early, approximate result which is the basis of Fig. 8.9 was obtained by van Krevelen and Hoftijzer [98], who made a very clever modification of the older

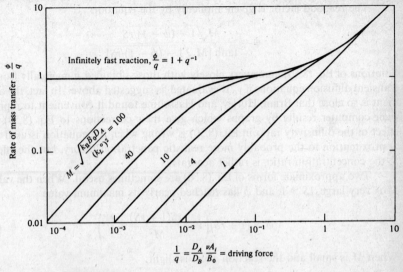

**FIGURE 8.10**
Effect of composition on rate of mass transfer into a liquid where a bimolecular reaction occurs.

film theory of Hatta [47]. Van Krevelen and Hoftijzer evaluated the film thickness $y_0$ from the rate of physical absorption of putting $k_L^\circ = D_A/y_0$, giving $ay_0$ in Eq. (8.12) equal to $(k_{II} B_i D_A/k_L^{\circ 2})$, or to $(\pi k_{II} B_i t/4)^{1/2}$ if $k_L^\circ$ is interpreted as the time-average value according to the penetration theory, Eq. (5.3). They noted, further, that $B_i$ could be found approximately by equating the rate of diffusion of $B$ toward the interface to the mole number times the *excess* rate of absorption of $A$, giving

$$v(\phi - 1) D_A \frac{C_{Ai}}{y_0} = D_B \frac{B_0 - B_i}{y_0} \tag{8.48}$$

or

$$B_i = B_0 - \frac{vC_{Ai}}{r}(\phi - 1) \tag{8.49}$$

The value of $ay_0$ therefore depends on the ratio of mass-transfer coefficients, $\phi$:

$$ay_0 = \sqrt{\frac{k_{II} B_i y_0^2}{D_A}} = \sqrt{\frac{k_{II} B_0 D_A}{k_L^{\circ 2}} \left[ 1 - \frac{vC_{Ai}}{B_0 r}(\phi - 1) \right]}$$

$$= M \left( 1 - \frac{\phi - 1}{rS} \right)^{1/2}$$

where $S = B_0/vC_{Ai}$ and $M = \sqrt{k_{II} B_0 D_A/k_L^{\circ 2}}$.

The reaction factor is given implicitly by the equation

$$\phi = \frac{M\sqrt{1 - (\phi - 1)/rS}}{\tanh\left[M\sqrt{1 - (\phi - 1)/rS}\right]} \tag{8.50}$$

Solutions of Eq. (8.50) agree very closely with those obtained numerically from the transient diffusion equations if $y_0$ is evaluated as suggested above. In fact, the agreement is so close that Brian, Hurley, and Hasseltine found it convenient to represent their computer results by graphs which give their corrections to Eq. (8.50). The effect of the diffusivity ratio in Eq. (8.50) is wrong when the equation is used as an approximation to the probably more realistic penetration theory, but the effect of $S$, the concentration ratio, is rather accurate.

Two approximate forms of Eq. (8.50) are sometimes useful. When the value of $M$ is very large, $rS > 1$, and $\phi$ has reached nearly its maximum value

$$\phi \approx (1 + rS)\left(1 - \frac{rS(1 + rS)}{M^2} + \cdots\right) \tag{8.51}$$

When $M$ is small and the reaction effect is slight,

$$\phi = 1 + \frac{M^2}{3} + \cdots \tag{8.52}$$

Finally, when $S$ is very large so that nearly first-order conditions prevail in the reaction zone and for $M$ sufficiently large (but $< rS$), this is approximated by

$$\phi = M\left(1 - \frac{M - 1}{2rS} + \cdots\right) \tag{8.53}$$

Equations (8.51) to (8.53) are sometimes useful for making small corrections to experimental data before attempting to plot them.

Only a few theoretical studies have been made of reversible reactions of the bimolecular type, despite the importance of such reactions in practice. Perry and Pigford [72] included a few numerical computations in their work, but the principal contribution to the subject is that of Olander [71], who gave several approximate solutions of the steady- and unsteady-state diffusion equations. Olander's method is subject to some criticism owing to the fact that by eliminating the reaction-rate terms from the simultaneous differential equations, he reduces the order of the equations and consequently solves a problem in which not all the boundary conditions can be satisfied. Some discretion must be exercised, therefore, in choosing the boundary values to be disregarded. One case in which an exact solution is available for direct comparison to test Olander's method is the one represented by Eq. (8.31). The reversible reaction which is first-order in both directions was considered. The limiting value of $\phi$ as the reaction-rate constant approaches infinity was given in Eq. (8.32); it is noteworthy that Olander's approximate method gives the same result, suggesting that similar results obtained with his method in other cases for more complicated reaction-rate expressions are very likely correct.

**Table 8.2 VALUES OF THE REACTION FACTOR, $\phi$, FROM EQ. (8.50)***

Stoichiometric ratio $(D_B/D_A)(B_0/\nu A_1) = rS$

| M | 1.0 | 2.0 | 3.0 | 5.0 | 7.0 | 10.0 | 20.0 | 30.0 | 50.0 | 70.0 | 100. | 200. | 300. | $\infty$ |
|---|-----|-----|-----|-----|-----|------|------|------|------|------|------|------|------|----------|
| 0.1 | 1.0033 | 1.0131 | 1.0033 | ... | ... | ... | ... | ... | ... | ... | ... | ... | ... | 1.0033 |
| 0.2 | 1.0131 | 1.0132 | ... | ... | ... | ... | ... | ... | ... | ... | ... | ... | ... | 1.0133 |
| 0.3 | 1.0290 | 1.0294 | 1.0295 | ... | ... | ... | ... | ... | ... | ... | ... | ... | ... | 1.0298 |
| 0.5 | 1.0759 | 1.0788 | 1.0798 | 1.0807 | 1.0810 | ... | ... | ... | ... | ... | ... | ... | ... | 1.0820 |
| 0.7 | 1.1371 | 1.1469 | 1.1505 | 1.1535 | 1.1548 | 1.1558 | 1.1570 | ... | ... | ... | ... | ... | ... | 1.1582 |
| 1.0 | 1.2410 | 1.2726 | 1.2849 | 1.2956 | 1.3004 | 1.3041 | 1.3085 | 1.3100 | 1.3112 | 1.3117 | ... | ... | ... | 1.3130 |
| 2.0 | 1.5444 | 1.7306 | 1.8211 | 1.9089 | 1.9516 | 1.9860 | 2.0287 | 2.0437 | 2.0559 | 2.0612 | ... | ... | ... | 2.0746 |
| 3.0 | 1.7218 | 2.0928 | 2.3051 | 2.5324 | 2.6504 | 2.7486 | 2.8749 | 2.9200 | 2.9572 | 2.9734 | ... | ... | ... | 3.0149 |
| 5.0 | 1.8748 | 2.5101 | 2.9601 | 3.5244 | 3.8523 | 4.1426 | 4.5374 | 4.6838 | 4.8065 | 4.8607 | 4.9020 | ... | ... | 5.0004 |
| 7.0 | 1.9313 | 2.7066 | 3.3262 | 4.2009 | 4.7618 | 5.2899 | 6.0518 | 6.3458 | 6.5966 | 6.7085 | 6.7942 | ... | ... | 7.0000 |
| 10.0 | 1.9649 | 2.8409 | 3.6098 | 4.8326 | 5.7143 | 6.6190 | 8.0475 | 8.6344 | 9.1489 | 9.3822 | 9.5623 | 9.7781 | 9.8514 | 10.0000 |
| 20.0 | ... | 2.9568 | 3.8869 | 5.6070 | 7.1143 | 8.9828 | 12.8035 | 14.7291 | 16.5913 | 17.4868 | 18.1990 | 19.0749 | 19.3777 | 20.0000 |
| 30.0 | ... | 2.9805 | 3.9481 | 5.8123 | 7.5559 | 9.9090 | 15.5953 | 18.9855 | 22.6070 | 24.4613 | 25.9836 | 27.9090 | 28.5874 | 30.0000 |
| 50.0 | ... | ... | 3.9810 | 5.9297 | 7.8284 | 10.5544 | 18.3161 | 24.0558 | 31.3471 | 35.5712 | 39.2808 | 44.2630 | 46.0896 | 50.0000 |
| 70.0 | ... | ... | ... | 5.9637 | 7.9106 | 10.7636 | 19.4551 | 26.6513 | 37.0174 | 43.7083 | 49.9933 | 58.9860 | 62.4239 | 70.0000 |
| 100.0 | ... | ... | ... | ... | ... | ... | ... | 28.5540 | 42.1267 | 52.0417 | 62.2497 | 78.3199 | 84.8770 | 100.0000 |
| 200.0 | ... | ... | ... | ... | ... | ... | ... | 30.3109 | 48.1071 | 63.3627 | 83.5489 | 124.0536 | 144.4678 | 200.0000 |
| 300.0 | ... | ... | ... | ... | ... | ... | ... | 30.6861 | 49.6315 | 67.4604 | 91.6641 | 150.5995 | 185.86 | 300.0000 |
| 500.0 | ... | ... | ... | ... | ... | ... | ... | 30.8855 | 50.4901 | 69.6420 | 97.2194 | 166.3921 | 234.83 | 500. |
| 700.0 | ... | ... | ... | ... | ... | ... | ... | ... | 50.7373 | 70.2941 | 98.9998 | 176.1710 | 259.71 | 700. |

*In the application of the table, it is customary to substitute $r^{1/2}$ for r, since this makes the film theory agree approximately with the transient diffusion theory.

EXAMPLE 8.6   Based on measured values of the reaction-rate constant for the bimolecular reaction

$$CO_2 + OH^- \longrightarrow HCO_3^-$$

in dilute solution at 20°C, compute the rates of absorption of $CO_2$ gas into (a) 0.01-normal KOH solution and (b) a buffered solution containing equal concentrations of $KHCO_3$ and $K_2CO_3$, from gases containing (1) pure $CO_2$ at 1 atm pressure and (2) 10 percent $CO_2$.

SOLUTION   According to Pinsent, Pearson, and Roughton [74], the bimolecular rate constant is 5,900 l/(g mole) (s); the Henry's-law coefficient for $CO_2$ in water, assumed equal to the value in the salt solutions, is $0.392 \times 10^{-4}$ g mole/(cm$^3$) (atm) at 20°C.

(a) The coordinates of Fig. 8.9 are

$$M = \left(\frac{k_{II} B_0 D}{k_L}\right)^{1/2} = \left[\frac{\pi k_{II} B_0 t}{4}\right]^{1/2} = \left[\frac{(3,14)(5900)(0.01)(0.05)}{4}\right]^{1/2}$$
$$= 1.57$$

$$\frac{B_0}{v A_i} = \frac{(0.01)}{(2)(0.0392Y)} = \frac{0.1273}{Y}$$

where $Y$ = mole fraction $CO_2$ in gas. The stoichiometric coefficient $v = 2$ because the primary rate-determining reaction with $OH^-$ is followed immediately by the fast reaction

$$HCO_3^- + HO^- \longrightarrow CO_3^{--} + H_2O$$

which consumes a second hydroxyl ion.

Based on the diffusivity of $CO_2$ in the liquid,

$$k_L^\circ = \sqrt{\frac{(4)(1.96 \times 10^{-5})}{(3.14)(0.05)}} = 0.0224 \text{ cm/s}$$

When $Y = 1.0$, the reaction factor is about $\phi = 1.13$; for $Y = 0.1$, it is about 1.5. The rates of absorption are $(1.13)(0.0224)(0.0000392) = 9.9 \times 10^{-7}$ and $(1.5)(0.0224)(3.92 \times 10^{-6}) = 1.3 \times 10^{-7}$ g mole/(cm$^2$) (s), respectively. Note that the rate is not proportional to the partial pressure of $CO_2$ in the gas.

(b) In the carbonate-bicarbonate buffer solution the ionic equilibrium can be represented by the reactions

1 Second ionization of carbonic acid: $HCO_3^- \rightarrow H^+ + CO_3^{--}$   $K_2$
2 Ionization of water: $H_2O \rightarrow H^+ + OH^-$   $K_w$

Writing the two expressions for the mass-action equilibrium and eliminating the concentration of hydrogen ion,

$$[OH^-] = \frac{K_w}{K_2} \frac{[CO_3^-]}{[HCO_3^-]}$$

Harned and Owen [45] give $K_2 = 4.20 \times 10^{-11}$ g ion/l and $K_w = 0.681 \times 10^{-14}$ (g ion/l)$^2$ at 20°C. For equal concentrations of the buffer ions, $[OH^-] = 1.62 \times 10^{-14}$ g ion/l and the abscissa for Fig. 8.9 is

$$\sqrt{k_{II}[OH^-]\frac{D}{k_L^{\circ 2}}} = 0.193$$

The upper envelope curve on the figure is used because very rapid buffer action causes the value of $[OH^-]$ to be nearly the same at the interface and in the bulk of the solution. It is difficult to read the figure in this range, but Eq. (8.26) can be used since it represents the envelope curve for pseudo first-order conditions. Since

$$kB_0 t = (5900)(1.6 \times 10^{-4})(0.05) = 0.0472$$

we have
$$\phi = 1 + \frac{0.0472}{3} - \frac{0.0472^2}{30} + \cdots = 1.016$$

The calculated rates of absorption are $(1.016)(0.0224)(0.392 \times 10^{-4}) = 8.9 \times 10^{-7}$ and $8.9 \times 10^{-8}$ g mole/(cm$^2$) (s) at 1 atm and 0.1 atm, respectively. Note that owing to the pseudo first-order conditions, the rate is proportional to the driving force. Actually, the solubility of free $CO_2$ in the solution may be smaller than in pure water, depending on the ionic strength of the solution.  ////

**EXAMPLE 8.7**[1]  Dissolved oxygen and carbon monoxide react with haemoglobin Hb (ca. 300 g/l, molecular weight $\approx$ 16,700) inside the red blood cells of the human body as follows.

*1*  $O_2 + Hb \xrightarrow{k_1} O_2 \cdot Hb$ (oxyhaemoglobin)

*2*  $CO + Hb \xrightarrow{k_2} CO \cdot Hb$

Reaction 1 is the normal reaction in a healthy human lung: oxygen is carried in the chemically combined form by blood circulation to the body tissues, where the reaction reverses. Reaction 2, if it occurs sufficiently to cause the absorption of about 10 cm$^3$ CO per kilogram of body weight, can cause carbon monoxide poisoning. Roughton and his coworkers [121, 122] have reported values of several of the physical constants associated with these systems, and the solubilities of oxygen and carbon monoxide are given by West [120] as $0.940 \times 10^{-6}$ and $0.695 \times 10^{-6}$ g mole/(cm$^3$) (atm), respectively at 37 °C. Using these values find (a) the number of standard cubic centimeters of oxygen absorbed per second from atmospheric air and (b) the number of cubic centimeters of carbon monoxide from air containing 100 cc CO per million cubic centimeters of gas in human lung (volume 107 cc of blood, surface of red blood cells 3,140 cm$^2$/cm$^3$ of blood).

SOLUTION  (a) For oxygen we assume that the reverse reaction is negligible in the blood freshly returned to the lung. We use the following physical data to evaluate the coordinates of Fig. 8.9:

$$D_{O_2} = 7.1 \times 10^{-6} \text{ cm}^2/\text{s}$$
$$D_{Hb} = 8.3 \times 10^{-8} \text{ cm}^2/\text{s}$$
$$k_1 = 1.8 \times 10^6 \, \text{l/(g mole) (s)}$$

The molar concentration of haemoglobin is

$$B_0 = \frac{300}{16,700} = 0.018 \text{ g mole/l}$$

[1] Based on a discussion by Ramachandran [77].

and the interfacial concentration of oxygen, neglecting the resistance to diffusion through cell walls, the blood plasma, and the arterial walls, is

$$C_{Ai} = 0.21 \times 0.940 \times 10^{-3} = 0.197 \times 10^{-3} \text{ g mole/1}$$

With these values the abscissa of Fig. 8.9 is

$$M = \sqrt{\frac{(1.8 \times 10^6)(0.018)(7.1 \times 10^{-6})}{(0.1)(0.1)}} = 4.8$$

based on an estimate [77] that $k_L^\circ = 0.1$ cm/s. The stoichiometric parameter is

$$\frac{B_0}{vC_{Ai}}\left(\frac{D_B}{D_A}\right)^{1/2} = \frac{0.018}{0.197 \times 10^{-3}}\left(\frac{8.3 \times 10^{-8}}{7.1 \times 10^{-6}}\right)^{1/2} = 9.88$$

This locates a point on Fig. 8.9 where $\phi = 4.0$, the rate being slightly reduced from its maximum value because of depletion of haemoglobin at the blood-cell walls. The mass-transfer coefficient is $k_L = 4.0 \times 0.1 = 0.4$ cm/s and the rate of absorption of oxygen is

$$\text{Rate} = (0.4)(0.197 \times 10^{-6})(22,400)(3,140)(107) = 593 \text{ cm}^3/\text{s} \qquad (0°\text{C, 1 atm})$$

(By allowing for additional mass-transfer resistances in the blood-cell walls, in the blood stream, and in blood-vessel walls, Ramachandran calculated a value of 274 cm³/s. Both the results represent maximum rates of transport based on the assumption that the breathing rate is fast enough to keep the lung cavity full of air.)

(b) For carbon monoxide a similar calculation can be based on

$$C_{Ai} = (100 \times 10^{-6})(0.695 \times 10^{-3}) = 0.695 \times 10^{-7} \text{ g mole/l}$$

$$M = 3.03$$

$$\frac{B_0}{vA_i}\left(\frac{D_B}{D_A}\right)^{1/2} = 2.7 \times 10^4$$

indicating that the conditions are those for a pseudo first-order reaction. The mass-transfer coefficient is $k_L = (k_2 B_0 D_A)^{1/2} = 0.30$ cm/s. (If oxygen absorption is taking place simultaneously, there will be some depletion of haemoglobin at the cell walls, causing the mass-transfer rate to be slightly less.) The rate of absorption of CO is

$$\text{Rate} = (0.3)(0.695 \times 10^{-10})(22,400)(3,140)(107) = 0.16 \text{ cm}^3/\text{s, or } 570 \text{ cm}^3/\text{h}$$

For a man weighing 175 lb the critical total uptake of CO will occur in 1.4 h total exposure.

////

# 8.10 BIMOLECULAR REACTION IN A TURBULENT FLUID NEAR A SOLID INTERFACE

Although the emphasis in this chapter has been on molecular diffusion as the means by which reactive molecules move through the fluid near an interface, another transport mechanism is well known and is likely to have a considerable bearing on the transfer coefficients. This is transfer of material by turbulent diffusion in which the random motion of molecules is that of the small fluid eddies in which they are present, as well as the motion of the individual molecules themselves.

The rates of such processes depend only weakly on the molecular diffusivities, as brought out in Chapter 4. Such a view is not at all like that used in the Danckwerts surface-replacement model, even though both represent ways of describing a random fluid motion at the interface. In the Danckwerts model, the eddies remain intact while they are at the interface and absorb gas which penetrates them gradually by molecular diffusion. However, if the interface region is filled with a fluid which has a very fine grained turbulent motion, turbulent diffusion may strongly influence mass transfer near the interface itself. No fluid particles may remain in place at the interface long enough for molecular diffusion to be effective.

An analysis of the role of a bimolecular reaction in a turbulent boundary layer has been carried out by Sherwood and Ryan [92] by assuming that molecular and turbulent transport rates are additive. The rate expressions in the Hatta film theory are then modified by introduction of the integral used in Eq. (5.19).

The rate of transport of $A$ through the turbulent fluid from the interface to the plane of the extremely fast reaction is given by

$$\frac{q}{k_A} = \frac{C_{Ai} - 0}{N_A} q = \Phi(y_{y'}^+, \sigma_A) \tag{8.54}$$

where the function $\Phi$ is given by Eq. (5.19). The other symbols are defined by $q = (\tau_w g_c/\rho)^{1/2} = u_0(f/2)^{1/2}$ (the friction velocity), $y_{y'}^+ = y'g\rho/\mu$ (the dimensionless distance of the reaction plane from the interface), and $\sigma_A = \mu/\rho D_A$ (the Schmidt number based on the molecular diffusivity of A). Similarly, for the turbulent mass transfer of $B$ from the inner edge of the film to the reaction plane,

$$\frac{(0 - B_0)}{N_B} q = \Phi(y_0^+, \sigma_B) - \Phi(y_{y'}^+, \sigma_B) \tag{8.55}$$

where $y_0^+$ refers to the value of the dimensionless distance corresponding to the total film thickness. Stoichiometry at the reaction plane requires that $vN_A = -N_B$, giving

$$\frac{k_A}{k_A^\circ} = \frac{\Phi(y_L^+, \sigma_A)}{\Phi(y_0^+, \sigma_B) - \Phi(y_{y'}^+, \sigma_B)} \frac{B_0}{vC_{Ai}} \tag{8.56}$$

The integrals designated by $\Phi$ have to be evaluated from the velocity distribution function as shown in Chapter 5.

Although Eq. (8.56) may appear somewhat complicated owing to the difficulty of carefully evaluating $\Phi$, it turns out that in some special cases the result is simple. If the two molecular diffusivities are equal, $\sigma_A = \sigma_B$ and Eq. (8.56) reduces to

$$\frac{k_A}{k_A^\circ} = \frac{(B_0/vC_{Ai})}{1 - [\Phi(y_{y'}^+, \sigma_A)/\Phi(y_0^+, \sigma_A)]} = \frac{(B_0/vC_{Ai})}{1 - (k_A^\circ/k_A)} \tag{8.57}$$

from which, solving for $k_A/k_A^\circ$, Hatta's result is obtained:

$$\phi = \frac{k_A}{k_A^\circ} = 1 + \frac{B_0}{vC_{Ai}} = 1 + S \tag{8.58}$$

A result identical with Eq. (8.45) can be obtained as a good approximation even when $D_B \neq D_A$. Here again we see that although they may have a strong influence on the physical coefficient, $k^\circ_A$, the details of the fluid motion near the interface are not very important for determining the *ratio* of mass-transfer coefficients.

Marangozis, Trass, and Johnson [64] found that values of $\Phi$ calculated for turbulent exchange mechanisms agree approximately with the estimates based on the Hatta equation. They also noted that variations in the estimates of turbulence intensities near the interface were capable of influencing $\Phi$. On the other hand, some realistic estimates led to very nearly the same $\phi$ values as did the laminar boundary-layer theory. The boundary-layer calculations were those of Meyerlink and Friedlander [65]. These results make it appear very unlikely that it will be possible to make a clear distinction among different assumptions about fluid motion at an interface based on measured values of $\phi$.

## 8.11 THE EFFECT OF REVERSIBILITY OF THE CHEMICAL REACTION ON THE MASS-TRANSFER RATE

When the chemical reaction can reach an equilibrium in the solution phase, the diffusion of both reactants and products must be calculated to compute the effect of the reaction rate on the rate of transport across the interface. As we have seen already in Sec. 8.6, for a reaction that is first-order in both directions, the presence of a reverse reaction not only limits the interfacial concentration of the volatile species at equilibrium with the gas phase, it also limits the maximum value of the reaction factor $\phi$ to the value $1 + K$, where $K$ is the mass-action equilibrium constant. Similar effects are found for more complex equilibrium reactions, but the calculations must be done numerically.

Secor and Beutler [86] have carried out such computations for several of the reactions symbolized by

$$\gamma_A A + \gamma_B B \;\rightleftharpoons\; \gamma_M M + \gamma_N N$$

having the rate expression

$$k_1 [A]^\alpha [B]^\beta = k_2 [M]^\mu [N]^\nu$$

with the condition that $\alpha/\gamma_A = \beta/\gamma_B = \mu/\gamma_M = \nu/\gamma_N$. They have reported values of the reaction factor for various combinations of the stoichiometric and rate constants and equilibrium constant. Their results qualitatively resemble those of Brian et al. [12] and are similar to Fig. 8.9, even though that figure applies to an irreversible bimolecular reaction. For the equilibrium reactions, $\phi$ tends toward a maximum value as $\theta = k B_0^\beta A_i^{\alpha-1} t$ increases. The limiting value depends both on the stoichiometric ratio $B_0/A_i$ and on the equilibrium constant. The smaller the equilibrium constant, the greater the reduction in $\phi$ from its value according to Fig. 8.9.

Although the complete computation of $\phi$ values for a reversible reaction is tedious, approximate calculations of the behavior at very large values of $\theta$, where the quilibrium constant has its greatest effect, can be made more readily by a local equilibrium method due to Olander [71]. Consider, for example, a reaction represented by the chemical equation

$$A + B \; \rightleftharpoons \; 2C$$

and let the diffusion of $A$, $B$, and $C$ be represented by the film-theory equations

$$D_A \frac{d^2 A}{dy^2} = r \tag{8.59a}$$

$$D_B \frac{d^2 B}{dy^2} = r \tag{8.59b}$$

$$D_C \frac{d^2 C}{dy^2} = -2r \tag{8.59c}$$

where $r$ represents the reaction rate per unit volume of fluid in the forward direction. If the reaction is very rapid, $C^2 = KAB$ at all values of $y$. Subtracting the equations from each other to eliminate $r$ and integrating twice with respect to $y$ gives

$$D_A A + \frac{D_C}{2} C = a_1 + a_2 y \tag{8.60a}$$

$$D_B B + \frac{D_C}{2} C = a_3 + a_4 y \tag{8.60b}$$

The constant $a_1$ can be evaluated from the interfacial concentrations, $A_i$ and $C_i$; $a_2$ depends on the total rate of transport of $A$ and its reaction product, $C$, across the interface:

$$N_A = -D_A \left( \frac{dA}{dy} \right)_{y=0} - D_C \left( \frac{dC}{dy} \right)_{y=0} \tag{8.61}$$

Thus, Eq. (8.60a) becomes

$$D_A(A_i - A_0) + \frac{D_C}{2}(C_i - C_0) = \frac{N_A D_A}{k_L^\circ} \tag{8.62}$$

where $A_0$ and $C_0$ are the bulk concentrations at $A$ and $C$.

The reaction factor is given by

$$\phi = \frac{N_A}{k_L^\circ(A_i - A_0)} = 1 + \frac{D_C}{2D_A} \frac{C_i - C_0}{A_i - A_0} \tag{8.63}$$

which shows that the computation of $\phi$ depends on the evaluation of the concentration of the equilibrium product at the interface.

We turn now to Eq. (8.60b) and note that $a_4$ must be zero if neither $B$ nor $C$ is volatile. Then, applying the equation at the interface and in the bulk of the liquid,

$$D_B(B_i - B_0) + \frac{D_C}{2}(C_i - C_0) = 0 \tag{8.64}$$

or, using the equilibrium relation,

$$D_B(B_i - B_0) + \frac{D_C}{2}\left(\sqrt{KA_iB_i} - \sqrt{KA_0B_0}\right) = 0 \tag{8.65}$$

This is a quadratic equation for $B_i^{1/2}$ (and therefore for $C_i$). It yields eventually

$$\frac{C_i - C_0}{K^{1/2}} = \sqrt{\left(\frac{D_C}{D_B}\right)^2 \frac{KA_i^2}{16} + A_iB_0 + \frac{D_C}{D_B}A_i(KA_0B_0)^{1/2}} \\ - \left[(A_0B_0)^{1/2} + \frac{D_C}{D_B}A_i\left(\frac{K}{16}\right)^{1/2}\right] \tag{8.66}$$

Combination of Eqs. (8.63) and (8.66) yields a rather involved expression for $\phi$, the value of which is always smaller than the result obtained from the simpler expression for the irreversible reaction.

If we assume that all the diffusivities are equal and that the bulk of the solution does not contain $A$, we obtain a simple result,

$$\phi = 1 + \left(\frac{K}{4}\right)^{1/2}\left(\sqrt{\frac{K}{4} + \frac{B_0}{A_i}} - \sqrt{\frac{K}{4}}\right) \tag{8.67}$$

In the limit $K \to \infty$, the equation agrees with the Hatta expression, Eq. (8.58). For finite $K$, the result shows that the rate of absorption is not a linear function of the concentration of reagent $B_0$ in the liquid, as it was for an irreversible reaction, but varies more slowly with $B_0$. If the interfacial concentration of $A$ is zero, as in desorption into an $A$-free gas, the local equilibrium theory gives

$$\phi = 1 + (KB_0/2C_0) = 1 + (KB_0/4A_0)^{1/2} \tag{8.68}$$

There is no comparable expression for an irreversible reaction.

Secor and Beutler [86] found from their extensive calculations that values of $\phi$ found from the equations for transient diffusion usually agreed with those according to the Olander theory within 5 percent at the limit of infinite $\theta$. Thus, substitution of the physically unrealistic film theory for the intuitively more appealing unsteady-state diffusion calculations gives nearly correct results for $\phi$.

EXAMPLE 8.8   Given an equilibrium reaction having the stoichiometry represented by

$$CO_2 + CO_3^- + H_2O \rightleftharpoons 2HCO_3^-$$

find (a) the maximum value of the reaction factor for absorption of pure $CO_2$ gas at 1 atm pressure into an aqueous solution of sodium carbonate containing $[CO_3^{--}]_0 = 1$ g ion/l at 25°C. (b) Also find the maximum value of $\phi$ for desorption of $CO_2$ from a solution containing $[CO_3^{--}]_0 = 0.5$ g ion/l and $[HCO_3^-]_0 = 0.5$ g ion/l into a gas containing no $CO_2$.

SOLUTION (a) At 25°C, based on equilibrium constant for the ionic reactions, $K = 7,700$. For absorption, Eq. (8.67) gives

$$\phi = 1 + \left(\frac{7,700}{4}\right)^{1/2} \left(\sqrt{\frac{7,700}{4} + \frac{1}{0.04}} - \sqrt{\frac{7,700}{4}}\right) = 13.5$$

(b) For desorption, Eq. (8.68) gives

$$\phi = 1 + \frac{K}{2} = 3,851 \qquad \qquad ////$$

It should be clear from the example that mass-transfer coefficients for absorption and desorption of gases can differ considerably, at least for very rapid reactions which are governed by nonlinear equilibrium relationships.

A practically important example of mass transfer with a chemical reaction that can be reversed occurs in the absorption of carbon dioxide into a buffered alkaline solution containing carbonate and bicarbonate ions, followed in the process by the regeneration of the solution in a $CO_2$ stripper, where dissolved gas is expelled and carbonate ion is regenerated. The net reaction is

$$CO_2 + CO_3^{--} + H_2O \;\rightleftharpoons\; 2HCO_3^-$$

and the reaction-rate expression is made up of two parts owing to the presence of parallel reactions with hydroxyl ion and with water.

$$r = k_1([CO_2][OH^-] - \frac{1}{K_A}[HCO_3^-]) + k_2\left([CO_2][H_2O] - \frac{1}{K_B[HCO_3^-][H^+]}\right) \quad (8.69)$$

The first expression on the right corresponds to the reaction of dissolved gas with hydroxyl ions that are formed very rapidly in the buffer; the second corresponds to the hydration of $CO_2$ followed by the very rapid ionization of the carbonic acid produced. By using equilibrium constants for the ionic reactions

$$HCO_3^- \;\rightleftharpoons\; CO_3^{--} + H^+ \qquad K_2$$
$$H_2O \;\rightleftharpoons\; OH^- + H^+ \qquad K_w$$

the rate expression is found to be equivalent to

$$r = k_1 \frac{K_w}{K_2}\left([CO_2]\frac{[CO_3]}{[HCO_3^-]} - K^{-1}[HCO_3^-]\right) + k_2[H_2O]\left([CO_2] - K^{-1}\frac{[HCO_3^-]^2}{[CO_3^{--}]}\right)$$

$$(8.70)$$

The second term on the right is more significant the lower the pH of the solution. At 25°C, the two terms are equal when the pH is about 8.5.

This chemical-rate expression can be incorporated into differential material balances for $CO_2$, $HCO_3^-$, and $CO_3^{--}$, and solutions of the three simultaneous partial differential equations can be obtained representing a semi-infinite region filled with solution. The result is a calculation of the rate of mass transfer of carbon dioxide across the interface. Such computations were carried out by Wall [102], using dilute-solution values of the various physical constants. Figure 8.11

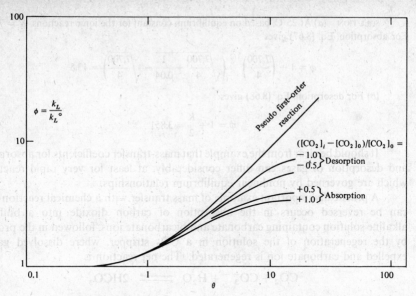

**FIGURE 8.11**
Reaction factors calculated by Wall [102] for the reversible reaction $CO_2 + CO_3^= + H_2O \rightleftharpoons 2HCO_3$.

shows some of his results for absorption and desorption from a solution of pH = 8 at 100°C.

Note that the calculated reaction factor varies with the $CO_2$ content of the solution, as expected from the example given above. The effects shown in the figure are smaller than those indicated in the example because of the finite reaction rates. Figure 8.12 shows the same results in another way. The ordinate is a quantity proportional to the rate of transfer, positive for absorption and negative for desorption; the abscissa is proportional to the driving force if the gas composition is varied. Clearly the rate is not proportional to the driving force expressed in concentrations of free $CO_2$, except in the limit of zero time of exposure of the interface, at which purely physical diffusion controls. In agreement with the qualitative result from the method of Olander [71], mass transfer during desorption is more rapid than that during absorption for the same driving force. The difference is smaller because the reactions are finite, not infinite, in speed.

A rather different kind of reversible reaction which is very much influenced by diffusion occurs in the manufacture of condensation polymers, such as polyamids or polyesters. The reaction occurs through successive chemical reactions between end groups, such as

$$-R_1-NH_2 + HOOC-R_2- \rightleftharpoons -R_1-(NHCO)-R_2- + H_2O$$

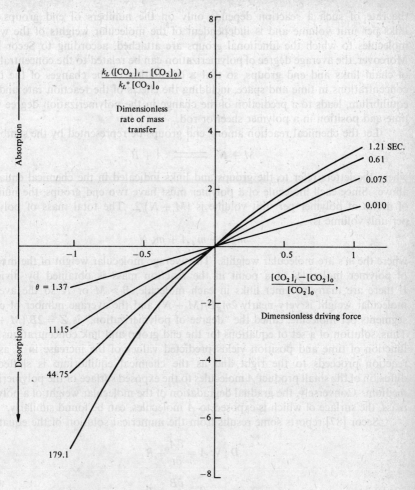

**FIGURE 8.12**
Calculated rates of absorption and desorption of $CO_2$ in buffered aqueous solutions for various times of exposure.

In the case of polyamids such as nylon, the reactants on the left are the amine and carboxylic acid ends of long molecules, and the first structure on the right is a polymer link. Polymerization can continue in place without diffusion of ends or links, but growth of polymer chains to very great length is inhibited by the accumulation of water, which needs to be removed from the reaction mass by diffusion to a surface where it can evaporate. Except near the beginning of a polymerization reaction where the active groups are attached to short molecules, the rate of such a reaction depends only on the numbers of end groups and links per unit volume and is independent of the molecular weights of the whole molecules to which the functional groups are attached, according to Secor [87]. Moreover, the average degree of polymerization can be related to the concentrations of chain links and end groups, so that a calculation of the changes of the three concentrations in time and space, including the effects of the reaction rate and the equilibrium, leads to a prediction of the change in the polymerization degree with time and position in a polymer sheet or rod.

Let the chemical reaction among end groups be represented by the symbols

$$M + N \;\rightleftharpoons\; A + B$$

where the letters refer to the groups and links indicated in the chemical equation above. Since each molecule of a polymer must have two end groups, the number of moles of polymer per unit volume is $(M + N)/2$. The total mass of polymer per unit volume is

$$m_B B + m_A A + m_N N$$

where the $m$'s are molecular weights. The average molecular weight of the mixture of polymer molecules at a point in the reaction mass is obtained by division. If there are many polymer links in each molecule, $B \gg M$ or $N$ so the average molecular weight is very nearly $2m_B B/(M + N)$ and the average number of chain segments per molecule, called the "degree of polymerization," is $Z = 2B/(M + N)$. Thus, solution of a set of equations for the end group and link concentrations as a function of time and position yields predicted values of the increase in $Z$ as the reaction proceeds to the right and as the chemical equilibrium is shifted by diffusion of the small product $A$ molecules to the exposed surface of the polymerizing medium. Conversely, the gradual degradation of the molecular weight of a polymer mass, the surface of which is exposed to $A$ molecules, can be found similarly.

Secor [87] reports some results from the numerical solution of the equations

$$D_A \nabla^2 A = \frac{\partial A}{\partial t} + R$$

$$0 = \frac{\partial B}{\partial t} + R$$

$$0 = \frac{\partial M}{\partial t} - R$$

$$0 = \frac{\partial N}{\partial t} - R \tag{8.71}$$

with boundary conditions representing initial chemical equilibrium in the reactive mass and slablike, rodlike, or spherical geometry. Figure 8.13 shows the variation across a flat slab of the local average degree of polymerization when the initial value of $Z$ was 100. By making the slab thinner, the rate of increase of $Z$ would be greater and the variation across the slab smaller.

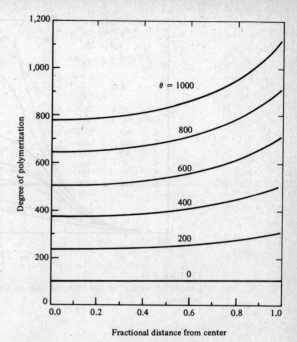

FIGURE 8.13
Calculated polymerization profiles in polymer slab (from Secor [87]).

These results suggest the importance of knowledge of diffusion phenomena for the rational design of condensation polymerization reactors and the need to distinguish between the chemical and diffusional rates.

## 8.12 COMPUTATIONS OF THE REACTION EFFECT FOR A FEW, MORE GENERAL CHEMICAL SITUATIONS

**Simultaneous reaction of two or more dissolved gases ($A$ and $B$) with a reagent ($R$) in the liquid:** The two gases compete for the same chemical reagent, according to the chemical equations

$$A + v_A R \longrightarrow \text{products}$$
$$B + v_B R \longrightarrow \text{products}$$

as in the simultaneous absorption of $CO_2$ and $H_2S$ into an alkaline reagent. Roper, Hatch, and Pigford [82] found the following expression for the reaction factor for any number of simultaneous reactions when all the reactions are

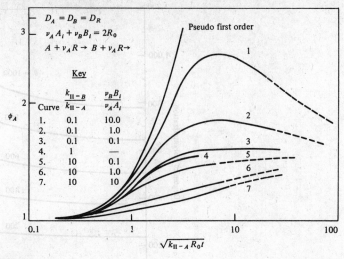

FIGURE 8.14
Calculated values of the reaction factor for two simultaneous bimolecular reactions.

instantaneous and occur at a common, moving reaction plane. The solution is similar to that given in Eqs. (8.42) and (8.43); an approximate expression of the complex result, exactly true when all the diffusivities are equal, is

$$\phi_A = r_A^{1/2} + \frac{r_A^{1/2}}{\sum r_i S_i^{-1}} \qquad (8.72)$$

where the summation is over all the reactions taking place. The result is therefore a generalization of Eq. (8.44). Since all the stoichiometric ratios, $S_i = R_0/v_i C_i$, affect the reaction factor for each substance, the presence of a second reactive gas can reduce the rate of absorption of a substance below the rate at which it would dissolve in the same solution if it were present alone. The reduction in rate depends primarily on the solubilities of the other substances taking part in the competition for reagent $C$.

When the reaction-rate constants $k_{II A}$ and $k_{II B}$ are finite and the reaction zone is diffuse, numerical solutions are necessary, as shown by Goettler and Pigford [42]. Figure 8.14 gives some of their results. The parameters were chosen so that $\phi_A \to 1.5$ at $t \to \infty$ for all the curves. It is striking to notice that the reaction factor for one of the soluble gases can be greater than its asymptotic value, at least temporarily.

Roper, Hatch, and Pigford [82] considered the problem of the irreversible reaction of the two gases $A$ and $B$ with each other after they have dissolved, as in the simultaneous solution of ammonia and carbon dioxide in water followed by their combination with water to make ammonium carbamate. Numerical solutions of the diffusion-plus-reaction equations were obtained and values of $\phi$ plotted versus the two different values $S_A$ and $S_B$. Experimental data for the $NH_3$–$CO_2$ reaction agreed approximately with the calculated results.

EXAMPLE 8.9   Assume that 1-$N$ aqueous solution of KOH at 25°C is used for absorption of carbon dioxide and hydrogen sulfide from a gas stream at 1 atm absolute pressure, containing 10 percent by volume of each substance. Find the ratio of the liquid-film mass-transfer coefficients and the ratio of the rates of absorption.

Assume that each dissolved gas reacts instantly with $OH^-$ present in the solution on contact. The solubilities of $CO_2$ and $H_2S$ in water are 0.0338 and 0.102 g mole/(l)(atm), respectively, and the diffusivities of $CO_2$, $H_2S$, and $OH$ ion in solution are $1.94 \times 10^{-5}$, $1.5 \times 10^{-5}$, and $2.5 \times 10^{-5}$ cm$^2$/s, respectively.

SOLUTION   Using Eq. (8.72) for the parallel reactions

$$CO_2 + 2OH^- \longrightarrow CO_3^{--} + H_2O$$
$$H_2S + 2OH^- \longrightarrow S^{--} + H_2O$$

we need to evaluate

$$S_{CO_2} = \frac{1.0}{(2)(0.00338)} = 148$$

$$S_{H_2S} = \frac{1.0}{(2)(0.0102)} = 49$$

$$r_{CO_2} = \frac{1.94}{2.5} = 0.775 \qquad r_{CO_2}^{1/2} = 0.88$$

$$r_{H_2S} = \frac{1.5}{2.5} = 0.60 \qquad r_{H_2S}^{1/2} = 0.77$$

$$\phi_{CO_2} = 0.88 + \frac{0.88}{(0.775/148) + (0.60/49)} = 51.4$$

$$\phi_{H_2S} = 0.77 + \frac{0.77}{(0.775/148) + (0.60/49)} = 43.0$$

The ratio of the mass-transfer coefficients becomes $k_{L,CO_2}/k_{L,H_2S} = 1.19$; the ratio of rates of absorption is $1.19(0.0338/0.102) = 0.39$.                                     ////

Brian and Beaverstock [14] solved the diffusion problem in which consecutive chemical reactions occurred such as

$$A + B \longrightarrow C \qquad C + B \longrightarrow \text{products}$$

Figure 8.15 shows some of their results, which indicate that the reaction factor for $A$, the gas which begins the reaction sequence, can exceed the value expected if the second reaction does not occur.

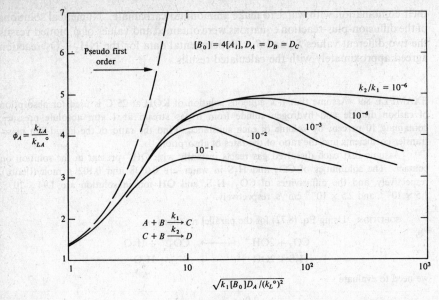

FIGURE 8.15
Reaction factor for consecutive reactions near an interface.

An extension of the van Krevelen-Hoftijzer method to single bimolecular reactions having chemical-rate expressions of the type

$$r = k_{II} C_A{}^n C_B{}^m \tag{8.73}$$

was carried out by Hikita and Asai [50] and by Secor and Beutler [86]. The exponents $m$ and $n$ are positive and $B$ is a nonvolatile reagent. Astarita [7] has presented some of the results.

## 8.13  THE REACTIONS OF $NO_x$ WITH WATER AND AQUEOUS SOLUTIONS

The absorption of nitrogen oxides into liquids is an industrial problem of major importance, not only for the manufacture of nitric acid but also for control of atmospheric pollution. That it is also a problem of considerable complexity is indicated by the enormous literature on the subject. Despite this great effort, the rate-determining steps in the reactions are still not wholly clear. As a result, the design of industrial equipment remains largely empirical.

The principal source of difficulty lies in the fact that there are several oxides, from $N_2O$ to $N_2O_5$, of which NO, $N_2O_3$, $NO_2$, and $N_2O_4$ play important roles in the reaction and diffusion processes in both gas and liquid phases. In addition,

$HNO_2$ and $HNO_3$ are found in both gas and liquid phases and may be formed in each. Only in recent years has it been possible by using the diffusion-reaction theory to interpret laboratory experiments so as to obtain some of the important physical properties, such as solubilities of the gases in water and the rates of reaction in the liquid phase. Reactions in the gas phase include the oxidation of NO, an unusual trimolecular reaction with an apparently negative temperature coefficient which was studied many years ago by Bodenstein [11]. The rates of formation of $HNO_3$ and $HNO_2$ in the gas and, especially, the rates of formation of acid mist in the gas and of its collection by the liquid remain poorly understood.

**a   Reactions of $NO_2(N_2O_4)$ with water**   The principal reaction by which nitric acid is formed in industrial absorbers can be represented by

$$2NO_2^{\cdot}(N_2O_4) + H_2O \longrightarrow HNO_3 + HNO_2 \qquad (a)$$

$$3HNO_2 \longrightarrow HNO_3 + H_2O + 2NO \qquad (b)$$

which are equivalent to the overall reaction

$$\tfrac{3}{2}N_2O_4(g) + H_2O(l) \longrightarrow 2HNO_3(l) + NO(g) \qquad (c)$$

The equilibrium in this reaction can be expressed in a practically useful way by

$$K_c = \frac{p_{NO}}{p_{N_2O_4}^{3/2}} \qquad (8.74)$$

where $K_c$, one factor in the mass-action equilibrium expression, is a function of acid strength. Its values have been measured by Abel and Schmidt, [1], Epshtein [36], Denbigh and Prince [32], and Chambers and Sherwood [18]. Interpretations of the data are given by Wenner [107] and Carberry [17], who found that $K_c$ is independent of temperature. Carberry's plot of the equilibrium data given in several references is shown in Fig. 8.16. It is this equilibrium that limits the concentration of acid manufactured from ammonia oxidation gases in cooled, high-pressure absorbers to about 68 weight percent $HNO_3$, even when a large interfacial surface is exposed on bubble trays to accommodate the low rates of the reactions. A few performance data on an absorber producing 55 tons/d of nitric acid are given by Graham, Lyons, and Faucett [118].

EXAMPLE 8.10   The gas mixture of nitrogen oxides entering the bottom plate of a gas absorber which produces nitric acid contains 0.1 atm partial pressure of NO and 0.25 atm partial pressure of $NO_2 + N_2O_4$. The total pressure is 1 atm. Find the maximum concentration of aqueous $HNO_3$ into which this gas will dissolve, increasing the $HNO_3$ concentration, if the temperature is 25°C.

SOLUTION   Initially we neglect the probably small partial pressures of $N_2O_3$ and $HNO_2$ in the following calculation. First we must find the partial pressure of $N_2O_4$ in the gaseous mixture of NO, $NO_2$, and $N_2O_4$. The equilibrium constant for the reaction $2NO_2 \rightleftharpoons N_2O_4$ is given by [15]

$$\log K_p = \frac{2993}{T°K} - 9.226$$

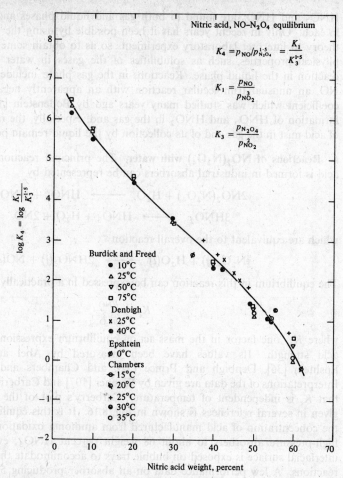

FIGURE 8.16
Equilibrium between phases for the reaction $\frac{3}{2}N_2O_4 + H_2O \rightleftharpoons 2HNO_3 + NO$.

which gives $K_p = 6.571$ atm$^{-1}$. At equilibrium we have

$$\frac{p_{N_2O_4}}{p_{NO_2}^2} = 0.25\frac{1-\alpha^2}{\alpha^2} = 6.571$$

where $\alpha$ is the fraction of the $N_2O_4$ which is dissociated. This quadratic equation has a root at $\alpha = 0.3634$. The partial pressure of $N_2O_4$ is $0.25(1-\alpha)/(1+\alpha) = 0.1166$ atm; that of $NO_2 = 0.25(2\alpha)/(1+\alpha) = 0.1333$ atm. To determine the acid strength, use Fig. 8.16 with

$$K_c = \frac{0.1}{0.1166^{3/2}} = 2.510$$

$$\log K_c = 0.3997$$

Figure 8.16 shows that this gas stream will be at equilibrium with 60 percent nitric acid. To obtain stronger acid, the pressure must be increased or the temperature must be reduced, causing more of the $NO_2$ to associate.

To fix the acid strength more precisely, an empirical equation for the line on Fig. 8.16 can be used [10],

$$\log K_c = 7.412 - 20.28921w + 32.47322w^2 - 30.87w^3$$

where $w$ is the weight fraction of $HNO_3$. This can be solved for $w$ by Newton's method, giving $w = 55.0$.

To determine the equilibrium partial pressures of the trace components, we take account of the additional reactions

$$NO + NO_2 = N_2O_3$$

having [10] $\log K_p = 2072/T - 7.234$, $K_p = 0.524$ atm$^{-1}$ at 25°C, and

$$NO + NO_2 + H_2O = 2HNO_2$$

having $$K_p = 1.67 \text{ at } 25°C \quad [10]$$

$K_p$ for the reaction varies slightly with $p_{NO_2}$, according to Beattie [10]. The empirical equation is based on values extrapolated to zero partial pressure.

Using $p_{NO} = 0.1$ and $p_{NO_2} = 0.0806$ atm as approximations, we have

$$p_{N_2O_3} \approx (0.524)(0.1)(0.1333) = 0.0070 \text{ atm}$$

Assuming that $p_{H_2O} = 23.76/760 = 0.312$ atm, the value at equilibrium with pure $H_2O$ at 25°C (the equilibrium partial pressure over 55 percent nitric acid is lower still), we find the approximate value of $p_{HNO_2}$ as

$$p_{HNO_2} \approx \sqrt{(1.67)(0.1)(0.1333)(0.0312)} = 0.0263 \text{ atm}$$

These values indicate that the calculated partial pressure of $NO_2$ and $N_2O_4$ were slightly too great and might be corrected iteratively. On the other hand, the computation of the equilibrium acid strength should not be changed, because Fig. 8.16, on which the value is based, was computed from experimental data, not allowing for pressure of the minor components, $N_2O_3$ and $HNO_2$. ////

**EXAMPLE 8.11**[1] The gas entering the bottom plate of a nitric acid absorber contains components flowing at the rates given in the following table.

| Component | lb moles/h Entering | Partial pressures, atm | |
|---|---|---|---|
| | | Entering | Leaving |
| $O_2$ | 60.0 | 0.4200 | 0.4253 |
| NO | 3.4 | 0.0238 | 0.0627 |
| $NO_2^* = NO_2 + 2N_2O_4$ | 68.2 | 0.4474† | 0.3573‡ |
| $N_2$ | 868.4 | 6.0788 | 6.1547 |
| | 1,000.0 | 7.000 | 7.0000 |

† $NO_2 = 0.1816$, $N_2O_4 = 0.1479$.

‡ $NO_2 = 0.1530$, $N_2O_4 = 0.1050$.

[1] Based on an example given by Chilton [20].

The pressure is 7 atm absolute and the acid leaving the plate is at 30°C, contains 60 percent by weight $HNO_3$, and flows at the rate of 9,570 lb/h. (The gaseous compounds listed in the table will be assumed to be virtually insoluble in the liquid.) The vertical spacing from the liquid surface on the tray to the floor of the next tray above is 12 in.

Compute the gas compositions (a) leaving the froth and (b) entering the next tray above.

SOLUTION  We use subscripts 1 and 2 to designate partial pressure entering and leaving the tray, respectively, and use $K_c$ and $K_{c3}$ to represent equilibrium constants for the main interphase reaction ($\frac{3}{2}N_2O_4 + H_2O \rightarrow 2HNO_3 + NO$) and the gas-phase association reaction ($2NO_2 \rightarrow N_2O_4$), respectively. Values of $K_c$ can be read from Fig. 8.16 or calculated from the equation quoted by Chilton [20], and used above.

For $w = 0.6$, $K_c = 1.823$ atm$^{-1/2}$. At 30°C, $K_p = 4.486$ atm$^{-1}$. Using the expression for association equilibrium,

$$2NO_2 \rightleftharpoons N_2O_4$$
$$2\alpha \qquad\qquad 1 - \alpha \qquad \text{sum} = 1 + \alpha$$

We have, for the entering conditions, since the sum of the partial pressures is $0.4774 \times [(1 + \alpha)/2]$,

$$4.486 = \frac{p_{N_2O_{4,1}}}{(p_{NO_{2,1}})^2} = \frac{1 - \alpha}{(0.4774)(2)\alpha^2}$$

from which $\alpha = 0.3803$, $p_{NO_{2,1}} = 0.1816$ atm, and $p_{N_2O_{4,1}} = 0.1479$ atm.

Since three moles of $NO_2^*$ are consumed for each mole of NO produced in the reaction,

$$3[K_c\, p_{N_2O_{4,2}}^{3/2} - 0.0238] = 0.4774 - [p_{NO_{2,2}} + 2K_{p3}\, p_{NO_{2,2}}^2]$$

If we let $x = p_{NO_{2,2}}$ and introduce the gas-phase equilibrium into the first term, we find

$$3(K_c K_{p3}^{3/2} x^3 - 0.0238) = 0.4774 - (x + 2K_{p3} x^2)$$

This is the cubic equation

$$51.96x^3 + 8.972x^2 + x - 0.5488 = 0$$

Solution of the equation by Newton's method gives $x = 0.1530$, the partial pressure of $NO_2$ in the gas leaving the liquid surface of the bottom tray. The other partial pressures are $p_{N_2O_4} = 4.486(0.1530)^2 = 0.1050$ and $p_{NO_2}^* = 0.1530 + 2(0.105) = 0.3529$ atm. The partial pressures of $O_2$ and $N_2$ are adjusted slightly, as shown in the table.

The cubic equation is based on a material balance that is not exact since it assumed no change in the total moles of gas reacting with the liquid. A precise calculation would have allowed for the change, but the results would not have been changed appreciably.

A material balance around the tray is based on $9,570 \times 0.60/63.02 = 91.14$ lb moles/h of $HNO_3$ and $9,560 \times 0.40/18.02 = 212.2$ lb moles/h of $H_2O$ leaving the tray. The number of moles of $NO_2^*$ removed from the gas is $68.2 - 1000(0.3528/7) = 17.79$; this implies that $91.14 - (2/3)(17.79) = 79.28$ lb moles/h of $HNO_3$ and $212.2 + (17.79/3) = 218.14$ lb moles/h of $H_2O$ enter the tray. The weight percent $HNO_3$ in this liquid is $100(79.28 \times 63.02)/(79.28 \times 63.02 + 218.4 \times 18.02) = 55.96$ percent. The quantity of heat transferred to a cooling coil on the tray needed to keep the temperature constant could be found from an energy balance.

In the space between the trays, containing 19.7 ft$^3$, the time of contact of the gas is 1.25 s. The NO in the gas undergoes an essentially irreversible reaction at 30°C, represented by the rate equation

$$-\frac{ap_{NO}}{dt} = kp_{NO}\,p_{O_2}$$

for which the rate data of Bodenstein [11] imply that

$$\log k = \frac{652.1}{T} - 0.7356$$

and $k$ is in units of s$^{-1}$ atm$^{-2}$, equals 26.1 at 30°C. (The equation given by Chilton [20] was in error, apparently.) Assuming plug flow of the gas stream between the trays, the fraction, $\alpha$, of the NO oxidized is given by [20]:

$$t = \frac{2}{k(2p_{O_2,2} - p_{NO,2})}\left\{\frac{\alpha}{p_{NO,2}(1-\alpha)} - \frac{1}{2p_{O_2,2} - p_{NO,2}} \ln\left[\frac{2p_{O_2,2} - p_{NO,2}\,\alpha}{2p_{O_2,2}(1-\alpha)}\right]\right\}$$

Trial solution using assumed values of $\alpha$ gives $\alpha = 0.696$. The partial pressures in the gas entering the next tray are given in the following table.

| Compound | Partial pressure, atm | |
| | Calculated | Adjusted |
| --- | --- | --- |
| $O_2$ | 0.4109 | 0.4117 |
| NO | 0.0339 | 0.0340 |
| $NO_2^*$ | 0.3861 | 0.3869 |
| $N_2$ | 6.1547 | 6.1674 |
| | 6.9856 | 7.0000 |

Note that the sum of the calculated partial pressures is not quite equal to the known total because of the assumption implied in the calculations of NO oxidation that there is no reduction in the total moles of gas. The last column gives the "adjusted" partial pressures, calculated by assuming that each compound's pressure is raised by the same factor.

It is seen, by comparing the two tables, that gaseous oxidation of NO has partially restored the original content of NO in the gas before it enters the second tray. The absorber's operation depends on repeated removal of $NO_2^*$, followed by reoxidation of the gaseous NO released. Each process is affected by the column pressure and the gas composition, especially the oxidation reaction, which is third-order. As shown by Chilton [20], calculations such as those illustrated indicate that the fraction of total gaseous nitrogen oxides in the tetravalent state and the fraction of total oxides removed by a tray become progressively smaller as the top of the absorber is approached.                                                                    ////

The rates of the heterogeneous reaction $a$ have been studied by Kramers, Blind, and Snoeck [59], Dekker, Snoeck, and Kramers [31], Gerstacker [44], Wendel and Pigford [106], Kameoka [55], and Corriveau [21], using short wetted-wall,

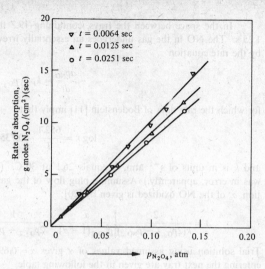

**FIGURE 8.17**
Absorption rates of $N_2O_4$ measured in a water-jet absorber at 20°C.

liquid-jet, and wetted-sphere absorbers. Each apparatus was designed so that the time of exposure of the liquid surface could be calculated from the flow rate and dimensions of the surface. The rates of absorption were computed from chemical analysis of the liquid for $NO_3^-$ and $NO_2^-$ ions. Figure 8.17 shows the results of Kramers et al. [59], who used a gas composed of pure $NO_2/N_2O_4$ mixtures and whose data are similar to those of the other investigators. (These two oxides are in chemical equilibrium in the gas. The proportions of the two vary with temperature and pressure. Figure 8.18 shows the effect of composition at 1 atm pressure). Note that the partial pressure of $N_2O_4$ is shown for the abscissa in Fig. 8.17 because the measured rate of absorption turns out to be proportional to $p_{N_2O_4}$ rather than $p_{NO_2}$. (The rate is proportional to the square of $p_{NO_2}$.) The result suggests that reaction $a$ actually takes place by the following steps.

1   $2NO_2(g) \longrightarrow N_2O_4(g)$
2   $N_2O_4(g) \longrightarrow N_2O_4(aq)$
3   $N_2O_4(aq) + H_2O \longrightarrow HNO_3 + HNO_2$

The pseudo first-order rate constant $k$, for reaction 3 is evidently large enough that the rate of absorption is little affected by $t$. Thus Eq. (8.27) can be used

$$N_{N_2O_4} = 2N_{NO_2} = H\sqrt{Dk}\left(1 + \frac{1}{2kt}\right)p_{N_2O_4} \qquad (8.27)$$

where $H$ is the Henry's-law coefficient for unreacted $N_2O_4$ in water and $D$ is $N_2O_4$'s diffusivity. By using a jet apparatus, Kramers et al. [59] were able to make

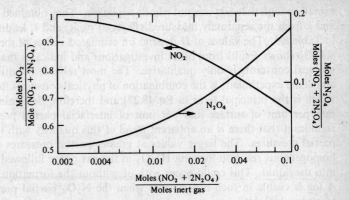

**FIGURE 8.18**

Equilibrium between $NO_2$ and $N_2O_4$ in gases at 1 atm, 30°C.

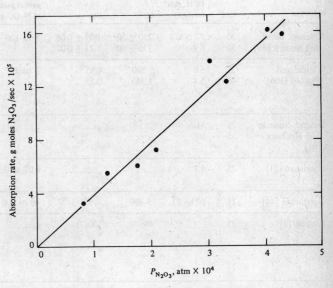

Interface partial pressure of $N_2O_3$

**FIGURE 8.19**

Rate of absorption of $N_2O_3$ into water.

$t$ small enough that $1/kt$ was not wholly negligible. They obtained values of $H(Dk)^{1/2}$ and $k$ from the separately measured effects of $p_{N_2O_4}$ and $t$, leading to values listed in Table 8.3. The values of $H$ depend on estimated values of the diffusivity. Table 8.3 also shows results from other investigations and indicates that agreement in the physical constants is only qualitative. The most reliable quantity to emerge from any of the experiments is the combination of physical constants for $N_2O_4$, $H(Dk)^{1/2}$, which is the principal term in Eq. (8.27) and therefore determines the absorption rate per unit of surface and per unit of interfacial partial pressure. It may be significant that there is an apparent trend of this quantity with the range of $N_2O_4$ partial pressures. The larger values at greater partial pressures may result from a homogeneous reaction forming $HNO_3$ in the gas phase followed by its absorption into the liquid. This could occur with or without the formation of nitric acid fog. A fog is visible in such experiments when the $N_2O_4$ partial pressure is permitted to go too high [18].

When the apparatus used for these rate measurements produces a larger surface exposure than that in the jet absorber, $kt$ becomes too great to use the parenthetical factor of Eq. (8.27) for determining $k$ and $HD^{1/2}$ separately. Then one must use some data of Abel and Schmidt [1] on the rates of homogeneous

**Table 8.3  VALUE OF PHYSICAL CONSTANTS FOR** $N_2O_4 + H_2O \rightarrow HNO_3 + HNO_2$

| Reference | $t$, °C | $10^5 H\sqrt{kD}$† for $N_2O$ | $k$‡ | $H$§ | Range of gas partial pressure of $N_2O_4$, mm | Method of resolution |
|---|---|---|---|---|---|---|
| Kramers, Blind, and Snoeck [59] | 20<br>30 | $7.7 \pm 0.2$<br>8.9 | $250 \pm 30$<br>$330 \pm 40$ | $1.39 \pm 0.08$<br>$1.23 \pm 0.07$ | 8 to 100 | Vary $t$ liquid jet |
| Wendel and Pigford [106] | 25<br>40 | 5.8<br>5.4 | 290<br>1,340 | 0.95<br>0.35 | 8 to 38 | Data [1] on decomposition of $HNO_2$; wetted-wall column |
| Dekker, Snoeck, and Kramers [31] | 25<br>35 | 11<br>10 | ...... | ...... | ...... | Not resolved |
| Corriveau [21] | 25 | 5.7 | ...... | ...... | 0.01 to 0.02 | Wetted-sphere absorber |
| Gerstacker [44] | 25 | 10 to 11 | $>490$ | $<1.4$ | 40 to 200 | Not resolved |
| Andrew [2] | 25 | ...... | 48 | 130(?) | ...... | Thermodynamic data plus $H(kD)^{1/2}$ from Kramers, Blind, and Snoeck [59] |
| Kameoka [55] | 25 | 6.85 | ...... | ...... | 0.3 to 0.9 | Wetted-sphere absorber |

† g mole/(cm²)(s)(atm)     § g mole/(l)(atm)
‡ s⁻¹

reaction $c$ in solution. The rate of formation of $HNO_3$ is equal to $k'$ $[HNO_2]^4/p_{NO}^2$, where $k' = 0.77$ $(l/g\ mole)^3\ atm^2/s$ at $25°C$. As this reaction evidently proceeds through reaction 3 as a rate-determining step, independent information is available from which the absorption rate data can be resolved, as shown by Wendel and Pigford [106].

**EXAMPLE 8.12**  A gas mixture of nitrogen and nitrogen dioxide containing 10% $NO_2 + 2N_2O_4$ is in contact with a surface of liquid water in a gas absorber. If the mass-transfer coefficient for the gas phase is $4 \times 10^{-5}$ g mole/(s)(cm$^2$)(atm), find the interfacial composition of the gas and the rate of solution of $NO_x$.

SOLUTION   (a) *Solution of diffusion plus instantaneous reaction problem for the gas film:*

$$N^{4+} = \frac{1}{RT}\left(D_{NO_2}\frac{dp_{NO_2}}{dy} + 2D_{N_2O_4}\frac{dp_{N_2O_4}}{dy}\right)$$

$$= \frac{1}{RT}(D_{NO_2} + 4D_{N_2O_4}K_p p_{NO_2})\left(\frac{dp_{NO_2}}{dy}\right)$$

$$= \frac{D_{NO_2}}{RTy_0}\left\{p_{NO_2} - p_{NO_2}^i + 2\left(\frac{D_{N_2O_4}}{D_{NO_2}}\right)K_p[(p_{NO_2})^2 - (p_{NO_2}^i)^2]\right\}$$

$$= k_{GNO_2}\left\{1 + 2\left(\frac{D_{N_2O_4}}{D_{NO_2}}\right)K_p(p_{NO_2} + p_{NO_2}^i)\right\}(p_{NO_2} - p_{NO_2}^i)$$

(b) *Determination of bulk gas composition:*   From Fig. 8.18,

$$\frac{p_{NO_2}}{0.1} = 0.92 \qquad \frac{p_{N_2O_4}}{0.1} = 0.04$$

(c) *Determination of interface partial pressure:*   Assuming that $D_{N_2O_4} \approx D_{NO_2}$ and taking $K_p = 6.57\ atm^{-1}$ at $25°C$,

$$4 \times 10^{-5}[1 + (2)(6.57)(0.092 + p_{NO_2}^i)](0.092 - p_{NO_2}^i) = 2(4.7 \times 10^{-5})K_p(p_{NO_2}^i)^2$$

This quadratic equation gives $p_{NO_2}^i = 0.0686$ atm and at equilibrium $p_{N_2O_4}^i = 6.57(0.0686)^2 = 0.0309$ atm. The rate of absorption is $(4.7 \times 10^{-5})(0.0309) = 1.45 \times 10^{-6}$ g moles $N_2O_4/$(s)(cm$^2$) or $2.91 \times 10^{-6}$ g moles/(s)(cm$^2$) of tetravalent nitrogen.     ////

**b   Reactions of $N_2O_3$ with water**   The partial pressure of $N_2O_3$ is very small in gas mixtures usually used experimentally or industrially because of the strong tendency of this oxide to dissociate into NO and $NO_2$. It has been realized for some time, however, that $N_2O_3$ should react so quickly with water,

$$N_2O_3 + H_2O \longrightarrow 2HNO_2$$

that the rate of solution might be very fast even at low interfacial partial pressures. Figure 8.19 shows a few data from Corriveau [21], who used a laboratory gas absorber containing five wetted spheres over which water flowed in a vertical direction. The gas mixtures were prepared by mixing NO and $NO_2$ and contained. $N_2O_4$, as well as small amounts of $N_2O_3$ and $HNO_2$. The rate of absorption of $N_2O_3$ was computed from chemical analyses of the effluent liquid for both $NO_2^-$ and $NO_3^-$, assuming that the decomposition of $HNO_2$ by reaction did not occur appreciably. Typical results showed that when the partial pressures of NO and of $NO_2 + 2N_2O_4$ were about equal, the rates of absorption of $N_2O_4$ and $N_2O_3$ were about equal also, despite the fact that the interfacial partial pressure of $N_2O_4$ was four times that of $N_2O_3$.

The value of $H\sqrt{Dk}$ for $N_2O_3$ corresponding to the slope of the straight line on Fig. 8.19 is $1.59 \times 10^{-4}$ g moles/$(cm^2)(s)(atm)$ at 25°C, a value to be compared with $0.57 \times 10^{-4}$ for $N_2O_4$. The values of $k$ and $H$ inferred from this result and from published equilibrium data [10] are $k = 1.2 \times 10^4$ s$^{-1}$ and $H = 0.39$ g mole/(1)(atm). Thus, $N_2O_3$ is only about 30 percent as soluble as $N_2O_4$ in $H_2O$, but it reacts 40 times as fast.

Corriveau's data indicate indirectly that the formation of gaseous $HNO_2$ at the interface and its solution in water did not contribute significantly to the rate of appearance of $HNO_3$ in the liquid. If the opposite had been true, the measured rate should have been proportional to $p_{HNO_2}$, which is proportional to $\sqrt{p_{N_2O_3}}$. The straight line on the figure indicates that trivalent nitrogen entered the liquid primarily as $N_2O_3$.

EXAMPLE 8.13 A gas mixture at 1 atm and 25°C contains 10 mole percent NO. In order to remove this unreactive oxide, $NO_2$ is added to the gas such that the moles of NO and $NO_2$ before reaction are in the ratio of 1 to 2.5. The mixture is sent to a gas absorber for scrubbing with a large volume of $H_2O$. Neglecting the gas-phase resistance to mass transfer, find the rates of removal of trivalent and tetravalent nitrogen from the gas at the bottom of the absorber.

SOLUTION The gas composition is that of Example 8.11, which showed that $p_{NO} \approx 0.1$, $p_{NO_2} \approx 0.133$, $p_{N_2O_4} \approx 0.117$, $p_{N_2O_3} \approx 0.0070$, and $p_{HNO_2} \approx 0.026$ atm. Thus, the rates of solution are

$$N^{++} = N_{N_2O_3} = (1.59 \times 10^{-4})(0.0070) = 1.1 \times 10^{-6} \text{ g moles/cm}^2 \text{ s}$$

$$N^{4+} = N_{N_2O_3} + 2N_{N_2O_4} = (1.59 \times 10^{-4})(0.0070) + 2(4.7 \times 10^{-5})(0.117)$$

$$= 1.2 \times 10^{-5} \text{ g moles/cm}^2 \text{ s}$$ ////

c **Absorption of NO/$NO_2$ gaseous mixtures into water** The rate of formation of nitric acid by the overall reaction

$$3NO_2(g) + H_2O(aq) \longrightarrow 2HNO_3(aq) + NO(g)$$

is known empirically from extensive tests in gas absorbers in commercial acid plants (Chilton [20]) to be increased by lowering the temperature and raising the pressure. In the light of what is known about the chemical mechanisms involved, this can be accounted for qualitatively. Reducing the temperature has several effects. It causes the gas-phase reaction of NO and oxygen to proceed more rapidly in the space between plates of a tower. It also increases the concentration of $N_2O_4$ at the expense of $NO_2$ in the gas and increases the concentration of $N_2O_4$ at the interface. These effects apparently override the smaller detrimental effects of lower reaction-rate constants. Raising the pressure has a strong effect on the equilibrium in the overall reaction and also shifts the association of $NO_2$ toward $N_2O_4$ and thereby promotes faster diffusion and reaction near the interface while increasing the rate of the gas-phase oxidation of NO.

As a result of experience and qualitative reasoning, modern plate-type nitric acid absorbers are fitted with tubular coolers for the removal of the exothermic heat of reaction by cooling the acid on each plate; in addition, corrosion resistant stainless steel alloys are used for construction so that high pressure can be employed [95].

In spite of the considerable effort that has been spent on the design and development of such gas absorbers, very little information has been available about their plate efficiencies and no attempts have been published, until recently, to make use of the information available from laboratory data such as those discussed above. It has been reported, however, that the departure from theoretical tray performance has been greatest on the top trays of an absorber, where most of the oxidized nitrogen that remains in the dilute gas stream is NO rather than $NO_2$. In view of what has been said earlier about the two significant mass-transfer processes, absorption of $N_2O_4$ and absorption of $N_2O_3$, and the vanishingly small concentrations of these substances in dilute gases, such trends in the plate efficiency are not surprising.

The first attempt to use existing knowledge of the chemical reaction and diffusion rates along with equilibrium data to correlate plate efficiency is that of Andrew and Hanson [2]. {See also Andrew [112].} They define the efficiency, $\eta$, as the ratio of the total higher oxide, $NO_2^*$, absorbed on a plate to the total higher oxide entering the plate—not quite the standard definition, in which the denominator of the ratio is the maximum removal of the components if the gas came to equilibrium with the liquid. The difference may be slight, however, if the $HNO_3$ concentration in the liquid does not become very large. Figure 8.20 shows Andrew and Hanson's results and indicates that measured efficiencies depend on the concentrations both of total higher oxides in the entering gas and of the lower oxides essentially NO. The number of moles of "higher oxide," used for finding the partial pressure, is computed from

$$NO_2^* = NO_2 + 2N_2O_4 + N_2O_3 + \tfrac{1}{2}HNO_2$$

The lines in Fig. 8.20 were calculated by Andrew and Hanson to represent the cumulative effect of four different mechanisms for introducing higher oxides into the liquid, as illustrated in Fig. 8.21. In strong gases, as on the bottom plate of an absorber, the efficiency is large, primarily because the reaction involving both diffusion and reaction of $N_2O_4$ is favored. For dilute gases, however, the partial pressure of $N_2O_4$ becomes so small that the second reaction path involving $N_2O_3$ produces $HNO_3$ more efficiently. The line representing the formation of $HNO_3$ by a homogeneous gas reaction was based on an approximate value of the bimolecular homogeneous reaction-rate constant of $4 \times 10^{-4}$ cm$^3$/(g mole)(s)

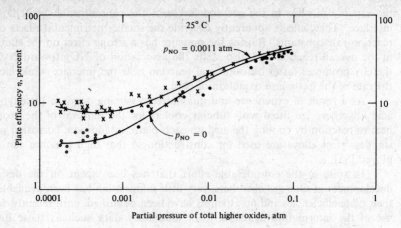

FIGURE 8.20
Comparison of measured and predicted plate efficiencies of $NO_2$ absorption.

computed from the observed effect of increasing the residence time available for this gas reaction before the gaseous mixture entered the absorber. The fourth mechanism, which becomes important at the lowest partial pressures of $NO_2^*$, involves the plain physical diffusion of $NO_2$ without reaction through gas and liquid films, followed by a slow reaction occurring throughout the volume of liquid held on the tray. The calculations were based on Andrew and Hanson's interpretation of the known rate of the homogeneous, liquid-mass reaction between $N_2O_4$ and water, plus their estimate of the solubility of $NO_2$ in water of 0.0021 g mole/(1)(atm) at 25°C. Had they not included the fourth mechanism, their calculated plate efficiencies would have continued to fall with decreasing $p_{NO_2^*}$.

Detailed studies of results such as these can lead to improved designs of absorbers in order to provide maximum opportunity for the reaction paths that are most efficient. Such studies have not been reported, however.

EXAMPLE 8.14   A gaseous mixture of an inert substance, $NO_2$, and NO is washed with a large volume of water in a packed column at 25°C. The partial pressures in the inlet gas are $p_{NO} = 0.1$ atm and $p_{NO_2^*} = p_{NO_2} + 2p_{N_2O_4} = 0.25$ atm. Calculate the partial pressure of the major components of the gas vs. the distance from the bottom of the packing. Assume that all diffusional resistance is in the water phase.

SOLUTION   The fluxes of tetravalent and trivalent nitrogen at the interface are

$$N^{4+} = 2[H\sqrt{kD}]_{N_2O_4} p_{N_2O_4} + [H\sqrt{kD}]_{N_2O_3} p_{N_2O_3}$$
$$= 2[H\sqrt{kD}]_{N_2O_4} K_4 p_{NO_2}^2 + [H\sqrt{kD}]_{N_2O_3} K_3 p_{NO_2} p_{NO}$$
$$N^{++} = [H\sqrt{kD}]_{N_2O_3} K_3 p_{NO_2} p_{NO}$$

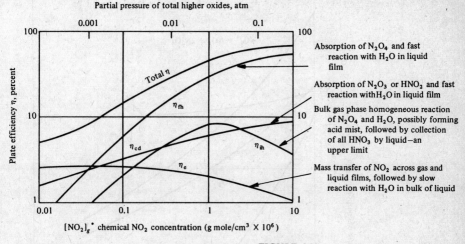

FIGURE 8.21
Predicted component and total plate efficiencies.

where $K_3$ and $K_4$ correspond to the reactions $NO + NO_2 \rightarrow N_2O_3$ and $NO_2 + NO_2 \rightarrow N_2O_4$, respectively. Material balances, neglecting the small partial pressure of $N_2O_3$ and $HNO_2$ in the bulk of the gas, give

$$- G_M \frac{d(p_{NO_2} + 2p_{N_2O_4})}{P} = N^{4+} a \, dx$$

$$- G_M \frac{d(p_{NO})}{P} = N^{++} a \, dx$$

where $G_M$ is the molar mass velocity of the gas mixture and $a$ is the interfacial area per unit of packed volume. If we let

$$n = \frac{Pa[H\sqrt{kD}]_{N_2O_4}}{G_M} x$$

$$\alpha = \frac{[H\sqrt{kD}]_{N_2O_3}}{[H\sqrt{kD}]_{N_2O_4}}$$

the equations above can be expressed as

$$\frac{dp_{NO_2}}{dn} = - \frac{p_{NO_2}(2K_4 p_{NO_2} + \alpha K_3 p_{NO})}{1 + 2K_4 p_{NO_2}}$$

$$\frac{dp_{NO}}{dn} = -\alpha K_3 p_{NO} p_{NO_2}$$

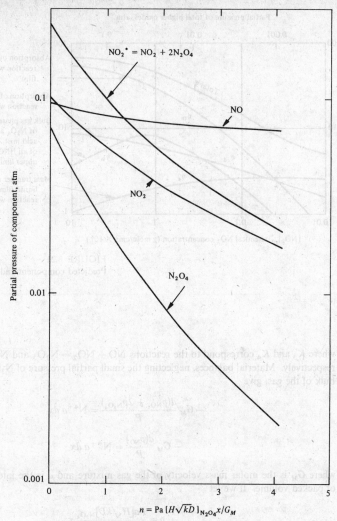

**FIGURE 8.22**
Changes in gas composition during absorption of $NO/NO_2$ into water.

A numerical solution of this set of equations using $K_4 = 6.53$ atm$^{-1}$, $K_3 = 0.52$ atm$^{-1}$, $[H\sqrt{kD}]_{N_2O_4} = 4.7 \times 10^{-5}$ g mole/(s)(cm$^2$)(atm), and $[H\sqrt{kD}]_{N_2O_3} = 1.59 \times 10^{-4}$ g mole/(s) (cm$^2$)(atm) is shown in Fig. 8.22. The curves start at $p_{NO_2} = 0.1052$, $p_{N_2O_4} = 0.0722$, and $p_{NO} = 0.1000$, as in Example 8.13. The figure shows that none of the profiles is straight on semilogarithmic coordinates, as would be true if there were no chemical reactions. $NO_2^*$ as well as $NO_2$ and $N_2O_4$ falls faster than NO because two reaction paths are effective for tetravalent nitrogen compounds but only one for divalent. The rate of reduction of nitric oxide's partial pressure approaches zero as $NO_2$ disappears from the gas.

Using the values of $G_M$ and $a$ that were employed by Andrew and Hanson [2] in their more involved computations (see also Andrew [112]), gas velocity = 10 cm/s, $a = 11$ $cm^2/cm^2$, $\rho_G = 4.1 \times 10^{-5}$ g mole/$cm^3$, the value of $n$ corresponding to one sieve tray is

$$n = \frac{(1)(11)(4.7 \times 10^{-5})}{(10)(4.1 \times 10^{-5})} = 1.26$$

Similarly, using a packed column filled with 1-in Raschig rings at a liquid flow of 0.6 $cm^3/(cm^2)(s)$, $a = 1.2$ cm/$cm^3$, and the value of $n$ for 1 ft of packing is

$$n = \frac{(1)(1.2)(4.7 \times 10^{-5})}{(30)(4.1 \times 10^{-5})(30.5)} = 1.40$$

and at a gas velocity of 30 cm/s 26 percent of the NO and 69 percent of total $NO_2$ would be removed from the gas in 1 ft of packing. ////

## 8.14 THE REACTIONS OF $CO_2$ WITH ALKALINE AQUEOUS SOLUTIONS

Removal of carbon dioxide from gas mixtures by washing with alkaline solutions is one of the most widely practiced industrial gas-absorption processes. In the manufacture of ammonia, for example, hydrogen is made from the reactions of hydrocarbons or coal with steam. The carbon dioxide which forms must be removed very completely from the gas mixture before it can be sent to the ammonia synthesis reactor. According to Danckwerts and Sharma [28], the quantity to be removed is about 1.2 to 2.2 tons for each ton of $NH_3$ made. Similar requirements apply in the manufacture of methanol from carbon monoxide and hydrogen and for the production of carbon dioxide itself as dry ice. Moreover, regeneration of the $CO_2$-rich solutions to reverse the reaction is also required and accounts for mass-transfer equipment as large or larger than that needed for absorption in the first place.

Owing to the industrial importance of the unit operations involved, considerable effort has been devoted to the collection of empirical equilibrium and rate data for a variety of aqueous solutions and a variety of processes exist. Many of these are described by Kohl and Riesenfeld [58], who outline empirical methods of process design. The monograph by Danckwerts and Sharma [28] discusses the use of the various diffusion-and-reaction theories for a few of the more important alkaline reagents.

**The chemical properties of the system $CO_2$-$H_2O$-$OH^-$**   In alkaline solutions both of the reactions

$$1 \quad CO_2 + H_2O \rightleftharpoons H_2CO_3 \longrightarrow HCO_3^- + H^+$$

and

$$2 \quad CO_2 + OH^- \longrightarrow HCO_3^-$$

proceed simultaneously. In the absence of $OH^-$ the extent of reaction *1* is limited by chemical equilibrium, the first ionization of carbonic acid being expressed by $[H^+][HCO_3^-]/([CO_2] + [H_2CO_3]) = 4 \times 10^{-7}$ g ion/l at 25°C. Less than 1 percent of the un-ionized carbon is in the hydrated form. In the presence of hydroxyl ion, however, the proton formed in *1* reacts instantaneously with $OH^-$ to form water. Thus the reaction products are the same by both paths but the rates are different, *2* being more important the higher the pH of the solution.

The rate of the hydration reaction is known rather accurately as a result of the work of Pinsent, Pearson, and Roughton [74]. The pseudo first-order rate constant can be expressed by

$$\log k_{H_2O} = 329.80 - 110.541 \log T - \frac{17,265.4}{T} \tag{8.75}$$

where $k_{H_2O}$ is in $sec^{-1}$ and $T$ is in °K. At 25°C, $k_{H_2O} = 0.028$ $s^{-1}$. The reaction is subject to catalysis by anions of weak acids or by molecules having a high affinity for protons [88, 89]. Among the most effective catalysts are arsenite ion, $As(OH)_2O^-$, the anion of formaldehyde hydrate, $CH_2(OH)_2$, and hypochlorite ion, $ClO^-$. Use of arsenite by adding arsenious oxide to a solution of potassium carbonate and bicarbonate is the basis of a patented process [28]. The rate of the reaction via the third, catalyzed reaction path can be expressed by the term $k_B[B][CO_2]$, where $B$ signifies the concentration of the nucleophilic anion. According to Sharma and Danckwerts [88], the available data for arsenious anion indicate that

$$\log k_B = 9.837 + \log T - \frac{2,967}{T} \tag{8.76}$$

At 25°C, $k_B = 229$ 1/(g mole)(s). The rate constant is greater at finite ionic strength.

In the absence of catalysis of reaction *1*, the second reaction is usually the faster of the two in the alkaline solutions usually employed. Its second-order rate constant has been measured in many studies, as indicated in Fig. 8.23, which shows most of the available data. The straight line on the graph is drawn through the data of Pinsent, Pearson, and Roughton [74]. Because it is a reaction involving an ion, the rate constant $k_{OH^-}$ is affected by the ionic strength of the solution. (The values of the constant in Fig. 8.23 are either extrapolated to zero strength or apply to such low values that the extrapolation is not required.) It has been suggested [15] that the ratio of the rate constants in solutions of finite and at zero strength should be a function of the "ionic strength,"

$$I = \tfrac{1}{2} \sum z_i^2 C_i \tag{8.77}$$

where $z_i$ is the number of positive or negative charges on an ion having molarity $C_i$. It is not possible using any ordinary means of measurement to determine $k_{OH^-}$ in concentrated solutions of hydroxyl ions, but Nijsing, Hendriksz, and Kramers [70] obtained values by measuring the rate of absorption of $CO_2$ into short liquid jets of potassium, sodium, and lithium hydroxides. Their results are shown in Fig. 8.24. Their extrapolated value at $I = 0$ agrees within 15 percent with the line on Fig. 8.23.

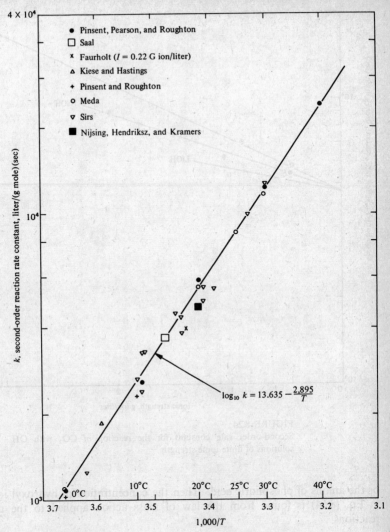

FIGURE 8.23
Second-order rate constant for the reaction of $CO_2$ with hydroxyl ion in aqueous solutions.

In general the total rate of reaction of dissolved $CO_2$ is found by adding the rates for the three parallel paths, giving

$$\text{Rate} = k_{H_2O} + k_B[B] + k_{OH^-}[OH^-][CO_2] \tag{8.78}$$

Often the reactive alkaline solution is buffered by the presence of several anions, as when carbonate–bicarbonate buffered solutions of sodium or potassium are used

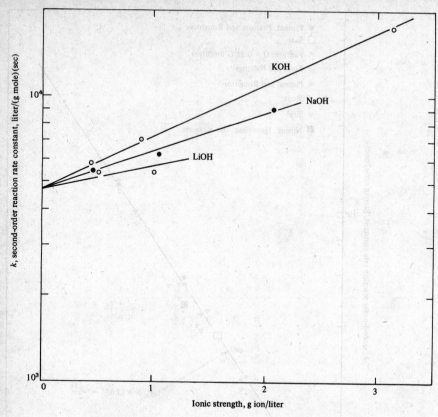

**FIGURE 8.24**
Second-order rate constant for the reaction of $CO_2$ with $OH^-$ in aqueous solutions of finite ionic strength.

or the anions of phosphoric acid. Then the concentration of hydroxyl ion for use in Eq. (8.64) is found from the law of mass action applied to the two rapid reactions

$$3 \quad HCO_3^- \; \rightleftharpoons \; H^+ + CO_3^{--} \qquad K_2$$

$$4 \quad H_2O \; \rightleftharpoons \; H^+ + OH^- \qquad K_W$$

$K_2$ is the second ionization constant for carbonic acid and $K_W$ is the ion product for water. Eliminating the concentration of $H^+$,

$$[OH^-] = \frac{K_W}{K_2} \frac{[CO_3^{--}]}{[HCO_3^-]}$$

at infinite dilution [45].

$$\log \frac{K_W}{K_2} = -\frac{1,568.60}{T} - 0.4105 + 0.00673T \tag{8.79}$$

giving the value of $2.16 \times 10^{-4}$ g ion/l at 25°C. The value of $k_{OH}(K_W/K_2)$ expected from the three constants at infinite dilution is $1.83$ s$^{-1}$ at 25°C. The equilibrium concentrations are affected by ionic strength, however. Data from Näsänen [68], obtained by adding inert KCl or NaCl to the solutions, were used by Roberts and Danckwerts [80] in combination with the values of $k_{OH^-}$ according to Fig. 8.23 to obtain Fig. 8.25. The figure gives the value by which the ratio $[CO_3^{--}]/[HCO_3^-]$ must be multiplied to obtain the term $k_{OH^-}[OH^-]$ in Eq. (8.78). According to Roberts and Danckwerts [80], the value of $k_{OH} - (K_W/K_2)$ is $0.86$ s$^{-1}$ at 25°C in solutions of ionic strength in the range from 1 g ion/l. Also shown on the figure are horizontal lines representing some values derived from absorption-rate data by Nijsing and Kramers [69]. The agreement is only approximate. There is a preference for potassium over sodium because higher concentrations of these compounds can be obtained before the solution becomes saturated and because the ionic strength effect on the rate and equilibrium constants is more favorable.

Rates of absorption depend on the solubility of the gas, a quantity that is affected by the concentrations of ions in the solution at the interface, as pointed out by van Krevelen and Hoftijzer [98, 99]. The Henry's-law coefficient for the solution is related to the value in water by the empirical equation

$$\log \frac{H}{H_0} = -hI \tag{8.80}$$

where $I$ is the ionic strength calculated from Eq. (8.77) and $h = h_+ + h_- + h_G$ where the $h$ values are taken from Tables 8.4 and 8.5.

EXAMPLE 8.15   Estimate the Henry's-law solubility coefficient of $CO_2$ in a solution having equal concentrations of carbonate and bicarbonate ions at 25°C, each concentration being 0.5 g ion/l, (a) for sodium and (b) for potassium cations.

**Table 8.4   CONSTANTS IN Eq. (8.80) FOR IONS**

| Cations | | Anions | |
|---|---|---|---|
| $H^+$ | $h_+ = 0.000$ l/g ion | $OH^-$ | $h_- = 0.066$ l/g ion |
| $N_a^+$ | 0.091 | $Cl^-$ | 0.021 |
| $K^+$ | 0.074 | $NO_3^-$ | −0.001 |
| $NH_4^+$ | 0.028 | $SO_4^{--}$ | 0.022 |
| $Mg^+$ | 0.051 | $Br^-$ | 0.012 |
| $Z_n$ | 0.048 | $CO_3^{--}$ | 0.021 |
| $Ca^{++}$ | 0.053 | $I^-$ | 0.005 |

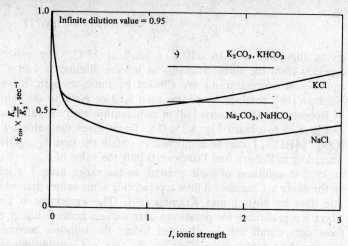

**FIGURE** 8.25

Effect of ionic strength on the value of $K_{OH} - K_w/K_2$ in carbonate-bicarbonate buffer solutions.

SOLUTION    The ionic strength of the solution is

$$I = \tfrac{1}{2}[(2^2)(0.5) + (1^2)(0.5) + (1^2)(1.5)] = 2.0$$

Note that the small concentrations of $OH^-$ and $H^+$ have a negligible effect.

(a) For sodium bicarbonate–carbonate,

$$\log \frac{H}{0.0339} = -(0.091 + 0.021 - 0.019)2.0$$

$$= -0.186$$

$$H = \frac{0.0339}{1.53} = 0.0221 \text{ g mole/(l)(atm)}$$

(It is assumed that the value of $h_-$ for $HCO_3^-$ is equal to that for $CO_3^{--}$.)

Table 8.5    CONSTANTS IN Eq. (8.80) FOR DISSOLVED GASES; VALUES OF $h_G$, 1/g ion*

| Gas | Temperature, °C | | | | |
|---|---|---|---|---|---|
| | 0 | 15 | 25 | 40 | 60 |
| $H_2$ | ...... | −0.008 | −0.002 | ...... | ...... |
| $O_2$ | ...... | 0.034 | 0.022 | ...... | ...... |
| $CO_2$ | −0.007 | −0.010 | −0.019 | −0.026 | −0.016 |
| $N_2O$ | ...... | 0.003 | 0.000 | ...... | ...... |
| $H_2S$ | ...... | ...... | −0.033 | ...... | ...... |
| $NH_3$ | ...... | ...... | −0.054 | ...... | ...... |
| $C_2H_2$ | ...... | ...... | −0.009 | ...... | ...... |
| $SO_2$ | ...... | ...... | −0.013 | ...... | ...... |

*See Mai and Babb [111] for solubilities of $CO_2$ and $H_2S$ in $Na_2CO_3/NaHCO_3$.

(b) For potassium bicarbonate–carbonate,

$$\log \frac{H}{0.0339} = -(0.074 + 0.021 - 0.019)2.0$$

$$= 0.152$$

$$H = \frac{0.0339}{1.42} = 0.0239 \text{ g mole/(l)(atm)}$$

Note that the solubility is greater in the solution containing potassium ions, one of the reasons for preferring it to sodium in industrial absorption processes. Note also that depletion of $CO_3^{--}$ near the interface, if it should occur because of slow diffusion of $HCO_3^-$ from a region of fast reaction, would raise the solubilities slightly.  /////

EXAMPLE 8.16  Assuming that the diffusivity of unreacted $CO_2$ in the buffer solutions of the previous problem is $1.3 \times 10^{-5}$ cm$^2$/s, estimate the rate of absorption per unit of interfacial partial pressure of $CO_2$ for the potassium and sodium solutions.

SOLUTION   From Fig. 8.24 the pseudo first-order reaction-rate constants at $I = 2.0$ g ion/l are 0.71 and 0.55 s$^{-1}$ for potassium and sodium solutions, respectively, based on the work of Nijsing, Hendriksz, and Kramers [70].

For the sodium buffer,

$$H\sqrt{kD} = 0.0221 \times 10^{-3} \sqrt{(1.3 \times 10^{-5})(0.55)}$$

$$= 5.9 \times 10^{-8} \text{ g mole/(cm}^2)(s)$$

For the potassium buffer,

$$H\sqrt{kD} = 0.0239 \times 10^{-3} \sqrt{(1.3 \times 10^{-5})(0.71)}$$

$$= 7.3 \times 10^{-8} \text{ g mole/(cm}^2)(s)$$

Because of the effect of $I$ on the rate constant, an 8 percent difference in the equilibrium solubility has become a 22 percent difference in the expected rate of absorption.  ////

**The rate of absorption of $CO_2$ into alkaline solutions**   Figure 8.26, taken from the work of Nijsing, Hendriksz, and Kramers [70], shows the results of their measurements of rates of absorption into aqueous solutions of KOH. The ordinate is the ratio of the observed rate to that expected in the same solution if the reaction were absent. The solubilities of free $CO_2$ in these hypothetical solutions were computed from Eq. (8.80), using $h = 0.113$ 1/g ion, and the theoretical rates were found from Eq. (8.47), using $D\mu^{0.85} = 1.68 \times 10^{-5}$ (cm$^2$/s)(Cp)$^{0.85}$ where $\mu$ is the viscosity of the solution in centipoise. The dimensionless abscissa is based on the calculated time of exposure of the surface of the liquid, the bulk concentration of $OH^-$, and the reaction-rate constants from Fig. 8.24. The parameter corresponds to $B_0/A_i$ in the theory of bimolecular reactions, as represented in Fig. 8.9.

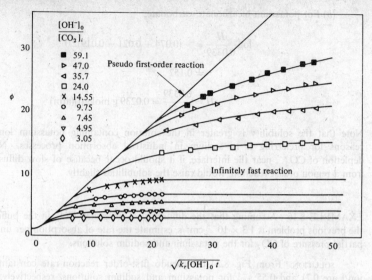

**FIGURE 8.26**
Reaction factor for absorption of $CO_2$ into KOH solutions.

Figures 8.26 and 8.9 closely resemble each other, giving qualitative support to the reality of the diffusion-plus-reaction theory. There is a clear transition from pseudo first-order conditions at high values of $[OH^-]_0/[CO_2]_i$ provided $k[OH^-]_0 t$ is not too great, i.e., if

$$\sqrt{kB_0 t} < \sqrt{\frac{D_A}{D_B}} \left(1 + \frac{D_B}{D_A}\frac{B_0}{A_i}\right) \tag{8.81}$$

(For Fig. 8.26, $D_{OH^-}/D_{CO_2^{--}} = 1.67$ is recommended [70].) At larger values of the abscissa, where the inequality is reversed, essentially instantaneous, diffusion-controlled reaction conditions occur and the ratio of the rates reaches a constant asymptotic limit. Figure 8.27 is a cross plot of Fig. 8.26 and shows the increase in the rate factor for constant-solution reactivity as $[OH^-]_0/[CO_2]_i$ is increased. It indicates that under conditions such that the chemical reaction is extremely rapid and diffusion of $CO_2$ and $OH^-$ to a thin reaction plane controls the rate,

$$\phi = 0.8 + 0.63\frac{[OH^-]_0}{[CO_2]_i} \tag{8.82}$$

The constants agree closely with the expected values according to Eq. (8.45). Figure 8.28 is another cross plot of the same data, the coordinate being chosen so that the ordinate is proportional to the rate of absorption if the hydroxyl ion concentration is fixed. The abscissa is proportional to the $CO_2$ driving force. The

[... text obscured ...] for this reason of the order conditions at low rates. Where [...] concentration is essentially constant throughout the liquid, to infinitely [...] conditions at this limit a [...] Here the rate is not proportional to [...] of the concentrations, $[OH^-]_0$.

[...] dioxide in aqueous solutions of sodium and potassium carbonate [...] the solution is used widely in industrially important gas absorption [...] has been studied by Nijsing and Kramers [60] [...] writers and Kramers [26] and Roberts and Danckwerts [60]. Table 8.6 lists [...] of their data. Note that because of the low hydroxyl-ion concentration in [...] solutions, the apparent rate constant is only about 0.1 to 1 s, as expected [...] indicating that the reaction factors for such solutions will be only moderately [...] than unity. It is expected that the apparent rate constant in such solutions [...]

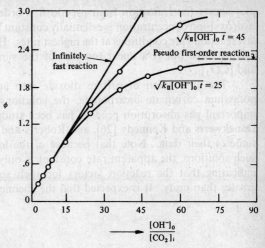

FIGURE 8.27
Effect of reagent concentrations on the reaction factor according to Nijsing, Hendriksz, and Kramers.

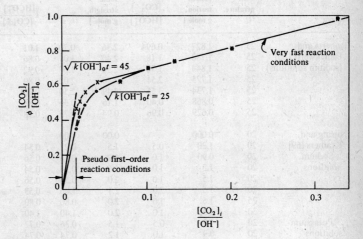

FIGURE 8.28
Effect of interface $CO_2$ concentration on the rate of absorption of $CO_2$ into KOH solution.

plot shows the transition from pseudo first-order conditions at low rates, where hydroxyl-ion concentration is essentially constant throughout the liquid, to infinitely fast reaction-rate conditions at the highest rates. Here the rate is not proportional to driving force but is a linear function of the sum of the concentrations, $[OH^-]_0$ and $[CO_2]_i$.

The absorption of carbon dioxide into aqueous solutions of sodium and potassium carbonate–bicarbonate, the solutions used widely in an industrially important gas absorption process, has been studied by Nijsing and Kramers [69], Danckwerts and Kennedy [26], and Roberts and Danckwerts [80]. Table 8.6 lists some of their data. Note that because of the low hydroxyl-ion concentration in such solutions, the apparent rate constant is only about 0.4 to 3 $s^{-1}$, as expected, indicating that the reaction factors for such solutions will be only moderately greater than unity. It is expected that the chemical-rate constant in such solutions

**Table 8.6** SUMMARY OF DATA ON THE RATE OF ABSORPTION OF CARBON DIOXIDE INTO CARBONATE–BICARBONATE BUFFER SOLUTIONS

| Source | Temperature, °C | Metal ion concentration, g mole/l | $\dfrac{[CO_3^-]}{[HCO_3^-]}$ | Ionic strength, g mole/l | $k$, $s^{-1}$ | $k\dfrac{[HCO_3^-]}{[CO_3^-]}$, $s^{-1}$ | $10^8(C^*/p)D^{1/2}$, g mole/(cm²)(atm)(s) |
|---|---|---|---|---|---|---|---|
| Roberts and | 25 | 1.827 | 0.694 | 2.36 | 0.70 | 1.01 | 7.3 |
| Danckwerts [80] | 25 | 1.780 | 1.18 | 2.40 | 1.01 | 0.86 | 7.3 |
| (sodium solutions) | 25 | 1.844 | 1.74 | 2.56 | 1.60 | 0.92 | 7.1 |
| | 25 | 1.775 | 2.51 | 2.50 | 2.02 | 0.80 | 7.4 |
| | 25 | 1.734 | 3.55 | 2.49 | 2.50 | 0.70 | 7.6 |
| | 25 | 0.868 | 0.40 | 1.16 | 0.35 | 0.88 | 10.05 |
| | 25 | 0.629 | 0.96 | 0.94 | 0.75 | 0.78 | 10.75 |
| | | | | | | | |
| Nijsing and | 20 | 0.000 | ... | 0.00 | 0.00 | ... | 15.83 |
| Kramers [69] | 20 | 1.20 | 0.5 | 1.5 | 0.27 | 0.54 | 9.87 |
| Sodium | 20 | 0.90 | 1.0 | 1.2 | 0.56 | 0.56 | 11.12 |
| solutions | 20 | 1.3 | 1.0 | 1.6 | 0.54 | 0.54 | |
| | 20 | 1.5 | 1.0 | 2.0 | 0.56 | 0.56 | 10.12 |
| | 20 | 1.5 | 2.0 | 2.1 | 1.06 | 0.53 | 8.50 |
| | 25 | 1.5 | 1.0 | 2.0 | 0.80 | 0.80 | 8.40 |
| | 30 | 1.5 | 1.0 | 2.0 | 1.40 | 1.40 | 8.05 |
| Potassium | 20 | 1.2 | 0.5 | 1.5 | 0.36 | 0.72 | 11.63 |
| solutions | 20 | 0.9 | 1.0 | 1.2 | 0.74 | 0.74 | 12.67 |
| | 20 | 1.2 | 1.0 | 1.6 | 0.74 | 0.74 | 11.66 |
| | 20 | 1.5 | 2.0 | 2.1 | 1.43 | 0.71 | 10.84 |
| | 20 | 1.5 | 1.0 | 2.0 | 0.72 | 0.72 | 10.84 |
| | | | | | | | |
| Danckwerts and | 25 | 1.795 | 0.601 | 2.29 | 0.70 | 1.16 | 6.7 |
| Kennedy [26] | 25 | 1.775 | 1.158 | 2.40 | 1.51 | 1.30 | 6.8 |
| (sodium solutions) | 25 | 1.760 | 1.70 | 2.44 | 1.45 | 0.85 | 6.8 |
| | 25 | 1.765 | 2.47 | 2.50 | 2.25 | 0.91 | 6.7 |
| | 25 | 1.755 | 3.40 | 2.52 | 2.55 | 0.75 | 6.8 |
| | 25 | 1.735 | 4.45 | 2.52 | 3.74 | 0.84 | 7.6 |

will be proportional to the carbonate:bicarbonate molar ratio, as indicated previously. The table shows that the numerical values of $k[HCO_3^-]/[CO_3^{--}]$ have average values of 0.55, 0.90, and 1.40 s$^{-1}$ for sodium buffer solutions at temperatures equal to 20, 25, and 30°C, respectively; the average value for potassium solutions is 0.73 at 20°C. The values in the last column of the table vary primarily because of the reduced solubility of carbon dioxide in the solutions of high ionic strength. The observed reduction factors agree closely with the estimates from Eq. (8.80), according to van Krevelen and Hoftijzer [99]. The reduction is greater for sodium ions than for potassium ions.

## 8.15   THE OXIDATION OF SULFITE ION BY DISSOLVED OXYGEN

Measurements of the rate of absorption of oxygen into aqueous solutions of sodium sulfite have been used widely for the determination of the mass-transfer characteristics of stirred-tank absorbers, as outlined in Sec. 10.6. The presence of the reducing agent and the low solubility of unreacted oxygen in the solution make it possible to conduct a convenient batch experiment over a long time. Two types of experiments are carried out: (1) those used to determine the physical mass-transfer coefficient per unit volume, $k_L^\circ a$, in order that the device can be characterized for use with an entirely different system, and (2) those for determination of the interfacial area of the bubbles in the dispersion.

For the former it is customary to assume that because of the irreversible reaction, very little dissolved free oxygen is present in the bulk of the liquid and the driving force is proportional to the oxygen partial pressure. The reaction rate in the whole volume of the liquid must be sufficient, even at the low concentration, to keep up with the rate of absorption. If the measurements are to yield $k_L^\circ a$, however, the homogeneous reaction rate must not be so great that the reaction factor $\phi$ is appreciably greater than unity. These are the conditions within a narrow range of the first-order rate constant on Fig. 8.1 at which the relative absorption rate has a nearly horizontal intermediate range. Depending on the values of $k_L^\circ$, $\theta$, and $a$, the range of $k_1$ does not extend over more than two powers of 10. In order to interpret such absorption rate data correctly, it is therefore necessary to have adequate information about the rate of the homogeneous reaction.

For measurements of the interfacial area, on the other hand, it is necessary to conduct the experiments under conditions such that the rate of adsorption per unit of surface can be expressed solely in terms of the chemical-rate constant and the diffusivity, as in Eq. (8.13). The rate must be independent of the time of exposure of the interface or of any aspect of the fluid motion around the bubbles. This is the condition existing at the upper right of Figs. 8.1 and 8.3, showing that the reactivity of the solution must be sufficiently great that the diffusing oxygen molecules disappear by reaction before they have moved very far from the interface.

It might be assumed that it is not necessary that the reaction-rate constant be known for area measurements if the rate of absorption can be measured for the same solution using an apparatus, such as a wetted-wall column, for which the surface area is known accurately. Then a direct comparison of Na determined for the gas dispersion in the stirred tank with N determined for an absorber of known area clearly permits $a$ to be computed by division. Such direct comparisons are only approximate, however, unless the conditions just stated are realized, for it may be that the values of $k_L^o$ for the two surfaces are so different that fluid motion has an effect at one surface and not at the other. Either it must be proved experimentally that the rate of absorption is independent of $t$ or $s$ for both surfaces or the homogeneous rate expression must be known so that an appropriate theory can be applied.

Unfortunately, the reaction with sulfite may not be first-order with respect to dissolved oxygen in the presence of the catalysts commonly used: copper [9] or cobalt ions. (Cobalt is the more effective of the two.) At concentrations of sulfite less than 0.06 molar, Chen and Barron [19] have reported data from rapid-mixing studies from which the rate of the homogeneous reaction can be represented by the equation

$$r = 6.6[SO_3^{--}]^{3/2} [Co]^{1/2} \qquad \text{g mole/(cm}^3\text{)(s)} \tag{8.83}$$

where the concentrations are in gram moles per liter and [Co] refers to all the cobalt ions present, regardless of their state of oxidation. Note that the rate is zero-order with respect to oxygen. (The rate is of course zero when the oxygen concentration is zero.) The reaction mechanism apparently involves the generation of small amounts of the free-radical ion, $\cdot SO_5^-$, in solution by steps that are initiated by an oxidized form of cobalt but are independent of oxygen concentration provided it is not zero. Sulfite ion is attacked by the free radical, creating a chain reaction. Under these conditions the steady-state rate of mass transfer, reached after prolonged exposure of the interface, might be expected to be represented by Eq. (8.36), giving

$$N_{Ox} = \sqrt{2D_{Ox} \cdot 6.6[O_2]_i [SO_3^{--}]_i^{3/2} [Co]^{1/2}} \tag{8.84}$$

Note that the absorption rate should be proportional only to the square root of oxygen partial pressure when this lower limit to $N_{Ox}$ has been reached.

Such results are not observed; however de Waal and Okeson [33] and Sawiki and Barron [85] using wetted-wall columns found under certain conditions that the rate of absorption was proportional to the first power of $p_{O_2}$, suggesting that the homogeneous rate may be first-order with respect to oxygen. Note, however, that such observations do not prove that the homogeneous reaction is first-order with respect to oxygen unless the reaction factor is considerably greater than unity, for the purely diffusion-controlled reaction is always first-order. Similar conclusions were reached, however, by Yagi and Inoue [110], who used a polarograph to follow the disappearance of dissolved oxygen in dilute solutions of sulfite and sulfate ions and cobalt catalyst.

On the other hand, several authors [61, 78, 81] have observed that the rate of oxygen absorption is proportional to the 3/2 power of the oxygen partial pressure. They infer from this that the rate of the homogeneous reaction is proportional to the square of dissolved oxygen concentration, as is implied by Eq. (8.35). When applied to a reaction that is second-order with respect to the molecules diffusing inward from the interface, the equation becomes

$$N_{Ox} = C_{i,Ox}^{3/2}\sqrt{\tfrac{2}{3}D_{Ox}k} \qquad (8.85)$$

where $k_{II}(= r/[O_2]^2)$ may also depend on the concentrations of other substances such as $SO_3^{--}$, catalyst, and hydrogen ion. Linek and Mayrhoferova [61] measured the rates of oxygen absorption through the upper surface of the pool of liquid in a stirred vessel to determine $N_{Ox}$ and concluded that Eq. (8.85) should be used when $C_{i,Ox}$ is smaller than about $6 \times 10^{-4}$ g mole/l, corresponding to about the concentration of oxygen in pure water at equilibrium with pure $O_2$ at 0.6 atm and room temperature. For greater oxygen concentrations the reaction rate was apparently first-ofder. The rate appears to be proportional to the first power of the cobalt catalyst concentration [33, 62, 78, 110] and the activation energy is about 12 kcal/g mole [33, 61, 85]. The rate also depends on the pH of the solution [33, 62, 110], which tends to fall as $SO_3^{--}$ is converted to $SO_4^{--}$. Trace impurities can have considerable effect. To obtain reproducible results in a series of experiments, it is desirable to use only one stock of reagents and to protect the solutions carefully from contaminants [33].

Thus, the mechanism and the chemical kinetic expression for the reaction rate is not known very well. Nevertheless, measurements of the bubble area are successful, as shown by de Waal and Okeson [33], Westerterp, van Dierendonck, and Kraa [109], and Linek [62]. Operating under conditions in which the reaction was apparently first-order in oxygen, based on wetted-wall column tests, de Waal and Okeson [33] applied Danckwerts' formula for the rate of absorption at the bubble surface of specific area $a$ in an agitated vessel:

$$N_{Ox} a = C_{i,Ox}\, a\sqrt{D_{Ox}(k_I + s)}$$

$$\left(\frac{N_{Ox}\, a}{C_{i,Ox}}\right)^2 = a^2 D_{Ox}(k_I + s) \qquad (8.86)$$

The rate constant $k_I$, calculated from the wetted-wall column measurements, was varied by changing the amount of catalyst. Figure 8.29 is taken from their paper. It shows how $a$ and $s$ can be found from the slope and the intercept, respectively. Moreover, measurements of the increase in liquid depth caused by the presence of the bubbles yield values of $\varepsilon$; values of bubble diameter and population can be computed from $a$ and $\varepsilon$, as shown on the figure. This is an example of the use of the "chemical method" for the determination of interfacial area, a method that has also been used for packed gas absorbers [27], for perforated plates [34], and for liquid-liquid extractors [67].

**EXAMPLE 8.17** Oxygen is absorbed from air into aqueous solutions of sodium sulfite at 30°C and 1 atm. The solution contains cobalt ion as a homogeneous catalyst. The absorber is an agitated vessel having dimensions and stirrer speed such that $k_L^\circ = 0.04$ cm/s and

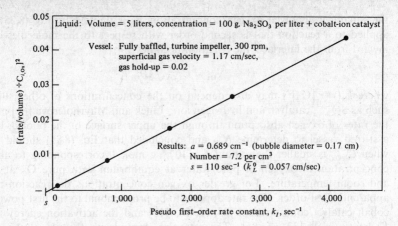

FIGURE 8.29
Evaluation of surface area and surface renewal rate in an agitated gas-liquid reactor (from de Waal and Okeson [33]).

$a = 0.1$ cm²/cm³. Sulfite concentration equal to 0.8 g mole/l is used and the pH is adjusted to 8.0.

(a) Find the catalyst concentrations such that the oxidation reaction throughout the bulk of the solution will cause the free oxygen concentration there to be only 1 percent of the concentration at the interface, at equilibrium with the bulk gas.

(b) Compare the steady-state rates of mass transfer per unit of surface with the rates for no reaction effect near the interface to determine whether essentially physical rates of solution of oxygen will be observed in tests of the absorber.

(c) What catalyst concentration is required to make the reaction factor equal to 10, such that the rate per unit of surface is essentially independent of the fluid motion around the gas bubbles?

Data [78]: The solubility of oxygen in water at 30°C is $H = 1.16 \times 10^{-6}$ g mole/(cm³) (atm); the value in the 0.8-molar sodium sulfite is $0.58 \times 10^{-6}$ [29]. The diffusivity of oxygen in the concentrated solution is $1.2 \times 10^{-5}$ cm²/s. The rate of the homogeneous reaction is $r = 2.5 \times 10^{16} [O_2]^2 [Co]$, where oxygen and cobalt concentrations are in gram moles per cubic centimeter.

SOLUTION (a) The steady-state material balance for a liquid volume $V$ in the absorber is

$$N_{Ox} aV = rV$$

The concentration of dissolved oxygen at the interface, neglecting the gas-phase resistance, is $0.21H = (0.21)(0.58 \times 10^{-6}) = 0.122 \times 10^{-6}$ g mole/cm³; that in the bulk liquid is $0.122 \times 10^{-8}$ g mole/cm³; the driving force for physical diffusion of oxygen across the liquid film is $(0.99) \times (0.21H)$; and the rate of physical mass transfer into the bulk liquid is $(0.04)(0.99)(0.21H) = 0.48 \times 10^{-8}$ g mole/(cm²)(s). The bulk reaction rate per unit volume of liquid must be one-tenth of this.

$$0.48 \times 10^{-9} = r = 2.5 \times 10^{16}(0.12 \times 10^{-6} \times 0.01)^2 [Co]$$

from which

$$[Co] = 1.33 \times 10^{-10} \text{ g mole/cm}^3 = 1.33 \times 10^{-7} \text{ g mole/l}$$

This value is smaller than the smallest concentration used by Reith, approximately $10^{-5}$ molar, indicating that the reaction is fully capable of keeping the bulk concentration of oxygen at a very low level.

(b) Using the low catalyst concentration just computed, the steady-state rate of solution of oxygen, according to Eq. (8.34), is

$$N_{Ox} = \sqrt{\frac{2}{2+1}(1.2 \times 10^{-5})(2.5 \times 10^{16})(0.12 \times 10^{-6})^3(1.33 \times 10^{-10})}$$

$$= 2.14 \times 10^{-10} \text{ g mole/(cm}^2)(s)$$

This lower limiting rate is much smaller than the purely physical rate, equal to $0.48 \times 10^{-8}$ g mole/(cm²)(s), indicating that at the low catalyst concentration a measured rate of solution in an agitated absorber would give very nearly the physical rate of solution.

At the lowest catalyst concentration used by Reith and Beek [78], the steady-state rate of combined diffusion and reaction near the interface would be $N_{Ox} = 1.86 \times 10^{-9}$ g mole/(cm²)(s), still only about one-third the rate of physical diffusion.

The comparison indicates that there is an appreciable range of catalyst concentrations over which it is possible to use the reaction to keep the dissolved oxygen at a comparative low level in a sulfite system without having a large effect on the process of physical mass transfer near the interface. The comparisons depend heavily on the assumed value of $k_L^\circ$, however, showing the importance of knowing $k_L^\circ$ and $a$ separately for such absorbers.

(c) If the reaction factor is to be equal to 10, the rate of absorption according to Eq. (8.83) must be nearly as great as $0.48 \times 10^{-7}$ g mole/(cm²)(s). Thus, using Eq. (8.35) to solve for the required catalyst concentration,

$$[Co] = \frac{(3)(0.48 \times 10^{-7})^2}{(2)(1.2 \times 10^{-5})(2.5 \times 10^{16})(0.122 \times 10^{-6})^3} = 0.63 \times 10^{-5} \text{ g mole/cm}^3$$

or $0.63 \times 10^{-2}$ g mole/l. This concentration is impractically high because the solubility limit for the cobalt compounds is apparently reached at somewhat lower values, especially at low sulfite-ion concentrations [33]. Thus, measurements of interfacial area using the chemical system described above can not be expected to be satisfactory.

An acceptable solution is provided, however, by using pure oxygen gas at 1 atm rather than air. Multiplying the cobalt concentration just computed by the cube of the ratio of gas partial pressures, we find

$$[Co] = 0.63 \times 10^{-2}\left(\frac{0.21}{1.00}\right)^3 = 0.58 \times 10^{-4} \text{ g mole/l}$$

This value is within the range which has been used successfully in area measurements. For these reaction conditions we can estimate the depletion of sulfite-ion concentration at the interface by equating an estimate of its rate of diffusion to half the inward flux of oxygen:

$$(0.04)([SO_3^{--}] - [SO_3^{--}]_i) = 0.24 \times 10^{-7}$$

from which

$$[SO_3^{--}] = [SO_3^{--}]_i = 0.6 \times 10^{-7} \text{ g mole/cm}^3 \quad \text{or} \quad 0.6 \times 10^{-4} \text{ g mole/l}$$

Thus the difference between interface and bulk concentrations is too small to have an appreciable effect on the rate of oxygen transfer.

////

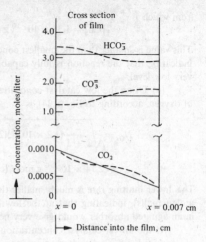

**FIGURE 8.30**
Concentration profiles in a bicarbonate–carbonate film across which a carbon dioxide concentration difference is maintained (from Ward and Robb [104]).

## 8.16 CARRIER-FACILITATED MASS TRANSFER THROUGH MEMBRANES

We have seen that the effect of a fast chemical reaction occurring near a gas-liquid interface is to increase the rate of absorption of a reactive gas. As indicated in Sec. 8.11, when the dissolved molecules, $A$, form a nonvolatile reaction product, $B$, by a reversible chemical reaction, both $A$ and $B$ can diffuse together through the liquid. If the reaction is very fast, the result is an increase in the rate of uptake of $A$ by the enhancement factor, $\phi = 1 + K$. Similarly, if carbon dioxide molecules, dissolving in an aqueous buffer solution, can undergo reaction to form $HCO_3^-$, a nonvolatile form of $CO_2$, by a reversible reaction we may think of the bicarbonate ion as a "carrier" for $CO_2$ in the solution. If its equilibrium concentration is large, as in solutions of high pH, the large difference in its concentration at two sides of a liquid film may create a large diffusion driving force across the film, even though the concentrations of dissolved $CO_2$ may be small everywhere in the film.

Figure 8.30 is a diagram [104] of a thin liquid film across which dissolved $CO_2$ is diffusing. The concentration of dissolved $CO_2$ at the left side is calculated to be at equilibrium with gas containing 34 mm partial pressure of $CO_2$; that on the right is lower because the gas there contains only 4 mm partial pressure; the driving force for free, dissolved $CO_2$ is only about $10^{-3}$ g mole/l because of the low solubility—not enough to produce a large diffusion flux. Because the water in the film is nearly saturated with $CeHCO_3$ and $Ce_2CO_3$, the concentrations of the ions are on the order of 1 g ion/l. If the reaction

$$CO_2 + CO_3^{--} + H_2O \;\rightleftharpoons\; 2HCO_3^-$$

is capable of forming large amounts of the carrier ion at equilibrium, there can be a significant difference in its concentration at left and right, as shown by the dashed lines at the top of the figure. The horizontal solid lines show the distribution of $HCO_3^-$ and $CO_3^{--}$ if no reaction occurs at all. If the reaction goes toward the right at the side where $CO_2$ enters and toward the left at the side where $CO_2$ leaves, nonvolatile $HCO_3^-$ will be created in the left and will be destroyed on the right as shown by its dashed line. $CO_3^{--}$, on the other hand, will diffuse in the opposite direction. However, because two $HCO_3^-$ are made from one $CO_3^{--}$, the net transport of $CO_2$ in carrier form is toward the right.

Ward and Robb [104] calculated the enhanced net-transfer rate for $CO_2$, assuming local equilibrium and using the method of Olander [71]. Their numerical results are the basis of Fig. 8.30. Their computed $CO_2$ transfer rate was $3.8 \times 10^{-7}$ g mole/(s)(cm$^2$), an increase by a factor of $\Phi = 29$ over the rate expected for $CO_2$ diffusing through pure water. The calculated permeability was $2.3 \times 10^{-4}$ (std. cm$^3$ $CO_2$)/(s)(cm$^2$)(mm Hg/cm). On the other hand, the rate of transfer of chemically inert oxygen through the same film was calculated to be only one-fortieth the value for oxygen in water, owing to the lowered solubility of $O_2$ in the concentrated salt solution, leading to the possibility of a method of separating $CO_2$ and $O_2$ in the life-support systems used aboard spacecraft. Ward and Robb measured the $CO_2$ permeability of a 3-mil-thick film of cellulose acetate containing $CeHCO_3/CO_3$. They found a value of $0.75 \times 10^{-6}$, much smaller than the value expected from the equilibrium theory and even smaller than the value for pure water. The result shows the critical importance of the rates of the reactions by which free $CO_2$ is converted into a carrier ion. By adding 0.5-$N$ sodium arsenite catalyst to the solution in the film, the experimental permeability was increased to $2.1 \times 10^{-6}$, much smaller than the value if chemical equilibrium occurred but 4,100 times the measured value for $O_2$.

Quinn and Otto [76] have made similar calculations for the passage of $CO_2$ into the surface of the ocean. They found that although the ionic equilibria are capable of causing facilitated transport, expected rates of reaction are too small to give facilitated transport unless some of the substances dissolved in sea water are homogeneous catalysts.

Biologists have been interested in a somewhat similar problem of facilitated transport, as discussed by Smith, Meldon, and Colton [93] and others. The problem concerns the possible increase in the rate of transfer of oxygen across living membranes owing to the reaction $O_2 +$ hemoglobin $\rightleftharpoons$ oxyhemoglobin as in Example 8.7. Using independently measured values of the rate and equilibrium constant for the reaction, Kutchai, Jacquez, and Mather [60] computed that the steady-state flux of oxygen across films about 1 $\mu$m thick may be about 30 percent greater than if the reaction did not occur.

From a theoretical point of view the increase in transport rate has been studied by Aris and Keller [3], Bailey and Luss [8], and Luss, Bailey, and Sharma [63], who were interested in the generation of carriers by enzyme-catalyzed reactions at the two surfaces of a film, and by Friedlander and Keller [40] and Smith, Meldon, and Colton [93] for homogeneous reactions inside a film. For the latter case the approach of $\phi$ toward its equilibrium value depends very much on the thickness of the film. For thick films the concentrations of reactants and products may be everywhere very nearly equal to their equilibrium values, making $\phi$ large.

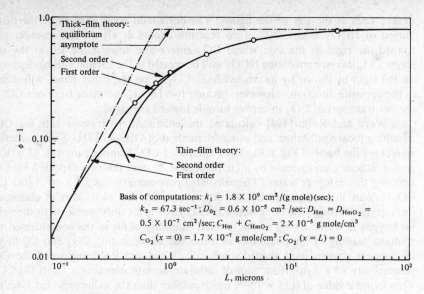

**FIGURE 8.31**
Calculated reaction factors for diffusion of oxygen through a film containing hemoglobin (from Smith, Meldon, and Colton [93]).

(Because of the large film thickness, the flux may be small, nevertheless.) For a thin film, however, only the center of the film may approach equilibrium concentrations, the speed of the forward and reverse reactions having a predominating influence near the two surfaces. For the reaction

$$A + B \underset{k_2}{\overset{k_1}{\rightleftharpoons}} AB$$

with $B$ and $AB$ representing the nonvolatile dissolved substances and $A$ the unreacted dissolved gas, Smith et al. [93] were able to show that for "thin" films the enhancement factor is given by

$$\phi = 1 + \frac{1}{12}\frac{k_2 C_B^* L^2}{D_A} \tag{8.87}$$

plus higher-order terms proportional to $L^4$. In Eq. (8.87) $L$ is the film thickness and $C_B^*$ is the concentration of $B$ at equilibrium with average of the two surface concentrations of $A$ and with the constant sum of the concentrations of $B$ and $AB$. Films are thin when $\phi \approx 1$.

For "thick" films, however, the series approximation leading to Eq. (8.87) are invalid and the method of matched asymptotic expansions in inverse powers of $k$ is needed. The results are complex, but a sample of them is shown in Fig. 8.31, which applies to the oxygen-hemoglobin system. The figure shows the results of the numerical calculations of Kutchai et al. [60], indicating that the series expansions and the general conclusions of Smith et al. are true.

## 8.17 DIFFUSION AND REACTION IN POROUS CATALYSTS

Solid catalysts are employed widely in the chemical and petroleum industries to promote many important chemical reactions. Porous catalysts are preferred since they can provide an enormous surface in a very small volume (up to several hundred square meters per gram). The porous materials are usually manufactured from powders, and are available as beads, pellets, and extrudates, commonly $\frac{1}{16}$ to $\frac{3}{8}$ in in diameter. Reactants diffuse into the porous structure, reaction takes place, and the products formed diffuse back out to the ambient fluid.

Diffusion into the pores involves a decrease in the concentration of the diffusing reactants, and the concentration effective in promoting the chemical reaction at the active sites is everywhere less than if diffusion were not involved. The catalyst, therefore, is less effective than if all the surface were in contact with the reactants at the concentrations maintained in the external or ambient fluid. This loss of catalyst effectiveness due to diffusion has been studied extensively, and is the subject of a very large literature.

The mechanism of diffusion and reaction within a porous bead is quite analogous to absorption and reaction in a liquid film, as discussed in the earlier sections of this chapter. The reactants must first overcome the resistance to diffusion from ambient fluid to the bead surface and diffuse into the porous structure with reaction occurring all along the diffusion path. Diffusion within the pores may be ordinary molecular diffusion, Knudsen diffusion, surface diffusion, or a combination of all three. The nature of these diffusion modes has been described in Chap. 2.

If the chemical reaction is slow, as in the case of a poor catalyst, the diffusion is accomplished with a small concentration gradient, and the concentration throughout is nearly the same as at the bead surface. Fast reactions, however, may take place in the pores very near the bead surface, and the internal pore surfaces contribute little. It is usual, therefore, for the effectiveness of a poor catalyst to be high and that of an excellent catalyst to be low. Here the standard of comparison in describing effectiveness is the chemical rate which would be attained if all the internal pore surface were in contact with reactants at the concentrations prevailing at the bead surface.

Consider a simple irreversible surface reaction involving a single reactant, the rate being expressed as $k_s c^m$ gram moles per second per square centimeter of surface. The catalyst is in the form of a porous spherical bead of radius $r_0$, and the effective diffusion coefficient for diffusion of reactant into the bead is $D_P$. A reactant balance on a differential shell at radius $r$ then leads to the differential equation

$$\frac{d^2 c}{dr^2} + \frac{2}{r}\frac{dc}{dr} = \frac{S_v k_s c^m}{D_P} \tag{8.88}$$

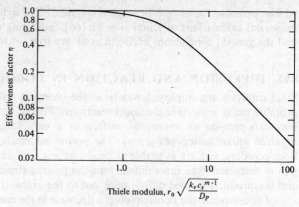

$$\text{Thiele modulus, } r_0 \sqrt{\frac{k_v c_s^{\,m-1}}{D_P}}$$

**FIGURE 8.32**
Effectiveness factor for spherical catalyst pellets.

where $S_v$ represents the pore surface per unit volume of the bead. This can be solved (the derivation is given in detail in [117]) to give

$$\text{Reaction rate} = 4\psi\pi r_0 D_P c_s \left[\frac{1}{\tanh\psi} - \frac{1}{\psi}\right] \tag{8.89}$$

where $c_s$ is the reactant concentration at the bead surface, and

$$\psi = r_0 \sqrt{\frac{S_v k_s c_s^{\,m-1}}{D_P}} = r_0 \sqrt{\frac{k_v c_s^{\,m-1}}{D_P}} \tag{8.90}$$

Equations (8.89) and (8.90) express the rate as gram moles per second for the single bead of radius $r_0$; $k_s$ is the intrinsic rate constant based on pore surface area and $k_v$ is per unit of gross volume of the bead. The quantity $\psi$ is known as the Thiele modulus, named after E. W. Thiele [113], who, along with Zeldowitsch [114] and Damköhler [115], pioneered in this analysis.

If the reactant concentration were $c_s$ throughout the bead, the rate would be $(\frac{4}{3}\pi)r_0^3 k_v c_s$. The actual rate as a fraction of this the limiting rate is known as the "effectiveness factor," represented by $\eta$:

$$\eta = \frac{3}{\psi}\left[\frac{1}{\tanh\psi} - \frac{1}{\psi}\right] \tag{8.91}$$

Figure 8.32 shows the nature of this relation.

The foregoing outlines the analysis for one particular case: a porous sphere, constant $D_P$ and $k_v$, and a surface reaction described by power-law kinetics, with a single reactant. Many other cases have been analyzed: various geometries, reaction mechanisms, and reactions described by Langmuir-Hinshelwood kinetics. Heat effects require allowance for the temperature gradients within the catalyst and for the resulting variation in the reaction-rate constant along the diffusion path. In the case of exothermic reactions, the interior of the porous mass becomes hotter than the ambient fluid, and the effectiveness factor, based on conditions at the sphere surface, may be greater than unity.

The effectiveness factor $\eta$ is a measure of the extent to which diffusion reduces the reaction rate within the porous catalyst. If the catalyst is highly active, $\eta$ can be increased by reducing the size of the particles, but this has the practical disadvantage of increasing the pressure drop for fluid flow in packed beds of catalyst particles.

Design optimization requires knowledge of the manner in which $\eta$ varies with $r_0$ for each catalyst. Laboratory tests provide data on reaction rate per unit of bed volume, but not on $k_v$. However, since the molar reaction rate per unit of bed volume is $\eta k_v (1 - \varepsilon) c_s{}^m$, it follows that $\Phi = \psi^2 \eta$ is given by

$$\Phi \equiv \psi^2 \eta = \frac{r_0{}^2}{D_P c_s} \left( - \frac{1}{V_R(1 - \varepsilon)} \frac{dn}{dt} \right) \tag{8.92}$$

where $-(dn/dt)$ is the molar conversion rate in the bed volume $V_R$, and $\varepsilon$ is the void volume fraction in the bed. All the quantities on the right with the exception of $D_P$ are obtained experimentally in the course of a catalyst test; $D_P$ must be measured separately or estimated by the methods of Chap. 2. Figure 8.32 can evidently be replotted as $\eta$ vs. $\Phi = \psi^2 \eta$, and $\eta$ determined from the test data without knowledge of $k_v$.

As Weisz [116] has noted, the curves of $\eta$ vs. $\Phi$ fall in a surprisingly narrow band for various catalyst geometries and reaction orders. The effectiveness factor $\eta$ is essentially unity for $\Phi < 1.0$ but starts to decrease appreciably as $\Phi$ becomes larger, reaching 0.1 at $\Phi$ of 8 to 15. The resistance of diffusion in the pores evidently becomes significant as $\Phi$ becomes greater than about 1.0. If $\eta$ is near unity, little increase in conversion can be accomplished by reducing catalyst particle size.

EXAMPLE 8.18  A hydrotreater is operated at 714°F and 1,800 psia to remove sulfur (as $H_2S$) by catalytic hydrogenation of an oil. The liquid hourly space velocity is 1.2 [volume of cold liquid feed/(h)(volume of bed)], and the hydrogen consumption is 800 SCF per barrel of feed.[1] The cobalt-molybdate catalyst is in the form of pellets equivalent to spheres having a radius of 1.4 mm. The void fraction of the bed is 0.38 and the pore volume of the pellets is 36 percent of the pellet volume. The flowing hydrocarbon is largely liquid, so hydrogen must diffuse into liquid-filled pores. Assuming the reactions to be irreversible and first-order in dissolved hydrogen, estimate the effectiveness factor for the catalyst pellets.

It is calculated that the concentration of hydrogen in the liquid at the pellet surface is 0.00069 g mole/cm$^3$, and the diffusion coefficient for hydrogen in the liquid is $37 \times 10^{-5}$ cm$^2$/s.

---

[1] *Note:* 1 bbl = 42 gal = 5.61 ft$^3$ = 159,000 cm$^3$; SCF refers to gas at 60°F and 1.0 atm.

SOLUTION  The reaction rate is

$$\frac{800}{359} \frac{492}{520} \frac{454 \times 1.2}{159,000 \times 3600} = 2.01 \times 10^{-6} \text{ g moles } H_2/(s)(cm^3 \text{ bed})$$

Assuming a tortuosity factor of 3 (see Chap. 2), $D_P$ is $37 \times 10^{-5} \times 0.36/3 = 4.44 \times 10^{-5}$ cm$^2$/s. Substitution in Eq. (8.92) gives

$$\Phi = \phi^2 \eta = \frac{0.14^2}{4.44 \times 10^{-5} \times 0.00069} \left( \frac{2.01 \times 10^{-6}}{1 - 0.38} \right) = 2.07$$

whence from Fig. 8.31, $\eta$ is 0.85.                                         ////

# REFERENCES

1  ABEL, E., and H. SCHMIDT: *Z. Phys. Chem.*, **143A:** 279 (1928).

2  ANDREW, S. P. S., and D. HANSON: *Chem. Eng. Sci.*, **14:** 105 (1961).

3  ARIS, R., and K. H. KELLER: *Proc. Nat. Acad. Sci.*, **69:** 777 (1972).

4. ASTARITA, G.: *Chem. Eng. Sci.*, **17:** 708 (1962).

5  ASTARITA, G., and G. MARRUCCI: *Ind. Eng. Chem. Fundam.*, **2:** 4 (1963).

6  ASTARITA, G., and F. GIOIA: *Ind. Eng. Chem. Fundam.*, **4:** 317 (1965).

7  ASTARITA, G.: "Mass Transfer with Chemical Reaction," Elsevier, Amsterdam, 1966.

8  BAILEY, J. E., and D. LUSS: *Proc. Nat. Acad. Sci.*, **69:** 1460 (1972).

9  BARRON, G. H., and H. A. O'HERN: *Chem. Eng. Sci.*, **21:** 397 (1966).

10  BEATTIE, I. R.: Dinitrogen Trioxide, in "Progress in Inorganic Chemistry," vol. 5, Wiley, New York, 1963.

11  BODENSTEIN, M.: *Z. Phys. Chem.*, **100:** 68 (1922).

12  BRIAN, P. L. T., H. F. HURLEY, and E. H. HASSELTINE: *AIChE J.*, **7:** 226 (1961).

13  BRIAN, P. L. T., J. E. VIVIAN, and A. G. HABIB: *AIChE J.*, **8:** 205 (1962).

14  BRIAN, P. L. T., and M. C. BEAVERSTOCK: *Chem. Eng. Sci.*, **20:** 47 (1965).

15  BRONSTED, J. N.: *Z. Phys. Chem.*, **102:** 169 (1922).

16  CALDERBANK, P. H.: *Trans. Inst. Chem. Eng. (London)*, **36:** 443 (1958).

17  CARBERRY, J.: *Chem. Eng. Sci.*, **9:** 189 (1959).

18  CHAMBERS, F. S., and T. K. SHERWOOD: *Ind. Eng. Chem.*, **29:** 1415 (1937); *Trans. AIChE*, **33:** 579 (1937).

19  CHEN, T.-T., and C. H. BARRON: *Ind. Eng. Chem. Fundam.*, **11:** 466 (1972).

20  CHILTON, T. H.: *Chem. Eng. Progr. Monog. Ser.* 3, **56** (1960).

21  CORRIVEAU, C. E.: Master's thesis in chemical engineering, U. Calif., Berkeley, 1971.

22  CRANK, J.: "The Mathematics of Diffusion," chap. 7, Clarendon, Oxford, 1956.

23  CRYDER, D. S., and J. O. MALONEY: *Trans. AIChE*, **37:** 827 (1941).

24  DANCKWERTS, P. V.: *Trans. Faraday Soc.*, **46:** 701 (1950).

25  DANCKWERTS, P. V.: *Trans. Faraday Soc.*, **46:** 300 (1950).

26  DANCKWERTS, P. V., and A. M. KENNEDY: *Chem. Eng. Sci.*, **8:** 201 (1958).

27  DANCKWERTS, P. V., and A. J. GILLHAM: *Trans. Inst. Chem. Engrs. (London)*, **44:** T42 (March 1966).

28  DANCKWERTS, P. V., and M. M. SHARMA: *Chem. Eng., Inst. Chem. Engrs. (London)*, CE244-280 (October 1966).

29  DANCKWERTS, P. V.: "Gas-Liquid Reactions," McGraw-Hill, New York, 1970.
30  DAVIDSON, J. F., and E. J. CULLEN: *Trans. Inst. Chem. Engrs. (London)*, **35:** 51 (1953).
31  DEKKER, W. A., E. SNOECK, and H. KRAMERS: *Chem. Eng. Sci.* **11:** 61 (1959).
32  DENBIGH, K. G., and A. J. PRINCE: *J. Chem. Soc.*, **59:** 316 (1937).
33  DE WAAL, K. J. A., and J. C. OKESON: *Chem. Eng. Sci.*, **21:** 559 (1966).
34  EBEN, C. D., and R. L. PIGFORD: *Chem. Eng. Sci.*, **20:** 803 (1965).
35  EMMERT, R. E., and R. L. PIGFORD: *AIChE J.*, **8:** 171, 702 (1962).
36  EPSHTEIN, D. A.: *J. Gen. Chem. (USSR)*, **9:** 792 (1939).
37  FAURHOLT, C.: *J. Chim. Phys.*, **21:** 400 (1925).
38  FAURHOLT, C.: *J. Chim. Phys.*, **22:** 1 (1925).
39  FIELD, J. H., H. E. BENSEN, G. E. JOHNSON, J. S. TOSH, and A. J. FORNEY: *U.S. Bureau of Mines* Bull. 597, 1962.
40  FRIEDLANDER, S. K., and K. H. KELLER: *Chem. Eng. Sci.,* **20:** 121 (1965).
41  FURNAS, C. C., and F. BELLINGER: *Trans. AIChE*, **34:** 251 (1938).
42  GOETTLER, L. A., and R. L. PIGFORD: *Inst. Chem. Engrs. (London) Symp.* Ser. 28, 1, 1968.
43  GOETTLER, L. A., and R. L. PIGFORD: *AIChE J.*, **17:** 793 (1971).
44  GERSTACKER: *Chem. Eng. Sci.*, **14:** 124 (1961).
45  HARNED, H. S., and B. B. OWEN: "The Physical Chemistry of Electrolyte Solutions," 2d ed., Reinhold, New York, 1950.
46  HATCH, T. F., and R. L. PIGFORD: *Ind. Eng. Chem. Fundam.*, **1:** 209 (1962).
47  HATTA, S: Tohoku Imperial U. Tech. Rept., **8:** 1 (1928); **10:** 119 (1932).
48  HEERTJES, M. H., M. H. VAN MENS, and M. BUTAYE: *Chem. Eng. Sci.*, **10:** 47 (1959).
49  HIGBIE, R.: *Trans. AIChE*, **31:** 365 (1935).
50  HIKITA, H., and S. ASAI: *Kagaku Kogaku*, **27:** 823 (1963).
51  HUANG, C.-J., and C.-H. KUO: *AIChE J.*, **11:** 901 (1965).
52  INGERSOLL, L. R., O. J. ZOBEL, and A. J. INGERSOLL: "Heat Conduction," pp. 190ff., McGraw-Hill, New York, 1949.
53  JENSEN, M. B., F. JORGENSON, and C. FAURHOLT: *Acta Chem. Scand.*, **8:** 1137 (1954).
54  JESSER, B. W., and J. C. ELGIN: *Trans. AIChE*, **39:** 277 (1943).
55  KAMEOKA, Y.: Master's thesis in Chemical Engineering, U. Calif., Berkeley, 1973.
56  KIESE, M., and A. B. HASTINGS: *J. Biol. Chem.*, **132:** 267 (1940).
57  KILPI, S., K. S. MIKKOLA, and M. K. VALANTI: *Soumal. Tiedeakat. Torim*, **A2:** 52 (1953); cf. Danckwerts and Sharma [28].
58  KOHL, A. L., and F. C. RIESENFELD: "Gas Purification," McGraw-Hill, New York, 1960; 2d ed., 1974.
59  KRAMERS, H., M. P. P. BLIND, and E. SNOECK: *Chem. Eng. Sci.*, **14:** 115 (1961).
60  KUCHAI, H., J. A. JACQUEZ, and F. J. MATHER: *Biophys. J.*, **10:** 38 (1970).
61  LINEK, V., and J. MAYRHOFEROVA: *Chem. Eng. Sci.*, **17:** 411 (1961); **24:** 481 (1969); **25:** 787 (1970).
62  LINEK, V.: *Chem. Eng. Sci.*, **21:** 77 (1966).
63  LUSS, D., J. E. BAILEY, and S. SHARMA: *Chem. Eng. Sci.*, **27:** 1555 (1972).
64  MARANGOZIS, J., O. TRASS, and A. I. JOHNSON: *Can. J. Chem. Eng.*, **41:** 195 (1963).
65  MEYERINK, E. S. C., and S. K. FRIEDLANDER: *Chem. Eng. Sci.*, **17:** 121 (1962).
66  MOORE, W. J., and J. K. LEE: *Trans. Faraday Soc.*, **47:** 501 (1951).
67  NANDA, A. K., and M. M. SHARMA: *Chem. Eng. Sci.*, **21:** 707 (1966).
68  NÄSÄNEN, R.: *Acta Chim. Fenn*, **B19:** 90 (1946); cf. P. V. Danckwerts and M. M. Sharma [28].
69  NIJSING, R. A. T. O., and H. KRAMERS: *Chem. Eng. Sci.*, **8:** 81 (1958).
70  NIJSING, R. A. T. O., R. H. HENDRIKSZ, and H. KRAMERS: *Chem. Eng. Sci.*, **10:** 88 (1959).

71 OLANDER, D. R.: *AIChE J.*, **6:** 233 (1960).

72 PERRY, R. H., and R. L. PIGFORD: *Ind. Eng. Chem.*, **45:** 1247 (1953).

73 PINSENT, B. R. W., and F. J. W. ROUGHTON: *Trans. Faraday Soc.*, **47:** 263 (1951).

74 PINSENT, B. R. W., L. PEARSON, and F. J. W. ROUGHTON: *Trans. Faraday Soc.*, **52:** 1512 (1956).

75 PORTER, K. E., M. B. KING, and K. C. VARSKNEY: *Trans. Inst. Chem. Eng. (London)*, **44:** T274 (1966).

76 QUINN, J. H., and N. C. OTTO: *J. Geophys. Res.*, **76:** 1539 (1971).

77 RAMACHANDRAN, P. A.: *Chem. Process. Eng.*, **3:** 23 (1969).

78 REITH, T., and W. J. BEEK: *Chem. Eng. Sci.*, **28:** 1331 (1973).

79 RIESENFELD, F. G., and J. F. MULLOWNEY: *Petr. Refiner*, **38:** 161 (1959).

80 ROBERTS, D., and P. V. DANCKWERTS: *Chem. Eng. Sci.*, **17:** 96 (1962).

81 ROBINSON, C. W., and C. R. WILKE: "Chemca 70," Butterworths, London, 1971; also Lawrence Radiation Lab. Rept. 20472 U. Calif., Berkeley, April, 1971.

82 ROPER, G. H., T. F. HATCH, and R. L. PIGFORD: *Ind. Eng. Chem. Fundam.*, **1:** 144 (1962).

83 SAAL, R. N. J.: *Rec. Trav.*, **47:** 73, 264 (1928).

84 SIRS, J. A.: *Trans. Faraday Soc.*, **54:** 201 (1958).

85 SAWIKI, J. E., and C. H. BARRON: *Chem. Eng. J.*, **5:** 153 (1973).

86 SECOR, R. M., and J. A. BEUTLER: *AIChE J.*, **13:** 365 (1967).

87 SECOR, R. M.: *AIChE J.*, **15:** 861 (1969).

88 SHARMA, M. M., and P. V. DANCKWERTS: *Chem. Eng. Sci.*, **18:** 729 (1963).

89 SHARMA, M. M., and P. V. DANCKWERTS: *Trans. Faraday Soc.*, **59:** 386 (1963).

90 SHERWOOD, T. K., and F. A. L. HOLLOWAY: *Trans. AIChE*, **36:** 21, 39 (1940).

91 SHERWOOD, T. K., and J. C. WEI: *AIChE J.*, **1:** 522 (1955).

92 SHERWOOD, T. K., and J. M. RYAN: *Chem. Eng. Sci.*, **11:** 81 (1959).

93 SMITH, K. A., J. H. MELDON, and C. K. COLTON: *AIChE J.*, **19:** 102 (1973).

94 SPECTOR, N. A., and B. F. DODGE: *Trans. AIChE*, **42:** 827 (1946).

95 TAYLOR, G. B., T. H. CHILTON, and S. L. HANDFORTH: *Ind. Eng. Chem.*, **23:** 860 (1931).

96 TEPE, J. B., and B. F. DODGE: *Trans. AIChE*, **39:** 255 (1943).

97 TOSH, J. H., J. H. FIELD, H. E. BENSON, and W. P. HAYNES: *U.S. Bur. Mines. Rept. of Investigations* 5484, 1959.

98 VAN KREVELEN, D. W., and P. J. HOFTIJZER: *Rec. Trav. Chim.*, **67:** 563 (1948).

99 VAN KREVELEN, D. W., and P. J. HOFTIJZER: "On the Solubility of a Gas in Aqueous Solutions," p. 168, 21st Int. Congr. Ind. Chem., Brussels, 1948.

100 VIVIAN, J. E., and R. P. WHITNEY: *Chem. Eng. Progr.*, **43:** 691 (1947); **44:** 54 (1948).

101 VIVIAN, J. E., and D. W. PEACEMAN: *AIChE J.*, **2:** 437 (1956).

102 WALL, H. H.: M.S. thesis, U. Del., June 1966.

103 WAGNER, C., and K. GRUNEWALD: *Z. Phys. Chem.*, **B40:** 455 (1938).

104 WARD, W. J., JR., and W. L. ROBB: *Sci.*, **156:** 1481 (16 June 1969).

105 WEBER, H. C., and K. NILLSON: *Ind. Eng. Chem.*, **18:** 1070 (1926).

106 WENDEL, M. M., and R. L. PIGFORD: *AIChE J.*, **4:** 249 (1958).

107 WENNER, R. R.: "Thermochemical Calculations," McGraw-Hill, New York, 1941.

108 WESTERTERP, K. R.: *Chem. Eng. Sci.*, **18:** 495 (1963).

109 WESTERTERP, K. R., L. L. VAN DIERENDONCK, and J. A. DE KRAA: *Chem. Eng. Sci.*, **18:** 157 (1963).

110 YAGI, S., and H. INOUE: *Chem. Eng. Sci.*, **17:** 411 (1962).

111 MAI, K. L., and BABB, A. L.: *Ind. Eng. Chem.*, **47:** 1749 (1955).

112 ANDREW, S. P. S.: *Corso Estivodi Chimica*, **5:** 153 (1961).

113  THIELE, E. W.: *Ind. Eng. Chem.*, **31**: 916 (1939).

114  ZELDOWITSCH, J. B.: *Acta Physicochim.* (*USSR*), **10**: 583 (1939).

115  DAMKÖHLER, G.: *Deut. Chem. Ing.*, **3**: 430 (1937).

116  WEISZ, P. B.: *Sci.*, **179**: 433 (1973).

117  SATTERFIELD, C. N.: "Mass Transfer in Heterogeneous Catalysis," M.I.T., Cambridge, Mass., 1970.

118  GRAHAM, H. G., V. E. LYONS, and H. L. FAUCETT: *Chem. Eng. Progr.*, **60**: 77 (July 1964).

119  DANCKWERTS, P. V.: *Ind. Eng. Chem.*, **43**: 1460 (1951).

120  WEST, E. S.: "Textbook of Biophysical Chemistry," Macmillan, New York, 1957.

121  LONGMUIR, I. S., and F. J. W. ROUGHTON: *J. Physiol.*, **118**: 264 (1952).

122  SIRS, J. A., and F. J. W. ROUGHTON: *J. Appl. Physiol.*, **18**: 158 (1963).

## PROBLEMS

*8.1*  A chemically inert film of thickness $X$ separates two well-stirred masses of fluid. Each contains a substance, called a substrate, in solution. The substrate molecules can react on the adjacent membrane surface, where an enzyme catalyst is located. The rate of the reaction per unit of surface is expressed by

$$r = \frac{k}{1 + (K_M/C) + (C/K_s)}$$

where  $K_M$ = Michaelis constant
  $K_s$ = inhibition constant
  $C$ = substrate concentration in adjacent fluid

Substrate left after the surface reaction is able to diffuse through the film at a rate dependent on the diffusion coefficient $D$.

Set up mass-balance equations representing the steady-state concentrations of substrate at the two film surfaces in terms of the specified bulk concentrations, the membrane thickness, the mass-transfer coefficients, and the various physical constants.

Are dynamically stable solutions of the equations possible for which the bulk concentrations are equal but there is nevertheless a mass flux of substrate through the film? Is it possible for there to be a flux from the mass of lower substrate concentration toward the one of higher concentration? (Such a situation would be called "active transport" in biology.) (Refs. 3, 63, 8).

*8.2*  The table shows some of the data reported by Andrew [*Chem. Eng. Sci.*, **3**: 279–86 (1954)] from measurements of the rate of absorption of carbon dioxide into an aqueous solution of 7.65-normal ammonia. Andrew found that his measured rates were very nearly proportional to the square root of the ammonia concentration; they were also approximately proportional to the gas-phase partial pressure of $CO_2$, but some deviations in this proportionality were noted. Presumably the small deviations were owing to the depletion of the ammonia concentration at the interface.

Based on the data cited below, compute an approximate correction to each of the measured absorption rates to adjust each for the small effect of ammonia depletion.

*Data:* The apparatus used by Andrew was a "disk column" in which small glass disks were strung on a vertical wire which passed through holes parallel to the faces of the disks. The disks were hung inside a tube through which the gas flowed. Liquid drained downward over the disks. Andrew probably used 40 disks in line, each 1.45 cm in diameter and 0.39 cm thick. Prior to Andrew's work, Stephens and Morris [*Chem. Eng. Progr.*, **47**: 232 (1951)] had used the same apparatus for measuring the rate of absorption of $CO_2$ into water and had obtained $k_L^\circ = 0.016$ cm/s.

Values of the second-order rate constant for the reaction,

$$CO_2 + 2NH_3 = NH_4COONH_2$$

are given by the formula

$$k_{II} = 10^{(11.13 - 2,530/T)} \ 1/(g \ mole)(s)$$

where $T$ is in degrees Kelvin, based on Faurholt [38] and Pinsent, Pearson, and Roughton [74].

| Driving force, atm | $CO_2$ absorption rate, standard liter/h |
|---|---|
| 0.106 | 10.6 |
| 0.213 | 20.0 |
| 0.320 | 30.0 |
| 0.427 | 38.0 |
| 0.533 | 47.0 |
| 0.634 | 54.5 |
| 0.634 | 54.5 |
| 0.741 | 63.5 |
| 0.832 | 70.0 |
| 0.927 | 76.5 |

*Data:* Liquid flow rate = 7.7 l/h; temperature = 23.9°C; average fraction of $NH_3$ converted in bulk liquid = 0.406.

**8.3** A mass-transfer coefficient has been measured for the *absorption* of carbon dioxide into 0.5-molar sodium carbonate solution in a column filled with 1-in Raschig rings at $L = 10,000$ lb/(h)(ft²). The observed value was $K_G a = 1.2$ lb mole/(h)(ft³)(atm) when the solution was partly converted to the bicarbonate form by the reaction

$$CO_2 + CO_3^{--} + H_2O = 2HCO_3^-$$

and 50 percent of the sodium ions were "attached" to bicarbonate.

Now it is desired to use these data in the design of a carbon dioxide *stripping* column which will operate at 150°C, with an essentially zero gas partial pressure of $CO_2$. The liquid rate and the type of packing will be the same as in the absorber.

Using the theory for a hypothetical reversible reaction that is first-order in each direction as a rough approximation, estimate the mass-transfer coefficient for the new conditions.

**8.4** Derive an expression for the time-dependence of the rate of absorption of a gas $A$ into a liquid where the irreversible reaction

$$2A \longrightarrow B$$

occurs. The rate of the homogeneous reaction is given by

$$r = k_{II} C_A^2$$

*Suggestion:* Find two approximate solutions, one in which the time-dependent terms in the equation have vanished and the steady-state rate of diffusion and reaction has set in, the other in which the reaction-rate term in the equation is very small and $\phi \approx 1.0$. (Reference: Roper, Hatch, and Pigford [82]).

8.5 In the manufacture of cellophane (cf. R. N. Shreve, "The Chemical Process Industries," McGraw-Hill, New York, 1945, pp. 724ff.) a sheet of cellulose xanthate is immersed in a bath of sulfuric acid, producing the following reaction:

$$\left( C_{18}H_{27}O_{12}(OH)_2OC \overset{\displaystyle S}{\underset{\displaystyle SNa}{\big/}} \right)_x + \left(\frac{x}{2}\right)H_2SO_4 \longrightarrow (C_6H_9O_4OH)_{3x} + xCS_2 + \left(\frac{x}{2}\right)Na_2SO_4$$

cellulose xanthate                                                                     regenerated

                                                                       cellulose

The reaction can be assumed to be irreversible and very fast, the speed of the process depending on the diffusion of hydrogen ions to the reaction site.

Plot a curve showing the depth of the regenerated cellulose layer near the surface of a very thick polymer film versus time for $0 < t < 1$ s. Assume that the diffusion coefficient for the ion is $D = 10^{-5}$ cm$^2$/s and that the polymer molecules are immobile.

8.6 In the absorption of carbon dioxide into sodium hydroxide solution in a packed column, there are conditions where the concentration of hydroxyl ion is so small that the rate of mass transfer is diffusion-controlled (cf. Fig. 8.33). Nevertheless, even in the absence of gas-phase resistance, the rate may depend not only on the interface concentration of $CO_2$ but also on the local bulk concentration of $OH^-$. Answer the following

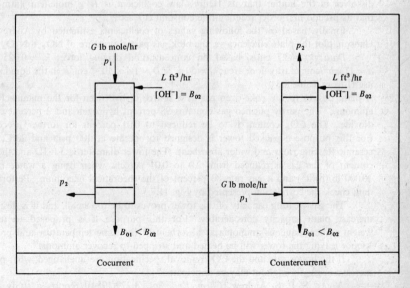

| Cocurrent | Countercurrent |

FIGURE 8.33

questions for a case in which $p_1 = 10p_2$ and half the entering $OH^-$ is converted by the reaction

$$CO_2 + 2OH^- \longrightarrow CO_3^{--} + H_2O$$

(a) How much error would be introduced by assuming that the total rate of solution is proportional to the log-mean partial pressure of $CO_2$?

(b) Is there any advantage in the required tower-packed height of countercurrent over cocurrent operation?

8.7 Andrew and Hanson [2] collected data on the rates of absorption of nitrogen dioxide into water using a sieve-tray apparatus. They also presented a theory leading to an estimate of the tray efficiency based on the computations of the liquid-phase diffusion-reaction effect. The results and the theory show that the efficiency is low when the partial pressure of $NO_2$ is low in the gas.

In accounting for the diffusional resistance of the gas phase, Andrew and Hanson assumed that the rate of transport of tetravalent nitrogen was proportional to the difference in the hypothetical partial pressures, defined by

$$p^* = p_{NO_2} + 2p_{N_2O_4}$$

Thus, they used the expression

$$\text{Rate} = k_g(p_1^* - p_2^*)$$

where the subscripts refer to the bulk gas and to the interface. Assuming that the gas-phase diffusivities of $NO_2$ and $N_2O_4$ are equal, show that their rate expression is correct.

Show also that the diffusion-plus-reaction theory for steady-state conditions can be modified to allow for gas-phase mass-transfer resistance. Assume that only $N_2O_4$ dissolves in the liquid, that its Henry's-law coefficient is $H$ g mole/($cm^3$)(atm), and that its pseudo first-order reaction-rate constant is $k$ $sec^{-1}$.

Finally, based on the following values of coefficients estimated by Andrew and Hanson, plot the plate efficiency vs. the bulk gas partial pressure of $NO_2 + N_2O_4$.

*Data:* $k_G = 3.7$ cm/s, based on concentration driving force; $k_L^\circ = 0.025$ cm/s, $a = 11$ $cm^2/cm^2$ of tray floor area; $k_1 = 50$ $s^{-1}$; $D = 1.4 \times 10^{-5}$ $cm^2/s$ in the liquid phase; use H from Table 8.3.

8.8 A plant desires to use coke-oven gas as a source of hydrogen for the manufacture of ammonia. The partly purified gas contains 98 percent hydrogen and 2 percent carbon dioxide. The $CO_2$ content is to be reduced to 0.01 percent for further processing. For this purpose a packed tower is designed to operate at 100 psia and 20°C, using ceramic Raschig rings and water absorbent. The tower diameter is 3 ft ID and the $CO_2$ content of the gas is reduced from 2.0 to 0.01 percent when using a water rate of 10,000 lb/(h)(ft²) and a gas rate 80 percent of the theoretical maximum. Performance data check the correlation presented by Fig. 11.7.

The gas-treating capacity of this tower proves to be too small, and it is desired to increase plant capacity considerably. For this purpose it is proposed to use 2.0 weight percent aqueous ammonia as absorbent at the same temperature and pressure. Liquor leaving the tower will be heated and stripped to recover ammonia.

If the liquor rate and the $CO_2$ removal specification are unchanged, what percent increase in gas throughput may be anticipated?

*Notes:* S. P. S. Andrew [*Chem. Eng. Sci.*, **3**:279(1954)] reports a study of the absorption of $CO_2$ from air by aqueous ammonia in a disk column, as described in

Prob. 8.2. Several general results of this work suggest the mechanism of the absorption process:

*1* The absorption is sufficiently slow so that variations in the gas-flow rate do not affect the rate of absorption.

*2* Except when the disks were completely wetted, liquor rate does not affect the rate of absorption.

*3* The rate is nearly proportional to the partial pressure of $CO_2$ in the gas.

*4* The reaction appears to be the rapid, irreversible combination of 1 mole of $CO_2$ with 1 mole of $NH_3$ [rate constant = $3.56 \times 10^5$ cm$^3$/(g mole)(s) at 20°C], followed by the very fast combination with a second mole of $NH_3$.

*5* In tests using very low carbonation ratios the rate was found to be proportional to the square root of the total ammonia content of the liquor. (Carbonation ratio is combined $CO_2$ divided by total $NH_3$.)

*6* Using liquors with carbonation ratios up to 0.54, the rate was proportional to the square root of the partial pressure of $NH_3$ over the liquor. This latter was measured and found to follow Henry's law.

*7* The solubility of $CO_2$ in the carbonated liquor was the same as in water.

*8* Values of liquid-phase diffusivity for $CO_2$ and $O_2$ are $1.8 \times 10^{-5}$ and $2.3 \times 10^{-5}$ cm$^2$/s, respectively.

*9* Make any reasonable assumption necessary about the extent of wetting of the packing.

8.9 Danckwerts and Gillham [27] measured the rate of absorption of carbon dioxide into an alkaline aqueous solution (0.6-$M$ $K_2CO_3$ and 0.2-$M$ $KHCO_3$) containing both $OH^-$ and $ClO^-$ ions. The latter catalyzes the hydration reaction of $CO_2$ and leads to a pseudo first-order rate constant given by

$$k_1 = 1.5 + 2,700[ClO^-] \qquad s^{-1}$$

at 20°C. By varying the hypochlorous-ion concentration, experimental data could be collected with solutions of different chemical reactivity.

The apparatus consisted of a tower filled with $6.04 \times 10^5$ cm$^3$ of 1.5-in Raschig-ring packing. The following absorption rate data were obtained at a liquid superficial velocity of 1.51 cm$^3$/(s)(cm$^2$).

| $k_1$, s$^{-1}$ | Absorption rate, cm$^3$/(s)(cm$^3$ of packed space) |
|---|---|
| 1.5 | $1.94 \times 10^{-2}$ |
| 7 | 2.38 |
| 23 | 2.68 |
| 37 | 2.96 |
| 55 | 3.21 |
| 130 | 3.95 |

The diffusion coefficient of unreacted $CO_2$ in these solutions is estimated to be $1.4 \times 10^{-5}$ cm$^2$/s; the solubility is 0.029 g mole/l at a $CO_2$ partial pressure of 1 atm.

From the data estimate the active interfacial area of the liquid phase per unit volume of packed space and the average exposure time of a liquid-surface fluid element.

8.10 According to T. L. Nunes and R. E. Powell [*Inorg. Chem.*, **9**:1916(1970)], nitric oxide reacts in aqueous solution with sulfite ion by two parallel mechanisms such that the pseudo first-order rate constant is given by

$$k_1 = 0.13 + 0.45[SO_3^{--}] \quad s^{-1}$$

at 25°C. $[SO_3^{--}]$ represents the molarity of the sulfite ion.

For a packed-column gas absorber filled with 2-in Raschig rings operating at $L = 5,000$ lb/(h)(ft$^2$) and $G = 500$ lb/(h)(ft$^2$), is the reaction fast enough to have an appreciable effect on the rate of absorption of NO at a point where the bulk liquid contains no dissolved NO? The solution circulated through the absorber contains 0.5 g mole/l each of $Na_2SO_3$ and $NaHSO_3$.

8.11 According to Danckwerts and Sharma [28] the rate of absorption of carbon dioxide into a buffered alkaline solution containing carbonate and bicarbonate ions can be computed from the theory of a pseudo first-order reaction, using the bulk concentrations of the ions, provided that

$$[CO_2]_i\left(\frac{1}{[CO_3^{--}]_b} + \frac{2}{[HCO_3^-]_b}\right)\left(\sqrt{1 + \frac{Dk_{OH^-}[OH^-]_b}{(k_L^\circ)^2}} - 1\right) \ll 1$$

The subscript $i$ refers to interface values; $b$, to bulk values of concentrations.

(a) Derive the inequality.

(b) For a buffer solution made from potassium carbonate and potassium bicarbonate in which the total potassium concentration is $K^+$, the "degree of carbonation" $\alpha$ can be expressed by

$$[HCO_3^-]_b = \alpha[K^+] \quad \text{and} \quad [CO_3^{--}]_b = \frac{1-\alpha}{2}[K^+]$$

Based on these definitions, express the above inequality in terms of the groups of variables, $A = Dk_{OH}(K_w/K_2)(k_L^\circ)^2$, $B = Hp_{CO_2,i}/[K^+]$, and $\alpha$.

(c) Plot lines on $A,B$ coordinates representing the above relationship as an equality, showing lines corresponding to $\alpha = 0.2$, 0.5, and 0.8. Locate a point on these coordinates for a packed-tower operation in which $k_L^\circ = 0.02$ cm/s, $p_{CO_2,i} = 0.7$ atm, temperature $= 60°C$, carbonate and bicarbonate concentrations of 1.5 and 0.05 molar, respectively. Determine whether the inequality is satisfied for these conditions.

8.12 In a study of the rates of absorption of carbon dioxide into hot solutions of potassium carbonate–bicarbonate, Field, Benson, Johnson, Toch, and Forney [39] reported the following values of the overall mass-transfer coefficient $K_G a$ based on gas-phase partial pressures of $CO_2$. The absorber was made from 6-in diameter, schedule-80 steel pipe and was filled with 41 in of $\frac{1}{2}$-in Raschig-ring packing, supported on an additional 3 in of 3/4-in rings. Initially the solution contained 40 percent by weight $K_2CO_3$, which had a density of about 1.4 g/cm$^3$. The gas pressure was 300 lb/in$^2$ gage. The composition of the entering gas varied from 1.3 to 29.0 percent carbon dioxide but only a slight decrease in $K_G a$ was observed to occur over this range. In all tests reported below, the gas-flow rate was 1,500 standard ft$^3$/h.

**EFEECT OF FRACTION CONVERSION TO BICARBONATE, $f$, AND LIQUID TEMPERATURE, $t$ (LIQUID-FLOW RATE = 123 gal/h)**

| $t$, °F | Average value of $f$ in packing | $K_G a$, lb/(h)(ft$^2$)(atm) |
|---|---|---|
| 231 | 0.23 | 0.40 |
|  | 0.315 | 0.34 |
|  | 0.45 | 0.25 |
|  | 0.615 | 0.22 |
|  | 0.65 | 0.19 |
| 199 | 0.175 | 0.43 |
|  | 0.315 | 0.26 |
|  | 0.44 | 0.21 |

**EFFECT OF LIQUID-FLOW RATE AND LIQUID TEMPERATURE (AVERAGE VALUE OF $f$ IN PACKING VARIED FROM 0.30 TO 0.39)**

| $t$, °F | Liquid-flow rate, gal/h | $K_G a$, lb/(h)(ft$^2$)(atm) |
|---|---|---|
| 230 | 31 | 0.20 |
|  | 46 | 0.17 |
|  | 61.5 | 0.25 |
|  | 93 | 0.27 |
|  | 123 | 0.34 |
|  | 260 | 0.40 |
| 260 | 31 | 0.29 |
|  | 61.5 | 0.36 |

Use the theory developed in Chap. 8 to interpret these data in order to answer the question are the observed effects of solution composition, temperature, and liquid-flow rate reasonable?

# 9

# DESIGN PRINCIPLES
# FOR MASS-TRANSFER EQUIPMENT

## 9.0  SCOPE

Elementary principles or techniques for the design of separation-process equipment
are presented for equilibrium stage and continuous differential contactors. Gas
absorption in plate columns and in packed towers is employed as the principal
vehicle for illustration of the basic concepts. The treatment is limited primarily to
binary systems at constant temperature and pressure. Multicomponent hydrocarbon
absorption and nonisothermal effects are considered briefly, along with some
general observations concerning application of the theory to distillation, extraction,
and stripping. Shortcut methods useful for preliminary design of plate and packed-
column absorbers are given, including procedures applicable to concentrated gases.
An approximate design theory is developed for multicomponent mass transfer in
absorption. A general computer program for rigorous design of isothermal
absorbers is presented. Plate efficiency definitions and mass-transfer relationships
are reviewed. It is intended that this basic discussion coupled with fundamental
concepts presented elsewhere in the book should enable the reader to analyze or
design more complex absorption systems and other operations. Detailed treatment
of the total field of separation-process-equipment design would be too voluminous

for inclusion in the present volume. References to more complete treatment are included. Prior knowledge of equipment performance characteristics, for which data sources and predictive methods are discussed in Chap. 11, is assumed in the design procedures.

## 9.1   PRINCIPAL SYMBOLS

| | |
|---|---|
| $a$ | Constant defined by Eq. (9.3c) |
| $a_v$ | Effective mass-transfer surface per unit packed volume |
| $A$ | van Laar constant, °K [Eq. (9.4)]; or absorption factor [Eq. (9.18)] |
| $b$ | Constant defined by Eq. (9.3c) |
| $B$ | van Laar constant, °K [Eq. (9.4)] |
| $c$ | Bulk liquid-phase solute concentration, lb mole/ft$^3$ |
| $c_i$ | Liquid-phase solute concentration at gas liquid interface, lb mole/ft$^3$ |
| $E$ | Overall plate efficiency [Eq. (9.164)] |
| $E_a$ | Murphree vapor efficiency corrected for entrainment |
| $E_{ML}$ | Murphree liquid efficiency [Eq. (9.174)] |
| $E_{MV}$ | Murphree vapor efficiency [Eq. (9.162)] |
| $E_{OG}$ | Murphree point efficiency [Eq. (9.165)] |
| $f$ | Fractional approach to equilibrium at concentrated end of tower, $Y_1^*/Y_1$ |
| $G$ | Gas-mass velocity, lb/(h) (ft$^2$) |
| $G_M$ | Molar gas-mass velocity, lb mole/(h) (ft$^2$) |
| $G_M'$ | Molar mass velocity of solute-free gas, lb mole/(h) (ft$^2$) |
| $G_M^\circ$ | Generalized gas-mass velocity, lb mole/(h) (ft$^2$) [Eq. (9.141)] |
| $h$ | Depth of packing, ft |
| $h_T$ | Total depth of packing in the tower, ft |
| $H$ | Henry's-law constant, defined by Eq. (9.2) |
| HETP | Height equivalent to a theoretical plate, ft |
| $H_G$ | Gas-phase height of a transfer unit, ft [Eq. (9.61)] |
| $H_L$ | Liquid-phase height of a transfer unit, ft [Eq. (9.69)] |
| $H_{OG}$ | Overall gas-phase height of a transfer unit, ft [Eqs. (9.85) and (9.125)] |
| $H_{OL}$ | Overall liquid-phase height of a transfer unit, ft [Eq. (9.91)] |
| HTU | Height of a transfer unit, ft |
| $J'$ | Ratio of equilibrium line slope to operating line slope [Eq. (9.187)] |
| $k_g$ | Gas-phase mass-transfer coefficient, lb mole/(h) (ft$^2$) (atm) |
| $K$ | Vapor-liquid equilibrium constant, $Y/X$ |
| $K_{OG}$ | Overall gas-phase mass-transfer coefficient, lb mole/(h) (ft$^2$) (atm) |
| $K_{OL}$ | Overall liquid-phase mass-transfer coefficient, lb mole/(h) (ft$^2$) (lb mole/ft$^3$) |
| $L$ | Liquid-mass velocity, lb/(h) (ft$^2$) |
| $L_M$ | Molar liquid-mass velocity, lb mole/(h) (ft$^2$) |
| $L_M^\circ$ | Generalized molar liquid-mass velocity, lb mole/(h) (ft$^2$) [Eq. (9.140)] |
| $L_0$ | Moles of solvent entering the column |
| $m$ | Equilibrium line slope, $Y$-vs.-$X$ graph |
| $\overline{m}$ | Effective average equilibrium line slope for the tower |
| $N_A$ | Mass-transfer flux of component $A$, lb mole/(h) (ft$^2$) |
| $N_G$ | Number of gas-phase transfer units [Eqs. (9.63) and (9.73)] |
| $N_H$ | Number of transfer units for heat transfer |

| | |
|---|---|
| $N_L$ | Number of liquid-phase transfer units [Eq. (9.70)] |
| $N_M$ | Number of transfer units for mass transfer |
| $N_{OG}$ | Number of overall gas-phase transfer units [Eqs. (9.86), (9.104), and (9.123)] |
| $N_p$ | Number of theoretical plates |
| $N_p'$ | Number of actual plates |
| $N_T$ | Number of overall gas-phase transfer units for a dilute gas [Eq. (9.104)] |
| $p$ | Partial pressure of a component in the vapor |
| $p_s$ | Vapor pressure of the pure component |
| $P$ | Total pressure |
| $-Q$ | Heat of solution of the absorbed component |
| $Q^s$ | Heat of solution of the solvent vapor |
| $r_A$ | Rate of absorption of component $A$ per unit packed volume, lb mole/(h) (ft$^3$) |
| $R$ | Molar flow ratio, $L_M/G_M$ |
| Sc | Schmidt number, dimensionless |
| $t$ | $1/\overline{\phi}_A$ |
| $T$ | Absolute temperature, °K |
| $V_{N+1}$ | Moles of vapor entering the column |
| $X$ | Mole fraction in the liquid |
| $\overline{X}$ | Effective average liquid-phase mole fraction for the tower |
| $X^\circ$ | Generalized mole fraction in liquid [Eq. (9.139)] |
| $X'$ | Mole fraction of a component in solute-free liquid, $X/(1-X)$ |
| $X^*$ | Liquid-phase mole fraction corresponding to equilibrium with the bulk-gas composition |
| $X_{BM}$ | Log-mean mole fraction of inert component in liquid [Eq. (9.68)] |
| $X_{BM}^*$ | Log-mean mole fraction of inert component in liquid [Eq. (9.89)] |
| $X_f$ | Film factor for a component [Eq. (9.145)] |
| $Y$ | Mole fraction in the gas |
| $\overline{Y}$ | Effective average gas-phase mole fraction for the tower |
| $Y'$ | Mole fraction of a component in solute-free vapor, $Y/(1-Y)$ |
| $Y^s$ | Mole fraction of solvent vapor in gas |
| $Y^*$ | Gas-phase mole fraction corresponding to equilibrium with the bulk-liquid composition |
| $Y_{BM}$ | Log-mean mole fraction of inert component in gas [Eq. (9.60)] |
| $Y_{BM}^*$ | Log-mean mole fraction of inert component in gas [Eq. (9.84)] |
| $Y_f$ | Film factor for a component [Eq. (9.118)] |
| $Y_f^*$ | Film factor, overall driving-force basis [Eq. (9.126)] |
| $Y_{in}$ | Mole fraction solute in gas entering a plate |
| $Y^\circ$ | Generalized mole fraction in the gas [Eq. (9.138)] |

## Subscript designations

| | |
|---|---|
| av | Average quantity |
| $A$ | Component $A$ |
| $B$ | Component $B$ |
| $C$ | Component $C$ |
| $D$ | Component $D$ |
| $e$ | An equilibrium composition or an "effective" absorption or stripping factor |
| $G$ | Gas |
| $i$ | Gas-liquid interface |

| $L$ | Liquid |
|---|---|
| $M$ | Molar quantity |
| $N$ | $N$th plate numbered from the top of a column |
| $n$ | $n$th plate |
| 1 | Concentrated end of a packed tower, or top plate of a plate tower |
| 2 | Dilute end of a packed tower, or plate 2 of a plate tower |

**Greak letters**

| $\gamma$ | Liquid-phase-activity coefficient [Eq. (9.4)] |
|---|---|
| $\Delta$ | Difference—as in difference of concentrations used for driving force, for example |
| $\Delta g$ | Difference in mass velocity of a component over the tower, lb mole/(h) (ft$^2$) |
| $\Delta g_A$ | Total moles of component $A$ absorbed over the tower, lb mole/(h) (ft$^2$) |
| $\Delta g_I$ | Net total moles of all components other than $A$ absorbed over the tower, lb mole/(h) (ft$^2$) |
| $\Delta N_{OG}$ | Correction to apparent number of transfer units [Eq. (9.108)] |
| $\varepsilon$ | Moles liquid entrainment per mole dry vapor |
| $\bar{\rho}$ | Average molal liquid density, lb mole/ft$^3$ |
| $\bar{\phi}_A$ | Average relative mass-transfer rate for component $A$ in the tower [Eq. (9.129)] |
| $\phi_A$ | Relative mass-transfer rate for component $A$ [Eq. (9.113)] |
| $\psi_A$ | Counterdiffusion factor [Eq. (9.122)] |
| $\psi_A^*$ | Counterdiffusion factor, overall driving-force basis [Eq. (9.124)] |

## 9.2   INTRODUCTION

Most separation processes involve the transfer of material from one phase to another. A few special types of separation, such as thermal diffusion, depend upon developing concentration differences within single fluid phases. Perhaps the most common mode of transfer encountered in chemical processing is that between gases and liquids, occurring typically in gas absorption, stripping, and distillation. Other modes include liquid-liquid transfer in solvent extraction and fluid-solid transfer in drying, leaching, and crystallization. The design engineer must normally select the type of mass-transfer equipment to be used and calculate the required size and number of units needed for a given separation operation. Conversely, the plant engineer might employ the same knowledge to establish operating conditions for a desired production with existing equipment, or to diagnose operating difficulties.

Chapter 11 describes the common types of mass-transfer equipment and indicates sources of information on their performance, flow capacity, and other operating characteristics. Chapter 5 discusses the mechanisms of mass transfer and the rate processes which govern the movement of molecules from one phase to another, and which are fundamental, therefore, to the overall mass-transfer behavior of a separation process.

The present chapter considers methods of engineering analysis and computational techniques which may be employed in design of mass-transfer equipment.

Because of its relative simplicity the process of steady-state continuous-gas absorption will be used to demonstrate the basic principles and methods of approach. Similar procedures may be applied to other processes such as distillation or extraction, and extended to unsteady-state operations.

**Gas-absorber design**  Absorption entails the removal of a substance from a gas by contacting it with liquid into which the desired component dissolves. Some typical examples of importance are the removal of sulfur dioxide from stack gases by absorption with alkaline solutions, absorption of carbon dioxide from combustion products into aqueous amine solutions, and the removal of propane or other heavier components from natural gas by absorption into hydrocarbon oil.

All absorption processes involve the following basic steps: (1) the gas and liquid are brought together in a suitable contacting apparatus, (2) the two phases are allowed to approach equilibrium, and (3) the gas-liquid phases are separated. The desired material is transferred from the gas to the liquid in step 2 at a rate which depends on its concentrations in the gas and liquid, the mass-transfer coefficients in each phase, the solubility of the material in the liquid, and the amount of gas-liquid interfacial area made available in the contactor.

Two basic types of contactors are employed: (1) stagewise and (2) continuous differential. The corresponding computational methods applicable to each type of contactor form a convenient basis for classification of design procedures.

## 9.3  PHASE EQUILIBRIA

Consider a component in a gas which is in contact with a liquid. At equilibrium the activities of the component in the gas and liquid phases will be equal. This criterion of equilibrium may be expressed in various ways.

For ideal solutions, Raoult's law applies:

$$p = p_s X \tag{9.1}$$

where $p$ = the partial pressure of the component in the gas phase

$p_s$ = vapor pressure of pure component

$X$ = mole fraction in the liquid

For moderately soluble gases with relatively little interaction between the gas and liquid molecules Henry's law is often applicable:

$$Y = \frac{p}{P} = \frac{HX}{P} \tag{9.2}$$

where $Y$ = mole fraction in the gas phase

$P$ = total pressure

$H$ = Henry's constant, here defined as $p/X$

Usually $H$ is dependent upon temperature but relatively independent of pressure at moderate levels. In solutions containing inorganic salts, $H$ will also be a function of the ionic strength [54]. Henry's constants are tabulated for many of the common gases in water [40].

A more general way of expressing solubilities is through the vapor-liquid equilibrium constant $K$, defined by

$$Y = KX \tag{9.3a}$$

A common alternate notation replaces $K$ by $m$:

$$Y = mX \tag{9.3b}$$

$K$ values are widely employed to represent hydrocarbon vapor-liquid equilibria in absorption and distillation calculations.

When Eqs. (9.1) or (9.2) are applicable at constant pressure and temperature [equivalent to constant $K$ or $m$ in Eqs. (9.3a) and (9.3b)], a plot of $Y$ vs. $X$ for a given solute will be linear from the origin. In other cases the $YX$ plot may be approximated by a linear relationship over limited regions of interest. More generally, for nonideal solutions or for nonisothermal conditions, $Y$ will be a curving function of $X$, which must be determined from experimental data or rigorous theoretical relationships. The $YX$ plot when applied to absorber design is commonly called the "equilibrium line."

Systems of moderate curvature in the equilibrium line may be represented conveniently for approximate design calculations by a relation similar to the Langmuir adsorption isotherm as expressed by

$$Y = \frac{aX}{1 + bX} \tag{9.3c}$$

Equation (9.3c) has the useful property of providing an essentially linear relation in the dilute region and of fitting an upward or downward curvature depending on the sign of the constant $b$. The constant $a$ is equal to the slope of the equilibrium line at the dilute end.

Vapor-liquid equilibria for binary solutions can often be satisfactorily correlated by the van Laar equation [55] which may be expressed for ideal vapor behavior as follows:

$$\log \gamma \equiv \log \left( \frac{YP}{Xp_s} \right) = \frac{A/T}{\left\{ 1 + \dfrac{AX}{B(1 - X)} \right\}^2} \tag{9.4}$$

where  $T$ = absolute temperature, °K
  $A, B$ = the van Laar constants, °K
  $\gamma$ = liquid-phase activity coefficient for component of mole fraction $X$

The van Laar constants for various systems are tabulated in Landolt-Börnstein [29], and methods for obtaining them from experimental data are described by a number of authors (8, 23, 46]. The van Laar equation often does not give an adequate representation of the temperature dependence of $\gamma$. It is perhaps most useful for the smoothing of isothermal data with the constants applicable to the particular temperature. A more satisfactory general equation is that based on the expression of Wilson for the excess Gibbs energy of a binary solution [37, 59, 42].

Gas solubility is treated extensively in the classical text of Hildebrand and Scott [21]. A comprehensive theoretical treatment of various methods for the thermodynamic prediction of phase equilibria is given by Prausnitz [42]. Computer techniques and programs have been prepared by Prausnitz and associates [44, 43]. Extensive graphical correlations of $K$ values for hydrocarbons have been developed by Benedict, Webb, Rubin, and Friend [3] and by Edmister [15].

## 9.4   ABSORPTION IN A SINGLE EQUILIBRIUM STAGE

Industrial absorption processes normally deal with flowing streams. An equilibrium stage is a hypothetical device in which the gas and liquid streams are brought together for a sufficient time to reach equilibrium and then separated. A stage may be depicted schematically by the simple mixer-separator arrangement shown in Fig. 9.1a. Gas and liquid streams enter the mixer at molal flows $G_{M2}$ and $L_{M0}$ moles per hour, respectively, containing the component to be absorbed at mole fractions $Y_2$ and $X_0$, and leave the separator at equilibrium. The outlet compositions, $Y_{1e}$ and $X_{1e}$, must lie on the equilibrium line. Composition changes over the stage may be computed by satisfaction of the material balance over the stage and the equilibrium relation. Thus,

$$G_{M2} Y_2 + L_{M0} X_0 = G_{M1} Y_{1e} + L_{M1} X_{1e} \qquad \text{material balance}$$

and
$$Y_e = f(X_e) \qquad \text{equilibrium line}$$

$$(9.5)$$

For the special case of absorption of a single component from a dilute gas in which the gas-liquid flows are changed negligibly by absorption, Eq. (9.5) becomes

$$\frac{Y_2 - Y_{1e}}{X_0 - X_{1e}} = -\frac{L_M}{G_M} \tag{9.6}$$

where $L_M$ and $G_M$ are the average flows (assumed constant). In this case the history of the absorption process may be determined graphically as shown in Fig. 9.1b. The composition changes, until the equilibrium line is reached, along the line $abc$, which has a slope equal to $-L_M/G_M$, as required by Eq. (9.6).

It should be noted at this point that in practice complete equilibrium can never be attained since infinite contact time would be necessary. Depending on the prevailing mass-transfer rates in the stage, the composition will move along the line $ac$ and stop at some point $b$ of compositions $X_1$, $Y_1$ short of reaching the equilibrium line. A "stage efficiency" $E$ may be defined as the ratio of the composition change actually obtained to that which would be attained at equilibrium. In the present case for either the gas or liquid phase,

$$E = \frac{Y_2 - Y_1}{Y_2 - Y_{1e}} = \frac{X_1 - X_0}{X_{1e} - X_0} = \frac{ab}{ac} \tag{9.7}$$

In many of the stagewise contactors used in practice, stage efficiencies close to unity can be realized so that the equilibrium-stage concept is useful in the mathematical

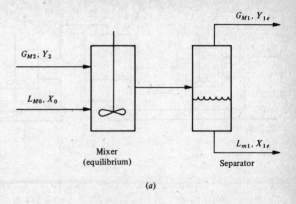

(a)

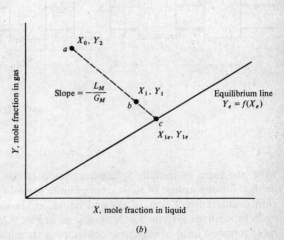

(b)

FIGURE 9.1
Absorption in a single equilibrium stage.

analysis of such devices. Also, knowledge of the stage efficiency which might be obtained in a given apparatus is usually available from experimental data, operating experience, or estimation procedures as discussed in Chapter 11.

## 9.5 MULTISTAGE COUNTERCURRENT ABSORPTION

Complete absorption in a single stage is theoretically possible only when the absorbed component has a negligible vapor pressure over the liquid, as in the case of absorption with an irreversible chemical reaction forming a nonvolatile species, or

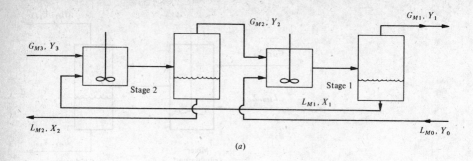

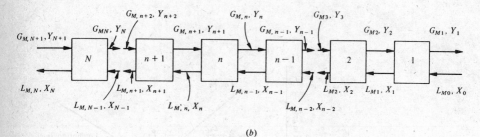

FIGURE 9.2
Countercurrent stagewise gas absorption.

when the ratio of initially solute-free liquid to gas is infinite. A more effective method for absorption of substances of finite solubility is to use two or more stages with countercurrent flow of gas and liquid, as shown in Fig. 9.2a for two stages, or in generalized systems for $N$ stages, as in Fig. 9.2b. For gas-liquid systems the most common form of stagewise contactor is the bubble-plate column shown schematically in Fig. 9.3 (see Chap. 11 for a detailed description). Each plate is assumed to act as a well-mixed stage with vapor-liquid disengagement in the space between plates. The liquid and gas leaving each plate are assumed to be in equilibrium.

### 9.5.A Graphical Design Procedure

Consider an $N$-stage countercurrent process as shown in Fig. 9.2b or 9.3b. A single component is to be absorbed from the gas at constant temperature and pressure. Operation may be assumed to be at steady state. Let $L'_M$ and $G'_M$ be the molal flows of solute-free liquid and gas, respectively. A material balance around plate $n$ gives

$$G'_{M, n+1} \frac{Y_{n+1}}{1 - Y_{n+1}} + L'_{M, n-1} \frac{X_{n-1}}{1 - X_{n-1}} = G'_{M, n} \frac{Y_n}{1 - Y_n} + L'_{M, n} \frac{X_n}{1 - X_n} \quad (9.8)$$

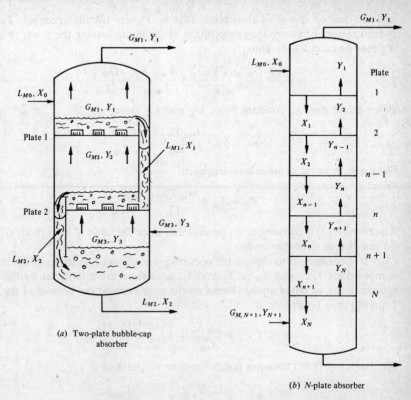

FIGURE 9.3
Schematic representations of the plate column as a multistage contactor.

Setting $Y' = Y/(1 - Y)$ and $X' = X/(1 - X)$ and noting that $L'$ and $G'$ are constant, Eq. (9.8) may be rearranged in the form

$$Y'_{n+1} - \frac{L'_M}{G'_M} X'_n = Y'_n - \frac{L'_M}{G'_M} X'_{n-1} \qquad (9.9)$$

where $Y'$ and $X'$ are molal "stoichiometric units." For a dilute gas for which $Y'$ and $X'$ may be assumed to be essentially equal to the corresponding mole fractions Eq. (9.9) becomes

$$Y_{n+1} - \frac{L_M}{G_M} X_n = Y_n - \frac{L_M}{G_M} X_{n-1} \qquad (9.10)$$

A graph of $Y_{n+1}$ vs. $X_n$, made in accordance with either Eq. (9.9) or (9.10), constitutes the "operating line" for the absorption process. To place the operating line the flows, the composition of the entering gas $Y_{N+1}$, the entering liquid $X_0$, and the desired degree of absorption, that is, $Y_1$, are usually specified. From an overall material balance the composition of the liquid leaving the tower (Fig. 9.3b) $X_N$ may be calculated. Thus,

$$X_N = \frac{G_{M,N+1} Y_{N+1}}{L_{M,N}} + \frac{L_{M0} X_0}{L_{M,N}} - \frac{G_{M,1} Y_1}{L_{M,N}} \tag{9.11}$$

or for dilute gas, i.e., constant flows, $G_M$ and $L_M$ assumed,

$$X_N = X_0 + \frac{G_M Y_{N+1}}{L_M} - \frac{G_M Y_1}{L_M}$$

Similarly, in terms of stoichiometric units,

$$X_N' = X_0' + \frac{G_M' Y_{N+1}'}{L_M'} - \frac{G_M' Y_1'}{L_M'} \tag{9.13}$$

Equation (9.13) is convenient for general use, since the flows of inert fluids $G'$ and $L'$ do not change over the tower.

It is convenient to express the operating-line equation in terms of the terminal compositions $Y_{N+1}$ and $X_N$ or $X_0$ and $Y_1$, since these are usually set by the design specifications, as noted above. Based on the conditions at the bottom of the tower, Eq. (9.9) may be written as

$$Y_{n+1}' = \frac{L_M' X_n'}{G_M'} + Y_{N+1}' - \frac{L_M' X_N'}{G_M'} \tag{9.14}$$

Similarly, Eq. (9.10) becomes (again for $Y \approx Y'$ and $X \approx X'$)

$$Y_{n+1} = \frac{L_M X_n}{G_M} + X_{n+1} - \frac{L_M X_N}{G_M} \tag{9.15}$$

The equilibrium line over the range of gas-liquid compositions of interest may be determined from knowledge of Henry's constant or other appropriate forms of the vapor-liquid equilibria for the system. A design diagram may be prepared by plotting the equilibrium and operating lines in $YX$ coordinates as shown in Fig. 9.4. Theoretical plates are stepped off by moving from the operating line vertically to the equilibrium line, and then horizontally to the operating line, starting at the point $X_N Y_{N+1}$ as determined from the overall material balance. Each such sequence of movement, or step, between the operating and equilibrium lines constitutes a theoretical plate. Four plates are required in the examples shown in Fig. 9.4. Alternatively, plates may be stepped off starting at the point $X_0$, $Y_1$ and moving upward between the lines from left to right.

The design diagram may also be prepared using stoichiometric units through Eq. (9.14) and an appropriately modified equilibrium relation. An identical step construction is used to determine the theoretical stages. This basis of construction may be useful with concentrated gas, which gives a curved operating line in mole fraction units, but the convenience of the straight operating line thus obtained may be offset by the curvature introduced in the equilibrium line.

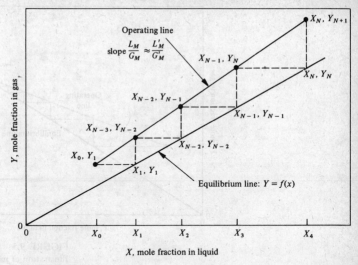

**FIGURE 9.4**
Graphical method for countercurrent gas absorption (four theoretical plates).

## 9.5.B   Minimum Liquid Rate

In the case of a straight or concave downward operating line and a straight or concave upward equilibrium line, the minimum liquid rate is that at which the operating line just touches the equilibrium line [Eq. (9.3a)] at the bottom of the tower, that is, $X_N = Y_{N+1}/K$. This situation is illustrated in Fig. 9.5, where it can be seen that an infinite number of steps (plates) are required (i.e., there is a pinch point) to reach the point $X_N$, $Y_{N+1}$. For the case of constant flows (i.e., straight operating line), the minimum-flow ratio $L_M/G_M$ is given by the slope of the line connecting the points $X_0$, $Y_1$ and $X_N$, $Y_{N+1}$.

In the case of an irregularly shaped equilibrium line or an operating line which is concave downward, a pinch point may occur elsewhere on the diagram where the equilibrium and operating lines intersect. If this latter type of behavior is suspected, it may be necessary to determine the minimum liquid rate by trial-and-error graphs of the design diagram for various flow rates. Alternatively, in this case trial-and-error graphs may be avoided by using stoichiometric units for the design diagram, since the operating line will then be straight and its point of intersection with the curved equilibrium line can easily be determined.

The actual liquid rate to be specified in a given design depends rigorously upon an optimization of all economic factors not affected by the absorption process. For example, use of a liquid rate far above the minimum value will result in a small number of plates and relatively low absorber cost, but at the expense of a

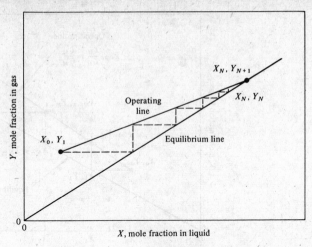

**FIGURE 9.5**
Illustration of minimum liquid rate.

higher cost for solvent circulation and subsequent removal of the solute from the relatively dilute solution. Conversely, a liquid rate very near minimum will require a large number of plates, but at a lower cost for solvent and solute recovery. Often an absorber will be operated in conjunction with a stripper, and it is the total cost of operating both units which must be optimized. Use of a liquid rate on the order of 1.5 times the minimum value is not unusual. A value of the group $L_M/KG_M$ (designated as the "absorption factor," see Sec. 9.5.D) of 1.4 is often specified in the absence of a detailed cost analysis [10].

**EXAMPLE 9.1**   Carbon dioxide evolved during the production of ethyl alcohol by fermentation contains 0.01 mole fraction of alcohol vapor. It is proposed to remove the alcohol by absorption into water in a bubble-plate tower 8 ft in diameter. Absorption may be assumed to occur isothermally at 40°C, 1 atm. The water for absorption is supplied from the subsequent distillation step for alcohol recovery and may be assumed to contain 0.0001 mole fraction alcohol. To be processed are 500 lb moles of gas per hour. Over the conditions of operation, the solubility of alcohol in water may be approximated satisfactorily by the relation $Y = 1.0682X$ (where $Y$ and $X$ are the gas-phase and liquid-phase mole fractions) based on a fit of the van Laar equation to isothermal vapor pressure data in the manner described by Hougen and Watson [23].

    (*a*) Calculate the minimum water rate for 98 percent absorption of the alcohol vapor (pound moles per hour).

    (*b*) Calculate the number of theoretical plates required for 98 percent absorption at a water rate of 1.5 times the minimum.

    (*c*) Calculate the percentage absorption which would be obtained in one equilibrium stage at the flow rates of part *b*.

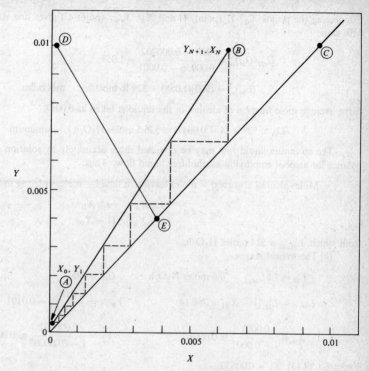

FIGURE 9.6
Graphical construction for Example 9.1

SOLUTION:

(a) Minimum liquid rate. The equilibrium line is drawn on the $XY$ diagram, Fig. 9.6, in accordance with the equation $Y = 1.0682X$. At minimum liquid rate the operating line, Eq. (9.9) or (9.11), will intersect the equilibrium line at $Y_{N+1} = 0.01$, or

$$X_N = \frac{Y_{N+1}}{1.0682} = \frac{0.01}{1.0682} = 0.009362 \qquad \text{point } C$$

Composition of the gas leaving the tower is obtained from a material balance.

Total gas entering = 500 lb moles/h

$g_{N+1}$ = alcohol entering = $500 \times 0.01 = 5$ lb moles/h

$\Delta g$ = alcohol removed = $5 \times 0.98 = 4.9$ lb mole/h

$g_1$ = alcohol leaving = $5.0 - 4.9 = 0.1$ lb mole/h

$G'_M$ = inert-gas flow = $500 - 5 = 495$ lb mole/h

$Y_1$ = mole fraction alcohol = $0.1/(495 + 0.1) = 0.202 \times 10^{-3}$ in leaving gas

$X_0$ = mole fraction alcohol = $0.0001$ in entering liquid

Connecting the points $X_0$, $Y_1$ (point $A$) and $X_N$, $Y_{N+1}$ (point $C$) gives line $AC$ (not shown) with slope

$$(L_M/G_M)_{av} = \frac{0.01 - 0.000202}{0.00936 - 0.0001} = 1.058$$

$$(L_M)_{av} = (500)(1.058) = 529 \text{ lb mole/h} \qquad \text{minimum}$$

If the average mole fraction of alcohol in the liquid is taken as 0.0047,

$$L'_M \approx 529 \times (1 - 0.0047) \approx 526.5 \text{ moles } H_2O/h \qquad \text{minimum}$$

The minimum liquid rate may be obtained more accurately by solution of the material balance for alcohol employing alcohol-free liquid flow. Thus,

Moles alcohol absorbed = moles leaving in liquid − moles entering in liquid

$$\Delta g = 4.9 = \frac{L'_M X_N}{(1 - X_N)} - \frac{L'_M X_0}{(1 - X_0)}$$

from which, $L'_{M_{min}} = 524$ moles $H_2O/h$.

(b) Theoretical stages

$$L'_M = 1.5 L'_{M_{min}} = 786 \text{ moles } H_2O/h \qquad G'_M = 495$$

$$L_{M,0} = L'_M/(1 - X_0) = 786.14 \qquad Y'_{N+1} = \frac{0.01}{1 - 0.01} = 0.0101$$

$$X'_0 = \frac{0.0001}{1 - 0.0001} \approx 0.0001 \qquad Y'_1 = \frac{0.000202}{1 - 0.000202} \approx 0.000202$$

From Eq. (9.13), $X'_N = 0.00633$.

Substitution of the above values in Eq. (9.14) gives the operating-line equation:

$$Y'_{n+1} = \frac{Y_{n+1}}{1 - Y_{n+1}} = \frac{1.588 X_n}{1 - X_n} + 4.8 \times 10^{-5}$$

Values of $Y_{n+1}$ vs. $X_n$ obtained from the equation above are plotted on line $AB$ on the $YX$ diagram. For example, at $X_n = 0.005$, $Y_{n+1} = 0.00796$. From the step construction starting at point $B$, that is, $Y_{N+1}$, $X_N$, it is found that nine theoretical stages are required to reach the desired final-gas composition of 0.0202 mole percent alcohol.

It should be noted that in the present example, curvature of the operating line is so slight that satisfactory results could be obtained by connecting points $A$ and $B$ directly with a straight line instead of calculating intermediate points via Eq. (9.14).

(c) Absorption in a single stage (see Fig. 9.1). From part $b$,

$$\frac{L'_M}{G_M} = \frac{L'_M/(1 - X_0)}{G'_M/(1 - Y_{N+1})} = 1.572$$

$$X_0 = 0.0001 \qquad Y_2 = 0.01$$

From the plot of Eq. (9.6), line $DE$, $Y_{1e} = 0.00408$; $X_{1e} = 0.0038$; $G'_M = 495$; $Y'_2 = 0.01010$; $Y'_{1e} = 0.00410$. Alcohol absorbed = $G'_M(Y'_2 - Y'_e) = 2.97$ moles/h; Percent absorption = $2.97 \times 100/5 = 59.4$ percent.

*Note:* The great advantage of countercurrent processing is brought out in the comparison of the separations obtained in parts $b$ and $c$, i.e., 98 percent vs. 59.4 percent.     ////

## 9.5.C  Algebraic Methods for Dilute Gases [51]

For a dilute gas, Eq. (9.8) becomes

$$G_M(Y_{n+1} - Y_n) = L_M(X_n - X_{n-1}) \tag{9.16}$$

For an ideal stage ("theoretical plate"), with the equilibrium line assumed linear from the origin, Eqs. (9.3a) or (9.3b) and (9.10) may be combined and rearranged to give

$$Y_n = \frac{Y_{n+1} + AY_{n-1}}{1 + A} \tag{9.17}$$

where $\qquad A = \dfrac{L_M}{KG_M} \qquad$ or $\qquad \dfrac{L_M}{mG_M} \equiv$ the absorption factor $\qquad$ (9.18)

For a single theoretical plate absorber,

$$Y_1 = \frac{Y_2 + AY_0}{1 + A} \tag{9.19}$$

where $Y_1$ is the mole fraction of absorbed component in gas leaving the plate, and $Y_0$ is the mole fraction of component in gas in equilibrium with entering liquid.

For a two-plate absorber,

$$Y_2 = \frac{Y_3 + AY_1}{1 + A} \tag{9.20}$$

Eliminating $Y_1$ by combining Eqs. (9.19) and (9.20),

$$Y_2 = \frac{(A + 1)Y_3 + A^2 Y_0}{A^2 + A + 1} \tag{9.21}$$

Proceeding in a similar manner to an $N$-plate absorber

$$Y_N = \frac{(A^N - 1)Y_{N+1} + A^N(A - 1)Y_0}{A^{N+1} - 1} \tag{9.22}$$

Eq. (9.22) may be combined with a material balance over the absorber to eliminate $Y_N$; and after setting $Y_0 = KX_0$,

$$\frac{Y_{N+1} - Y_1}{Y_{N+1} - KX_0} = \frac{A^{N+1} - A}{A^{N+1} - 1} \tag{9.23}$$

where $N =$ number of theoretical plates
$\quad Y_{N+1} =$ mole fraction in entering gas
$\quad Y_1 =$ mole fraction in leaving gas
$\quad X_0 =$ mole fraction of solute in liquid entering the tower
Equation (9.23) is a form of the Kremser equation [28], which is quite useful for

quick, approximate estimates of the relationship between the extent of absorption and the number of theoretical plates.

When $A = 1$, Eq. (9.23) is indeterminate. It may be shown, however, that for this condition,

$$\frac{Y_{N+1} - Y_1}{Y_{N+1} - KX_0} = \frac{N}{N+1} \tag{9.24}$$

An alternate result which gives the number of plates $N_p$ directly as suggested by Colburn [10] is given by Eq. (9.25), in which $K$ is replaced by $m$, the slope of the equilibrium line, and subscripts 1 and 2 refer to the concentrated and dilute terminal compositions, respectively:

$$N_P = \frac{\ln\{(1 - mG_M/L_M)[(Y_1 - mX_2)/(Y_2 - mX_2)] + mG_M/L_M\}}{\ln(L_M/mG_M)} \tag{9.25}$$

The left-hand side of Eq. (9.23) represents the actual change in composition of the gas flowing through a column having $N$ theoretical plates, divided by the change which would occur if the gas came to equilibrium with the liquid entering, i.e., in an infinite column. A high degree of recovery may be obtained either by using a large number of plates or by employing a large liquid-gas ratio.

Equation (9.25) is the preferred form when the number of plates is the unknown quantity in the problem. Figure 9.7 provides a convenient graphical solution for Eq. (9.24) or (9.25), which is useful for visualizing the effects of varying $mG_M/L_M$ or the number of plates on the separation which can be obtained.

When $mG_M/L_M$ is greater than unity, complete removal of the solute from the gas in an absorber is impossible, no matter how many plates are available and even if there is no solute in the entering liquid. This is indicated by the fact that Eq. (9.23) reduces to

$$\frac{Y_1 - Y_2}{Y_1 - Y_{e2}} = \frac{L_M}{mG_M} \tag{9.26}$$

when $N_P$ is infinite and $L_M/mG_M$ is a fraction.

EXAMPLE 9.2   For the conditions of Example 9.1$b$ compute the following:

(a) The percentage of alcohol recovery which would be obtained with four theoretical plates

(b) The number of theoretical plates required for 98 percent recovery

SOLUTION   Calculate $G'_M$ and $L'_M$ as in Example 9.1.

(a) Recovery with four theoretical plates. Since the absorption factor needed in Eq. (9.23) varies through the column, one should use an average value. In order to estimate the absorption factor at the ends of the tower, assume 100 percent absorption.

$$G_{M,1} \approx G'_M = 495 \text{ lb moles/h}$$

$$L_{M,N} = L_{M,0} + g_{N+1} = 786 + 5 = 791 \text{ lb moles/h}$$

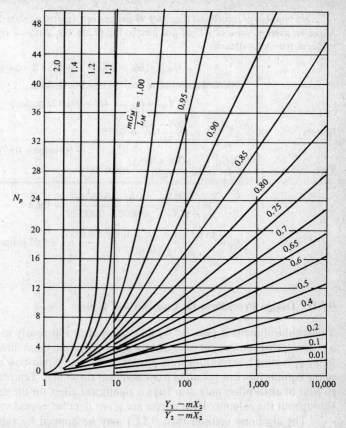

**FIGURE 9.7**
Number of theoretical plates in an absorption column, condition of constant $mG_m/L_m$.

Bottom:   $A = \dfrac{L_{M,N}}{KG_{M,N+1}} = \dfrac{791}{(1.068)(500)} = 1.481$

Average 1.484

Top:   $A = \dfrac{L_{M,0}}{KG_{M,1}} = \dfrac{786}{(1.068)(495)} = 1.487$

$$\frac{A^{N+1} - A}{A^{N+1} - 1} = 0.922 = \frac{0.01 - Y_1}{0.01 - 1.068 \times 0.0001} \qquad Y_1 = 0.000874$$

Alcohol recovery:   $1 - \dfrac{0.000874}{(1 - 0.000874)} \dfrac{495}{5} = 0.913$   or   91.3 percent

(b) Number of theoretical stages for 98 percent recovery. Calculate $Y_1$ as in Example 9.1. Again, an average value of $KG_M/L_M$ is used in Eq. (9.25), but it is now based on the specified values in the outlet streams.

$$G_{M,1} = G'_M + 0.02 \times 5 = 495.1 \text{ lb moles/h}$$

$$\text{Alcohol absorbed} = 5 \times 0.98 = 4.9 \text{ lb moles/h}$$

$$L_M = L_{M,0} + 4.9 = 791.0 \text{ lb moles/h}$$

Bottom: $\quad \dfrac{KG_{M,N+1}}{L_{M,N}} = \dfrac{1.068 \times 500}{791} = 0.6751$

$$\text{Average} = 0.6739$$

Top: $\quad \dfrac{KG_{M,1}}{L_{M,0}} = \dfrac{1.068 \times 495.1}{786.1} = 0.6726$

$$\frac{Y_{N+1} - KX_0}{Y_1 - KX_0} = \frac{0.01 - 0.000107}{0.0002 - 0.000107} = 106.4$$

$$N_P = \frac{\ln\left[(1 - 0.6739) \times 106.4 + 0.6739\right]}{-\ln(0.6739)} = 9.03 \text{ plates} \qquad ////$$

## 9.5.D  Design Procedures for Concentrated Gases

The graphical method (Sec. 9.5.A) may be applied rigorously to the absorption of a concentrated component provided the design diagram takes into account curvature in the operating line resulting from changes in the gas-liquid-flow ratio and curvature in the equilibrium line resulting from nonideal solubility. Temperature changes due to heat of absorption may also have a significant effect on the equilibrium relation throughout the column. Heat effects are given detailed consideration in Sec. 9.5.G.

The algebraic method (Sec. 9.5.C) may be applied by substitution of molal stoichiometric units $Y'$ and $X'$ [see (Eq. 9.9)] and solute free flows $L'_M$ and $G'_M$ in Eq. (9.23), provided the equilibria may be expressed satisfactorily in the form $Y' = m'X'$. Figure 9.8 illustrates the behavior of Eq. (9.23) expressed in this manner.

Horton and Franklin [22] have developed an algebraic method for the case when the absorption factor varies throughout the column. An absorption factor is defined for each plate $n$ as

$$A_n = \frac{L_n}{K_n V_n} \qquad (9.27)$$

where $L_n$ and $V_n$ are the molal liquid and vapor flows from the plate. For an absorber of $N$ plates, the fractional absorption of any component is given by the equation

$$\frac{Y''_{N+1} - Y''_1}{Y''_{N+1}} = \left(\frac{A_1 A_2 A_3 \cdots A_N + A_2 A_3 \cdots A_N + \cdots + A_N}{A_1 A_2 A_3 \cdots A_N + A_2 A_3 \cdots A_N + \cdots + A_N + 1}\right)$$

$$- \frac{L_0 X'_0}{V_{N+1} Y''_{N+1}} \left(\frac{A_2 A_3 \cdots A_N + A_3 \cdots A_N + \cdots + A_N + 1}{A_1 A_2 A_3 \cdots A_N + A_2 A_3 \cdots A_N + \cdots + A_N + 1}\right) \qquad (9.28)$$

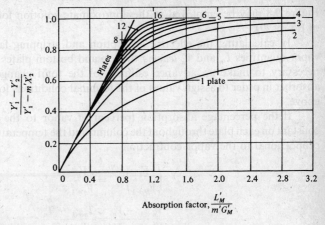

FIGURE 9.8
Relation between column performance and number of theoretical plates based on
Eq. (9.23), with molal stoichiometric variables.

where $L_0$ = moles solvent entering the column

$\quad X'_0$ = moles of absorbed component per mole of solvent entering

$V_{N+1}$ = moles of vapor entering the column

$Y''_{N+1}$ = moles of absorbed component entering the column per mole of entering vapor

$\quad Y''_1$ = moles of absorbed component leaving the column per mole of entering vapor

Edmister [14] defines average effective absorption factors $A'$ and $A_e$ to replace the values for each plate to obtain a solution to Eq. (9.28) in a form somewhat similar to the Kremser equation:

$$\frac{Y''_{N+1} - Y''_1}{Y''_{N+1}} = \left(1 - \frac{L_0 X'_0}{A' V_{N+1} Y''_{N+1}}\right)\left(\frac{A_e^{N+1} - A_e}{A_e^{N+1} - 1}\right) \tag{9.29}$$

where $A'$ and $A_e$ are defined as follows:

$$\frac{A_e^{N+1} - A_e}{A_e^{N+1} - 1} = \left(\frac{A_1 A_2 A_3 \cdots A_N + A_2 A_3 \cdots A_N + \cdots A_N}{A_1 A_2 A_2 \cdots A_N + A_2 A_3 \cdots A_N + \cdots + A_N + 1}\right) \tag{9.30}$$

$$\frac{1}{A'}\left(\frac{A_e^{N+1} - A_e}{A_e^{N+1} - 1}\right) = \left(\frac{A_2 A_3 \cdots A_N + A_3 \cdots A_N + \cdots A_N + 1}{A_1 A_2 A_3 \cdots A_N + A_2 A_3 \cdots A_N + \cdots + A_N + 1}\right) \tag{9.31}$$

Edmister has shown that as a good approximation for design purposes, $A_e$ and $A'$ are independent of the number of plates and can hence be expressed in terms of the terminal conditions of the absorber, as

$$A_e = \sqrt{A_N(A_1 + 1) + 0.25} - 0.5 \tag{9.32}$$

$$A' = \frac{A_N(A_1 + 1)}{A_N + 1} \tag{9.33}$$

Figure 9.8 may be employed in the approximate solution for the right-hand term of Eq. (9.29).

In calculating the effective absorption and stripping factors, the liquid and vapor quantities $L_n$ and $V_n$ *leaving* the top and bottom plates should be used. It is necessary to make an advance estimate of the total change occurring over the absorber in order to assign values of the terminal conditions for use in the equations above.

If the percentage absorption (percent of vapor to the plate in question) is constant on each plate throughout the column and the temperature change is assumed proportional to the vapor contraction,

$$\left(\frac{V_1}{V_{N+1}}\right)^{1/N} = \frac{V_n}{V_{n+1}} \tag{9.34}$$

$$\frac{V_{n+1} - V_n}{V_{N+1} - V_1} = \frac{T_{n+1} - T_n}{T_N - T_0} \tag{9.35}$$

$L_n$ can be calculated from a molal material balance

$$L_n = L_0 + V_{n+1} - V_1 \tag{9.36}$$

$T_N$ can be calculated from a heat balance over the column.

Equation (9.28) may be used by calculating conditions on each plate through Eqs. (9.34) to (9.36) and determining absorption factors on each plate. This is one of the procedures suggested by Horton and Franklin.

**EXAMPLE 9.3** From its use as a solvent for cottonseed-oil extraction, excess pentane is removed from the solid phase by a vent gas consisting of 79 percent nitrogen and 21 percent carbon dioxide. The gas leaves the extraction process at 100°F, 1 atm, at a flow rate of 1.224 lb mole/min and is 52.77 percent saturated with pentane.

The solvent is to be recovered by absorption into a hydrocarbon oil of molecular weight 160 and density 0.84 g/cm³ in a bubble-tray column at 100°F and 1 atm. Isothermal operation is maintained by internal cooling.

Oil enters the column from a stripper containing 0.005 mole fraction of pentane at a flow rate equal to 170 percent of the minimum required for 99 percent recovery. The $K$ value [Eq. (9.3a)] for pentane may be taken as 1.0.

Estimate the number of trays required for the desired recovery.

SOLUTION  The vapor pressure of pentane at 100°F is 806.5 mm Hg.

$Y_{N+1} = \frac{1}{760} 806.5 \times 0.5277 = 0.56$

$G_{M,N+1} = 1.224$ lb moles/min, which contains $0.56 \times 1.224 = 0.685$ lb moles pentane/min

$G'_M = 1.224 \times 0.44 = 0.538$ lb moles inert gas/min

Pentane absorbed: $0.685 \times 0.99 = 0.678$ lb mole/min

$G_{M,1} = 1.224 - 0.6784 = 0.545$

$Y_1 = (0.685 \times 0.01)/0.545 = 0.0126$; $Y'_1 = 0.0128$

Minimum oil rate:

$$X_{max} = \frac{0.56}{1} = 0.56 \qquad \text{equilibrium at rich end}$$

$$0.678 = L'_M \left( \frac{0.56}{0.44} - \frac{0.005}{0.995} \right) \qquad \text{whence } L'_{min} = 0.535 \qquad \text{and}$$

$$L'_M = 0.5352 \times 1.7 = 0.909 \text{ lb moles oil/min}$$

Computation of $X_N$:

$$0.678 = 0.909(X'_N - X'_0) \qquad \text{whence } X'_N = 0.751 \qquad \text{and} \qquad X_N = 0.429$$

*Rigorous solution*   The rigorous solution can be obtained graphically in the same way as in Example 9.1:

$$\frac{L'_M}{G'_M} = 1.689 \qquad Y'_{N+1} - \frac{L'_M}{G'_M} X'_N = 0.004222$$

| $X_n$ | $X'_n$ | $Y'_{n+1}$ | $Y_{n+1}$ |
|-------|--------|------------|-----------|
| 0.4   | 0.6666 | 1.130      | 0.5307    |
| 0.3   | 0.4286 | 0.7283     | 0.4214    |
| 0.2   | 0.25   | 0.4266     | 0.2990    |
| 0.1   | 0.1111 | 0.1919     | 0.1610    |
| 0.005 | 0.005  | 0.0128     | 0.0126    |

A $YX$ diagram is set up as shown in Fig. 9.9. The step construction indicates that the number of theoretical trays required is 7.95.

*Estimation by Edmister's equation*   In order to determine the effective absorption factors through Eqs. (9.32) and (9.33), values have to be assigned to all streams leaving the top and the bottom plate: $G_{M,1} = 0.545$ lb mole/min, and $L_{M,N} = L'_M(X'_N + 1) = 0.909 \times 1.751 = 1.592$ lb mole/min. $G_{M,N}$ and $L_{M,1}$ could be estimated using Eq. (9.34) as suggested by Horton and Franklin. Due to the presence of an inert gas, however, the direct application of the equation to this case proves to yield an incorrect result. Horton and Franklin's assumptions of constant fractional overall absorption on each tray would incorrectly imply an increase of fractional absorption of pentane toward the dilute end of the tower and also an increase in the driving forces in the same direction. A much better result is therefore obtained by assuming that

$$\frac{Y_n G_{M,n}}{Y_{n+1} G_{M,n+1}} = \left( \frac{Y_1 G_{M,1}}{Y_{N+1} G_{M,N+1}} \right)^{1/N} = \left( \frac{Y'_1}{Y'_{N+1}} \right)^{1/N}$$

Rough estimate of $N$:

$$\frac{Y_{N+1} - KX_2}{Y_1 - KX_2} = 73 \qquad \left( \frac{L_M}{G_M} \right)_{av} \approx 1.5 \qquad \left( \frac{mG_M}{L_M} \right) \approx 0.67$$

whence from Fig. 9.7 or Eq. (9.25), $N \approx 9$.

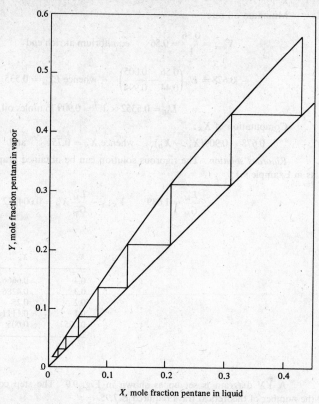

FIGURE 9.9
Design diagram for Example 9.3.

$$\frac{Y_n G_{M,n}}{Y_{n+1} G_{M,1}} = (0.01)^{1/9} = 0.599$$

$$Y_N G_{M,N} = 0.6853 \times 0.599 = 0.410 \text{ lb mole/min}$$

$$Y_2 G_{M,2} = \frac{0.00685}{0.599} = 0.0114$$

$$G_{M,N} = G'_M + Y_N G_{M,N} = 0.410 + 0.538 = 0.948 \text{ lb mole/min}$$

$$A_N = \frac{1.592}{0.948} = 1.678$$

$$L_1 = L_{M,0} + (Y_2 G_{M,2} - Y_1 G_{M,1}) = \frac{0.909}{0.995} + (0.0114 - 0.00685) = 0.918$$

and

$$A_1 = \frac{0.918}{0.545} = 1.685$$

Now the effective absorption factors may be determined.

$$A_e = \sqrt{1.678 \times 2.685 + 0.25} - 0.5 = 1.680$$

$$A' = \frac{1.6779 \times 2.6851}{2.6779} = 1.682$$

If we introduce the following notation:

$$R = \frac{(Y''_{N+1} - Y''_1)/Y''_{N+1}}{1 - L'_0 X'_0 / A' V_{N+1} Y''_{N+1}}$$

Eq. (9.29) may be expressed in the following form:

$$N = \frac{\ln\left[(A_e - R)/(1 - R)\right]}{\ln A_e} - 1$$

$$R = \frac{0.99}{[1 - (0.9098 \times 0.005025)/(1.682 \times 0.685)]} = 0.994$$

$$N = \frac{\ln(0.6866/0.006)}{\ln 1.680} - 1 \qquad N = 8.1 \qquad\qquad ////$$

## 9.5. E  Design Procedures for Multicomponent Absorption Graphical

**Design method**  The graphical design method described in Sec. 9.5.A was first applied to multicomponent systems by Lewis [30]; it is a slight modification of Lewis's method that is outlined below. For convenience, concentrations will be expressed in the following special units; $X'$ represents the moles of one solute per mole of solute-free solvent fed at the top of the countercurrent absorption column, and $Y''$ represents the moles of solute in the gas phase per mole of rich feed gas to be treated. The liquid rate will be represented by $L'$, the moles of *solvent* per unit time and the gas rate by $G''$, the moles of gas to be treated per unit time. The subscripts 1 and 2 will be used in referring to the rich and lean ends of the column, respectively. The material balance for any one component may be written

$$L'(X' - X'_2) = G''(Y'' - Y''_2) \tag{9.37}$$

$$L'(X'_1 - X') = G''(Y''_1 - Y'') \tag{9.38}$$

where $X'$ and $Y''$ represent the concentrations in the liquid and gas phases flowing across any imaginary horizontal plane located between any two trays of the column. In terms of these special units, the equilibrium relationship, Eq. (9.3a), becomes

$$\frac{Y''}{\sum Y''} = \frac{KX'}{1 + \sum X'} \tag{9.39}$$

$$Y'' = KX' \frac{\sum Y''}{1 + \sum X'} \tag{9.40}$$

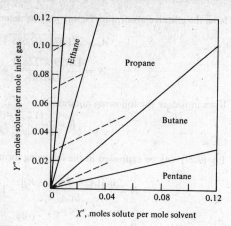

FIGURE 9.10
Graphical construction for absorption from a lean gas containing several soluble constituents present in small amounts.

where $\sum Y'' =$ sum of values of $Y''$ for components present
$\quad \sum X' =$ sum of values of $X'$ for each component, except the solvent, present in the liquid phase

In the special case of absorption from gases with large amounts of solvent, Eq. (9.40) reduces to

$$Y'' = KX' \qquad (9.41)$$

The operation of a column treating a lean gas may be represented by a plot of $Y''$ vs. $X'$, with straight equilibrium lines for each component. Each equilibrium line will pass through the origin and have a slope $K$. Each component will have its own straight operating line, and these lines will be parallel, since the slope of each is $L'/G''$. The required value of $L'/G''$ is determined by the solubility of the least soluble compound present in the gas which it is desired to absorb completely.

A typical diagram for absorption from a lean gas is shown as Fig. 9.10. The oil used as a solvent is assumed to be completely denuded ($X'_2 = 0$), and the lower ends of the dashed operating lines, at their intersections with the ordinate scale, indicate the composition of the treated gas with respect to each component. The percentage of each component absorbed may be determined by comparison of the values of $Y''$ at the two ends of each operating line. Thus 50 percent of the butane is absorbed, and the oil leaves the tower containing 0.05 mole butane per mole of solute-free oil. The operating line for propane is drawn parallel to that for butane and is placed so that the change in $Y''$, that is, $Y''_1 - Y''_2$, divided by the mean driving force $\Delta Y''$ (mean vertical distance between operating and equilibrium lines) is the same as for butane. In the case illustrated by Fig. 9.10 the equilibrium lines are straight, and a logarithmic mean $\Delta Y''$ may be employed. This may be obtained for butane by calculating the logarithmic mean of the values of $\Delta Y''$ at the two ends of the butane operating line. The propane line is then placed by trial so that $(Y_1'' - Y_2'')/\Delta Y_M''$ for propane is equal to the corresponding ratio for butane.

It is evident from the plot that the upper end of the operating line for propane must be very near the propane equilibrium curve, since the large slope of the latter curve reduces the amount absorbed and at the same time increases the driving force at the lean end of the column. In other words, the oil leaving the column is very nearly saturated with propane. In the case of ethane, the equilibrium curve is still steeper, and the oil leaving will be even more nearly saturated with ethane. The slope of the equilibrium curve for pentane is less than the slope of the operating line, so that the driving force is largest at the rich end of the column. The driving force at the rich end is fairly large compared with the total amount of pentane absorbed, and the operating and equilibrium curves will approach each other at the lean end, i.e., the pentane content of the gas leaving the column is essentially in equilibrium with the oil entering; substantially all the pentane is removed from the gas.

The "key component," in this case the butane, is defined as that component absorbed in appreciable amount whose equilibrium curve falls most nearly parallel to the operating line, i.e., the component having a value of $K$ most nearly equal to $L'/G''$. In general, the composition of the gas with respect to components more volatile than the key component approaches equilibrium with the liquid phase at the rich end of the column, and the composition of the gas with respect to components less volatile than the key component approaches equilibrium with the oil entering. Thus, components heavier than the key component are absorbed nearly completely if the fresh solvent contains none of these components. Varying the oil-gas ratio clearly changes the nature of the key component; the optimum ratio is the one which will allow substantially complete absorption of the desired components.

The theoretical-plate method may be employed [51] to fix the operating lines, as indicated by Fig. 9.11. This plot is quite similar to Fig. 9.10, with the addition of the steps used in counting theoretical plates. Assuming any one operating line to be fixed, as may be done by fixing the oil-gas ratio and fraction of the butane absorbed, the other operating lines must be drawn with the same slope and placed to give the same number of theoretical plates. Except for the method of placing the various operating lines, the construction is the same as that of Fig. 9.10.

Figure 9.11 serves to illustrate the differences in absorption distribution through the column for the various components. Starting at the ordinate scale at the left (representing the top of the column) and counting off one step for each component, the points $c$, $c'$, $c''$, etc., on the various operating lines represent the composition of the gas entering and of the liquid leaving the top plate of the column. For ethane the composition of the gas entering the top plate (point $c$) is nearly equal to the composition of the gas entering the column, $Y_1'$. For pentane it is evident that no appreciable change occurs in the gas composition in passing through the top plate. It follows that the absorption of the more volatile components takes place in the upper part of the column, and the less volatile components are absorbed in the lower section. The analysis of the oil-solute mixture on any plate may be determined by reading from the plot the values of $X'$ for each of the various components on that plate.

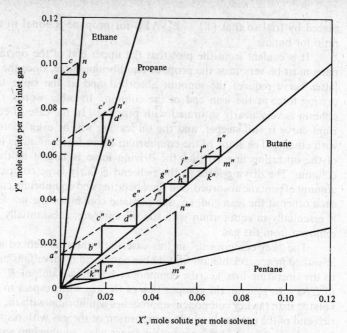

**FIGURE 9.11**
Construction for counting theoretical plates in multicomponent absorption. Solid lines are equilibrium lines; dashed lines are operating lines.

**Algebraic method** The method of calculation outlined above based on the theoretical-plate concept, may be followed algebraically, using the Kremser formula discussed in Sec. 9.5.C. In terms of the molal stoichiometric units, this equation becomes

$$\frac{Y''_1 - Y''_2}{Y''_1 - m'X'_2} = \frac{(L'/m'G'')^{N+1} - (L'/m'G'')}{(L'/m'G'')^{N+1} - 1} \tag{9.42}$$

where, as before, $N$ represents the number of theoretical plates. It applies only when the equilibrium curve is a straight line through the origin with a slope $m'$. In the case of lean-gas absorption with a large amount of solvent, Eq. (9.41) applies, and $m' \approx K = Y/X$. The left-hand side represents the ratio of the amount of solute absorbed to the amount which would be absorbed were the gas to leave in equilibrium with the entering oil, and consequently represents the efficiency of absorption of any one component. If the oil enters solute-free, as in Figs. 9.10 and 9.11, this ratio becomes equal to the fraction of the component entering the column which is absorbed. Where $N$, $L$, and $G''$ are fixed, the fraction absorbed may be computed directly for each component, and the operating lines need not be placed by trial and error.

It is apparent from Eq. (9.42) that where $m'$ is greater than $L'/G''$ and $N$ is larger, the result is

$$\frac{Y''_1 - Y''_2}{Y''_1 - m'X'_2} = \frac{L'}{m'G''} \tag{9.43}$$

This is also evident from Fig. 9.8, which represents Eq. (9.42). The same conclusion may be reached by referring to Fig. 9.10 or 9.11, from which it follows easily by simple geometry that in cases where equilibrium is reached at the rich end of the column, the change in $Y''$, divided by $Y''_1$, is equal to the ratio of slopes of operating and equilibrium curves. It is usually convenient to estimate the fraction absorbed for each of the more volatile components by use of Eq. (9.43). The procedure is satisfactory in the case of components for which the value of $m'$ is three to four times the value of $m'$ for the key component.

Furthermore, it is apparent that where $m'$ is small compared with $L'/G''$, and $N$ is large, the right-hand side of Eq. (9.42) reduces to unity, which means that the gas leaves in equilibrium with the entering oil. This is also shown by Fig. 9.8 and is apparent from inspection of Figs. 9.10 and 9.11. For the least volatile components, therefore, equilibrium at the lean end of the column may be assumed. If the oil used is solute-free, this corresponds to complete absorption of these components.

Where $L'/m'G'' = 1$, the right side of Eq. (9.42) becomes indeterminate. It may be shown, however, that for this condition,

$$\frac{Y''_1 - Y''_2}{Y''_1 - m'X'_2} = \frac{N}{N + 1} \tag{9.44}$$

It is seen that the computations are simple for the relatively volatile and nonvolatile components. It is only for the key component, and for the components on either side of the key component with respect to volatility, that the graphical method or the use of Eq. (9.42) is necessary.

Precise plate-by-plate computations for multicomponent systems in which the dilute-gas assumption cannot be justified, or for which isothermal absorption cannot be assumed, are best conducted by digital computer. A range of computer programs applicable to absorption and other separation processes have been prepared by Hanson, Duffin, and Somerville [18]. Because of the voluminous and specialized nature of the subject, computer techniques will be omitted in the present discussion.

For cases in which the absorption factor may be assumed to vary in a fairly regular manner throughout the column, i.e., no pinch points between the top and bottom plates, the Horton, Franklin, and Edmister algebraic method described in Sec. 9.5.D may be employed for approximate estimates.

### 9.5.F   Application of the Graphical Method to the Absorption of Rich Gases

**Isothermal Column**   Because of the variation through the column of the amount of solutes carried by the oil, the equilibrium lines are curved, their curvature depending on the fraction of the gas absorbed and on the relative amounts of gas and liquid.

This is evident from the equilibrium relation

$$Y'' = KX' \frac{\sum Y''}{1 + \sum X'} \tag{9.40}$$

For lean gases and high oil rates, the fractional term may be assumed to approximate unity, as was done in the preceding section; but for rich gas or for low oil rates, the variations in the terms $\sum Y''$ and $1 + \sum X'$ must be allowed for, unless the two flow rates on a molar basis should be equal at some level in the absorber. Where the column is supplied with oil containing no dissolved solute, the equilibrium line may be located by two facts: (1) the slope at the origin is $K\sum Y_2''$, as determined by differentiating Eq. (9.40), and (2) a point on the equilibrium curve for conditions at the rich end of the column is determined by substituting $Y_1''$ or $X_1'$, $\sum Y_1''$, and $\sum X_1'$ in Eq. (9.40). $\sum Y_1''$ is unity by definition, and $\sum X_1'$ may be calculated if the total moles of all components absorbed in the column is known. In many practical problems, the curvature of the equilibrium lines is not serious, and the lines may be placed with sufficient accuracy if the slope at the origin and the location at the rich end are known in each case. The error involved in locating these lines is negligible except for the key component, for which certain additional calculations described below may be necessary. The equilibrium lines are straight over the entire range, even for rich gas or low oil rates, providing the total moles of liquid flowing is equal to the total moles of gas passing any point in the column. (If the flow rates are equal at any point, they are equal at all points in the column.)

The exact location of the equilibrium lines is determined by the total absorption, and the amount absorbed is dependent on the equilibrium relations. Consequently, the procedure is necessarily by trial and error, although it is possible to reduce the necessary computations greatly if a logical method is employed for the first estimates. For the volatile components, the absorption is determined by the location of the equilibrium curve at the rich end of the column, and the shape and exact location of the remainder of the curve are unimportant. For the absorption of these volatile components from a lean gas, the performance of the column was shown above to be

$$\frac{Y_1'' - Y_2''}{Y_1'' - m'X_2'} = \frac{L'}{m'G''} \tag{9.43}$$

In Eq. (9.43), $m'$ represents the slope of a straight line through the origin and through a point on the equilibrium curve at the rich end of the column. If the solvent supplied is solute-free, $X_2'$ is zero, and the relation gives the fraction absorbed. Allowing for the solvent effect of dissolved solute and neglecting the solute content of the liquid, this becomes

$$\frac{Y_1'' - Y_2''}{Y_1''} = \frac{L'(1 + \sum X_1')}{K_1 G'' \sum Y_1''} = \frac{L'(1 + \sum X_1')}{K_1 G''} \tag{9.45}$$

where $K_1 \sum Y_1''/(1 + \sum X_1')$ has replaced $m'$ in Eq. (9.43), $K_1$ referring to the value of $K$ at the bottom of the absorber. The moles of total gas to be treated, $G''$, is usually specified, but the moles absorbed, $L' \sum X_1'$, must first be estimated. This is best done by assuming straight equilibrium curves of slope $K$ and using Eq. (9.42) to estimate the fraction absorbed for each component. If the preformance of the tower with respect to one component is specified, Eq. (9.42) or Fig. 9.8 may be used first to estimate the required number of perfect plates and then to estimate the absorption of the other components. This procedure leads to an estimate of the total absorption of all components, and consequently of the value of $\sum X'_1$, which may be employed to repeat the calculation, this time using $L'(1 + \sum X'_1)/KG''$ in place of $L'/m'G''$.

The second calculation will be very nearly correct, except for the key component and possibly one component on either side of the key component. For these it is necessary to construct a graph, placing the equilibrium lines from the known slope at the origin and from the known location of the curve at the rich end, and placing the operating lines by trial until the correct number of plates is counted. A new value of the total absorption is obtained by addition. If this is appreciably different from that obtained by the first trial, a third calculation should be made, using the corrected value of $\sum X'_1$.

In many cases it is possible to make a rough guess of the absorption by inspection from the gas analysis. Where this can be done, the first trial calculation using a straight equilibrium curve may be omitted. More than three trials are very seldom required, and in cases where the total moles of oil leaving the column is approximately the same as the total moles of the gas entering, the equilibrium curves are nearly straight, and the first trial is sufficient.

The use of the inlet gas as a "basis" for all gas composition is simple and convenient. The distance from equilibrium at any point in the tower expressed as a $\Delta Y''$ is not, however, strictly proportional to the diffusional driving force called for by the diffusion equations. The plate concept is commonly used in this type of absorber, however, and the choice of basis for expressing concentrations has no effect on the calculation of the number of the theoretical plates.

The design calculation outlined above for multicomponent systems may be summarized in the form of a list of steps given below in order. This tabulation assumes that the problem is to calculate the performance of a column, given the operating conditions, solvent and gas rates, and composition of the feed gas. Minor modifications of this procedure allow the calculation of the number of theoretical plates when the performance of the column is specified, etc.

*1*  Determine from suitable sources the value of $K$ for each component for the conditions prevailing in the column.

*2*  Calculate $L'$ and $G''$, the moles of fresh solvent and inlet gas per unit time.

*3*  Make a rough estimate of the total absorption, as moles per 100 moles feed gas.

*4*  Calculate $\sum X'_1$, based on this estimate of the total absorption. Calculate values of $L'(1 + \sum X'_1)/KG''$, and employ Eq. (9.42), using these ratios in place of $L'/m'G''$, to obtain a second estimate of the absorption of the very volatile and the nonvolatile components.

*5*  For the key component (or components), construct a plot of $Y''$ vs. $X'$, and draw a line with a slope $K\sum Y''_2$ through the origin. Locate the point

$Y_1''$, $X_1'$ (equilibrium) from Eq. (9.40), and draw the equilibrium curve in its approximate position.

6  Locate the operating line for the key component with a slope $L/G''$, and in a position above the equilibrium curve corresponding to the proper number of theoretical plates. From this line read off $Y_2''$ for the key component, and calculate the moles of that component absorbed.

7  Add the moles of the components absorbed to obtain a third estimate of the total absorption. If this differs appreciably from the value obtained under step 4, repeat steps 4, 5, and 6.

The procedure just outlined is not difficult to follow, but it is suggested that the following illustrative problem be studied carefully if a clear understanding of the method of calculation is desired.

EXAMPLE 9.4  A 24-plate absorber is designed to operate at 470 psig, treating 18,780,000 ft$^3$/24 h (1 atm, 60°F) of a gas containing 83.02 percent CH$_4$, 8.41 percent C$_2$H$_6$, 4.76 percent C$_3$H$_8$, 0.84 percent i-C$_4$H$_{10}$, 1.66 percent n-C$_4$H$_{10}$, 0.61 percent i-C$_5$H$_{12}$, 0.16 percent n-C$_5$H$_{12}$, and 0.54 percent C$_6$H$_{14}$ and higher. The tower will be supplied with 111,840 gal/24 h of a denuded oil having an average molecular weight of 161 and a specific gravity of 0.8363 at the temperature at which it is metered. The tower will operate with an average oil temperature of 87°F. Assuming the Atkins and Franklin [1] value of 18 percent for the overall plate efficiency (see below), calculate the composition of the treated gas.

SOLUTION

$$\text{Inlet gas} = \frac{18,780,000}{359}\frac{492}{520}\frac{1}{(24)(60)} = 34.4 \text{ lb moles/min}$$

$$\text{Inlet oil} = \frac{111,840}{(24)(60)} \text{ gal/min} \frac{(8.33)(0.8363)}{161} = 3.36 \text{ lb moles/min}$$

$$\text{Operating pressure} = 1 + \frac{470}{14.7} = 33 \text{ atm}$$

Because of the high proportion of methane and ethane in the gas, the total absorption will not be large, even at this relatively high pressure. Assume 10 moles absorbed per 100 moles inlet gas. Values of $K$ are obtained from Ref. 48 by interpolation (graphically) for 87°F at 33 atm. The tabulation of the values of $K$ completes steps 1, 2, and 3.

$$\sum X_1' = \frac{10}{100} 34.4 \frac{1}{3.36} \text{ moles absorbed/min} = 1.02 \text{ moles/mole oil}$$

$$\frac{L'(1 + \sum X_1')}{G''} = \frac{3.36(1 + 1.02)}{34.4} = 0.197$$

Values of $L'(1 + \sum X_1')/KG''$ are tabulated and used in place of $L'/m'G''$ in Eq. (9.42) to calculate the absorption of each component. The number of theoretical plates, $N$, is taken as 4.3, corresponding to the assumed plate efficiency of 18 percent. For n-C$_4$H$_{10}$, for example,

$$\text{Fraction absorbed} = \frac{0.94^{5.3} - 0.94}{0.94^{5.3} - 1.0} = 0.79$$

| Component | Mole percent | $K$ | $[L'(1 + \Sigma X_i')]/K G''$ | Percent absorbed | Moles absorbed |
|---|---|---|---|---|---|
| $CH_4$ | 83.02 | 5.7 | 0.035 | 3.5 | 2.9 |
| $C_2H_6$ | 8.41 | 1.2 | 0.164 | 16.4 | 1.4 |
| $C_3H_8$ | 4.76 | 0.51 | 0.386 | 38.6 | 1.8 |
| $i\text{-}C_4H_{10}$ | 0.84 | 0.27 | 0.73 | 67 | 0.6 |
| $n\text{-}C_4H_{10}$ | 1.66 | 0.21 | 0.94 | 79 | 1.3 |
| $i\text{-}C_5H_{12}$ | 0.61 | 0.09 | 2.19 | 100 | 0.6 |
| $n\text{-}C_5H_{12}$ | 0.16 | 0.072 | 2.74 | 100 | 0.2 |
| $C_6H_{14}$ | 0.54 | ...... | ..... | 100 | 0.5 |
| | | | | | 9.3 |

Since the total absorption checks closely the assumed value of 10 moles, the calculation may be continued with step 5, a more careful treatment of the key component. The previous calculation showed the key component to be $n$-butane, for which it is necessary to plot equilibrium and operating lines. The slope of the equilibrium curve at the origin is

$$K(\textstyle\sum Y_2'') = (0.21)(1 - 0.093) = 0.19$$

The dotted line $OB$ of Fig. 9.12 is drawn with this slope. The value of $Y''$ in equilibrium with $X_1'$ is obtained from Eq. (9.40):

$$Y''(\text{equilibrium}) = 0.21 \frac{(1.3)(34.4)}{(100)(3.36)[1 + (9.3/100)(34.4/3.36)]} = 0.0144$$

This locates one point on the equilibrium curve near the rich end of the column, as shown by the point $A$ on Fig. 9.12. The equilibrium curve is drawn in its approximate position, starting at the origin with the slope $OB$ and curving off to pass through $A$.

The slope of the operating line is $L'/G'' = 3.36/34.4 = 0.0975$. A straight line with this slope is placed by trial in a position such that 4.3 steps corresponding to theoretical plates can be stepped off between operating and equilibrium curves.

The $n$-butane content of the gas leaving the column is read as the intercept of the operating line with the $Y''$ axis as $Y_2'' = 0.0045$. The moles of $n$-butane absorbed is consequently $1.66 - 0.45 = 1.21$, vs. 1.3 previously estimated. The correct total absorption is 9.2 moles, checking the original estimate sufficiently closely so that further computations are not necessary.

The composition of the gas leaving the column may now be calculated.

| Component | Feed, mole percent | Moles absorbed | Moles in exit gas | Moles percent in exit gas | |
|---|---|---|---|---|---|
| | | | | Calculated | Observed |
| $CH_4$ | 83.02 | 2.9 | 80.1 | 88.4 | 88.7 |
| $C_2H_6$ | 8.41 | 1.4 | 7.0 | 7.7 | 6.9 |
| $C_3H_8$ | 4.76 | 1.8 | 3.0 | 3.3 | 3.5 |
| $i\text{-}C_4H_{10}$ | 0.84 | 0.56 | 0.28 | 0.31 | 0.42 |
| $n\text{-}C_4H_{10}$ | 1.66 | 1.21 | 0.45 | 0.49 | 0.47 |
| $i\text{-}C_5H_{12}$ | 0.61 | 0.6 | 0.00 | | |
| $n\text{-}C_5H_{12}$ | 0.16 | 0.2 | 0.00 | | |
| $C_6H_{14}$ | 0.54 | 0.5 | 0.00 | | |
| | | | 90.83 | | |

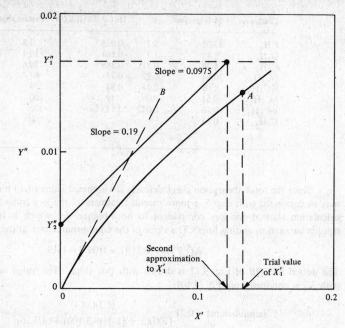

**FIGURE 9.12**
Curved equilibrium line for key component obtained in case of low oil rate (Example 9.4).

The conditions described in Example 9.4 are those of a test reported by Brown and Souders [6] in which an actual natural gasoline absorber was tested under the conditions described. The measured analysis of the exit gas is given in the last column of the table for comparison with the calculated composition. The agreement between calculated and observed compositions of the treated gas is seen to be quite good. The calculations show 73 percent absorption for $n$-butane, whereas the actual absorption was about 69 percent. This discrepancy is probably within the accuracy of the analysis for the small quantity of $n$-butane in the lean gas.

Similar calculations for hydrocarbon absorbers have been presented by Jackson and Sherwood [25] who found that overall tower performance in several tests was consistent with plate-efficiency data obtained in small laboratory equipment.

Available data on plate efficiencies for absorbers are presented in Chap. 11. The low value found in the example above is consistent with other data. Distillation columns usually give plate efficiencies which are considerably higher than 18 percent, but it must be remembered that simple liquid mixtures flowing through such columns are at their boiling points and therefore have viscosities of the order of 0.2 to 0.3 cP, whereas the viscosity of the liquid phase in an absorber may be several times greater, and the plate efficiency correspondingly lower.

Although the plate efficiency may be somewhat different for the different hydrocarbon gases, as discussed in Chap. 11, the numbers of moles of the gases heavier or lighter than the key component which are absorbed are almost independent of the plate efficiencies of these components, since the number of theoretical plates is usually sufficient to absorb substantially all the heavier gases or to allow the lighter ones to reach almost complete equilibrium with the exit oil. Thus, only the recovery of the key and adjacent components is affected by the value of plate efficiency; conversely, test data for an absorber can be expected to give reliable plate-efficiency data only for the key component.

The effect of changes in the equilibrium data on the computed plate efficiency is brought out by the fact that Brown and Souders [6] calculate an overall plate efficiency of 50 percent from these data using different values of $K$ [7].

The equilibrium curve for the key component may be placed much more accurately if an additional point, corresponding to the middle of the column, is located. The absorption of the components more volatile than the key component will take place in the upper part of the column; the absorption of the less volatile will take place on the lower plates. Consequently $1 + \sum X'$ for the middle plate may be estimated closely; and by using Eq. (9.40), the coordinates of a point on the equilibrium curve at about the middle of the column can be found. This additional calculation removes some of the uncertainty in placing the curved equilibrium line for the key component and improves the accuracy of the method. If the key component is one of the least volatile constituents, almost all the curvature will occur near the origin (see Fig. 9.12); if the key component is one of the most volatile constituents, the equilibrium line will follow the original slope ($OB$ of Fig. 9.12) for a considerable distance from the origin.

### 9.5.G  Allowance for Thermal Effects

In an adiabatic absorber almost all the heat of solution of the dissolved gases must be compensated for by an increase in the sensible heat of the liquid stream. If the resulting temperature rise of the solvent as it flows through the tower becomes too great, the vapor pressure of the dissolved hydrocarbons over the exit oil may become excessively high. This situation resembles that described in Sec. 9.5.B in connection with the determination of the minimum solvent-flow rate, and similar considerations apply.

The latent heats of condensation of hydrocarbons are relatively small, however; and as the gases dissolve in liquids of very similar molecular structure, the heats of mixing are negligibly small. Consequently, the temperature rise of the solvent passing through the column is not large unless the quantity of solvent is small, as in absorbers which operate at high pressure, or unless a large fraction of the rich gas is recovered.

Commercial natural-gasoline absorbers are sometimes operated so that the stream of oil is diverted through a heat exchanger, or "intercooler," and then returned to the column, in order to remove the heat of solution. This sometimes proves to be economically attractive. In one series of commercial tests, the use of

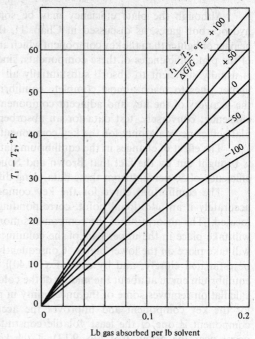

FIGURE 9.13

Temperature rise due to absorption of dissolved solutes; oil absorption of light hydrocarbons. The ordinate $T_1 - T_2$ is the rise in temperature of the liquid, top to bottom. The inlet gas temperature is $t_1$, and $\Delta G/G$ is the ratio of total absorption to gas fed, pound per pound.

an intercooler increased the absorption from 25.1 to 36.8 percent of the total moles of gas entering [25]. The rate of absorption was more than twice as great in the bottom half of the column when an intercooler was used as when the column was operated adiabatically.

Figure 9.13 may be used for approximate predictions of the temperature rise of the solvent in a natural-gasoline absorber which operates at a pressure lower than about 10 atm. This chart is based on a latent heat of 160 Btu/lb for all hydrocarbons condensed and on heat capacities of 0.5 and 0.46 Btu/(lb)(°F) for rich oil and rich gas, respectively. The lean gas is assumed to leave at the temperature of the lean oil, and the sensible heat gained or lost by the gas is allowed for by means of the parameter $(t_1 - T_2)/(\Delta G/G)$. (All flow rates are expressed in pounds rather than in pound moles.) It might be expected that the temperature rise of the solvent in a high-pressure absorber would be relatively large, owing to the low oil-gas ratio. The effect of pressure on the enthalpy of hydrocarbon vapors, however, is such that the heat of solution is smaller. At 33 atm, the temperature rise may be only 40 percent as great as indicated by the curve [7]. The temperature rise for such cases must be computed from an enthalpy balance involving actual thermal data for the hydrocarbons under pressure. Modification of the design diagram to include heat effects is discussed in Sec. 9.6.I.

## 9.6  CONTINUOUS DIFFERENTIAL-CONTACT PACKED-COLUMN DESIGN

The stagewise contactor, such as the bubble-plate column or mixer, provides interfacial surface for mass transfer by forming an intimate dispersion of gas in the liquid in the form of bubbles or foam. An alternate method of providing mass-transfer surface is to allow the liquid to flow over a subdivided solid material or packing and to pass the gas through the solid-liquid matrix. In such a device the gas and liquid phases are more or less continuously in contact as they pass through the apparatus, and hence it may be termed a "continuous differential contactor." Commonly, in absorption the gas and liquid are passed countercurrent to each other in such a tower filled with packing. Packed columns and typical packings are described in Chap. 11.

This section will consider various methods for estimating the required tower height, or depth of packing, for absorption of one component from a gas stream into a nonvolatile liquid. In Secs. 9.6.A, 9.6.B, and 9.6.C methods of rigorous calculations are described. In Secs. 9.6.D through 9.6.G simplified procedures are presented for shortcut estimates which are often useful for preliminary design purposes.

A rigorous method of computation employing a digital computer is presented in the appendix. An alternate design procedure (based on the height equivalent to a theoretical plate) of intermediate difficulty and accuracy is described in Sec. 9.8.

Consider the packed absorption tower shown schematically in Fig. 9.14. The tower has a total cross-sectional area $S$ perpendicular to the direction of flow. The packing provides an effective mass-transfer area $a_v$ per unit packed volume, and the total height of the packed section is $h_T$. Absorption is assumed to occur at constant temperature and pressure.

Gas enters the bottom of the tower at a molal mass velocity $G_{M1}$ moles per hour per square foot based on the tower cross section perpendicular to the flow, and liquid enters at the top of the tower at mass velocity $L_{M2}$ moles per hour per square foot. Mole fractions of the absorbed component are $Y$ and $X$ in the gas and liquid phases, respectively. The gas and liquid are assumed to pass through the tower in plug flow, i.e., with uniform composition and velocity over any horizontal cross section. (Effects of nonuniform flow distribution and axial mixing are discussed in Chap. 4 for single-phase flow and in Chapter 11 for a countercurrent two-phase-flow system.)

At any height $h$ in the tower the component being absorbed passes from the gas into the liquid at a rate which depends upon the mass-transfer coefficients and concentrations in each phase, the interfacial area and the gas solubility. As in the

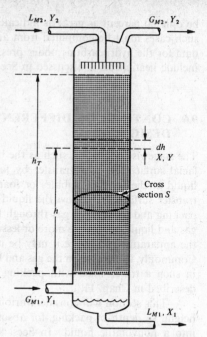

**FIGURE 9.14**
Continuous differential contactor.

case of the plate column, the process may be represented by operating and equilibrium lines on a $YX$ diagram. It may be assumed that the terminal flows and compositions are determined by an overall material balance analogous to Eq. (9.8):

$$G'_M \frac{Y_1}{1 - Y_1} + L'_M \frac{X_2}{1 - X_2} = G'_M \frac{Y_2}{1 - Y_2} + L'_M \frac{X_1}{1 - X_1} \tag{9.46}$$

where $G'_M$ and $L'_M$ are the molal mass velocities of solute-free gas and liquid respectively. From knowledge of the flow rates and composition of the entering streams and of the specified outlet gas composition $Y_2$, the outlet liquid composition $X_1$ may be obtained from Eq. (9.46).

The equation for the operating line, representing the relation between $Y$ and $X$ at any horizontal section is obtained by a material balance around the bottom (or top) to the section considered:

$$\frac{Y}{1 - Y} = \frac{L'_M}{G'_M} \frac{X}{1 - X} + \frac{Y_1}{1 - Y_1} - \frac{L'_M}{G'_M} \frac{X_1}{1 - X_1} \tag{9.47}$$

If the streams are sufficiently dilute to permit the assumption of constant flows, Eq. (9.47) may be written

$$Y = \frac{L_M X}{G_M} + Y_1 - \frac{L_M X_1}{G_M} \tag{9.48}$$

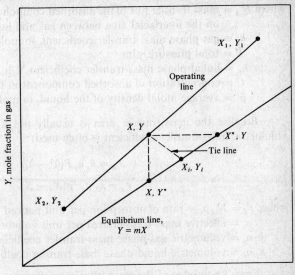

FIGURE 9.15
Graphical design method based on single-phase coefficients.

The equilibrium line is obtained from the gas solubility, which may be represented in the general form

$$Y = f(X) \tag{9.49}$$

where $f(X)$ is given by Eq. (9.3), (9.4), or other suitable expressions.

Figure 9.15 illustrates the equilibrium and operating lines for the case when $f(X)$ is linear in the form of Eq. (9.3$b$).

## 9.6.A   Design Procedure Based on Single-Phase Coefficients

Consider the absorption process as it occurs at any point $X$, $Y$ on the operating line for the tower, as shown in Fig. 9.15. The phases are assumed to be in equilibrium at the gas-liquid interface at compositions $X_i$, $Y_i$. In steady state, the mass-transfer flux may be expressed in terms of the mass-transfer coefficients and driving forces for each phase:

$$N_A = k_g P(Y - Y_i) \tag{9.50}$$

$$N_A = k_L(C_i - C) = k_L \bar{\rho}(X_i - X) \tag{9.51}$$

where $N_A$ = mass-transfer flux of the absorbed component $A$, lb mole/(h) (ft$^2$) based
    on the interfacial area between gas and liquid
 $k_g$ = gas-phase mass-transfer coefficient, lb mole/(h) (ft$^2$) (atm)
 $P$ = total pressure, atm
 $k_L$ = liquid-phase mass-transfer coefficient, ft/h
 $C$ = concentration of absorbed component in the liquid, lb mole/ft$^3$
 $\bar{\rho}$ = average molal density of the liquid, lb mole/ft$^3$ (assumed constant)

Because the mass-transfer area is usually not known in packed towers, a
volumetric mass-transfer coefficient is often used:

$$r_A = k_g a_v P(Y - Y_i) \tag{9.52}$$

or
$$r_A = k_L a_v \bar{\rho}(X_i - X) \tag{9.53}$$

where $r_A = N_A a_v$ = rate of absorption per unit packed volume
 $a_v$ = effective mass-transfer area per unit volume
 $k_g a_v$ = volumetric gas-phase mass-transfer coefficient, moles/(h) (ft$^3$) (atm)
 $k_L a_v$ = volumetric liquid-phase mass-transfer coefficient, h$^{-1}$

From knowledge of the mass-transfer coefficients (Chap. 11), Eqs. (9.50) and
(9.51) or (9.52) and (9.53) may be used to determine the compositions at the interface and the mass-transfer flux. Equating the rates of transfer in the gas and liquid
phases gives

$$\frac{Y - Y_i}{X - X_i} = -\frac{k_L \bar{\rho}}{k_g P} = -\frac{k_L a_v \bar{\rho}}{k_g a_v P} \tag{9.54}$$

Thus, according to Eq. (9.54), the interfacial composition lies on a line of negative
slope $k_L \bar{\rho}/k_g P$ connecting the operating line point $X$, $Y$ with the equilibrium line.
Values of the interfacial composition obtained in this manner may be substituted
into Eq. (9.50) or (9.51) to give the rate of absorption corresponding to any composition on the operating line.

Point values of the rate may be coupled with a material balance to calculate
the packed height required to effect a desired composition change over the
tower. Referring to Fig. 9.14, consider a differential element of packed height $dh$
over which the gas-stream composition changes by an amount $dY$ with a point
value mass-transfer flux $N_A$. A material balance for the absorbed component gives

$$-d(G_M Y) = N_A a_v \, dh = -G_M \, dY - Y \, dG_M \tag{9.55}$$

If only one component is transferred,
$$dG_M = -N_A a_v \, dh \tag{9.56}$$

Combining Eqs. (9.55) and (9.56) gives

$$dh = -\frac{G_M \, dY}{N_A a_v (1 - Y)} \tag{9.57}$$

Substituting for $N_A$ from Eq. (9.50) and integrating over the tower gives

$$h_T = \int_1^2 dh = \int_{Y_2}^{Y_1} \frac{G_M \, dY}{k_g P a_v (1 - Y)(Y - Y_i)} \qquad (9.58)$$

A similar expression is obtained using $k_L$ and $X$.

Equation (9.58) may be integrated graphically by evaluating its component terms at a series of points on the operating line. Alternatively, the integration may be easily performed numerically by digital computer (see the end of chapter).

### 9.6.B  Design Procedure Based on the Height of a Transfer Unit

Mass-transfer data for absorption are not usually available in a form most suitable for direct use in Eq. (9.58). As discussed in Chap. 5, for mass transfer of one component, the group $G_M/k_g P Y_{BM}$ or $G_M/k_g' P$ is theoretically independent of concentration and pressure. Therefore, if Eq. (9.58) is multiplied and divided by the factor $Y_{BM}$,

$$h_T = \int_{Y_2}^{Y_1} \left( \frac{G_M}{k_g P Y_{BM} a_v} \right) \left[ \frac{Y_{BM} \, dY}{(1 - Y)(Y - Y_i)} \right] \qquad (9.59)$$

$$Y_{BM} = \frac{(1 - Y) - (1 - Y_i)}{\ln [(1 - Y)/(1 - Y_i)]} \qquad (9.60)$$

The left-hand term under the integral has the dimensions of length (or height) and is designated the gas-phase "height of a transfer unit," $H_G$. Thus,

$$H_G = \frac{G_M}{k_g a_v P Y_{BM}} \qquad (9.61)$$

As noted above, $H_G$ is theoretically independent of concentration and pressure.

Gas-phase absorption data are often reported in the form of $H_G$. Unless the changes in flow rate and in $a_v$ are large over the tower, it is satisfactory to choose an average value of $H_G$ and remove it from the integral, so that Eq. (9.59) becomes

$$h_T = (H_G)_{\text{av}} \int_{Y_2}^{Y_1} \frac{Y_{BM} \, dY}{(1 - Y)(Y - Y_i)} \qquad (9.62)$$

The integral of Eq. (9.62) is designated as the "number of gas-phase transfer units" $N_G$. Thus,

$$N_G = \int_{Y_2}^{Y_1} \frac{Y_{BM} \, dY}{(1 - Y)(Y - Y_i)} \qquad (9.63)$$

The tower height is therefore equal to the product of the number of transfer units and the height of a transfer unit:

$$h_T = (H_G)_{\text{av}} N_G \qquad (9.64)$$

In a qualitative sense, the number of transfer units required represents the difficulty of the separation to be made. A high degree of absorption requires a large number of transfer units. Similarly, the height of a transfer unit indicates inversely the relative ease with which a given tower and packing can accomplish mass transfer. As can be seen from Eq. (9.61), a large mass-transfer coefficient and large interfacial area per unit volume will tend to give a low value of $H_G$. A quantitative significance can be given to $N_G$ in special cases. Consider, for example, a dilute-gas system for which the operating and equilibrium lines are straight and parallel so that $(Y - Y_i)$ is approximately constant. Equation (9.63) gives

$$N_G \approx \frac{\int_2^1 dY}{(Y - Y_i)_{av}} = \frac{Y_1 - Y_2}{(Y - Y_i)_{av}} \tag{9.65}$$

Therefore, in this case, one transfer unit corresponds to the height of packing over which the gas-stream composition changes by an amount equal to the average mass-transfer driving force.

An alternate set of design equations may be formulated based on use of the liquid-phase mass-transfer coefficient [Eq. (9.51)] to express the rate of absorption. The material balance for the absorbed component is given by

$$d(L_M X) = N_A a_v \, dh = L_M \, dX + X \, dL_M$$

Setting
$$dL_M = N_A a_v \, dh$$

and substituting $N_A$ from Eq. (9.51) gives

$$h_T = \int_1^2 dh = \int_{X_1}^{X_2} \frac{L_M \, dX}{k_L \bar{\rho} a_v (X_i - X)(1 - X)} \tag{9.66}$$

multiplying and dividing by $X_{BM}$

$$h_T = \int_{X_1}^{X_2} \left( \frac{L_M}{k_L \bar{\rho} X_{BM} a_v} \right) \left[ \frac{X_{BM} \, dX}{(1 - X)(X_i - X)} \right] \tag{9.67}$$

where
$$X_{BM} = \frac{(1 - X) - (1 - X_i)}{\ln \left[ (1 - X)/(1 - X_i) \right]} \tag{9.68}$$

The liquid-phase height of a transfer unit is defined by the left-hand term under the integral:

$$H_L = \frac{L_M}{k_L \bar{\rho} X_{BM} a_v} \tag{9.69}$$

The number of liquid-phase transfer units is given by the right-hand term under the integral:

$$N_L = \int_{X_1}^{X_2} \frac{X_{BM} \, dX}{(1 - X)(X_i - X)} \tag{9.70}$$

If the flows, temperature, and physical properties of the liquid do not change appreciably over the tower, an average value of $H_L$ may be used, and the tower height is given by

$$h_T = (H_L)_{av} N_L \qquad (9.71)$$

Evaluation of the integrals in Eqs. (9.62), (9.63), (9.66), or (9.67) requires knowledge of the interfacial composition $X_i$ or $Y_i$. These compositions may be related to the gas- and liquid-phase heights of a transfer unit through an expression equivalent to Eq. (9.54):

$$\frac{Y - Y_i}{X - X_i} = -\frac{L_M}{G_M} \frac{H_G}{H_L} \frac{Y_{BM}}{X_{BM}} = -\frac{L'_M}{G'_M} \frac{(1 - Y)}{(1 - X)} \frac{H_G}{H_L} \frac{Y_{BM}}{X_{BM}} \qquad (9.72)$$

For a dilute gas Eq. (9.63) becomes

$$N_G = \int_{Y_2}^{Y_1} \frac{dY}{Y - Y_i} \qquad (9.73)$$

In certain cases the interfacial composition $Y_i$ may be constant, as in the case of evaporation of a liquid of constant vapor pressure, or $Y_i$ may be essentially zero, as in the case of a highly soluble gas. In such cases Eq. (9.73) becomes

$$N_G = \ln \left( \frac{Y_1 - Y_i}{Y_2 - Y_i} \right) \qquad (9.74)$$

Under similar assumptions such that $X_i$ is constant or zero, Eq. (9.74) reduces to

$$N_L = \ln \left( \frac{X_i - X_1}{X_i - X_2} \right) \qquad (9.75)$$

In the case of $X_i = Y_i = 0$, Eq. (9.74) indicates that the fraction absorbed in one transfer unit is $1 - e^{-1}$, or 63.2 percent.

### 9.6.C   Design Procedure Based on Overall Coefficients

Employment of the single-phase coefficients is sometimes inconvenient or impractical because of the necessity of obtaining the interfacial composition $X_i$ or $Y_i$. This may be particularly the case if graphical integration is used or if the individual phase coefficients $k_L$, $k_L a_v$ or $k_G$, $k_G a_v$ are not well known. Accordingly, it has become customary to use overall coefficients, which are based on hypothetical interfacial compositions, and thus hypothetical driving forces for mass transfer.

The overall gas-phase coefficient is based upon the driving force $(Y - Y^*)$, where $Y^*$ is the gas-phase mole fraction corresponding to equilibrium with the bulk liquid composition $X$ as shown in Fig. 9.15. In terms of the overall gas-phase mass-transfer coefficient $K_{OG}$, the absorption flux corresponding to any point $X$, $Y$ on the operating line is given by

$$N_A = K_{OG} P(Y - Y^*) \qquad (9.76)$$

Similarly, an overall liquid-phase driving force $(X^* - X)$ may be defined, where $X^*$ is the liquid-phase composition which would be in equilibrium with gas of mole fraction $Y$ as shown in Fig. 9.15. The mass-transfer flux is given by

$$N_A = K_{OL}\bar{\rho}(X^* - X) \tag{9.77}$$

where $K_{OL}$ is the overall liquid-phase mass-transfer coefficient.

From Eqs. (9.50), (9.51), and (9.76), it may be shown that

$$\frac{1}{K_{OG}} = \frac{1}{k_g} + \frac{Y_i - Y^*}{X_i - X}\frac{P}{\bar{\rho}}\frac{1}{k_L}$$

or

$$\frac{1}{K_{OG}} = \frac{1}{k_g} + m\frac{P}{\bar{\rho}}\frac{1}{k_L} \tag{9.78}$$

where

$$m = \frac{Y_i - Y^*}{X_i - X} = \text{slope of equilibrium line over the interval } X, Y^* \text{ to } X_i, Y_i \tag{9.79}$$

Similarly from Eqs. (9.50), (9.51), and (9.77) the overall liquid-phase coefficient is given by

$$\frac{1}{K_{OL}} = \frac{1}{k_L} + \frac{\bar{\rho}}{Pm''}\frac{1}{k_g} \tag{9.80}$$

where

$$m'' = \frac{Y - Y_i}{X^* - X_i} = \text{slope of equilibrium line over the interval } X^*, Y \text{ to } X_i, Y_i \tag{9.81}$$

Equations (9.78) and (9.80) show that the overall mass-transfer coefficients depend upon the slope of the equilibrium line, so that their use is most satisfactory for systems in which this slope remains nearly constant over the concentration range involved.

Combining Eqs. (9.57) and (9.76) gives the following expression for the tower height:

$$h_T = \int_{Y_2}^{Y_1} \frac{G_M\, dY}{K_{OG}\, Pa_v(1 - Y)(Y - Y^*)} \tag{9.82}$$

The right-hand side of Eq. (9.82) may be multiplied and divided by the factor $Y^*_{BM}$ to give

$$h_T = \int_{Y_2}^{Y_1} \left(\frac{G_M}{K_{OG}\, PY^*_{BM}\, a_v}\right)\left[\frac{Y^*_{BM}\, dY}{(1 - Y)(Y - Y^*)}\right] \tag{9.83}$$

where

$$Y^*_{BM} = \frac{(1 - Y) - (1 - Y^*)}{\ln\left[(1 - Y)/(1 - Y^*)\right]} \tag{9.84}$$

The left-hand term under the integral defines the overall gas-phase height of a transfer unit $H_{OG}$:

$$H_{OG} = \frac{G_M}{K_{OG} \, P Y^*_{BM} \, a_v} \tag{9.85}$$

Similarly, the right-hand term under the integral defines the number of overall gas-phase transfer units $N_{OG}$:

$$N_{OG} = \int_{Y_2}^{Y_1} \frac{Y^*_{BM} \, dY}{(1 - Y)(Y - Y^*)} \tag{9.86}$$

If an average value of $H_{OG}$ may be used the tower height is given by

$$h_T = (H_{OG})_{av} N_{OG} \tag{9.87}$$

In a similar manner, design equations may be derived based on the overall liquid-phase coefficient. Thus,

$$h_T = \int_{X_2}^{X_1} \left( \frac{L_M}{K_{OL} \, \bar{\rho} X^*_{BM} \, a_v} \right) \left[ \frac{X^*_{BM} \, dX}{(1 - X)(X^* - X)} \right] \tag{9.88}$$

where

$$X^*_{BM} = \frac{(1 - X) - (1 - X^*)}{\ln \left[ (1 - X)/(1 - X^*) \right]} \tag{9.89}$$

In terms of the overall liquid-phase height of a transfer unit $H_{OL}$ and the number of overall liquid-phase transfer units $N_{OL}$,

$$h_T = (H_{OL})_{av} N_{OL} \tag{9.90}$$

where

$$H_{OL} = \frac{L_M}{K_{OL} \, \bar{\rho} X^*_{BM} \, a_v} \tag{9.91}$$

$$N_{OL} = \int_{X_2}^{X_1} \frac{X^*_{BM} \, dX}{(1 - X)(X^* - X)} \tag{9.92}$$

Overall heights of a transfer unit may be related to single-phase values in a manner analogous to the relations for the mass-transfer coefficients, Eqs. (9.78) to (9.81). Thus,

$$H_{OG} = \frac{Y_{BM}}{Y^*_{BM}} H_G + \frac{m G_M}{L_M} \frac{X_{BM}}{Y^*_{BM}} H_L \tag{9.93}$$

$$H_{OL} = \frac{X_{BM}}{X^*_{BM}} H_L + \frac{L_M}{m'' G_M} \frac{Y_{BM}}{X^*_{BM}} H_G \tag{9.94}$$

When $H_{OG}$ or $H_{OL}$ vary appreciably over the tower, it is necessary to retain them within the integral in Eq. (9.83) or (9.88) for accurate results, rather than to use average values.

As discussed in Sec. 9.6.D, if graphical procedures are employed and $mG_M/L_M$ is less than unity, particularly in the dilute region of the tower, it will usually be more convenient and more accurate to use $N_{OG}$ and $H_{OG}$ as the basis for design rather than to use $N_{OL}$ and $H_{OL}$.

**EXAMPLE 9.5** The gas stream described in Example 9.1 is to be processed by absorption at 40°C and 1 atm in a column packed with 1-in Raschig rings. Water containing 0.0001 mole fraction alcohol is to be used as the absorbent to obtain 98 percent recovery of alcohol vapor.

On the basis of considerations of column capacity and the criterion that $mG_M/L_M$ should be approximately 0.7 at the top of the tower, the liquid and gas molar mass velocities are to be 47.0 and 30.8 lb moles/(h)(ft$^2$) at the top of the column. At these conditions, based on absorption data from the literature, the height of a gas-transfer unit may be taken as 1.79 ft and the height of a liquid-transfer unit may be taken as 0.98 ft. Absorption may be assumed isothermal at 1 atm pressure. Due to the very low concentrations the change of these two parameters over the column is negligible.

(a) Employing a graphical integration based on the gas-phase mass-transfer coefficient, calculate the height of packing required to obtain the desired recovery.

(b) Assuming the heights of transfer units to be constant over the tower conditions, calculate by suitable graphical construction the number of gas-phase transfer units and the corresponding tower height.

(c) Repeat the calculation of part b above employing the overall height of a gas-phase transfer unit to obtain the number of transfer units and the corresponding tower height.

SOLUTION   First determine the conditions at the bottom of the tower from $G_{M,2} = 30.8$ lb moles/(h)(ft$^2$). A combined inert gas and alcohol balance over the gas phase gives

$$\frac{Y_2}{1-Y_2} = 0.02\,\frac{Y_1}{1-Y_1} = 0.02\,\frac{0.01}{1-0.01} \quad \text{and} \quad Y_2 = 0.000202$$

The amount of alcohol fed to the absorber in the gas stream follows, therefore, as

$$0.000202 \times 30.8\,\frac{1}{0.02} = 0.311 \text{ lb moles alcohol/(h) (ft}^2)$$

$$G_{M,1} = 31.1 \text{ lb moles/(h) (ft}^2)$$

$$\text{Alcohol absorbed} = 0.311 - 0.0002 \times 30.8 = 0.30478 \text{ lb moles/(h) (ft}^2)$$

$$X_1 = \frac{0.30478 + 0.0001 \times 47}{47 + 0.30478} = 0.006542$$

A $YX$ diagram is set up as shown in Figs. 9.16 and 9.17. The operating line is calculated by Eq. (9.47).

$$\frac{L'_M}{G'_M} = \frac{47 \times 0.9999}{30.8(1 - 0.000202)} = 1.5264$$

$$\frac{Y}{1-Y} - \frac{L'_M}{G'_M}\frac{X}{1-X} = 4.9369 \times 10^{-5}$$

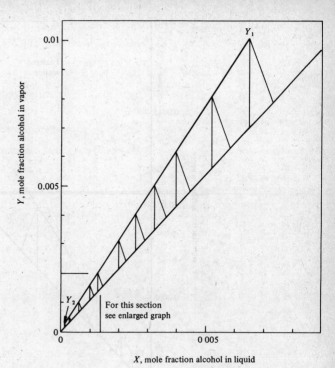

FIGURE 9.16
$YX$ graph for Example 9.5.

For differential conditions in the absorber, one may find the interfacial concentrations on the equilibrium line $Y = 1.0682X$ (see Example 9.1) by means of Eq. (9.72). Assume $Y_{BM}/X_{BM} = 1$ as a first approach. Tie-line construction locates $Y_i$ and $X_i$, as explained in Sec. 9.6.A. Alternatively, Eq. (9.54) may be used, with $Y_i = 1.0682X_i$.

| $X$ | $\dfrac{X}{1 - X}$ | $\dfrac{Y}{1 - Y}$ | $Y$ | $\dfrac{L_M}{G_M}\dfrac{H_G}{H_L}$ |
|---|---|---|---|---|
| 0.00654 | 0.006585 | 0.0101005 | 0.009999 | 2.778 |
| 0.004 | 0.004016 | 0.0061794 | 0.0061414 | 2.782 |
| 0.002 | 0.002004 | 0.0031082 | 0.0030986 | 2.785 |
| 0.001 | 0.001001 | 0.001577 | 0.001575 | 2.786 |
| 0.0001 | 0.00010001 | 0.000202 | 0.0002020 | 2.788 |

*Check*  At the bottom, $Y_{BM}/X_{BM} = 0.9911/0.9931 = 0.998 \approx 1$.

From the $YX$ diagram, the following table is prepared containing all information needed to solve the problem in the three different ways.

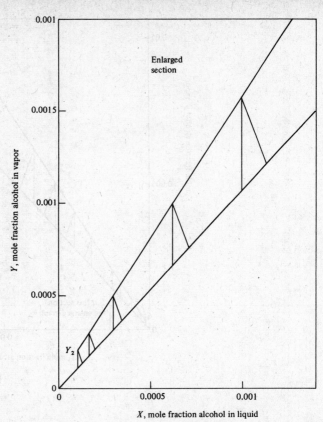

FIGURE 9.17
$YX$ graph for Example 9.5, dilute range.

## CALCULATED QUANTITIES—EXAMPLE 9.5

| $Y$ | $1 - Y$ | $Y_i$ | $Y_{BM}$ | $Y - Y_i$ |
|---|---|---|---|---|
| 0.01 | 0.99 | 0.00782 | 0.9911 | 0.00218 |
| 0.008 | 0.992 | 0.00624 | 0.99287 | 0.00176 |
| 0.006141 | 0.9938 | 0.004775 | 0.9945 | 0.001360 |
| 0.005 | 0.995 | 0.00388 | 0.99555 | 0.00112 |
| 0.004 | 0.996 | 0.00310 | 0.99645 | 0.0009 |
| 0.003099 | 0.9969 | 0.0024 | 0.99725 | 0.00061 |
| 0.002 | 0.998 | 0.00152 | 0.99824 | $4.8 \times 10^{-4}$ |
| 0.001575 | 0.9984 | 0.001209 | 0.9986 | $3.66 \times 10^{-4}$ |
| 0.001 | 0.999 | 0.00076 | 0.9991 | $2.40 \times 10^{-4}$ |
| 0.0005 | 0.9995 | 0.000366 | 0.9996 | $1.34 \times 10^{-4}$ |
| 0.0003 | 0.9997 | 0.00020 | 0.9997 | $1.0 \times 10^{-4}$ |
| 0.000202 | 0.9998 | 0.000133 | 0.9998 | $6.90 \times 10^{-5}$ |

(a) Gas-phase mass-transfer coefficient. In order to integrate Eq. (9.58), which is based on the gas-phase mass-transfer coefficient, the function under the integral is evaluated at different flow conditions according to the definition of $H_G$ [Eq. (9.61)]:

$$\frac{G_M}{k_G P a_v (1 - Y)(Y - Y_i)} = H_G \frac{Y_{BM}}{(1 - Y)(Y - Y_i)}$$

The numerical values of this function are given in the table. The integration of Eq. (9.58) can then be performed graphically as shown in Fig. 9.18 to give

$$h_T = 28.3 \text{ ft}$$

(b) Height of a gas-phase transfer unit. The number of gas-phase transfer units is obtained by a graphical integration as shown in Fig. 9.19 following Eq. (9.63):

$$N_G = 15.75$$
$$h_T = 15.75 \times 1.79 = 28.2 \text{ ft}$$

Because of the negligible variation of $H_G$ and $G_M$, methods a and b should give the same result. If, however, the concentrations were higher, method a, retaining $H_G$ under the integral, would allow for this variation and thus yield a better result.

(c) Overall height of a gas-phase transfer unit. Again, the number of overall gas-phase transfer units is computed by integrating Eq. (9.86) graphically as shown in Fig. 9.20:

$$N_{OG} = 11.46$$

The overall height of a transfer unit at the top of the packing follows from Eq. (9.93):

$$H_{OG} = \frac{0.9998}{0.9998} 1.79 + \frac{0.99989}{0.9998} \frac{1.0682 \times 30.8}{47.0} 0.98 = 2.48 \text{ ft}$$

Even though $H_G$ and $H_L$ have been assumed to be constant over the tower conditions, $H_{OG}$ varies slightly because of differences in $Y_{BM}/Y_{BM}^*$, $X_{BM}/X_{BM}^*$, and $(mG_M/L_M)$. In this case, however, the variation would affect $H_{OG}$ only in the fourth significant figure and we may accept $H_{OG} = 2.48$ ft.

$$h_T = 11.46 \times 2.48 = 28.4 \text{ ft} \qquad ////$$

| $\dfrac{Y_{BM}}{(1-Y)(Y-Y_i)}$ | $H_G \dfrac{Y_{BM}}{(1-Y)(Y-Y_i)}$ | $Y^*$ | $Y_{BM}^*$ | $Y - Y^*$ | $\dfrac{Y_{BM}}{(1-Y)(Y-Y^*)}$ |
|---|---|---|---|---|---|
| 459.2 | 822.0 | 0.00698 | 0.9915 | 0.00302 | 331.6 |
| 568.6 | 1,017.9 | 0.00558 | 0.9932 | 0.00242 | 413.7 |
| 732.3 | 1,310.9 | 0.00427 | 0.9948 | 0.00187 | 534.9 |
| 893.3 | 1,599.0 | 0.00347 | 0.9958 | 0.00153 | 654.0 |
| 1,111.5 | 1,989.7 | 0.00277 | 0.9967 | 0.00123 | 816.0 |
| 1,431.9 | 2,563.1 | 0.00213 | 0.9974 | 0.000969 | 1,032.9 |
| 2,083.3 | 3,729.2 | 0.001365 | 0.9983 | $6.35 \times 10^{-4}$ | 1,574.8 |
| 2,733.7 | 4,893.4 | 0.001069 | 0.9986 | $5.058 \times 10^{-4}$ | 1,977.1 |
| 4,166.6 | 7,458.3 | 0.000662 | 0.9992 | $3.38 \times 10^{-4}$ | 2,958.6 |
| 7,462.7 | 13,358.0 | 0.000313 | 0.9996 | $1.87 \times 10^{-4}$ | 5,347.6 |
| 10,000.0 | 17,900.0 | 0.000175 | 0.9998 | $1.25 \times 10^{-4}$ | 8,000.0 |
| 14,494.0 | 25,946.0 | 0.000107 | 0.9998 | $9.50 \times 10^{-5}$ | 10,526.0 |

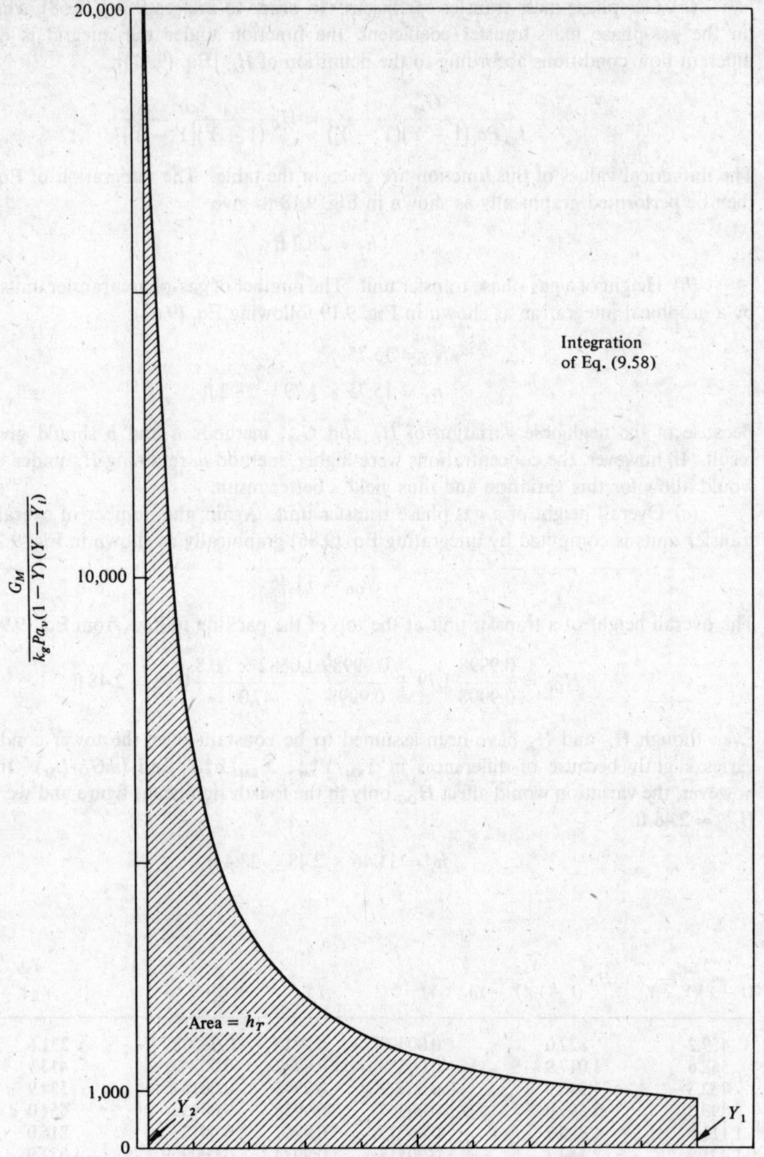
20,000
10,000
1,000
0
$\frac{G_M}{k_g P a_v (1 - Y)(Y - Y_i)}$
Integration
of Eq. (9.58)
Area = $h_T$
$Y_2$
$Y_1$
0
0.005
0.01
Y

FIGURE 9.18
Graphical integration for packed height, Example 9.5.

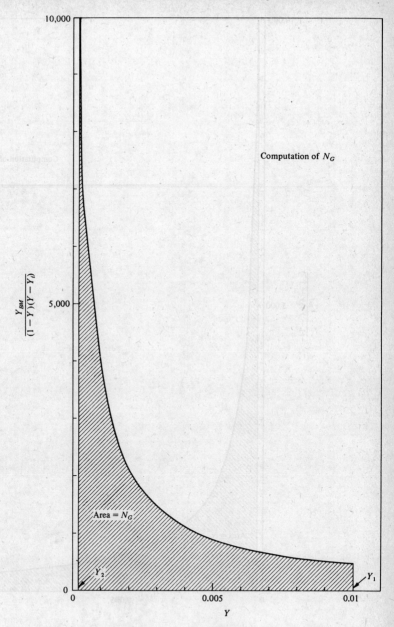

FIGURE 9.19
Graphical integration for $N_G$, Example 9.5.

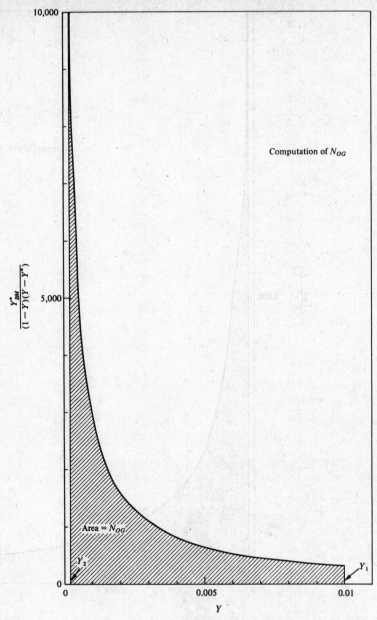

FIGURE 9.20
Graphical integration for $N_{OG}$, Example 9.5.

### 9.6.D  Simplified Design Procedures for Dilute Gases with Straight Equilibrium Line

**Logarithmic-mean driving force**   Although the graphical or analytical procedure outlined above must by employed in many practical design problems, it is frequently possible to use a simple mean driving force or potential and thereby obviate the need for the integration. Where it is possible to assume that the equilibrium curve and the operating line are linear over the range in which they are to be used, it can be shown that the logarithmic mean of the terminal potentials is theoretically correct. When the overall gas-film coefficient is used to express the rate of interphase transfer, the calculation then reduces to the solution of the equation

$$L_M(X_1 - X_2) = G_M(Y_1 - Y_2) = K_{OG} a_v Ph_T(Y - Y^*)_{av} \tag{9.95}$$

$$(Y - Y^*)_{av} = \frac{(Y - Y^*)_1 - (Y - Y^*)_2}{\ln[(Y - Y^*)_1/(Y - Y^*)_2]} \tag{9.96}$$

Equation (9.96) may be derived as follows: Assume the equilibrium curve to be linear in the range $X_1$ to $X_2$. Then, since the operating line is also linear, the difference in ordinates of the two lines must also be linear in $X$ or $Y$; that is, $Y - Y^*$, designated by $\Delta$, is linear in $Y$, and

$$\frac{d\Delta}{dY} = \frac{\Delta_1 - \Delta_2}{Y_1 - Y_2} \tag{9.97}$$

Substituting this relation into Eq. (9.82) for a dilute gas (i.e., $1 - Y \approx 1$) gives, upon rearrangement,

$$\frac{h_T K_{OG} a_v P}{G_M} = \int_{Y_2}^{Y_1} \frac{dY}{\Delta} = \frac{Y_1 - Y_2}{\Delta_1 - \Delta_2} \int_{Y_2}^{Y_1} \frac{d\Delta}{\Delta} = \frac{Y_1 - Y_2}{\Delta_1 - \Delta_2} \ln \frac{\Delta_1}{\Delta_2} \tag{9.98}$$

whence, from Eq. (9.95),

$$(Y - Y^*)_{av} = \Delta_{av} = \frac{\Delta_1 - \Delta_2}{\ln(\Delta_1/\Delta_2)} \tag{9.99}$$

In general, the logarithmic mean applies wherever the potential is linear in the variable defining the amount transferred: $(Y - Y^*)$ linear in $Y$; $\Delta t$ linear in $t$; etc.

A similar procedure may be followed to show that when the overall liquid-film coefficient is used,

$$G_M(Y_1 - Y_2) = L_M(X_1 - X_2) = \bar{\rho}_M K_{OL} a_v h_T(X^* - X)_{av} \tag{9.100}$$

$$(X^* - X)_{av} = \frac{(X^* - X)_1 - (X^* - X)_2}{\ln[X^* - X)_1/(X^* - X)_2]} \tag{9.101}$$

It may be noted that with gas film controlling, it is necessary that the equilibrium curve be linear between $X_2$ and $X_1$ if the logarithmic mean is to be used. With liquid film controlling, the logarithmic mean applies if the equilibrium curve is linear over the range $Y_2$ to $Y_1$.

**Height-of-a-transfer-unit methods for dilute gases**  For relatively dilute systems the ratios involving $Y_{BM}$, $Y^*_{BM}$, $X_{BM}$, and $Y_{BM}$ approach unity, so that Eqs. (9.93) and (9.94) become

$$H_{OG} = H_G + \frac{mG_M}{L_M} H_L \tag{9.102}$$

$$H_{OL} = H_L + \frac{L_M}{m'G_M} H_G \tag{9.103}$$

Similarly, Eqs. (9.86) and (9.92) become

$$N_{OG} \approx N_T = \int_{Y_2}^{Y_1} \frac{dY}{Y - Y^*} \tag{9.104}$$

where $N_T$ = number of transfer units for equimolar counterdiffusion.

$$N_{OL} \approx \int_{X_1}^{X_2} \frac{dX}{X^* - X} \tag{9.105}$$

As discussed in Secs. 9.6.F and 9.6.H, the equations above are also applicable if mass transfer of components other than the absorbed component is such that the gas and liquid flows (molal) do not change appreciably over the tower.

For cases in which the equilibrium and operating lines may be assumed straight, with slopes $L_M/G_M$ and $m$, respectively, Colburn [10] has developed algebraic expressions for the integrals of Eqs. (9.104) and (9.105). For the number of overall gas-phase transfer units,

$$N_{OG} \approx N_T = \int_2^1 \frac{dY}{Y - Y^*} = \frac{1}{1 - mG_M/L_M} \ln \left[ \left( 1 - \frac{mG_M}{L_M} \right) \left( \frac{Y_1 - mX_2}{Y_2 - mX_2} \right) + \frac{mG_M}{L_M} \right] \tag{9.106}$$

Figure 9.21 may be used for numerical evaluation of Eq. (9.106).

The corresponding equation for the number of overall liquid-phase transfer units is

$$N_{OL} = \frac{1}{[1 - L_M/mG]} \ln \left[ \left( 1 - \frac{L_M}{mG_M} \right) \left( \frac{X_1 - Y_2/m}{X_2 - Y_2/m} \right) + \frac{L_M}{mG_M} \right] \tag{9.107}$$

It may be noted that Eq. (9.107) is identical in form to Eq. (9.106), and that, therefore, Fig. 9.21 may be employed for solving the equation if the abscissa and ordinate are considered to represent $[(X_1 - Y_2/m)/(X_2 - Y_2/m)]$ and $N_{OL}$, respectively, and the parameter is considered to be $L_M/mG_M$.

Whether to use $N_{OG}$ and $H_{OG}$, or to use $N_{OL}$ and $H_{OL}$, as a basis for design will depend on the value of $mG_M/L_M$ for the system. Inspection of Fig. 9.21 shows that for values of $mG_M/L_M$ greater than 1.0, $N_{OG}$ changes so rapidly with small changes in composition that the accuracy obtainable in using the chart is impaired. Therefore, when $mG_M/L_M$ exceeds unity, better results will be obtained by

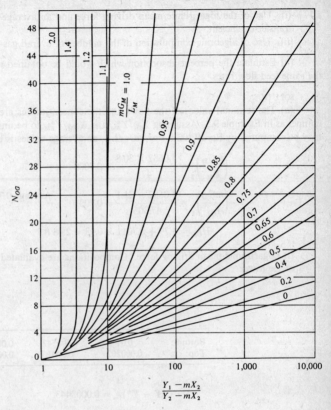

**FIGURE 9.21**
Number of transfer units in an absorption column, condition of constant $mG_M/L_M$.

employing $N_{OL}$ and $H_{OL}$ in the form of Eq. (9.107). Except for this consideration of graphical accuracy, there is no difference between Eqs. (9.106) and (9.107), since they lead to mathematically equivalent results under the assumptions made in their derivation. As a practical matter, however, the economic absorber design [12] will normally lead to specification of a value for $mG_M/L_M$ less than unity, usually in the vicinity of 0.7. It is apparent from Fig. 9.21 that for values of $mG_M/L_M$ greater than 1.0, a very large number of transfer units and corresponding large tower height will be required to give a reasonably high level of absorption. A more detailed discussion of economic design is presented in Sec. 9.6.K.

**EXAMPLE 9.6**

(a) For the flows and conditions of Example 9.5, calculate the tower height by the following approximate methods:

(*i*)  Use of the logarithmic mean driving force and an average overall gas-phase mass-transfer coefficient.

(*ii*)  Use of algebraic computation of the number of overall gas-phase transfer units.

(*b*) Estimate the percent absorption which would be obtained with 12 ft of packing at the same feed flow rates.

SOLUTION

(*a*) The terminal values of the flows and the concentrations are calculated in the same manner as in Example 9.5. Assuming $Y_{BM}/Y^*_{BM}$ and $X_{BM}/Y^*_{BM}$ to be unity and $m = 1.0682$, $H_{OG}$ may be computed by Eq. (9.102). For $mG_M/L_M$, an average values is used:

$$\left(\frac{mG_M}{L_M}\right)_2 = \frac{1.0682 \times 30.8}{47} = 0.7000$$

$$\left(\frac{mG_M}{L_M}\right)_1 = \frac{1.0682 \times 31.1}{47.305} = 0.7023 \qquad \text{Average } 0.7011$$

$$H_{OG} = 1.79 + 0.7011 \times 0.98 = 2.48 \text{ ft}$$

(*i*) Logarithmic-mean driving force. Compositions are evaluated at the top and bottom of the tower:

|          | $Y$      | $X$       | $Y^*$       | $Y - Y^*$   |
|----------|----------|-----------|-------------|-------------|
| Bottom:  | 0.01     | 0.006542  | 0.00698816  | 0.0030118   |
| Top:     | 0.000202 | 0.0001    | 0.00010682  | 0.00009518  |

$$(Y - Y^*)_{lm} = 0.0008443$$

The height of the packing follows from Eq. (9.95):

$$h_T = \frac{Y_1 G_{M,1} - Y_2 G_{M,2}}{K_{OG} a_v P (Y - Y^*)_{lm}}$$

The term $K_{OG} a_v P$ can be determined from $H_{OG}$ by applying its definition, Eq. (9.85):

$$K_{OG} a_v P = \frac{G_M}{H_{OG} Y^*_{BM}} = \frac{30.95}{2.48 \times 0.9956} = 12.53 \text{ lb moles/(h) (ft}^3)$$

which represents an average value over the tower.

$$h_T = \frac{31.1 \times 0.01 \times 0.98}{12.53 \times 0.0008443} = 28.8 \text{ ft}$$

(*ii*) Analytical solution

$$\frac{Y_1 - mX_2}{Y_2 - mX_2} = \frac{0.01 - 0.00010682}{0.000202 - 0.00010682} = 103.94$$

According to Eq. (9.106),

$$N_{OG} = \frac{\ln\left[(1-0.7011) \times 103.94 + 0.7011\right]}{1 - 0.7011} = 11.57$$

$$h_T = 11.57 \times 2.48 = 28.7 \text{ ft}$$

The same result may be obtained using Fig. 9.21.

(b)  The computation is performed by means of Eq. (9.106). In order to determine $(mG_M/L_M)_{av}$ assume a recovery of 90%.

Alcohol absorbed: $0.311 \times 0.9 = 0.2799$ lb mole/(h) (ft$^2$)

$L_{M,1} = 47.0 + 0.2799 = 47.2799$ lb moles/(h) (ft$^2$)

$G_{M,2} = 31.1 - 0.2799 = 30.82$ lb moles/(h) (ft$^2$)

$$\left(\frac{mG_M}{L_M}\right)_1 = \frac{1.0682 \times 31.1}{47.2799} = 0.7026$$

$$\left(\frac{mG_M}{L_M}\right)_2 = \frac{1.0682 \times 30.82}{47.0} = 0.7005$$

Average 0.7015

The needed value of $N_{OG}$ may be calculated from the available height of packing and $H_{OG}$. From Eq. (9.102),

$$H_{OG} = 1.79 + 0.7015 \times 0.98 = 2.477 \text{ ft}$$

$$N_{OG} = \frac{h_T}{H_{OG}} = \frac{12}{2.477} = 4.844$$

Equation (9.106) is rearranged in the following manner:

$$\frac{Y_1 - mX_2}{Y_2 - mX_2} = \frac{\exp\left[N_{OG}(1 - mG_M/L_M)\right] - mG_M/L_M}{1 - mG_M/L_M}$$

It follows that $(Y_1 - mX_2)/(Y_2 - mX_2) = 11.87$, and since $Y_1 - mX_2 = 0.009893$, $Y_2 - mX_2 = 0.0008333$ and $Y_2 = 0.00094$.

$$\text{Recovery fraction} = 1 - \frac{0.00094/(1 - 0.00094)}{0.01/0.99} = 0.907 \quad \text{or} \quad 90.7\%$$

*Comment*  The estimated tower heights in *a*, steps *i* and *ii*, agree closely with that obtained by more rigorous graphical integration in Example 9.5. This result would be expected in view of the slight departure of the operating line from linearity.    ////

## 9.6.E  Approximate Design Procedure for Concentrated Gas with Straight Equilibrium Line

For absorption from a concentrated gas stream, the operating line will be curved to an extent depending on the variation of the $L_M/G_M$ ratio over the tower. Values of $N_{OG}$ obtained from Eq. (9.86) will differ from those obtained from Eq. (9.104) because of the $Y_{BM}^*/(1 - Y)$ term in the former equation. A convenient approach for purposes of approximate design is to define a correction term $\Delta N_{OG}$ which can be added to Eq. (9.104). Thus

$$N_{OG} = \int_{Y_2}^{Y_1} \frac{dY}{Y - Y^*} + \Delta N_{OG} \tag{9.108}$$

Wiegand [56] has shown that for cases in which $Y^*_{BM}$ may be represented by an arithmetic mean,

$$\Delta N_{OG} = \frac{1}{2} \ln \frac{1 - Y_2}{1 - Y_1} \tag{9.109}$$

Equation (9.109) will be sufficiently accurate for most situations.

Some perspective concerning the significance of the correction term may be obtained by examining the solution of Eq. (9.86) for the limiting case in which $Y^* = 0$ [13] and comparing this solution with the corresponding solution of Eq. (9.108) to evaluate $\Delta N_{OG}$. This should closely represent the maximum value of $\Delta N_{OG}$ for a given extent of absorption. Thus,

$$[\Delta N_{OG}]_{Y^* \to 0} = \ln \left[ \frac{\ln(1 - Y_2)}{\ln(1 - Y_1)} \right] - \ln \frac{Y_1}{Y_2} \tag{9.110}$$

Figure 9.22 shows values of $\Delta N_{OG}$ from Eq. (9.110) as a function of the inlet-gas composition $Y_1$ and the apparent number of transfer units $N_T$ obtained from the solution of Eq. (9.104) for varying degrees of absorption. It is apparent from Fig. 9.22 that $\Delta N_{OG}$ will not usually be an important term for cases involving a substantial number of transfer units.

Figure 9.23 shows results of a study by the authors in which $\Delta N_{OG}$ was evaluated for a large number of absorption cases covering a wide range of gas and liquid concentrations without restriction on $Y^*$. The results were obtained by precise numerical integration of Eqs. (9.86) and (9.104) by digital computer. All the calculated values fall in a rather narrow band on or under the line corresponding to Eq. (9.109).

As a practical procedure for design, $\Delta N_{OG}$ may be evaluated from Eq. (9.109) or the upper curve of Fig. 9.23 to within 0.1 transfer unit. This procedure is believed conservative in accordance with the test cases shown in Fig. 9.23.

For approximate design purposes it is useful to have a method of evaluating the integral of Eq. (9.108) without resort to graphical integration of lengthy numerical computation. This is not generally possible for cases when $L'_M/G'_M$ is much less than unity, because the operating line becomes concave upward and may intersect the equilibrium line somewhere between $Y_1$ and $Y_2$ if the gas is very concentrated and the separation is close. Such cases must be assessed by plotting the design diagram or evaluating the integral by numerical means. However, when $L'_M/G'_M$ is in the vicinity of, or greater than, unity, an empirical procedure developed by the authors may be employed as described below.

A computational study was made covering several hundred hypothetical absorber designs for concentrated gas streams containing up to 80 mole percent

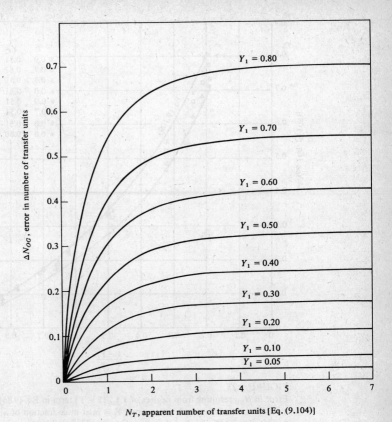

**FIGURE 9.22**
Error in neglecting the $Y_{BM}^*/(1 - Y)$ term in the number of transfer-unit calculations for gas absorption when $Y^* = 0$.

of the absorbed component for recoveries ranging from 81 to 99.9 percent. By numerical integration with a digital computer, precise values were obtained for $N_{OG}$ and $N_T$. It was also possible to compute the effective average value of the flow ratio, $R = L_M/G_M$, which when used in conjunction with Eq. (9.106) gave satisfactory values of $N_T$. It was found that the effective average flow ratio $[R_{av} = (L_M/G_M)_{av}]$ could be correlated satisfactorily as a function of the terminal values $R_1$ and $R_2$, of the change in the mole fraction of the absorbed component over the tower, and of the fractional approach to equilibrium $(f = Y_1^*/Y_1)$ between the concentrated gas entering the tower and the liquid leaving. Figure 9.24 shows typical results for $R_{av}$ at $f = 0.7$ for various test cases. The cases shown embraced a range of inert liquid-to-gas flow ratios, $L_M'/G_M'$ from 10 to 1.1, of inlet-gas mole fractions from 0.81 to 0.22 and of outlet mole fractions from 0.05 to 0.001.

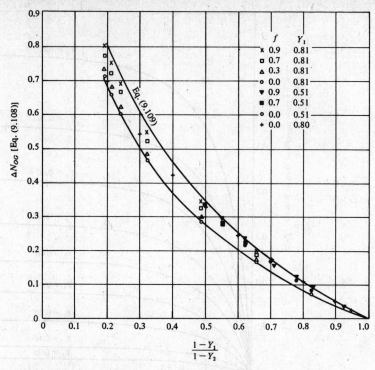

**FIGURE 9.23**
Error in $N_{OG}$ resulting from neglect of $Y^*_{BM}/(1 - Y)$ term in Eq. (9.86). $Y_2$ = outlet mole fraction of absorbed component; $Y_1$ = inlet mole fraction of absorbed component; $f = Y^*_1/Y_1$, fractional approach to equilibrium between entering gas and leaving liquid.

A generalized graph covering a wide range of saturation ratios is given in Fig. 9.25 for use in approximate design calculations. This graph may be used for values of $L'_M/G'_M$ in the vicinity of 1.0 and greater. For low percentages of absorption when $L'_M/G'_M$ is near unity, the correlation is less satisfactory; but the variation in $R$ over the tower is not large in such cases either, so that the error generally will not be serious.

It was found, further, that the value of $R_{av}$ predicted from this correlation yields a suitable average value of $mG_M/L_M$ in Eq. (9.93) to give an average overall height of a gas-phase transfer unit.

In the application of this principle to Eq. (9.93) it will generally suffice to estimate the corresponding average composition ratios from a rough sketch of the design diagram. The point of "average" composition will be at the intersection of the operating line and a line of slope equal to $R_{av}$ which passes through the

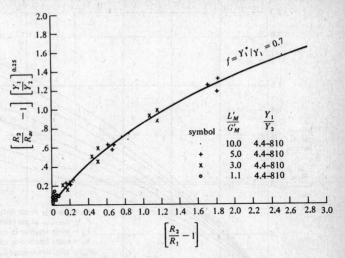

**FIGURE 9.24**
Correction of average-flow ratio for curved operating line at $f = 0.7$.

terminal point of the operating line at the dilute end of the tower. Alternatively, if the exit gas and entering liquid are dilute, the liquid mole fraction $\overline{X}$ corresponding to the average composition is given approximately by

$$\overline{X} = \frac{R_2/R_{av} - 1}{R_2 - 1} \tag{9.111}$$

From the point $\overline{X}$ on the operating line, the tie line corresponding to Eq. (9.72) is placed and the compositions $X$, $X_i$, $Y$, $Y^*$, and $Y_i$ are read from the graph to provide all necessary quantities for the composition ratios.

It may be possible to determine by inspection that the composition ratios in Eq. (9.93) may be assumed unity or estimated satisfactorily by assuming the gas-phase resistance to be controlling, i.e., by placing the tie line vertically. Examination of various cases for which $m$ was greater than unity showed that the term $Y_{BM}/Y_{BM}^*$ varied inversely with the term $X_{BM}/Y_{BM}^*$ in a manner such that use of the simpler Eq. (9.102) was preferable to using the vertical tie line. In any case, a high degree of precision in the computation of the ratios is usually not warranted since correspondingly accurate values of $H_G$ and $H_L$ are rarely available.

The tower height is then calculated from Eq. (9.87).

**EXAMPLE 9.7**   The absorption process described in Example 9.3 is to be performed by means of a packed column. The tower has an inner diameter of 2.5 ft and is packed with 2-in Raschig rings. Enough cooling surface is provided to ensure isothermal operation.

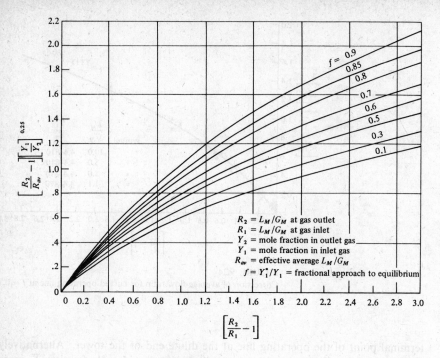

**FIGURE 9.25**
Design chart for estimation of average-flow ratio in absorption.

Since the mass velocities may vary considerably through the tower, the dependence of the height of a transfer unit on the mass velocities may not be neglected. Based on the data of Fellinger [49] for $H_G$ in absorption of ammonia (corrected for variation in physical properties) and based on correlations for $H_L$ [49], the variation of the gas-phase and liquid-phase height of a transfer unit with $G$ and $L$ may be described as

$$H_G = 5.3 \frac{G^{0.395}}{L^{0.417}} \tag{a}$$

$$H_L = 0.2723 \, L^{0.22} \tag{b}$$

(a) By means of the simplified design procedure proposed above estimate

(i) The number of overall gas-phase mass-transfer units required for the specified recovery
(ii) The tower height

(b) Check both results by suitable graphical integrations for the number of overall gas-phase transfer units and for the tower height.

SOLUTION   (a)   Estimation by means of simplified design procedure.

(i)   The cross-sectional area of the column is $2.5^2 \times \pi/4 = 4.909$ ft$^2$. Designate the bottom and top of the packing by subscripts 1 and 2, respectively:

$G_{M,1} = 1.224$ lb moles/min (see Example 9.3)
$\quad\quad = 14.958$ lb moles/(h) (ft$^2$)

$L'_M = 0.9098$ lb mole oil/min (see Example 9.3)
$\quad\quad = 11.12$ lb moles oil/(h) (ft$^2$) and, because $X_2 = 0.005$,

$L_{M,2} = 11.1759$ lb moles/(h) (ft$^2$)

Pentane absorbed: $14.958 \times 0.56 \times 0.99$ (see Example 9.3) $= 8.295$ lb moles pentane/(h) (ft$^2$)

$G_{M,2} = 14.958 - 8.295 = 6.663$ lb moles/(h) (ft$^2$)

$L_{M,1} = 11.1759 + 8.295 = 19.471$ lb moles/(h) (ft$^2$)

$$\frac{L_{M,1}}{G_{M,1}} = 1.3017 = R_1 \quad\quad \frac{L_{M,2}}{G_{M,2}} = 1.677 = R_2$$

$$\left(\frac{R_2}{R_1} - 1\right) = 0.288 \quad\quad f = \frac{X_1 K}{Y_1} = 0.766$$

where $K = 1.0$, $X_1 = 0.4288$, and $Y_1 = 0.56$ (see Example 9.3).

$$\left[\frac{Y_1}{Y_2}\right]^{0.25} = \sqrt[4]{\frac{0.56}{0.01257}} = 2.583 \quad\quad \text{see Example 9.3}$$

Reading the chart (Fig. 9.25):

$$\left.\begin{array}{r}\left(\dfrac{R_2}{R_1} - 1\right) = 0.288 \\[2mm] f = 0.766\end{array}\right\} \rightarrow \left(\frac{R_2}{R_{av}} - 1\right) 2.583 = 0.34$$

whence $R_{av} = 1.482$; and $(mG_M/L_M)_{av} = 0.6748$. Substituting in Eq. (9.106), $N_T = 9.84$ units. This result has to be corrected for the factor $Y^*_{BM}/(1 - Y)$ in Eq. (9.86), which has been neglected thus far. Reading Fig. 9.23,

$$\frac{1 - Y_1}{1 - Y_2} = \frac{0.44}{0.9874} = 0.4456$$

$$\Delta N_{OG} = 0.39$$

$$\frac{Y^*_1}{Y_1} = 0.766$$

$$N_{OG} = 9.84 + 0.39 = 10.23$$

(ii)   An average value for $H_{OG}$ is now sought. From Eq. (9.111),

$$\overline{X} = \frac{0.1316}{0.677} = 0.1944$$

A mass balance [Eq. (9.47)] gives

$$\overline{Y}' = \frac{Y_1}{1 - Y_1} + \frac{11.12}{G'_M}\left(\frac{0.1944}{1 - 0.1944} - \frac{X_1}{1 - X_1}\right)$$

where (see Example 9.3) $Y_1 = 0.56$, $X_1 = 0.4288$, and $G'_M = 14.958(1 - 0.56) = 6.581$ lb moles/ (h)(ft$^2$). Whence $\bar{Y}' = 0.412$, and $\bar{Y} = \bar{Y}'/(1 + \bar{Y}') = 0.2917$.

*Check:*

$$\frac{L_M}{G_M} = \frac{11.12}{6.581} \frac{(1 - 0.2917)}{(1 - 0.1944)} = 1.485$$

which is close enough to $R_{av}$.

$H_G$ and $H_L$ will now be determined:

$$G = G'_M \left( 72 \frac{\bar{Y}}{1 - \bar{Y}} + 31.4 \right) = 402 \text{ lb/(h)(ft}^2)$$

where 31.4 is the mean molecular weight of the vent gas.

$$L = L'_M \left( 72 \frac{\bar{X}}{1 - \bar{X}} + 160 \right) = 1,972 \text{ lb/(h)(ft}^2)$$

It follows from Eqs. (a) and (b) that

$$H_G = 5.3 \frac{402^{0.395}}{1972^{0.417}} = 2.39 \text{ ft}$$

$$H_L = 0.2723 \times 1972^{0.22} = 1.45 \text{ ft}$$

The interfacial concentrations are now determined on a design diagram (see Fig. 9.26) through Eq. (9.72). As a first approach, $Y_{BM}/X_{BM}$ is assumed to be unity and the interfacial concentrations $Y_i$ and $X_i$ ($Y_i = 1.0X_i$) are found by substitutions in Eq. (9.72) to be 0.222. Therefore, as the first guess,

$$Y_{BM} = \frac{0.222 - 0.292}{\ln\left[(1 - 0.292)/(1 - 0.222)\right]} = 0.742 \qquad X_{BM} = \frac{0.222 - 0.194}{\ln\left[(1 - 0.194)/(1 - 0.222)\right]} = 0.792$$

Substituting these values into Eq. (9.72) and repeating the procedure once or twice, one finds as the final values

$$X_i = Y_i = 0.224 \qquad Y_{BM} = 0.742 \qquad X_{BM} = 0.791$$
$$Y^* = \bar{X} = 0.1944 \qquad Y^*_{BM} = 0.756$$

The height of an overall gas-phase mass-transfer unit may now be computed according to Eq. (9.93).

$$(H_{OG})_{av} = 2.39 \frac{0.742}{0.756} + 1.45 \frac{1.0}{1.485} \frac{0.791}{0.756} = 3.37 \text{ ft}$$

Tower height [Eq. (9.87)]:

$$h_T = 10.23 \times 3.37 = 34.47 \text{ ft}$$

A simple way of finding $(H_{OG})_{av}$ would have been to assume $Y_{BM}/Y^*_{BM} = 1$ and $X_{BM} = 1 - \bar{X}$, which would hold true for a completely gas-film-controlled situation. The height of an overall transfer unit becomes

$$(H_{OG})_{av} = 2.39 + \frac{1.0}{1.48} \frac{0.806}{0.756} 1.45 = 3.43 \text{ ft}$$

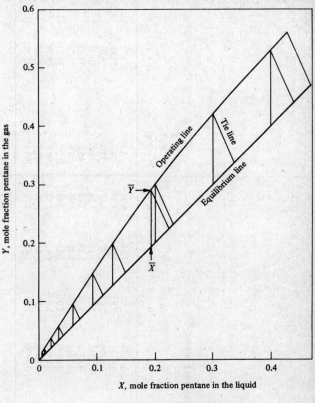

FIGURE 9.26
Design diagram for Example 9.7.

Equation (9.102) for dilute solutions yields a better result:

$$(H_{OG})_{av} = 2.39 + 1.45\,\frac{1.0}{1.485} = 3.37 \text{ ft}$$

which is in agreement with the procedures suggested in the text above for cases when $m > 1.0$.

The corresponding values for the required height of the packing are $h_T = 35.08$ ft and $h_T = 34.47$ ft, respectively.

(b) Because of the anticipated variations in the heights of a transfer unit, the rigorous solutions for the number of overall gas-phase transfer units and for the tower height require two separate integrations. They will be performed through Eqs. (9.86) and (9.59), respectively.

For each one of a set of $Y$ values, the corresponding $X$ value is found through a mass balance, and the point located on a $YX$ diagram (see Fig. 9.26). In the same way as outlined in part $a$, $G$ and $L$ and hence the gas-phase and liquid-phase heights of a transfer unit are found. By the trial-and-error procedure described in part $a$ the interfacial concentrations are determined as the intersection of the tie line with the equilibrium line.

## CALCULATED QUANTITIES—EXAMPLE 9.7

| $Y$ | $X(=Y^*)$ | $Y_i(=X_i)$ | $Y_{BM}$ | $H_G$, ft | $H_L$, ft | $H_{OG}$, ft | $\dfrac{H_G Y_{BM}}{(1-Y)(Y-Y_i)}$ | $\dfrac{Y^*_{BM}}{(1-Y)(Y-Y^*)}$ |
|---|---|---|---|---|---|---|---|---|
| 0.56 | 0.4288 | 0.4696 | 0.4838 | 2.92 | 1.51 | 4.07 | 35.5 | 8.71 |
| 0.5307 | 0.4 | 0.4405 | 0.5131 | 2.85 | 1.50 | 3.99 | 34.5 | 8.68 |
| 0.4214 | 0.3 | 0.3371 | 0.6198 | 2.63 | 1.47 | 3.68 | 33.4 | 9.07 |
| 0.3 | 0.2007 | 0.2308 | 0.7341 | 2.41 | 1.45 | 3.38 | 36.5 | 10.77 |
| 0.2 | 0.1270 | 0.1490 | 0.8252 | 2.24 | 1.43 | 3.16 | 45.3 | 14.31 |
| 0.1 | 0.0595 | 0.0717 | 0.9141 | 2.08 | 1.42 | 2.96 | 74.6 | 25.24 |
| 0.06 | 0.0341 | 0.0419 | 0.9490 | 2.01 | 1.42 | 2.88 | 112.5 | 39.11 |
| 0.03 | 0.0156 | 0.0199 | 0.9750 | 1.97 | 1.42 | 2.82 | 196.6 | 69.78 |
| 0.02 | 0.0095 | 0.0127 | 0.9840 | 1.95 | 1.41 | 2.80 | 267.7 | 95.67 |
| 0.01257 | 0.005 | 0.0073 | 0.9900 | 1.94 | 1.41 | 2.78 | 369.8 | 132.82 |

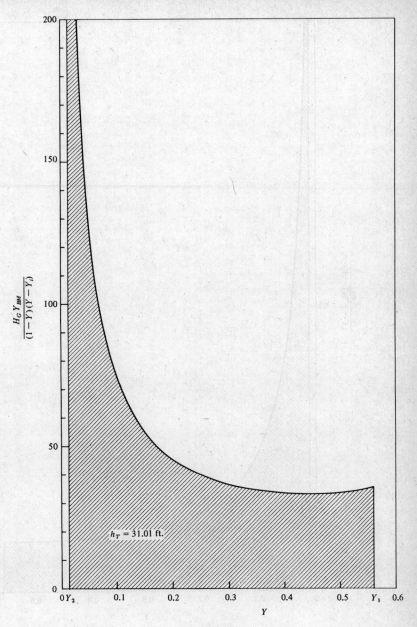

**FIGURE** 9.27
Graphical integration for tower height, Example 9.7.

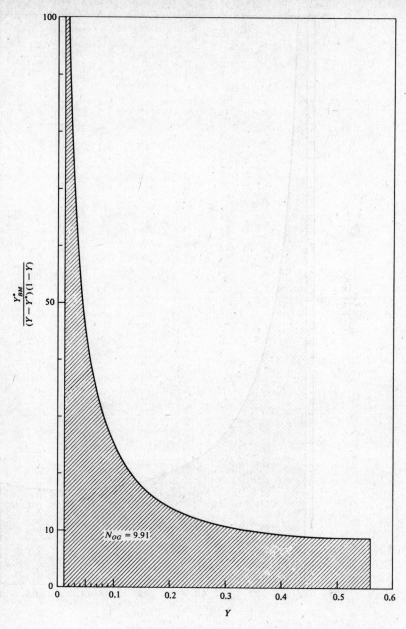

FIGURE 9.28
Graphical integration for $N_{OG}$, Example 9.7.

Based on this information, the function $H_G\{Y_{BM}/[(1 - Y)(Y - Y_i)]\}$ is evaluated and Eq. (9.59) is integrated graphically to give the depth of the packing (see Fig. 9.27). The number of overall gas-phase transfer units is found by computing the function $Y_{BM}^*/[(1 - Y)(Y - Y^*)]$ and integrating Eq. (9.86) graphically as shown in Fig. 9.28.

A summary of these calculations is given in the table of calculated quantities, together with the values of $H_{OG}$.

The total required height of the column from Fig. 9.27 is $h_T = 31.01$ ft. This is in good agreement with the shortcut procedure of part $a$, which appears to yield a conservative result.

In this example, the mass velocities change drastically through the column and range in the case of the gas stream from $G = 810$ lb/(h)(ft$^2$) at the concentrated end down to 210 lb/(h)(ft$^2$) at the dilute end. This causes an increase of the height of a transfer unit toward the bottom of the column which manifests itself in a slight increase of the function $H_G\{Y_{BM}/[(1 - Y)(Y - Y_i)]\}$ toward the concentrated end (see Fig. 9.27).

The number of required overall transfer units is given by Fig. 9.28, as $N_{OG} = 9.91$, which is in good agreement with the result in part $a$.

· *Comment*: An exact solution by computer (see appendix) gave the following results: $N_T = 9.52$, $N_{OG} = 9.91$, $\Delta N_{OG} = 0.39$, and $h_T = 31.18$ ft. The value for $\Delta N_{OG}$ is in excellent agreement with that obtained by Fig. 9.23. The correct average $H_{OG}$ is 31.18/9.91 or 3.14 compared with 3.37 ft by the shortcut method. The corresponding error in $N_{OG}$ is 3.2 percent. In view of the high pentane content of the gas and resulting large variation in flows and $H_{OG}$ over the tower, the results of the shortcut procedure are considered satisfactory for preliminary design. Excellent agreement between the graphical integration method and the exact solution was obtained. ////

### 9.6.F  Approximate Design Procedure for Straight Operating Line and Curved Equilibrium Line

An approximately straight operating line is obtained when (1) the gas stream is dilute or (2) when evaporation of the solvent counter to the direction of absorption, or mass transfer of other components, is such that the total molal gas and liquid flows do not vary appreciably over the tower. Equations (9.102), (9.103), (9.104), and (9.105) are applicable.

Colburn [11] has shown that when the equilibrium line is straight near the origin but curved slightly at its upper end, the value of $N_{OG}$ is given approximately by

$$N_{OG} = \int_{Y_2}^{Y_1} \frac{dy}{Y - Y^*}$$

$$= \frac{1}{1 - (mG_M/L_M)_2} \ln \left\{ \frac{[1 - (mG_M/L_M)_2]^2 (Y_1 - Y_2^*)}{[1 - (Y_1^*/Y_1)](Y_2 - Y_2^*)} + \left( \frac{mG_M}{L_M} \right)_2 \right\} \quad (9.112)$$

Othmer and Scheibel [38] have derived an equation for the same case which is somewhat less convenient in application. It should be apparent that Fig. 9.21 may be used instead of Eq. (9.112) if the abscissa is chosen as

$$\frac{(Y_1 - Y_2^*)}{(Y_2 - Y_2^*)} \frac{[1 - (mG_M/L_M)_2]}{[1 - (Y_1^*/Y_1)]}$$

and the parameter is taken to be $(mG_M/L_M)_2$.

Equation (9.112) is not satisfactory when the curvature of the equilibrium line is large. Alternate algebraic formulas are given by Onda and Sada [35].

The authors have considered the more general case in which the equilibrium line can be represented satisfactorily by Eq. (9.3c).

$$Y^* = \frac{aX}{1 + bX} \tag{9.3c}$$

A large number of absorption cases, in which $a$, $b$, and the value of $mG_M/L_M$ at the dilute end of the tower were varied over wide ranges, were studied. Values of $N_{OG}$ were computed precisely by numerical integration. It was found that an effective average equilibrium line slope, $\bar{m}$, could be determined which when used in conjunction with Eq. (9.106) gave satisfactory values of $N_T$. Graphical correlations for $\bar{m}$ are presented in Figs. 9.29 and 9.30 as a function of the initial slope $m_2$, of the slope $m_c$ of the chord connecting $Y_1^*$ and $Y_2^*$, and of the concentration change over the tower expressed as $Y_1/Y_2$ for various degrees of approach to equilibrium between the concentrated gas entering the tower and the liquid leaving. Figure 9.29 applies when the equilibrium line is concave upward, that is, $m_c > m_2$; and Fig. 9.30 applies when the curvature is concave downward, $m_c < m_2$.

It should be emphasized that this method assumes that the equilibrium data can be fitted by Eq. (9.3c). Unless the equilibrium line is quite irregular in slope, this will be a satisfactory assumption for most purposes of approximate design, provided care is taken to use the correct slope at the dilute end. Moderate deviation from this function in the more concentrated region of the tower will not have a large effect on the number of transfer units.

Having obtained the number of transfer units from Eq. (9.106) by employing the average value of $m$ obtained via Fig. 9.29 or 9.30, the tower height may be obtained from Eq. (9.87). To estimate an average $H_{OG}$ the same average value of $m$ may be used in Eq. (9.102) as in Eq. (9.106).

**EXAMPLE 9.8** It is desired to recover alcohol vapor from an air stream by absorption into water at 40°C and 1 atm in a column packed with 1-in Raschig rings. Recovery of 99 percent of the alcohol entering is to be obtained. Water enters the tower at 40°C containing 0.03 percent alcohol by weight.

The airstream is dry and contains 6.7 mole percent alcohol at 40°C. In order to conserve heat in the subsequent distillation process for alcohol recovery, the liquid rate in the absorber will be set at a rather low value, namely 110 percent of the minimum rate required for the desired recovery. The mass velocity of the gas stream entering the tower is to be set at 33.5 lb moles/(h)(ft$^2$), corresponding to a superficial velocity of 3.58 ft/s (at standard conditions). The design will be based on isothermal operation as a conservative assumption, since under actual adiabatic operation evaporative cooling of the liquid would be expected to exceed the effect of the heat of absorption of the alcohol.

For this preliminary design, constant gas-phase and liquid-phase heights of a transfer unit of 1.8 ft and 1.0 ft, respectively, may be assumed. Furthermore, the pressure may be assumed throughout the column to be the same as at the gas outlet, namely 1 atm. The

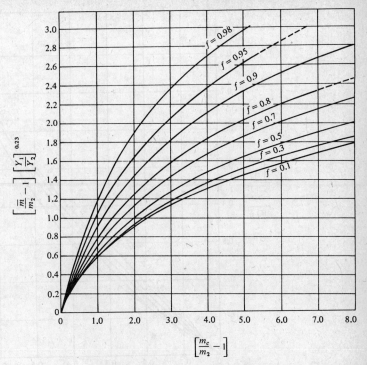

**FIGURE 9.29**
Correlation of average equilibrium line slope (concave upward).

calculation will thus be performed at a pressure value somewhat too low for the lower part of the tower and will therefore yield a conservative result.

Vapor pressure of alcohol over aqueous solution at 40°C may be computed by Eq. (9.4) fitted to isobaric vapor pressure data [24] at 40°C.

Estimate the depth of packing required.

SOLUTION Since water will evaporate in this column to an amount approximately equal to the absorption of alcohol, constant flows will be assumed. The required depth of packing will therefore be estimated by the design procedure suggested for straight operating line and curved equilibrium line.

$$Y_1 = 0.067 \qquad Y_2 = (0.1)(0.067) = 0.00067 \qquad \text{and}$$

$$X_2 = \frac{0.0003/46}{0.0003/46 + 0.9997/18} = 0.0001174$$

Because the equilibrium line is curved downward, the minimum water rate is determined not by the equilibrium at the bottom of the tower but rather by a pinch within the column. The minimum slope of the operating line must be established graphically.

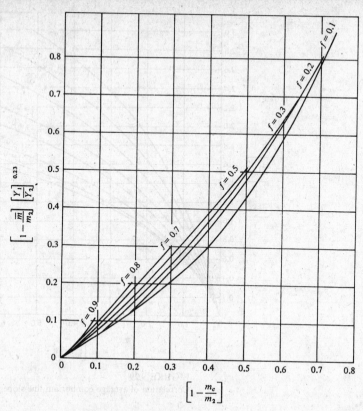

**FIGURE 9.30**
Correlation of average equilibrium line slope (concave downward).

Setting $p_s$ at 135.3 mm and constants $A$ and $B$ in Eq. (9.4) at 253.6°K and 122.1°K, respectively, the equilibrium line points are calculated as follows:

| X | Y | X | Y |
|---|---|---|---|
| 0.001 | 0.00114 | 0.01 | 0.01066 |
| 0.002 | 0.00226 | 0.014 | 0.01448 |
| 0.004 | 0.00446 | 0.02 | 0.01981 |
| 0.006 | 0.00659 | 0.04 | 0.03457 |
| 0.008 | 0.00865 | 0.06 | 0.04571 |

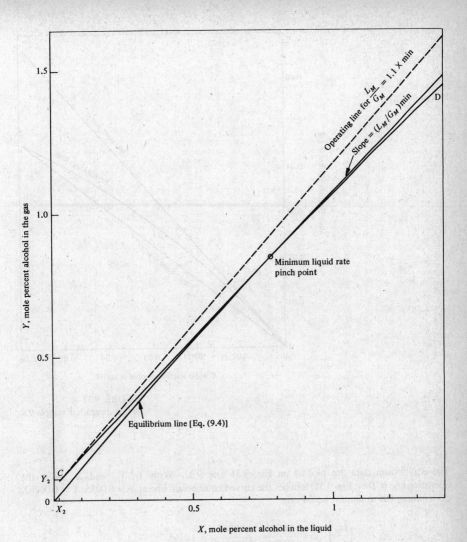

**FIGURE 9.31**
Design diagram for minimum liquid rate, Example 9.8.

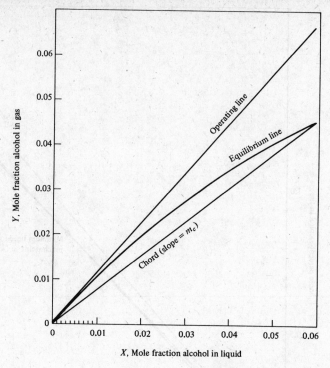

FIGURE 9.32
Design diagram, Example 9.8.

The equilibrium data are plotted on Fig. 9.31 and 9.32. With $Y_1$, $Y_2$, and $X_2$ fixed, the operating line ($CD$ on Fig. 9.31) touches the curved equilibrium line at $Y = 0.0085$, $X = 0.00782$, determining the minimum liquid rate.

$$\left(\frac{L_M}{G_M}\right)_{min} = \frac{0.0085 - 0.00067}{0.00782 - 0.0001174} = 1.0165$$

$$\left(\frac{L_M}{G_M}\right) = (1.016)(1.1) = 1.1182$$

$$L_M = 1.1182 \times 33.5 = 37.459 \text{ lb moles/(h)(ft}^2)$$

$X_1$ follows from an alcohol balance:

$$X_1 = \frac{G_M}{L_M}(Y_1 - Y_2) + X_2 = 0.05943$$

In order to apply the suggested method, a few parameters have to be computed. The slope of the equilibrium line at its dilute end, $m_2$, follows from constant $A$ of the van Laar equation:

$$m_2 = 10^{0.81} \frac{135.3}{760} = 1.1494 = \left(\frac{dY}{dX}\right)_{X \to 0}$$

$Y_1^*$ and $Y_2^*$ are read from the $YX$ diagrams (Figs. 9.31 and 9.32):

$$Y_1^* = 0.0457 \qquad Y_2^* = 0.0002 \qquad m_c = \frac{0.0457 - 0.0002}{0.05943 - 0.0001174} = 0.7671$$

$$1 - \frac{m_c}{m_2} = 0.3326 \qquad f = \frac{Y_1^*}{Y_1} = \frac{0.0457}{0.067} = 0.682$$

$$\left(\frac{Y_1}{Y_2}\right)^{0.23} = 100^{0.23} = 2.884$$

According to the chart in Fig. 9.30,

$$\left(1 - \frac{\overline{m}}{m_2}\right)2.884 = 0.31 \qquad \overline{m} = \left(1 - \frac{0.31}{2.884}\right)1.1494 = 1.0259$$

$$\overline{m}\,\frac{G_M}{L_M} = 0.9175 \qquad \frac{Y_1 - \overline{m}X_2}{Y_2 - \overline{m}X_2} = \frac{0.067 - 0.0001204}{0.067 - 0.0001204} = 121.7$$

Either Fig. 9.21 or Eq. (9.106) may now be used to determine $N_T$, which is equal to $N_{OG}$ for this case of equimolar counterdiffusion.

$$N_T = 29.02 = N_{OG} \qquad \text{by Eq. (9.106)}$$

Because of the low concentration and also because $\overline{m} > 1$, Eq. (9.102) is justified and may be used in order to obtain $(H_{OG})_{av}$:

$$(H_{OG})_{av} = 1.8 + 0.92 \times 1.0 = 2.72 \qquad \text{and } h_T = 29.02 \times 2.72 = 78.93 \text{ ft}$$

*Comment*   An exact solution by computer (see appendix) Example 9.8a gave $N_{OG} = 29.01$ and $h_T = 77.38$ ft. The corresponding average value of $H_{OG} = 77.38/29.01$ or 2.67 ft. In view of the more than twofold variation in $m$ over the tower, and the closeness of the liquid rate to the minimum value, the shortcut procedure appears quite satisfactory with respect to the calculation of both $(H_{OG})_{av}$ and $N_{OG}$.

Use of Eq. (9.3c) to express the equilibrium was tested by computer (see appendix, Example 9.8b). When the constants in Eq. (9.3c) were determined from the initial slope and a single experimental point ($X = 0.042$, $Y = 0.036$) of composition near to that at the bottom of the tower, good agreement with the result of Eq. (9.4) was obtained.     ////

## 9.6.G   Approximate Design Procedures for Concentrated Gas with Curved Equilibrium Line

For this case Eqs. (9.106) and (9.108) may be employed to estimate the number of transfer units based on an average value of $L_M/G_M$ from Fig. 9.25 and an average value of $m$ from Fig. 9.29 or 9.30. Similarly, Eq. (9.93) may be employed with the same average value of $mG_M/L_M$ to obtain $(H_{OG})_{av}$. The tower height is then obtained from Eq. (9.87). However, in cases in which the operating line is concave upward and the equilibrium line is concave downward; or if the curvatures are such that a pinch point could exist, a more rigorous graphical or numerical computation may be necessary. A graph of the design diagram should be made in doubtful cases to assess the applicability of this simplified procedure. A rough sketch of the design diagram may be employed, as suggested in the preceding sec. 9.6.E, to estimate the composition ratios in Eq. (9.93), since these will generally not be far from unity and do not affect $H_{OG}$ greatly. It should suffice in this case to base the tie-line construction on the composition corresponding to the average value of $m$ or of $L_M/G_M$, whichever appears to vary more over the tower.

This method is illustrated in the solution of Example 9.10.

### 9.6.H   Multicomponent Mass-Transfer Effects in Gas Absorption

In some situations it may be necessary to consider the effects of simultaneous mass transfer of other components on that for the component of primary interest. For example, if the solvent is volatile and the entering gas is unsaturated with the solvent, the component being absorbed must diffuse counter to the diffusion of the evaporating solvent. Multicomponent systems may involve simultaneous transfer of several components in either direction. While a rigorous treatment of such cases has not been well developed and is beyond the scope of the present discussion, it will be useful to consider in an approximate manner how such effects might enter into the design equations. The following discussion is based in part upon an approximate treatment of multicomponent mass transfer suggested by one of the authors [57].

Consider the absorption of component $A$ in a mixture of components $A$, $B$, $C$, $D$, ..., at a point in the tower the steady-state molar mass-transfer fluxes of component $A$ and of the mixture as a whole, including $A$, are $N_A$ and $N_t$, respectively. Over a differential packed height $dh$, a material balance for the gas phase gives for a unit cross section of tower, according to Eq. (9.55),

$$N_A a_v \, dh = -G_M \, dY - Y \, dG_M \qquad (9.55)$$

where $a_v$ = surface area of the packing per unit volume

$Y$ = mole fraction of component $A$

Defining the ratio of the flux of $A$ to the net flux of the entire mixture in the same coordinate direction as

$$\phi_A = \frac{N_A}{N_t} \qquad (9.113)$$

where $N_t = N_A + N_B + N_c + \cdots$. Then,

$$dG_M = -N_t a_v \, dh = -\frac{N_A}{\phi_A} a_v \, dh \qquad (9.114)$$

Equations (9.55) and (9.114) may be combined and rearranged to give

$$\frac{N_A a_v\, dh}{G_M} = -\frac{dY}{1 - Y/\phi_A} \tag{9.115}$$

$N_A$ may be written in terms of the mass-transfer coefficient according to Eq. (9.50); and upon rearrangement, Eq. (9.115) becomes

$$h_T = \int_1^2 dh = \int_2^1 \frac{G_M}{k_g a_v P} \frac{dY}{(1 - Y/\phi_A)(Y - Y_i)} \tag{9.116}$$

where the integration covers the terminal compositions of the tower. If the film factor $(Y_f)_A$ is now introduced in Eq. (9.116),

$$h_T = \int_2^1 \frac{G_M}{k_g a_v P(Y_f)_A} \frac{(Y_f)_A}{(1 - Y/\phi_A)} \frac{dY}{(Y - Y_i)} \tag{9.117}$$

where

$$(Y_f)_A = \frac{1}{\phi_A}(\phi_A - Y)_{lm} \tag{9.118}$$

and the log mean average of $(\phi_A - Y)$ is taken from $Y$ to $Y_i$.

The first quantity under the integral of Eq. (9.117) is the general expression for the height of a gas-phase transfer unit $(H_G)_A$, and the remainder is the differential number of transfer units. Thus,

$$(H_G)_A = \frac{G_M}{k_g a_v P(Y_f)_A} \tag{9.119}$$

$$N_G = \int_2^1 \frac{(Y_f)_A}{1 - Y/\phi_A} \frac{dY}{Y - Y_i} \tag{9.120}$$

Equation (9.120) may be simplified to give

$$N_G = \int_2^1 \psi_A \frac{dY}{(Y - Y_i)} \tag{9.121}$$

where

$$\psi_A = \frac{(\phi_A - Y)_{lm}}{(\phi_A - Y)} \tag{9.122}$$

and the log-mean average of $(\phi_A - Y)$ is taken from $Y$ to $Y_i$. The factor $\psi_A$ contains the effect of diffusion of other components on the number of transfer units for absorption of component $A$. $(H_G)_A$ is independent of these effects provided that $k_g(Y_f)_A$ is constant as supported by theory (see Chap. 5) and that the Schmidt number is not affected by changes in the "effective" composition of the mixture [57]. This latter effect will be negligible in most cases of practical interest and will not be considered in the present development. It is apparent from the form of the equation that $\psi_A$ will not deviate far from unity unless the log-mean average of $(\phi_A - Y)$ between the main stream and the gas-liquid interface differs substantially from $\phi_A - Y$. As a practical matter, therefore, it is often satisfactory to ignore the $\psi$ factor or use a simple average over the tower in the computation of the number of transfer units by Eq. (9.121). Usually a more important consideration than the effect of the $\psi$ factor will be the effect of transfer of the various components on the gas and liquid flows, and the resultant influence on the shape of the operating line, which determines the term $(Y - Y_i)$ in Eq. (9.121). Approximate methods for evaluating the foregoing effects in certain limiting situations are described later in this section.

It should be noted that for mass transfer of one component only $N_t = N_A$ and $\phi_A = 1$, and Eq. (9.120) becomes identical to Eq. (9.63).

For equimolal counterdiffusion, which is a satisfactory assumption for most cases of distillation, $N_t = 0$ and $\phi_A = \infty$, so that $(Y_f)_A$ and $\psi_A$ become equal to 1.0.

The analogous expressions for the number of overall gas-phase transfer units are

$$N_{OG} = \int_2^1 \psi_A^* \frac{dY}{Y - Y^*} \tag{9.123}$$

where

$$\psi_A^* = \frac{(\phi_A - Y)_{lm}}{\phi_A - Y} \tag{9.124}$$

and the log-mean average is taken over the compositions $Y$ and $Y^*$.

The corresponding overall height of a transfer unit is defined by the equation

$$H_{OG} = \frac{G_M}{K_{OG} Y_f^* P} \tag{9.125}$$

where

$$Y_f^* = \frac{1}{\phi_A} (\phi_A - Y)_{lm} \tag{9.126}$$

and the log-mean average is taken from $Y$ to $Y^*$.

**Generalized design procedure [58]**  The factor $\psi_A^*$ in the general case is analogous to the term $Y_{BM}^*/(1 - Y)$ in Eq. (9.86) for transfer of one component only. An assessment of the importance of $\psi^*$ in Eq. (9.123) can be made by considering the special case of absorption of a component of solubility such that its concentration at the gas-liquid interface is always zero, and with the further restriction that a constant average value of $\phi_A$ may be used over the integral. For this case Eq. (9.123) may be solved analytically to give

$$N_{OG} = \ln \left\{ \frac{\ln \left[ (\overline{\phi}_A - Y_1)/\overline{\phi}_A \right]}{\ln \left[ (\overline{\phi}_A - Y_2)/\overline{\phi}_A \right]} \right\} \equiv \ln \left[ \frac{\ln (1 - tY_1)}{\ln (1 - tY_2)} \right] \tag{9.127}$$

where $\overline{\phi}_A$ = average value of $\phi_A$

$t = 1/\overline{\phi}_A$

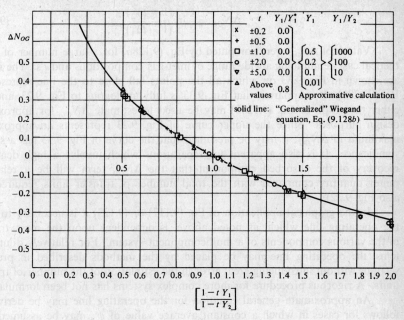

FIGURE 9.33
Correction for counterdiffusional effects on $N_{OG}$.

$N_{OG}$ may be expressed also in the form given previously:

$$N_{OG} = \int_2^1 \frac{dY}{Y - Y^*} + \Delta N_{OG} \tag{9.108}$$

and

$$N_T = \int_2^1 \frac{dY}{Y - Y^*} \tag{9.104}$$

For the present case ($Y^* = 0$), Eq. (9.104) gives

$$N_T = \ln \frac{Y_1}{Y_2}$$

and, therefore,

$$\Delta N_{OG} = \ln \left[ \frac{Y_2 \ln (1 - tY_1)}{Y_1 \ln (1 - tY_2)} \right] \tag{9.128a}$$

For the special case in which the log-mean average term of Eq. (9.124) can be represented by the arithmetic mean, Eq. (9.123) can be solved analytically to give an equation analogous to Eq. (9.109):

$$\Delta N_{OG} = \frac{1}{2} \ln \frac{(1 - tY_2)}{(1 - tY_1)} \qquad (9.128b)$$

Values of $\Delta N_{OG}$ were computed by Eq. (9.128a) for a large number of hypothetical cases embracing a wide range of inlet-gas compositions and $t$'s. The results are shown in Fig. 9.33 on which all the points fall close to the line representing Eq. (9.128b). It is apparent that Fig. 9.33 is fully analogous to Fig. 9.23, and that either the graph or Eq. (9.128b) may be used to estimate $\Delta N_{OG}$ for approximate design purposes. Since the upper curve of Fig. 9.23 represents an approximate maximum in $\Delta N_{OG}$, it may be presumed that the curve of Fig. 9.33 has a similar significance. It is also apparent for separations which involve a high degree of recovery of the absorbed component that the $\Delta N_{OG}$ term will be a relatively minor quantity compared with the total number of transfer units required (see also Fig. 9.22).

In the graphical application of Eq. (9.117) or (9.123) it is necessary to place the operating line which, as noted above, is dependent upon the mass transfer of the various components of a multicomponent system. For relatively dilute systems, the operating line may be placed by the methods described in previous sections without introducing great error in the estimation of the number of transfer units. A rigorous procedure for more complex systems has not been formulated.

An approximate general equation for the operating line may be derived as follows for cases in which a constant average value of $\phi_A$ may be assumed over the entire column. Consider a mixture of components $A$, $B$, $C$, $D$, etc. Define

$\Delta g_A$ = total moles of component $A$ absorbed over the tower, lb moles/(h) (ft$^2$)

$\Delta g_I = \Delta g_B + \Delta g_C + \Delta g_D + \cdots$

$\Delta g_I$ = net total moles of all components other than $A$ absorbed or desorbed over the tower, lb moles/(h) (ft$^2$)

The average value of $\phi_A$ is given by

$$\bar{\phi} = \frac{\Delta g_A}{\Delta g_A + \Delta g_I} \qquad (9.129)$$

It will generally suffice to estimate $\Delta g_A$ and $\Delta g_I$ from preliminary estimates of the total mass transfer of each component over the tower. Considerations such as those specified in Sec. 9.5.E for multicomponent hydrocarbon absorption may be useful for this purpose.

From material balances on the gas and liquid streams around the dilute end (compositions $X_2$, $Y_2$) and any point of compositions $X$, $Y$ within the tower, it can be shown that

$$G_M = G_{M,2} \frac{\bar{\phi}_A - Y_2}{\bar{\phi}_A - Y} \equiv G_{M,2} \frac{(1 - tY_2)}{(1 - tY)} \qquad (9.130)$$

and

$$L_M = L_{M,2} \frac{\bar{\phi}_A - X_2}{\bar{\phi}_A - X} \equiv L_{M,2} \frac{(1 - tX_2)}{(1 - tX)} \qquad (9.131)$$

The flow $G_{M,2}$ can be estimated from the relations

$$G_{M,2} = G_{M,1} \frac{\bar{\phi}_A - Y_1}{\bar{\phi}_A - Y_2} \equiv G_{M,1} \frac{(1 - tY_1)}{(1 - tY_2)} \tag{9.132}$$

or

$$G_{M,2} = G_{M,1} - (\Delta g_A + \Delta g_I) \tag{9.133}$$

The material balance [analogous to Eq. (9.47)] may be expressed in total molal flows as follows:

$$Y = \frac{L_M}{G_M} X + \frac{G_{M,2}}{G_M} Y_2 - \frac{L_{M,2}}{G_M} X_2 \tag{9.134}$$

Equations (9.130) and (9.131) may be combined with Eq. (9.134) to give after some rearrangement explicit expressions for $Y$ or $X$.

$$Y = \frac{L_{M,2}(1 - tX_2)X + (G_{M,2}Y_2 - L_{M,2}X_2)(1 - tX)}{G_{M,2}(1 - tX) + tL_{M,2}(X - X_2)} \tag{9.135}$$

where, as defined above, $t \equiv 1/\phi_A$ and

$$X = \frac{G_{M,2}(1 - tY_2)Y + (L_{M,2}X_2 - G_{M,2}Y_2)(1 - tY)}{L_{M,2}(1 - tY) + tG_{M,2}(Y - Y_2)} \tag{9.136}$$

Alternatively, Eq. (9.135) can be expressed by the following system of equations:

$$Y^0 = \frac{L_M^0}{G_M^0}(X^0 - X_2^0) + Y_2^0 \tag{9.137}$$

where

$$Y^0 = \frac{Y}{1 - tY} \tag{9.138}$$

$$X^0 = \frac{X}{1 - tX} \tag{9.139}$$

$$L_M^0 = L_{M,2}(1 - tX_2) = L_{M,1}(1 - tX_1) \tag{9.140}$$

$$G_M^0 = G_{M,2}(1 - tY_2) = G_{M,1}(1 - tY_1) \tag{9.141}$$

Equation (9.137) is useful for graphical methods because of its linear form. For example, the minimum liquid rate in case of a pinch point can readily be estimated graphically from the intersection of an equilibrium line in $X^0$, $Y^0$ coordinates with a corresponding operating line of slope $L_M^0/G_M^0$. Also, theoretical plates can easily be stepped off from a design diagram in this form.

A generalized design procedure may be developed in terms of the equations above which permits the handling of all cases with a common set of equations. This is particularly useful in permitting a single computer program to cover a range of cases, as described in the appendix.

An expression for the tower height may be obtained by combining Eqs. (9.123) and (9.125) and rearranging the result:

$$h_T = \int_{Y_2}^{Y_1} H_{OG} \frac{Y_f^* \, dY}{(1 - tY)(Y - Y^*)} \tag{9.142}$$

$$Y_f^* = \frac{(1 - tY) - (1 - tY^*)}{\ln\left[(1 - tY)/(1 - tY^*)\right]} \tag{9.143}$$

In performing numerical or graphical integration of Eq. (9.142), a value of $X$ corresponding to each value of $Y$ is found from Eq. (9.136), or alternatively, by solving Eq. (9.137) for $X^0$ and obtaining $X$ from Eq. (9.139). $Y^*$ is then obtained by substitution of $X$ in an appropriate equilibrium relation such as Eq. (9.3) or (9.4).

The height of a transfer unit may be expressed by a generalized form of Eq. (9.93):

$$H_{OG} = \frac{Y_f}{Y_f^*} H_G + \frac{mG_M}{L_M} \frac{X_f}{Y_f^*} \cdot H_L \tag{9.144}$$

where

$$X_f = \frac{(1 - tX) - (1 - tX_i)}{\ln\left[(1 - tX)/(1 - tX_i)\right]} \tag{9.145}$$

and

$$Y_f = \frac{(1 - tY) - (1 - tY_i)}{\ln\left[(1 - tY)/(1 - tY_i)\right]} \tag{9.146}$$

The flow ratio is given by

$$\frac{G_M}{L_M} = \frac{G_{M,2}}{L_{M,2}} \frac{(1 - tY_2)}{(1 - tX_2)} \frac{(1 - tX)}{(1 - tY)} = \frac{G_M^0}{L_M^0} \frac{(1 - tX)}{(1 - tY)}$$

To evaluate $Y_f$ and $X_f$ it is necessary to calculate the interfacial composition $X_i$, $Y_i$ at each $Y$ value used in the calculation of $H_{OG}$. For this purpose Eq. (9.72) may be expressed in the generalized form

$$\frac{Y - Y_i}{X - X_i} = -\frac{L_M}{G_M} \frac{H_G}{H_L} \frac{Y_f}{X_f} \tag{9.147}$$

**Application of short-cut design procedures to multicomponent systems**  For approximate estimates of tower height, the simplified procedures described in Secs. 9.4.D through 9.4.G for transfer of a single component should be applicable also to isothermal multicomponent systems, since the diffusional effects are handled through the appropriate $\Delta N_{OG}$ correlation and the integral to be solved for $N_T$ is the same in all cases. Example 9.10 below provides a rather severe test of the foregoing hypothesis, since the system involves substantial curvature of both the operating and equilibrium lines with varying degrees of counterdiffusion of alcohol and water vapor. In questionable cases, the tower height may be obtained by graphical integration of Eq. (9.142) in the manner described previously for simple cases or by the HETP method (Sec. 9.8), or by computer as described in the appendix.

**Discussion**   Employment of an average value of $t$ or $\phi$ as defined by Eq. (9.129) in the foregoing development is based on the principle that a satisfactory representation of the operating line will result provided that the flows vary through the tower in a relatively smooth manner. The value of $\bar{\phi}_A$ so defined will not necessarily correspond closely with local values of $\phi_A$ at various heights in the tower. For example, if component $A$ is absorbed from the gas while another component $B$ is stripped from the liquid, $\phi_A$ can vary from a positive value near unity at the bottom of the tower to a small negative value at the top. If the quantities of $A$ and $B$ transferred are equal, the operating line equation can be based satisfactorily on a $\bar{\phi}_A$ value of $\infty$. On the other hand, if the solvent vaporizes at a rate in constant ratio to the rate of absorption of $A$, $\bar{\phi}_A$ will be constant and equal to local values of $\phi_A$ over the tower. The present procedure is supported by the conclusion reached in the analysis leading to development of Fig. 9.33, namely that $\phi_A$, as expressed by the $\psi_A^*$ factor, has only a small effect on the integral of Eq. (9.123) for $N_{OG}$. Equation (9.123) depends primarily on the term $(Y - Y^*)$ and hence upon proper placement of the operating line. Therefore, the main objective is to select a value of $\bar{\phi}_A$ which satisfies the operating-line equation.

In case there are strong heat effects in the column, the operating line (and equilibrium line) may be distorted by local vaporization or condensation of the solvent sufficiently to invalidate the method. In such cases more rigorous design procedures, as discussed in Sec. 9.6.I below, may be necessary.

**EXAMPLE 9.9**   Hot, dry air at 490°F, 1 atm, containing ammonia, is to be cooled with simultaneous absorption of 99 percent of the ammonia before it is discharged to the atmosphere. This will be done by contact with slightly acidified ammonia-free water in an adiabatic column packed with 1-in Berl saddles.

Ammonia-air mixture enters the column at a molar mass velocity of 30 lb moles/(h) (ft²). Water enters the column at 50 lb moles/(h)(ft²) at 122°F, approximately the adiabatic saturation temperature of the entering air, therefore, the liquid temperature will be assumed constant at 122°F.

At these flow rates the gas-phase height of a transfer unit for $NH_3$ absorption may be taken as 1.6 ft. The equilibrium water partial pressure at 122°F is 92.51 mm Hg.

(a)   Estimate the required depth of packing and the effect of counterdiffusion thereupon for inlet ammonia mol fractions of 0.05, 0.01, and 0.001.

(b) Could the tower height be reduced substantially if the evaporation of the water vapor could be suppressed?

SOLUTION

(a) Because of the similarity in the diffusional properties of ammonia and water vapor, it will be assumed that the air leaving the tower is 99 percent saturated with water vapor, corresponding to the 99 percent absorption of the ammonia. In this situation a constant average value of $\phi$ may be determined by Eq. (9.129).

*Inlet ammonia mole fraction* = 0.01 (case 2). Designate ammonia, water, and air by the subscripts $A$, $B$, and $C$, respectively, and the top and bottom of the tower by subscripts 2 and 1, respectively.

$$g_{A,1} = (30)(0.01) = 0.3 \text{ lb mole NH}_3 \text{ entering/(h)(ft}^2)$$

$$\Delta g_A = (0.99)(0.3) = 0.297 \text{ lb mole NH}_3 \text{ absorbed/(h)(ft}^2)$$

$$g_{B,1} = 0$$

$$g_{C,1} = 29.7 \text{ lb moles air entering/(h)(ft}^2)$$

$$g_{A,2} = (0.3)(0.01) = 0.003$$

$$-\Delta g_B = \frac{(92.51)(0.99)}{760 - (92.51)(0.99)}(29.7 + 0.003) = 4.07 \text{ lb moles/(h)(ft}^2)$$

By Eq. (9.127),

$$\overline{\phi}_A = \frac{0.297}{0.297 - 4.07} = -0.0787 \quad \text{and} \quad t = -12.71$$

$$G_{M,2} = 0.003 + 4.07 + 29.7 = 33.773 \text{ lb moles/(h)(ft}^2)$$

$$Y_1 = 0.01 \quad \text{and} \quad Y_2 = \frac{0.003}{33.773} = 0.8883 \times 10^{-4}$$

$$N_{OG} = \ln\left\{\frac{\ln\left[1 + (12.71)(0.01)\right]}{\ln\left[1 + (12.71)(0.8883 \times 10^{-4})\right]}\right\} = 4.664$$

$$h_T = 4.664 \times 1.60 = 7.46 \text{ ft}$$

*Results for other concentrations* (cases 1 and 3). Following the procedure just given, results were computed for ammonia mole fractions of 0.05 and 0.001. Results for all three cases are summarized in the table below.

|  | $Y_1$ | $t$ | $Y_2 \times 10^{-4}$ | $N_{OG}$ | $h_T$, ft |
|---|---|---|---|---|---|
| Case 1 | 0.05 | −1.882 | 4.627 | 4.638 | 7.42 |
| Case 2 | 0.01 | −12.71 | 0.8883 | 4.664 | 7.46 |
| Case 3 | 0.001 | −136.3 | 0.08803 | 4.668 | 7.47 |

*Comment:* The results given in the table indicate that counterdiffusion of the solvent at rates in excess of equimolal counterdiffusion in the concentration range has essentially no effect on the number of transfer units required for a given percentage absorption, even though the mass-transfer flux of solvent was on the order of 130 times greater than the opposite flux of the component being absorbed.

(b) For the three cases of part *a* assume that the air is precooled and humidified by addition of steam [approximately 4.1 lb moles/(h)(ft$^2$)] sufficient to prevent solvent evaporation. In this case $t = 1.0$ in Eq. (9.127).

Following the procedure of part *a* the following results are obtained.

| | $Y_1$ before humidification | Steam injected, lb moles/ (h)(ft$^2$) | $Y_1$ after humidification | $Y_2 \times 10^4$ | $t$ | $N_{OG}$ | $h_T$, ft | $h_T$ from part *a*, ft |
|---|---|---|---|---|---|---|---|---|
| Case 1 | 0.05 | 3.95 | 0.04418 | 4.620 | 1.0 | 4.58 | 7.33 | 7.42 |
| Case 2 | 0.01 | 4.12 | 0.008792 | 0.8870 | 1.0 | 4.60 | 7.36 | 7.46 |
| Case 3 | 0.001 | 4.15 | $0.8784 \times 10^{-3}$ | 0.08792 | 1.0 | 4.60 | 7.36 | 7.47 |

*Comment:* It is apparent from comparison of the results above with those of part *a* that there would be little mass-transfer advantage in suppression of solvent evaporation in this way, and that additional facilities for air-cooling and humidification could not be justified. In fact, it will generally be desirable to allow the solvent to vaporize when circumstances permit, to obtain the beneficial effect of evaporative cooling on the gas solubility.          ////

**EXAMPLE 9.10** Alcohol vapor is to be recovered from a hot gas stream by absorption into water. Because of the availability of a large packed tower, rough cost estimates for the entire system, including the distillation column, indicated that it might be preferable in this case to run the absorption at a higher than usual temperature. It was therefore decided to design the column to be packed with 2-in Raschig rings for isothermal operation at 60°C and 1 atm to obtain 99 percent recovery of the alcohol.

The gas enters the tower at a rate of 30 lb moles/(h)(ft$^2$) and contains 0.2 mole fraction of alcohol and is 50 percent saturated with water. The subsequent distillation column supplies the absorption water with an alcohol concentration of 0.0003 mole fraction at 60°C. The amount of water is to be chosen in such a manner that the stripping factor will be 0.7 at the dilute end of the tower. Enough heat-transfer area will be provided to ensure isothermal operation.

The gas-phase and liquid-phase heights of a transfer unit may be assumed constant at 1.8 and 0.8 ft, respectively. The equilibrium data are given in the table. $p_A$ and $p_B$ denote the partial pressures of alcohol and water, respectively. The constants for the van Laar equation are $A = 312.0°K$ and $B = 126.6°K$.

| $X_A$ | $p_A$, mm Hg | $p_A + p_B$ | $Y_A$ |
|---|---|---|---|
| 0 | 0 | 149.38 | 0 |
| 0.051 | 69.2 | 219.0 | 0.091 |
| 0.086 | 97.8 | 249.0 | 0.1287 |
| 0.197 | 154.1 | 298.0 | 0.2028 |
| 0.375 | 193.7 | 325.0 | 0.2549 |
| 1.0 | 352.7 | 352.7 | |

(a) Determine by a suitable graphical integration the number of overall gas-phase transfer units required for this process. Use the method suggested in Sec. 9.6.G in order to estimate an average value of $H_{OG}$ and calculate the required depth of packing.

(b) Determine $N_{OG}$, $(H_{OG})_{av}$, and $h_T$ following the approximate design procedure outlined in Sec. 9.6.G for the situation where the inlet-gas stream is saturated with water. The flow rates and all the other specifications are to remain the same. Repeat this calculation for 50 percent saturation of the inlet-gas stream and for a dry feed gas.

(c) Repeat all calculations rigorously on a digital computer using the computer program shown in the appendix.

SOLUTION

(a) $g_{A,1} = (0.2)(30) = 6.0$ lb moles/(h)(ft$^2$)

$\Delta g_A = (0.99)(6.0) = 5.94$ lb moles/(h)(ft$^2$)

$g_{A,2} = 0.06$ lb mole/(h)(ft$^2$)

In order to determine $\Delta g_B$, the saturation concentrations of water at the bottom and at the top of the column, $Y_{B,1}{}^{sat}$ and $Y_{B,2}{}^{sat}$, respectively, have to be computed.

$$Y_{B,1}{}^{sat}$$

From a rough sketch of the equilibrium data given in the problem description, one finds that at $Y = 0.2$, the equilibrium concentration of alcohol in the liquid, $X_{A,1}^*$, is approximately 0.187. At $Y = 0.2028$, the equilibrium partial pressure of water is $298 - 154 = 144$ mm Hg. Linearizing the equilibrium curve for water one obtains

$$p_B = 144 \frac{1 - 0.187}{1 - 0.197} = 145.8 \text{ mm Hg} \quad \text{or} \quad Y_{B,1}{}^{sat} = 0.1918$$

$$Y_{B,2}{}^{sat}$$

According to the van Laar equation [(9.4)], at the dilute end,

$$\log \gamma = \frac{(312.0/333.2)(0.9997)^2}{[0.9997 + (312.0/126.6)0.0003]^2} = 0.9346$$

$$K = 10^{0.9346}(352.7/760) = 3.992$$

In order to obtain the equilibrium alcohol concentration in the liquid, a guess of $Y_{A, 2}$ is needed. The gas concentration leaving the column, $Y_{A, 2}$, will be between $0.2(1 - 0.99) = 0.002$, which represents its value at equimolar counterdiffusion, and $0.06/(30 - 5.94) = 0.002494$, which represents the value without any solvent evaporation. Let us therefore assume that $Y_{A, 2} = 0.00228$. Then, $X^*_{A, 2} = 0.00228/3.992 = 0.00057$. This allows an estimation of $Y_{B, 2}{}^{sat}$ on the basis of Raoult's law:

$$Y_{B, 2}{}^{sat} = \frac{149.38}{760} (1 - 0.00057) = 0.19644$$

It follows for part $a$ that

$$Y_{B, 1} = (0.5)(0.1918) = 0.0959$$

The assumption of equal heights of a transfer unit for the solute and the solvent suggests a saturation of the leaving gas stream with water of 99.5 percent. Thus

$$Y_{B, 2} = (0.995)(0.19644) = 0.1955$$

Gas streams entering the tower:

$g_{A, 1} = 6.0$ lb moles/(h)(ft$^2$)
$g_{B, 1} = 30(0.0959) = 2.877$ lb moles/(h)(ft$^2$)
$g_C\ \ = 30 - 6 - 2.877 = 21.123$ lb moles/(h)(ft$^2$)

Gas streams leaving the tower:

$g_C\ \ \ = 21.123$ lb moles/(h)(sqft)
$g_{A, 2} = 0.06$ lb mole/(h)(ft$^2$)
$g_{B, 2}\ = [Y_{B, 2}/(1 - Y_{B, 2})](g_{A, 2} + g_C) = 5.147$ lb moles/(h)(ft$^2$)
$G_{M, 2} = g_C + g_{A, 2} + g_{B, 2} = 26.33$ lb moles/(h)(ft$^2$)
$\Delta g_B\ \ = 2.877 - 5.147 = -2.27$ lb moles/(h)(ft$^2$)

The average value of $t$ $(= 1/\bar{\phi})$ is now estimated by Eq. (9.129):

$$t = 1 - \frac{2.27}{5.94} = 0.6178$$

$$Y_{A, 2} = \frac{0.06}{26.33} = 0.00228$$

which confirms the above assumption. Since $m_2$ will be approximately 4.0 and $(mG_M/L_M)_2 = 0.7$, $L_M/G_M$ will be set to 5.715 and $L_{M, 2}$ to 150.48 lb moles/(h)(ft$^2$).

*Graphical integration:*
The next step is the evaluation of the integral

$$N_T = \int_{0.00228}^{0.2} \frac{dY}{(Y - Y^*)}$$

It is most convenient to ignore the function $\psi$ during the integration and to correct the result for its influence afterwards according to the graph in Fig. 9.33.

The evaluation of the function under the integral is formally very similar to the earlier demonstrated cases. For each one of a set of $Y$ values, the corresponding $X$ value must be found through a mass balance. This will be done by means of the following form of Eq. (9.137):

$$X^\circ = \frac{G_M^\circ}{L_M^\circ} Y^\circ - \frac{G_M^\circ}{L_M^\circ} Y_2^\circ + X_2^\circ$$

$L_M^\circ = 150.48(1 - 0.6178 \times 0.0003) = 150.45 \text{ lb moles/(h)(ft}^2)$
$G_M^\circ = 30(1 - 0.6178 \times 0.2) = 26.29 \text{ lb moles/(h)(ft}^2)$
$Y_2^\circ = 0.00228/(1 - 0.6178 \times 0.00228) = 0.0022832$
$X_2^\circ = 0.0003/(1 - 0.6178 \times 0.0003) = 0.00030006$

$G_M^\circ/L_M^\circ = 0.1747; \quad X_2^\circ - \frac{G_M^\circ}{L_M^\circ} Y_2^\circ = -9.896.10^{-5}$

Example: $Y = 0.16$:

$Y^\circ = 0.16/(1 - 0.6178 \times 0.16) = 0.17755$
$X^\circ = 0.1775 \times 0.1747 - 9.896 \times 10^{-5} = 0.03093$
$X = 0.0309/(1 + 0.6178 \times 0.0309) = 0.03035$

$Y^*$ can now be found by the van Laar equation:

$$\log \gamma = \frac{(0.936)(0.96965)^2}{[0.96965 + (312/126.6)0.03035]^2} = 0.8065$$

$$K = 10^{0.8065} \frac{352.7}{760} = 2.974 \qquad Y^* = (2.974)(0.03035) = 0.090272$$

$$\frac{1}{Y - Y^*} = \frac{1}{0.16 - 0.090272} = 14.3$$

The rest of the calculations are summarized in the table below.

| $Y$ | $Y^\circ$ | $X^\circ$ | $X$ | $K$ | $Y^*$ | $\dfrac{1}{Y - Y^*}$ |
|------|------|------|------|------|------|------|
| 0.20 | 0.22819 | 0.03978 | 0.03883 | 2.759 | 0.10715 | 10.8 |
| 0.16 | 0.17755 | 0.03093 | 0.03035 | 2.974 | 0.90272 | 14.3 |
| 0.14 | 0.15325 | 0.02668 | 0.02625 | 3.088 | 0.08106 | 17.0 |
| 0.12 | 0.12961 | 0.02255 | 0.02224 | 3.205 | 0.07130 | 20.5 |
| 0.08 | 0.08416 | 0.01461 | 0.01448 | 3.455 | 0.0500 | 33.3 |
| 0.04 | 0.04101 | 0.007069 | 0.007038 | 3.722 | 0.02619 | 72.4 |
| 0.01 | 0.010062 | 0.00166 | 0.001658 | 3.935 | 0.006524 | 287.7 |
| 0.006 | 0.006022 | 0.000954 | 0.000953 | 3.964 | 0.003778 | 450.1 |
| 0.004 | 0.004010 | 0.000602 | 0.000602 | 3.977 | 0.002394 | 622.7 |
| 0.00228 | 0.002283 | 0.000300 | 0.000300 | 3.992 | 0.001198 | 923.9 |

The function $1/(Y - Y^*)$ is integrated graphically in the usual way. The result is $N_T = 12.31$. This is now corrected for the so far neglected influence of $\psi$. Reading the chart, Fig. 9.33,

$$\frac{1 - tY_1}{1 - tY_2} = \frac{0.8764}{0.9987} = 0.878 \qquad \Delta N_{OG} = 0.06 \qquad N_{OG} = 12.4$$

*Average value of the height of an overall gas-phase transfer unit:*

An average value of $(mG_M/L_M)$ is now determined according to the method proposed in Sec. 9.6.G:

$$G_{M,2} = 26.33 \text{ lb moles/(h)(ft}^2)$$
$$L_{M,1} = L_{M,2} + \Delta g_A + \Delta g_B = 150.48 + 5.94 - 2.27 = 154.15 \text{ lb moles/(h)(ft}^2)$$
$$R_2 = 150.48/26.33 = 5.715; \; R_1 = 154.15/30.0 = 5.138$$
$$Y_1^* = 0.107 \text{ (see table above)}; \; f = 0.107/0.2 = 0.535$$

$$\left(\frac{Y_1}{Y_2}\right)^{0.25} = \left(\frac{0.2}{0.00228}\right)^{0.25} = 3.06$$

Reading the chart, Fig. 9.25:

$$\left.\begin{array}{c}\left(\dfrac{R_2}{R_1} - 1\right) = 0.1123 \\[2mm] f = 0.535\end{array}\right\} \left(\dfrac{R_2}{R_{av}} - 1\right)3.06 = 0.13 \qquad R_{av} = 5.477$$

$$m_c = \frac{Y_1^* - Y_2^*}{X_1 - X_2} = \frac{0.107 - 0.0012}{0.0388 - 0.0003} \qquad \text{See table above}$$

$$m_c = 2.748 \qquad m_2 = 3.992 \qquad \left[1 - \frac{m_c}{m_2}\right] = 0.312$$

$$\left(\frac{Y_1}{Y_2}\right)^{0.23} = \left(\frac{0.2}{0.00228}\right)^{0.23} = 87.7^{0.23} = 2.798$$

Reading the chart, Fig. 9.30:

$$\left.\begin{array}{c}\left(1 - \dfrac{m_c}{m_2}\right) = 0.312 \\[2mm] f = 0.535\end{array}\right\} \left(1 - \dfrac{\overline{m}}{m_2}\right)2.798 = 0.29 \qquad \overline{m} = 3.578$$

$$\left(\frac{mG_M}{L_M}\right)_{av} = \frac{3.578}{5.477} = 0.6533$$

Since $m > 1$ and because of the relatively small value of $t$, the film factors may be omitted and Eq. (9.102) will yield very nearly the correct result

$$H_{OG} = 1.8 + (0.6533)(0.8) = 2.32 \text{ ft} \qquad h_T = 28.77 \text{ ft}$$

A $YX$ diagram for this process is shown in Fig. 9.34. The middle of the three operating lines represents this case and has been constructed using the values of the table given above.

(b) $\qquad Y_{B,1} = (1.0)(0.1918) = 0.1918 \qquad Y_{B,2} = (1.0)(0.1964) = 0.1964$

Because $Y_{A,2}$ is such a small quantity, the computation $Y_{B,2}$ may be based on the same guess of $Y_{A,2}$ as in part $a$.

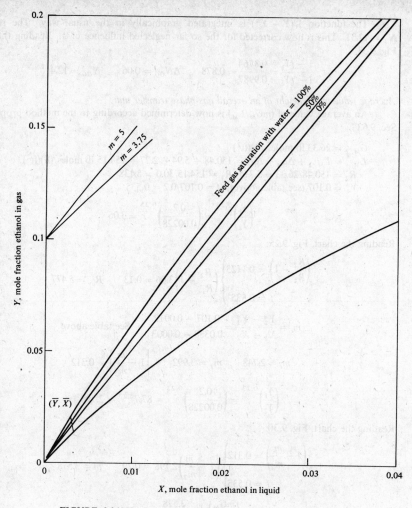

**FIGURE** 9.34
Design diagram for Example 9.10 illustrating displacement operating line by solvent evaporation.

A short algebraic manipulation of material balances shows that $\Delta g_B$ can be evaluated conveniently as

$$\Delta g_B = G_{M,1} \frac{Y_{B,1} - Y_{B,2}}{1 - Y_{B,2}} + \frac{\Delta g_A Y_{B,2}}{1 - Y_{B,2}} = 30 \frac{0.1918 - 0.1964}{0.8036} + 5.94 \frac{0.1964}{0.8036}$$

$$= 1.28$$

$$\Delta g_B = +1.28 \text{ lb moles/(h)(ft}^2)$$

Because of the considerable decrease in $G_M$, a condensation of water will take place rather than an evaporation.

$$G_{M, 2} = 30 - 5.94 - 1.28 = 22.78 \text{ lb moles/(h)(ft}^2)$$

[Check: $(30)(0.1918) - (22.78)(0.1964) = 1.28 = +\Delta g_B$]

$$t = 1 + \frac{1.28}{5.94} = 1.2155 \quad \text{by Eq. (9.129)}$$

$$Y_{A, 2} = \frac{0.06}{22.78} = 0.002634$$

$$L_{M, 1} = 150.48 + 5.94 + 1.28 = 157.7 \text{ lb moles/(h)(ft}^2)$$

$$X_{A, 1} = \frac{5.94 + (0.0003)(150.48)}{157.7} = 0.03795$$

The $YX$ diagram (Fig. 9.34) shows that $Y_1^* = 0.106$. Therefore, $f = 0.53$.

Determination of an average $(mG_M/L_M)$ value:

$$R_1 = \frac{157.7}{30} = 5.257 \qquad R_2 = \frac{150.48}{22.78} = 6.606$$

$$\frac{Y_1}{Y_2} = \frac{0.20}{0.002634} = 75.93 \qquad \left(\frac{Y_1}{Y_2}\right)^{0.25} = 2.952 \qquad \left(\frac{Y_1}{Y_2}\right)^{0.23} = 2.707$$

$$m_c = \frac{0.106 - 0.0012}{0.03795 - 0.0003} = 2.7835 \qquad m_2 = 3.992$$

$$\left. \begin{array}{l} \left(\dfrac{R_2}{R_1} - 1\right) = 0.2566 \\[2mm] f = 0.53 \\[2mm] R_{av} = 6.090 \end{array} \right\} \left(\frac{R_2}{R_{av}} - 1\right)2.952 = 0.25 \quad \text{see Fig. 9.25}$$

$$\left. \begin{array}{l} \left(1 - \dfrac{m_c}{m_2}\right) = 0.303 \\[2mm] f = 0.53 \\[2mm] \overline{m} = 3.57 \end{array} \right\} \left(1 - \frac{\overline{m}}{m_2}\right)2.707 = 0.285 \quad \text{see Fig. 9.30}$$

$$\left(\frac{mG_M}{L_M}\right)_{av} = \frac{3.57}{6.090} = 0.5862$$

*Calculation of* $N_T$ *and* $N_{OG}$:

$$\overline{m}X_2 = 3.57 \times 0.0003 = 0.001071 \qquad \frac{Y_1 - \overline{m}X_2}{Y_2 - \overline{m}X_2} = \frac{0.2 - 0.001071}{0.002634 - 0.001071} = 127.27$$

Substituting into Eq. (9.106), $N_T = 9.61$. Again, the correction for the influence of $\psi$ is read from Fig. 9.33.

$$\frac{1 - tY_1}{1 - tY_2} = \frac{1 - 1.2155 \times 0.2}{1 - 1.2155 \times 0.002634} = 0.759 \qquad \Delta N_{OG} = 0.14 \qquad N_{OG} = 9.75$$

Determination of an average value of $H_{OG}$:

Because $m > 1$, Eq. (9.102) is expected to yield very nearly the same result as the expressions including the film factors

$$H_{OG} = 1.8 + 0.5862 \times 0.8 = 2.27 \text{ ft}$$

The application of the simple Eq. (9.102) shall now be tested using the more rigorous expression (9.144). In order to evaluate the film factors and thus the interfacial concentrations, a point $\overline{X}, \overline{Y}$ is to be found by trial and error for which $(mG_M/L_M) = (mG_M/L_M)_{av} = 0.5862$. The following preparations have to be done:

$$L_M^\circ = 150.48(1 - 0.0003 \times 1.2155) = 150.425 \text{ lb moles/(h)(ft}^2)$$
$$G_M^\circ = 30(1 - 0.2 \times 1.2155) = 22.707 \text{ lb moles/(h)(ft}^2)$$

$$\frac{L_M}{G_M} = \frac{L_M^\circ}{G_M^\circ} \frac{1 - t\overline{Y}}{1 - t\overline{X}} = 6.625 \frac{1 - t\overline{Y}}{1 - t\overline{X}}$$

An operating line for this case is now inserted into the $YX$ diagram shown in Fig. 9.34 (left line). It is calculated in exactly the same way as in part $a$; the calculations are not shown here. In this case with $t \approx 1$, it can also be approximated by the conventional Eq. (9.47).

First trial. As a first approximation to $\overline{X}$, the point where $m = \overline{m} = 3.57$ is sought graphically on the design diagram and the corresponding $\overline{Y}$ read from the operating line. $L_M/G_M$ is then calculated as indicated above:

$$\overline{X} = 0.005 \qquad \overline{Y} = 0.033 \qquad \frac{L_M}{G_M} = 6.625 \frac{0.9598}{0.9939} = 6.398$$

A new value of $m$ is obtained as follows:

$$m = 6.398 \left( \frac{mG_M}{L_M} \right)_{av} = 6.398 \times 0.5862 = 3.750$$

Second trial. The procedure is repeated with $m = 3.75$.

$$\overline{X} = 0.0030 \qquad \overline{Y} = 0.020 \qquad \frac{L_M}{G_M} = 6.488 \quad m = 3.8$$

Third trial.

$$\overline{X} = 0.00308 \qquad \overline{Y} = 0.0206 \qquad \frac{L_M}{G_M} = 6.483 \qquad m = 3.8$$

The next step is the calculation of the corresponding interfacial concentrations. This can be done by trial and error working on an enlarged graph of Fig. 9.34. The procedure is strictly analogous to the one outlined in Example 9.7, but now using Eq. (9.147),

$$\frac{Y - Y_i}{X - X_i} = -\frac{L_M}{G_M} \frac{H_G}{H_L} \frac{Y_f}{X_f} = -6.483 \frac{1.8}{0.8} \frac{Y_f}{X_f} = -14.59 \frac{Y_f}{X_f}$$

$$\overline{X} = 0.00308 \qquad \overline{Y} = 0.0206$$

Assuming $Y_f/X_f = 1$ as a first approach, the interfacial concentrations are found graphically as

$$X_i = 0.003552 \qquad Y_i = 0.0137$$
$$1 - tX_i = 0.09568 \qquad 1 - tY_i = 0.9833$$
$$1 - t\overline{X} = 0.99626 \qquad 1 - t\overline{Y} = 0.97496$$
$$X_f = \frac{0.99626 - 0.99568}{\ln (0.99626/0.99568)} \qquad Y_f = \frac{0.9833 - 0.97496}{\ln (0.9833/0.97496)}$$

$$X_f = 0.99598 \qquad Y_f = 0.9791$$

$$\frac{Y - Y_i}{X - X_i} = -14.59 \qquad \frac{0.9791}{0.9960} = -14.34$$

The next trial yields

$$X_i = 0.00356 \qquad Y_i = 0.01378$$
$$X_f = 0.99597 \qquad Y_f = 0.9791$$

$$\frac{Y - Y_i}{X - X_i} = -14.343$$

$Y_f^*$ is now calculated as follows:

$$\overline{Y}^* = 0.01194 \qquad 1 - tY^* = 0.98549$$

$$Y_f^* = \frac{0.98549 - 0.97496}{\ln (0.98549/0.97496)} = 0.9802$$

$(H_{OG})_{av}$ may now be computed according to Eq. (9.144), which is analogous to Eq. (9.93):

$$H_{OG} = \frac{0.9791}{0.9802} (1.8) + (0.5862) \frac{0.99597}{0.9802} (0.8) \qquad H_{OG} = 2.274 \text{ ft}$$

which is almost the same as before.

$$h_T = (9.75)(2.274) = 22.17 \text{ ft}$$

*Calculations for 50 percent saturation with water vapor in the inlet gas*
$(mG_M/L_M)_{av}$ has already been determined in part *a*. It remains to compute $N_T$ according to Eq. (9.106):

$$\frac{Y_1 - \overline{m}X_2}{Y_2 - \overline{m}X_2} = \frac{0.2 - 3.578 \times 0.0003}{0.00228 - 3.578 \times 0.0003} = 164.86$$

Substituting this and $(mG_M/L_M)_{av} = 0.6533$ into Eq. (9.106),

$$N_T = 11.70 \qquad \Delta N_{OG} = 0.06 \qquad \text{from Fig. 9.33} \qquad N_{OG} = 11.8 \qquad h_T = 27.83 \text{ ft}$$

*Calculations for dry feed gas*
   The calculations are performed as shown in parts *a* and *b*. An almost equimolar counter-diffusion takes place in this case and $t = 0.022$. Since $L_M/G_M$ is almost constant, the arithmetic average was taken as $R_{av}$. The results are shown in the table below.
   (*c*)  All input variables for the rigorous computation have already been determined. The rigorous results are included in the table of results below.

| Saturation and design procedure used | $Y_{B,1}$ | $t$ | $Y_{A,2} \times 10^3$ | $X_{A,1} \times 10^2$ | $R_1$ | $R_2$ | $R_{av}$ | $\overline{m}$ | $\left(\dfrac{mG}{L}\right)_{av}$ | $\Delta N_{OG}$ | $N_{OG}$ | $(H_{OG})_{av}$, ft | $h_T$, ft |
|---|---|---|---|---|---|---|---|---|---|---|---|---|---|
| 100% according to Sec. 9.6.G | 0.1918 | 1.2155 | 2.634 | 3.795 | 5.257 | 6.606 | 6.09 | 3.570 | 0.5862 | 0.14 | 9.75 | 2.274 | 22.17 |
| 100% rigorous | 0.1918 | 1.2155 | 2.634 | 3.795 | 5.257 | 6.606 | ... | ... | ... | 0.14 | 9.91 | 2.24 | 22.20 |
| 50% graphical integration | 0.0959 | 0.6178 | 2.28 | 3.883 | 5.138 | 5.715 | ... | ... | ... | 0.06 | 12.4 | 2.32 | 28.77 |
| 50% according to Sec. 9.6.G | 0.0959 | 0.6178 | 2.28 | 3.883 | 5.138 | 5.715 | 5.477 | 3.578 | 0.6533 | 0.06 | 11.8 | 2.32 | 27.38 |
| 50% rigorous | 0.0959 | 0.6178 | 2.28 | 3.883 | 5.138 | 5.715 | ... | ... | ... | 0.06 | 12.08 | 2.29 | 27.71 |
| Dry, according to Sec. 9.6.G | 0.0 | 0.022 | 2.0087 | 3.974 | 5.0203 | 5.0378 | 5.0291 | 3.6248 | 0.7208 | 0.00 | 14.72 | 2.38 | 34.98 |
| Dry, rigorous | 0.0 | 0.022 | 2.0087 | 3.974 | 5.0302 | 5.0378 | ... | ... | ... | 0.00 | 15.35 | 2.35 | 36.05 |

*Comment:* When the feed gas is dry, an almost equimolar counterdiffusion results due to solvent evaporation, and $t$ is in the vicinity of zero. $L_M/G_M$ stays practically constant (compare $R_1$ and $R_2$) and the operating line remains straight.

With a saturated inlet gas the absorption process resembles an absorption through a stagnant gas. Therefore, $t$ is almost unity and the operating line shows considerable curvature.

The effect of counterdiffusion on the rate of mass transfer, represented by the $\psi$ function and by $\Delta N_{OG}$, is very small. $\Delta N_{OG}$ can be precisely estimated using Fig. 9.33. The main effect of counterdiffusion consists in its influence on the mass balance and reflects itself in a displacement of the operating line which overrides the effect of $\Delta N_{OG}$. The lowest $L_M/G_M$ ratio and thus the highest number of transfer units is realized with dry inlet gas. Some of the solvent will evaporate in this case and will therefore not be available for absorption.

The calculations demonstrate the validity of the approximate methods regardless of the $t$ values and in situations where both the operating and the equilibrium lines are curved. The agreement with the rigorous solution is good in all cases.

The trial-and-error calculations employed to calculate the film factors in Eq. (9.144) could have been avoided by simple inspection of the design diagram, since the effective average $mG_M/L_M$ occurs in the dilute region of the diagram, so that the film factor ratios could have been assumed unity without significant error.    ////

## 9.6.I  Heat Effects in Gas Absorption

If the absorbed component has an appreciable heat of solution and the amount absorbed is sufficiently large, the solvent may undergo an appreciable temperature rise as it passes through the tower. (See Sec. 9.3.E for hydrocarbon absorption.)

Since gas solubility depends on the temperature of the solvent, the equilibrium line cannot be located for the whole tower until the temperature of the solvent is known at every value of the solvent concentration. When a very dilute gas is contacted with a large quantity of solvent, the heat effects accompanying solution of the soluble material may be so small in comparison with the sensible-heat capacity of the liquid that substantially isothermal operation will result. In actual practice, however, there are many cases in which the solvent temperature rises appreciably. Examples are the drying of air by contact with strong sulfuric acid, the absorption of sulfur trioxide in sulfuric acid, and the solution of hydrochloric acid gas in water to make concentrated muriatic acid. In the last case, the quantity of heat liberated when HCl dissolves is so great that its removal becomes a controlling factor in determining the concentration of acid which can be obtained. In practice the absorption of HCl is often conducted without cooling and the liquid is allowed to boil. In this type of operation the acid strength is limited to about 38 percent HCl, although a high degree of absorption is possible [27].

In most cases it is a relatively simple matter to correct for the change in temperature due to the heat of solution or condensation of the vapor. This correction is possible because the heat liberated is a function of the change in composition of the liquid, so if the heat capacity of the solvent is known, the relation between temperature rise and concentration may be calculated easily. An equilibrium curve of gas composition vs. liquid composition may then be constructed, which takes into account the temperature variation through the column.

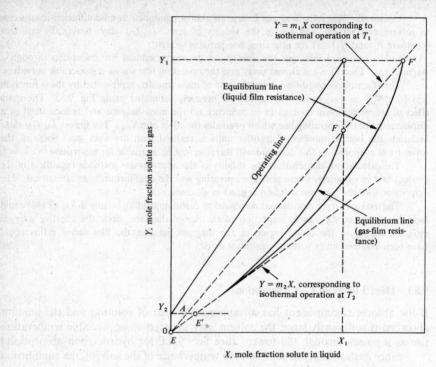

FIGURE 9.35
Effect of temperature on the equilibrium line.

The rate of heat liberation is usually largest in the bottom of the tower, where the solute is absorbed more rapidly. This has the effect of lifting the upper end of the equilibrium curve, as indicated by Fig. 9.35. The lower end of the curve, corresponding to the upper, dilute end of the absorber, may remain straight, the slope corresponding to the initial solvent temperature.

The equilibrium curve $EF$, corresponding to the actual temperatures existing in the column, is located by first calculating the temperature rise for each value of $X$ between $X_2$ and $X_1$. The curve $EF$ is then used in the design calculations. If the liquid-film resistance controls the rate of absorption but the temperature of the liquid is the same at the interface as in the bulk of the liquid, the equilibrium line will be $E'F'$; but the procedure will be the same. The two points $F$ and $F'$ come together as the solvent rate approaches its minimum value. The variation of the liquid-film resistance with temperature may be allowed for in the graphical integration for tower height, since the relation between temperature and liquid concentration is obtained in order to place the equilibrium curve.

Allowance for cooling of the solvent due to heat loss from the column is more complicated, because the cooling is a function of column length rather than of liquid composition. In theory, the equilibrium curve may be placed by trial, however, until graphical integration shows a relation between length and concentration which leads to a temperature-concentration relation agreeing with the assumed location of the equilibrium curve.

Since the minimum rate of liquid circulation required for good recovery of the solute gas is determined by the location of the upper end of the operating line, it is sometimes economical to provide special means for cooling the liquid near the base of the column. This may consist of coils laid on the bottom trays and supplied with cooling water or a refrigerant. Alternatively, it may be more desirable to divert the liquid stream through an external heat exchanger, or intercooler, and then to return it to the next lower tray or section of the column.

If the rich gas enters the absorber cold, the liquid may be cooled just before it leaves by transferring sensible heat to the gas. A similar effect occurs if the solvent is volatile and the entering gas is not saturated with respect to the solvent. When water is the absorbent liquid, evaporation of water into the unsaturated entering gas may have a large cooling effect owing to the large latent heat. In either of these cases, the liquid will be warmer at a point within the tower than at either end. Occasionally, the heat effects of this sort may be so large that the operating and equilibrium lines for the solvent may cross, with the result that the solvent may be evaporated from the liquid in a lower part of the tower and condensed higher up.

**EXAMPLE 9.11**   A gas containing 41.6 percent $NH_3$ by volume at 68°F is to be scrubbed with water to recover 99 percent of its $NH_3$ content. If the water enters at 68°F and if heat loss from the absorber is assumed to be negligible, what will be the minimum water rate, expressed as pounds water per pound gas mixture treated? What will be the corresponding maximum liquor strength obtainable? For heats of solution, see the International Critical Tables, vol. V, page 213 [24].

SOLUTION

$$NH_3(g) + nH_2O(l) \longrightarrow (NH_3 + nH_2O)(l)$$

At 68°F the heat of solution $(-Q)$ is related to the heat of formation $(a' + nb')$ of solution by

$$45.8676 + n(286.103) = a' + nb' + Q \qquad kJ/g \text{ mole}$$

| $n$, g moles $H_2O$ | $a'$ | $b'$ | $-Q$, kJ/g mole $NH_3$ | $-Q$, Btu/lb mole $NH_3$ | $X$, mole fraction $NH_3$ | Adiabatic temperature rise, °F | Liquid temperature, °F |
|---|---|---|---|---|---|---|---|
| 1 | 66.12 | 293.4 | 27.55 | 11,847 | 0.5 | 328.7 | 396.7 |
| 2.33 | 76.17 | 287.13 | 32.69 | 14,057 | 0.3 | 234.3 | 302.3 |
| 4 | 78.09 | 286.46 | 33.66 | 14,474 | 0.2 | 160.6 | 228.6 |
| 9 | 79.68 | 286.17 | 34.41 | 14,796 | 0.1 | 82.1 | 150.1 |
| 19 | 80.31 | 286.11 | 34.64 | 14,895 | 0.05 | 41.3 | 109.3 |
| 49 | 80.64 | 286.11 | 35.11 | 15,093 | 0.02 | 16.8 | 84.8 |

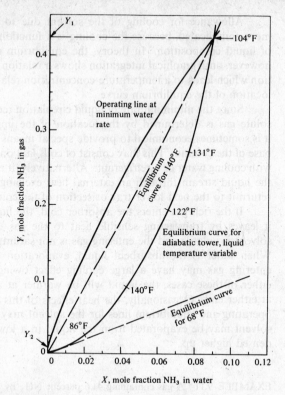

**FIGURE 9.36**
Equilibrium curve modified to allow for heat of solution of solute.

Neglecting the change in sensible heat of the air and gaseous $NH_3$, and assuming that no water is vaporized and that the heat capacity of the ammonia solution is the same as that of liquid water, a heat balance shows that $\Delta T = -Q/(18)(1 + n)$. The liquor temperature is obtained by adding the temperature rise to the inlet temperature of 68°F. The mole fraction of $NH_3$ in the gas at equilibrium is then plotted vs. liquor strength, each point corresponding to the actual temperature resulting from the heat of solution. The result is shown in Fig. 9.36.

$$Y_1' = \frac{0.416}{1 - 0.416} = 0.711 \text{ mole } NH_3/\text{mole air}$$

$$Y_2' = (0.01)(0.711) = 0.00711 \ , \quad Y_2 = \frac{0.00711}{1 + 0.00711} = 0.00710$$

Reading from the equilibrium curve of Fig. 9.36 where $Y_e = Y_1 = 0.416$, $X_1 = 0.088$, which is the maximum liquor strength. The corresponding mole ratio is $X_1' = 0.0965$ lb mole $NH_3/\text{lb}$

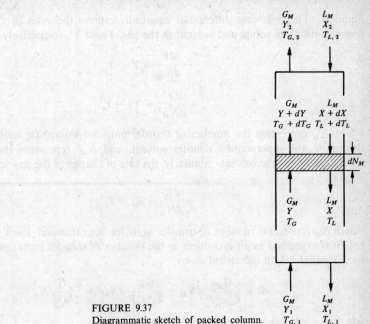

FIGURE 9.37
Diagrammatic sketch of packed column.

mole $H_2O$. The minimum water rate $= (0.99)(0.711/0.0965) = 7.3$ lb moles water per pound mole of air or $(7.3)(1 - 0.416) = 4.26$ moles/mole of entering gas mixture, corresponding to the maximum liquor strength (and equilibrium at gas inlet).

Maximum liquor strength $= 0.088$ mole fraction, or 8.2 weight percent ammonia.    ////

**Simultaneous mass and heat transfer in packed absorbers**  The effects of absorption of the solute, vaporization of the solvent, and sensible-heat transfer between the gas and the liquid, all taking place simultaneously, may be allowed for by a step-by-step procedure if necessary. The calculations involved are tedious, requiring a digital computer, however, and are mainly of theoretical interest. Rigorous design methods have been proposed recently for plate column absorption [4, 52, 45] which may provide useful insight into the corresponding analysis of heat effects in packed towers. An approximate method for packed-tower design will be developed in this section.

Consider a differential section of the column, as shown by Fig. 9.37. For the sake of simplicity, assume that the gas and liquid solutions are dilute so that $G_M$ and $L_M$ remain substantially constant. Assume also, for simplicity, that all the resistance to diffusion of both solute and evaporating solvent is in the gas phase, and that the temperature of the interface is the same as that of the bulk of the liquid, $T$. The following differential equations express the rates of change of the concentrations of solute and solvent in the gas, $Y$ and $Y^s$, respectively,

$$-\frac{dY}{dN_M} = Y - Y_e \tag{9.148}$$

$$-\frac{dY^s}{dN_M{}^s} = Y^s - Y_e{}^s \tag{9.149}$$

where $N_M$ represents the number of transfer units for interphase mass transfer of the solute, the superscript $s$ denotes solvent, and $N_M{}^s$ represents the number of transfer units for the solvent. Similarly, the rate of change of the gas temperature $T$ is given by

$$-\frac{dT_G}{dN_H} = T_G - T_L \tag{9.150}$$

where $N_H$ represents number of transfer units for heat transfer. Each increment of height is expressed as an increment in the number of transfer units; each is related to the actual height increment $dh$ by

$$dh = \frac{G_M}{k_G a_v P} dN_M = \frac{G_M}{k_G{}^s a_v P} dN_M{}^s = \frac{C_p G}{U a_v} dN_H \tag{9.151}$$

where $G$ = gas mass velocity, lb/(h)(ft$^2$)
$U$ = gas-phase to liquid heat-transfer coefficient, Btu/(h)(ft$^2$)(°F)
$C_p$ = gas-heat capacity, Btu/(lb)(°F)
$s$ = solvent

The transfer coefficients are related by the principle of similarity between mass and heat transfer. Thus

$$\frac{k_G{}^s a_v}{k_G a_v} = \left(\frac{D^s}{D}\right)^n = \alpha \tag{9.152}$$

and where $D$ and $D^s$ are the diffusion coefficients of the solute and solvent vapors, respectively,

$$\frac{U a_v / G}{k_G a_v P / G_M} = C_p \left(\frac{k}{C_p \rho_G D}\right)^n = \beta C_p \tag{9.153}$$

showing that where $k$ = gas thermal conductivity and $\rho_G$ = gas density,

$$dN_M{}^s = \alpha dN_M \qquad dN_H = \beta dN_M \tag{9.154}$$

and

$$-\frac{dY}{dN_M} = Y - Y_e \tag{9.155}$$

$$-\frac{dY^s}{dN_M} = \alpha(Y^s - Y_e{}^s) \tag{9.156}$$

$$-\frac{dT_G}{dN_M} = \beta(T_G - T_L) \tag{9.157}$$

A heat balance equating the heat leaving the differential section to that arriving leads to

$$\frac{dT_L}{dN_M} = \frac{G_M}{L_M C_L}\left(C_G \frac{dT_G}{dN_M} + Q \frac{dY}{dN_M} + Q^s \frac{dY^s}{dN_M}\right) \tag{9.158}$$

where $Q$ and $Q^s$ are the molal heats of solution of the solute and solvent vapors and $C_G$ and $C_L$ are the molar heat capacities of the liquid and gas, respectively. A material balance leads to

$$\frac{dX}{dN_M} = \frac{G_M}{L_M} \frac{dY}{dN_M} \tag{9.159}$$

Once all the concentrations and temperatures are known at one end of the packing, it is possible in principle to apply Eqs. (9.155) to (9.159) to successive thin slices of the packing, and thus to establish the profiles of gas and liquid concentrations or temperatures through the packing. If some of the concentrations or temperatures are not known at the ends, they must be estimated by overall heat and material balances. Sometimes numerical integration of the simultaneous differential equations can be avoided by calculating only the rates of change of the temperatures and concentrations at the top and bottom of the packing; then estimating the complete profiles can be done very roughly.

**Minimum liquid rate for an adiabatic tower**  The competing heat effects due to absorption of the solute and evaporation of the solvent may cause the liquid temperature to reach a maximum value at an intermediate point in the absorption tower. "Hot spots" have been observed by noting the temperature of the metal shell enclosing the packing; they were recently measured in plate-column studies by Bourne, von Stockar, and Coggan [5]. If the liquid is heated too much, the equilibrium and operating lines will touch, causing the equipment to become inoperable. Since the only way of removing the heat of solution of the solute in an adiabatic tower is by allowing the solvent to evaporate or to rise in temperature, it is obvious that an inoperable condition must be avoided by providing sufficient solvent to absorb all the solute and, in addition, to absorb the heat of solution.

Whereas in an isothermal tower, the minimum solvent rate is found as discussed in Sec. 9.5.B by considering the solubility of the solute at the exit liquid temperature only, in an adiabatic tower the operating and equilibrium lines may touch at an intermediate point. Even though the operating and the equilibrium lines are separated at the ends and an overall heat balance is satisfied, the absorber will not have been proved operable for the specified separation until an intermediate equilibrium point has been proved not to exist.

The determination of the minimum solvent rate of an adiabatic tower does not appear to have been worked out in simple terms. At present one must resort to step-by-step calculations of fluid temperatures and compositions, starting from the base or the top of the packing, to determine whether the opposite end of the packing can be reached, using an arbitrarily chosen liquid rate.

In order to bring out these principles more clearly, the effects of changes in the liquid-gas ratio have been calculated for a typical case of adiabatic absorption of ammonia in water (Example 9.12). The results show that the liquid rate must be 19 percent greater than would be estimated by considering the terminal conditions alone.

**EXAMPLE 9.12**  A packed absorption tower is to be designed for the removal of ammonia from air by adiabatic absorption in water, which is available at 61.3°F. The gas entering the absorber (at 1 atm and 77°F) is dry and contains 0.0890 mole fraction $NH_3$. It is required that 94.97 percent of the entering ammonia be removed from the gas.

SOLUTION  The composition of the exit gas is estimated by assuming that it leaves at a temperature 5°F higher than the inlet water and that the gas is 85 percent saturated at this temperature. On the basis of 1 lb mole/h of air in the feed gas, we calculate the quantities of the constituents in the exit gas as follows:

| Constituent | lb moles/h | Mole fraction |
|---|---|---|
| Air | 1 | 0.9768 |
| Ammonia | $(0.089/0.911) \times 0.0503 = 0.0049$ | 0.0048 |
| Water vapor | 0.0188 | 0.0184 |
| | 1.0237 | 1.0000 |

An approximate overall energy balance, based on Eq. (9.158), shows

$$T_{L,1} - 61.3 = \frac{G_M}{L_M}\left[\frac{7.0}{18}(77 - 66.3) + \frac{15380}{18}(0.0890 - 0.0048) + \frac{19250}{18}(0 - 0.0184)\right]$$

$$= 56.42 \frac{G_M}{L_M}$$

An overall material balance shows

$$X_1 = \frac{G_M}{L_M}(0.0890 - 0.0048) = 0.0842 \frac{G_M}{L_M}$$

Neither balance is precise. Changes in the enthalpy of solution are disregarded in the first and changes in the flow rate of inert air in the second. Changes for the sake of exactness would not modify the results significantly.

(a) *Approximate determination of minimum liquid rate*  Assume that equilibrium between phases for ammonia is reached at the bottom of the packing. Then $Y_1 = Y_{e1} = (p_{e1}/X_1)(X_1/P)$. For dilute ammonia solutions, the Henry's-law coefficient is given as a function of temperature by the following data:

| $T$, $F$ | $p_e/X$, atm/mole fraction |
|---|---|
| 50 | 0.64 |
| 68 | 1.03 |
| 86 | 1.65 |
| 104 | 2.54 |
| 122 | 3.80 |

The tabulated values can be expressed approximately by the formula

$$\frac{p_e}{X} = \exp\left(0.5008 + 0.4436q - 0.0122q^2\right) \qquad q = \frac{T - 86}{18}$$

The limiting gas-liquid ratio at which equilibrium is obtained is found by a cut-and-try calculation as follows:

| Assumed $G_M/L_M$ | $T_{L,1}$ | $\dfrac{Y_{e1}}{X_1}$ | $X_1$ | $Y_{e1}$ | $Y_1 - Y_{e1}$ |
|---|---|---|---|---|---|
| 0.45 | 86.69 | 1.678 | 0.03789 | 0.06358 | $+0.02542$ |
| 0.50 | 89.51 | 1.798 | 0.04210 | 0.07570 | $+0.01330$ |
| 0.55 | 92.33 | 1.926 | 0.04631 | 0.08919 | $-0.00019$ |
| 0.60 | 95.15 | 2.055 | 0.05052 | 0.10382 | $-0.01482$ |

By interpolation, $Y_{e1} = 0.0890 = Y_1$ when $G_M/L_M = 0.4593$. The true value is smaller, as shown below.

(b)  *Determination of typical profiles in the packing*  A simultaneous solution of the five differential Eqs. (9.155) through (9.159) is begun at the bottom of the packing, where all the dependent variables except $T_1$ are known. Different values are assumed for it until, at each assumed $G_M/L_M$, the integration yields $T_2 = 61.3$ when $X_2 = 0$. Using Sc = 0.74 and 0.60 for ammonia and water vapor in air, respectively, and assuming $n = 0.5$, we find $\alpha = 1.11$ and $\beta = 1.00$. The five equations are

$$\frac{dY}{dN_M} = Y_e(T_L, X) - Y$$

$$\frac{dY^s}{dN_M} = 1.11[Y_e^s(T_L, X) - Y^s]$$

$$\frac{dT_G}{dN_M} = T_G - T_L$$

$$\frac{dT_L}{dN_M} = \frac{G_M}{L_M}\left(0.3888\,\frac{dT_G}{dN_M} + 854.4\,\frac{dY}{dN_M} + 1{,}069.4\,\frac{dY^s}{dN_M}\right)$$

$$\frac{dX}{dN_M} = \frac{G_M}{L_M}\frac{dY}{dN_M}$$

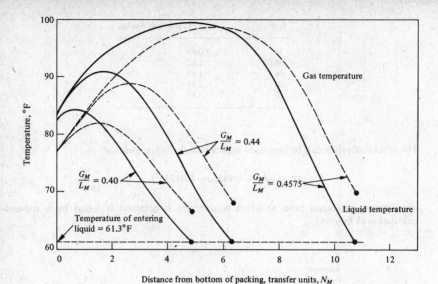

FIGURE 9.38
Calculated temperatures in adiabatic ammonia absorber.

The vapor pressure of water was approximated from Raoult's law, using $Y_e^s = (1 - X)$ $\exp(-3.1731 + 0.58212q - 0.03145q^2 + 0.00233q^3)$.

Figure 9.38 shows calculated values of liquid and gas temperatures as a function of $N_M$ at three values of $G_M/L_M$, each smaller than the value found above. In each case the liquid temperature goes through a maximum value inside the packing. The dotted lines, showing the gas temperature, indicate that it tends to follow the liquid temperature, the values of $T_G$ lying below $T_L$ at the bottom of the packing and above $T_L$ at the top. As the liquid rate is reduced, the height of the internal temperature maximum becomes greater and more packing depth is needed. Figure 9.39 shows the calculated values of the gas and liquid compositions at the highest of the values of $G_M/L_M$ appearing in Fig. 9.38.

Figure 9.40 shows the same results on a graph of $Y$ vs. $X$. The operating lines have slightly decreasing slopes as $G_M/L_M$ increases while the equilibrium lines move toward the operating lines because of the temperature rises just shown. At $G_M/L_M = 0.4575$, the greatest value for which a numerical solution could be obtained, the operating and equilibrium lines nearly touch. No solutions could be obtained at $G_M/L_M = 0.46$. Evidently the true minimum value lies between these values, very likely close to 0.4594.

Comparing the values found in this and the previous section, we see that the false assumption that phase equilibrium occurs first at the bottom of the packing results in a minimum liquid rate that is $100(1/(0.4594 - 1/0.5493)/(1/0.4594) = 16.4$ percent too small.

(c) *Calculation of $G_M/L_M$ when $dT_L/dX = 0$ at bottom of packing*  As shown in part b, the close approach of the ammonia driving force inside the packing is caused by the existence of a maximum liquid temperature inside the column. A safe estimate of the minimum liquid

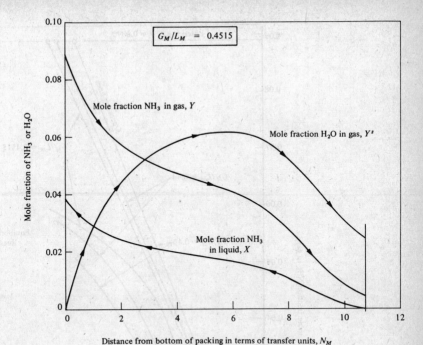

**FIGURE 9.39**
Calculated composition profiles in adiabatic ammonia absorber.

rate might be based on a calculation of the conditions required to prevent a maximum temperature, i.e., the value of $G_M/L_M$ at which $dT/dN_M$ is zero at the base of the column. Using the fourth differential equation above, we must find $G_M/L_M$ such that

$$\left.\frac{dT_L}{dN_M}\right|_{N_M=0} \equiv F\left(\frac{G_M}{L_M}\right) = 0.3889(T_{L,1} - 77) + 854.4(Y_{e1} - 0.0890) + 1069.4\alpha(Y_{e1}{}^s - 0) = 0$$

The values of $T_{L,1}$, $X_1$, $Y_{e1}$, and $Y_{e1}{}^s$ are found from overall energy and material balances. The following numerical values are found for trial values of $G_M/L_M$:

| Assumed $G_M/L_M$ | $X_1$ | $T_{L1}$ | $Y_{e1}$ | $Y_{e1}{}^s$ | $F(G_M/L_M)$, °F/transfer unit |
|---|---|---|---|---|---|
| 0.32 | 0.02694 | 79.35 | 0.03767 | 0.03272 | −2.703 |
| 0.34 | 0.02863 | 80.48 | 0.04128 | 0.03392 | +2.297 |
| 0.36 | 0.03031 | 81.61 | 0.04492 | 0.03516 | +7.371 |

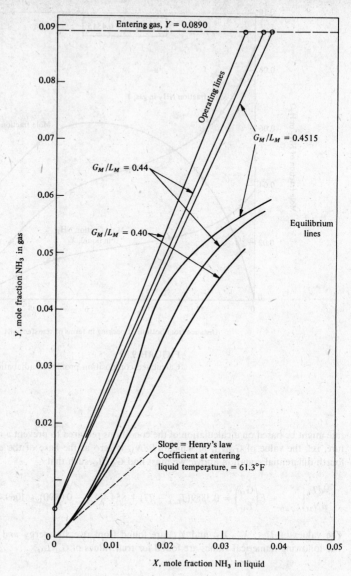

FIGURE 9.40
Operating and equilibrium lines in adiabatic ammonia absorber.

By interpolation, $G_M/L_M = 0.3308$ will yield a zero value for the rate of change of liquid temperature at the bottom of the packing.

The various estimates of the critical value of $G_M/L_M$ are as follows:

| Basis of estimate | Maximum $G_M/L_M$ |
|---|---|
| Phase equilibrium at bottom of packing | 0.5493 |
| True value, allowing for maximum internal liquid temperature | 0.4575 |
| "Safe" value, providing $dT/dN_M = 0$ at bottom of packing, i.e., no internal maximum $T$ | 0.3308 |

In this example the true value, found by rather tedious calculation, 4 percent above the arithmetic mean of the other values, which can be calculated quickly without solving the simultaneous differential equations.                                                   ////

In a water-cooling tower, where the only solute present in the gas is the vaporized solvent and the inert gas is insoluble, Eqs. (9.155) to (9.158) may be simplified considerably provided $\beta = 1$, and provided the ratio of $h$ to $k'$ is equal to the specific heat of the gas. These conditions have been discussed in detail in Chap. 7.

**General observations concerning heat effects**  Bourne, von Stockar and Coggan [4, 5] have studied the effect of nonisothermal operation in a wide range of cases for plate-column absorption. Several general conclusions based on this work which apply similarly to packed towers are summarized below.

The solvent content of the inlet gas can have an important influence on the performance of an absorber. A low inlet solvent concentration will permit substantial adiabatic cooling of the liquid and generally improve the degree of absorption. In the case of water as solvent, it is sometimes worthwhile to dehumidify the inlet gas by cooling to permit greater evaporation of solvent within the tower.

In the case of a nonvolatile solvent, use of a hot inlet gas may lead to a crossing of the solute operating and equilibrium lines within the tower resulting in absorption on the upper section and desorption within the lower section.

Inlet liquid temperature is often not a highly important variable in cases where heat effects occur within the tower as a result of heat of solution or solvent vaporization. In such cases the temperature of the liquid is determined primarily by the internal heat effects.

## 9.6.J  Stripper Design

Stripping involves the desorption of a component from a liquid by contact with an inert gas. Steam is frequently used as a stripping agent because it can be condensed and separated from the stripped component. Stripping is often conducted in conjunction with absorption to recover the absorbed substance and to purify the solvent for reuse. For example, in a common method for removal of hydrogen sulfide from natural gas, the sulfide is absorbed into an aqueous solution of an alkylamine at moderate temperature and pressure [50] in a packed or bubble-plate tower. The amine solution is subsequently heated to 250°F to make the equilibrium favorable for desorption and contacted countercurrently with steam which strips the hydrogen sulfide from the solution. Since stripping is simply the reverse of absorption, similar design procedures and types of contactors may be used for both processes. On the design diagram ($Y$ vs. $X$) for a stripper, the operating line must lie below the equilibrium line in order that mass transfer will occur from liquid to gas. If the flows and temperatures change significantly, rigorous graphical or plate-to-plate computation procedures may be necessary. In many cases, however, approximate design equations analogous to those described above for isothermal absorption into dilute gases will be adequate. Stripping conditions usually require that the solubility of the stripped component be low, i.e., a high value of $m$, and that the "stripping factor," $KG_M/L_M$ or $mG_M/L_M$, be greater than unity, usually around 1.4. Under these conditions the liquid-phase resistance will normally be dominant so that it is preferable to base the design calculations on liquid-phase mass-transfer coefficients. For the calculation of packed-tower height, Eqs. (9.107), (9.94), and (9.90) employing the overall liquid phase height of a transfer unit may be used when $mG_M/L_M$ may be assumed constant. For computation of theoretical plates, Eq. (9.160) analogous to Eq. (9.23) for absorption may be used [51]:

$$\frac{X_0 - X_N}{X_0 - X_N^*} = \frac{S^{N+1} - S}{S^{N+1} - 1} \qquad (9.160)$$

where  $N$ = number of theoretical plates, counting from the top
$X_0$ = concentration of the stripped component in the entering liquid
$X_N$ = concentration of the stripped component in the liquid leaving the tower
$X_N^*$ = concentration of the stripped component which would be in equilibrium with the entering gas = $Y_{N+1}/K$
$S$ = the stripping factor = $KG_M/L_M$
$K$ = the vapor-liquid equilibrium constant, $Y_e/X$.

If conditions are such that the stripping factor may be assumed to vary in a fairly regular manner over the tower, an "effective" stripping factor $S_e$ may be employed in Eq. (9.160) as defined by Edmister [14] for approximate estimates:

$$S_e = \sqrt{S_N(S_1 + 1) + 0.25} - 0.5 \qquad (9.161)$$

where $S_N$ and $S_1$ are the stripping factors for plates $N$ and 1, respectively; Eq. (9.161) is analogous to Eq. (9.32) for absorption.

**EXAMPLE 9.13** Absorption oil is to be removed from the mixture leaving the pentane absorbed described in Example 9.3. Part of the pentane will be flashed off by heating the mixture to 225°F at 1 atm. The subsequent stripping tower operates isothermally at 1 atm and at the same temperature, at which the $K$ value for pentane may be assumed to be 5.8.

Steam enters the tower at a rate of 150 percent of the minimum required to reduce the mole fraction of pentane in the oil to 0.005, so that the oil can be recycled to the absorber.

Estimate the required number of theoretical trays.

SOLUTION
Concentration of pentane after the flash:

$$X_0 = \frac{1}{5.8} = 0.1724 \qquad X'_0 = \frac{0.1724}{1 - 0.1724} = 0.2082$$

$$L'_M = 0.9098 \text{ lb moles oil/min}$$

Pentane balance:

$$G'_M Y'_1 = L'_M(X'_0 - X'_N) = 0.9098(0.2082 - 0.005025)$$
$$= 0.1849 \text{ lb mole pentane/min}$$

Minimum steam mass velocity:

If the gas leaving the stripper were in equilibrium with the entering liquid, the former would consist of pure pentane ($Y_1 = 1$). This indicates that the actual steam mass velocity will have an intrinsically smaller value than $L'$, thus causing a strong downward curvature in the operating line on the $XY$ diagram. The limiting minimum $G'$ value will therefore not be determined by the equilibrium at the top of the tower but rather by a pinch somewhere inside the column which will develop between the equilibrium and the operating lines due to the curvature of the latter.

In order to establish the maximum $L_M/G_M$ ratio and the minimum steam rate, the exact location of the point of contact of the two lines must be determined. This task is accomplished much more easily in terms of mole ratios instead of mole fractions, because on an $X'Y'$ diagram, as shown in Fig. 9.41, the operating line appears straight whereas the curvature is shown by the (otherwise straight) equilibrium line. It is established by computations of the following type.

| $X'$ | $X$ | $Y$ | $Y'$ |
|---|---|---|---|
| 0.010101 | 0.01 | 0.058 | 0.0616 |
| 0.02040 | 0.02 | 0.116 | 0.1311 |
| 0.04 | 0.03845 | 0.2231 | 0.2872 |
| 0.08 | 0.0741 | 0.430 | 0.755 |
| 0.1 | 0.0909 | 0.5275 | 1.117 |

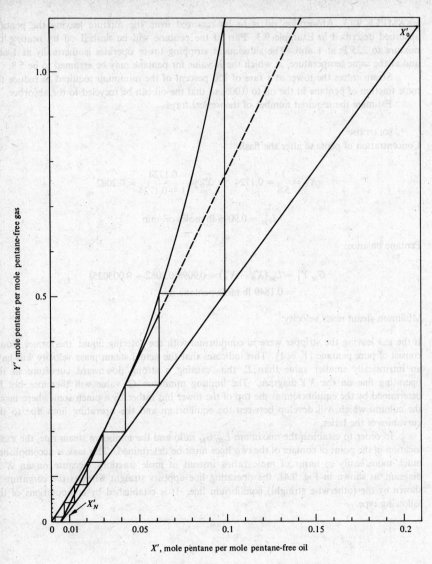

FIGURE 9.41
Design diagram for Example 9.13.

The dashed line on Fig. 9.41 represents the operating conditions at the minimum steam rate. Its slope is read from the diagram as

$$\frac{L'_M}{G'_{Mmax}} = \frac{0.2105}{0.03092 - 0.005025} = 8.125$$

$$G'_{M\,min} = \frac{0.9098}{8.125} = 0.1120 \qquad G'_M = 0.1120 \times 1.5 = 0.1680 \text{ lb mole steam/min}$$

$$Y'_1 = \frac{0.1849}{0.1680} = 1.101 \qquad X'_0 = 0.2082$$

The two last values constitute the upper end of the operating line, which may now be inserted into the diagram as a straight line. The step construction begins at the upper end of the operating line. It indicates the number of required theoretical trays as $N = 7.5$. //// 

**EXAMPLE 9.14**  Absorption oil is to be recovered from the mixture leaving the pentane absorption process described in Examples 9.3 and 9.7. Part of the pentane will be flashed off by heating the mixture up to 225°F. The subsequent stripping tower operates isothermally at 1 atm and at the same temperature at which the $K$ values for pentane may be assumed at 5.8.

Steam enters the tower at a rate of 150 percent of the minimum required to reduce the mole fraction of pentane in the oil to 0.005, so that the oil can be recycled to the absorber.

Estimate for the system described above in Example 9.13 the tower height which is required to perform the operation in a packed column with the same packing and the same inner diameter as the absorber described in Example 9.7. At the high operation temperature, the liquid-phase height of a transfer unit will have dropped to 0.3 ft and can be assumed constant, since owing to the flash, $L_M$ will remain fairly constant. A strong variation in $G_M$ is, however, anticipated; and it is therefore necessary to allow for the dependence of the gas-phase height of a transfer unit on the gas mass velocity according to the function given in the problem description of Example 9.7.

SOLUTION

$$L'_M = 11.12 \text{ lb moles oil/(h)(ft}^2) \qquad \text{See Example 9.7}$$

$$G'_M = \frac{11.12}{8.125} \, 1.5 = 2.0529 \text{ lb moles steam/(h)(ft}^2)$$

It is apparent from Eq. (9.72) and from the values of $L'_M$, $G'_M$, and $H_L$ that stripping process will be gas film controlled. It will therefore be designed on a gas-phase driving force basis, employing a modified version of Eq. (9.59).

$$h_T = \int_{Y_2}^{Y_1} H_G \frac{Y_{BM} \, dY}{(1 - Y)(Y_i - Y)}$$

For each one of a set of $Y$ values, ranging from $Y_2$ to $Y_1$, the corresponding $X$, $H_G$, and the interfacial concentrations must be found. The subscripts 1 and 2 denote the concentrated and the dilute end of the packing, respectively. $Y_1$ is found as follows.

Pentane balance:

$$\Delta g = g_1 = L'_M \left( \frac{1}{1 - X_1} - \frac{1}{1 - X_2} \right)$$

$$= 11.12 \left( \frac{1}{1 - 0.1724} - \frac{1}{0.995} \right)$$

$$= 2.260 \text{ lb moles/(h)(ft}^2)$$

$$Y_1 = \frac{2.260}{2.260 + 2.0529} = 0.5240$$

A material balance around the bottom of the stripper gives

$$X' = \frac{G'_M}{L'_M} Y' + X'_2 \qquad X' = 0.1846 Y' + 0.005025$$

Example:

$$Y = 0.5 \qquad Y' = 1 \qquad X' = (0.1846)(1.0) + 0.005025 = 0.1896$$
$$X = 0.1594$$

In order to find the interfacial concentrations for this point, $H_G$ has to be evaluated first:

$$G = G'_M (Y' \times 72 + 18)$$

where 72 and 18 are the molecular weights of pentane and steam, respectively.

$$G = 2.0529(1 \times 72 + 18) = 184.76 \text{ lb/(h)(ft}^2)$$
$$L = L'_M (X' \times 72 + 160) = 11.12(0.18964 \times 72 + 160) = 1,931 \text{ lb/(h)(ft}^2)$$
$$H_G = 5.3 \frac{(184 \times 76)^{0.395}}{(1,931)^{0.417}} = 1.775 \text{ ft}$$

The interfacial concentrations are now determined through Eq. (9.72) by a trial-and-error procedure similar to the one used in absorption calculations:

$$\frac{Y - Y_i}{X - X_i} = -\frac{11.12}{2.0529} \frac{(1 - 0.5)}{(1 - 0.15941)} \frac{1.775}{0.3} \frac{Y_{BM}}{X_{BM}} \qquad \frac{Y - Y_i}{X - X_i} = -19.071 \frac{Y_{BM}}{X_{BM}}$$

First trial:

Assume $Y_{BM} = X_{BM} = 1$. Then,

$$\frac{Y - Y_i}{X - X_i} = 19.07$$

The interfacial concentrations are found graphically on a $YX$ diagram (see Fig. 9.42) as

$$Y_i = 0.8256 \qquad X_i = 0.1423$$

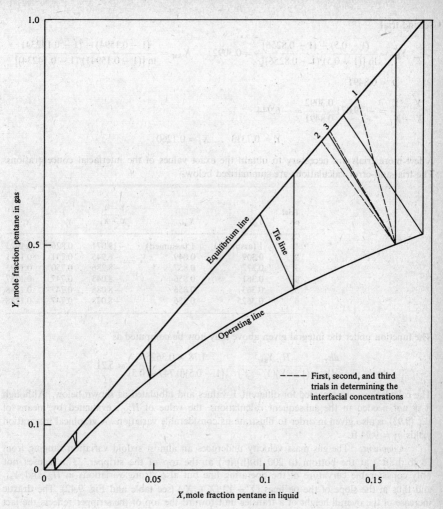

**FIGURE 9.42**
Design diagram for Example 9.14.

Second trial:

$$Y_{BM} = \frac{(1 - 0.5) - (1 - 0.8256)}{\ln\left[(1 - 0.5)/(1 - 0.8256)\right]} = 0.3092 \qquad X_{BM} = \frac{(1 - 0.1594) - (1 - 0.14234)}{\ln\left[(1 - 0.15941)/(1 - 0.14234)\right]}$$

$$= 0.8491$$

$$\frac{Y - Y_i}{X - X_i} = -19.071 \frac{0.3092}{0.8491} = -6.944$$

$$Y_i = 0.7313 \qquad X_i = 0.1260$$

A few more trials are necessary to obtain the exact values of the interfacial concentrations. The trial-and-error calculations are summarized below.

| Trial no. | $Y_{BM}$ | $X_{BM}$ | $\dfrac{Y - Y_i}{X - X_i}$ | $Y_i$ | $X_i$ |
|-----------|----------|----------|----------------------------|-------|-------|
| 1 | 1 (assumed) | 1 (assumed) | −19.071 | 0.826 | 0.1423 |
| 2 | 0.309 | 0.849 | −6.945 | 0.731 | 0.1261 |
| 3 | 0.372 | 0.857 | −8.286 | 0.750 | 0.1293 |
| 4 | 0.361 | 0.856 | −8.043 | 0.747 | 0.1287 |
| 5 | 0.363 | 0.856 | −8.085 | 0.747 | 0.1288 |
| 6 | 0.3625 | 0.856 | −8.078 | 0.747 | 0.1288 |

The function under the integral given above may now be computed as

$$\frac{dh_T}{dY} = \frac{H_G Y_{BM}}{(1 - Y)(Y_i - Y)} = \frac{1.78 \times 0.3625}{(1 - 0.5)(0.747 - 0.5)} = 5.21$$

The calculations are repeated for different $Y$ values and tabulated as shown below. Although it is not needed in the subsequent calculations, the value of $H_{OG}$, evaluated by means of Eq. (9.93), is also given in order to illustrate its considerable variation. A graphical integration yields $h_T = 9.94$ ft.

*Comment:* The gas mass velocity undergoes an almost sixfold variation ranging from 36 lb/(h)(ft$^2$) at the bottom to 200 lb/(h)(ft$^2$) at the top of the stripper. This causes not only considerable curvature of the operating line but also strong variations in $H_G$ and $Y_{BM}$ and thus in the slope of the tie lines $(Y - Y_i)/(X - X_i)$ (see table and Fig. 9.42). The drastic increase of the overall height of a transfer unit toward the top of the stripper reflects the fact that $Y_{BM}^*$ becomes very small as $Y^*$ approaches unity.

| $Y$ | $X$ | $\dfrac{Y - Y_i}{X - X_i}$ | $Y_i$ | $X_i$ | $Y_{BM}$ | $H_G$, ft | $H_{OG}$, ft | $\dfrac{H_G Y_{BM}}{(1 - Y)(Y_i - Y)}$, ft |
|------|------|------|------|------|------|------|------|------|
| 0.5240 | 0.1724 | −7.30 | 0.789 | 0.1360 | 0.326 | 1.83 | — | 4.71 |
| 0.5 | 0.15941 | −8.08 | 0.747 | 0.1288 | 0.3625 | 1.78 | 4.93 | 5.21 |
| 0.4 | 0.11355 | −11.08 | 0.570 | 0.0982 | 0.510 | 1.59 | 2.69 | 7.96 |
| 0.3 | 0.07761 | −13.58 | 0.405 | 0.0698 | 0.646 | 1.42 | 2.11 | 12.47 |
| 0.2 | 0.04869 | −15.54 | 0.260 | 0.0488 | 0.769 | 1.27 | 1.77 | 20.33 |
| 0.1 | 0.02490 | −16.89 | 0.1331 | 0.02294 | 0.883 | 1.12 | 1.51 | 33.24 |
| 0.05 | 0.01453 | −17.31 | 0.0756 | 0.01304 | 0.937 | 1.05 | 1.40 | 40.23 |
| 0.01 | 0.00684 | −17.50 | 0.0323 | 0.00557 | 0.979 | 0.99 | 1.32 | 43.77 |
| 0.001 | 0.00518 | −17.52 | 0.0228 | 0.00394 | 0.988 | 0.97 | 1.30 | 44.11 |
| 0.00 | 0.005 | −17.52 | 0.0218 | 0.00376 | 0.989 | 0.97 | 1.30 | 44.13 |

The calculation of the same problem on a digital computer (see appendix) yielded a result which agreed very closely: $h_T = 9.97$ ft. The required number of overall gas-phase transfer units turned out to be 5.94, which indicates an average value for $H_{OG}$ of 1.68 ft.     ////

### 9.6.K Optimum Absorber Design

The final design of an absorber will normally be based on an economic optimization of the absorption operation in relation to the total process of which it is a part. Most commonly an absorber will operate in conjunction with a stripper as discussed in Sec. 9.6.J, or in conjunction with a distillation tower, for recovery of the absorbed material and purification of the solvent for reuse. It is not feasible to completely generalize the design rules for optimum absorber design because of the great diversity of potential processing situations.

Absorber dimensions and operating conditions are properly determined by minimizing total annual costs. Many cost items must be included: power, water, steam, labor, chemicals, maintenance, etc. Investment costs are translated into annual costs by using appropriate percentages for depreciation, interest, insurance, taxes, etc. Location of the minimum annual cost fixes the optimum values of the design and operating variables.

In some cases it may be apparent that only two or three of the cost items dominate, and the optimization is made simpler. For such cases Colburn [12, 39] has suggested a mathematical approach to cost analysis and derived relationships for the optimum gas velocity in an absorber and for the optimum $mG_M/L_M$ for an absorber-stripper combination. While these relationships are based on overly simplified models, they may be useful in giving the designer a starting point for a more accurate economic analysis, or for specifying a very preliminary design.

In some cases the cost of power for liquid circulation, omitted in the Colburn treatment, may be much larger than that for gas flow. This is the case, for example, in the Stretford absorption of $H_2S$, as applied to the scrubbing of Claus-plant tail gas.

An alternate approach for establishing a starting point in the design is to set the ratio $L_M/G_M$ at a reasonable value based on the very general criterion that $mG_M/L_M$ should be in the vicinity of 0.7 at the dilute end of the tower for absorption, or 1.4 for stripping, and then to set the gas rate at about 50 percent of the estimated flooding velocity at the point of maximum flow. Actually, the permissible range of values of $mG_M/L_M$ is surprisingly narrow. For example, Fig. 9.21 or Eq. (9.106) shows that with $Y_1$, $Y_2$, $m$, and $X_2$ specified, $N_{OG}$ corresponding to 95 percent of the maximum possible absorption (i.e., equilibrium between entering liquid and leaving gas) varies as shown in Fig. 9.43. A similar trend of $N_P$ with $mG_M/L_M$ is observed. Evidently there is little reduction in $N_{OG}$ to be gained by reducing $mG_M/L_M$ below about 0.6; and at values above 0.9, $N_{OG}$ increases extremely rapidly. As a further illustration of this point, reducing $mG_M/L_M$ from 0.4 to 0.2 at a given $m$ and gas rate would mean doubling the required liquid rate and subsequent stripping costs while decreasing the absorber packing only 10 percent. Incorporation of the effect of varying the gas-liquid ratio in the optimization of total annual costs often suggests that $L_M/G_M$ should be 1.2 to 1.5 times the theoretical minimum corresponding to equilibrium at the rich end (infinite tower height), provided that flooding is avoided.

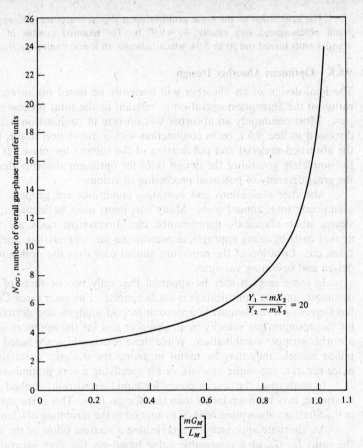

**FIGURE 9.43**
Effect of $mG_M/L_M$ on required $N_{OG}$ at 95 percent of maximum recovery.

**Absorber-stripper combination** The liquid leaving the absorber is usually distilled, either to recover the absorbed solute or to regenerate the solvent for reuse. A typical absorber-stripper arrangement is illustrated diagrammatically by Fig. 11.1. The optimization of the costs for the combined system is not generally amenable to an analytical approach; it must be done by what amounts to trial and error, i.e., by estimating the costs for a variety of cases.

The stripper and absorber costs are not independent and cannot be optimized separately. Both distillation and absorption costs, including stripper fixed charges, steam, and pumping, will depend on $X_2$, total liquid recycle, and $X_1$. If the solute recovery in the absorber is to be large, the liquid concentration $X_2$ will be perhaps the most important variable. A close approach to an optimum design can be made by calculating total annual costs for a number of "cases" involving combinations of $L_M/G_M$ and $G$ in the absorber, $X_2$, and steam rate or reflux ratio in the stripper. This means lengthy and tedious calculations, but they are not difficult if performance data on the absorber and distillation equipment, together with the needed cost data, are available.

## 9.7 RELATION BETWEEN ACTUAL AND THEORETICAL PLATES; PLATE EFFICIENCY

The calculation of the number of theoretically perfect plates required for an absorber or stripper (see Sec. 9.5) closely parallels the calculation of transfer units as already discussed. The actual height of a packed column is not known until some knowledge of the specific rate of interphase transfer is available, expressed in terms of an absorption coefficient or HTU. Similarly, the number of actual plates or trays cannot be estimated until the rate of interphase transfer on each actual plate is known. This information is usually expressed in terms of a "plate efficiency."

The simplest definition of plate efficiency from the point of view of application to column design is that of the *overall plate efficiency E*. This is equal to the ratio of the number of theoretical plates required for a given separation to the number of actual plates required. Once the number of theoretical plates has been calculated, the number of actual plates is found by dividing by $E$. The overall plate efficiency is less than 100 percent except under special circumstances.

The Murphree plate efficiency [34] is related more closely to the resistance to interphase diffusion on the plates.

Consider plate $N$ of an absorption or distillation column, Fig. 9.44. In the case of absorption it will be assumed that the gas stream is dilute to that both operating line and equilibrium line may be assumed straight. The corresponding assumption for distillation is that the mass transfer is equimolar. Liquid enters from above and flows across the plate, leaving at mole fraction $X_N$ of the absorbed component. Gas enters the plate well mixed from below at composition $Y_{N+1}$ and leaves the plate with a mixed average mole fraction $Y_N$.

The Murphree vapor efficiency [34] for the plate is defined by the equation

$$E_{MV} = \frac{Y_{N+1} - Y_N}{Y_{N+1} - Y_N^*} \tag{9.162}$$

where $Y_N^*$ is the gas-phase mole fraction corresponding to equilibrium with the liquid leaving the plate. In the present case,

$$Y_N^* = KX_N$$

where $K$ is the equilibrium constant.

If $E_{MV}$ is the same for each plate in a column and if both the operating and equilibrium lines are straight, the number of actual plates $N_P'$ may be calculated from the equation [32]

$$N_P' = \frac{\ln\{(1 - mG_M/L_M)[(Y_1 - KX_2)/(Y_2 - KX_2)] + mG_M/L_M\}}{-\ln[1 + E_{MV}(mG_m/L_M - 1)]} \tag{9.163}$$

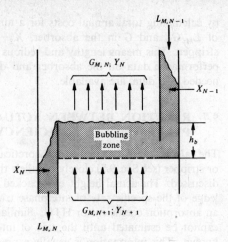

**FIGURE 9.44**
Flows and compositions across a bubble
plate; $h_b$ = depth of bubbling zone.

where subscripts 1 and 2 refer to the concentrated and dilute terminal compositions, respectively. Under these conditions, therefore, the relation between the overall
and Murphree vapor efficiencies is

$$\frac{N_P}{N_P'} = E = \frac{\ln\{1 + E_{MV}[(mG_M/L_M) - 1]\}}{\ln(mG_M/L_M)} \tag{9.164}$$

where $N_P$ = number of theoretical plates [Eq. (9.25)].

Figure 9.45 shows this relationship graphically. When operating and equilibrium lines are parallel, the two efficiencies are equal; if $E_{MV}$ is not equal to 1,
$E$ may be greater or less than $E_{MV}$ as $mG_M/L_M$ is greater or less than unity.
No exact relationship has been developed between the two efficiencies for cases
where either the operating line or the equilibrium line is curved, though the
efficiencies may be expected to be nearly equal if the two lines are parallel on the
average.

**EXAMPLE 9.15**  Carbon dioxide is being absorbed from hydrogen gas under pressure in a
plate tower using an aqueous solution of diethanolamine as absorbent. The temperature,
pressure, and degree of conversion of the amine solution are such that there is no appreciable
vapor pressure of $CO_2$ over the liquid at equilibrium. Determine the number of actual plates
required to reduce the carbon dioxide in the hydrogen from 10 to 1 percent by volume. The
Murphree vapor efficiency of each plate is assumed to be 25 percent. Solve the problem
both (a) graphically and (b) analytically.

SOLUTION  The conditions are such that the equilibrium curve coincides with the $X$ axis.
The liquid-gas ratio need not be specified, as shown by Eq. (9.163), since $L_M/G_M$ enters only
in combination with $m$ and its effect therefore disappears in this problem.

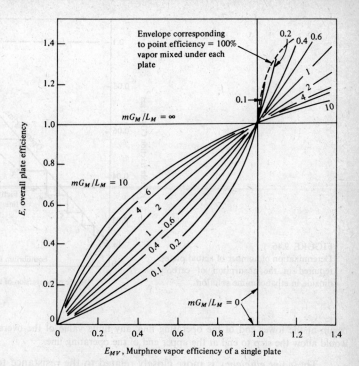

**FIGURE 9.45**
Relation of average plate efficiency to Murphree vapor efficiency.

(a) Graphical solution. An $XY$ diagram is constructed as shown by Fig. 9.46. A step corresponding to one actual plate is begun at the upper end of the operating line, where $Y = 0.1$. The vertical step $AB$ ends at 25 percent of the way toward the equilibrium curve (the $X$ axis in this case) and is followed by a horizontal step $BC$ ending at the operating line. In this way the composition of the gas entering the second plate above the bottom of the column is found to be 0.075 mole fraction. The required value $Y_2$ is reached after eight steps.

The lower end of each step ends on a straight line which passes through the points $0.75Y_2$ and $0.75Y_1$. This line may be thought of as a fictitious equilibrium curve which may be used as though the actual plates are theoretical plates. The same type of construction, using a fictitious equilibrium curve may be used even though both operating and equilibrium lines are curved [2].

(b) Analytical solution. Using Eq. (9.187),

$$N'_p = \frac{\ln (0.10/0.01)}{-\ln (1 - 0.25)} = 8.0 \text{ actual plates}$$

It should be noted that the concept of the theoretical plate and that of the overall plate efficiency have no significance here. One theoretical plate would be more than sufficient to accomplish the desired absorption. If the step corresponding to a theoretical plate were

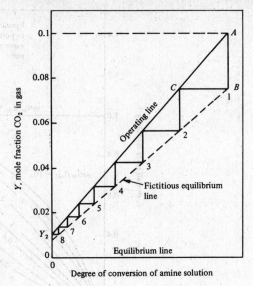

**FIGURE 9.46**
Determination of number of actual plates required in the absorption of carbon dioxide in ethanolamine solution.

begun at the lower end of the operating line, any finite value of the overall plate efficiency would allow the step to end at the upper end of the operating line. ////

The *point efficiency* is more closely related to the resistance to the interphase diffusion than any other plate efficiency. It is defined by the equation

$$E_{OG} = \frac{Y_{p, N+1} - Y_{p, N}}{Y_{p, N+1} - Y_{p, N}^*} \tag{9.165}$$

where $Y_p$ is taken above and below a single point on the plate.

Assume that the liquid is completely mixed in the vertical direction, that is, $Y_{p, N}^* = \text{constant} = KX_{p, N}$. Gas enters the section at a molar mass velocity $G_M$. As the gas passes upward through the liquid covering a small area of the plate, mass transfer from gas to liquid occurs in a manner similar to a packed tower of height $h_b$, the depth of the bubbling area. Thus,

$$-d(G_M Y) \approx -G_M \, dY = P K_{OG} a_v (Y_p - Y_p^*) \, dh \tag{9.166}$$

where $K_{OG}$ = overall gas-phase mass-transfer coefficient between gas and liquid
$\quad a_v$ = interfacial area per unit volume
$\quad P$ = total pressure
$\quad Y_p$ = mole fraction of absorbed component in the gas at position $h$ within the height, $h_b$

Equation (9.166) may be rearranged to give

$$-\int_{N+1}^{N} \frac{dY_p}{Y_p - Y_p^*} = \int_0^{h_b} \frac{K_{OG} a_v P \, dh}{G_M} = \frac{K_{OG} a_v P h_b}{G_M} \tag{9.167}$$

Integration of the middle term depends on the assumption that the product of the transfer coefficient per unit area and the interface area per unit volume is constant along the path of integration.

For the present case of a dilute gas absorption or (distillation with equimolar transfer) the $Y_{BM}^*$ term of Eq. (9.85) becomes unity so that

$$H_{OG} = \frac{G_M}{K_{OG} a_v P} \tag{9.168}$$

The left-hand integral of Eq. (9.167) may be recognized as the number of overall gas-phase transfer units $N_{OG}$ for the single plate. For $Y_p^* = \text{constant} = Y_{p,N}^*$,

$$N_{OG} = \ln \frac{Y_{p,N+1} - Y_{p,N}^*}{Y_{p,N} - Y_{p,N}^*} \tag{9.169}$$

Equation (9.165) may be combined with Eq. (9.168) to give

$$E_{OG} = 1 - e^{-N_{OG}} \tag{9.170}$$

If resistance to transfer is present in both phases, the overall gas-phase coefficient may be related to the single-phase coefficients in the manner of Eq. (9.78), and

$$N_{OG} = \frac{h_b/G_M}{1/Pk_g a_v + (m/\bar{\rho})(1/k_L a_v)} \tag{9.171}$$

or

$$\frac{1}{N_{OG}} = \frac{1}{N_G} + \frac{mG_M}{L_M}\frac{1}{N_L} \tag{9.172}$$

where $N_G = k_G a_v Ph/G_M$, the number of transfer units based on the gas-film resistance, and $N_L = \bar{\rho} k_L a_v h/L_M$, the number of transfer units based on the liquid-film resistance. Equation (9.172) may be combined with Eq. (9.170) to give

$$\frac{1}{-\ln(1 - E_{OG})} = \frac{1}{N_G} + \frac{mG_M}{L_M}\frac{1}{N_L} \tag{9.173}$$

As shown in Fig. 9.47 based on Eq. (9.173), the point efficiency is greater, the smaller the resistance of each phase.

In the case of a gas of low solubility, the value of $m$ in Eq. (9.173) is very large, and the second term dominates the right side of the equation. The point efficiency is extremely small under these circumstances, inasmuch as the vapor composition does not change appreciably as the gas flows through the tray. The liquid may come close to equilibrium, however. It is more logical in this case to reverse the roles of liquid and vapor, keeping in mind the progressive change in liquid composition toward equilibrium with the vapor as the liquid stream flows across the plate, It is logical, therefore, to define a new Murphree plate efficiency $E_{ML}$, in terms of liquid composition, by means of the equation

$$E_{ML} = \frac{X_{N-1} - X_N}{X_{N-1} - Y_N/K} \tag{9.174}$$

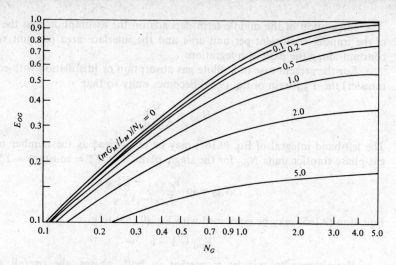

**FIGURE 9.47**

Relationship between $E_{OG}$, $N_G$, and $N_L$, according to Eq. (9.173). $E_{OG}$ = local or transfer efficiency on a gas basis, and $N_G$ and $N_L$ are numbers of gas- and liquid-transfer units.

where $X_{N-1}$ = mole fraction of absorbed component in liquid entering the tray

　$X_N$ = mole fraction of absorbed component in liquid leaving the tray

　$Y_N/K$ = composition of the liquid which would be in equilibrium with the average vapor stream from the tray

This efficiency is similar to the Murphree vapor efficiency $E_{MV}$ of the whole tray, across which a drop in liquid concentration takes place, rather than to the point efficiency. The relation between the two is

$$E_{ML} = \frac{E_{MV}}{E_{MV} + (1 - E_{MV})/(mG_M/L_M)} \tag{9.175}$$

Figure 9.48 shows the dependence of the Murphree liquid and vapor efficiencies, as well as the overall efficiency for an entire column, on the number of transfer units for each phase, $N_G$ and $N_L$, and on $mG_M/L_M$. When $m$ and also $mG_M/L_M$ are very large, as in oxygen stripping from water, the value of $E_{ML}$ is very nearly equal to $1 - e^{-(N_L)}$. Experimental measurements of $E_{ML}$ for absorption or desorption of slightly soluble gases may be used, therefore, for computing $N_L$.

　The composition of the liquid is usually assumed to be uniform along any vertical line perpendicular to the plate surface; i.e., it is assumed that the liquid is thoroughly mixed in the vertical direction. By definition, the point efficiency cannot exceed 100 percent.

　Particularly on a broad plate, where the horizontal path along which the liquid flows is several feet, the concentration of solute in the liquid may change

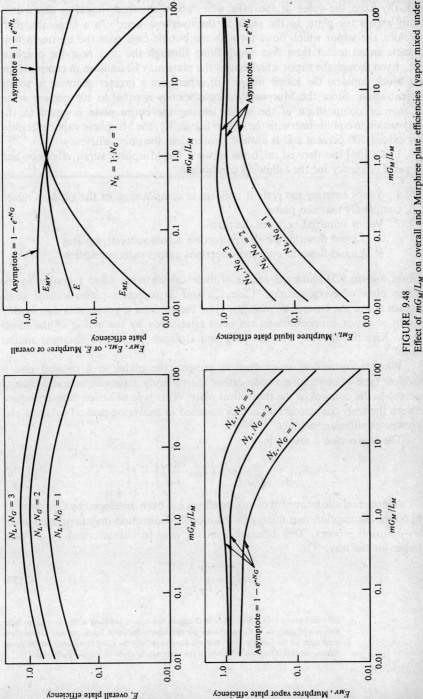

FIGURE 9.48
Effect of $mG_M/L_M$ on overall and Murphree plate efficiencies (vapor mixed under each plate).

gradually from the value at the inlet weir, where the fresh liquid is distributed evenly across the plate, to the value at the overflow weir.[1] In a large absorber, therefore, the vapor which flows through the bubble caps near the upstream weir contacts liquid leaner than that which flows through the caps near the overflow weir. Even though the vapor which enters the plate may be uniform in composition, that which contacts the leaner liquid will experience a greater decrease in solute concentration. Since the Murphree vapor efficiency is equal to 100 percent when the *average* composition of the vapor leaving the entire plate is equal to the composition in equilibrium with the *richest* liquid, $Y_N^*$, the Murphree vapor efficiency may exceed 100 percent and is always greater than the point efficiency.

Lewis [31] has derived relations between the Murphree vapor efficiency and the point efficiency for the following conditions:

*1* Vapor entering the plate is uniform in composition, or the vapor is mixed completely between plates
*2* Vapor unmixed between the plates
    *a* Liquid flows in the same direction across successive plates
    *b* Liquid flows in opposite directions across successive plates

Figures 9.49 to 9.51 show the results of these calculations. Case 1 applies to the bottom plate of every column. Cases 2*a* and 2*b* represent plates located some distance above the base of the column. For these plates a permanent distribution pattern of vapor concentrations has been established by the action of the lower plates. Kirschbaum [26], Hausen [19], and Colburn [10], have obtained similar results.

When a concentration gradient is present, the plates in a column give a crossflow type of contacting action rather than a truly stagewise action, which is assumed in the concept of the theoretical plate. This type of action is intermediate between the truly countercurrent action assumed in analyzing packed columns and stagewise equilibrium contact.

The Lewis case 1 results in

$$E_{MV} = \frac{L_M}{mG_M} \left[ \exp\left( E_{OG} \frac{mG_M}{L_M} \right) - 1 \right] \tag{9.176}$$

A more realistic treatment of mixing effects has been developed by Gerster et al. [16] on the assumption that mixing in the crossflow direction may be treated as an eddy-diffusion process. This diffusional mixing may be characterized by a Peclet number for the tray:

$$N_{Pe} = \frac{Z_1^2}{D_E t_L} \tag{9.177}$$

---

[1] On small plates only a few inches in diameter, the violent bubbling action may cause large amounts of spray to be thrown backward opposite to the main liquid current. Under these conditions all the liquid on the plate will have essentially the same composition as the liquid leaving, and the Murphree efficiency of the plate will be equal to the point efficiency.

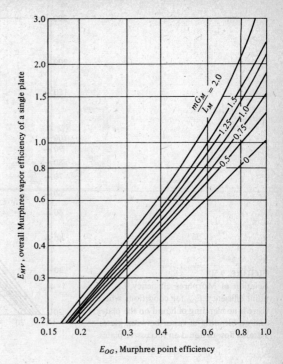

FIGURE 9.49
Relation of point efficiency to overall Murphree efficiency; vapor mixed under each plate.

where $D_E$ = the eddy diffusivity for the system
  $Z_1$ = distance of liquid travel across the tray
  $t_L$ = liquid contact time on the tray

The Murphree vapor efficiency is related to the point efficiency, Peclet number, and $mG_M/L_M$ by the equation

$$\frac{E_{MV}}{E_{OG}} = \frac{1 - e^{-(\eta + N_{Pe})}}{(\eta + N_{Pe})[(1 + \eta + N_{Pe})/\eta]} + \frac{e^{\eta} - 1}{\eta[1 + \eta/(\eta + N_{Pe})]} \qquad (9.178)$$

where

$$\eta = \frac{N_{Pe}}{2}\left[\left(\sqrt{1 + \frac{4\lambda E_{OG}}{N_{Pe}}}\right) - 1\right]$$

$$\lambda = \frac{mG_M}{L_M}$$

Numerical values for Eq. (9.178) are given in Table 11.1.

Methods for prediction of point efficiencies, Peclet numbers, and other factors required for estimation of plate efficiencies are described in Chap. 11.

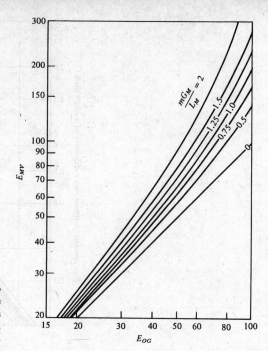

**FIGURE 9.50**
Relation of Murphree efficiency $E_{MV}$ to point efficiency $E_{OG}$ for conditions where there is no blending of liquid on the plates and no blending of vapor between plates. Parallel flow of liquid on successive plates.

An alternate definition of plate efficiency similar to the Murphree plate efficiency but based on the average of the inlet and outlet liquid compositions has been suggested by Onda et al. [36].

When liquid is carried by entrainment from one tray to the one next above, the strong liquid from the lower plate is mixed with the weaker liquid. This action tends to destroy the desired countercurrent contact, with the result that the apparent Murphree vapor efficiency $E_a$ of a tray in the presence of entrainment is lower than it would otherwise be. Colburn [9] has derived a simple approximate relation showing the effect of the amount of entrainment $\varepsilon$ and the gas-liquid ratio on the apparent vapor efficiency. The equation is

$$E_a = \frac{E_{MV}}{1 + (\varepsilon G_M/L_M)E_{MV}} \tag{9.179}$$

A plot of this equation is shown by Fig. 9.52 (see also Sherwood and Jenny [47]). This effect is discussed further in Chap. 11 (see Example 11.4).

A procedure for computing directly the actual number of plates required for an absorption or distillation column has been proposed recently by Mostafa (33a). The method includes the effects of entrainment, liquid back-mixing, and point efficiency for each tray.

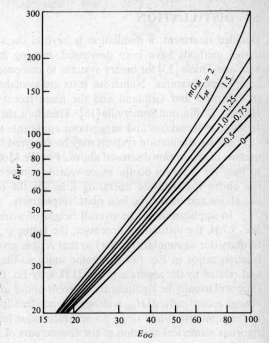

**FIGURE 9.51**
Relation of Murphree efficiency $E_{MV}$ to point efficiency $E_{OG}$ for conditions where there is no blending of liquid on the plates and no blending of vapor between plates. Flow of liquid in opposite direction on successive plates.

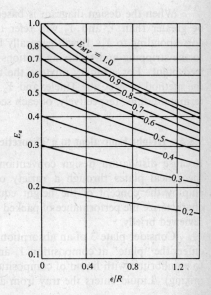

**FIGURE 9.52**
Relation of apparent Murphree plate vapor efficiency $E_a$ to dry vapor efficiency $E_{MV}$, to moles entrainment $\varepsilon$ per mole dry vapor, and to the reflux ratio $R = L_M/G_M$, according to Eq. (9.179).

## 9.8 DISTILLATION

Detailed treatment of distillation is beyond the scope of the present text. Many design methods have been developed, ranging from the well-known method of McCabe-Thiele [33] for binary systems to computer techniques for complex, multi-component mixtures. Numerous texts are available, including the pioneering book of Robinson and Gilliland and the more recent texts of Hengstebeck [20], and Hanson, Duffin, and Somerville [18]. Therefore, the present discussion will be limited to some observations and suggestions applicable to the design of packed towers.

Binary distillation systems may be analyzed by the $YX$ design diagram in the manner of absorption discussed above. For the McCabe-Thiele graph it is customary to base the diagram on the more volatile component so that the equilibrium line lies above the separate operating lines for the rectifying and stripping sections, i.e., above and below the feed plate, respectively.

In application of the overall height of a transfer unit method, as noted in Sec. 9.7.H, for distillation processes, the factor $\psi_A$ may usually be assumed equal to unity (for equimolar transfer) so that $N_{OG}$ is given by Eq. (9.104). Also the mole fraction ratios in Eq. (9.93) become unity so that $H_{OG}$ is defined by Eq. (9.168) and related to the separate phase HTUs by Eq. (9.102). Since the liquid-gas flow ratio will usually be significantly different above and below the feed plate, $H_{OG}$ will change accordingly. If $H_{OG}$ is strongly dependent upon concentration due to variation in the equilibrium-line slope, an accurate value for the packed height may require rigorous numerical solution of the counterpart of Eq. (9.83), namely,

$$h_T = \int_{Y_1}^{Y_2} \frac{G}{K_{OG} P a_v} \frac{dY}{(Y^* - Y)} \equiv \int_{Y_1}^{Y_2} H_{OG} \frac{dY}{Y^* - Y} \tag{9.180}$$

When the design diagram is based on the more volatile component, $Y^*$ will be greater than $Y$, and $Y_2$ will refer to a point in the column above the point corresponding to $Y_1$. It will generally be desirable to apply Eq. (9.180) separately to the stripping and rectifying sections since the flows will change sharply at the feed point. For a binary mixture the operating lines can be placed according to the method of McCabe-Thiele, and $Y_2$ and $Y_1$ will represent the upper and lower terminal points, respectively, of each section.

### 9.8.A Height Equivalent to a Theoretical Plate

Because distillation design conventionally involves computation of numbers of theoretical plates through a variety of techniques, it is a common practice to employ the concept of the height equivalent to a theoretical plate (HETP) to characterize the performance of packed towers for distillation. This concept will be reviewed briefly.

Consider plate 3 of an absorption column as illustrated in Fig. 9.53. (a) Gas enters the "plate" at composition $Y_4$ and leaves at composition $Y_3$, corresponding to equilibrium with liquid of composition $X_3$ leaving the tray (assuming complete mixing). Liquid enters the tray from above at composition $X_2$. (b) represents a

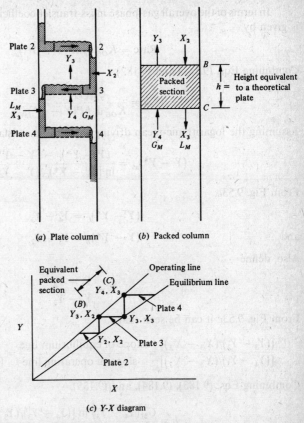

(a) Plate column          (b) Packed column

(c) Y-X diagram

**FIGURE 9.53**
Illustration of height equivalent to a theoretical plate concept.

section $BC$ of height $h$ of a packed column over which the composition changes of the gas and liquid streams for the component under consideration are identical to those occurring across the plate. It will be assumed also that the streams are dilute so that the molal flow rates over the section are constant. The situation described will apply equally to distillation in which constant molal overflow may be assumed. (c) is an $YX$ diagram showing the step construction for the plate; points $B$ and $C$ represent the terminal operating line points for the equivalent packed section.

The rate of absorption is given by a material balance over the plate section

$$\text{Rate} = G_M(Y_4 - Y_3) = L_M(X_3 - X_2) \quad \text{moles/(h)(ft}^2) \quad (9.181)$$

In terms of the overall gas-phase mass-transfer coefficient, the rate of absorption is given by

$$\text{Rate} = K_{OG} a_v hP(Y - Y^*)_{\text{av}} \tag{9.182}$$

Combining Eqs. (9.181) and (9.182),

$$h = \frac{G_M(Y_4 - Y_3)}{K_{OG} a_v hP(Y - Y^*)_{\text{av}}} \tag{9.183}$$

Assuming the logarithmic-mean driving force to be satisfactory,

$$(Y - Y^*)_{\text{av}} = \frac{(Y - Y^*)_C - (Y - Y^*)_B}{\ln [(Y - Y^*)_C/(Y - Y^*)_B]} \tag{9.184}$$

From Fig. 9.53c,

$$(Y - Y^*)_C = Y_4 - Y_3 \tag{9.185a}$$

and

$$(Y - Y^*)_B = Y_3 - Y_2 \tag{9.185b}$$

Also, define

$$J' = \frac{Y_3 - Y_2}{Y_4 - Y_3} \tag{9.186}$$

From Fig. 9.53c it can be seen that

$$\frac{[(Y_3 - Y_2)/(X_3 - X_2)]}{[(Y_4 - Y_3)/(X_3 - X_2)]} = \frac{\text{slope of equilibrium line}}{\text{slope of operating line}} = \frac{m}{(L_M/G_M)} = J' \tag{9.187}$$

Combining Eqs. (9.183), (9.184), and (9.185),

$$h = \frac{G_M(Y_4 - Y_3) \ln [(Y_4 - Y_3)/(Y_3 - Y_2)]}{K_{OG} a_v P[(Y_4 - Y_3) - (Y_3 - Y_2)]} \tag{9.188}$$

and

$$J' - 1 = \frac{Y_3 - Y_2}{Y_4 - Y_3} - 1 = \frac{(Y_3 - Y_2) - (Y_4 - Y_3)}{(Y_4 - Y_3)} \tag{9.189}$$

Combining Eqs. (9.188) and (9.189) gives

$$h = \text{HETP} = \frac{G_M \ln J'}{K_{OG} a_v P(J' - 1)} \tag{9.190}$$

For a dilute gas or for constant flows, $Y_{BM}^*$ approaches unity and

$$H_{OG} = \frac{G_M}{K_{OG} a_v P} \tag{9.168}$$

Equations (9.168) and (9.190) may be combined to give

$$h = \text{HETP} = \frac{H_{OG} \ln (mG_M/L_M)}{(mG_M/L_M - 1)} \tag{9.191}$$

It can be shown also that if the arithmetic mean of $(Y - Y^*)$ is satisfactory in Eq. (9.182),

$$\text{HETP} = \frac{2H_{OG}}{(mG_M/L_M + 1)} \tag{9.192}$$

Equation (9.191) or (9.192) may be used to estimate the HETP of a packed section for distillation or dilute absorption systems from the corresponding value of $H_{OG}$. Although Eq. (9.191) is more rigorous, Eq. (9.192) will be satisfactory in most cases.

If the operating and equilibrium lines are straight and parallel, $mG_M/L_M = 1$, and by Eq. (9.192) $\text{HETP} = H_{OG}$.

### 9.8.B  Use of HETP for Distillation

A relatively simple design procedure can be employed with the HETP as follows. A $YX$ diagram is constructed in the usual manner for distillation design and the theoretical plates are stepped off in the rectifying and stripping sections. From the compositions at each step on the graph the flows, individual phase HTUs, equilibrium line slope, $H_{OG}$, and HETP can be calculated. The height of packing required for each step is summed for all steps to give the required depth of packing in each section. It appears preferable to step off the plates in each section starting at the ends of the tower and proceeding toward the feed plate. If the column has a reboiler, the vapor will enter the packed section with a composition corresponding to the reboiler vapor as determined from the $YX$ diagram. Example 9.16 illustrates the design procedure for the case of a saturated liquid feed. The method can readily be modified to accommodate other situations following the principles outlined above.

EXAMPLE 9.16   A solution at its boiling point containing 15 mole percent methanol in water is to be fractionated in a tower packed with 1-in Raschig rings at 1 atm. It is desired to obtain an overhead product containing 98 mole percent methanol and a bottom product containing 98 mole percent water. Equilibria are given by Perry et al. [41] as shown in Fig. 9.54. A reflux rate of 0.252 moles per mole of feed is to be used (corresponding to 150 percent of minimum) and the column is operated with an externally heated reboiler. A total feed rate of 23,000 lb/h (1,144 lb moles) is to be distilled in a column 6 ft in diameter. Estimate the required tower height employing the HETP concept with the assumption of constant molal overflow.

SOLUTION   The $YX$ diagram based on methanol is plotted as shown in Fig. 9.54 according to the method of McCabe and Thiele [33]. At the specified reflux rate and separation, the slope of the operating line $L_M/G_M$ is 0.65 in the rectifying section and 3.24 in the stripping section. Feed is introduced to the column at a point corresponding to the intersection of the operating lines at $X = 0.15$.

Theoretical plates in the stripping section are stepped off starting at the bottom product composition ($X = 0.02$ on the 45° line) and moving up the tower until a value of $Y$ corresponding to the intersection of the operating lines for the stripping and rectifying sections is

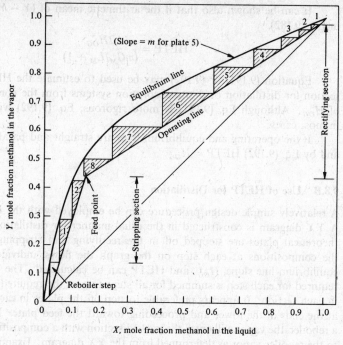

**FIGURE 9.54**
Design diagram for Example 9.16.

reached. A total of 2.18 plates is required in addition to the reboiler. Similarly, rectifying section plates are stepped off starting at the top product composition ($X = 0.98$ on the 45° line) and moving down the tower until a value of $Y$ corresponding to the operating-line intersection is reached. A total of 7.88 plates is required.

The overall gas-phase height of a transfer unit corresponding to the composition and flows over the tower section corresponding to each theoretical plate can now be calculated. Employing the methods suggested in Chap. 11, the gas-phase and liquid-phase heights of a transfer unit are estimated from the flows and properties of the fluids. In the present case, average HTU values for each phase will be employed in each section since the HTUs are not well known and do not vary greatly within each section.

Average gas-phase heights of a transfer unit $H_G$ of 0.87 ft in the stripping section and 2.16 ft in the rectifying section are assumed. Similarly average values of $H_L$ of 0.27 and 0.29 ft are assumed for the stripping and rectifying sections, respectively. $H_G$ is considerably higher in the rectifying section because of the low liquid rate. Values of the equilibrium-line slope $m$ [as defined above by Eq. (9.187)] are determined from the diagram graphically for each step. It is assumed that the reboiler acts as one theoretical plate and that, therefore, the gas stream enters the packed section from the reboiler at a composition $Y_1 = 0.134$ and the liquid leaves the section at a mole fraction $X_1 = 0.055$, corresponding to the first step above the reboiler on the operating line. Based on the corresponding average values of $m$, $L_M/G_M$, $H_G$ and $H_L$, $H_{OG}$ are calculated for each step according to Eq. (9.102).

The HETP is calculated for each step according to Eq. (9.192). The packed height corresponding to each step is calculated as the product of the HETP and the number of plates required in the step. Fractional plates are specified adjacent to the feed plate in both sections, since the feed composition is not reached with an integral step from either direction.

Key quantities in the calculation are summarized in the table below.

| | Step | $Y_{in}$ | $H_G$, ft | $H_L$, ft | $m$ | $\dfrac{L_M}{G_M}$ | $H_{OG}$, ft | HETP, ft | Plates required | Increment of packed height, ft |
|---|---|---|---|---|---|---|---|---|---|---|
| Stripping | 1 | 0.134 | 0.87 | 0.27 | 3.04 | 3.24 | 1.12 | 1.16 | 1.0 | 1.16 |
| section | 2 | 0.286 | 0.87 | 0.27 | 2.00 | 3.24 | 1.04 | 1.29 | 1.0 | 1.29 |
| | 3 | 0.424 | 0.87 | 0.27 | 1.36 | 3.24 | 0.98 | 1.38 | 0.18 | 0.25 |
| Total for stripping section | | | | | | | | | 2.18 | 2.70 |
| Rectifying | 1 | 0.960 | 2.16 | 0.29 | 0.41 | 0.65 | 2.34 | 2.87 | 1.0 | 2.87 |
| section | 2 | 0.929 | 2.16 | 0.29 | 0.43 | 0.65 | 2.35 | 2.83 | 1.0 | 2.83 |
| | 3 | 0.886 | 2.16 | 0.29 | 0.435 | 0.65 | 2.36 | 2.81 | 1.0 | 2.81 |
| | 4 | 0.824 | 2.16 | 0.29 | 0.44 | 0.65 | 2.36 | 2.81 | 1.0 | 2.81 |
| | 5 | 0.731 | 2.16 | 0.29 | 0.44 | 0.65 | 2.36 | 2.81 | 1.0 | 2.81 |
| | 6 | 0.610 | 2.16 | 0.29 | 0.49 | 0.65 | 2.38 | 2.71 | 1.0 | 2.71 |
| | 7 | 0.495 | 2.16 | 0.29 | 1.69 | 0.65 | 2.47 | 2.40 | 1.0 | 2.40 |
| | 8 | 0.441 | 2.16 | 0.29 | 1.13 | 0.65 | 2.66 | 1.80 | 0.88 | 1.58 |
| Total for rectifying section | | | | | | | | | 7.88 | 20.82 |
| Total for tower | | | | | | | | | 10.06 | 23.52 ft |

*Comment*  The procedure above provides a relatively convenient method for approximate design of packed columns for binary distillation. Use of constant average single-phase HTUs in each section may not be appropriate in general, but the method can easily be extended to include calculation of the flows and fluid properties needed to estimate the HTUs at each step with greater accuracy. A more precise result can be obtained by graphical integration of Eq. (9.180), which gives a tower height of 23.7 ft for the present example.

Inspection of the HETP values in the table above illustrates the strong variation of this quantity which can occur over a tower. Use of an overall average HETP, obtained by dividing the packed height by the number of theoretical plates, which is sometimes used to report packed-column performance, may have limited general value.          ////

## 9.8.C  Use of HETP for Absorber Design

The method can be applied directly to absorption in dilute systems in the manner described above. However, for a concentrated gas a slight correction is required because of the simplifications introduced in deriving the log-mean driving force relation, Eq. (9.99), via Eq. (9.98). The latter equation is based on Eq. (9.86) with the $Y_{BM}^*/(1 - Y)$ term neglected. Therefore, a more rigorous relation for the HETP for absorption can be shown to be

$$\text{HETP} = \left(\frac{Y_{BM}^*}{1 - Y}\right)_{\text{av}} \left[\frac{H_{OG} \ln (mG_M/L_M)}{mG_M/L_M - 1}\right] \tag{9.193}$$

Similarly, if the arithmetic-mean driving force is used

$$\text{HETP} = \left(\frac{Y_{BM}^*}{1 - Y}\right)_{\text{av}} \left(\frac{2H_{OG}}{mG_M/L_M + 1}\right) \tag{9.194}$$

For most purposes Eq. (9.193) should be satisfactory using the arithmetic mean over the plate for $Y_{BM}^*$ and for the group $Y_{BM}/(1 - Y)$. Solution of Example 9.7 by this procedure gives a packed height of 32.4 ft compared with 31.2 ft by exact calculation.

As an alternate to calculating $[Y_{BM}^*/(1 - Y)]_{\text{av}}$ at each plate Equation (9.192) can be applied to the entire column, and the resulting column height multiplied by the factor $(N_P + \Delta N_{OG})/N_P$, where $N_P$ is the number of theoretical plates required by the $YX$ diagram. This procedure should be adequate as a first-order correction since the $\Delta N_{OG}$ term is generally small compared with $N_{OG}$ as noted previously, and since $N_P$ is generally similar in magnitude to $N_{OG}$ as demonstrated by the similarity of Figs. 9.7 and 9.21. Application of the latter procedure to Example 9.7 gave results identical to those obtained above with Eq. (9.194).

## 9.9 SOLVENT EXTRACTION

Solvent extraction involves the contacting of immiscible liquids under conditions such that a desired component will be transferred from one phase (raffinate) to the second phase (extract or solvent). Countercurrent flow in packed or plate columns, or in mixer-settler combinations, is commonly employed.

Extraction problems in which the component of interest is in dilute concentration and in which the raffinate and solvent phases are highly immiscible may be solved by the methods described above for absorption. The phase of lower density may be viewed as equivalent to the gas phase in absorption and an $XY$ diagram may be constructed as a basis for design.

As in the case of distillation, a detailed treatment of solvent extraction lies beyond the scope of this text. The comprehensive book by Treybal [53] and the more recent review volume of Hanson [17] provide extensive coverage of the subject. Digital computer techniques and programs applicable to multistage extraction are given by Hanson et al. [18].

# APPENDIX

# A DIGITAL COMPUTER PROGRAM FOR PACKED-COLUMN ABSORPTION AND STRIPPING [58]

The following FORTRAN IV program is an illustrative application of the more important basic design equations for packed columns. The program was written for use with the CDC 6400 computer at Berkeley, so that some modifications may be required for use with other equipment. A central core memory of 2,071 words is required. A typical tower-height calculation uses 3 CPU seconds on the CDC 6400. All equations are used in their generalized form described in Sec. 9.4.H and the program follows the design procedure outlined therein. It will compute the number of transfer units and the required tower height rigorously for isothermal absorption and stripping processes involving one solute, an inert gas and a solvent. The latter may be volatile, provided that $\phi$ can be assumed constant through the tower and its value is known. Heat effects leading to nonisothermal conditions can be taken into account in certain cases.

**Program Organization** The main program organizes and performs the integration, while the material balance calculations and the evaluation of the equilibrium relationship are done in the subroutines MATBAL and PHYSIC. The former uses Eq. (9.136) and thus allows for the displacement of the operating line due to solvent evaporation effects.

Since not every value of $\phi$ is normally compatible with the rest of the specifications, a subroutine TEST is called once at the beginning of the calculations which tests for physically impossible specifications and diagnoses them.

The integration is performed by two nested loops: The outer loop breaks the integration range down into a number (JE-1) of intervals and calculates an average value of the overall gas-phase height of a transfer unit for each interval by means of Eq. (9.144). The interfacial concentrations are found iteratively, using Eq. (9.147). The exact number of those intervals depends on the specific case and will typically range from 17 to 22.

The inner loop computes for each interval of the outer loop both the number of apparent transfer units defined by Eq. (9.104) (TNT) and the number of overall gas-phase transfer units defined by Eq. (9.123) (TNU). The numerical integration employs the trapezoidal rule. Three different values for the step length are used for the integration which are chosen in such a way that the first 60 percent of the total integration range on the concentrated side is covered by approximately 50 steps, the next 20 percent by again 50 steps, and the final 20 percent next to the dilute end by 100 to 200 steps. Test calculations employing various numbers of steps were conducted to ensure that the procedure above gives adequate accuracy.

The increment of tower height corresponding to each interval is calculated according to Eq. (9.87), and the sum of the heights for all the outer loop intervals gives the total for the tower.

**Important Variable Names Used in the Program**

| | |
|---|---|
| KY | Integer number for identification of the calculated case |
| ISTRIP | Flag labeling stripper calculations: ISTRIP $=O$ indicates absorber, any other (integer) value means stripper |
| Y1, Y2 | Mole fractions $Y$ of solute in the gas at the concentrated end and dilute end of the packing, respectively |
| X1, X2 | Mole fractions $X$ of solute in the liquid at the concentrated end and dilute end of the packing, respectively |
| G1, G2 | $G_{M,1}$ and $G_{M,2}$, molar mass velocity of gas at the concentrated and dilute end of the packing, respectively |
| L1, L2 | $L_{M,1}$ and $L_{M,2}$, molar mass velocity of liquid at the concentrated and dilute end of the packing, respectively |
| S | Slope of the linearized equilibrium line ($\equiv Y^*/X$) |
| HG | $H_G$, gas-phase height of a transfer unit |
| HL | $H_L$, liquid-phase height of a transfer unit |
| HOG | $H_{OG}$, overall gas-phase height of a transfer unit |
| RPHI | $t$ ($\equiv 1/\phi$) [see Eq. (9.129)] |
| YB1, YB2 | $Y_{B,1}$ and $Y_{B,2}$, mole fractions of solvent vapors in the gas at the concentrated and dilute end of the packing, respectively |
| GO, LO | $L_M^\circ$ and $G_M^\circ$ [see Eqs. (9.140) and (9.141)] |
| GC | Molar inert-gas mass velocity, $G'_M$ |
| DY, DELY | Step lengths of outer and inner loops, respectively |
| Y(J) | $Y$ value of the beginning of the $j$th DY interval |
| YH, XH | Mole fractions at which $H_{OG}$ is evaluated |
| YI, XI | Mole fractions at gas-liquid interface |
| TNT | Value of $N_T$ up to the $j$th interval |
| TNU | Value of $N_{OG}$ up to the $j$th interval |
| TL | Value of $h_T$ up to the $j$th interval |

**Input requirements and use of the program**  For each case, three data cards with the values of the independent variables listed below must be supplied.

For absorption towers,

| | | |
|---|---|---|
| *1* | Run number KY | Format: I3 |
| *2* | $Y_1, Y_2, X_2, G_{M,1}, L_{M,2}$ | Format: 5F10.4 |
| *3* | $S, H_G, H_L, t, Y_{B,1}$ | Format: 5F10.4 |

For strippers,

| | | |
|---|---|---|
| *1* | Run number KY, ISTRIP | Format: 2I3 |
| *2* | $X_1, X_2, Y_2, L_{M,1}, G_{M,2}$ | Format: 5F10.4 |
| *3* | $S, H_G, H_L, t, Y_{B,2}$ | Format: 5F10.4 |

After completion of the calculations, control is transferred back to statement 1 in the program, which reads in KY of the next case. Whenever a KY is encountered being zero or negative, the program stops.

**Counterdiffusional effects**  If the solvent may be regarded as nonvolatile and only one component is absorbed, as is the case for instance in Examples 9.5 and 9.7, the value of $t$ is set to unity. The calculations will then correspond strictly to the principles for gas absorption (or stripping) through a stagnant gas, as presented in Secs. 9.6.B and 9.6.C. If, on the other hand, the molar flows $G_M$ and $L_M$ may be assumed constant due to equimolar solvent evaporation, as in the Example 9.8, $t$ is set to zero and the program will yield a straight operating line. In other, more general cases, where an unspecified amount of solvent evaporation occurs, the average value of $t$ must be estimated according to the guidelines described in Sec. 9.6.H.

**Equilibrium relationship**  The two subroutines PHYSIC and HEIGHT compute the slope of the linearized equilibrium line $S$ and the heights of a transfer unit $H_G$ and $H_L$ if those parameters cannot be assumed constant. The appropriate functions, being different from case to case, are to be supplied by the user in the form of FORTRAN statements. The subroutines will then override whatever values have been read in for $S$, $H_G$, and $H_L$ by the main program.

In Examples 9.5, 9.7, and 9.9, for instance, the assumption of a linear equilibrium line was justified. In calculating those examples using the program, the value of $S$ can therefore be read directly (data card 3 above) and the subprogram PHYSIC remains empty. However, in order to run the program for Examples 9.8 and 9.10, the van Laar equation or another suitable expression for $S$ with the correct parameter values has to be substituted into PHYSIC. A listing of the corresponding statements for Example 9.8, employing the van Laar equation in the form of Eq. (9.4) is provided at the end of this section.

**Heat effects**  For nonvolatile solvents it is possible to include the effect of the heat of absorption by neglecting the heat capacity of the gas stream. This is normally a good assumption at atmospheric pressure. In this case, the subprogram has to be provided with a statement computing the temperature through a heat balance and the value of $S$ must be evaluated at that temperature. The calculation procedure is illustrated in Example 9.11.

When the liquid-film resistance cannot be neglected, special care has to be given to the heat balance in order to make sure that it is based on the bulk liquid concentration and not on the liquid concentration at the interface. Therefore, the bulk liquid concentration must be made available through a COMMON statement in the subroutine PHYSIC and a corresponding one in the main program, where the two names XH and XA are used for the bulk concentration.

This technique will direct the program to evaluate the equilibrium concentration $Y^*$, at the *same* temperature as at the interface, and to substitute a value of $m$ into Eq. (9.144), which represents the average slope of the *isothermal* equilibrium line at that temperature between the two points $Y^*$, $X$ and $Y_i$, $X_i$. Thus, the correct overall driving force and a correspondingly adapted value for $H_{OG}$ is used. But it should be noted that in this case, the line connecting the interfacial concentrations will not coincide with the locus of the equilibrium concentrations $(Y^*, X)$.

The results of this procedure will be much less accurate when solvent evaporation occurs; but if the solvent is only slightly volatile, it will still be possible to obtain an approximate estimation of the required depth of packing. As an illustration of the computation procedure suggested here, Example 11.2 was recalculated on the computer. The application of the procedure to this example is discussed in more detail in the paragraph on program listing and illustrative examples.

The subroutine PHYSIC made up for this purpose is shown following the complete program listing.

**Heights of a transfer unit**    The values for the gas-phase and liquid-phase heights of a transfer unit may often be read in as constants. This is the case for all examples of this chapter except for Example 9.7. In order to run the program for this example, a few statements have to be substituted into the subroutine HEIGHT so that $H_G$ and $H_L$ are computed according to the functions given in the problem description.

**Application to stripper design**    The program may be applied to stripper calculations by setting the flag ISTRIP to any value different from zero. This will direct the program to skip the subsequent read statements in the main program. Since other variables are normally specified for stripper design than for the design of absorbers, the specified variable values for strippers are read in by a special subroutine called STRIP, which also applies Eq. (9.135) in order to perform a material balance around the whole column.

No equivalent to the subprogram TEST has been provided for stripper calculations, as $t$ will normally remain in the vicinity of one during the course of stripping processes. If, however, the program is to be used to calculate stripping processes with unusually high $t$ values, the input data should be checked for physically impossible specifications in order to avoid the possibility of an abnormal termination of the program execution.

**Program listing and illustrative examples**    The program is shown below in the form used to check the graphical integration of Example 9.7, and the appropriate statements to calculate $H_G$ and $H_L$ are therefore included in the subroutine HEIGHT. The complete program listing is followed by a listing of a few statements illustrating the use of the van Laar equation (9.4) in the subroutine PHYSIC. The first group of statements refers to the application of the equation to the isothermal process described in Example 9.8, while the second group illustrates the computation of the nonisothermal equilibrium line of Example 11.2.

The results printed out by the computer for the Example 9.7 are also shown below. The first two lines of data, being unlabeled, reflect the input data, with the exception of the last figure on the first line, which represents $X_1$. The third unlabeled data line shows the different step lengths used.

```
      PROGRAM ISGA    (INPUT, OUTPUT)
C
C
C
C
C         GASABSORPPTION IN PACKED COLUMNS
C         ================================
C
C
C         A = SOLUTE, B = SOLVENT,  C = INERT GAS.
C         1 = CONCENTRATED END,  2 = DILUTE END OF PACKING
C
C
      DIMENSION Y(100),DEL(100)
      REAL NT, NU, L2, LO, L1
      COMMON/CONSTS/RPHI, GO, LO, GC
      COMMON/READIN/KY,Y1,Y2,X1,X2,G1,G2,L1,L2, S, HG, HL, YB1, YB2
    1 READ 901, KY, ISTRIP
      IF (KY .LE. 0)      STOP
      IF (ISTRIP .NE. 0)     CALL STRIP
      IF (ISTRIP .NE. 0)     GO TO 2
C
      READ 902, Y1, Y2, X2, G1, L2
      READ 902, S, HG, HL, RPHI, YB1
      GC   = (1. - Y1 - YB1) * G1
      GO   = (1.-Y1*RPHI) * G1
      LO   = (1.-X2*RPHI) * L2
      CALL TEST(IFLAG)
      IF (IFLAG) 1, 1, 2
C
    2 CONTINUE
C
C     SET THE GRID FOR THE INTEGRATION
C     --------------------------------
C
      YDEL1= (Y1-Y2)*0.4 + Y2
      YDEL2= Y2 + (Y1-Y2)/5.0
      DY   = (Y1-Y2)/15.0
      DO 40 J=1,15
      DY   = 10.*DY
      IF(DY .GT. 1.0)  GO TO 41
   40 CONTINUE
   41 IDY   = DY
      DY   = IDY/(10.0**J)
      STEP1= DY/5.0
      STEP2= DY/10.0
      STEP3= DY/50.0
C
      JE=(Y1-Y2)/DY
      AJE=(Y1-Y2)/DY
      IF( AJE.GT. JE)  JE = JE+1
      YZ=Y1
      DO 550 J=1,JE
      Y(J)=YZ
      YZ=YZ-DY
      IF (Y(J).GT.YDEL1) DEL(J)=STEP1
```

```
      IF (Y(J).LT.(YDEL1+DY).AND.Y(J).GT.YDEL2) DEL(J)=STEP2
      IF (Y(J).LT.(YDEL2+DY)) DEL(J)=STEP3
  550 CONTINUE
      JT=JE+1
      Y(JT)=Y2
      PRINT 900
      PRINT 908, KY
      IF (ISTRIP .NE. 0)      GO TO 552
      PRINT 909 , Y1, Y2, X2, G1, L2, X1
      PRINT 909, S, HG, HL, RPHI, YB1
      GO TO 555
  552 PRINT 909, X1, X2, Y2, L1, G2
      PRINT 909, S, HG, HL, RPHI, YB2
  555 CONTINUE
      PRINT 909, DY, YDEL1, YDEL2, STEP1, STEP2, STEP3
      PRINT 905
      PRINT 906
      PRINT 905
C
C
      TL=0
      TNT=0
      TNU=0
      DO 200 J=1,JE
C
C     COMPUTATION OF HCG
C     ------------------
C
      YH= (Y(J)+Y(J+1))/2
      CALL MATBAL(XH, YH)
C     DETERMINATION OF HG AND HL
      CALL HEIGHT(HG, HL, XH, YH)
      CALL PHYSIC(S, XH)
      YQH=S*XH
C
C     DETERMINATION OF INTERFACIAL CONCENTRATIONS BY TRIAL AND ERROR
C
      YHT  = 1. - YH*RPHI
      XHT  = 1. - XH*RPHI
      YQT  = 1. - YQH*RPHI
      XP   = XH
      YF   = 1.0
      XF   = 1.0
      IF (HL.EQ. 0.0)  GO TO 253
      BL   = HG*LO*YHT/(HL*GO*XHT)
C
  230 CONTINUE
      BR   = BL * YF/XF
  240 XI   = (XH*BR + YH)/(BR + S)
      CALL PHYSIC(S, XI)
      YI   = S*XI
      CRIT = ABS(1. - XP/XI)
      IF (CRIT .LT. 0.0001) GO TO 254
      XP   = XI
      IF(RPHI .EQ. 0.0)      GO TO 240
C     COMPUTATION OF THE FILM FACTORS
```

```
      XIT  =  1. - XI*RPHI
      YIT  =  1. - YI*PPHI
      XF   =  (XHT - XIT)/ALOG(XHT/XIT)
      IF (YHT.EQ. YIT) GO TO 240
      YF   =  (YHT - YIT)/ALOG(YHT/YIT)
      GO TO 230
C
C     SPECIAL CASES
C
C     THE SPECIAL CASES RPHI = 0 (YF=XF=YSF=1), HG = 0 (YF=1) AND YI = 0
C     (YF=YSF, HOG=HG) ARE HANDLED BY THE ABOVE 230-LOOP. HL = 0 (YF=1, HOG=HG)
C     NEEDS SPECIAL TREATEMENT.
  253 HOG  =  HG
      GO TO 260
C
  254 YSF  =  1.0
      IF (RPHI . EQ. 0.0)  GO TO 256
      YSF  =  (YHT - YQT)/ALOG(YHT/YQT)
  256 AM   =  (YI - YQH)/(XI - XH)
      HGP=HG*YF/YSF
      HLP=HL*XF/YSF
      HOG  =  HGP  +  HLP*AM*GO*XHT/(LO*YHT)
  260 CONTINUE
      PRINT 903, J, Y(J), YH, XH, YI, XI, HG, HL, HOG,      TNT, TNU, TL
C
C
C     INTEGRATION OF NOG OVER INTERVALL DY
C     ------------------------------------
C
      DELY=DEL(J)
      ME   =  (Y(J)-Y(J+1)) / DELY
      AME  =  (Y(J)-Y(J+1)) / DELY
      IF(AME .GT. ME)   ME = ME + 1
      YU=Y(J)
      NT=0
      NU=0
      DO 100 N=1,ME
      IF (N .EQ. ME) DELY =  YU - Y(J+1)
      YA=YU-DELY/2
      CALL MATBAL(XA, YA)
      CALL PHYSIC(S, XA)
      YQ=S*XA
      IF (YA.LE.YQ.AND.ISTRIP.EQ.0.OR.YA.GE.YQ.AND.ISTRIP.NE.0) GO TO 60
      GO TO 50
   60 PRINT 61, KY, J, Y(J), YA, XA
   61 FORMAT(/, 1X, *OPERATING AND EQUILIBRUIM LINES CROSS*, /,
     1  2I3, 6F10.6, /)
      GO TO 1
   50 CONTINUE
      G    =  1.0
      IF(RPHI .EQ. 0.0)  GO TO 70
      YBU = 1. - YA*RPHI
      YBQ = 1. - YQ*RPHI
      R=ALOG(YBU/YBQ)
      YBM=(YBU-YBQ)/R
      G=YBM/YBU
```

```
     70 DNT = DELY/ABS(YA-YQ)
        NT=NT+DNT
        DNU=G*DNT
        NU=NU+DNU
        YU=YU-DELY
    100 CONTINUE
        DH = HOG * NU
        TL=TL+DH
        TNT=TNT+NT
        TNU=TNU+NU
    200 CONTINUE
        PRINT 904, JT, Y(JT), TNT, TNU, TL
        PRINT 905
        GO TO 1
C
C
C
    900 FORMAT(1H1)
    901 FORMAT(10I3)
    902 FORMAT(8F10.4)
    903 FORMAT(1X, I3, 6X, 5F10.6, 6F10.2)
    904 FORMAT(1X, I3, 6X, F10.6, 70X, 3F10.2)
    905 FORMAT( /, 1X, 120(1H=), /)
    906 FORMAT( 1X, * J *, 6X, 6X, *Y(J)*,  6X, *YH *, 6X, *XH *,
       1 6X, *YI *, 6X, *XI *, 6X, *HG *, 6X, *HL *, 6X, *HOG *,
       2 6X, *TNT *, 6X, *TNU *, 6X, *TL*)
    908 FORMAT(     20X, *RUN NUMBER   *, I3, /,  20X, 16(1H=), /)
    909 FORMAT( 20X, 6F10.5)
        END




        SUBROUTINE STRIP
C
C       CALCULATES BASIC QUANTITIES IF COLUMN IS A STRIPPER
C
        REAL L2, LO, L1
        COMMON/CONSTS/RPHI, GO, LO, GC
        COMMON/READIN/KY,Y1,Y2,X1,X2,G1,G2,L1,L2, S, HG, HL, YB1, YB2
        READ 902, X1, X2, Y2, L1, G2
        READ 902, S, HG, HL, RPHI, YB2
        GC  = (1. - Y2 - YB2) * G2
        GO  = (1. - Y2*RPHI) * G2
        LO  = (1. - X1*RPHI) * L1
        L2  = LO/(1. - X2*RPHI)
C
C       MATERIAL BALANCE FOR A
C
        XT  = 1. - RPHI*X1
        Y1  = LO*X1 + G2*Y2*XT - L2*X2*XT
        Y1  = Y1/(G2*XT + L2*RPHI*(X1-X2))
        RETURN
C
    902 FORMAT(8F10.4)
        END
```

```
      SUBROUTINE TEST(IFLAG)
C
C     TESTS FOR UNMEANINGFUL SPECIFICATIONS OF T
C     ********************************************
C
      REAL L2, LO, L1
      COMMON/CONSTS/RPHI, GO, LO, GC
      COMMON/READIN/KY,Y1,Y2,X1,X2,G1,G2,L1,L2, S, HG, HL, YB1, YB2
      IFLAG = 1
      IBRAN = 1
C
      IF (RPHI*X2 .EQ. 1.0)  IBRAN = 2
      IF (RPHI*Y2 .EQ. 1.0)  IBRAN = 3
      IF (RPHI*Y1 .EQ. 1.0)  IBRAN = 3
      IF (IBRAN .GT. 1)  GO TO 91
C
C     NORMAL CASES
C     ------------
C
    1 CONTINUE
      D   = LO/GO
      G2  = GO/(1. - Y2*RPHI)
    3 CALL MATBAL(X1, Y1)
      IF (X1 .GT. 1.0)       IBRAN = 4
      IF (X1 .LT. 0.0)       IBRAN = 4
      IF (G2 .LT.  GC/(1.-Y2)) IBRAN = 5
      IF (IBRAN .LT. 4)      RETURN
C
C     FATAL CASES
C     -----------
C
   91 CONTINUE
      IFLAG = -1
      PRINT 900
      PRINT 908, KY
      PRINT 909 , Y1, Y2, X2, G1, L2
      PRINT 909, S, HG, HL, RPHI, YB1
      PRINT 910
      IBRAN = IBRAN - 1
      GO TO (92, 93, 94, 95), IBRAN
C
   92 CONTINUE
      PRINT 911
      PRINT 909, RPHI
      PRINT 917
      RETURN
C
   93 CONTINUE
      PRINT 911
      PRINT 909, RPHI
      PRINT 912
      RETURN
```

```
      C
         94 CONTINUE
            PRINT 913
            PRINT 915, PPHI, X1, G2
            PRINT 914
            RETURN
      C
         95 CONTINUE
            PRINT 913
            PRINT 915, RPHI, X1, G2
            PRINT 916
            RETURN
      C
      C
        900 FORMAT(1H1)
        908 FORMAT(      20X, *RUN NUMBER    *, I3, /,  20X, 16(1H=), /)
        909 FORMAT( 20X, 6F10.5)
        910 FORMAT(///, 20X, *UNMEANINGFUL SPECIFICATION OF T*, /, 20X,
           1 31(1H*), /)
        911 FORMAT(26X, *T    *)
        912 FORMAT(10X, *PHI IS EQUAL TO Y1 OR Y2.*, /, 10X, *THIS MAKES IT *,
           1 *IMPOSSIBLE TO SPECIFY Y1 AND Y2 INDEPENDENTLY, FOR Y WOULD*,
           2 * STAY CONSTANT.*)
        913 FORMAT(26X, *T    *, 6X, *X1  *, 7X, *G2*)
        914 FORMAT(10X, *T IS TOO SMALL. *, /, 10X, *THIS WOULD CAUSE MORE S*,
           1*OLVENT TO EVAPORATE THAN INITIALLY PRESENT IN THE LIQUID PHASE.*)
        915 FORMAT(20X, 2F10.5, F10.2)
        916 FORMAT(10X, *T IS TOO LARGE.*, /, 10X, *THIS WOULD CAUSE MORE SO*,
           1 *LVENT TO CONDENSE THAN INITIALLY PRESENT IN THE GAS PHASE.*)
        917 FORMAT(10X, *PHI IS EQUAL TO X2.*, /, 10X, *X WOULD STAY CONSTA*,
           1 *NT. THE ABSORBER WOULD BEHAVE LIKE  A WELL MIXED TANK.*)
            END



            SUBROUTINE MATBAL(X, Y)
      C
      C     PERFORMS A MATERIAL BALANCE FOR A
      C
            REAL LO, L2, L1
            COMMON/CONSTS/RPHI, GO, LO, GC
            COMMON/READIN/KY,Y1,Y2,X1,X2,G1,G2,L1,L2, S, HG, HL, YB1, YB2
            YT   = 1. - RPHI*Y
            X    = GO*Y - G2*Y2*YT + L2*X2*YT
            X    = X/ (L2*YT + G2*RPHI*(Y-Y2))
            RETURN
            END
```

```
      SUBROUTINE   PHYSIC(S, X)
C
C     COMPUTES THE SLOPE OF THE EQUILIBRIUM LINE
C
      RETURN
      END




      SUBROUTINE HEIGHT(HG, HL, X, Y)
C
C     CALCULATES THE GAS AND LIQUID HEIGHTS OF A TRANSFER UNIT
C
      REAL L,  LLB,  MWA,  MWB,  MWC,  LO
      COMMON/CONSTS/RPHI, GO, LO, GC
C     INSERT HERE VALUES FOR MWA, MWB AND MWC (MOL. WEIGHTS)
      MWA =  72.0
      MWB =  160.0
      MWC =  31.4
      G   =  GO /(1. - Y*RPHI)
      GLB =  G*Y*MWA + (G-G*Y-GC)*MWB + GC*MWC
      L   =  LO /(1. - X*RPHI)
      LLB =  L * (X*MWA + (1.-X)*MWB)
      HG  =  5.3 * GLB**0.395/LLB**0.417
      HL  =  0.2723*LLB**0.22
      RETURN
      END




STATEMENTS USED IN SUBROUTINE PHYSIC FOR COMPUTATION OF EXAMPLE 9.8
------------------------------------------------------------------

      GAMLOG = 253.66/313.16/ (1.0 + 253.66*X/(122.132*(1-X)))**2
      S    =  10.0**GAMLOG * 135.3/760.0




STATEMENTS USED IN SUBROUTINE PHYSIC FOR COMPUTATION OF EXAMPLE 11.1
-------------------------------------------------------------------

C     THIS IS FOR NON-ISOTHERMAL EQUILIBRIUM LINE
      COMMON XB
      T   =  XB*7656.0/((1.0-XB)*18.0)
      T   =  T + 273.16  + 15.0
      PO  =  EXP(-3794.06/T  + 18.1594)
      A   =  2.39333*T - 454.43
      B   =  600.7 - 1.403*T
      GAMLOG = A/T /(1.0 + A*X/(B*(1-X)))**2.0
      S    =  10.0**GAMLOG * PO/760.0
```

```
 .56000     .01257     .00500   14.95780   11.17590    .42882
1.00000   72.00000  160.00000    1.00000   -0.        -.00060
 .03000     .23154     .12206    -.00600     .00300
```

| J | Y(J) | YH | XH | YI | XI | HG | HL | HDG | TNT | TNU | TL |
|---|------|------|------|------|------|------|------|------|------|------|------|
| 1 | .560000 | .545000 | .413976 | .454624 | .454624 | 2.88 | 1.50 | 4.03 | 0. | 0. | 0. |
| 2 | .530000 | .515000 | .384977 | .425130 | .425130 | 2.82 | 1.49 | 3.94 | .23 | .26 | 1.05 |
| 3 | .500000 | .485000 | .356859 | .396269 | .396269 | 2.76 | 1.48 | 3.85 | .46 | .52 | 2.08 |
| 4 | .470000 | .455000 | .329582 | .368014 | .368014 | 2.70 | 1.48 | 3.77 | .69 | .78 | 3.09 |
| 5 | .440000 | .425000 | .303111 | .340340 | .340340 | 2.64 | 1.47 | 3.69 | .93 | 1.05 | 4.09 |
| 6 | .410000 | .395000 | .277409 | .313223 | .313223 | 2.58 | 1.46 | 3.61 | 1.18 | 1.32 | 5.09 |
| 7 | .380000 | .365000 | .252443 | .286639 | .286639 | 2.52 | 1.46 | 3.54 | 1.43 | 1.60 | 6.10 |
| 8 | .350000 | .335000 | .228182 | .260566 | .260566 | 2.47 | 1.45 | 3.47 | 1.70 | 1.89 | 7.13 |
| 9 | .320000 | .305000 | .204598 | .234984 | .234984 | 2.42 | 1.45 | 3.40 | 1.98 | 2.19 | 8.18 |
| 10 | .290000 | .275000 | .181661 | .209871 | .209871 | 2.36 | 1.44 | 3.33 | 2.28 | 2.51 | 9.26 |
| 11 | .260000 | .245000 | .159346 | .185207 | .185207 | 2.31 | 1.44 | 3.26 | 2.60 | 2.85 | 10.40 |
| 12 | .230000 | .215000 | .137627 | .160973 | .160973 | 2.26 | 1.43 | 3.19 | 2.95 | 3.22 | 11.61 |
| 13 | .200000 | .185000 | .116481 | .137151 | .137151 | 2.21 | 1.43 | 3.13 | 3.34 | 3.63 | 12.91 |
| 14 | .170000 | .155000 | .095886 | .113721 | .113721 | 2.16 | 1.43 | 3.07 | 3.78 | 4.09 | 14.34 |
| 15 | .140000 | .125000 | .075821 | .090667 | .090667 | 2.12 | 1.42 | 3.01 | 4.29 | 4.61 | 15.95 |
| 16 | .110000 | .095000 | .056265 | .067969 | .067969 | 2.07 | 1.42 | 2.95 | 4.90 | 5.24 | 17.85 |
| 17 | .080000 | .065000 | .037199 | .045611 | .045611 | 2.02 | 1.42 | 2.89 | 5.68 | 6.04 | 20.19 |
| 18 | .050000 | .035000 | .018604 | .023574 | .023574 | 1.98 | 1.42 | 2.83 | 6.78 | 7.15 | 23.40 |
| 19 | .020000 | .016285 | .007236 | .009983 | .009983 | 1.95 | 1.41 | 2.79 | 8.69 | 9.08 | 28.86 |
| 20 | .012570 | | | | | | | | 9.52 | 9.91 | 31.18 |

The following additional problem was also solved and included here in order to illustrate the application of the program to processes which involve heat effects and lead to nonisothermal operation.

**EXAMPLE 9.17** *Illustration of a nonisothermal case* (*see Chap. 11*)

(*a*) The process described in Example 11.2, involving the absorption of acetone vapor into water, is to be redesigned using the program proposed here.

(*b*) Rigorous temperature and concentration profiles, which take into account all heat effects including the effect of solvent evaporation and transfer of sensible heat to the gas phase, can be obtained for plate columns by application of the computational design procedure for plate absorbers proposed by Bourne, von Stockar, and Coggan [4]. Based on data obtained in such a manner, use the HETP method described in Sec. 9.8 to compute a new value for the required depth of packing which reflects the influence of all heat effects.

SOLUTION

(*a*) Although water is a more than slightly volatile solvent, the application of the procedure for nonisothermal absorption processes suggested in the respective paragraph of this appendix, which neglects the influence of solvent evaporation, is justified here because the anticipated temperature rise is comparatively low and the entering gas is already saturated with water.

An expression for the equilibrium vapor concentration of acetone as a function of the liquid composition and also of temperature must be found for use in subroutine PHYSIC. The vapor pressure of pure acetone in the temperature range considered here is well represented by the equation

$$P_A{}^s = \exp\left(-\frac{3794.06}{T} + 18.1594\right)$$

where $T$ denotes the absolute temperature in °K and $P_A{}^s$ the vapor pressure in mm Hg.

The constants for the van Laar equation (9.4) may be found by plotting the vapor pressure data of Othmer et al. in the following form (Ref. 81 of Chap. 11). A separate line is obtained for each of the three temperatures, 15°C, 35°C, and 45°C.

$$\frac{1}{\sqrt{T \log \gamma}} = \frac{\sqrt{A}}{B}\frac{X}{1-X} + \frac{1}{\sqrt{A}}$$

From such a plot it becomes obvious that due to the strong nonideality of this system, $A$ and $B$ show considerable variation with temperature. Values for $A$ and $B$ were therefore determined at each temperature using three different least-square fits of the above equation. The temperature variation was then found to be

$$A = 2.3933T - 454.43°K \qquad B = 600.7°K - 1.403T$$

With this information the subroutine PHYSIC may be set up. Based on the bulk concentration (XB), the temperature rise is first computed by a heat balance and added to the temperature of the entering liquid (see listing). Based thereupon, the vapor pressure of pure acetone and the activity coefficients are evaluated by means of the above relations and Eq. (9.4). From the value of $\gamma$ at each value of $X$, the corresponding value of $Y$ is obtained from the equilibrium line.

According to Example 11.2 the gas- and liquid-phase heights of a transfer unit may be assumed constant as 1.37 and 0.99 ft, respectively, and can therefore be read into the program from the data cards.

The results obtained by computer were

$$N_{OG} = 4.01 \qquad h_T = 7.82 \text{ ft}$$

*Comment* The required depth of packing is markedly lower than the one obtained graphically in Chap. 11, because the line connecting the equilibrium concentrations $(Y^*, X)$ was used as an equilibrium line in Chap. 11 in order to simplify the calculations. According to the section on heat effects of this appendix, this line does not coincide with the line connecting the actual concentrations at the interface $(Y_i, X_i)$. Both lines, labeled $b$ and $a$, respectively, are shown in the design diagram, Fig. 9.55. Using the former line $b$ as an equilibrium line implies the assumption that the interfacial temperatures may be calculated by a heat balance based on the liquid interfacial concentration itself. The temperatures and $h_T$ are thereby somewhat overestimated.

The more rigorous computer analysis uses correctly evaluated interfacial concentrations (line $a$) for the computation of $H_{OG}$. The overall driving force, however, must be taken at the same temperature but at the bulk liquid concentration, and $Y^*$ is thus represented by the line $b$.

Furthermore, the equilibrium line used in the graphical solution in Chap. 11 differs slightly from the data given by the subroutine PHYSIC at the same concentrations and temperatures (line $b$). This explains the difference in $N_{OG}$, which otherwise would agree.

(b) The procedure of Bourne, von Stockar, and Coggan [4] computes a rigorous operating line and equilibrium line by simulating a column with an integer number of theoretical plates. Therefore, if a given degree of absorption is specified in a design problem, it is necessary to make the column computation for at least two sets of plate specifications which give a solute recovery above and below that specified, and to interpolate between them. One procedure is to apply the HETP principle to each set of plate specifications to estimate the depth of packing required for each case, then to interpolate for the depth of packing which gives the desired performance. It would be conservative, of course, to specify that integer number of plates which gives a separation greater than and closest to that desired. In the present example the desired degree of acetone recovery is obtained closely with four theoretical plates, i.e., 90.4 percent vs. the 90 percent recovery specified. The computed temperatures and concentrations around each theoretical plate are shown in the table below.

| Number of plate $j$ | Temperature of liquid, °C (leaving plate) | $X_j$ | Temperature of gas, °C (leaving plate) | $Y_{A,j}$ | $Y_{B,j}$ | $\left(\dfrac{mG}{L_M}\right)_{\text{isoth}}$ |
|---|---|---|---|---|---|---|
| 0 | 15.00 | 0.0 | (entering liquid) | | | |
| 1 | 18.34 | 0.004 | 18.34 | 0.00604 | 0.0198 | 0.585 |
| 2 | 21.69 | 0.00896 | 21.69 | 0.01590 | 0.0243 | 0.654 |
| 3 | 23.91 | 0.01438 | 23.91 | 0.02794 | 0.0276 | 0.693 |
| 4 | 23.61 | 0.02233 | 23.61 | 0.04093 | 0.0268 | 0.627 |
| 5 | | (entering gas) | 15.00 | 0.06 | 0.0168 | |

$H_{OG}$ is computed for each plate according to Eq. (9.103) on the assumption that the concentration ratios in the more rigorous Equation (9.144) may be neglected for the relatively dilute system employed.

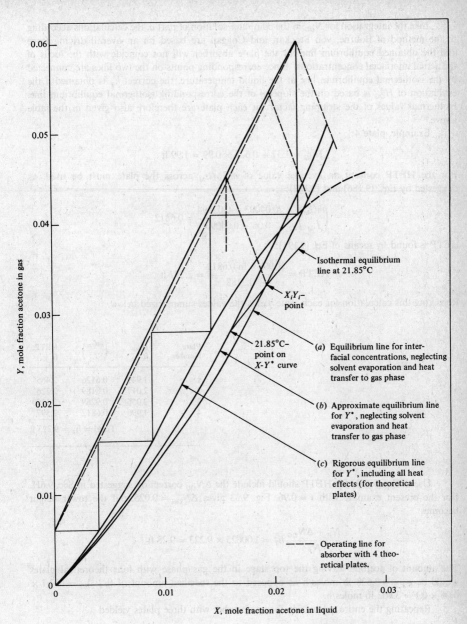

Isothermal equilibrium line at 21.85°C

$X_iY_i$-point

21.85°C-point on $X$-$Y^*$ curve

(a) Equilibrium line for inter-facial concentrations, neglecting solvent evaporation and heat transfer to gas phase

(b) Approximate equilibrium line for $Y^*$, neglecting solvent evaporation and heat transfer to gas phase

(c) Rigorous equilibrium line for $Y^*$, including all heat effects (for theoretical plates)

– – – – Operating line for absorber with 4 theoretical plates.

$Y$, mole fraction acetone in gas

$X$, mole fraction acetone in liquid

**FIGURE 9.55**
Design diagram for Example 9.17.

Like the integration for $N_{OG}$ in the computer solution of part $a$, the calculations according to the method of Bourne, von Stockar, and Coggan are based on an overall driving force and the obtained equilibrium line for the plate absorber will not coincide with the locus of the actual interfacial concentrations. Since corresponding points on the two lines are connected by the isothermal equilibrium line at the liquid temperature, the correct $h_T$ is obtained if the evaluation of $H_{OG}$ is based on the slope $m$ of the corresponding isothermal equilibrium line. Isothermal values of the stripping factor for each plate are therefore also given in the table above.

Example, plate 4:

$$H_{OG} = 1.37 + 0.627 \times 0.99 = 1.99 \text{ ft}$$

For the HETP concept an average value of $mG_M/L_M$ across the plate must be used, as suggested by Eqs. (9.186) and (9.187):

$$\left(\frac{mG_M}{L_M}\right)_4 = \frac{0.04093 - 0.02794}{0.06 - 0.04093} = 0.6812$$

HETP is found by means of Eq. (9.191):

$$\text{HETP} = \frac{(1.99) \times \ln 0.6812}{-0.3188} = 2.397 \text{ ft}$$

Repeating this calculation for each plate yields the values summarized below.

| Plate Number | $H_{OG}$, ft | $\left(\dfrac{mG_M}{L}\right)_{av}$ | HETP, ft |
|---|---|---|---|
| 1 | 1.949 | 0.6126 | 2.465 |
| 2 | 2.017 | 0.8189 | 2.226 |
| 3 | 2.056 | 0.9269 | 2.1350 |
| 4 | 1.990 | 0.6812 | 2.397 |
| | | Total $= h_T =$ | 9.223 ft |

Use of Eq. (9.191) for HETP should include the $\Delta N_{OG}$ correction suggested in Sec. 9.6H. For the present example with $t = 0.96$, Fig. 9.33 gives $\Delta N_{OG} = 0.025$ and the tower height becomes

$$\frac{N_P + \Delta N_{OG}}{N_P} h_T = 1.00625 \times 9.223 = 9.28 \text{ ft}$$

The amount of acetone leaving the top stage in the gas phase with four theoretical plates would be $g_{A, 2} = 3.628$ lb moles/h as opposed to the specified amount of $(g_{A, 2})_{SP} = 633.7 \times 0.06 \times 0.1 = 3.802$ lb moles/h.

Repeating the entire calculation for an absorber with three plates yielded

$$h_T = 6.99 \text{ ft} \qquad \text{with } g_{A, 2} = 5.128 \text{ lb moles/h}$$

A linear interpolation for the tower height suggests

$$h_T = 9.0 \text{ ft}$$

*Comment* A design diagram with the nonisothermal equilibrium lines used in part *a* is shown in Fig. 9.55. The lower line *a* connects the interfacial concentrations while the upper line *b* was used to evaluate the overall driving force. The step construction and the corrected equilibrium line *c*, based on the rigorous solution of a plate absorber, are also shown for comparison. They again represent the overall driving force. Use of 0.97 plates for plate 1 would give exactly the specified acetone recovery.

The calculations in part *b* suggest that the effect of solvent evaporation is to cause not only an internal temperature maximum and a bulge in the equilibrium line but also higher average temperatures and thus to increase the required tower height. (This accounts for most of the difference in the results of part *a* and part *b*, although the method of part *b* tends to yield somewhat conservative results.) This surprising condition derives from the fact that the gas is cooled on the two top plates by the cold liquid stream (see table at beginning of part *b*). Water is therefore condensed on those plates and thereby causes the liquid to be heated up more than would be the case by the absorption alone. This gives rise to higher temperatures than would be observed with a nonvolatile solvent.

**Summary of computer results for example problems** The various packed-column design problems described in Secs. 9.4 to 9.6 were solved by means of the program shown above. Results for each case are summarized in Table 9.1 to serve as a standard of comparison for the results obtained by graphical integration and the shortcut methods.

**Table 9.1 SUMMARY OF COMPUTER RESULTS FOR DESIGN CASES DISCUSSED IN TEXT**

| Example | $t$ | Subroutines PHYSIC | Subroutines HEIGHT | $\Delta N_{OG}$ | $N_{OG}$ | $(H_{OG})_{av}$, ft | $h_T$, ft |
|---|---|---|---|---|---|---|---|
| 9.5 | 1.0 | .......... | ............ | 0.01 | 11.57 | 2.48 | 28.64 |
| 9.7 | 1.0 | .......... | Statements | 0.39 | 9.91 | 3.15 | 31.18 |
| 9.8a | 0.0 | van Laar equation (see listing) | shown above | 0 | 29.01 | 2.67 | 77.38 |
| 9.8b | 0.0 | Eq. (9.3) | | 0 | 28.18 | 2.66 | 75.06 |
| 9.9a | | | | | | | |
| Case 1 | −1.882 | .......... | ............ | −0.46 | 4.64 | 1.60 | 7.42 |
| Case 2 | −12.71 | .......... | ............ | −0.06 | 4.66 | 1.60 | 7.46 |
| Case 3 | −136.3 | .......... | ............ | −0.06 | 4.67 | 1.60 | 7.47 |
| 9.9b steam injection | | | | | | | |
| Case 1, | 1.0 | .......... | ............ | 0.02 | 4.58 | 1.60 | 7.33 |
| Case 2 | 1.0 | .......... | ............ | 0.00 | 4.60 | 1.60 | 7.36 |
| Case 3 | 1.0 | .......... | ............ | 0.00 | 4.60 | 1.60 | 7.37 |
| 9.10, 100% saturation | 1.2155 | van Laar equation | ............ | 0.14 | 9.91 | 2.24 | 22.20 |
| 9.10, 50% saturation | 0.6178 | van Laar equation | ............ | 0.06 | 12.08 | 2.29 | 27.71 |
| 9.10, dry feed gas | 0.022 | van Laar equation | ............ | 0.00 | 15.35 | 2.35 | 36.05 |
| 9.14 (stripper) | 1.0 | .......... | See text | 0.29 | 5.95 | 1.68 | 9.97 |
| 9.17 | 1.0 | van Laar equation and heat balance (see listing) | ............ | 0.02 | 4.01 | 1.95 | 7.82 |

Since in this case the effect of water evaporation is contrary to what is expected normally, the calculations demonstrate the difficulty in predicting the effect of solvent evaporation. Neglecting solvent evaporation and transfer of sensible heat to the gas phase will generally not be a safe assumption in the case of water.

A still more rigorous procedure for design would be to place the equilibrium line as described in sec. 9.6.I in which the relative rates of heat and mass transfer are considered. However, the above procedure should be adequate for most situations.     /////

# REFERENCES

1   ATKINS, G. T., and W. B. FRANKLIN: *Refiner Natur. Gasoline Mfg.*, **15**(1): 30 (1936).
2   BAKER, T., and J. S. STOCKHARDT: *Ind. Eng. Chem.*, **22**: 376 (1930).
3   BENEDICT, M., G. B. WEBB, L. C. RUBIN, and L. FRIEND: *Chem. Eng. Progr.*, **47**: 571, 609 (1951).
4   BOURNE, J. R., U. VON STOCKAR, and G. C. COGGAN: *Ind. Eng. Chem. Process Des. Dev.*, **13**: 115 (1974).
5   BOURNE, J. R., U. VON STOCKAR, and G. C. COGGAN: *Ind. Eng. Chem. Process Des. Dev.*, **13**: 124 (1974).
6   BROWN, G. G., and M. SOUDERS: *Oil Gas J.*, **31** (5): 34 (1932); compare also Brown and Souders, "The Science of Petroleum," vol. 2. p. 1557, Oxford, New York, 1938; *Refiner*, **11**: 376 (1932).
7   BROWN, G. G., and M. SOUDERS: Separation of Petroleum Hydrocarbons by Distillation, in "The Science of Petroleum," vol. 2, sec. 25, Oxford, New York, 1938.
8   CARLSON, H. C., and A. P. COLBURN: *Ind. Eng. Chem.*, **34**: 581 (1942).
9   COLBURN, A. P.: *Ind. Eng. Chem.*, **28**: 526 (1936).
10   COLBURN, A. P.: *Trans. AIChE*, **35**: 211 (1939).
11   COLBURN, A. P.: *Ind. Eng. Chem.*, **33**: 459 (1941).
12   COLBURN, A. P.: Collected papers on the teaching of chemical engineering, American Society for Engineering Education. Summer School for Teaching of Chemical Engineering, Pennsylvania State College, 1936.
13   DREW, T. B.: *Trans. AIChE*, **36**: 679 (1940).
14   EDMISTER, W. C.: *Ind. Eng. Chem.*, **35**: 837 (1943).
15   EDMISTER, W. C.: "Applied Hydrocarbon Thermodynamics," Gulf, Houston, 1961.
16   GERSTER, J. A., A. B. HILL, N. F. HOCHGRAF, and D. G. ROBINSON: Tray Efficiencies in Distillation Columns. Final report from the University of Delaware. *AIChE* (1958).
17   HANSON, C. (ed.): "Recent Advances in Liquid-Liquid Extraction," Pergamon, New York, 1971.
18   HANSON, D. N., J. H. DUFFIN, G. S. SOMERVILLE: "Computation of Multistage Separation Processes," Reinhold, New York, 1962.
19   HAUSEN, E.: *Z. Ges. Kälte-Ind*, **32**: 93, 114 (1928); *Forsch. Gebeite Ingenieurw.*, **7**: 177 (1936).
20   HENGSTEBECK, R. J.: "Distillation Principles and Design Procedures," Reinhold, New York, 1961.
21   HILDEBRAND, J. H., and R. L. SCOTT: "Regular Solutions," Prentice-Hall, Englewood Cliffs, N.J., 1962.
22   HORTON, G., and W. B. FRANKLIN: *Ind. Eng. Chem.*, **32**: 1384 (1940).
23   HOUGEN, O. A., and K. M. WATSON: "Chemical Process Principles," pt 2, p. 653, Wiley, New York, 1947.
24   International Critical Tables, vol. 3, McGraw-Hill, New York, 1928.
25   JACKSON, R. M., and T. K. SHERWOOD: *Trans. AIChE*, **37**: 959 (1941).

26  KIRSCHBAUM, E.: *Forsch. Gebiete Ingenieurw.*, **5**: 245 (1934); *Z. Ver. Deut. Ing.*, **80**(1): 633 (1936).

27  KOLBE, E., and F. BRANDMAIR: *Chem. Ind. Technol.*, **35**: 262 (1963).

28  KREMSER, A.: *Nat. Petrol. News*, **22** (21): 42 (May 21, 1930).

29  LANDOLT-BÖRNSTEIN: Tables, "Zahlenwerte und Funktionen," 6th ed., band II/2a, pp. 336–709, Springer-Verlag, Berlin, 1960.

30  LEWIS, W. K.: *Trans AIChE*, **20**: 1 (1927).

31  LEWIS, W. K., JR.: *Ind. Eng. Chem.*, **28**: 399 (1936).

32  MARSHALL, W. R., and R. L. PIGFORD: "Application of Differential Equations to Chemical Engineering Problems," Univ. of Delaware, 1947.

33  MCCABE, W. L., and E. W. THIELE: *Ind. Eng. Chem.*, **17**: 605 (1925).

33a MOSTAFA, H. A.: *Chem. Eng. Sci.*, **29**: 1997 (1974).

34  MURPHREE, E. V.: *Ind. Eng. Chem.*, **17**: 747 (1925).

35  ONDA, KAKUSABURO, and EIZO SADA: Method of Calculating N.T.U. for Large Transfer Amounts, *Memoirs of the Faculty of Engineering*, Nagoya Univ., vol. 10, no. 2, November, 1958.

36  ONDA, K., E. SADA, K. TAKAHASHI, and S. P. MAKLOV: *AIChE J.*, **17**: 1141 (1971).

37  ORYE, R. V., and J. M. PRAUSNITZ: *Ind. Eng. Chem.*, **57**: 19 (1965).

38  OTHMER, D. F., and E. G. SCHEIBEL: *Trans. AIChE*, **38**: 339 (1942).

39  PERRY, J. H., "Chemical Engineers' Handbook," 3d ed., p. 708, McGraw-Hill, New York, 1950.

40  PERRY, J. H.: "Chemical Engineers' Handbook," 4th ed., McGraw-Hill, New York, 1963.

41  PERRY, R. H., C. H. CHILTON, and S. D. KIRKPATRICK (EDS.): "Chemical Engineers' Handbook," 4th ed., p. 13–5, McGraw-Hill, New York, 1963.

42  PRAUSNITZ, J. M.: "Molecular Thermodynamics of Fluid Phase Equilibria," Prentice-Hall, Englewood Cliffs, N.J., 1969.

43  PRAUSNITZ, J. M., and P. L. CHUEH: "Computer Calculations for High Pressure Vapor-Liquid Equilibria," Prentice-Hall, Englewood Cliffs, N.J., 1968.

44  PRAUSNITZ, J. M., C. A. ECKERT, R. V. ORYE, and J. P. O'CONNELL: "Computer Calculations for Multicomponent Vapor-Liquid Equilibria," Prentice-Hall, Englewood Cliffs, N.J., 1967.

45  RAAL, D. J., and M. K. KHURANA: *Can. J. Chem. Eng.*, **51**: 162 (1973).

46  REID, R. C., and T. K. SHERWOOD: "The Properties of Gases and Liquids," p. 323, McGraw-Hill, New York, 1958.

47  SHERWOOD, T. K., and F. J. JENNY: *Ind. Eng. Chem.*, **27**: 265 (1935).

48  SHERWOOD, T. K., and R. L. PIGFORD: "Absorption and Extraction," pp. 190-192, McGraw-Hill, New York, 1952.

49  SHERWOOD, T. K., and R. L. PIGFORD: "Absorption and Extraction," pp. 282, 288, McGraw-Hill, 1952.

50  SHREVE, R. N.: " Chemical Process Industries," p. 87, McGraw-Hill, New York, 1967.

51  SOUDERS, M., and G. G. BROWN: *Ind. Eng. Chem.*, **24**: 519 (1932).

52  STRICHLMAIR, J., and A. MERSMANN: *Chem-Ing. Tech.*, **43**: 17, (1971).

53  TREYBAL, R. E.: "Liquid Extraction," 2d ed., McGraw-Hill, New York, 1963.

54  VAN KREVELIN, D. W., and P. J. HOFTIJZER: *Chem. Eng. Progr.* **44**: 532 (1948).

55  VAN LAAR, J. J.: *Z. Phys. Chem.*, **83**: 599 (1913).

56  WIEGAND, J. H.: *Trans. AIChE*, **36**: 679 (1940).

57  WILKE, C. R.: *Chem. Eng. Progr.*, **46**: 95 (1950).

58  WILKE, C. R., and URS VON STOCKAR: Private communication, University of California, Berkeley, March 1974.

59  WILSON, G. M.: *J. Amer. Chem. Soc.*, **86**: 129 (1964).

# PROBLEMS

**9.1**  In a manufacturing plant, electronic equipment is assembled in a room ventilated and cooled by circulation of air drawn in from the outside. Because of atmospheric contamination from surrounding industrial plants, the air on certain occasions contains 0.1 mole percent of $SO_2$. To avoid possible damage to sensitive devices, it is necessary to keep the $SO_2$ level at or below 0.001 mole percent. Fresh air is drawn in from the outside at the rate of 100 $ft^3$/min (0.259 moles/min) at 760 mm pressure. It is proposed to remove the $SO_2$ by scrubbing the air with water in a 1.5-ft-diameter bubble-tray tower. Although the temperatures of the air and water change somewhat in the scrubbing tower, for purpose of preliminary design the scrubbing process may be assumed to occur at 760 mm pressure and 68°F, which is the temperature at which the water enters the tower. The air may be assumed saturated with water vapor at 68°F (humidity = 0.01475 lb water/lb dry air). The water entering the tower contains no $SO_2$. Equilibria for $SO_2$-water may be assumed to follow Henry's law according to the equation

$$p = 8,900X$$

where $p$ = partial pressure of $SO_2$, mm Hg
  $X$ = mole fraction $SO_2$ in water

(a) What would be the minimum water rate required to accomplish the desired absorption?

(b) How many theoretical plates would be required at a water rate of 130 percent of minimum?

**9.2**  The absorption specified in Prob. 9.1 is to be accomplished in a 1.5-ft-diameter tower packed with 1-in Raschig rings. At the specified flows, the gas-phase and liquid-phase heights of a transfer unit may be assumed constant at 0.7 and 1.0 ft, respectively.

(a) What depth of packing will be required to accomplish the desired absorption at a water feed rate of 4,700 lb/h?

(b) What would be the effect on the required tower height of increasing the tower pressure at the same mass-flow rates as part a?

**9.3**  It is desired to evaluate the performance of a packed tower by experimental tests from which the gas-and liquid-phase heights of a transfer unit can be deduced. The test tower is 6 ft in diameter packed to a depth of 10 ft with 1-in Raschig rings. Test runs are made for absorption of hexane vapor from air into a nonvolatile hydrocarbon oil. In the tests, gas and liquid enter the tower at 140°F, and absorption may be assumed to occur isothermally at this temperature (presumably the heat of absorption is offset by heat losses from the tower). The equilibrium constant, $K = Y/X$ (mole fraction basis) for hexane in the oil at 140°F, 1 atm is 0.79. The molecular weight of the oil is 200.

|  | Test 1 | Test 2 |
|---|---|---|
| Gas rate to tower, lb mole/h | 493 | 493 |
| Liquid rate to tower, lb mole/h | 540 | 540 |
| Pressure at top of tower (absolute) (assumed constant over the tower) | 14.7 psia | 29.4 psia |
| Mole fraction hexane in entering gas | 0.01 | 0.01 |
| Mole fraction hexane in leaving gas | 0.00208 | 0.00065 |
| Mole fraction hexane in entering liquid | 0.0004 | 0.0004 |

On the basis of the foregoing test data,

(a) Calculate the height of an overall gas-phase transfer unit $(HTU)_{OG}$, for the system at 140°F, 1 atm

(b) Calculate the height of a gas-phase transfer unit $(HTU)_G$ at 140°F, 1 atm

(c) Calculate the height of a liquid-phase transfer unit $(HTU)_L$ at 140°F, 1 atm

(d) Estimate the gas-phase mass-transfer coefficient with the same flows at 20 psig, lb mole/(h)(ft²) (mole fraction)

9.4 In a process using methanol as a solvent, it is desired to recover methyl alcohol vapors from 940 ft³/min of exhaust air by absorption in water.

The air may be assumed to have an average alcohol content of 30 mole percent. Humidity of the air (lb $H_2O$/lb dry air) may be assumed constant throughout the process at a value corresponding to saturation with respect to liquid entering the tower.

Absorption is to be carried out at 39.9°C in a 2.25-ft-ID column packed randomly with 1-in Raschig rings. 99 percent removal of methanol from the air is to be obtained in the absorber. The pressure is maintained at 1 psig at the top of the tower. The prevailing barometric pressure is 1 atm. Absorption may be assumed to occur at constant pressure of 1 psig and constant temperature of 39.9°C. (Note that these assumptions are not generally valid.)

The water leaving the absorber is distilled for alcohol recovery and returned to the absorber for reuse. The absorber water may be assumed to enter the absorber at 39.9°C containing 1.0 percent methanol by weight.

The ratio of gas flow to liquid flow is 1.2 lb gas/lb of liquid. This corresponds closely to the usual economic criterion that $mG/L$ should be approximately 0.7 at the dilute end.

Vapor-liquid equilibrium data for the system at 39.9°C are as follows (based on International Critical Tables, vol. III, p. 290 [24]):

| X alcohol | Partial pressure of alcohol, mm Hg | Partial pressure of water, mm Hg |
|---|---|---|
| 0 | 0 | 54.7 |
| 0.1499 | 66.1 | 39.2 |
| 0.1785 | 75.5 | 38.5 |
| 0.2107 | 85.2 | 37.2 |
| 0.2731 | 100.6 | 35.8 |
| 0.3106 | 108.8 | 34.9 |

At the operating conditions, the gas- and liquid-phase heights of a transfer unit may be assumed constant at 2.5 and 0.64 ft, respectively.

(a) Calculate the minimum liquid rate, lb/(h)(ft²).

(b) Calculate the required tower height by rigorous graphical integration.

(c) Compare the above result with that obtained by the approximate method of Colburn [Eq. (9.106)].

(d) Compare the result of part b with that obtained by the procedure of Sec. 9.6.G.

(e) Estimate the required height if the gas entered the tower at zero humidity, but still assuming isothermal absorption at 39.9°C.

9.5 The absorption of methanol specified in Prob. 9.4 is to be conducted in a bubble-cap-tray tower of 2.50 ft inside diameter. Isothermal operation at 39.9°C, 1 psig may be assumed.

Estimate the following items:

(a) Number of theoretical plates required
(b) Actual number of plates required assuming a constant average Murphree point efficiency $E_{OG}$ of 60 percent with complete mixing
(c) Actual number of plates for $E_{OG}$ of 60 percent, but with no mixing in the cross-flow direction.

9.6 Consider the following alternative for the recovery of methyl alcohol as described in Prob. 9.4. The air-alcohol stream is first cooled to 20°C and then fed to the packed column, where it will be absorbed by pure water, which is available at 20°C.

Vapor-liquid equilibrium data for the system are to be calculated using the liquid-activity coefficient $\gamma_1$, which may be determined from the relation

$$\ln \gamma_1 = \frac{A/T}{(1 + AX_1/BX_2)^2}$$

where $A = 296°K$
$B = 159.5°K$
$T$ = absolute temperature, °K
$X_1$ = mole fraction of methanol
$X_2$ = mole fraction of water
Assume that the gas phase is ideal.

Heats of solution at 20°C are as follows (based on International Critical Tables, vol. V, p. 213 [24]):

| Mole percent methanol | 5 | 10 | 15 | 20 | 25 |
|---|---|---|---|---|---|
| $\Delta H_s$ (kJ/mole methanol) | −6.838 | −5.989 | −5.114 | −4.357 | −3.667 |

Estimate the minimum liquid rate required to remove 98 percent of the methanol.

9.7 A plastic coating is applied to cardboard containers by spraying them with resin containing hexane as a solvent. The containers are passed through a curing oven where the resin sets and hexane is removed by a current of warm nitrogen. Under normal processing conditions 10,000 ft³/min of air at 140°F, 1 atm containing 0.30 mole fraction hexane is discharged from the oven.

To permit recirculation of gas to the oven, it is necessary to remove hexane from the nitrogen. One method to be investigated is absorption of the hexane in a nonvolatile hydrocarbon oil followed by recovery in a stripping column. As part of a study of the process, it is desired to estimate the number of theoretical plates required for absorption and stripping, cooling water requirements, and steam requirements.

98 percent removal of hexane from the nitrogen is accomplished in the absorber. 99 percent removal of hexane from the oil is accomplished in the stripper.

Air enters the absorber at 140°F, 1 atm. Lean oil enters the absorber at 140°F, 1 atm at a rate equal to 150 percent of minimum based on conditions on the bottom tray. Nitrogen may be assumed to leave the absorber at the same temperature as the oil entering.

The stripping column is operated at flow rates corresponding to $KV/L = 1.5$ at 230°F, 1 atm at flows prevailing at the bottom of the column. ($V$ = molal vapor rate, $L$ = molal liquid rate, and $K$ = equilibrium constant $Y/X$). The oil feed to the stripper actually enters at a temperature sufficiently above 230°F, so that the heat requirement for stripping can be met by cooling of the liquid stream. To simplify the calculations, stripping may be assumed to occur isothermally at 230°F, 1 atm throughout the entire column. Liquid may be assumed to leave the column at 230°F. Hexane is recovered by condensation of the stripper vapors.

(a) Calculate the number of theoretical plates required assuming the absorption to occur isothermally at the temperature of the liquid leaving the absorber.

(b) Calculate the number of theoretical plates for nonisothermal absorption by graphical construction and compare the result with the algebraic method.

(c) Calculate the number of theoretical plates for the stripper by graphical construction and check the result with the algebraic method.

(d) Calculate the pounds of stripping steam required per pound of hexane recovered.

Basic data:

Properties of absorption oil: Boiling point 500°F (mean average); molecular weight 200; gravity 40° API; heat capacity at 140°F, 0.547 Btu/(lb) (°F); heat capacity at 230°F, 0.592 Btu/(lb) (°F) (may be assumed linear in temperature)

Equilibrium constants at one atmosphere are as follows:

| Temperature, °F | $K$ |
| --- | --- |
| 100 | 0.35 |
| 150 | 0.90 |
| 200 | 1.90 |
| 300 | 6.00 |

Properties of nitrogen: Heat capacity of nitrogen may be assumed constant at 6.98 cal/(g) (mole) (°C).

Simplifying assumptions: Enthalpies of all components, liquid or gas, may be assumed additive; heats of mixing may be assumed negligible; thermal equilibrium may be assumed to be reached between gas and liquid on each tray.

# 10

# DESIGN OF FIXED-BED SORPTION AND ION EXCHANGE DEVICES

## 10.0  INTRODUCTION AND SCOPE

The adsorption of molecules from a gas mixture or a liquid solution onto the internal surface of a porous solid offers many opportunities for the purification of process streams, as in the drying of gases, and for the recovery of valuable components, as in the treatment of solutions with "molecular sieves." Although it is sometimes possible to use countercurrent flow of fluid and solid phases [65], more often the solid particles are held in a fixed bed, the fluid feed being run through until the bed becomes nearly saturated and small quantities of the adsorbate begin to "break through." Then the bed must be regenerated to restore its adsorptive capacity and to recover the adsorbed material. Similarly, synthetic ion-exchange resins [42] or some naturally occurring clays will adsorb ions from aqueous solutions, these ions displacing other ions originally in the resin matrix until the resin becomes nearly saturated with the feed stream. Regeneration then follows.

In both operations, generally referred to as "sorption," the concentrations in fluid and solid phases inside the bed depend both on position and on time. A steady state is never reached in practical conditions because the bed loses its sorptive ability and permits feed to pass through unchanged in composition when equilibrium is achieved. Thus, the design of fixed-bed devices is mathematically more difficult than the design of continuous, steady-state absorption or distillation equipment. Nevertheless, a rather complete theory of fixed-bed phenomena exists. It requires a knowledge of phase equilibria and of interphase mass-transfer rates, just as steady-state devices do. The object of this section is to describe such information and to show how the theory can be used in typical design situations.

The fixed-bed operation known as "chromatography" can be dealt with by the existing sorption theory if the phase equilibrium can be represented by Henry's law. Largely for laboratory use as a convenient analytical instrument, a fixed bed is fed with a continuous stream of a sweep fluid into which a short pulse of an unknown mixture is injected. The components form different adsorbed bands on the solid, these bands moving at different speeds toward the outlet depending on their adsorption coefficients. The theory shows how to interpret the shapes and the times of appearance of the output pulses in terms of the phase equilibria and the rate phenomena.

For more complete discussions of these subjects reference may be made to Vermeulen, Hiester, and Klein in a recent edition of Perry's "Handbook" [65], to a collection on "Chromatography" by Heftman [73], and to discussions of ion-exchange phenomena by Kunin and Myers [42] and by Helfferich [32].

## 10.1 PRINCIPAL SYMBOLS

| | |
|---|---|
| $a$ | Interfacial area of solid phase per volume of bed |
| $b$ | Correction factor to obtain kinetic coefficients, equal to $\kappa[k_f{}^{-1} + (c_0/\rho_B q_m)k_p{}^{-1}]$ |
| $c$ | Concentration in fluid, moles per volume; $c_0$ is feed concentration for adsorption or concentration at equilibrium with initial solid for desorption; $c_i$ is interface value |
| $D_f$ | Fluid-phase diffusivity, cm$^2$/s |
| $D_p$ | Particle-phase diffusivity, cm$^2$/s |
| $d_p$ | Diameter of solid particles, cm |
| $f(c)$ | Equilibrium value of $q$; $f'(c) = dq^*/dc$ |
| $F$ | General driving force in Eq. (10.11) |
| $J(\alpha, \beta)$ | Function used in solution of the breakthrough function, defined by Eq. (10.34) |
| $K_A$ | Adsorption equilibrium constant |
| $K_f$ | Overall coefficient for fluid phase, cm/s |
| $k$ | General mass-transfer coefficient, moles per area per time, in Eq. (10.11) |
| $k_f$ | Mass-transfer coefficient for fluid phase, cm/s |
| $k_p$ | Mass-transfer coefficient for solid phase, cm/s |
| $m$ | limiting slope of adsorption isotherm, $dq^*/dc$, or $q_m K_A/kT$, cm$^3$ g$^{-1}$ |
| $n$ | Distance or number of transfer units, $\kappa a x/v\varepsilon$ |
| $nT$ | Dimensionless time, $\kappa a c_0 t/q_m \rho_B$ |
| Pe | Peclet number, $d_p U/D_f$ |
| $p$ | Partial pressure of adsorbable component of gas |
| $q$ | Solid-phase concentration of adsorbed material, moles per mass; $q_m$ is the maximum value corresponding to complete coverage of surface by a monolayer; $q_0$ is value of $q$ at equilibrium with feed fluid concentration $c_0$; $q^*$ in equilibrium with concentration $c$ or partial pressure $p$; $q_i$ is interface value |

| | |
|---|---|
| $r$ | Equilibrium constant, equal to $(k + K_A p_0)^{-1}$ for gas adsorption or to $K^{-1}$ for ion exchange |
| $T$ | $v \varepsilon c_0 \hat{t} / q_0 \rho_B x$ equal to moles introduced with feed $\div$ moles needed to saturate bed to distance $x$ |
| $t$ | Time, s |
| $\hat{t}$ | Time following the arrival of a fluid particle, equal to $t - (x/v)$, s |
| $U$ | Superficial fluid velocity, $v \varepsilon$, cm/s |
| $u$ | Velocity of propagation of a constant pattern through the bed, cm/s |
| $v$ | Velocity of fluid in interstices of bed; note that $\varepsilon v$ is superficial fluid velocity, cm/s |
| $v^*$ | Concentration propagation velocity, cm/s |
| $x$ | Axial distance along bed, cm; $x_2$ is bed length |

## Greek Symbols

| | |
|---|---|
| $\varepsilon$ | Fraction void space in packed bed |
| $\theta$ | Fraction of surface covered by adsorbed molecules |
| $\kappa$ | Kinetic mass-transfer coefficient in Thomas equation, (10.27); equal to $K_f$ when $K = 1$, cm/s |
| $\rho_B$ | Bulk density, mass of solid phase $\div$ volume of bed, g/cm$^3$ |

## 10.2   FLUID-SOLID EQUILIBRIA

Many solid adsorbents have a limited total internal surface area on which volatile molecules can be held, limiting the total amount of material that can be held on the surface if only a single layer of adsorbed molecules can form. (This is especially true for ion-exchange resins, which contain only a fixed number of chemically active functional groups per unit mass.) For a solid surface that has uniform properties and therefore holds each adsorbed molecule as tightly as all the others present, the equilibrium between adsorbed molecules and *gas* molecules can be found, following Langmuir [44], by equating the rate of capture of molecules from the gas to the rate of escape of molecules from the surface. The capture rate is assumed proportional to the fraction $1 - \theta$ of the area that is not covered:

$$kp(1 - \theta) = k'\theta \tag{10.1}$$

where $k$ and $k'$ are forward and reverse rate constants and $p$ is the gas partial pressure of the adsorbate molecules. Solving for $\theta$ and dividing by the surface area occupied by one molecule give the number of molecules adsorbed per unit total surface. From this the equilibrium number of moles of adsorbate per unit mass of solid, $q^*$, is easily found.

$$q^* = q_m \frac{K_A p}{1 + K_A p} \tag{10.2}$$

where $q_m$ represents the capacity of the solid phase, corresponding to complete coverage of the surface. Equation (10.2) is represented by a curved line on $q^*p$ coordinates, which approaches a horizontal tangent when $p$ becomes very great, as in Fig. 10.1. The initial slope of the curve is $K_A q_m$. When multilayer adsorption

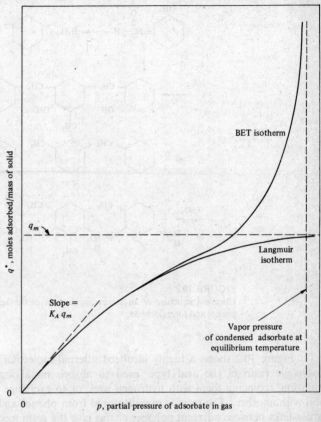

**FIGURE 10.1**
Types of gas adsorption equilibrium isotherms.

is assumed to occur, the second and subsequent layers being assumed to have the same properties that a condensed liquid phase would have in the absence of the solid, the adsorption isotherm behaves very differently at high partial pressures, as shown by Brunauer, Emmett, and Teller (BET) [13, 12] and as illustrated in Fig. 10.1. For many adsorption design problems, however, the simpler Langmuir formula is used very successfully.

Langmuir isotherms are sometimes applied to *liquid-solid* equilibria as well. Under such conditions the solvent liquid may fill the pores in the solid phase, contributing to the mass of the phase. Usually experimental measurements of $q$ express the number of moles of solute contained within the exterior outlines of the solid phase, whether the solute molecules are held on the solid's internal surface or are in solution in the liquid filling the pores. Likewise, the mass of the solid phase used for calculating $q$ may include the mass of liquid in the pores.

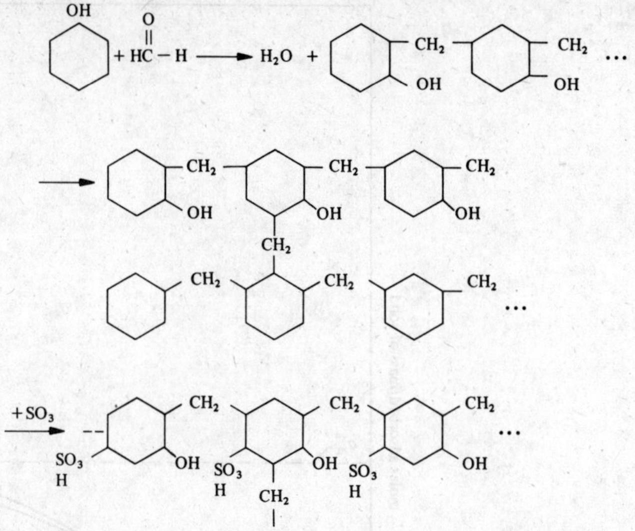

FIGURE 10.2
Chemical structure of an ion-exchange resin of the acid type, synthesized from phenol and formaldehyde.

Figure 10.2 shows a highly idealized internal molecular structure of an ion-exchange resin of the acid type, used to absorb metal ions from an aqueous solution, replacing them with hydrogen ions, or to exchange one type of positive ion with another. The resin is manufactured from phenol and formaldehyde. The cross-links between adjacent polymer chains give the resin mechanical stability and low solubility in most solvents. The acid functional groups are introduced by sulfonation of the polymer.

The sulfonic acid groups ionize easily, but being anchored to the polymer cage, they are immobile and hold their hydrogen ions nearby by electrostatic attraction. Many other chemical types are used as discussed by Kunin and Myers [42] and Helfferich [31]; these include basic resins using amine groups and weak-acid resins having carboxylic groups.

The ion-exchange reaction of a resin can be represented typically by the chemical equation

$$Na^+ + RH \longrightarrow H^+ + RNa$$

where R means the polymer structure associated with one acid-type functional group. Since the resin has a finite capacity, i.e., a fixed total number of acid sites per gram, the sum of the concentrations of RH and RNa is equal to a constant,

$$[RH] + [RNa] = q_m \tag{10.3}$$

Similarly, in the solution phase outside the resin, the total molar concentration of hydrogen and sodium is constant:

$$[H^+] + [Na^+] = c_0 \tag{10.4}$$

At equilibrium,

$$K = \frac{[H^+][RNa]}{[Na^+][RH]} = \frac{(c_0 - c)q^*}{c(q_m - q^*)} \tag{10.5}$$

where $c$ and $q^*$ represent the equilibrium concentrations of sodium in the fluid outside the resin particles and inside the particles, in moles or equivalents per unit volume and per unit mass, respectively. Solving Eq. (10.5) for $q^*$,

$$q^* = q_m \frac{Kc}{c_0 + (K - 1)c} \tag{10.6}$$

Note that Eq. (10.6) is remarkably similar to the Langmuir isotherm, Eq. (10.2). In fact, both equations can be expressed by the same formula. In Eq. (10.2) let $q_0^*$ be the value of $q^*$ corresponding to a fluid partial pressure $p_0$, selected in a particular design situation to correspond to some standard concentration such as that of a feed stream. Then from Eq. (10.2),

$$q_0^* = q_m \frac{K_A p_0}{1 + K_A p_0} \tag{10.7}$$

and Eq. (10.2) becomes

$$\frac{q^*}{q_0^*} = \frac{(1 + K_A p_0)(p/p_0)}{(1 + K_A p)} \tag{10.8}$$

and if we let $r = 1 + K_A p_0$, we can put the Langmuir formula in the same form as the formula for ideal-solution binary vapor-liquid equilibrium:

$$\frac{q^*}{q_0} = \frac{r(p/p_0)}{1 + (r - 1)(p/p_0)} \tag{10.9}$$

Obviously $r > 1$ for the Langmuir formula; the value of $r$ changes as the standard partial pressure varies. Moreover, Eq. (10.6) is of exactly the same form with $r = K$. Because of this similarity, many design problems involving ion exchange can be based on the same mass-transfer theory as problems in gas adsorption.

Expression of the equilibrium relationship according to the law of mass action in Eq. (10.5) assumes ideal-solution behavior in both fluid and resin phases and neglects any differences between ion concentrations inside and outside the particles. Such assumptions are shown to be very crude by Michaeil and Katchalsky [51], mostly because of strong effects of ionic charges in the resin structure. For polyvalent ions, even the mass action expression must be modified, as shown, for example, by Vasisth and David [63] for the exchange of ferric ion with hydrogen.

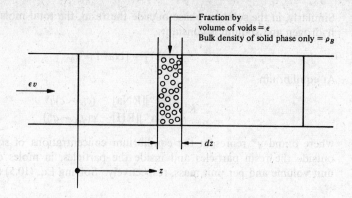

FIGURE 10.3
Diagram of a fixed-bed adsorber.

Nevertheless, use of Eq. (10.9) proves satisfactory in many design problems, probably because the value of $r$ is almost always found empirically and in many cases the equilibrium line need not be exact over more than a part of the full range, $0 \leq c \leq c_0$.

## 10.3 THE EQUATIONS OF TRANSPORT

Referring to Fig. 10.3, we see that a material balance on the fluid and solid phases contained within the section $dx$ shows that

$$\varepsilon \frac{\partial c}{\partial t} + \rho_B \frac{\partial q}{\partial t} + \varepsilon v \frac{\partial c}{\partial x} = 0 \qquad (10.10)$$

where $c$ and $q$ are fluid and solid concentrations, respectively, $\varepsilon$ is the fraction of fluid-filled space outside the particles, and $\rho_B$ is the bulk density of solid particles in the tube, including any fluid which the particles may contain in their pores. Note that $v$ is the average fluid velocity in the interstices between the particles; $\varepsilon v$ is the superficial velocity based on an empty tube. Longitudinal diffusion along the axis of the tube is neglected and "piston flow" of the fluid is assumed.

Equation (10.10) must be coupled with another equation representing the behavior of the solid phase alone:

$$\rho_B \frac{\partial q}{\partial t} = kaF(c, q) \qquad (10.11)$$

This equation expresses the rate of addition of solute to the solid phase in terms of the interphase mass-transfer coefficient, $ka$, and a driving force, $F(q, c)$, to be selected. Following the procedure used for packed columns operating at steady state, $k$ is based on a unit of exterior particle surface and $a$ is the total surface in a unit volume of packed space.

Design of a fixed-bed device for sorption or ion exchange requires solution of these two equations, given the fluid concentration at the entrance to the bed and the initial state of the fluid and solid in the bed. Solutions have been obtained for a variety of situations, as described by Vermeulen, Klein, and Hiester [65]. They are moderately complex, however. Before we outline some of the most important results, it will be helpful to think about the behavior of the packed bed in an extreme case in which $k$ is infinitely great. Then *local equilibrium* exists at all points and all times between the particles and the adjacent fluid. Under these conditions, $q = q^* = f(c)$, as expressed, for example, by Eq. (10.9). Then Eq. (10.11) is not needed and Eq. (10.10) becomes

$$\left[1 + \frac{\rho_B}{\varepsilon} f'(c)\right] \frac{\partial c}{\partial t} + v \frac{\partial c}{\partial x} = 0 \qquad (10.12)$$

## 10.4 THE LOCAL-EQUILIBRIUM THEORY OF FIXED-BED DEVICES

Equation (10.12) is a first-order partial differential equation that is linear in the derivatives but has a variable coefficient because of $f'(c)$. Such equations have remarkably simple geometrical properties as expressed by a solution procedure known as the "method of characteristics" [10]. Suppose we assume that a solution $c = c(x, t)$ has been found and ask ourselves if we can compute values of the two partial derivatives from Eq. (10.12) and from the solution itself. For this purpose we would want to solve (10.12) simultaneously with

$$\left(\frac{\partial c}{\partial t}\right) dt + \left(\frac{\partial c}{\partial x}\right) dx = dc \qquad (10.13)$$

The solution would be expressed by two equations, such as

$$\frac{\partial c}{\partial t} = \frac{\begin{vmatrix} 0 & v \\ dc & dx \end{vmatrix}}{\begin{vmatrix} 1 + \dfrac{\rho_B}{\varepsilon} f'(c) & v \\ dt & dx \end{vmatrix}} \qquad (10.14)$$

and a similar equation for $\partial c/\partial x$. Note that there are certain ratios of $dx$ to $dt$, that is, certain characteristic directions in the $xt$ plane, which make the denominator equal to zero. When this occurs, the numerator must also be zero if $\partial c/\partial t$ is to be finite. Then $dc$ will be zero. Thus, $c$ is constant along the "characteristic lines" corresponding to

$$\frac{dx}{dt} = \frac{v}{1 + (\rho_B/\varepsilon)f'(c)} \tag{10.15}$$

Since $c$ is constant, $f'(c)$ is constant and the characteristics are straight lines in the $xt$ plane. When such a line crosses the line $x = 0$, for example, the specified concentration at the bed's entrance is maintained for all other values of $x$ and $t$ along the same characteristic.

The method of characteristics brings out an important aspect of fixed-bed behavior: at least when there is efficient mass transfer between phases, fluid marked by a certain composition travels through the packing without a change in the concentration at a velocity $v^* = v/[1 + (\rho_B/\varepsilon)f'(c)]$, called the concentration wave velocity. The solvent fluid travels at velocity $v$, which is greater. The slower progress of the adsorbable molecules is owing to the fact that they spend a certain fraction of their time in a fixed state while the fluid is moving past.

If, depending on the shape of the equilibrium relationship and the range of concentrations involved, $f'(c)$ is nearly constant, the velocities will be equal for all concentrations and a wavefront will maintain its shape as it moves through the bed. On the other hand, if $v^*(c)$ varies, some concentrations will move faster than others and distortion of waveshapes will occur [28].

Consider the movement of a wavefront of concentration $c_0$ through a fixed adsorbent bed which is initially free of solute, where $c_0$ corresponds to a step change to concentration $c_0$ at the bed's entrance at time zero. Figure 10.4 shows the characteristic lines corresponding to an equilibrium curve similar to the Langmuir curve on Fig. 10.1. The horizontal axis represents the conditions at the entrance to the bed, where the concentration is constant $(c = c_0)$ after $t = 0$. The characteristics, such as $CD$ crossing the axis, all have the same steep slope corresponding to $v^* = v/[1 + (\rho_B/\varepsilon)f'(c_0)]$. [Note that $f'(c_0)$ is smaller than $f'(0)$ on Fig. 10.1.]

The line $OA$ has a slope equal to the fluid velocity $v$. To the left of this line, fluid initially in the packing has not yet been displaced by feed. Immediately to the right, the feed fluid has arrived but the concentration has not yet changed and is still equal to zero. All the characteristic lines, such as $AB$, crossing $OA$ have a smaller slope, $v^* = v/[1 + (\rho_B/\varepsilon)f'(0)]$, because $f'(0)$ is relatively large on Fig. 10.1.

The two families of characteristics cross each other on Fig. 10.4a. At such a point of intersection, the concentration has *two* values, $c_0$ and zero, and the solution of Eq. (10.12) is discontinuous. The concentration wave moves through the bed as a "shock wave" as in the upper part of Fig. 10.4a. The velocity of translation of this wave is found from a material balance. During the time $t$ required for the wave to emerge from the packing of depth $x$, the total amount of solute flowing in is $c_0 v\varepsilon t$; none flows out. The quantity that accumulates in the bed is $\varepsilon c_0 x + \rho_B q_0^* x$. Equating the two quantities gives

$$\frac{x}{t} = \frac{v}{1 + (\rho_B/\varepsilon)(q_0^*/c_0)} \tag{10.16}$$

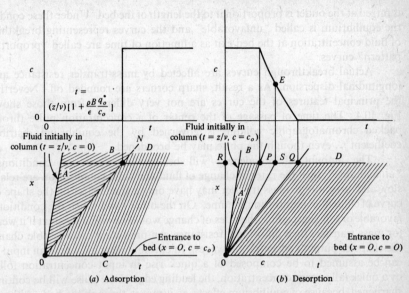

FIGURE 10.4

Graphical solution of fixed-bed sorption problems using the Langmuir equilibrium.

On the other hand, for desorption the slopes of characteristics crossing the two boundaries differ in the opposite direction, as shown on Fig. 10.4b. Then the regions of constant fluid concentration extend only from $R$ to $P$ and to the right of $Q$, as indicated in the figure. For times between points $P$ and $Q$, characteristics, such as $OS$, have slopes lying between the extremes. The concentrations at corresponding points, such as $E$, are found from the slope of lines such as $OS$, using

$$\frac{x}{t} = \frac{v}{1 + (\rho_B/\varepsilon)f'(c)} \tag{10.17}$$

to solve for $c$.

The properties of Eq. (10.12) have been explained in detail because they bring out an important feature of fixed-bed behavior and its relation to the shape of the equilibrium curve. Whenever the initial-fluid concentration in the bed and the feed concentrations are such that $f'(c_{feed}) < f'(c_{initial})$, the action of the bed on the concentration wave is to make it always steeper. Such behavior is called "self-sharpening" and the type of equilibrium line that permits this is called "favorable" or "constant-pattern." Then the concentration at the bed outlet varies abruptly from $c_{initial}$ to $c_{feed}$ at a time given by Eq. (10.16). When $f'(c_{initial}) > f'(c_{initial})$, as in Fig. 10.4b, the volume of fluid leaving the bed while the concentration is changing may be considerable. A concentration front that was sharp when it was introduced as a step function at $x = 0$ becomes diffuse and extended by mass transfer with the solid before the front leaves. As the construction shows, the volume of fluid which is mixed at the outlet is proportional to the length of the bed. Under these conditions the equilibrium is called "unfavorable" and the curves representing breakthrough of fluid concentration at the bed exit as a function of time are called "proportional-pattern" curves.

Actual breakthrough curves are affected by mass-transfer resistance and by longitudinal dispersion. As a result, sharp corners are rounded off. Nevertheless, the principal features of the curves are not very different from those shown in Fig. 10.4. The time of passage of the center of a concentration pulse through a packed chromatographic column is governed by the equilibrium distribution coefficient $f'$, even though the pulse may be broadened.

The amount of broadening will be smaller when the conditions are "unfavorable"; then the rates of change of fluid and solid composition are relatively slow and resistance to mass transfer may have only a small effect on the shape of the curve of effluent concentration vs. time. On the other hand, when the conditions are favorable or self-sharpening, the rates of change would be infinitely great if it were not for mass transfer; then diffusional resistance can produce a considerable change in the shape of output concentration waves. In chromatography, where an input pulse can be assumed to be composed of a quick rise in input concentration followed by a quick fall to zero concentration, the leading edge of the pulse will be continually sharpened, because of equilibrium effects, as it moves through the bed. Although it will be rounded by mass transfer, the leading edge of the pulse will emerge with a steep slope at the average concentration. The trailing edge, on the other hand, will be self-broadening and will trail out behind the emerging pulse for a long time, relatively. Such distortions of waveshape are the result of nonlinear response of the column arising from curvature of the equilibrium line and from large changes in composition. All abrupt changes are smoothed by diffusion, either between phases or in the fluid itself in the flow direction.

Adsorption responses tend to be much sharper than desorption. The slight broadening of a self-sharpening front depends on the mass-transfer coefficient but for a fixed value of the coefficient, the width of the front tends toward a constant value as the wave moves through the bed, called constant-pattern behavior. Even with mass-transfer resistance, the width of the diffuse part of a desorption curve is nearly proportional to the time of passage through the packing, retaining the proportional-pattern behavior expected if mass transfer were perfectly efficient. When the equilibrium line is straight ($f'(c) =$ constant), the width of the transition region of a breakthrough curve increases as the square root of the time of transit.

We shall see that these results are understandable in the following section, in which we return to the equations of transport and solve them using realistic expressions for the driving force, $F(c, q)$.

EXAMPLE 10.1   Estimate the values of $q^*$ as a function of $c$ for the absorption of $CO_2$ on porous carbon, using the desorption breakthrough data of Wicke [70]. The dimensions of the bed are length = 72 cm; bulk density = 0.36 g/cm$^3$; inside cross-sectional area of tube = 0.405 cm$^2$; fraction voids in bed = 0.345. The feed gas stream flowed at the rate 1.85 cm$^3$

(NTP)/s. The temperature was $0°C$ and the pressure was approximately 1 atm. The bed had been brought to equilibrium with a feed gas containing 100 mm partial pressure of $CO_2$. The coordinates of the breakthrough curve, abbreviated from a table given by Wicke [70], were as shown in the table.

| $t$, s | $c/c_0$ |
|---|---|
| 0 | 1.000 |
| 240 | 1.000 |
| 360 | 0.906 |
| 450 | 0.790 |
| 570 | 0.610 |
| 660 | 0.493 |
| 840 | 0.320 |
| 1,020 | 0.203 |
| 1,320 | 0.103 |
| 1,710 | 0.043 |

Figure 10.5a is a graph of the breakthrough data.

SOLUTION  Using the local-equilibrium theory for the desorption (unfavorable equilibrium) curve, the method of characteristics shows that at each point

$$\frac{t}{x} = \frac{1}{v^*} = \frac{1}{v}\left[1 + \frac{\rho_B}{\varepsilon}\frac{dq^*}{dc}\right]$$

or

$$\frac{dq^*}{dc} = \frac{\varepsilon}{\rho_B}\left(\frac{vt}{x} - 1\right)$$

$$= \frac{0.345}{0.36}\left[\frac{1.85}{(0.345)(0.405)(72)}t - 1\right]$$

$$= 0.1762t - 0.96 \text{ (g mole)}(cm^3)/(g) \text{ (g mole)}$$

As a sample calculation, take the point at $t = 840$ s, where $c = 0.320c_0 = (0.320)$ $\times (100/760)/(82.06)(273.1) = 0.188 \times 10^{-5}$ g mole/cm$^3$; $dq^*/dc = 147.0$ cm$^3$/g. Similar values of the slope of the equilibrium line can be found at all the concentrations. Starting at $c = 0$, where $q^* = 0$, values of $q^*(c)$ can be found by numerical integration. The following table shows the results.

| $c$, g mole/cm$^3$ | $q^*$, g mole/g |
|---|---|
| 0 | 0 |
| $0.025 \times 10^{-5}$ | $0.081 \times 10^{-3}$ |
| $0.060 \times 10^{-5}$ | $0.185 \times 10^{-3}$ |
| $0.119 \times 10^{-5}$ | $0.304 \times 10^{-3}$ |
| $0.188 \times 10^{-5}$ | $0.415 \times 10^{-3}$ |
| $0.289 \times 10^{-5}$ | $0.548 \times 10^{-3}$ |
| $0.358 \times 10^{-5}$ | $0.618 \times 10^{-3}$ |
| $0.464 \times 10^{-5}$ | $0.712 \times 10^{-3}$ |
| $0.532 \times 10^{-5}$ | $0.760 \times 10^{-3}$ |
| $0.587 \times 10^{-5}$ | $0.789 \times 10^{-3}$ |

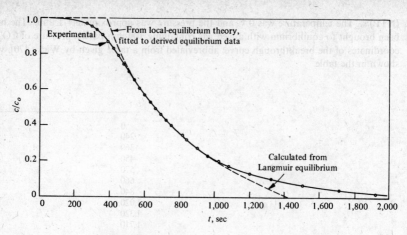

**FIGURE 10.5a**
Desorption breakthrough curve for $CO_2$ on carbon; data of Wicke [70].

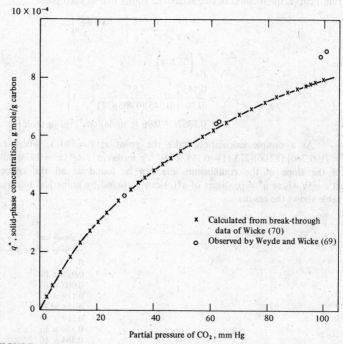

**FIGURE 10.5b**
Equilibrium adsorption of $CO_2$ on carbon. Comparison of values directly observed at 0°C with those derived from desorption breakthrough data.

Figure 10.5b is a plot of the derived equilibrium data also showing a few points measured directly by Weyde and Wicke [69], using conventional phase-equilibrium methods. The agreement between the calculated and observed values is remarkably good at low values of $c$, corresponding to $c/c_0$ smaller than about 0.5. The deviations at the larger concentrations is due to the failure of the local equilibrium theory near the point on Fig. 10.5a at which the breakthrough curve first turns downward. If the calculated equilibrium data are fitted by a Langmuir formula,

$$q^* = 1.54 \times 10^{-3} \frac{0.955 \times 10^{-2}p}{1 + 0.955 \times 10^{-2}p}$$

and the local-equilibrium breakthrough curve is computed, the dashed line in Fig. 10.5a is found. Note the very sudden change in slope of the calculated curve near $t = 400$ s, which is in contrast with the rounded experimental curve in the same region. The gradual changes in slope observed are due to the effect of interfacial mass-transfer resistance. However, when the equilibrium theory predicts that fluid and solid compositions should change gradually, mass transfer between phases can take place sufficiently fast and the local-equilibrium theory is more accurate.

The failure of the Langmuir isotherm to represent the data at long times is probably owing to deficiency of the Langmuir formula at the lowest values of $c$. The inference is that another formula, such as that of Freundlich or a combination of the two, may be superior [69].

////

## 10.5 THE EFFECTS OF MASS-TRANSFER RESISTANCE BETWEEN FLUID AND SOLID PHASES

Although we shall want to investigate the solutions of Eqs. (10.10) and (10.11) as written, it will be helpful first to think about the shape of a breakthrough curve after the mutual interaction of self-sharpening and of mass-transfer resistance have led to a constant pattern. Figure 10.6 shows some experimental adsorption breakthrough curves observed by Weyde and Wicke [69] using a bed of carbon particles to adsorb carbon dioxide from nitrogen. Note that following the introduction of a positive step change in gas concentration at the entrance to the tube, the patterns become virtually constant after bed lengths of about 18 cm. (The lower part of the figure shows similar data for desorption under the same conditions, demonstrating the effects of an unfavorable equilibrium situation and proportional-pattern behavior.)

Constancy of the breakthrough pattern implies that the concentration can be expressed as a function of the single variable $\hat{t} = t - x/u$, where $u$ is the velocity at which the pattern moves along the bed. Under these conditions Eq. (10.10) reduces to the ordinary differential equation

$$\left(1 - \frac{v}{u}\right)\frac{dc}{d\hat{t}} + \frac{\rho_B}{\varepsilon}\frac{dq}{d\hat{t}} = 0 \tag{10.18}$$

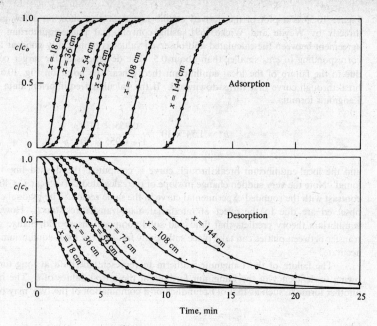

**FIGURE 10.6**
Breakthrough curves for carbon dioxide on solid carbon. Effect of depth of packing on adsorption (above) and desorption (below) using $v\varepsilon = 4.26$ cm/s (from Weyde and Wicke [69]).

Noting that $u < v$, this implies that

$$\left(\frac{v-u}{u}\right)c - \frac{\rho_B}{\varepsilon} q = \text{constant} = 0 \qquad (10.19)$$

where the last equality follows from the fact that initial values of $c$ and $q$ are both zero. Moreover, the equation must apply when the bed is saturated and $q_0 = q^*$ with $c = c_0$. Thus, if constant-pattern conditions exist,

$$\frac{q}{q_0} = \frac{c}{c_0} \qquad (10.20)$$

and the rate equation, Eq. (10.11), becomes

$$\rho_B \frac{q_0}{c_0} \frac{dc}{d\hat{t}} = kaF\left(c, \frac{q_0}{c_0} c\right) \qquad (10.21)$$

from which the breakthrough curve $c(t - x/u)$ can be found once the driving force $F$ is specified.

As an example, consider the calculations when the major resistance to mass transfer is in the fluid phase outside the solid particles and $kaF(q, c) = ka(c - c^*)$ where $c^*$ is computed from the Langmuir isotherm:

$$\frac{c^*}{c_0} = \frac{q/q_0}{1 + K_A p_0(1 - q/q_0)} = \frac{c/c_0}{1 + K_A p_0(1 - c/c_0)} \tag{10.22}$$

$$\frac{\rho_B q_0}{c_0} \frac{dc}{d\hat{t}} = ka \frac{K_A p_0}{1 + K_A p_0(1 - c/c_0)} c(1 - c/c_0) \tag{10.23}$$

Integrating between limits starting at the point where $c = c_0/2$ where $\hat{t} = \hat{t}_{1/2}$, we have

$$\frac{1}{K_A p_0} \ln\left(\frac{c}{c_0 - c}\right) + \ln\left(\frac{2c}{c_0}\right) = \frac{kac_0}{\rho_B q_0}(\hat{t} - \hat{t}_{1/2}) \tag{10.24}$$

Equation (10.24) is similar to one given by Glueckauf and Coates [26]. Note that the right side of the equation can be written as

$$\frac{kac_0(\hat{t} - \hat{t}_{1/2})}{\rho_B q_0} = \frac{kax_2}{\varepsilon v} \frac{t - t_{1/2}}{(\rho_B q_0 x_2/\varepsilon v c_0)} \tag{10.25}$$

The first factor, $kax_2/\varepsilon v$, is the number of transfer units in the bed; the denominator, $\rho_B q_0 x_2/\varepsilon v c_0$, is the time required for the input flow to supply the bed with an amount of adsorbate equal to the bed's capacity. From Eq. (10.23), the slope of the breakthrough curve at the midpoint, where $c = c_0/2$, is

$$\left.\frac{d(c/c_0)}{dt}\right|_{c=c_0/2} = \left(\frac{kax}{\varepsilon v}\right)\left(\frac{\varepsilon v c_0}{\rho_B q_0 x}\right)\frac{(K_A p_0/4)}{1 + (K_A p_0/2)} \tag{10.26}$$

**EXAMPLE 10.2**  Find the shape of a constant-pattern breakthrough curve for the adsorption of benzene vapor from nitrogen on charcoal at atmospheric pressure and 20°C. The partial pressure of benzene with the feed gas is 0.9 mm. The adsorption equilibrium constant is 0.33 mm$^{-1}$ and the maximum amount of benzene corresponding to a monolayer on the internal carbon surface is 0.007 g mole/g carbon. The bed is composed of porous carbon particles 2 mm in diameter, is 100 cm long, and has a bulk density of 0.36 g/cm$^3$ and a porosity of 0.4. The superficial velocity of fluid passing through the bed is 15 cm/s. Find the time at which breakthrough of benzene occurs at a partial pressure equal to one-twentieth of the feed value.

SOLUTION  At $p_0 = 0.9$ mm, $K_A p_0 = 0.3$, Eq. (10.24) can be used to find the shape of the curve, as shown in Fig. 10.7.

To relate the dimensionless horizontal coordinate on Fig. 10.7 to time, we need values of the mass-transfer coefficient. The Reynolds number based on the particle diameter is

$$\frac{d_p \varepsilon v}{v} = \frac{(0.2)(15)}{(0.148)} = 20.3$$

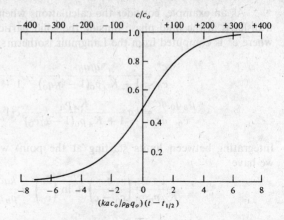

**FIGURE 10.7**
Calculated constant-pattern breakthrough. Langmuir isotherm with $K_A p_0 = 0.3$ and mass-transfer resistance in fluid phase.

From Fig. 6.17, $j_D$ is $(k/\varepsilon v)(Sc)^{2/3} = 0.35$. From Table 2.4 the Schmidt number is

$$\frac{v}{D} = \frac{0.148}{0.102}\left(\frac{293}{311.3}\right)^{1.75} = 1.61$$

and

$$\frac{k}{\varepsilon v} = \left(\frac{0.35}{1.61}\right)^{2/3} = 0.255$$

The particle surface area per unit volume of solid is $\pi d_p^2/(\pi d_p^3/6) = 6/d_p = 30 \text{ cm}^2/\text{cm}^3$ and $a = 30(1 - 0.4) = 24 \text{ cm}^2/\text{cm}^3$, giving

$$\frac{kax_2}{\varepsilon v} = (0.255)(24)(100) = 612 \text{ transfer units}$$

The time required to saturate the bed with the input stream requires a knowledge of

$$q_0^* = 0.007 \frac{0.3}{1 + 0.3} = 0.00161 \text{ g mole/g carbon}$$

and

$$c_0 = \frac{p_0}{RT} = \frac{0.9}{(82.06)(293)(760)} = 0.492 \times 10^{-7} \text{ g mole/cm}^3$$

Thus,

$$\frac{\rho_B q_0^* x_2}{\varepsilon v c_0} = \frac{(0.36)(1.61 \times 10^{-3})(100)}{(15)(0.492 \times 10^{-7})} = 78,500$$

and one unit on the abscissa of Fig. 10.7 corresponds to $78,500/612 = 128$ s. Any decrease in the overall mass-transfer coefficient because of diffusional resistance inside the solid particles will cause this figure to increase.

The time corresponding to $c = \frac{1}{2}c_0$ is very nearly equal to the time required to saturate the bed because the emission of benzene before $t_{1/2}$ is nearly equal to the deficiency of benzene in the bed near the exit. Thus $t_{1/2} \approx 78,500$ s. The time at which benzene vapor appears in the exit stream at a partial pressure of 0.045 mm (one-twentieth the partial pressure in the feed) is $78,500 - (128)(5.2) = 77,830$ s. (The abscissa on Fig. 10.7 is $-5.2$ at $c/c_0 = 0.05$.)    ////

## 10.6 THE THOMAS SOLUTION FOR SORPTION BREAKTHROUGH CURVES

The most useful treatment of the sorption design problem is that of Thomas [61]. It takes into account the curved shape of the equilibrium relationship, which has an important bearing on the shapes of breakthrough curves, as we have seen. Thomas assumed that the rate of adsorption on the right of Eq. (10.11) can be represented by an expression suggested by the stoichiometry of the monovalent ion-exchange reaction

$$Na^+ + RH \longrightarrow H^+ + RNa$$
$$c \;, q_m - c \qquad\qquad c_0 - c \;, q$$

Equation (10.11) then becomes

$$\rho_B \frac{\partial q}{\partial t} = \kappa a \left[ c \left( 1 - \frac{q}{q_m} \right) - \frac{1}{K} (c_0 - c) \frac{q}{q_m} \right] \qquad (10.27)$$

The expression in square brackets is called the "kinetic driving force" [65] and $\kappa$ is the corresponding "kinetic coefficient." The rate expression includes the factor $a$, because despite the implication of the terms in the square bracket, actual rates are almost always diffusion-controlled and must depend on the surface area available for interphase transfer, as Thomas recognized. Although it is possible to treat the motion of ions inside resin particles [31] or adsorbed molecules on solid surfaces or in pores by logically correct diffusion equations [22, 45, 55], solution of the resulting equations is much more difficult that for equations such as (10.27) in which an overall driving force is used. Moreover, Eq. (10.27) has in its favor the fact that the phase equilibrium is represented accurately either for Langmuir adsorption or for the simplest cases of an ion exchange. As will be shown, it is possible to relate $\kappa$ to individual mass-transfer coefficients representing diffusional resistances in fluid and solid phases.

   Equation (10.27) is simplified for further use by redefining the time scale by

$$\hat{t} = t - \frac{x}{v} \qquad (10.28)$$

such that $\hat{t}$ is the elapsed time at a point $x$ after a fluid particle has arrived at the point, having started at the entrance to the bed at $t = 0$. The initial condition of the fluid and solid phases in the column is easily specified in terms of $\hat{t}$ rather than $t$. Equations (10.10) and (10.27) become

$$\frac{\rho_B}{\varepsilon} \frac{\partial q}{\partial \hat{t}} + v \frac{\partial c}{\partial x} = 0 \tag{10.29}$$

$$\rho_B \frac{\partial q}{\partial \hat{t}} = \kappa a \left[ c \left( 1 - \frac{q}{q_m} \right) - \frac{1}{K} \frac{q}{q_m} (c_0 - c) \right] \tag{10.30}$$

Consider an operation in which a bed of ion-exchange resin, initially in the hydrogen form throughout, that is, $q(x, 0) = 0$, is fed at $t = 0$ with a solution containing $c_0$ moles per cubic centimeter of NaCl. Eventually the whole bed will come to equilibrium with the feed solution, which contains no $H^+$. Then the resin will contain $q_m$ moles per gram of $Na^+$, equal to the "capacity" of the resin. The resin will have been converted completely to the sodium form. Then the value of $c/c_0$ in the effluent will be equal to unity. The gradual breakthrough of sodium ion, i.e., the function $c(x, t)/c_0$, and the gradual accumulation of $Na^+$ on the resin, that is, $q(x, t)$, are the aims of the mathematical development which follows.

Thomas [61] was able to reduce Eqs. (10.29) and (10.30) to a linear equation by introducing a transformation of the dependent variables. His solution for adsorption boundary conditions on $c(x, \hat{t})$ and $q(x, \hat{t})$, such that $c(0, \hat{t}) = c_0 = $ constant and $q(x, 0) = 0$, can be expressed as

$$\frac{c}{c_0} = \frac{J(n/K, nT)}{J(n/K, nT) + [1 - J(n, nT/K)] \exp{[(1 - K^{-1})(n - nT)]}} \tag{10.31a}$$

$$\frac{q}{q_m} = \frac{1 - J(nT, n/K)}{J(n/K, nT) + [1 - J(n, nT/K)] \exp{[(1 - K^{-1})(n - nT)]}} \tag{10.31b}$$

where

$$n = \frac{\kappa a x}{v \varepsilon} = \text{dimensionless distance or number of transfer units to position } x \tag{10.31c}$$

$$T = \frac{v \varepsilon c_0 \hat{t}}{q_m \rho_B x} \tag{10.32}$$

$$nT = \frac{\kappa a c_0 \hat{t}}{q_m \rho_B} = \text{dimensionless time} \tag{10.33}$$

$$J(\alpha, \beta) = 1 - e^{-\beta} \int_0^\alpha e^{-\xi} I_0(2\sqrt{\beta \xi}) \, d\xi \tag{10.34}$$

$T$ is the ratio of elapsed time $\hat{t}$ to the time needed for the input flow to saturate the bed of the depth $x$. Table 10.1 [74] lists values of the $J$ function.

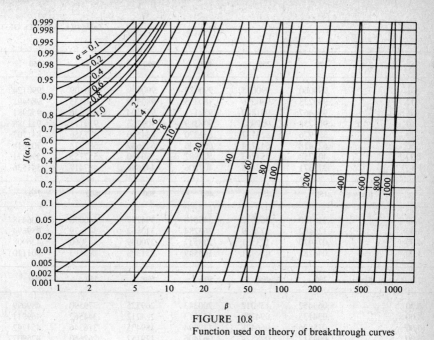

**FIGURE 10.8**

Function used on theory of breakthrough curves

According to Thomas [61], values of $J$ can be found from an asymptotic series when $\alpha$ and $\beta$ are large. The first two terms are

$$J(\alpha, \beta) \approx \frac{1}{2}\left[1 - \text{erf}\left(\sqrt{\alpha} - \sqrt{\beta}\right)\right] + \frac{\exp\left[-(\sqrt{\alpha} - \sqrt{\beta})^2\right]}{2\pi^{1/2}[(\alpha\beta)^{1/4} + \beta^{1/2}]} \quad (10.35)$$

The error is less than 1 percent when $\alpha\beta > 36$. When $\alpha\beta > 3{,}600$, the second term in Eq. (10.35) can be dropped and the Klinkenberg approximation [40], can be used:

$$J(\alpha, \beta) \approx \begin{cases} \frac{1}{2}\,\text{erfc}\left(\sqrt{\alpha} - \sqrt{\beta}\right) & \text{if } \beta < \alpha \quad (10.36a) \\ \frac{1}{2}\left[1 + \text{erf}\left(\sqrt{\alpha} - \sqrt{\beta}\right)\right] & \text{if } \beta > \alpha \quad (10.36b) \end{cases}$$

Figure 10.8, from Vermeulen, Klein, and Hiester [65], shows the behavior of the $J$ function over the full range of $n$ and $\theta$. Numerical values of $J$ have been listed by Schumann [58], by Furnas [23], and by Hougen and Marshall [48]. Figure 10.9 shows breakthrough curves calculated from Eq. (10.35) for a column having $\kappa ax/v\varepsilon = 40$ transfer units, the different lines corresponding to different values of the equilibrium constant $K$. The tendency for the breakthrough to be self-sharpening when the equilibrium is "favorable" and $K > 1$ is apparent; conversely, when $K < 1$ the duration of the flow from the bed while the concentration lies between $c_0$ and 0 is greater, the equilibrium being "unfavorable." Gilliland and Baddour [24] have shown mathematically that Eq. (10.31) tends toward constant-pattern or proportional-pattern behavior in the limit as $\kappa x/v\varepsilon$ becomes very large. Figure 10.8 [65]

Table 10.1   VALUES OF THE

| $\alpha$ | $\beta/\alpha =$ | .100 | .250 | .400 | .500 | .600 | .750 |
|---|---|---|---|---|---|---|---|
| .01 | | .990060 | .990075 | .990089 | .990099 | .990109 | .990124 |
| .02 | | .980238 | .980296 | .980355 | .980394 | .980433 | .980491 |
| .05 | | .951467 | .951820 | .952171 | .952404 | .952636 | .952981 |
| .10 | | .905738 | .907073 | .908389 | .909256 | .910115 | .911388 |
| .20 | | .821976 | .826737 | .831370 | .834390 | .837356 | .841707 |
| .50 | | .621418 | .642714 | .662846 | .675649 | .687978 | .705618 |
| 1.00 | | .403758 | .454263 | .500982 | .530130 | .557755 | .596471 |
| 1.50 | | .272371 | .342458 | .407806 | .448686 | .487417 | .541536 |

| $\alpha$ | $\beta/\alpha =$ | .150 | .250 | .400 | .500 | .600 | .750 |
|---|---|---|---|---|---|---|---|
| 2.00 | | .216160 | .269012 | .345570 | .394297 | .440890 | .506438 |
| 3.00 | | .123462 | .177849 | .263284 | .320862 | .377679 | .459694 |
| 4.00 | | .074470 | .123381 | .208537 | .270039 | .333060 | .426908 |
| 5.00 | | .046288 | .087791 | .168569 | .231308 | .298193 | .401110 |

| $\alpha$ | $\beta/\alpha =$ | .250 | .400 | .500 | .600 | .750 | .900 |
|---|---|---|---|---|---|---|---|
| 6.00 | | .063452 | .138012 | .200343 | .269525 | .379590 | .489099 |
| 8.00 | | .034135 | .094790 | .153513 | .224212 | .344562 | .469911 |
| 10.00 | | .018826 | .066478 | .119794 | .189435 | .316346 | .454742 |
| 15.00 | | .004517 | .028824 | .067406 | .129152 | .262660 | .425907 |
| 20.00 | | .001137 | .013028 | .039345 | .090853 | .223017 | .403988 |

| $\alpha$ | $\beta/\alpha =$ | .400 | .500 | .600 | .700 | .800 | .900 |
|---|---|---|---|---|---|---|---|
| 30.00 | | .002838 | .014187 | .047206 | .116138 | .226828 | .370308 |
| 40.00 | | .000647 | .005334 | .025436 | .080791 | .188063 | .344034 |
| 50.00 | | .000151 | .002055 | .014001 | .057202 | .157981 | .322131 |
| 60.00 | | .000036 | .000804 | .007817 | .040984 | .133892 | .303199 |
| 80.00 | | .000002 | .000127 | .002506 | .021542 | .097905 | .271432 |
| 100.00 | | .000000 | .000021 | .000823 | .011563 | .072748 | .245286 |

| $\alpha$ | $\beta/\alpha =$ | .700 | .750 | .800 | .850 | .900 | .950 |
|---|---|---|---|---|---|---|---|
| 150.00 | | .002574 | .011025 | .036085 | .093132 | .195121 | .341156 |
| 200.00 | | .000599 | .003993 | .0185 4 | .062408 | .158496 | .315230 |
| 300.00 | | .000035 | .000558 | .005167 | .029348 | .108222 | .274405 |
| 400.00 | | .000002 | .000081 | .001502 | .014297 | .075885 | .242506 |

| $\alpha$ | $\beta/\alpha =$ | .880 | .900 | .920 | .940 | .960 | .980 |
|---|---|---|---|---|---|---|---|
| 500.00 | | .026089 | .054077 | .101132 | .171744 | .266664 | .381352 |
| 600.00 | | .016588 | .038984 | .080789 | .149020 | .246575 | .369321 |
| 800.00 | | .006874 | .020684 | .052553 | .113935 | .213151 | .348469 |
| 1000.00 | | .002913 | .011201 | .034783 | .088342 | .186127 | .330608 |

\* Note: Mathematical properties of the $J$ function include: $(a)$ $\partial J(\alpha, \beta)/\partial \alpha = -e^{-(\alpha+\beta)}I_0(2\sqrt{\alpha\beta})$;

## FUNCTION $J(\alpha, \beta)$ FROM VERMEULEN [74]*

| $\beta/\alpha =$ | 1.000 | 1.300 | 1.600 | 2.000 | 3.000 | 5.000 | 10.000 |
|---|---|---|---|---|---|---|---|
| | .990148 | .990178 | .990207 | .990246 | .990342 | .990533 | .990992 |
| | .980587 | .980702 | .980816 | .980968 | .981341 | .982065 | .983756 |
| | .953550 | .954225 | .954890 | .955761 | .957866 | .961782 | .970051 |
| | .913469 | .915903 | .918268 | .921319 | .928457 | .940850 | .963242 |
| | .848701 | .856689 | .864257 | .873731 | .894633 | .926654 | .970404 |
| | .732880 | .762353 | .788641 | .819310 | .878175 | .945102 | .992819 |
| | .654254 | .713597 | .763436 | .817415 | .906137 | .976650 | .999427 |
| | .621500 | .701797 | .766968 | .834124 | .932268 | .990212 | .999952 |

| $\beta/\alpha =$ | 1.000 | 1.300 | 1.600 | 2.000 | 3.000 | 4.000 | 5.000 |
|---|---|---|---|---|---|---|---|
| | .603501 | .699994 | .776415 | .851936 | .951231 | .985277 | .995835 |
| | .583329 | .705249 | .797972 | .882809 | .974356 | .995245 | .999218 |
| | .571716 | .714027 | .818302 | .906894 | .986272 | .998423 | .999848 |
| | .563917 | .723595 | .836450 | .925608 | .992550 | .999468 | .999970 |

| $\beta/\alpha =$ | 1.000 | 1.250 | 1.500 | 1.750 | 2.000 | 2.250 | 2.500 |
|---|---|---|---|---|---|---|---|
| | .558213 | .707848 | .818671 | .893427 | .940254 | .967862 | .983331 |
| | .550272 | .722888 | .844408 | .919352 | .960987 | .982228 | .992320 |
| | .544890 | .737078 | .865779 | .938399 | .974205 | .990024 | .996401 |
| | .536573 | .768291 | .905528 | .967693 | .990498 | .997545 | .999434 |
| | .531639 | .794327 | .932278 | .982604 | .996385 | .999374 | .999907 |

| $\beta/\alpha =$ | 1.000 | 1.100 | 1.200 | 1.300 | 1.400 | 1.500 | 1.600 |
|---|---|---|---|---|---|---|---|
| | .525806 | .670489 | .788471 | .874175 | .930376 | .964009 | .982544 |
| | .522337 | .688375 | .817845 | .904254 | .954501 | .980334 | .992221 |
| | .519972 | .704361 | .841896 | .926278 | .969819 | .989065 | .996465 |
| | .518228 | .718806 | .861967 | .942753 | .979768 | .993846 | .998372 |
| | .515782 | .744095 | .893450 | .964861 | .990710 | .998002 | .999645 |
| | .514114 | .765715 | .916758 | .978073 | .995649 | .999337 | .999921 |

| $\beta/\alpha =$ | 1.000 | 1.050 | 1.100 | 1.150 | 1.200 | 1.250 | 1.300 |
|---|---|---|---|---|---|---|---|
| | .511521 | .675900 | .808830 | .900000 | .953594 | .980854 | .992955 |
| | .509977 | .697983 | .841489 | .929457 | .973369 | .991445 | .997650 |
| | .508145 | .734039 | .887913 | .963487 | .990799 | .998196 | .999722 |
| | .507053 | .762988 | .918911 | .980505 | .996701 | .999603 | .999966 |

| $\beta/\alpha =$ | 1.000 | .1020 | 1.040 | 1.060 | 1.080 | 1.100 | 1.120 |
|---|---|---|---|---|---|---|---|
| | .506309 | .629451 | .739535 | .829058 | .895458 | .940491 | .968486 |
| | .505759 | .640225 | .758137 | .850439 | .915146 | .955892 | .979005 |
| | .504987 | .659266 | .789456 | .883925 | .943116 | .975267 | .990465 |
| | .504461 | .675850 | .815068 | .908753 | .961252 | .985872 | .995580 |

(b) $\partial J(\alpha, \beta)/\partial \beta = e^{-(\alpha+\beta)}(\alpha/\beta)^{1/2}I_1(2\sqrt{\alpha\beta})$; (c) $J(\alpha, \beta) + J(\beta, \alpha) = 1 + e^{-(\alpha+\beta)}I_0(2\sqrt{\alpha\beta})$; (d) $J(\alpha, 0) = e^{-\alpha}$; (e) $J(0, \beta) = 1$.

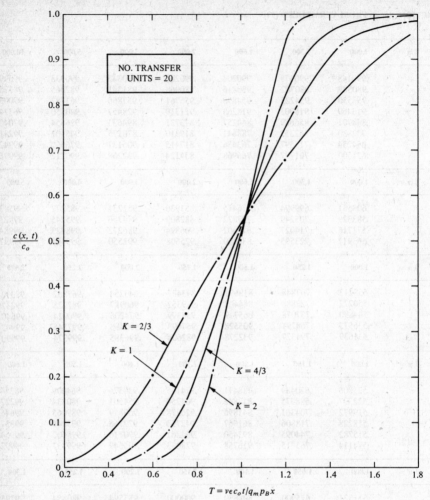

FIGURE 10.9
Effect of equilibrium constant on calculated breakthrough curves for 40 transfer units [61].

shows the varying shapes of breakthrough curves on log-probability coordinates for linear phase equilibrium. Elution of adsorbed material from a bed requires a change in the boundary conditions to $c(0, t) = 0$ and $q(x, 0) = q_m$ and yields

$$\frac{c}{c_0} = \frac{1 - J(n/K, nT)}{1 - J(n/K, nT) + [J(n, nT/K)] \exp\left[(K^{-1} - 1)n(T - 1)\right]} \quad (10.37)$$

These results are not limited to ion adsorption. They apply equally to the elution of $Na^+$ from a bed by feeding a solution of HCl to it, or to the adsorption or desorption involving any uniform initial solid loading and any constant-feed concentration. Gaseous and liquid adsorption and desorption are also included because of the similarity of many adsorption equilibrium isotherms, as in Eq. (10.2), to ion-exchange equilibria, as in Eq. (10.6). It is necessary in order to preserve the similarity of the formulas for the isotherms to replace $K$ in Eq. (10.27) by $r = 1 + K_A p_0$ and $q_m$ by $q_0$, the solid loading at equilibrium with the feed gas.

## 10.7 BREAKTHROUGH CURVES FOR LINEAR PHASE EQUILIBRIA AND CHROMATOGRAPHY

Inspection of the right side of Eq. (10.27) will show that if $K = 1$, the driving force becomes the linear expression $c - (c_0/q_0)q$. The solution, Eqs. (10.31) and (10.32), is much simplified, becoming

$$\frac{c}{c_0} = J(n, nT) \tag{10.38}$$

$$\frac{q}{q_0} = 1 - J(T, n) \tag{10.39}$$

If the value of $n$ is large and the Klinkenberg approximation [40] to the $J$ function is used, we find that

$$\frac{c}{c_0} \approx \frac{1}{2} \operatorname{erfc} \left( \sqrt{\frac{\kappa a x}{v \varepsilon}} - \sqrt{\frac{\kappa a c_0 \hat{t}}{\rho_B q_m}} \right) \tag{10.40}$$

It follows, from tables of the error function, that the time interval $\Delta t$ during which $c$ is changing from $0.01 c_0$ to $0.99 c_0$ is

$$\Delta t = t_{0.99} - t_{0.01} = 3.2894 \left( \frac{q_m \rho_B}{c_0} \right) \sqrt{\frac{x}{v \varepsilon \kappa a}} \tag{10.41}$$

or
$$\frac{t_{0.99} - t_{0.01}}{\hat{t}_{0.5}} = 3.2894 n^{-1/2} \tag{10.42}$$

In general, for any values of $n$ and $T$, the slope of the breakthrough curve is found by differentiation to be

$$\frac{\partial (c/c_0)}{\partial t} = \frac{\kappa a c_0}{q_m \rho_B} e^{-n(T+1)} \frac{I_1(2n\sqrt{T})}{\sqrt{T}} \tag{10.43}$$

When $n$ is sufficiently large, this can be approximated, using the asymptotic series for the Bessel function, by

$$\frac{\partial (c/c_0)}{\partial t} = \frac{1}{2\sqrt{\pi}} \frac{\kappa a c_0}{q_m \rho_B} \frac{e^{-(\sqrt{n} - \sqrt{nT})^2}}{(n^2 T^3)^{1/4}} \tag{10.44}$$

and when $T = 1$ (where $c$ is very nearly $c_0/2$), this reduces to

$$\frac{\partial(c/c_0)}{\partial t}\bigg|_{\bar{t}=q_0\rho_B x/v\varepsilon c_0} = \frac{v\varepsilon c_0}{2\sqrt{\pi}\rho_B q_0 x}\sqrt{n} = 0.28209\frac{c_0}{\rho_B q_0}\sqrt{\frac{v\varepsilon \kappa a}{x}} \qquad (10.45)$$

Note that the slope never becomes independent of $x$, as it would if $K > 1$ and a constant pattern developed, nor does it become inversely proportional to $x$, as if $K < 1$ and there were a proportionate pattern at large $x$. The conditions existing for a straight-line phase equilibrium are therefore rather special. They will occur rarely in real ion-exchange or sorption problems, most equilibria being curved. As a practical matter, therefore, behavior according to Eq. (10.38) is realized only when $c_0$ and $q_0$ are small enough for the equilibrium between phases to behave essentially according to Henry's law, or when the fluid causes only a small change in $q$.

EXAMPLE 10.3   An experiment has been conducted in which a gas mixture containing trace quantities of a radioactive component was passed at a constant rate through a column filled with activated carbon adsorbent particles. The level of radioactivity in the effluent gas was measured as a function of time, yielding the data given in the table.

| Time, s | Ratio, effluent-gas radioactivity: feed-gas radioactivity |
|---------|-----------------------------------------------------------|
| 540 | 0.041 |
| 615 | 0.134 |
| 689 | 0.303 |
| 763 | 0.518 |
| 838 | 0.719 |
| 912 | 0.862 |
| 987 | 0.943 |

The superficial fluid velocity was 2.1 cm/s. The carbon bed was 100 cm long and had a bulk density of 0.48 g/cm$^3$ and a void fraction of 0.41 cm$^3$ gas/cm$^3$ bed.

Estimate the equilibrium distribution coefficient and the mass-transfer coefficient between phases.

SOLUTION   Since the gas contains only trace quantities of the substance being measured, the changes in solid composition will all be very small and the equilibrium compositions will be confined to a short, nearly straight segment of the adsorption isotherm. Then the theory of the breakthrough curve for a straight equilibrium line can be used, i.e., Eq. (10.38). Computation of $t_{1/2}$ from the data will yield the distribution coefficient, $q_0^*/c_0$.

*First approximation, assuming $n \to \infty$.* In this limit, as Table 10.1 shows, the breakthrough curve crosses $c/c_0 = \frac{1}{2}$ when $T = 1$. From the graph of the data (Fig. 10.10), this occurred at

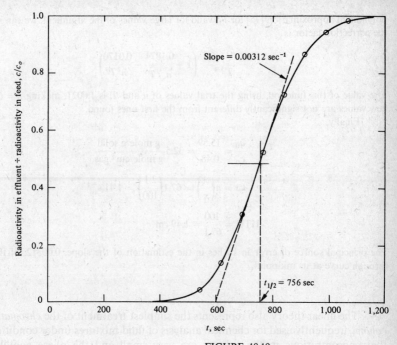

**FIGURE 10.10**
Plot of breakthrough data for Example 10.3.

$t_{1/2} = 756$ s. Since $x/v = 100/(2.1/0.41) = 19.5$ s, $\hat{t}_{1/2} = 756 - 19.5 = 736$ s. The solid–gas distribution ratio is found from

$$1 = T_{1/2} = \frac{v\varepsilon c_0 \hat{t}_{1/2}}{\rho_B q_0 x} = \frac{(2.1)(736)}{100} \frac{c_0}{\rho_B q_0}$$

from which $\rho_B q_0/c_0 = 15.47$ (g mole/cm³)/(g mole/cm³). Using this value in Eq. (10.45) for the slope at $\hat{t}_{1/2}$, we have

$$0.00312 = (0.28209)\left(\frac{2.1}{(100)(15.47)}\right)\sqrt{n}$$

for which $n = 66.3$ transfer units.

*Second approximation, using equations for finite n.* By interpolation in Table 10.1 at $n = 66.3$, we find that $T_{1/2} = 0.9921$ rather than unity. This makes $\rho_B q_0/c_0 = 15.59$. The revised value of $n$ is found from Eq. (10.43). After arranging it in the form

$$\frac{\partial(c/c_0)}{\partial t} = \frac{v\varepsilon c_0}{q_0 \rho_B x} \frac{1}{2}\sqrt{\frac{n}{\pi}}\left[\frac{2\sqrt{\pi}e^{-n(T+1)}I_1(2n\sqrt{T})}{\sqrt{T/n}}\right]$$

The factor in brackets is equal to unity when $n$ is infinite. Its numerical value for $n = 66.3$ and $T = 0.9921$ will provide a correction value for the previous estimate of $n$. Using a polynomial approximation [75] for $I_1$, valid for large values of the argument, we can show that the correction factor is

$$\frac{e^{-n(1-\sqrt{T})^2}}{T^{3/4}}\left(1 - \frac{0.1874}{n\sqrt{T}} - \frac{0.0170}{n^2 T}\right)$$

The value of this function, using the trial values of $n$ and $T$, is 1.0021, making $n = 67.1$. The new values are not significantly different from the first ones found.

Finally,

$$\frac{q_0}{c_0} = \frac{15.59}{0.48} = 32.5 \ \frac{\text{g mole/g solid}}{\text{g mole/cm}^3 \text{ gas}}$$

$$ka = n\left(\frac{v\varepsilon}{x}\right) = 67.1\left(\frac{2.1}{100}\right) = 1.41 \text{ s}^{-1}$$

$$\text{HTU} = \frac{100}{67.1} = 1.49 \text{ cm}$$

The principal source of error in $ka$ lies in the estimation of the slope, 0.00312, of the breakthrough curve at its midpoint.                    /////

The linear theory also represents the simplest treatment of the *chromatographic column*, frequently used for chemical analysis of fluid mixtures under conditions such that concentrations in the column are very small and the phase equilibrium is represented by its behavior near the origin, $q^* = q_m K_A p$.

A chromatograph consists of a long, slender tube filled with a powdered solid adsorbent through which a stream of weakly adsorbed carrier gas, such as helium, passes continuously. The operator of this analytical device quickly injects a small quantity of an unknown mixture into the entering gas stream. The different components of the mixture are carried through the column by the helium stream; but if they are reversibly adsorbed on the solid they are held in the packing for different times depending on the equilibrium adsorption coefficients. The result is a time-varying output composition with peaks corresponding to the different components. When the mass-transfer rate between phases is very large and there is not too much longitudinal diffusion, the peaks representing the different components are separated, and comparison of their measured areas obtained from a chart recorder yields the mixture composition.

The influence of the mass-transfer rate between phases on the width of a peak and the relation between its time of appearance in the effluent gas and its adsorption coefficient on the solid can be found from mass-balance and -rate equations which include transient terms. The analysis can be based on the assumption that only one adsorbable component is present, because if the mass-transfer rate is sufficiently great, different components with different adsorption equilibrium distributions will be separated along the column. (Hence the name "chromatograph," derived from the fact that sometimes the different adsorbed bands in the column of solid are colored.)

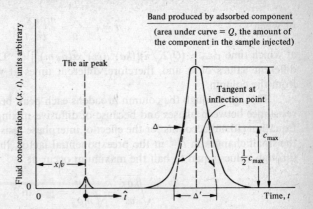

**FIGURE 10.11**
Variation of composition of gas stream leaving a chromatographic column.

Figure 10.11 shows the qualitative behavior of the fluid concentration at the exit from such a column. The first peak to emerge is caused by traces of air that inevitably get into the system when the sample is injected. Its presence is useful in interpreting the data because the air peak occurs at a time equal to the time of passage of the fluid itself through the column, $x/v$. Then times occurring in the theory, $\hat{t} = t - x/v$, can be measured from a vertical line drawn through the air peak. Any component that is adsorbed exhibits a broadened peak at a later time. To separate the peaks, it is desirable to design the column such that the width of a peak, either $\Delta$ or $\Delta'$, is smaller than the time interval between successive peaks. This requires an understanding of the mass-transfer behavior of the bed of packing.

The calculated time-dependence of the component concentration in the stream leaving a column of length $x$ is given by Eq. (10.44), with the quantity $Q$ of material injected quickly in the sample per unit of column cross-sectional area replacing $c_0$ on the left. The simplified result is

$$c(x, t) = \frac{Q}{2\sqrt{\pi}} \left[ \frac{(ka)^2}{(v\varepsilon)^3 x \hat{t} (\rho_B m)^3} \right]^{1/4} \exp\left[ -\left( \sqrt{\frac{kax}{v\varepsilon}} - \sqrt{\frac{ka\hat{t}}{m\rho_B}} \right)^2 \right] \qquad (10.46)$$

where $m = q_m^*/c_0 = q_m K_A RT$ is the distribution coefficient between phases, a function of temperature. The different components have different values of $m$, which is the source of their separation into successive peaks by the column.

When the mass transfer is very easy between phases ($kax \gg v\varepsilon$), the term in Eq. (10.46) that varies most rapidly is the exponential. The maximum peak concentration therefore occurs when

$$\frac{kat}{m\rho_B} = \frac{kax}{v\varepsilon}$$

or when

$$\hat{t} = t - \frac{x}{v} = \frac{m\rho_B x}{v\varepsilon} \tag{10.47}$$

at which time $c_{max} \approx (Q/2\sqrt{\pi})[(ka)^2/(v\varepsilon)^3 x\hat{t}(\rho_B m)^3]^{1/4}$. Different components have different values of $m$ and, therefore, different times at which their peaks emerge from the column.

Passage through the column broadens each peak because of the mass-transfer resistance between phases and because of diffusive mixing in the direction of flow. Equation (10.46) accounts for the effect of interphase resistance. Again disregarding the small changes in $\hat{t}^{1/4}$ in the preexponential factor, the times at which $c$ passes through values equal to half the maximum occur at

$$\sqrt{\frac{kax}{v\varepsilon}} - \sqrt{\frac{ka\hat{t}}{m\rho_B}} = \pm\sqrt{\ln 2} \tag{10.48}$$

from which the width of the peak, $\Delta$, at its half-maximum concentration is

$$\Delta = 4\rho_B m \sqrt{\ln 2 \frac{x}{v\varepsilon ka}} = \hat{t}_{max}\left(4\sqrt{\ln 2 \frac{\varepsilon v}{kax}}\right) \tag{10.49}$$

In terms of the number of transfer units in the bed,

$$n = \frac{kax}{v\varepsilon} = 16 \ln 2 \left(\frac{\hat{t}_{max}}{\Delta}\right)^2 \tag{10.50}$$

which provides an easy way experimentally to determine the mass-transfer characteristics of a chromatographic column. Alternatively, if the bandwidth is defined by drawing tangents to the curve representing the rounded peak at its inflection points and extending these to the time axis, the constant changes from $16 \ln 2 = 11.90$, to 32. Glueckauf [25] obtained an equivalent result with the constant equal to 16 instead of 32. He assumed that the column consists of a large number of stirred, equilibrium stages in series. Thus the height of a theoretical plate is apparently twice the length of a transfer unit.

The concept of a series of continuous, well-stirred equilibrium stages has also been used to calculate breakthrough curves. Such a sequence of stages will be governed by the difference-differential equation

$$\text{HETP}\left(\varepsilon + \rho_B \frac{q_m}{c_0}\right)\frac{dc_n}{dt} + \varepsilon v c_n = \varepsilon v c_{n-1} \tag{10.51}$$

with $c_n(0) = 0$ for all $n$ and $c_0 = $ constant for $t > 0$. The left side represents the time rate of change of the total moles of sodium ion in the fluid and in the resin within one equilibrium stage of bed length equal to the bed length equivalent to a theoretical stage, HETP. According to Said [76], the time-dependence of the effluent concentration from the $N$th stage turns out to be given by the summation

$$\frac{c}{c_0} = e^{-\alpha} \sum_{k=1}^{N} \frac{\alpha^{N-k}}{(N-k)!} \tag{10.52}$$

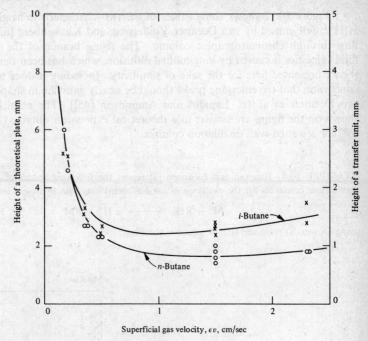

Superficial gas velocity, $\epsilon v$, cm/sec

FIGURE 10.12

Height of a theoretical plate in a gas chromatographic column. Data of van Deemter, Zuiderweg, and Klinkenberg [62] for a tube 18 cm long and 6 mm inside diameter filled with Celite particles (around 80 microns average diameter) wetted with a hydrocarbon oil of molecular weight 312. For $i$-butane, HETP = $0.8/(\epsilon v) + 0.8 + 0.08(\epsilon v)$; for $n$-butane, HETP = $0.3 + 0.9/(\epsilon v) + 0.5(\epsilon v)$.

where $\alpha = \epsilon v \hat{t}/[\epsilon + (\rho_B q_m/c_0)]$ HETP, or the ratio of the cumulative liquid flow past the last stage to the sum of void volume per stage plus resin mass per stage times the distribution coefficient, $q_m/c_0$. The function on the right of Eq. (10.52) is the Poisson exponential distribution function used in statistics, which is available in tables.

Both the mass-transfer-coefficient approach and the equilibrium-stage approach yield S-shaped breakthrough curves as the number of stages or the number of transfer units grow larger. The differences in shape become smaller. Both distributions approach the gaussian error function. Thus the choice between the two points of view may be unimportant if mass transfer is very efficient; if it is not, the point of view embodied in the mass-transfer-rate equations will certainly be preferred if one wants to relate apparatus performance to elementary processes; the imaginary and physically artificial equilibrium stage view may be acceptable if one is content with the purely empirical scaleup of model experiments.

Figure 10.12 shows some values of the two characteristic heights, HTU and HETP, determined by van Deemter, Zuiderweg, and Klinkenberg [62] for slow gas flow through chromatographic column. The rising branch of the curves at low fluid velocities is caused by longitudinal diffusion, which has been omitted from the theory presented here for the sake of simplicity. Including it does not change the conclusion that the emerging peaks should be nearly gaussian in shape as shown by van Deemter et al. (cf. Lapidus and Amundson [45]). The empirical equations shown on the figure are similar to a theoretical expression obtained by Westhaver [68] for a wetted-wall distillation column.

EXAMPLE 10.4   Bauman and Eichhorn [3] report the following values of the mass-action equilibrium constants for the exchange of various metal ions with hydrogen ion:

$$M^+ + RH \longrightarrow H^+ + RM$$

using Dowex-50 resin and 0.01-$N$ aqueous solutions:

| Metal ion | $K$, in Eq. (10.5) |
|-----------|--------------------|
| $Li^+$    | 0.61 |
| $Na^+$    | 1.20 |
| $K^+$     | 1.50 |
| $Cs^+$    | 2.04 |
| $Rb^+$    | 2.22 |
| $Tl^+$    | 8.60 |
| $Ag^+$    | 8.70 |

An ion-exchange column having a fractional void space of 0.4, a bulk density of 0.4 g/cm$^3$ dry resin, and a resin with an adsorptive capacity of 2.8 mequiv/g dry resin in the $H^+$ form is used as a chromatographic column for the analysis of an unknown mixture of these cations in a sample of the aqueous solutions. The chromatographic pattern is obtained by using 0.01-$N$ HCl as a sweep fluid. At time zero, a small sample of the solution of ions is injected quickly into the sweep fluid as it enters the bed.

Estimate the values of the ratio of cumulative volume of feed to volume of fluid in the bed at which each of the ions will emerge from the column.

SOLUTION   The velocities of the concentration waves along the column for the various ions are dependent on their equilibrium properties through

$$v^* = \frac{v}{1 + (\rho_B/\varepsilon)(dq^*/dc)}$$

The limiting value of the derivative at zero concentration will be used. This is calculated from Eq. (10.6):

$$q^* = q_m \frac{K(c/c_0)}{1 + (K-1)(c/c_0)} \qquad \frac{dq^*}{dc} = \frac{q_m c_0 K}{[c_0 + (K-1)c]^2} \rightarrow \frac{q_m K}{c_0} \qquad \text{as } c \rightarrow 0$$

Use is made of the concentration velocity for an ion $i$ by setting $v_i^* = x/t_i$, from which

$$\frac{\text{Volume fluid passed through column}}{\text{Volume fluid in column}} = \frac{\varepsilon v t_i}{\varepsilon x} = \frac{v}{v_i^*} = 1 + \frac{\rho_B}{\varepsilon} K_i \frac{q_m}{c_0}$$

$q_m$ represents the mequiv of metal plus hydrogen ions adsorbed per gram of dry resin at equilibrium with the fluid having total ion concentration $c_0$. Since the different ions are separated into chromatographic bands as they pass through the column, there is only one ion other than $H^+$ present in the solution at any point away from the bed's entrance. For the numerical values given

$$\frac{\text{Volume fluid passed through}}{\text{Volume fluid in column}} = 1 + \frac{0.4}{0.4} K_i \frac{2.8}{0.01}$$

and the following values of the ratio are computed.

| Ion | Ratio of fluid volumes |
|---|---|
| $Li^+$ | $1.70 \times 10^3$ |
| $Na^+$ | $3.36 \times 10^3$ |
| $K^+$ | $4.20 \times 10^3$ |
| $Cs^+$ | $5.71 \times 10^3$ |
| $Rb^+$ | $6.22 \times 10^3$ |
| $Tl^+$ | $24.1 \times 10^3$ |
| $Ag^+$ | $24.4 \times 10^3$ |

Note that separation of the pulses representing silver and thallium would require a column having extremely efficient mass transfer between the phases to avoid overlay of pulses.     ////

## 10.8  THE RESISTANCE TO MASS TRANSFER BETWEEN PHASES FOR ION EXCHANGE AND SORPTION

The precise design of fixed-bed units should allow for several sources of mass-transfer resistance, as discussed by Vermeulen, Klein, and Hiester [65]. These resistances include that in the fluid outside the particles, that in the particles owing to diffusion in pores and through the solid phase itself, and any resistance to the adsorption reaction at the surface. Among these, the last is likely to be insignificant, the major resistance being due to diffusion in one phase or the other. In addition, breakthrough patterns may be affected by longitudinal dispersion [32], as represented by the eddy diffusivities discussed in Chap. 4.

In view of the importance of diffusion phenomena, it is perhaps surprising that the major design method in use today is based on the Thomas solution of Eq. (10.31), which contains a kinetically derived rate expression. If diffusion were the major source of resistance, the rate expression should have been either

$$\rho_B \frac{\partial q}{\partial t} = k_f a(c - c_i) \tag{10.53}$$

if the dominating resistance is in the fluid phase surrounding the particles, or

$$\frac{\partial q}{\partial t} = k_p a(q_i - q) \tag{10.54}$$

if the main resistance is inside the particles.

Actually, Eq. (10.54) is an approximation because the distribution of concentrations inside a particle must be governed by an unsteady-state diffusion equation, as in Chap. 3. The true rate expression would make the instantaneous rate dependent on the particles' recent history and not simply on the instantaneous interface and bulk concentrations, as in Eq. (10.54). A true solution of the combined resistance problem has been given by Rosen [55], but his results apply only to linear equilibria with $K = 1$. Helfferich [33] has found that the electrostatic interactions in ion-exchange resin particles lead to a variety of diffusion patterns, some involving sharp reaction boundaries. Fleck, Kirwan, and Hall [22] have shown from their accurate calculations of transient pore diffusion using curved equilibrium relationships that the shapes of breakthrough curves differ somewhat from those found more simply using Eq. (10.31). The differences are usually not serious for design calculations, however, because the errors in computing the breakthrough times are not great. Thus, as a practical compromise and because of the simplicity of the equations obtained, the Thomas solution, Eq. (10.31), is often used, even though its rate expression can hardly be expected to allow precisely for the unsteady-state-diffusion situation inside the particles. As an alternative, any other rate equations involving driving forces can be used rather easily, provided constant-pattern conditions apply, as in Sec. 10.5. Moreover, there is empirical evidence [24, 52] that the kinetic-rate expression actually works very well despite its questionable form. Figure 10.13 shows some data reported by Gilliland and Baddour [24] from measurements of the uptake of sodium ion from an aqueous salt solution by ion-exchange particles of the acid type. The straight lines shown in the figure are expected from the rate expression in Eq. (10.27).

For Eq. (10.27) to be employed in design calculations two problems occur: (1) how can the individual-phase mass-transfer coefficients $k_f$ and $k_p$ be estimated, and (2) how can the Thomas kinetic coefficient $\kappa$ be found from the two mass-transfer coefficients?

The fluid-side coefficient $k_f$ can be estimated from a correlation of data for beds of small particles given in Chap. 5. Vermeulen, Klein, and Hiester [65] recommend the use of

$$k_p = \frac{10D_p}{d_p(1 - \varepsilon)} \tag{10.55}$$

and

$$a = \frac{6(1 - \varepsilon)}{d_p} \tag{10.56}$$

or $k_p a = 60D_p/d_p^2$ for estimating the solid-phase coefficient per unit volume. $D_p$ is the diffusion coefficient inside the particle, as measured, for example, by Boyd and Soldano [11], and $d_p$ is the particle diameter. Based on an empirical correlation of

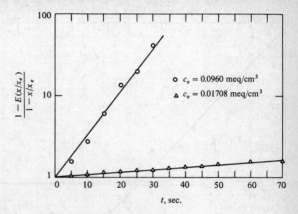

FIGURE 10.13
Batch adsorption of sodium ion from aqueous solution of NaCl using Dowex-50 ion-exchange-resin particles [24].

$$E = \left[ \frac{K}{K-1} \left( 1 + \frac{c_0}{\rho_B q_m} \right) - 1 \right]^{-1}$$

where $\rho_B$ = g resin/cm³ of resin-fluid mixture
$x = c_0 - c(t)$
$x_e = \rho_B q_m$
$c_0$ = initial concentration of Na⁺ in solution

$$\text{Slope} = \left[ \left( 1 - \frac{c_0}{\rho_B q_m} \right)^2 + \frac{4}{K} \frac{c_0}{\rho_B q_m} \right]^{1/2} \kappa a$$

nearly constant-pattern breakthrough data from a variety of sources, shown in Fig. 10.14, Hiester et al. [36] find a result equivalent to

$$\frac{1}{k_f} + \frac{c_0/q_m \rho_B}{k_p} = \frac{1}{(3.45 D_f/d_p)(\text{Pe})^{1/2}} + \frac{c_0}{q_m \rho_B} \frac{1}{16.7 D_p/d_p} \qquad (10.57)$$

where the Peclet number Pe is equal to $d_p \varepsilon v/6(1 - \varepsilon) D_f$ or $\varepsilon v/a D_f$. Figure 10.14 is a graph showing the data on which Eq. (10.55) is based. The solid-state diffusivities used in Eq. (10.57) were equal to $0.168 D_f$, based on several measurements for ion-exchange resins. The values of $k_f$ implied by the first term of Eq. (10.55) are not very different from estimates based on Fig. 6.18. The last term implies that the factor 10 appearing in Eq. (10.55) is somewhat too small.

How can the kinetic coefficient be found from $k_p$ and $k_f$, or from the sum of reciprocals in the empirical equation? The question can be answered by considering the three different driving forces appearing in Eqs. (10.27), (10.53), and (10.54). Each expression must represent the flux of material into the solid phase.

$$\kappa \left[ c \left( 1 - \frac{q}{q_m} \right) - \frac{1}{K} \frac{q}{q_m} (c_0 - c) \right] = k_f (c - c_i) = \frac{k_p}{\rho_B} (q_i - q) \qquad (10.58a,b)$$

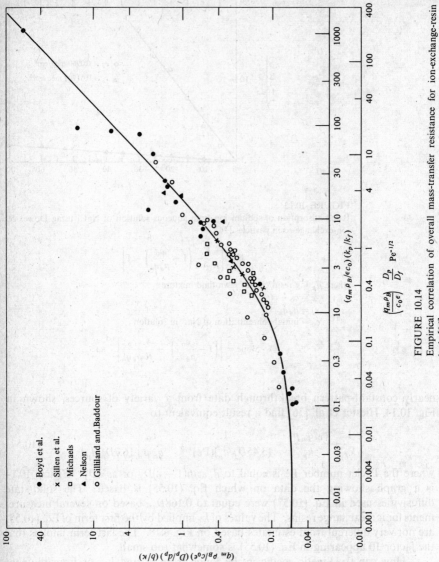

**FIGURE 10.14**

Empirical correlation of overall mass-transfer resistance for ion-exchange-resin beds [36].

using

$$q_i = q_m \frac{K(c_i/c_0)}{1 + (K - 1)(c_i/c_0)} \qquad (10.59)$$

It is clear that the relationship among the three coefficients must depend on the bulk concentrations $c$ and $q$ as well as the equilibrium constant $K$. Only if $K = 1$ will $1/\kappa$ be equal to the sum on the left of Eq. (10.57) at all compositions.

While it is not possible to find a simple relationship among the three $k$ values in general terms, a result applicable to constant-pattern conditions ($c/c_0 = q/q_m$) is possible. Moreover, the principal use of the overall coefficient $\kappa$ is for finding the shape of a breakthrough curve near the point $c = \frac{1}{2}c_0$. In this special condition, Eqs. (10.58$a$, $b$) reduce to

$$\kappa\left[\frac{c_0}{2}\left(1 - \frac{1}{2}\right) - \frac{1}{K}\frac{1}{2}\left(c_0 - \frac{c_0}{2}\right)\right] = k_f\left(\frac{c_0}{2} - c_i\right) = (k_p \rho_B)\left(q_i - \frac{q_m}{2}\right) \qquad (10.60a,b)$$

The second of these two equations can be solved simultaneously with Eq. (10.59) to find $c_i$, given $K$, $c_0$, and $q_m k_p/\rho_B c_0 k_f$. Then the result can be used to find either $\kappa/k_f$ or $\kappa/(k_p q_m/\rho_B c_0)$. Figure 10.15 shows the results of such computations. It indicates that if $K > 1$, the sum of reciprocals on the left of Eq. (10.57) must be divided by a factor $b$, which exceeds unity:

$$\frac{1}{\kappa} = \frac{1}{b}\left(\frac{1}{k_f} + \frac{c_0/q_m \rho_B}{k_p}\right) \qquad (10.61)$$

Thus, for constant-pattern conditions, $\kappa$ should somewhat exceed the value estimated by simple addition of the resistance of the phases.

**EXAMPLE 10.5**  Estimate the kinetic mass-transfer coefficient $\kappa$ for flow of a 0.01-$M$ water solution of NaCl through a bed of ion-exchange particles having $\varepsilon = 0.4$. The superficial fluid velocity is 1.0 cm/s and the temperature is 25°C. The particles are 2 mm in diameter and the diffusion coefficient of sodium ion is $1.2 \times 10^{-5}$ cm²/s in the fluid phase (cf. Table 2.11) and $9.4 \times 10^{-7}$ cm²/s inside the particles [11]. The bulk density of the bed is 0.7 g dry resin/cm³ and the capacity of the resin is 4.9 mequiv/g dry resin. The mass-action equilibrium constant is 1.2.

SOLUTION
Estimate $k_f$ using Fig. 6.18:

$$\text{Re} = \frac{d_p U \rho}{\mu} = \frac{(0.2)(1.0)}{(0.00913)} = 21.9$$

From Fig. 6.18,

$$j_D = \frac{k_f}{U}(\text{Sc})^{2/3} = 0.35$$

Using $\text{Sc} = \mu/\rho D_f = 0.00913/1.2 \times 10^{-5} = 761$,

$$k_f = \frac{(0.35)(1.0)}{761^{2/3}} = 0.0042 \text{ cm/s}$$

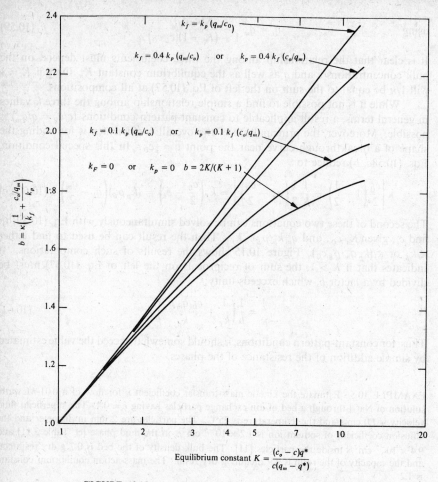

**FIGURE 10.15**
Correction to kinetic mass-transfer coefficient when it is calculated from addition of resistances in fluid and solid phases; constant-pattern conditions; mass-action equilibrium.

Estimate $k_p$ using Eq. (10.55):

$$k_p = \frac{10 D_p}{d_p(1 - \varepsilon)} = \frac{(10)(9.4 \times 10^{-7})}{0.2 \times 0.6} = 7.83 \times 10^{-5} \text{ cm/s}$$

The equilibrium between sodium and hydrogen ions

$$K = \frac{q(0.01 - c)}{(4.9 - q)c} = 1.2$$

where $c$ and $q$ are in mequiv/cm$^3$ and mequiv/g dry resin, respectively.

Combine resistance to find $\kappa$: Equating the rates of mass transfer,

$$\kappa\left[c\left(1 - \frac{q}{q_m}\right) - \frac{1}{K}\frac{q}{q_m}(c_0 - c)\right] = k_f[c - c_i] = k_p[q_i - q]\rho_B$$

Since we are primarily interested in conditions applying to constant-pattern behavior near the point $c = c_0/2 = 0.01/2$ mequiv/cm$^3$, where $q = 4.9/2 = 2.45$ mequiv/g we can find $c_i$ and $q_i$ by simultaneously solving

$$4.2 \times 10^{-3}(0.005 - c_i) = (7.83 \times 10^{-5})(0.7)\frac{q_i - 2.45}{1,000} \qquad \text{mequiv/(cm)}^2\text{(s)}$$

and

$$q_i = 4.9\frac{1.2c_i}{0.01 + 0.2c_i}$$

The result is $c_i = 0.0046065$ mequiv/cm$^3$ and $q_i = 2.4801$ mequiv/g. Now using the first equality,

$$\kappa = 0.0042\frac{0.005 - 0.0046065}{0.005(1 - 0.5) - 1/1.2(0.5)(0.01 - 0.0005)} = 3.46 \times 10^{-3} \text{ cm/s}$$

Alternatively, $\kappa$ can be found using the $b$ factor given by Fig. 10.15, after finding the apparent sum of resistances,

$$\frac{1}{k_f} + \frac{(c_0/\rho_B q_m)}{k_p} = \frac{1}{4.2 \times 10^{-3}} + \frac{0.01}{(0.7)(4.9)(7.83 \times 10^{-5})}$$

$$= 238 + 37.2 = \frac{1}{3.63 \times 10^{-3}}$$

From Fig. 10.15, $b = 1.09$ and $\kappa = (3.63 \times 10^{-3})(1.091) = 3.96 \times 10^{-3}$ cm/s. (The $b$ factor is a correction to the straightforward addition of resistances to allow for the curved equilibrium relationship.)

Estimate $\kappa$ from the empirical correlation of ion-exchange-rate data in Fig. 10.14. Using Eq. (10.57),

$$\text{Pe} = \frac{d_p \varepsilon v}{6(1 - \varepsilon)D_f} = \frac{(0.2)(1.0)}{(6)(0.6)(1.2 \times 10^{-5})} = 4,629$$

$$\frac{1}{k_f} + \frac{(c_0/q_m \rho_B)}{k_p} = \frac{0.2}{(3.45)(1.2 \times 10^{-5})(4.629^{1/2})} + \frac{(0.01)(0.2)}{(4.9)(0.7)(16.7)(9.4 \times 10^{-7})}$$

$$= 71 + 37.1 = \frac{1}{9.25 \times 10^{-3}}$$

and $\kappa = (9.25 \times 10^{-3})(1.091) = 1.01 \times 10^{-3}$ cm/s. The larger value of the transfer coefficient is owing primarily to the larger fluid-phase mass-transfer coefficient represented by the correlation.

The interfacial area per unit of bed volume is

$$a = \frac{\pi d_p^2(1 - \varepsilon)}{d_p^3/6} = \frac{6(1 - \varepsilon)}{d_p} = \frac{(6)(0.6)}{0.2} = 18 \text{ cm}^2/\text{cm}^3 \qquad \text{////}$$

## 10.9 THE DESIGN OF FIXED-BED ADSORBERS FOR REPEATED, CYCLIC USE

EXAMPLE 10.6 Design a fixed-bed adsorber for the removal of methane from a methane-hydrogen mixture at 10 atm and 25°C, containing 10 mole percent $CH_4$. The flow rate of the feed mixture is 100 $ft^3$/min at NTP. The maximum $CH_4$ content of the $H_2$ stream leaving the bed is not to exceed 0.1 mole percent after a 10-min adsorption cycle. Activated carbon particles having a bulk density of 0.509 $g/cm^3$, a particle density of 0.777 $g/cm^3$, and a true solid density of 2.178 $g/cm^3$ will be used as an adsorbent. Adsorption equilibrium data are reported by Grant, Manes, and Smith [72] and can be represented by the equation

$$q = 3.0 \times 10^{-3} \frac{K_A p}{1 + K_A p} \quad \text{g mole/g solid}$$

where $p$ is the $CH_4$ particle pressure in atmospheres and $K_A = 0.346 \exp [2,200(T^{-1} - 298.1^{-1})]$, $atm^{-1}$. Find a few acceptable values of bed volume, length, and cross-sectional area for solid particle sizes corresponding to 20-mesh and 60-mesh screen sizes.

SOLUTION   Physical properties of solid phase. From the densities reported, the fraction of void space outside the particles is

$$\varepsilon = \frac{0.509^{-1} - 0.777^{-1}}{0.509^{-1}} = 0.345$$

The fraction void space inside the porous particles is

$$\chi = \frac{0.777^{-1} - 2.178^{-1}}{0.777^{-1}} = 0.643$$

From work on diffusion inside catalyst particles [56], the diffusivity in the gas-filled pores of the solid particles can be estimated from [57]:

$$\frac{1}{D_p} = \frac{\tau}{\chi}\left(\frac{1}{D_K} + \frac{1}{D_f}\right) \qquad D_K = \frac{19,400\chi}{S_g \rho_p}\sqrt{\frac{T}{M}}$$

where $D_K$ = Knudsen diffusion coefficient
  $\tau$ = tortuoisty factor $\approx 4.0$
  $S_g$ = surface area per gram of solid = $1.1 \times 10^7$ $cm^2$/g
  $\rho_p$ = particle density.

$$D_p = \frac{0.643}{4.0}\left[\frac{(1.1 \times 10^7)(0.777)}{(1.94 \times 10^4)(0.643)}\sqrt{\frac{16.04}{298.1}} + \frac{1}{0.0742}\right]^{-1}$$

$$= 0.932 \times 10^{-3} \text{ cm}^2/\text{s}$$

The value of $D_f = 0.0742$ $cm^2$/s is taken from Table 2.4 and has been adjusted to 25°C and 10 atm. Using Eq. (10.57)

$$k_p = \frac{16.7 D_p}{d_p} = \frac{(16.7)(0.932 \times 10^{-3})}{d_p}$$

According to Eq. (10.55), we find the values of the solid-phase mass-transfer coefficient listed in the table below.

| Screen size | Sieve opening, $d_p$, cm | $k_p$, cm/s | $a$, cm$^2$/cm$^3$ |
|---|---|---|---|
| 4-mesh | 0.746 | 0.0208 | 5.26 |
| 10-mesh | 0.168 | 0.0927 | 23.4 |
| 20-mesh | 0.0841 | 0.185 | 46.6 |

Fluid-phase properties (hydrogen-methane):

$$D_f = 0.0742 \text{ cm}^2/\text{s} \quad \text{as above}$$

$$\mu/\rho = 0.1083 \text{ cm}^2/\text{s}$$

$$\text{Sc} = \nu/D_f = 1.46 \quad (\text{Sc})^{2/3} = 1.287$$

Use superficial fluid velocities, $U = \varepsilon v$, of 3 and 30 cm/s, corresponding to Reynolds numbers and fluid-side mass-transfer coefficients (from Fig. 6.17) as in the following table.

| $U$, cm/s | Screen size | $\text{Re} = d_p U/\nu$ | $k_f = j_d U/(\text{Sc})^{2/3}$, cm/s |
|---|---|---|---|
| 3 | 4 | 13.2 | 1.06 |
| | 10 | 4.65 | 2.02 |
| | 20 | 2.33 | 3.05 |
| 30 | 4 | 132 | 3.4 |
| | 10 | 46.5 | 5.4 |
| | 20 | 23.3 | 13.2 |

Overall mass-transfer coefficient for the Thomas equations. During adsorption,

$$c_0 = p_0/RT = \frac{(0.1)(10.0)}{(82.06)(298.1)} = 4.09 \times 10^{-5} \text{ g mole/cm}^3$$

$$K_A = 0.346 \text{ atm}^{-1}$$

$$q_0 = \frac{(3.0 \times 10^{-3})(0.346)(1.0)}{1.0 + (0.346)(1.0)} = 0.771 \times 10^{-3} \text{ g mole/g}$$

$$\frac{c_0}{\rho_B q_0} = \frac{4.09 \times 10^{-5}}{(0.509)(0.771 \times 10^{-3})} = 0.1042$$

Phase equilibrium is represented in the adsorber by

$$\frac{q^*}{q_0} = \frac{K(c/c_0)}{1 + (K-1)(c/c_0)}$$

with $K = 1 + K_A p_0 = 1.346$.

Having found the individual phase coefficients $k_f$ and $k_p$, their combination to find $\kappa$ is needed. Note that $k_p$ has been found on the assumption that mass transfer inside the particles occurs by diffusion along the gas-filled pores rather than through the solid phase itself. Moreover, surface diffusion in the adsorbed layer has been neglected. There is no need to introduce the adsorption equilibrium to relate interface compositions. The overall driving force should be simply

$$c - c^* = c - c_0 \frac{(q/q_0)}{K - (K - 1)(q/q_0)}$$

not the kinetic driving force in the Thomas equation, Eq. (10.27). Nevertheless, we need to use the Thomas solution, Eq. (10.31), to find effluent compositions because it is the only available general solution that allows for the nonlinear equilibrium relationship.

Thus, we must find a way to compute $\kappa$ in Eq. (10.30) from $k_f$ and $k_p$. The calculation is similar to that leading to Eq. (10.59) but the value of $b$ is different from that in Fig. 10.15. The development is left to the student in Prob. 10.3 at the end of the chapter. The result is the simple formula $b = 2K/(K + 1)$. The following table lists the results.

| $U$, cm/sec | Mesh size | $k_f$ | $k_p$ | $\kappa$, cm/s | $a$, cm$^2$/cm$^3$ | $\kappa a$, s$^{-1}$ |
|---|---|---|---|---|---|---|
| 3 | 4 | 1.06 | 0.0327 | 0.0364 | 8.26 | 0.300 |
| | 10 | 2.02 | 0.927 | 0.1016 | 23.4 | 2.38 |
| | 20 | 3.05 | 0.185 | 0.200 | 46.6 | 9.33 |
| 30 | 4 | 3.4 | 0.0327 | 0.0371 | 8.26 | 0.307 |
| | 10 | 5.4 | 0.0927 | 0.1045 | 23.4 | 2.44 |
| | 20 | 8.0 | 0.185 | 0.207 | 46.6 | 9.67 |

**Calculation of size and weight carbon bed.** The dimensionless coordinates in Eq. (10.31a) are

$$n = \frac{\kappa a x}{U}$$

$$nT = \frac{\kappa a x}{U} \frac{c_0 U(t - x\varepsilon/U)}{\rho_B q_0 x} = 0.1042\kappa a \left( 600 - \frac{0.345x}{U} \right)$$

For each value of $U$, $\kappa a$ is available from the table above and $x$ can be found by solution of Eq. (10.31) using $c/c_0 = 0.01$.

It is convenient to rearrange the equation in the form

$$F(x) \equiv J\left(\frac{n}{K}, nT\right) - \frac{c}{c_0 - c}\left[1 - J\left(\frac{nT}{K}, n\right)\right] \exp\left[(1 - K^{-1})(n - nT)\right] = 0$$

and to solve the equation for $x$ using the Newton-Raphson method. This requires the evaluation of the derivative $dF/dx$. If $n$ is sufficiently large, values of $F$ and $dF/dx$ can be computed numerically using the approximation, Eq. (10.35). The computations are easily done in a digital computer. Results are listed as in the following table.

| $U$, cm/s | Particle size | $n$ | $nT$ | $x$, ft | Volume, ft$^3$ | $\Delta p$, in H$_2$O |
|---|---|---|---|---|---|---|
| 3 | 4-mesh | 30.6 | 17.65 | 10.05 | 185.7 | 0.42 |
| | 10-mesh | 160.9 | 143.0 | 6.63 | 122.5 | 2.24 |
| | 20-mesh | 579.9 | 562.4 | 6.12 | 113.1 | 8.26 |
| 30 | 4-mesh | 31.1 | 18.1 | 99.7 | 184.3 | 83.1 |
| | 10-mesh | 164.4 | 146.9 | 66.2 | 122.4 | 313 |
| | 20-mesh | 600.5 | 582.9 | 61.1 | 112.9 | 824 |

The table shows that there is no advantage in using the larger fluid velocity in the bed. The volume of the bed is reduced insignificantly (because of the negligible mass transfer in the fluid phase), but the bed is made very long and the pressure drop is much greater. At the lower fluid velocity, there is some advantage in bed volume when the particles are reduced from 4- to 10-mesh, but little further reduction is obtained by using still finer carbon particles.

Obviously the bed needs to be regenerated before it can be used for another 10-min adsorption cycle. Can you propose ways to accomplish this?

What would be the effect on the design of the adsorber if the pressure had been increased to 100 atm.? ////

# REFERENCES

1 ANDERSON, J. S.: *Z. Phys. Chem.*, **88**: 212 (1914).
2 ANZELIUS, A.: *Z. Angew. Math. Mech.*, **6**: 291 (1926).
3 BAUMAN, W. C., and J. EICHHORN: *J. Amer. Chem. Soc.*, **69**: 2830 (1947).
4 BEATON, R. H., and C. C. FURNAS: *Ind. Eng. Chem.*, **33**: 1500 (1941).
5 BOHART, G. S., and E. Q. ADAMS: *J. Amer. Chem. Soc.*, **42**: 523 (1920).
6. BOYD, G. E., J. SCHUBERT, and A. W. ADAMSON: *J. Amer. Chem. Soc.*, **69**: 2818 (1947).
7 BOYD, G. E., A. W. ADAMSON, and L. S. MYERS, JR.: *J. Amer. Chem. Soc.*, **69**: 2836 (1947).
8 BOYD, G. E., L. S. MYERS, JR., and A. W. ADAMSON: *J. Amer. Chem. Soc.*, **69**: 2854 (1947).
9 BOYD, G. E., L. S. MYERS, JR., and A. W. ADAMSON: *J. Amer. Chem. Soc.*, **69**: 2849 (1947).
10 ARIS, R., and N. R. AMUNDSON: "Mathematical Methods in Chemical Engineering." vol. 2, Prentice Hall, Englewood Cliffs, N.J., 1973.
11 BOYD, G. E., and B. A. SOLDANO: *J. Amer. Chem. Soc.*, **74**: 6091 (1953).
12 BRUNAUER, S.: "The Adsorption of Gases and Vapors," Princeton, Princeton, N.J., 1945.
13 BRUNAUER, S., P. H. EMMETT, and E. TELLER: *J. Amer. Chem. Soc.*, **60**: 309 (1938).
14 CASSIDY, H. G., and S. E. WOOD: *J. Amer. Chem. Soc.*, **63**: 2628 (1941).
15 CONWAY, D. E., J. H. S. GREEN, and D. REICHENBERG: *Trans. Faraday Soc.*, **50**: 511 (1954).
16 COOLIDGE, A. S.: *J. Amer. Chem. Soc.*, **49**: 708 (1927); **46**: 609 (1924).
17 COONEY, D. O., and E. N. LIGHTFOOT: *Ind. Eng. Chem. Fundam.*, **4**: 233 (1965).
18 DEVAULT, D.: *J. Amer. Chem. Soc.*, **65**: 532 (1943).
19 DRANOFF, J. S., and L. LAPIDUS: *Ind. Eng. Chem.*, **50**: 1648 (1958); **53**: 71 (1961).
20 DUNCAN, J. F., and B. A. LISTER: *J. Chem. Soc.*, **1949**: 3285
21 EMMETT, P. H., and S. BRUNAUER: *J. Amer. Chem. Soc.*, **59**: 1553 (1937).
22 FLECK, R. D., JR., D. J. KIRWAN, and K. R. HALL: *Ind. Eng. Chem. Fundam.*, **12**: 95 (1973).
23 FURNAS, C. C.: *Trans. AIChE*, **24**: 142 (1930).
24 GILLILAND, E. R., and R. F. BADDOUR: *Ind. Eng. Chem.*, **45**: 330 (1953).

25 GLUECKAUF, E.: In "Ion Exchange and Its Applications," p. 34, Soc. Chem. Ind., London, 1955.

26 GLUECKAUF, E., J. I. COATES: *J. Chem. Soc.*, **1947**: 1315; *Trans. Faraday Soc.*, **51**: 1540 (1955).

27 GOLDMAN, F., and M. POLANYI: *Z. Phys. Chem.*, **132**: 321 (1928).

28 GOLDSTEIN, S.: *Proc. Roy. Soc. (London)*, **A219**: 151, 171 (1953).

29 GREGOR, H. P., J. J. BREGMAN, F. GUTOFF, R. D. BROADLEY, D. E. BALDWIN, and C. G. OVERBERGER: *J. Colloid Sci.*, **6**: 20 (1951).

30 HALE, D. K., and D. REICHENBERG: *Discuss. Faraday Soc.*, **7**: 79 (1949).

31 HELFFERICH, F.: "Ion Exchange," McGraw-Hill, New York, 1962.

32 HELFFERICH, F.: Ion Exchange Kinetics, chap. 2 in J. A. Marinsky (ed.), "Ion Exchange—A Series of Advances," Dekker, New York, 1966.

33 HELFFERICH, F.: *J. Phys. Chem.*, **69**: 1178 (1965).

34 HOUGHTON, G.: *J. Phys. Chem.*, **67**: 84 (1963).

35 HIESTER, N. K., E. F. FIELDS, JR., R. C. PHILLIPS, and S. B. RADDING: *Chem. Eng. Progr.*, **50**: 139 (1954).

36 HIESTER, N. K., S. B. RADDING, R. L. NELSON, and T. VERMEULEN: *AIChE J.*, **2**: 404 (1956).

37 HIESTER, N. K., and T. VERMEULEN: *Chem. Eng. Progr.*, **48**: 505 (1952).

38 HUANG, T.-C, and K.-Y LI: *Ind. Eng. Chem. Fundam.*, **12**: 50 (1973).

39 KETTELE, B. E., and G. E. BOYD: *J. Amer. Chem. Soc.*, **69**: 2800 (1947).

40 KLINKENBERG, A.: *Ind. Eng. Chem.*, **40**: 1970 (1948).

41 KUNIN, R., and R. J. MYERS: *J. Phys. Chem.*, **51**: 5 (1947).

42 KUNIN, R., and R. J. MYERS: "Ion Exchange Resins," Wiley, New York, 1950.

43 LAMBERT, B., and A. M. CLARK: *Proc. Roy, Soc. (London)*, **A122**: 497 (1929).

44 LANGMUIR, I.: *J. Amer. Chem. Soc.*, **38**: 2221 (1916).

45 LAPIDUS, L., and N. R. AMUNDSON: *J. Phys. Chem.*, **56**: 984 (1952).

46 MACDOUGALL, F. H.: *J. Phys. Chem.*, **38**: 945 (1938).

47 MARINSKY, J. A. (ED.): "Ion Exchange—A Series of Advances," vol. 1, Dekker, New York, 1966.

48 HOUGEN, O. A. and W. R. MARSHALL, JR.: *Chem. Eng. Progr.*, **43**: 197 (1947).

49 MCBAIN, J. W.: "The Sorption of Gases and Vapors by Solids," Routledge, London, 1932.

50 MCBAIN, J. W., and G. T. BRITTON: *J. Amer. Chem. Soc.*, **52**: 2217, 2218, 2220 (1930).

51 MICHAEIL, I., and A. KATCHELSKY, A.: *J. Polymer Sci.*, **23**: 683 (1957).

52 MICHAELS, A. S.: *Ind. Eng. Chem.*, **44**: 1922 (1952).

52a NACHOD, F. C., and W. WOOD: *J. Amer. Chem. Soc.*, **66**: 1380 (1944).

53 MCGAVACK, J., and W. A. PATRICK: *J. Amer. Chem. Soc.*, **42**: 946 (1920).

54 HARRIS, F. E., and S. A. RICE: *J. Chem. Phys.*, **24**: 1258 (1956).

55 ROSEN, J. B.: *J. Chem. Phys.*, **20**: 387 (1952); *Ind. Eng. Chem.*, **46**: 1590 (1954).

56 SATTERFIELD, C. N., and T. K. SHERWOOD: "The Role of Diffusion in Catalysis," Addison-Wesley, Reading, Mass., 1963.

57 SATTERFIELD, C. N.: "Mass Transfer in Heterogeneous Catalysis," M.I.T., Cambridge, Mass., 1970

58 SCHUMANN, T. E. W.: *J. Franklin Inst.*, **208**: 405 (1929).

59 SELKE, W. A., and H. BLISS: *Chem. Eng. Progr.*, **46**: 509 (1950).

60 SOLDANO, B. A., and G. E. BOYD: *J. Amer. Chem. Soc.*, **75**: 6099 (1953).

61 THOMAS, H.: *J. Amer. Chem. Soc.*, **66**: 1664 (1944).

62 VAN DEEMTER, J. J., F. J. ZUIDERWEG, and A. KLINKENBERG: *Chem. Eng. Sci.*, **5**: 271 (1956).

63 VASISTH, R. C., and M. M. DAVID: *AIChE J.*, **5**: 391 (1959).

64 VERMEULEN, T.: *Ind. Eng. Chem.*, **45**: 1664 (1953).

65 VERMEULEN, T., G. KLEIN, and N. K. HIESTER: Sec. 16 in J. H. Perry (ed.), "Chemical Engineers' Handbook," McGraw-Hill, New York, 1973.

66 VERMEULEN, T., and N. K. HIESTER: *Ind. Eng. Chem.*, **44**: 636 (1952).

67 WALTER, J. E.: *J. Chem. Phys.*, **13**: 229 (1945).

68 WESTHAVER, J.: *Ind. Eng. Chem.*, **34**: 681 (1942).

69 WEYDE, E., and E. WICKE: *Kolloid Z.*, **90**: 156 (1940).

70 WICKE, E.: *Kollid Z.*, **86**: 167 (1939); **86**: 295 (1939).

71 YOUNG, D. M., and A. D. CROWELL: "Physical Adsorption of Gases," Butterworths, Washington, 1962.

72 GRANT, R. J., M. MANES, and S. B. SMITH: *AIChE J.*, **8**: 403 (1962).

73 HEFTMANN, E. (ED.): "Chromatography," Reinhold, New York, 1961.

74 VERMEULEN, T.: Private communication, 1972.

75 ABRAMOWTIZ, A., and I. A. STEGUN: "Handbook of Mathematical Functions," p. 378, Dover, New York, 1965.

76 SAID, A. S.: *AIChE J.*, **2**: 477 (1956).

## PROBLEMS

*10.1* Derive a relationship between the overall mass-transfer coefficient $K_f$ defined by

$$\text{Rate, m mole/(s)(cm}^2) = K_f[c - c^*]$$

for ion exchange and the individual-phase coefficients $k_f$ and $k_p$. Assume that the equilibrium relationship is of the mass action form

$$c = c_0 \frac{q/q_m}{K - (K-1)(q/q_m)}$$

a relationship that applies both to the bulk particle-phase concentration $c^* = c_0 F(q/q_m)$ and to the interface concentrations $c_i = c_0 F(q_i/q_m)$. Evaluate the bulk-fluid and particle-phase concentrations by assuming constant-pattern behavior and use $c = c_0/2$. Compare the results of your numerical calculations for selected values of $k_p \rho_B q_m / k_f c_0$ and $K$ with Fig. 16.12 of Vermeulen, Klein, and Hiester [65].

*10.2* Extend the lines of Fig. 10.15 to values of $K < 1$, that is, to breakthrough curves of the unfavorable equilibrium type by assuming near-equilibrium behavior. Evaluate the bulk-fluid and particle-phase concentrations at $T = \varepsilon v c_0 (t - x/v)/x \rho_B q_m = 1.0$.

*10.3* When the resistance to mass transfer inside the solid particles of the adsorbent is due to gaseous diffusion inside pores, the rate of mass transfer can be written as $k_p (c_i - c^*)$ where $c^*$ is the gas concentration at equilibrium with the bulk solid composition, $q$. Show that under these circumstances and constant-pattern conditions the kinetic coefficient in the Thomas formula is given by

$$\frac{1}{\kappa} = \frac{K+1}{2K} \left( \frac{1}{k_p} + \frac{1}{k_f} \right)$$

where $K = 1 + K_A p_0$

$K_A$ = Langmuir adsorption coefficient

$p_0$ = partial pressure of the adsorbate in the feed-gas mixture

*10.4* Selke and Bliss [59] reported the following equilibrium data for the exchange reaction:

$$Cu^{++} + RMg \longrightarrow Mg^{++} + RCu$$

where $R$ signified Amberlite IR-120 resin. The constant sum of the aqueous-phase concentrations of the cupric and magnesium ions was 21.6 mequiv/l.

| $c/c_0$ | $q$, mequiv/g dry resin |
|---------|-------------------------|
| 0.148 | 1.00 |
| 0.288 | 1.66 |
| 0.53 | 2.58 |
| 0.61 | 2.95 |

Find the resin capacity in mequiv/g dry resin and the mass-action equilibrium constant for the above reaction. Which ion is more strongly held by the resin particles?

*10.5* By comparing the Thomas solutions for $c$ and $q$ values during the adsorption break-through curve, find an approximate expression for the ratio $(c/c_0)/(q/q_0)$ in terms of $n$, $T$, and $K$. (Note that the ratio will be equal to unity when constant-pattern conditions exist.) Show that the ratio is precisely unity if the Klinkenberg estimate of the $J$ function is valid. Using the better approximation given by Eq. (10.35), find expressions for the ratio for three points on the breakthrough curve: $c \ll c_0$, $c = c_0/2$, and $c \approx c_0$; show that the ratio is very nearly equal to unity when $n$ is large.

# THE PERFORMANCE OF
# MASS-TRANSFER EQUIPMENT

## 11.0 SCOPE

Chapters 9 and 10 describe the computational procedures employed in the design of mass-transfer equipment. Perry's "Chemical Engineers' Handbook" and various published reviews summarize the data needed for quantitative design problems. These sources, however, offer little insight into the nature of the operation of typical equipment and the factors which must be considered by the design engineer. The present chapter is intended to fill this gap by providing a qualitative description of the performance characteristics of several of the more popular mass-transfer devices.

The purpose of the sections which follow is to show the design engineer how the empirical correlations available in Perry can be combined with the material of Chap. 9 to arrive at reasonable equipment designs. The illustrative examples are important for this purpose and should be studied as part of the text.

## 11.1  PRINCIPAL SYMBOLS

| | |
|---|---|
| $a$ | Area of contact of phases, $ft^2/ft^3$ |
| $a_p$ | Area of dry packing, $ft^2/ft^3$ |
| $A$ | Interfacial area of gas-liquid contact, $ft^2/ft^2$ tray floor area |
| $A_0$ | Total area of orifices in sieve tray, $ft^2$ |
| $A_c$ | Total active area of sieve tray, $ft^2$ |
| $B$ | Tray width normal to liquid flow, ft |
| $C_r$ | "Packing factor," $a_p/\varepsilon^3$ |
| $d_b$ | Diameter of bubble, cm |
| $d_p$ | Nominal packing size |
| $D_E$ | Eddy-diffusion coefficient for dispersion in froth, $ft^2/s$ |
| $D_i$ | Impeller diameter |
| $E_a$ | Axial dispersion coefficient, $cm^2/s$ |
| $E_{MG}$ | Murphree efficiency for whole tray, in terms of gas-phase compositions |
| $E_{MGE}$ | Effective gas-phase efficiency with liquid entrainment |
| $E_{ML}$ | Murphree efficiency of whole tray in terms of liquid-phase compositions |
| $E_{OG}$ | Local or "point" tray efficiency in terms of gas-phase compositions |
| $F$ | $U_G(\rho_G)^{1/2}$, $(ft/s)(lb/ft^3)^{1/2}$. |
| $g$ | Local acceleration due to gravity, $ft/(s)^2$ |
| $g_c$ | Conversion factor, $(lbm)(ft)/(lbf)(s)^2$ |
| $G$ | Mass velocity of gas, $lb/(h)(ft^2)$ |
| $G_M$ | Molal mass velocity of gas, lb moles/(h)$(ft^2)$ |
| $h$ | Packed height, ft |
| $h_{DP}$ | Friction for gas flow through dry plates, $(ft)(lb/f)/lbm$. |
| $h_L$ | Liquid head over weir |
| $H$ | Henry's law coefficient, $X^*/p$, liquid mole fraction/atm, or lb moles/$(ft^3)$(atm); liquid depth in stirred vessel, ft |
| $H_L$ | Height of an individual liquid-phase transfer unit, ft |
| $H_G$ | Height of an individual gas-phase transfer unit, ft |
| $H_{OG}$ | Height of an overall gas-phase transfer unit, ft |
| $H_{OL}$ | Height of an overall liquid-phase transfer unit, ft |
| $k_G a$ | "Gas-side" mass-transfer coefficient, lb moles/(h)$(ft^3)$(atm) |
| $k_G RT$ | Gas-phase mass-transfer coefficient, or ft/h |
| $k_L a$ | "Liquid-side" mass-transfer coefficient, $h^{-1}$ |
| $K$ | Overall mass-transfer coefficient; ratio $Y/X$ at equilibrium; constant in Eq. (11.4) |
| $Ka$ | Overall mass-transfer coefficient on a volume basis |
| $K_L a$ | Overall mass-transfer coefficient, $h^{-1}$ |
| $L$ | Mass velocity of liquid, $lb/(h)(ft^2)$ |
| $L'$ | Liquid flow, gal/(min) (ft tray width) |
| $L_M$ | Molal mass velocity of liquid, lb moles/(h)$(ft^2)$ |
| $m$ | Slope of equilibrium line, $dY^*/dX$ |
| $M$ | Molecular weight |
| $n$ | Rotational speed, rps |
| $N_G$ | Number of gas-transfer units, no liquid-phase resistance |
| $N_L$ | Number of liquid-phase transfer units, no gas-phase resistance |
| $N_{OG}$ | Number of transfer units, gas phase, with both gas and liquid resistances involved |
| $p$ | Partial pressure, atm or mm Hg |
| $p^*$ | Equilibrium partial pressure, atm or mm Hg |

| | |
|---|---|
| $P$ | Total pressure, atm or psia |
| Pe | Peclet number for dispersion in froth |
| $P_w/V$ | Power input per unit volume, ft-lb/(s)(ft$^3$) |
| $R$ | Gas constant |
| $S'$ | Distance from top of froth to tray above, in |
| Sc | Schmidt number for gas mixture |
| $T$ | Absolute temperature, °K; diameter of agitated vessel; tray spacing |
| $U_F$ | Linear velocity of froth across tray, ft/s |
| $U_G, U_s$ | Superficial velocity, ft/s |
| $U_i$ | Interstitial velocity, ft/s, or cm/s· |
| $W$ | Weir height above tray floor, in |
| $X$ | Mole fraction in liquid |
| $Y$ | Mole fraction in gas mixture; $Y^*$ in equilibrium with liquid |
| $Z_c$ | Equivalent depth of clear liquid on plate, in |
| $Z_F$ | Depth of foam on tray, in |
| $Z_L$ | Length of liquid travel across plate, ft |
| $\Delta$ | Difference in clear-liquid depth, liquid inlet to liquid outlet on tray, in |
| $\Delta P$ | Pressure drop, in water or lbf/ft$^2$ |
| $\varepsilon$ | Void fraction |
| $\varepsilon_F$ | Volumetric fraction gas in froth |
| $\varepsilon_T$ | Entrainment ratio, lb liquid/lb gas |
| $\mu$ | Viscosity, poise (cP where indicated) |
| $\mu_L$ | Liquid viscosity, cP |
| $\sigma$ | Surface tension |
| $\nu$ | Kinematic viscosity, cm$^2$/s |
| $\rho_c$ | Density of continuous phase, lb/ft$^3$ |
| $\rho_G$ | Gas density, lb/ft$^3$ |
| $\rho_L$ | Liquid density, lb/ft$^3$ |
| $\rho_{mL}$ | Molal density of liquid, lb moles/ft$^3$ |

## 11.2 MASS-TRANSFER EQUIPMENT

Many industrial separation processes depend on mass transfer between a gas and a liquid, gas or liquid and a solid, or between two liquids. Such processes include distillation, gas absorption, liquid extraction, adsorption, partial condensation, and ion exchange. Drying, humidification, dehumidification, and water cooling might also be properly described as separation processes involving mass transfer. Heterogeneous catalysis requires the mass transfer of the reactants and products to and from the solid surface at which a chemical reaction takes place.

Numerous types of mass-transfer equipment have been used successfully. In most cases each is designed for a particular application and not purchased from a manufacturer, as are pumps and blowers. The designer's goal is to arrive at a proper economic balance of investment and operating costs, since one usually increases as the other is decreased.

The investment in a given type of equipment to be designed for a specific operation is roughly proportional to its size and inversely proportional to the rate of mass transfer. The latter, in turn, depends on the nature of the phase equilibria and is proportional to the mass-transfer coefficient and to the surface of contact between the two phases. Large coefficients can be obtained by increasing the turbulence of one or both phases and designing for high relative velocities of the two. The advantage is gained at the expense of an increase in the cost of power for operation. In many designs the power is supplied by external pumps and blowers, though there has been a trend in recent years to develop more devices which employ moving parts within the equipment.

High flow rates not only tend to increase $K$ but often the contact area as well. Up to a point the length of the contacting device in the direction of flow is not increased as much as the flow cross section is reduced, so the volume required is less, and the greater power cost may be offset by the reduced invest-ment. High throughput is desirable but often limited, as in packed columns, because flooding makes it impossible to maintain countercurrent flow.

Perhaps the most effective way to reduce the size of the equipment is to provide a very large surface of contact between phases. The surface-to-volume ratio $a$ can often be increased more easily than can the $K$ in $Ka$. Gas-liquid contactors can be made small by dispersing one phase as small drops or bubbles, thus providing a very large contact area per unit volume. The use of porous catalysts accomplishes the same thing in catalytic reactors. In many instances, more can be accomplished by designing for larger surface than for an enhanced transfer coefficient.

Most mass-transfer equipment is operated with continuous flow of both phases, though batch operation is often employed, especially where a chemical reaction is being carried out, as by hydrogenation in an autoclave. Countercurrent operation is to be preferred if the equilibrium tends to become less favorable as mass transfer proceeds in batch or cocurrent operation. Reflux is required in most distillation equipment, including "reboiled absorbers," and is sometimes used in liquid extrac-tion and moving-bed adsorbers.

Mass transfer between phases is normally accompanied by a heat effect, usually exothermic. The adiabatic absorption of hydrochloric acid by water, for example, results in a substantial increase in the temperature of the liquid and so increases the vapor pressure of HCl over the solution. This adverse effect on the equilibrium reduces the strength of the acid which it is possible to attain. A similar problem is encountered in the absorption of $NO_2$ to produce nitric acid. Cooling coils may be installed within the absorber, or the liquid may be removed, cooled, and returned. This second alternative is used in oil absorption of refinery gases rich in light hydrocarbons. Internal cooling coils can sometimes be designed to serve as packing to provide a large surface of contact between the phases.

The equilibrium properties of the system are of basic importance and the choice of absorbent, solvent, or adsorbent is critical. It is usually the intention to treat a stream to remove an impurity or to recover a product of value. The re-moval of one or a few constituents, but not others, may be required, so the treating agent must have equilibrium properties which make it selective. It must be cheap if it is to be discarded. Aqueous caustic can be used to remove traces of $CO_2$

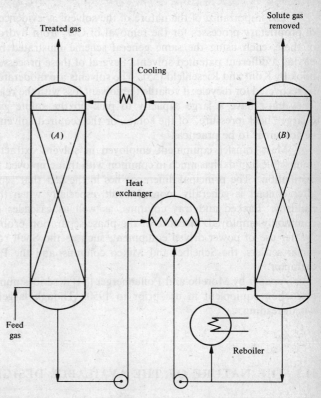

**FIGURE 11.1**
Gas absorber using a solvent regenerated by stripping. (a) Absorber; (b) stripper.

from hydrogen, but since the carbonate formed has little value, caustic cannot be used economically to remove large amounts of $CO_2$ from gas streams.

In most cases it is an economic necessity to recover and recycle the solvents employed in gas absorption and liquid extraction, or the carbon or other solids used in adsorption. An important example of this is the use of aqueous ethanol-amines for the absorption of $CO_2$ in hydrogen manufacture. These solutions form loose chemical compounds with $CO_2$ which can be broken up by heat or reduced pressure to release the $CO_2$ and permit the solvent to be reused. Figure 11.1 shows a common arrangement of absorber and distillation column for this purpose. The same system is employed with a regenerable absorbent to remove $SO_2$ from power-plant stack gases and Claus-plant tail gases, using either a "pressure swing" or "temperature swing," or both. In a pressure swing, the absorber operates at a much higher pressure but at about the same temperature as the stripper. In a temperature swing, the two may be at about the same pressure, but the absorber is cold and the stripper heated.

The importance of the nature of the solvent is evidenced by the proliferation of proprietary processes for the removal of $CO_2$ from hydrogen. There are many of these, each using the same general scheme illustrated by Fig. 11.1, but each having a different patented solvent. Several of these processes are described in the book by Kohl and Riesenfeld [55]. The solvents are moderately expensive and must be recovered for recycle; if volatile the solvent loss with the rejected $CO_2$ is excessive. They must have a large capacity to pick up the solute gas without developing a large "back pressure" of the solute, or the required solvent circulation rate may be too great to be practical.

Mass-transfer equipment employed in solvent extraction (transfer between immiscible liquids) has much in common with that employed for gas absorption and distillation. The principal difference lies in the fact that separation of the phases after contact is generally more difficult, especially when there is a tendency to emulsify. Packed and tray columns, as well as cascades of stirred vessels, are commonly employed. Because of the phase-separation problem, however, there is greater use of power-driven equipment, such as the Shell rotating-disc contactor, mixer settlers, the Scheibel and Mixco columns, and the Podbielniak centrifugal extractor.

A paper by Morello and Poffenberger [74] describes numerous types of solvent extraction equipment in use prior to 1950. Though largely obsolete, it is well worth reading.

## 11.3   THE NATURE OF THE AVAILABLE DESIGN DATA

The design engineer is usually required to determine the type and detailed specifications of the equipment to be installed for some well-defined purpose. That is, the amount and composition of the feed and the required degree of separation are stipulated. He must choose between many alternatives to develop the most economic system. Alternatively, he must be able to predict the effect of changes in load or operating variables on the performance of existing equipment. For these purposes he needs to understand the theory of mass transfer and phase equilibria, as well as the nature of the phenomena involved in the operation. Finally, he must have access to empirical design and cost data.

The early chapters represent an attempt to describe the existing relevant theory, phenomena, and calculation methods for design. Perry's "Chemical Engineers' Handbook" [83, 4th ed., Secs. 14 to 18 and 20] provides an extensive review of most of the published design data. Since this is readily available to most chemical engineers, no attempt will be made to summarize the same material. Rather, the sections which follow will try to indicate how handbook and other published data may be used for design purposes. This rather cursory treatment of the operating characteristics of mass-transfer equipment is intended to bridge the gap between the design procedures presented in Chap. 9 and the published data, so that the student may be better prepared to tackle quantitative design problems.

## 11.4  PACKED COLUMNS

Packed columns, or "towers," are widely used for gas absorption, and sometimes for liquid extraction. The column is vertical and usually cylindrical. Fresh or regenerated liquid absorbent is fed at the top and distributed over the packing, through which it falls by gravity. The solid packing is designed to distribute the liquid over a large surface, thereby providing a large gas-liquid contact area. The gas to be treated usually enters at the bottom in order to have countercurrent flow of the two streams. There are no moving internal parts; power is needed only for the gas blower and the pump which lifts the liquid to the top.

The early packing materials were sized stone or coke, but these were superseded by various packings manufactured for the purpose: Raschig rings, Berl saddles, and other proprietary types, including Intalox saddles, Tellerettes, and Pall rings; several are shown in Fig. 11.2. These packings may be of acid-proof ceramic, carbon, plastic or steel; Tellerettes are plastic and Pall rings are usually made of steel. The dry surface of "dumped" packings ranges from about 20 ft$^2$/ft$^3$ for the 3-in sizes to more than 200 ft$^2$/ft$^3$ for $\frac{1}{4}$-in sizes. Ceramic packings are used for the absorption of acid gases; Pall rings are common in the petroleum industry for use in noncorrosive hydrocarbon systems. The objective in designing a packing is to provide a large surface and yet have a low pressure drop for gas flow. The "packing factor," defined as the ratio of the dry surface area of the packing to the cube of the bed void volume fraction $(a_p/\varepsilon^3)$, provides a rough measure of how well this objective has been met. In general, a large surface means a high pressure drop, so on these counts there is not a wide spread among the various packing materials of a given nominal size. The main virtue of the manufactured packings is the size uniformity of the pieces, which makes for greater uniformity of gas and liquid flow through the bed.

The packed volume required is small if the gas-side mass-transfer resistance is controlling, in which case a packing of low surface area and low pressure drop may be used. This may consist of vertical sheets, wood grids, or plastic eggcrate forms. The splash decks used in cooling towers are examples, though these have very low wetted-surface areas and rely largely on the splash and drops of water to provide the gas-liquid contact surface needed.

Recent variations of the standard packed column are the Aerotec Floating Bed Scrubber, and the Turbulent Contact Absorber developed by Universal Oil Products Co. These employ a fluidized bed of plastic spheres (Ping Pong balls) constrained between coarse horizontal screens several feet apart. The upward flow of gas lifts and agitates the light spheres, which are well-wetted by the downflow of liquid. Flow rates are high but limited to the range in which the spheres neither rest on the lower screen nor are held against the upper screen. Absorption data are reported by Douglas [23] and by Wozniak and Østergaard [122].

Gas flow through most beds with fixed packings is uniform over the cross section of the column, but flow along the wall may be excessive if the ratio of column diameter to packing size is less than 8 to 10. Though the liquid may be distributed uniformly over the packing at the top of the column, it tends to

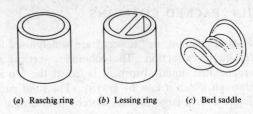

(a) Raschig ring      (b) Lessing ring      (c) Berl saddle

**FIGURE 11.2**
Individual pieces of manufactured packings.

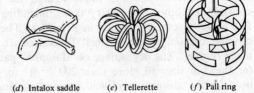

(d) Intalox saddle      (e) Tellerette      (f) Pall ring

gravitate toward the wall and should be collected and redistributed if the column is tall. This practice is much more common in Britain than in the United States. Maldistribution of liquid is partially offset by the greater transfer coefficients attained in regions where the local flow rate is greater than the average flow, particularly when the liquid-side resistance is controlling [100].

Since blower power represents a significant operating cost, it is important to have data on *pressure drop* in packings. Such data are available on ceramic rings, Berl saddles, and various other packing materials. The pressure drop $\Delta P$, usually given as inches of $H_2O$ per foot of packed height, increases steeply with increase in gas-flow rate and somewhat less rapidly with increase in liquid rate. For air and water at $G = 200$ lb/(h)(ft$^2$) and $L = 1,000$ lb/(h)(ft$^2$), $\Delta P$ is 0.17 in $H_2O$/ft in $\frac{1}{2}$-in ceramic rings but only 0.013 in/ft in 2-in rings. If the blower efficiency is 65 percent, the power in this case for the $\frac{1}{2}$-in rings is (0.0018)(packed height in feet)(cross section in ft$^2$), hp, or 1.3 hp for a tower 8 ft in diameter with a packed height of 15 ft. The pressure drop per foot is lower for the larger packing sizes, but the height must be greater because the mass-transfer coefficient on a volume basis is smaller. Pressure drop is also smaller for the packings of a given size but larger void fraction. Sheet, grid, and eggcrate packings have very low $\Delta P$ per foot.

Figure 11.3 illustrates the nature of the pressure-drop data widely available in handbooks and manufacturers' technical bulletins. At low gas rates, $\Delta P$ increases approximately as the square of the velocity, even at high liquid rates. At high gas rates, however, the curves bend upward steeply because the column tends to flood. The vertical asymptote indicates the gas rate corresponding to *flooding* at the indicated liquid rate. Under flooding conditions the liquid cannot flow downward

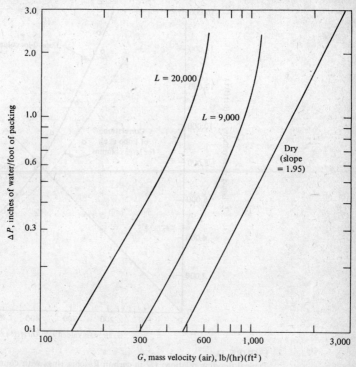

**FIGURE 11.3**

Pressure drop for airflow at 1.0 atm through 1-in ceramic Intalox saddles [83]. $L$ = countercurrent water flow, lb/(h)(ft$^2$).

over the packing; it is pushed up and out the top of the packing by the gas flowing upward at a high rate. The column is then inoperable as an absorber.

The *loading point* has been defined as the flow condition at which the curve of $\Delta P$ vs. $G$ begins to deviate appreciably from the straight line having a slope of a little less than 2. Referring to Fig. 11.3, one can see that for this packing with a liquid rate of 9,000, the air rate at flooding is about 1,100 lb/(h)(ft$^2$) and the loading point is in the vicinity of 700 lb/(h)(ft$^2$). The loading point is difficult to specify and it is more important to know the flooding rates; in any case, the entire pressure-drop curve is needed for the economic design of a gas absorber.

There are very few data on pressure drop in the flow of two liquids through packings, as in liquid extraction, but this is not often important except as it relates to flooding. For beds of small particles, such as catalyst pellets or extrudates in catalytic reactors, the Kozeny-Carman equation or the Ergun correlation (Reference 28) is used to obtain values of $\Delta P$ in single-phase flow. Few data are available, however, on pressure drop for cocurrent flow of both gas and liquids through beds

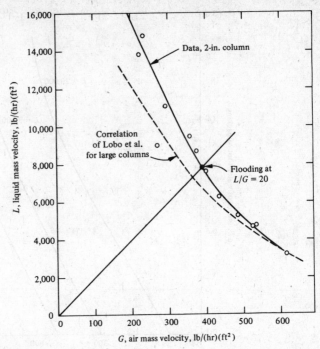

**FIGURE 11.4**
Typical flooding relation: 1/2-in carbon Raschig rings with countercurrent flow of air and water in a small laboratory column.

of small particles, as encountered in trickle-bed reactors. See, however, Weekman and Myers [119], Charpentier, Prost, and LeGoff [12], and Larkins, White, and Jeffrey [56].

The flooding characteristics of a packing are of obvious importance, since they determine the allowable gas and liquid flow rates, i.e., the minimum cross section of the column for any assumed or stipulated values of $L$ or $G$. This assumes that flooding results from the flow in the packing, and is not due to a poorly designed packing support plate. Furthermore, competition among vendors of columns is keen, and construction contracts are often awarded the company which can guarantee the performance of a slightly smaller column than those offered by competitors. In some cases the column capacity is limited by the development of excessive entrainment of liquid at flow rates well below those corresponding to flooding. Though flooding normally determines the limiting capacity, the actual flow rates to be used are properly determined by economic considerations, basically an economic balance of the costs of power and investment charges.

As suggested by Fig. 11.3, the limiting gas velocity decreases as the liquid rate increases. Figure 11.4 illustrates the nature of the flooding curve. The data shown were obtained in a small laboratory column; similar curves are obtained with various packings in larger columns. If the ratio $L/G$ is determined by design considerations fixing the operating line, then a straight line through the origin of slope $L/G$ will intersect the flooding curve at the maximum allowable $L$ and $G$. The economic balance of costs often leads to a choice of $L$ and $G$ in the vicinity of 70 to 80 percent of the flooding rates.

Methods of correlating flooding data with allowance for the physical properties of the fluids, together with the packing factor $a_p/\varepsilon^3$, have been developed and permit approximate predictions. One of the early correlations [103] was in the form of a graph of

$$\frac{\bar{U}_s{}^2 a_p \rho_G}{g \varepsilon^3 \rho_L} \mu^{0.2} \quad \text{vs.} \quad \frac{L}{G} \sqrt{\frac{\rho_G}{\rho_L}}$$

where $U_s$ = superficial velocity of the gas
$\quad g$ = acceleration due to gravity
$\rho_G$ and $\rho_L$ = densities of gas and liquid
$\quad \mu$ = viscosity of the liquid, cP

The general correlation obtained was later modified by Lobo, Friend, Hashmall, and Zenz [62] and by Zenz [124], using the same coordinates. The dashed curve on Fig. 11.4 represents the correlation of Lobo et al. for columns of large cross-section. These correlations have been used widely, though there is evidence that flooding rates may be considerably lower than indicated if the liquid surface tension is low, as for water containing a surfactant [75]. It appears that the liquid flooding rate, other quantities fixed, is inversely proportional to the cube of the surface tension in the range 32 to 72 dyn/cm, so that incorporation of $(\sigma_w/\sigma)^3$ in the abscissa of the usual correlation [103] brings the data on liquids of low surface tension into line with the data on water. The basic mechanism of flooding of packed and wetted-wall columns is discussed by Hutton, Leung, Brooks, and Nicklin [49a].

The possibility of foaming, which can be serious, is not covered by the flooding correlation. In one instance a large high-pressure absorber designed to remove methane from hydrogen by oil absorption performed poorly. The difficulty was corrected by changing to another suitable oil which did not foam, as determined by simply shaking samples in a glass graduate.

Most of the reported flooding data are for gas-liquid systems, but there are several publications dealing with limiting flows in packed columns used for liquid extraction, e.g., Ref. 17.

The amount of liquid held by the packing is known as the "liquid holdup," usually expressed as cubic feet of liquid per cubic feet of packed volume. In liquid extraction, the total holdup of the two liquid phases is the packing void fraction. In gas-liquid systems, the liquid draining after flow is stopped is the "dynamic" or "operating" holdup. The liquid not draining but held by the packing is the "static" holdup. The total holdup is the sum of the two. The total holdup increases with liquid flow rate but is nearly independent of gas flow below the loading point. Figure 11.5 illustrates the nature of the data available for counter-current flow of gas and liquid in various packings [106, 111]. Mohunta and

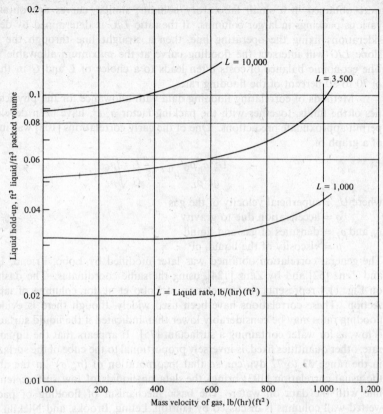

**FIGURE 11.5**
Total hold-up in 1.0-in unglazed white porcelain rings [106].

Laddha [73] have developed a correlation of such data for small rings and spheres. There are limited data on holdup with cocurrent flow [41, 91, 44, 22].

The liquid holdup may be important in cases where the packed column is intended to operate both as a gas absorber and a chemical reactor. The gas passage is reduced by the presence of liquid, and the residence time for a gas-phase chemical reaction is correspondingly less than would be calculated from the dry bed void volume.

A large liquid holdup is desirable if absorbed gases are to react in the liquid phase, though this may not be so if the reaction is fast (see Chap. 8). In the production of acetaldehyde by the Oxo process, for example, CO, $H_2$, and $C_2H_4$ are absorbed simultaneously at high pressure to produce the aldehyde by reaction in liquid carrying a soluble catalyst. Holdup data show, however, that with presently available catalysts the production capacity using ordinary packings would be small because the liquid volume is not large. One might say that for this application a packed column would have excessive capacity for gas absorption in relation to its performance as a chemical reactor.

Even with fairly high liquid-flow rates, the packing surface is not entirely wetted, as shown by the early work of Mayo, Hunter, and Nash [69] and of Shulman et al. [105]. Even wet surface may be inactive if the liquid is held up and is stagnant. The total surface of the dry packing, therefore, is but a rough index of the surface available for mass transfer. Stagnant liquid elements equilibrate with the flowing gas and no mass transfer occurs. In this situation the active surface may be only 25 to 50 percent of that of the dry packing, as found by de Waal and Van Mameren [21] for 1-in ceramic Raschig rings (see also Chap. 8). However, all the wetted surface, including the stagnant liquid elements, is active for the evapora-tion of a pure liquid into a gas. In experiments with a wetted-wall absorption column, Bond and Donald [6] found that the water flow required for complete wetting of the surface had to be increased as the rate of exothermic gas absorption was increased by feeding gas richer in ammonia. The effect was related to the difference in surface tension between the bulk and the surface of the liquid.

A recent paper by Puranik and Vogelpohl [83a] reviews the published data on the effective contact areas of gas on liquid for counterflow in packed columns. These authors propose a general correlation, distinguishing between dry surface and areas available for vaporization and for gas absorption with and without simultaneous reaction. Their correlation appears to be good to $\pm 20$ percent for a wide variety of systems, packings, and flow rates.

Data on flooding velocities provide a basis for choosing a suitable cross section of a mass-transfer device, but the height or length in the direction of flow must also be specified. For this purpose it is essential to have data on the HTU or $Ka$ to be expected under the operating conditions. Chapter 9 describes the methods of calculating the height or packed volume required to meet design criteria. The mass velocities of the streams are fixed when the cross section is selected, and pressure drops and power are then calculated from these and the packed heights.

There are literally hundreds of published investigations reporting *rate co-efficients* on packed columns used for gas absorption and solvent extraction, with and without simultaneous chemical reaction. The large majority of these describe laboratory studies with small columns, and the data are of questionable value for use in the design of large equipment. Data on columns 10 in or more in di-ameter are relatively scarce, but much of the data suitable for design purposes is summarized in the Perry handbook [83] and will not be reproduced here. The nature of the available correlations will be described briefly, with examples of how they can be used.

For gas absorption or desorption it may be expected that there will be resis-tances to mass transfer on both sides of the gas-liquid interface. In many cases, however, one of the two is negligible in comparison with the other, and either the gas-side or liquid-side resistance is "controlling." Consequently, much of the data pertains to systems in which $K_G a \approx k_G a$, or $K_L a \approx k_L a$. In principal, such data can be employed in intermediate cases by adding the two resistances, as shown

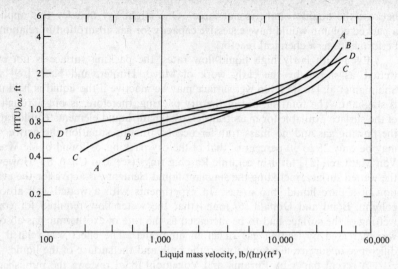

FIGURE 11.6

Data on desorption of oxygen from water, with countercurrent flow in ceramic Raschig rings. Ring sizes, in: A-A, 0.5; B-B, 1.0; C-C, 1.5; D-D 2.0 [101].

in Chaps. 5 and 9. It may be noted that the absorption of a solute of low solubility requires a large ratio $L/G$, giving a large $k_L a$ and small $k_G a$. The situation is reversed if the gas is highly soluble. The compensating effects of solubility and flow rates tend toward making the resistances in both phases important.

Figure 11.6 illustrates the nature of the data on rates of mass transfer where the liquid-side resistance is controlling. The curves represent Holloway's data [101] on desorption of oxygen from water at 25°C in a 20-in column packed to depths of 16 to 49 in with Raschig rings of various sizes. The curves of $(HTU)_{OL}$ $(= L/\rho_L K_L a)$ vs. $L$ turn up in the vicinity of the loading point at the modest gas rates employed ($G = 100$ to 230); at lower liquid rates, variation in $G$ had no affect on $(HTU)_{OL}$. These and other data on standard packings are given in Perry [83].

Holloway measured desorption coefficients for hydrogen and $CO_2$ as well as for oxygen. Vivian and King [115], in a similar study, desorbed helium and propylene in addition to the three gases used by Holloway, using $\frac{1}{2}$-in rings in a 12-in column. The data show convincingly that $k_L a$ is proportional to the square root of the molecular diffusion coefficient of the solute gas in water ($HTU_L$ proportional to $D^{-1/2}$). Since $D$ for oxygen in water at 25°C is $2.41 \times 10^{-5}$ cm²/s, the data shown in Fig. 11.6 may be used to predict $k_L a$ or $HTU_L$ for another solute in water at 25°C by introducing the square root of the ratio of the new $D$ to $2.41 \times 10^{-5}$ cm²/s. Holloway obtained a correlation of his data (see Perry), indicating $HTU_L$ below the loading point to be proportional to $(Sc)^{1/2}$ and to $(L/\mu)^n$, where $n$ fell in the range 0.22 to 0.35 for the different packings. However, since only water

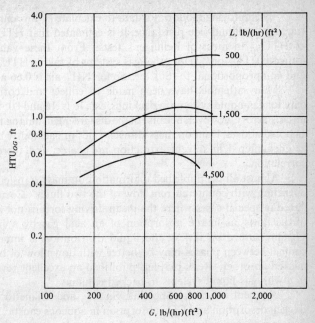

**FIGURE 11.7**
Data of Fellinger [30] on absorption of ammonia by water in 1.0-in Raschig rings.

was used, the effect of varying liquid viscosity was not established. Perry quotes other correlations of these and similar data.

The situation with regard to systems where the gas-side resistance is controlling is poor, in that published values are in disagreement. Lynch and Wilke [63] illustrate the wide scatter of the published data by a graph of $HTU_G$ vs. $G$ at $L = 1,575$; at $G = 200$ to $800$, reported values of $HTU_G$ vary some threefold for the same packing and same gas-flow rate. Perhaps the main reason for this is that equilibrium is approached in only a few inches in such systems, as brought out by Example 6.4, so that driving forces are difficult to measure accurately, and end effects cannot be easily eliminated or estimated. Relatively few data have been obtained in columns 10 in or more in diameter.

An estimate of $HTU_G$ for spiral tile, Raschig rings, and Berl saddles may be obtained from the extensive data of Fellinger [30], who employed an 18-in column for the absorption of ammonia from air by water. Figure 11.7 is a sample of the several graphs of Fellinger's data to be found in the third edition of Perry. It is seen that $HTU_{OG}$ ($\approx H_G$) increases with gas-flow rate, but passes through a maximum in the vicinity of the loading point, at a value of $G$ some 70 percent of flooding. Beyond the loading point, $K_G a$ tends to increase rapidly with increase in $G$. By contrast, it follows from Fig. 11.6 that $K_L a$ tends to become insensitive to changes in $L$ in the same region.

By employing Holloway's data to calculate $k_L$ for ammonia in water and subtracting the liquid-side resistance, it is estimated that $HTU_G$ was 85 to 95 percent of $HTU_{OG}$ in most of Fellinger's tests. From these values of $HTU_G$ one may estimate $HTU_G$ or $k_G a$ for other gas systems by taking $HTU_G$ proportional to $(Sc)^{1/2}$ and $k_G a$ proportional to $(Sc)^{-1/2}$; Sc for $NH_3$–air is 0.66 at 25°C.

Many attempts have been made to collect and correlate mass-transfer-rate data for gas-liquid systems in packings, e.g., Refs. 16 and 80. The results are generally unsatisfactory, however, because of the disagreement among the various investigators, as noted above. Data on mass-transfer rates in packings with liquid-liquid systems, as encountered in solvent extraction are scarce, and there is no reliable general correlation.

Almost all the published information on mass transfer in packings pertains to operation with countercurrent flow of the two fluids. Cocurrent flow is to be preferred in special cases where the mean driving force is not affected by the direction of the flows, as in the absorption of an acid gas by a highly alkaline solution. The pressure drop is less, the liquid distribution is improved, and the effective contact between phases may be better with downflow of both gas and liquid in a packed absorber. Reiss [85] has published an excellent review of cocurrent operation with gas-liquid systems in packed columns.

A recent article by Specchia, Sicardi, and Gianetto [109a] reports data on oxygen desorption and $CO_2$ absorption in aqueous caustic, using cocurrent upward flow in a packed column 8.0 cm in diameter. In another study, Gianetto, Specchia, and Baldi [39a] report data on ammonia absorption and oxygen desorption with cocurrent downward flow in an 8-cm column, using three different 6-mm packings.

Gas-side coefficients $k_G a$ for the absorption of $SO_2$ in aqueous solutions of sodium hydroxide are reported by Shende and Sharma [99a], using $\frac{5}{8}$-in polypropylene and stainless steel Pall rings in 4- and 8-in-ID columns. High gas- and liquid-flow rates were attained in cocurrent downflow. The effective interfacial area increased sharply with liquid flow, and to a lesser degree with gas flow. The increase in $k_G a$ with liquid-flow rates was explained as being entirely due to the greater effective contact area. Ufford and Perona [112a] absorbed $CO_2$ from air in cocurrent downflow using a 10-cm glass column and small Raschig rings and Berl saddles. Gas rates were low, but water rates up to 60,000 lb/(h)(ft²) were employed. An appreciable effect of gas rate was noted, and $K_L a$ was found to be substantially lower than reported in similar studies using $CO_2$ and water in countercurrent flow.

It is usual to operate packed columns with steady flow of both phases. Cannon in 1956 [11], and various others in recent years, however, have shown that under some conditions the mass-transfer performance and capacity can be improved by intermittent flow of one of the streams. In countercurrent rectification or gas absorption, for example, the liquid flow is stopped, the liquid holdup allowed to approach equilibrium with the flowing gas, and the liquid flow then continued. The development of the concept is reviewed by Schrodt [97]. See also Ref. 33. As applied to rectification, the vapor flow may be interrupted for a brief time while the liquid on each plate is drained to the plate below. This method of operation is not common, but the mathematical analysis of its possibilities presents a problem which should interest students. A worked example showing one approach is to be found in a book by Jensen and Jeffries [50]. A recent article [87] describes the application of the same idea to improve selectivity in a series of well-stirred chemical reactors.

Intermittent liquid flow has sometimes been used to improve the wetting of the packing in countercurrent gas absorption. If the solute is highly soluble in the liquid, the indicated $L/G$ may be so small that the liquid-mass velocity is inadequate to wet the packing well. If the liquid is stored and dumped intermittently, as by a laboratory Soxhlet device, the packing is thoroughly wetted and the performance improved. Pulsed flow of packed and plate columns has been shown to improve the performance for gas absorption, distillation, and liquid-liquid extraction.

In designing packed columns for gas-liquid, liquid-liquid, or fluid-solid contacting, it is common practice to assume "piston," or "plug," flow, ignoring axial dispersion. This can lead to unsafe designs, since axial dispersion decreases the mean driving force for solute transfer, as illustrated diagrammatically by Fig. 11.8. There is a sudden concentration change at each inlet, and the concentration slope is zero at each outlet. The mean potential is much less for curves $B$ than for the piston-flow case, illustrated by curves $A$.

The effect of axial dispersion, or "back-mixing," is especially serious in solvent extraction when unpacked "spray" columns are employed, since small buoyant liquid drops readily mix and reverse their direction of flow. In extreme cases the mixing end to end is so effective that the column operates as a well-stirred vessel. The effect is usually less pronounced in packed gas-liquid contactors, though an absorber or stripper can be a spectacular failure if back-mixing of low-velocity gas is ignored in the design. High $L$ and small $G$ can lead to "pumping" of the gas from top to bottom of a countercurrent absorber, destroying the benefit of countercurrent operation. In stripping $CO_2$ from water by air in a column packed with 2-in steel rings, Cooper, Christl, and Peery [14] obtained values of $(HTU)_{oL}$ two to three times greater than measured by Holloway in similar equipment. In these tests $L$ was very large, ranging from 13,200 to 56,000 lb/(h)(ft$^2$), while the superficial gas velocities were only 0.08 to 1.3 ft/s ($19 < G < 368$). These rather extreme conditions might be encountered in a high-pressure absorber designed to remove methane from reformer off-gas by a large flow of oil, in which methane is not very soluble.

The word back-mixing is used somewhat loosely in the literature. As Klinkenberg [54] points out, back-mixing implies the actual backflow in a direction opposite to that of the main flow. This is observed in liquid-liquid spray columns, in bubble columns, and in packed columns at large values of $L/G$, as noted above. However, the word is also used to describe the effect of axial dispersion by eddy diffusion in spreading the residence-time distribution of a stream. Hiby [43] shows a photograph of the spread of a dyed liquid issuing from a point source in the single-phase flow of liquid in a packing; no color appears upstream from the source, as called for by the diffusion model describing eddy diffusion, and there is evidently no back-mixing in the literal sense.

Section 4.10 discusses axial dispersion in single-phase flow at some length, and

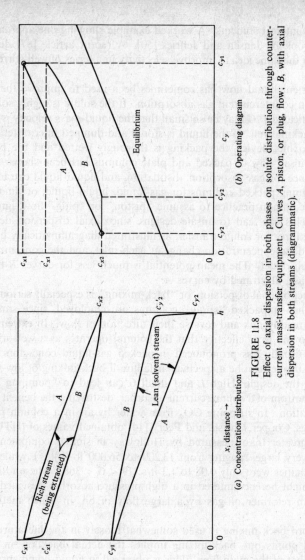

**FIGURE 11.8**

Effect of axial dispersion in both phases on solute distribution through counter-current mass-transfer equipment. Curves $A$, piston, or plug, flow; $B$, with axial dispersion in both streams (diagrammatic).

describes two of the models used in studying the phenomena. The diffusion model is used in most of the literature dealing with the effect of axial dispersion in mass-transfer equipment, and will be used here. Forcing the data to fit the diffusion model is somewhat of an empirical approach, though axial dispersion in open conduits, due only to velocity gradients, was treated successfully by Taylor, using the diffusion equation, as shown in Sec. 4.11.

For axial dispersion in steady two-phase flow, as in a packed gas absorber or liquid-liquid extraction column, Eq. (4.47) is assumed to apply to each phase, and radial concentration gradients are ignored:

$$E_x \frac{d^2 c_x}{dx^2} - \bar{U}_x \frac{dc_x}{dx} - K_x(c_x - c_x^*) = 0 \tag{11.1}$$

$$E_y \frac{d^2 c_y}{dx^2} + \bar{U}_y \frac{dc_y}{dx} + K_y(c_x - c_x^*) = 0 \tag{11.2}$$

The subscripts $x$ and $y$ refer to the phase being extracted and to the solvent phase, respectively. Thus, in a gas absorber, $x$ refers to the gas and $y$ to the liquid. $\bar{U}_x$ and $\bar{U}_y$ are the superficial velocities, and $E_x$ and $E_y$ are the dispersion coefficients based on the total cross section of the column. The symbol $c_x^*$ represents the concentration of solute in the $x$ phase in equilibrium with $c_y$ at the same level in the column.

These equations have been solved by Sleicher [108], Miyauchi and Vermeulen [72], and by Hartland and Mecklenburgh [42], using the linear equilibrium relation $c_x^* = q + mc_y$. Each uses the same boundary conditions (see Fig. 11.8a):

$$\bar{U}_x c_{x1} - \bar{U}_x c_{x1}' = E_x \frac{dc_x}{dx} \quad \text{and} \quad \frac{dc_y}{dx} = 0 \quad \text{as } x \to 0 \quad \text{(just inside the column)}$$

$$\bar{U}_y c_{y2}' - \bar{U}_y c_{y2} = E_y \frac{dc_y}{dx} \quad \text{and} \quad \frac{dc_x}{dx} = 0 \quad \text{as } x \to h$$

The analytical solution to this set of equations is complex and not easily used. However, McMullen, Miyauchi, and Vermeulen [65] have tabulated numerical results for a large number of cases. Watson and Cochran [117] have obtained a simple empirical equation to represent computer solutions, but their nomenclature seems to be ambiguous. Sleicher and Miyauchi and Vermeulen offer relatively simple approximate equations, and Rod [94] describes a graphical method. A simplified model of the effect of axial mixing involving the concept of a cascade of ideal mixers in each phase is described by Kerkhof and Thijssen [53a].

In many practical cases it is the dispersion in but one phase that is important. This is obvious where there is but a single phase flowing, as in an ion-exchange column, but it holds also in two-phase flow when there is little concentration change in one phase. In the desorption of oxygen from water into air, for example, the dispersion in the gas phase is of little consequence. The analytical solution describing the effect of *single-phase* axial dispersion on column performance has been presented by several writers [18, 120, 72]:

$$\psi_1 = \frac{4b \exp(P_{cx}/2)}{(1 + b)^2 \exp(P_{cx} b/2) - (1 - b)^2 \exp(-P_{cx} b/2)} \tag{11.3}$$

where $\psi_1$ = fraction unaccomplished approach of the $x$ phase concentration to equilibrium with the $y$ phase feed = $[c_{x2} - (q + mc_{y2})]/[c_{x1} - (q + mc_{y2})]$ = $e^{-N_{oxP}}$

$P_{cx}$ = column Peclet number for the $x$ phase = $\overline{U}_x h/E_x$

$b = (1 + 4N_{ox}/P_{cx})^{1/2}$

$c_x^* = q + mc_y$

$N_{ox} = h/H_{ox}$, where $H_{ox}$ is the "true" height of a transfer unit $\equiv \overline{U}_x/K_{ox} a$

$N_{oxP}$ = number of transfer units ("exterior-apparent") based on solute concentrations in inlet and outlet streams = $-\ln \psi_1$

The situation considered is one where the absorption factor $m\overline{U}_x/\overline{U}_y$ is essentially zero, as in typical cases of desorption of a relatively insoluble gas, where $m$ is very small. $H_{ox}$ is the true HTU, related to the mass-transfer coefficient $K_{ox}$ by definition. $N_{ox}$ is $h/H_{ox}$, and $N_{oxP}$ is $h/N_{oxP}$, the apparent number of transfer units as calculated from overall column performance. Because of the axial dispersion, $H_{oxP}$ is larger than $H_{ox}$, and $N_{ox}$ is greater than $N_{oxP}$. For gas absorption, $m\overline{U}_x/\overline{U}_y$ is the $mG_M/L_M$ of Chap. 9.

A column "efficiency" can be defined as $H_{ox}/H_{oxP}$, which is one measure of the effect of axial dispersion. Figure 11.9 shows this as a function of $N_{oxP}$ and $P_{cx}$, as found from Eq. (11.3). If $P_{cx}$ were 1.0, 1 ft of packing found to provide one transfer unit by the usual calculation ($N_{oxP} = H_{oxP} = 1$) would have an efficiency of 71 percent, that is, $H_{ox} = 0.71$ ft. If the same packing were used with the same flow rates to approach equilibrium more closely, requiring, say, $N_{oxP} = 8$, then $P_{cx}$ would increase in proportion to the column length $h$, which is $N_{ox} H_{ox}$. Using Fig. 11.9 it is found by trial that $P_{cx}$ is 10, $N_{ox}$ is 14, and $h = 10$ ft. Ignoring axial dispersion, $h$ would be calculated to be $N_{oxP} H_{oxP} = 8$ ft.

Data on axial dispersion for the flow of a single phase through a packed bed are summarized and discussed in Sec. 4.10. A limited amount of data on dispersion in two-phase flow in liquid-liquid contacting devices has been collected by Vermeulen, Moon, Hennico, and Miyauchi [114]. This reference includes data on packings with steady flow, pulsed packed columns, unpacked "spray" columns, Shell RDC extractors, and Mixco extractors. Bibaud and Treybal [5] have also reported data on axial dispersion in both phases in a Mixco (or Oldshue-Rushton) extractor.

Data on axial dispersion in packings with countercurrent flow of gas and liquid are quite scarce, though there are several reports of studies in small laboratory columns [34, 96, 8]. The only data on commercial packings in a large column appear to be those of Dunn et al. [26] and Woodburn [121a], who used a tracer technique to obtain values of $E_a$ in both liquid and gas phases. Three packings were studied, using a column 2 ft in diameter packed to a depth of 6 ft. Figure 11.10 represents Dunn's correlation of his data for both air and water. The "packing Peclet number" $P_p$ is used instead of $P_c$, since the packing size is more

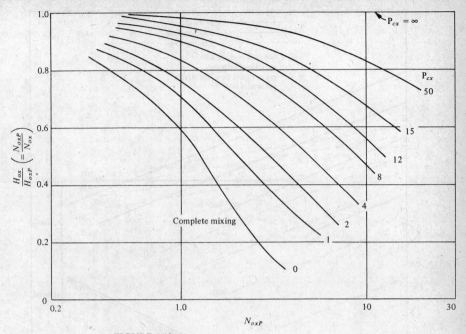

**FIGURE 11.9**
Effect on column performance of axial dispersion in phase being extracted; $mU_x/U_y \to 0$ (negligible concentration change in solvent phase).

relevant to dispersion than the packing height $h$. In the gas phase, $E_a$ evidently increases some threefold as the liquid rate is increased from zero to $L = 12,000$ at constant $G$, and in the 1-in packings there is a similar increase in $E_a$ at constant $L$ as $G$ is increased from 300 to 1,200. In the liquid, however, $E_a$ increased only moderately with increase in $L$ and was essentially independent of gas-flow rate in the range studied.

Woodburn measured only axial dispersion of the gas, with countercurrent flow of air and water in a 1-ft-ID column packed to a depth of 36 ft with 1-in stoneware Raschig rings. Measured values of $E_a$ were somewhat larger than those reported by Dunn, but the range of flow rates much greater. At $L = 7,600$ and $G = 300$, $P_p$ was found to be about 0.6, whereas Dunn found $P_p = 1.14$ at the same flow rates, as seen from Fig. 11.10.

Data on small columns agree roughly with Fig. 11.10. Sater and Levenspiel [96], using $\frac{1}{2}$-in rings and saddles in a 4-in column, obtained values of $P_p$ for liquid flow falling 20 to 40 percent below the lowest line on Fig. 11.10. Furzer and Michell [34], who employed $\frac{1}{4}$-in rings in a 2-in column, obtained values of $P_p$ nearly twice as large as those reported by Dunn for liquid flow in 1-in rings.

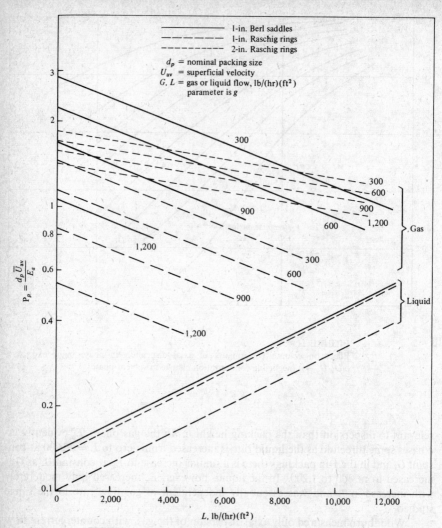

FIGURE 11.10
Axial dispersion coefficients for countercurrent flow of air and water in packings (data of Dunn, Vermeulen, Wilke, and Word [26]).

Hochman and Effron [44] report data on axial dispersion of liquid in trickle-bed operation of a column containing $\frac{3}{16}$-in glass beads, from which it would appear that axial dispersion of liquid has little effect on conversion in a reactor containing catalyst in the form of small pellets.

An illustration of the use of experimental values of $E_a$ to estimate the effect of axial dispersion on column performance is given in Example 11.1, which follows.

This pertains to the relatively simple case of dispersion in but one of the two flowing streams and a negligible solute concentration change in the second phase.

Miyauchi and Vermeulen's case 5 (their Table 1) explains the calculation procedure when dispersion in but one phase is important but where the concentration change in the second phase is substantial ($m\overline{U}_x/\overline{U}_y$ finite). These authors provide a numerical example of the use of their solution of the general case, with dispersion and concentration change in both phases.

## Trickle Bed Reactors

Fixed-bed reactors are widely employed in the chemical and petroleum industries, using solid catalysts to promote various chemical reactions. In many cases, as in the hydrosulfurization of medium to heavy oils, the feed is only partially vaporized. This feed is mixed with hydrogen and the gas and liquid flow in cocurrent fashion down through a bed of small pellets or extrudates, sometimes only $\frac{1}{32}$ in in diameter.

Much of the material in this section is related to mass transfer in such reactors. However, most of the published data is on packings designed for gas absorption, ranging in size from $\frac{1}{4}$ to 3 in, and much less is known about flow and mass transfer for beds of small particles. However, Satterfield's recent review [96a] provides a summary of information published through 1974 on pressure drop, mass transfer, holdup, flow patterns, and axial dispersion in such systems.

In most applications of trickle beds the reactants are present in both gas and liquid, and all reactants must reach the active surface of the porous catalyst particles. The liquid film is very thin and the resistance to transfer from gas to outer surface of the particle is negligible unless the catalyst is exceptionally active (low effectiveness factor). The major mass transfer resistance is that encountered by reactants and products diffusing in the liquid-filled pores, but even this may be negligible if the chemical conversion rate is low (high effectiveness factor). Overall conversion is frequently reduced appreciably by uneven liquid flow and the resulting spread of residence times of different elements of the flowing liquid.

EXAMPLE 11.1    *Axial Dispersion in a Packed Desorber*    Holloway's data included one test (run 158) in which oxygen was desorbed from water into air flowing at 1.0 atm in a 20-in-ID column packed to a depth of 6.1 in with dumped 2-in Raschig rings. The water and air rates $L$ and $G$ were 4,000 and 230 lb/(h)(ft$^2$), respectively, and the reported $(HTU)_{OL}$ at 25°C was 0.96 ft, calculated assuming piston flow. The void fraction in the bed was 0.83; and the dry surface of the packing, 29 ft$^2$/ft$^3$.

Does it appear that the assumption of piston flow of both air and water introduced an appreciable error in the reported $(HTU)_{OL}$? If a desorber were to be designed to bring the oxygen content of the water to within 1.0 percent of equilibrium with air, how much would the calculated packed height be in error if axial dispersion were ignored? Assume $L$ and $G$ to be 4,000 and 230, as in the test.

SOLUTION    This is clearly a case where essentially all the resistance to mass transfer is in the water phase. Oxygen solubility is low, and the oxygen concentration in the air increased by a negligible amount as a result of the desorption from the water. Axial dispersion in the gas phase could not have affected the mean driving force, and attention can be focused on the liquid.

Using the nomenclature of the preceding section, $h = 0.508$ ft; $\overline{U}_x = 4{,}000/(62.3 \times 3{,}600)$ $= 0.0178$ ft/s; $\overline{U}_y = 230 \times 359 \times 298/(29 \times 3{,}600 \times 273) = 0.863$ ft/s; $N_{oxP} = 0.508/0.96 = 0.529$; $\psi_1 = e^{-0.529} = 0.589$; $H_{oxP} = (HTU)_{OL} = 0.96$ ft.

The Henry's-law constant for oxygen in water at 25°C is $4.38 \times 10^4$ atm/mole fraction, so

$$m = \frac{82.07 \times 298}{4.38 \times 10^4 \times 18} = 0.031$$

whence the absorption factor $m\overline{U}_x/\overline{U}_y = 0.00064 \approx 0$.

From Fig. 11.10, $P_p$ is 0.21 and $P_{cx}$ is $0.21 \times 0.508 \times 12/2 = 0.64$. Substituting in Eq. (11.3),

$$\psi_1 = 0.589 = \frac{4be^{(0.64/2)}}{(1 + b)^2 e^{(0.32b)} - (1 - b)^2 e^{(-0.32b)}}$$

with $b = (1 + 4N_{ox}/0.64)^{1/2}$. Solving this by trial, $N_{ox}$ is found to be 0.659, whence $H_{ox} = 0.508/0.659 = 0.77$ ft. This indicates that the true $H_{ox}$ was only $(0.77/0.96) \times 100$, or 80 percent of the $(HTU)_{OL}$ value reported (and shown by Fig. 11.6).

If the new tower is to desorb oxygen to within 1.0 percent of equilibrium, $\psi_1 = 0.01$. The new height $h$ is $H_{ox}N_{ox}$, or $0.77N_{ox}$ ft, since $H_{ox}$ is presumed to remain constant, as is $P_p$. The column Peclet number is $0.21h \times 12/2 = 1.26h = 0.97N_{ox}$, and $b = (1 + 4N_{ox}/0.97N_{ox})^{1/2}$ $= 2.263$. Substituting in Eq. (11.3),

$$0.01 = \frac{4 \times 2.263 e^{(0.97N_{ox}/2)}}{(1 + 2.263)^2 e^{(0.97 \times 2.263N_{ox}/2)} - (1 - 2.263)^2 e^{(-0.97 \times 2.263N_{ox}/2)}}$$

whence $N_{ox} = 7.25$, and $h = 7.25 \times 0.77 = 5.6$ ft.

Had axial dispersion been ignored, the height of packing needed would have been calculated to be $N_{oxP}H_{oxP} = (-\ln \psi_1)(0.96) = 4.4$ ft. The simple design method based on the assumption of piston flow would evidently lead to an unsafe design. Conversely, had the experimental data been obtained on the taller column, the design of the shorter column would be overly conservative.

The correction to be applied to allow for axial dispersion in the gas stream would appear to be small in most practical cases of gas absorption in packings. Experimental values of $P_p$ are generally greater than for liquids, but $E_a$ is usually larger because the typical linear flow velocities of gases are much larger. There is a smaller residence time for the axial dispersion to affect the concentration.

For example, assume $H_{ox}$ to be 1.0 ft for the absorption of ammonia from air by acid in 1.0-in-ring packing with $L = 1{,}000$ and $G = 600$. From Fig. 11.10, $P_p$ is about 1.1 (the gas is the $x$ phase; dispersion in the liquid is unimportant). Then $P_{cx}$ is $13.2N_{ox}$ and $b$ is 1.142. Substituting in Eq. (11.3), it is found that the efficiency $N_{oxP}/N_{ox}$ is essentially constant at 0.94 over a wide range of values of packed heights.

*Note:* In many practical cases the second term in the denominator of Eq. (11.3) is negligible. The use of Eq. (11.3) is greatly simplified: $N_{oxP}$ becomes linear in $N_{ox}$.     ////

EXAMPLE 11.2  *Acetone Recovery*  It is proposed to recover acetone from an acetone–air stream by absorption in water, using a countercurrent packed scrubber. The gas is saturated with water at 15°C and contains 6.0 mole percent acetone. The flow rate is 4,000 ft³/min at

15°C and 1.0 atm. Suggest a workable design employing 1.0-in Berl saddles, with water supplied at 15°C. As a sample calculation, estimate the column diameter and packed height required to recover 90 percent of the acetone.

SOLUTION: The vapor pressure of water at 15°C is 12.8 mm Hg, so the composition of the gas to be treated is 0.0168 water, 0.06 acetone, and 0.923 air (mole fractions) with an average molecular weight of 30.5. The gas-feed rate is

$$\frac{4,000}{359} \frac{273}{288} \times 60 = 633.7 \text{ lb moles/h} \qquad \text{or} \qquad 19,328 \text{ lb/h}$$

and the acetone recovered as an aqueous solution is

$$633.7 \times 0.06 \times 0.9 = 34.2 \text{ lb moles/h} \qquad \text{or} \qquad 1,984 \text{ lb/h}$$

The acetone absorption can be expected to cause the water temperature to rise appreciably, so it is necessary to prepare an equilibrium curve allowing for this temperature change. Othmer, Kollman, and White [81] give data on the partial pressure of acetone over aqueous solutions at 15° and 30°C, and values for other temperatures can be obtained by interpolation assuming $\log p^*$ to be linear in $1/T$.

Lacking an enthalpy-concentration graph for the system, it will be assumed that the temperature rise from $X = 0$ to $X = X$ can be calculated on the assumption that the heat delivered to the water corresponds to the enthalpy of condensation of the acetone vapor—about 132 g cal/g, or 13,780 Btu/lb mole, and that the heat capacity of the solution is that of its water content. For example, when the mole fraction acetone has increased from zero (water) to 0.02, the temperature rise will be

$$\frac{0.02 \times 13,780}{0.98 \times 18 \times 1.8} = 8.7°C$$

and the temperature of the solution will have risen to $15 + 8.7 = 23.7°C$. Repeating this calculation for other values of $X$, and interpolating the limited equilibrium data as suggested, the $Y \approx X$ equilibrium curve shown in Fig. 11.11 is obtained.

From this curve it is seen that equilibrium with the feed ($Y = 0.06$) corresponds to 0.0265 mole fraction acetone in the liquid. If the column were of infinite height, this concentration could be reached in the effluent liquid. Since there seems to be no possibility of a "pinch" between the ends of the column, the corresponding (minimum) water rate (see Chap. 9) is

$$34.2 \frac{1 - 0.0265}{0.0265} 18 = 22,615 \text{ lb/h}$$

The actual water rate must be greater than this, and for the purpose of the ultimate design optimization it is desirable to determine the column size for a number of different assumed water rates. The column will be smaller if the water rate is large, but the solution obtained will then be more dilute and its concentration for acetone recovery more expensive. By way of example, assume that the water rate will be 20 percent greater than the theoretical minimum, i.e., $22,600 \times 1.2$ or $27,120$ lb/h.

The operating line can now be placed on Fig. 11.11. At the lean end (top of the column),

$$Y = Y_2 = \frac{633.7 \times 0.06 \times 0.1}{633.7 - 34.2} = 0.00634$$

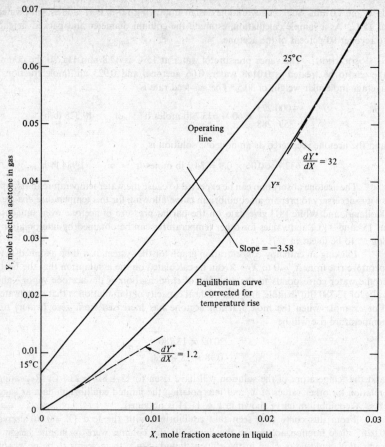

FIGURE 11.11
Operating diagram for acetone absorption—Example 11.2.

The equation for the operating line is then

$$\frac{27{,}120}{18}\frac{(X-0)}{1-X} = 633.7 \times (1-0.06)\left(\frac{Y}{1-Y}-\frac{0.00634}{0.994}\right)$$

which is represented by the upper nearly straight line on Fig. 11.11.

The required number of overall transfer units $N_{OG}$ is now obtained by a graphical integration. Reading values of $(Y - Y^*)$ from Fig. 11.11, a graph of $1/(Y - Y^*)$ vs. $Y$ is prepared from which the integral is found:

$$\int_{0.00634}^{0.06} \frac{dY}{Y - Y^*} = 4.19 = N_{OG}$$

The column cross section and packed height remain to be determined.

The liquid and gas mass velocities $L$ and $G$ will be arbitrarily specified to be 70 percent of the flooding rates. Figure 18.51 in Perry, 4th ed., will be used to estimate the flooding rates. Table 18.5 of the same source lists the packing factor $C_f$ ($= a_p/\varepsilon^3$) for 1.0-in Berl saddles as 82.5. The gas density $\rho_G$ is 0.081 and $\rho_L = 62$ lb/ft$^3$. The ratio $L/G$ is $27,120/(19,328 - 1,984) = 1.56$ at the lean end, and $(27,120 + 1,984)/19,328 = 1.51$ at the rich end. Using the maximum value, the abscissa of Fig. 18.51 is then

$$\frac{L}{G}\sqrt{\frac{\rho_G}{\rho_L}} = 1.56\sqrt{\frac{0.081}{62}} = 0.056$$

and the ordinate is approximately 0.17. Hence, with $\psi \approx 1$, and $\mu \approx 1$ cP,

$$\frac{G^2 a \psi^2 \mu^{0.2}}{3,600^2 \rho_G \varepsilon^3 \rho_L g_c} = 0.17 = \frac{82.5 G^2}{3,600^2 \times 0.081 \times 62 \times 32.2}$$

from which $G$ at flooding is 2,078 lb/(h)(ft$^2$). Using 70 percent of this, $G = 1,455$ and $L = 1.56 \times 1,455 = 2,270$ lb/(h)(ft$^2$). From Perry's Fig. 18.53, it is seen that these flow rates are slightly higher than the loading velocities.

The column cross section for $G = 1,455$ is $19,328/1455 = 13.28$ ft$^2$, corresponding to a diameter of 4.11 ft.

In order to estimate the packed volume required, and hence the packed height, it is necessary to estimate $K_G a$ or $H_{OG}$. From Fig. 28, p. 690, of Perry's 3d ed., it is estimated that $H_G \approx 0.9$ ft for the absorption of ammonia by water at room temperature. By the methods of Chap. 2 [Eq. (2.18)], $D_{AB}$ at 30°C is 0.256 cm$^2$/s for ammonia-air and 0.111 for acetone-air. Taking $H_G$ to inversely proportional to $D_{AB}^{1/2}$, $H_G$ for acetone-air at the chosen flow rates is then 1.37 ft, and

$$k_G a = \frac{G_M}{H_G P} = \frac{1,455}{30.5 \times 1.37 \times 1} = 34.8 \text{ lb mole/(h)(ft}^3)(\text{atm})$$

which at 1 atm has the same value when the driving force is in units of $Y$ in place of partial pressure.

The liquid-side coefficient can be estimated from Holloway's data. From Fig. 18.71$b$ in Perry's 4th ed., $H_L$ is 0.71 ft for oxygen desorption from water at 25°C. The molecular diffusivities are $2.41 \times 10^{-5}$ and $1.25 \times 10^{-5}$ cm$^2$/s for oxygen-water and acetone-water, so $H_L$ for acetone is 0.99 ft, and $k_L a = L_M/\rho_M H_L \approx 2,270/(62 \times 0.99) = 36.9$ lb moles/(h)(ft$^3$)(lb mole/ft$^3$). The overall resistance is given approximately by Eq. (9.78):

$$\frac{1}{K_G a P} = \frac{1}{k_G a P} + \frac{1}{k_L a \rho_M}\left(\frac{dY^*}{dX}\right) = \frac{1}{34.9} + \frac{1}{36.9}\frac{18}{62}\left(\frac{dY^*}{dX}\right)$$

$$= 0.0286 + 0.0079\left(\frac{dY^*}{dX}\right)$$

From Fig. 11.11 it is seen that $dY^*/dX$ varies from 1.2 at $X = 0$ to 3.2 at $Y = 0.06$. It follows that both gas-side and liquid-side resistances are of the same order of magnitude, and that $K_G a$ varies appreciably through the column. The construction illustrated by Fig. 9.15 will be employed to handle this complication.

The construction lines used to locate $Y_i$ have the negative slope

$$-\frac{\rho_M k_L}{k_G P} \approx -\frac{62}{18}\frac{36.9}{34.8} = -3.6$$

One such line is shown in Fig. 11.11, from which $Y - Y_i$ is $0.04 - 0.031 = 0.009$, whereas $Y - Y^*$ is $0.04 - 0.0245 = 0.0155$. From a number of such constructions $Y - Y_i$ is found as a function of $Y$ and the packed height calculated by a graphical integration, using the relation

$$h = \frac{G_M}{k_G aP} \int_{0.00634}^{0.06} \frac{dY}{Y - Y_i} = \frac{1,455}{30.5 \times 34.9 \times 1} \int_{0.00634}^{0.06} \frac{dY}{Y - Y_i}$$

In this way the packed height is found to be 8.9 ft.

If the $Y_{BM}$ is included in the estimate of $G_M/k_G aP = H_G Y_{BM}$, the graphical integration gives a packed height of $h \approx 8.6$ ft. Application of the short-cut procedure (Sec. 9.6F) gives $N_{OG} = 4.04$ and a height of 8.2 ft, which is in satisfactory agreement with the above.

The pressure drop for gas flow through the packing may be estimated by the use of Fig. 18.47b of Perry, 4th ed.:

$$\phi = \sqrt{\frac{\rho_G}{0.075}} = \sqrt{\frac{0.081}{0.075}} = 1.039 \qquad \frac{G}{\phi} = \frac{1,455}{1.039} = 1,400$$

whence the pressure drop per foot is about 0.9 in water, or 8.1 in for the 9 ft of packing. Assuming a blower efficiency of 65 percent, the power required is

$$\text{Power} = \int v\, dP = 4,000 \frac{0.9 \times 62 \times 9.0}{12 \times 0.65 \times 33,000} = 7.8 \text{ hp}$$

The illustrative calculation described above introduces several approximations, including neglect of the $Y_{BM}$ effect on the gas side, failure to allow for the decreasing mass velocity of the gas as acetone is absorbed, and others. The result, which suggests that the column should be 4.1 ft in diameter with a packed height of 9.0 ft, is probably within 20 percent, perhaps somewhat better than the precision of the data. Discussion of the column construction, including packing support, liquid distribution, foundation, etc., has been omitted. The construction of packed columns and column internals is described in various manufacturers' bulletins, and in Ref. 126. Example 9.17 is a more precise calculation of this case.

It is evident that the calculation should be repeated for other water rates and other packing materials in order that an optimized design may be specified by considering the total of fixed and operating costs, including the value of the acetone and the cost of its recovery by distillation. There are many questions which have not been discussed. Can a plastic packing be used with aqueous acetone? Does the design permit underload or overload? Is the operation inherently stable? Is it safe to handle an acetone–air mixture containing 6.0 mole percent acetone? Is it worthwhile to recover acetone so diluted by water, if pure acetone must be then obtained by distillation? (In the example the effluent contained only 0.0221 mole fraction or 6.8 weight percent acetone.)

It is of interest how best to employ a safety factor in the design. Increase in $L/G$ would make the effluent more dilute, increase the pressure drop per foot, and perhaps cause flooding. If the column diameter were increased, from 4.1 to 5.0 ft, the reduction in $L$ and $G$ would mean that both $k_G a$ and $k_L a$ would be smaller, and the required packed volume larger. Would the increase in diameter (for the same packed height) provide the additional packed volume needed? ////

**EXAMPLE 11.3** *Air Pollution Abatement* A 1,000-MW power plant is faced with a government requirement that it remove 90 percent of the sulfur ($SO_2$ and $SO_3$) from its stack gas.

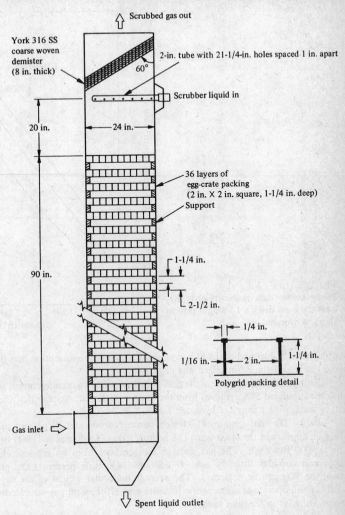

FIGURE 11.12

Pilot scrubber used in studies of the absorption of $SO_2$ from power-plant stack gas by lime slurries.

It is considering the use of countercurrent scrubbers, employing an aqueous slurry of finely ground lime or limestone as absorbent.

In order to obtain needed design data, pilot tests have been carried out in a small scrubber 2 ft in diameter packed with plastic polygrid eggcrate packing. The test device is illustrated by Fig. 11.12. Figure 11.13 shows the data obtained using an aqueous slurry containing 1.5 weight percent hydrated lime. This slurry was alkaline, and the values of $K_G a$ were

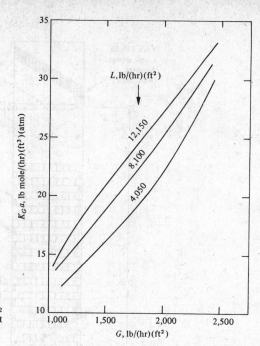

FIGURE 11.13
Pilot-plant-test data on absorption of $SO_2$ from stack gas using a 1.5 weight percent slurry of hydrated lime.

calculated from inlet and outlet $SO_2$ concentrations on the assumption that the partial pressure of $SO_2$ at the gas-liquid interface was negligible.

Estimate the dimensions of the grid-packed section of a countercurrent scrubber to accomplish the stipulated $SO_2$ removal from the plant's stack gas. Assume that the eggcrate packing will be used, with a slurry of hydrated lime.

*Data* The coal contains 71.0 weight percent carbon and 2.1 percent sulfur, essentially all of which appears in the stack gas. The heating value of the coal is 11,000 Btu/lb, and the heat rate is 9000 Btu/kWh. The gas goes to the scrubber from an efficient electrostatic precipitator and contains little fly ash. It contains 14.8 mole percent $CO_2$ plus CO on a dry basis; the $SO_3$ can be neglected. The average molecular weight of the dry gas is 30.4. The scrubber will operate at atmospheric pressure and 120°F, with gas saturated with water at that temperature, as in the pilot tests.

The test scrubber showed no tendency to flood at the flow rates employed, but mist entrainment appeared to become excessive at the highest liquid rate at a gas-mass velocity of about 2,500 lb/(h)(ft²). The pressure drop in the 7.5-ft packed section was low, being about 3.5 in. water at a gas rate of 2,000 SCFM and roughly proportional to the square of the gas flow rate.

SOLUTION   The coal rate for the 1,000-MW plant is $1,000 \times 1,000 \times 9,000/11,000 = 818,200$ lb/h, and the dry-gas rate is

$$\frac{818,200 \times 0.71 \times 100}{12 \times 14.8} = 327,100 \text{ lb moles/h}$$

Since the vapor pressure of water at 120°F is 88 mm Hg, the wet-gas flow through the scrubber is $327,100 \times 760/(760 - 88)$, or 369,930 lb moles/h (note that this is 2,609,000 ft$^3$/min). The average molecular weight of the wet gas is $[88 \times 18 + (760 - 88) \times 30.4]/760 = 28.96$, so the total mass flow of wet gas is $28.96 \times 369,930 = 10,713,000$ lb/h.

Since the pressure drop in the grid packing is low, a high gas velocity can be used. Assume $G = 2,200$, somewhat less than that at which entrainment becomes serious. The total cross section of the scrubber is then $10,713,000/2,200 = 4,870$ ft$^2$.

Figure 11.13 shows that the liquid rate has but a minor effect on $K_G a$, so it can be held low to minimize pumping costs. Assume $L = 8,100$ lb/(h)(ft$^2$), whence $K_G a$ is seen to be 28 lb moles/(h)(ft$^3$)(atm) at $G = 2,200$.

Since the SO$_2$ concentration is so low, there is no need for $Y_{BM}$ correction, and the molal flow rate is essentially constant. The relation giving the packed height is

$$-G_M\, dY = K_G aP(Y - Y^*)dh$$

The molal mass velocity is $2,200/28.96 = 76$ lb moles/(h)(ft$^2$) and $Y^*$ is assumed to be zero. For 90 percent reduction in $Y$,

$$\int_{Y_1}^{Y_2} \frac{dY}{Y} = \ln \frac{100}{10} = \frac{K_G ah}{G_M} = \frac{28h}{76}$$

whence $h = 6.25$ ft (height of packing).

The total height of the scrubber will be perhaps 12 ft to provide for gas and liquid distribution and space for a demister. The overall dimensions of the scrubber indicated by these calculations are $50 \times 100$ ft in plan and 12 ft high. For counterflow operation, how might one expect to supply the gas uniformly over such a large cross section? Crossflow would be simpler, but the eggcrate packing could not be used. What type of packing might give similar results with crossflow?

The SO$_2$ concentration of the gas is $2.1/71 \times 12/32 \times 0.148 \times 10^6 = 1,642$ ppm dry and 1,452 ppm wet. The stoichiometric lime (molecular weight 74) equivalent of the SO$_2$ absorbed is $(76 \times 0.001452 \times 0.9 \times 74 \times 100)/8,100 = 0.091$ weight percent of the slurry. The 1.5 percent in the slurry feed provides a large excess.

Since the gas-flow pressure drop in 7.5 ft of packing was 3.5 in water at 2,000 SCFM $(G = 2,916)$, $\Delta P$ for 6.25 ft at $G = 2,200$ is $3.5 \times 6.25 \times 2,200^2/7.5 \times 2,916^2 = 1.66$ in water, and the power required using a blower with an efficiency of 65 percent is

$$\left(\frac{1.66 \times 62.3}{12 \times 0.65}\right)(2,609,000)\left(\frac{0.746}{33,000}\right) = 782 \text{ kW}$$

which is only 0.08 percent of the power output of the plant. This is for flow through the packing only; there is additional pressure drop in flow through the entrainment separator and ducts.

The test data on the scrubber shown in Fig. 11.12 indicated that 70 to 80 percent of the fly ash not caught by the precipitator would be removed in the slurry. Though the quantity is not large, both this ash and precipitated calcium sulfite and sulfate must be removed by centrifuging a slipstream from the recycled slurry. This also rejects unreacted lime, but the lime loss can be minimized by maintaining a high total-solids' content of ash, sulfite, and sulfate in the circulating slurry (10 to 15 weight percent).

Though the lime slurry is quite alkaline, the assumption that $Y^*$ is zero under dynamic absorption conditions is suspect. However, the $K_G a$ used was calculated from the data with this assumption, and the results of the design calculation essentially duplicate scrubber performance as actually observed in the course of the tests. Could the interpretation of the pilot-plant data be improved and the design made more reliable by an analysis of the rate processes taking place in the liquid phase? What additional pilot-plant or laboratory data is needed to do this?

That there was an appreciable liquid-side resistance is suggested by tests of the same scrubber with other absorbents. At $L = 8,100$ and $G = 2,200$, $K_G a$ was found to be 37 for sodium carbonate solution at pH 9, 28 for the 1.5 weight percent lime slurry, and 20 for a 1.5 weight percent slurry of limestone. It is speculated that $K_G a$ for the soda ash is close to $k_G a$, and the lower values for the slurries are explained by the finite rate of solution of the suspended particles. The latter rate was evidently not quite sufficient to maintain the alkalinity at the liquid surface as the $SO_2$ was absorbed. It is relevant that the ground limestone was somewhat coarser than the lime. Students interested in developing the theory of this case of absorption and simultaneous chemical reaction in a slurry should read the articles by Ramachandran and Sharma [84] and Juvekar and Sharma [52].                    ////

## 11.5   TRAY OR "PLATE" COLUMNS

### Introduction

By far the most frequently used type of gas-liquid mass-transfer equipment, especially for distillation, is the multiple tray or plate column. This consists of a series of horizontal metal plates arranged one above another in a vertical cylindrical column. Figure 11.4 is a view of a commercial installation of this type. A single tray equipped with the popular bubble caps is illustrated in Fig. 11.16; representative bubble-cap designs are illustrated in Fig. 11.17. Gas flow is upward, with liquid passing downward from tray to tray. The gas stream flows through chimneys set in holes under each bubble cap and out through the serrated cap rim, producing a mass of bubbles in the pool of liquid on the tray. An alternate design is the sieve tray, pictured in Fig. 11.18. The bubble caps are omitted, a very large number of small holes in the plate serving to disperse the upflowing gas as bubbles in the liquid on each tray. Liquid is held on the tray by a barrier or weir at the entrance to a downcomer, which is a liquid-sealed conduit carrying liquid to the plate below.

The principal reason for preferring trays over packing in certain applications is that trays provide positive contact between the two streams, avoiding some of the problems associated with bypassing and back-mixing. At each plate all the gas is dispersed through all the liquid; at the discharge from each tray all the liquid is collected, mixed, and redistributed on the next tray below. The more positive control of countercurrent action and more certain performance than packed columns is especially important in the distillation of systems having low relative volatility and low driving forces and requiring a great many countercurrent equilibrium contacts. The offsetting disadvantage is the greater pressure drop for gas flow in the tray column, though this may not be important in some cases, as in operation

FIGURE 11.14
Plate-type distillation columns in a commercial ethylene plant. (*M. W. Kellogg Co., Houston, Texas.*)

at elevated pressures. The expense of the greater pressure drop is discussed by Teller, Miller, and Scheibel in Perry [110], and by Norman [77]. Additional practical information is given by Nielsen [76].

Sieve trays provide good mass-transfer efficiency with generally lower pressure drop than bubble-cap trays. They have good liquid-handling capacity because of the absence of obstructions to the flow of liquid across the plate, from downcomer to overflow weir. Little liquid "weeps" through the perforations when the column is operating. However, their region of stable operation without liquid leakage is somewhat smaller than for bubble caps.

### Fluid-Handling Characteristics of Trays

The fluid-flow capacity of a tray may be limited by any one of three principal factors, as illustrated by Fig. 11.19. This figure explains some of the nomenclature which will be used later. (1) At high gas-flow rates the drop in gas pressure across

**FIGURE 11.15**
Installation of a large distillation tower for fractionation of propylene at Amoco Chemicals Corp., Texas City. The tower is 265 ft tall and weighs 455 tons. (*M. W. Kellogg Co., Houston, Texas.*)

**FIGURE 11.16**
Photograph of a single bubble-cap tray. (*F. W. Glitsch and Sons, Dallas, Texas.*)

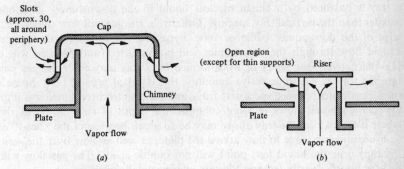

**FIGURE 11.17**
Devices for dispersing gas into a liquid on a plate. (*a*) Bubble cap; (*b*) valve cap, partly open.

FIGURE 11.18
Seven-foot-diameter single perforated sieve tray. The downcomer to the next tray is located at the left. (*F. W. Glitsch and Sons, Dallas, Texas.*)

a tray is balanced by a height of clear liquid in the downcomer. This must be smaller than the vertical tray spacing. Otherwise, the gas will flow under the lower edge of the downcomer baffle or tube, bypassing the liquid on the tray above. Liquid flow through the downcomer will be restricted, and *flooding* will occur. (2) Alternatively, the spray or froth formed on a tray may enter the gas passages into the tray above, in excessive amounts; this is called *priming* [29]. Some liquid then recirculates between successive trays, and the countercurrent action is impaired. (3) At high liquid flows in large columns, the depth of clear liquid at the side where it enters from the tray above may be so great, because of the "head" needed to overcome resistance to flow across the plate, as well as flow over the weir, that the caps near the liquid feed point will not bubble at all. The gas flow will then be very poorly distributed and the tray efficiency will suffer.

A widely used correlation of the maximum allowable vapor velocity in bubble-cap columns is that of Souders and Brown [109]. This was based on an analysis of which gas velocities are needed to suspend droplets and cause entrainment of

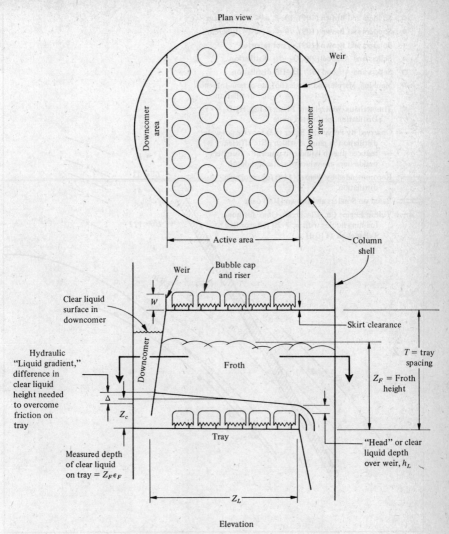

FIGURE 11.19
Illustration of bubble-cap-tray design, with various terms and dimensions.

liquid up the column. (See also Ref. 92.) This led to the expression

$$U_G(\text{max}) = K\left(\frac{\rho_L - \rho_G}{\rho_G}\right)^{1/2} \tag{11.4}$$

where $U_G$ = allowable superficial gas velocity based on the column cross section
$\rho_L$ and $\rho_G$ = densities of liquid and gas
$K$ = empirical constant

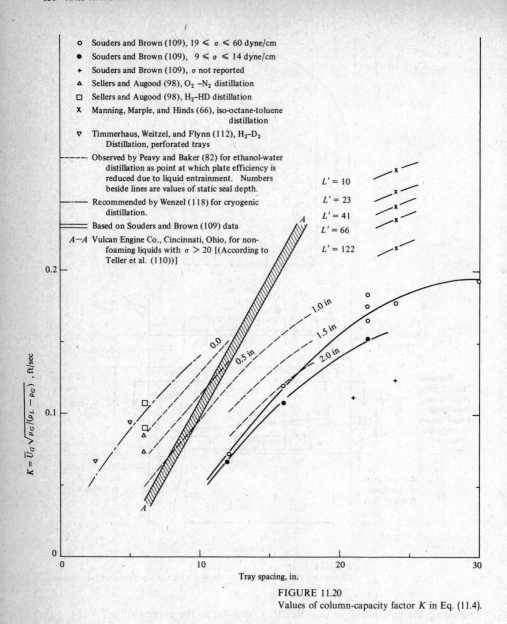

○ Souders and Brown (109), $19 \leqslant \sigma \leqslant 60$ dyne/cm
● Souders and Brown (109), $9 \leqslant \sigma \leqslant 14$ dyne/cm
+ Souders and Brown (109), $\sigma$ not reported
△ Sellers and Augood (98), $O_2$–$N_2$ distillation
□ Sellers and Augood (98), $H_2$–HD distillation
X Manning, Marple, and Hinds (66), iso-octane-toluene distillation
▽ Timmerhaus, Weitzel, and Flynn (112), $H_2$-$D_2$ Distillation, perforated trays
---- Observed by Peavy and Baker (82) for ethanol-water distillation as point at which plate efficiency is reduced due to liquid entrainment.  Numbers beside lines are values of static seal depth.
—— Recommended by Wenzel (118) for cryogenic distillation.
═══ Based on Souders and Brown (109) data
$A$–$A$ Vulcan Engine Co., Cincinnati, Ohio, for non-foaming liquids with $\sigma > 20$ [(According to Teller et al. (110))]

$L' = 10$
$L' = 23$
$L' = 41$
$L' = 66$
$L' = 122$

$K = \overline{U}_G \sqrt{\rho_G(\rho_L - \rho_G)}$ , ft/sec

Tray spacing, in.

FIGURE 11.20
Values of column-capacity factor $K$ in Eq. (11.4).

The last will depend on the tray spacing, the physical properties of the fluids, and perhaps on the dimensions of the caps and tray.

Figure 11.20 shows the manner in which $K$ may be expected to vary with tray spacing, depth of liquid seal, and the surface tension of the liquid.  Most of the data are quite old, but a few newer values are available.  The Souders-Brown

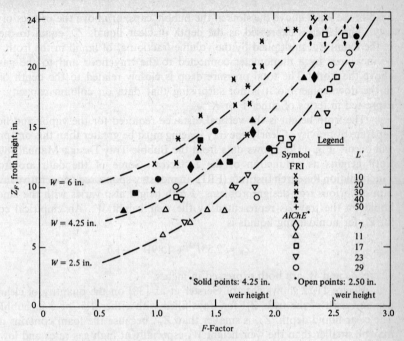

**FIGURE 11.21**
Variation of froth height with $F$ factor in an acetone-benzene distillation, with tower at 7 to 80 psia total.

curves indicate values of $K$ which are believed to be conservative, being perhaps 20 to 50 percent low. Note that for the air-water system at atmospheric pressure and ambient temperature, $U_G(\text{max})$ varies from 1 to 5 ft/s ($270 \leq G \leq 1,350$) as the tray spacing is increased and the liquid seal depth is reduced (the seal depth is the depth of clear liquid above the tops of the slots or serrations in the bubble caps). Note also that as the total pressure is increased, the allowable linear gas velocity will decrease and the allowable mass velocity will increase. According to Huang and Hodson [47], the allowable velocities for sieve trays are somewhat greater than indicated by Fig. 11.20 for bubble-cap trays.

The gas density $\rho_G$ is normally small compared with $\rho_L$ in the numerator of Eq. (11.4). The gas density in the denominator gives the effect of gas properties, but $\rho_L$ is only one of several liquid properties which influence $K$. Accordingly it has become customary to compare bubble trays at constant $F$ factor, defined by

$$F = U_G \rho_G^{1/2} \tag{11.5}$$

where $U_G$ is in feet per second and $\rho_G$ in pounds per cubic feet. The pressure drop through a *dry* tray is proportional to $F^2$; in operation there is the additional head of the foam above the slots in the bubble caps, or above the orifices of a sieve tray. The latter is expressed as the depth of clear liquid, $Z_c$, equal to the depth of the foam, $Z_F$, multiplied by the volume fraction $\varepsilon_F$ of liquid in the froth. It can be measured by a manometer connected to the tray floor and to the gas space above the froth. The total pressure drop is closely related to the depth of liquid in the downcomer, so it is not surprising that data on column capacity can be expressed in terms of either $F$ or $K$.

The froth height is the vertical distance required for the vapor and liquid to separate by gravity. Clearly, the tray spacing must be greater than this or "priming" will occur. Fig. 11.21 shows data from the Bubble Tray Design Manual [89], with froth heights as a function of the $F$ factor. Some of the data are from the Fractionation Research Institute (FRI), an industry-sponsored cooperative investigation of various tray-design problems. Froth height also varies with the amount of liquid on the tray, as represented by the weir height $W$. An empirical equation for $Z_F$ for nonfoaming liquids is

$$Z_F = 2.53F^2 + 1.89W - 1.6 \qquad (11.6)$$

where $Z_F$ and $W$ are both expressed in inches.

Figure 11.22 shows data of Gerster et al. [38] on the quantity of clear liquid per unit area of the tray, $Z_c$, for trays filled with either 1.5- or 3-in bubble caps. The clear liquid depth, $Z_c$, is smaller than $Z_F$, because the foam contains gas and may be smaller than the weir height $W$, especially at high gas rates and low liquid rates. According to Gerster et al. these and similar data can be represented by the empirical equation

$$Z_c = 1.59 + 0.29W - 0.65F + 0.020L' \qquad (11.7)$$

where $Z_c$ and $W$ are in inches, and $L'$ is the liquid-flow rate expressed as gallons per minute per foot of tray width.

As noted earlier, the pressure drop across each tray $\Delta P$ has a direct bearing on the possibility of flooding, since the depth of liquid in each downcomer must be sufficient to balance the pressure drop, i.e., be equal to or greater than $\Delta P/\rho_L$. It has also been noted that the $\Delta P$ is composed of (1) $\Delta P$ across the tray alone, and (2) that corresponding to the weight of the foam above the tray. The former is approximately equal to $\Delta P$ across the dry tray—although generally somewhat greater because the liquid tends to restrict the gas flow through the chimneys and caps, or through the holes in sieve trays.

Figure 11.23 shows representative data on $\Delta P$ for dry sieve trays, as reported by Hunt, Hanson, and Wilke [49], who employed several gases of different densities. Similar results were obtained with trays having different thicknesses and different ratios of total hole area ($A_o$) to active tray area, $A_c$. (Units of feet × pound force, per pound mass were used for $h_{DP}$; $P$ in pounds force per square foot is obtained by multiplying $h_{DP}$ by the gas density in pounds mass per cubic feet.)

Figure 11.24 compares these data for flow through dry sieve trays with the observed total $\Delta P$ minus that equivalent to the weight of liquid on the tray.

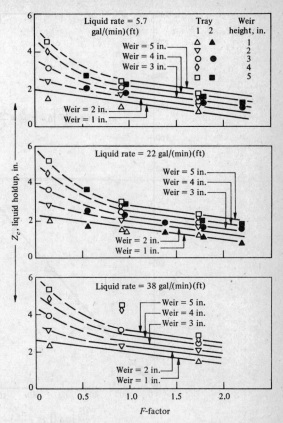

FIGURE 11.22
Liquid holdup on operating bubble-cap trays. System ammonia-air-water at 1.0 atm, 20°C.

Adopting an average value for the small difference (in inches water) between the two lines, Hunt, Hanson, and Wilke recommended the following equation for the total pressure drop with operating *sieve trays*:

$$\Delta P = 1.14 \frac{\rho_G U_G^2}{2g_c} \left[ 0.4 \left( 1.25 - \frac{A_o}{A_c} \right) + \left( 1 - \frac{A_o}{A_c} \right)^2 \right] + \frac{\rho_L g Z_c}{12 g_c} + 2.5 \text{ lbf/ft}^2 \quad (11.8)$$

where $Z_c$ is again in inches. (1.0 lbf/ft$^2$ is equivalent to 0.016 in water.) Additional data are given by Mayfield et al. [68], Jones and Pyle [51], Arnold et al. [2], Hughmark and O'Connell [48], and Manning et al. [66].

Pressure drop across bubble-cap trays is likely to be greater than for sieve trays handling the same total vapor flows. Because of the different designs of slots

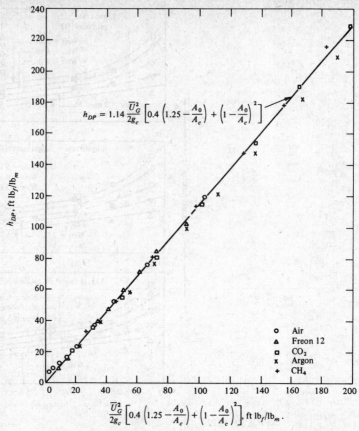

$$h_{DP} = 1.14 \frac{\overline{U}_G^2}{2g_c} \left[ 0.4 \left( 1.25 - \frac{A_0}{A_c} \right) + \left( 1 - \frac{A_0}{A_c} \right)^2 \right]$$

Air ○
Freon 12 ▲
$CO_2$ □
Argon ×
$CH_4$ +

$h_{DP}$, ft $lb_f$/$lb_m$

$$\frac{\overline{U}_G^2}{2g_c} \left[ 0.4 \left( 1.25 - \frac{A_0}{A_c} \right) + \left( 1 - \frac{A_0}{A_c} \right)^2 \right], \text{ft } lb_f/lb_m.$$

Figure 11.23
Pressure drop for flow of various gases through a dry sieve plate.

and chimneys, the data on dry trays, corresponding to the first term in Eq. (11.8), are difficult to correlate. The dry tray $\Delta P$ has been expressed as $k\rho_G U_{Gs}^2/2g_c$ lbf/ft², where $U_{Gs}$ is the gas velocity through the slots. The value of $k$ is 3 to 6 [2, 48].

*Entrainment* of liquid from one tray to the next tray above has an adverse effect both on the composition change across the tray, as expressed by the tray efficiency, and on the fluid-handling capacity. Various data showing these effects have been published, but the results are inconsistent and no generally accepted correlation is available [89]. This is not surprising in view of the variations in tray designs, and especially the apparent large effects of interfacial phenomena, not always measured or understood.

Entrainment can be measured either by catching the entrained liquid or by injecting a nonvolatile tracer, such as an inorganic salt, into the liquid on a tray

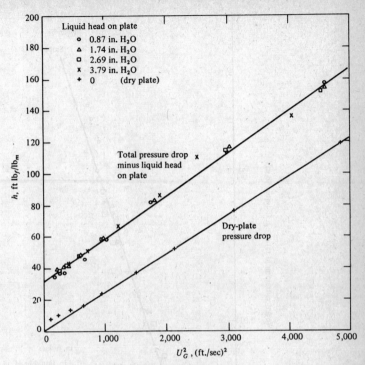

**FIGURE 11.24**
Pressure drop with liquid on a sieve plate. Air-water system.

and measuring the tracer concentration in the liquid on the tray above. Using the first method, Hunt, Hanson, and Wilke [49] obtained data on sieve trays with various gases and liquids which they expressed by the equation

$$\varepsilon_T = 0.22\left(\frac{73}{\sigma}\right)\left(\frac{U_G}{S'}\right)^{3.2} \tag{11.9}$$

where $\varepsilon_T$ is the pounds of liquid carried per pound of gas, $\sigma$ is the interfacial tension, dyne per centimeter, and $S'$ is the "effective" tray spacing, in inches. The last is defined as the actual tray spacing, less $Z_F$. Figure 11.25 illustrates the data obtained, these being for the air-water system. $\varepsilon_T$ was found to be independent of the gas velocity through the holes and independent of gas and liquid densities.

Other data on entrainment with different trays are reported by Holbrook and Baker [45], Sherwood and Jenny [102], Jones and Pyle [51], Ashrof et al. [3], Atterig et al. [4], and the Bubble Tray Design Manual [89].

A *hydraulic gradient* of the clear liquid depth develops because of the frictional resistance to flow of liquid across the tray. The effect is larger for bubble-cap than

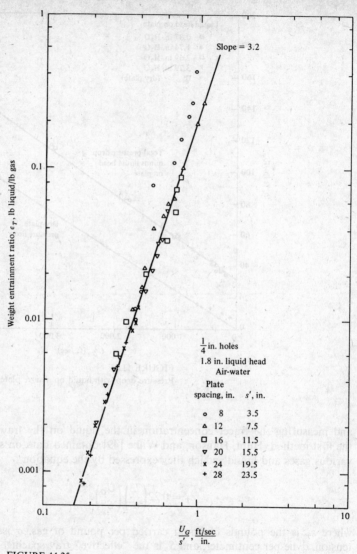

Slope = 3.2

Weight entrainment ratio, $\epsilon_T$, lb liquid/lb gas

$\frac{1}{4}$ in. holes
1.8 in. liquid head
Air-water

| Plate spacing, in. | $s'$, in. |
|---|---|
| ○ 8 | 3.5 |
| △ 12 | 7.5 |
| □ 16 | 11.5 |
| ▽ 20 | 15.5 |
| × 24 | 19.5 |
| + 28 | 23.5 |

$\dfrac{U_G}{s'}$, $\dfrac{\text{ft/sec}}{\text{in.}}$

**FIGURE 11.25**
Liquid entrainment as a function of gas velocity and "effective" tray spacing on a sieve tray.

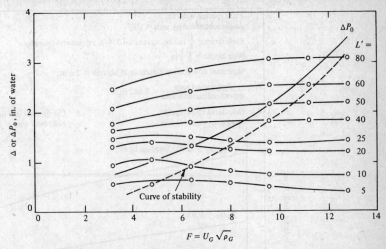

**FIGURE 11.26**

Sample graph showing the variation of the hydraulic gradient with the $F$ factor for a bubble-cap tray. Sixteen rows of 3-in caps, flush, on 4 1/4-in equilateral triangular centers; minimum depth of clear liquid above top of slots = 1.0 in. $\Delta P_0$ is the total pressure drop minus the weight of liquid above the top of the slots in the caps; all caps bubble in the region to the right of the curve of stability.

for sieve trays, since the caps obstruct the flow. It is also greater in columns of larger diameter and increases with liquid-flow rate. This liquid gradient may limit the width of the tray, since the liquid depth at the upstream side may become so great that some caps do not bubble. Columns larger than 6 ft in diameter sometimes employ downcomers alternately on both sides and at the center, so the liquid flow is split into two streams, each of which flows only half as far [89].

The variation of the hydraulic gradient with gas- and liquid-flow rates is illustrated by data of Kemp and Pyle [53], shown in Fig. 11.26. Here $\Delta$ is the difference in depth of clear liquid between liquid inlet and outlet, and $F$ is the $F$ factor defined by Eq. (11.5). Though $\Delta$ increases somewhat with increase in gas rate, it is evident that it is affected much more by liquid-flow rate and plate width. Davies [19] has developed a correlation of such data, based largely on the results of a study by Good, Hutchinson, and Rousseau [40].

Figure 11.26 shows a dashed "curve of stability" separating a region on the right, where all the bubble caps are active, from the region at the left, where some caps are idle, and the tray action is said to be "unstable." An increase in the gas-flow pressure drop always improves the uniformity of gas distribution among the caps. Even when $\Delta$ is large, as at $L' = 80$ gal/(min)(ft), an increase in $F$ to the point where $\Delta$ is smaller than $\Delta P_o$ will cause all caps to bubble. ($\Delta P_o$ is the pressure drop when the clear liquid level is just at the top of the slots.) Smaller gas flows may lead to danger of flooding, because of excessive entrainment and poor efficiency.

Bubble-cap columns do not leak liquid from one tray to the tray below, because

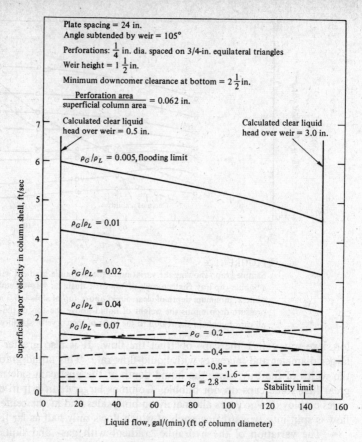

Plate spacing = 24 in.
Angle subtended by weir = 105°
Perforations: $\frac{1}{4}$ in. dia. spaced on 3/4-in. equilateral triangles
Weir height = $1\frac{1}{2}$ in.
Minimum downcomer clearance at bottom = $2\frac{1}{2}$ in.
$\dfrac{\text{Perforation area}}{\text{superficial column area}}$ = 0.062 in.

Calculated clear liquid head over weir = 0.5 in.

Calculated clear liquid head over weir = 3.0 in.

$\rho_G/\rho_L$ = 0.005, flooding limit

$\rho_G/\rho_L$ = 0.01

$\rho_G/\rho_L$ = 0.02

$\rho_G/\rho_L$ = 0.04

$\rho_G/\rho_L$ = 0.07

$\rho_G$ = 0.2

0.4

0.8

1.6

$\rho_G$ = 2.8

Stability limit

Superficial vapor velocity in column shell, ft/sec

Liquid flow, gal/(min) (ft of column diameter)

FIGURE 11.27
Operating limits of a sieve tray.

the chimneys under the caps form dams. Sieve trays, however, depend on the pressure drop due to gas flow to prevent leakage. The sieve tray, therefore, has a smaller range of $F$ factors within which it operates properly. Figure 11.27 shows ranges of acceptable gas velocities for a typical sieve tray. These are bounded above by the solid lines, representing flooding, and below by the dashed lines, which represent leakage due to insufficient gas pressure drop.

## Plate Efficiencies

The resistance to mass transfer between gas and liquid on a tray is conveniently expressed in terms of the ratio of the change in gas composition to the maximum possible change; this ratio is the *plate efficiency*. Several ways to define the efficiency are described in Chap. 9, where the relation between the efficiency and the mass-transfer-rate coefficients is also brought out.

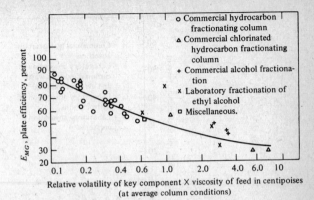

FIGURE 11.28
Variation of plate efficiencies in distilling columns with relative volatility and liquid viscosity.

Tray efficiencies vary widely, depending on the physical properties and flow rates of the fluids and, to a lesser degree, on the mechanical design of the tray. In many distilling columns the efficiencies, as defined in terms of change in vapor composition, are in the vicinity of 50 percent. In these systems the resistance to mass transfer is largely in the gas phase. In many absorbers and strippers, however, efficiencies may be only a few percent if the component being exchanged is but slightly soluble and the resistance in the liquid phase becomes more important than that in the gas. Some of these effects were reported many years ago by Walter and Sherwood [116] in experiments with a small single-tray laboratory apparatus. It was observed that efficiency was depressed by increased liquid viscosity and decreased gas solubility. Other studies by Drickamer and Bradford [24], O'Connell [79], and others have led to widely used empirical relationships, such as those illustrated by Figs. 11.28 for distilling columns and Fig. 11.29 for tray absorbers.

Since the liquid viscosities of most substances are about 0.3 cP at their atmospheric boiling points, the point efficiencies, $E_{OG}$, will lie between 50 and 90 percent for distillation if the relative volatilities do not exceed about 3. Efficiencies for vacuum distillation may be lower because of the greater viscosity at the lower temperature.

Figure 11.29, by contrast with Fig. 11.28, shows that $E_{MG}$ for tray absorbers may fall below 10 percent because the viscosity of the colder liquid is greater than in distillation, and because the gas solubility may be small.

Although such empirical results have proved to have practical value for rough estimates, they do not account for several significant variables, including fluid velocities, tray dimensions, and such physical properties of the system as molecular-diffusion coefficients. The careful investigation of these factors is made difficult by the fact that experiments must be conducted on equipment at least 1 or 2 ft in diameter if the results are to have practical value for design purposes. Such studies are too costly to be carried out in university laboratories without financial support.

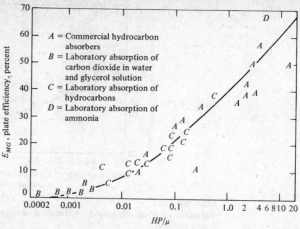

**FIGURE 11.29**

Plate efficiencies in absorbers as a function of gas solubility, pressure, and liquid viscosity. $H$ = Henry's-law constant, lb moles/(ft$^3$)(atm partial pressure); $P$ = total pressure, atm; $\mu$ = liquid viscosity, cP.

Fortunately, some of the problems were solved by the Research Committee of the American Institute of Chemical Engineers that with financial assistance from more than 50 chemical and petroleum companies supported extensive studies in three universities during the period 1952 to 1957. The results have been summarized in the Bubble Tray Design Manual [89], and reported in more detail elsewhere [38]. They provide much further understanding of the quantitative relations, so that even though experimental tests may still be needed in critical cases, many of the relationships between variables can now be understood. Several summaries of these and other data have been published—notably those by Perry [83] and Norman [77]. The section which follows presents a brief account of the underlying principles; reference should be made to sources previously mentioned for details.

The relation between the point-tray efficiency, $E_{OG}$, and the design parameters is conveniently expressed by

$$\frac{1}{-\ln(1 - E_{OG})} = \frac{1}{N_{OG}} = \frac{1}{N_G} + \frac{mG_M}{L_M}\frac{1}{N_L} = \frac{U_G}{k_G RTA} + \frac{mG_M}{L_M}\frac{(L/\rho_L)}{k_L A} \quad (11.10)$$

where $E_{OG}$ is the local efficiency, in terms of gas compositions, at a point on a tray. The $N$'s represent numbers of transfer units per tray (see Chap. 9), and $A$ is the total interfacial area of gas-liquid contact in the froth per unit of active area of the tray floor. As before, $U_G$ is the superficial gas velocity based on the active tray area, and $L_M$ and $G_M$ represent the corresponding molal mass velocities.

The behavior of the froth is much too complicated to permit the use of exact diffusion calculations for the prediction of $k_G$ and $k_L$. The attempts to do this

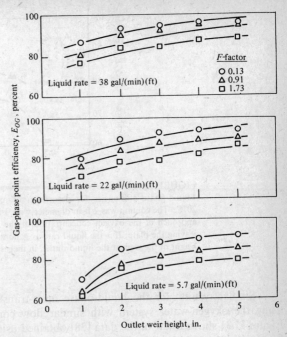

FIGURE 11.30
Gas-phase-point efficiencies for the absorption of ammonia by water in a tray
fitted with 1.5-in bubble caps (1.0 atm, 20°C).

(Ref. 36) lead to relations which involve gas-bubble size and other quantities
which are not known. Nevertheless, the effects on $k_G RTA$ and $k_L A$ of such factors
as physical properties and fluid velocities can be found empirically.

In a study of the gas-phase resistance, Gerster et al. [38, 37, 88] measured
plate efficiencies for the evaporation of water into air, the absorption of ammonia
by water, and the distillation of acetone-benzene mixtures. A column 2 ft in diameter
was employed, with several tray designs and operating pressures. Figure 11.30 shows
results obtained for the absorption of ammonia from air using a plate filled with
small 1.5-in-diameter bubble caps. This shows $E_{OG}$ to increase as either liquid-flow
rate or depth of liquid on the tray was increased. However, $E_{OG}$ decreased with
increase in gas-flow rate, because $k_G RTA$ increases with $U_G$ less rapidly than $U_G$
increases. The results for this and eleven other bubble-cap trays were correlated by
the empirical equation

$$N_G = -\ln\left(1 - E_{OG}\right) = \frac{k_G RTA}{U_G} = (0.776 + 0.116W - 0.290F + 0.0217L')(\mathrm{Sc})^{-1/2}$$

$$(11.11)$$

where Sc is the Schmidt number for the gas mixture.

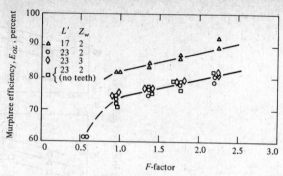

**FIGURE 11.31**
Liquid-phase efficiencies for desorption of oxygen from water at 1.0 atm and 20°C. The column was 2 ft in diameter with 3-in bubble caps. $E_{OL}$ is the fractional approach of the oxygen concentration in the water to equilibrium with the gas leaving the plate; $L$ is the liquid rate as gallons per minute per foot, and $Z_w$ is the height of the weir at the liquid outlet, in inches.

Similar results for the liquid-phase mass-transfer resistance were obtained using the oxygen-water system with similar flow conditions on the same trays. Figure 11.31 shows a few of the data [38] obtained using a column 2 ft in diameter fitted with 3-in bubble caps. The ordinate $E_{OL}$ is the change in liquid composition across the plate divided by the maximum possible change corresponding to the liquid coming to equilibrium with the gas leaving the tray (in these tests, the change in gas composition was negligible). The principal effect of increasing flow rates and weir height is to increase the area $A$ for contact between the phases.

Gerster et al. [38] obtained the following empirical correlation of their data for several tray designs:

$$N_L = \frac{k_L A}{(L/\rho_L)} = \left(\frac{D_L}{2.41 \times 10^{-5}}\right)^{1/2} \frac{(9.7F + 5.6)Z_c Z_L}{L'} \tag{11.12}$$

where $Z_c$ is in inches and $Z_L$ in feet (see Fig. 11.19). This is based on tests using oxygen desorption from water at 25°C, in which case $D_L = 2.41 \times 10^{-5}$ cm$^2$/s. The height of clear liquid, $Z_c$, can be estimated by the use of Eq. (11.7). Note that $Z_c Z_L$ represents the volume of liquid per unit of tray width, and that $Z_c Z_L/L'$ is proportional to the time required for the liquid to flow across the tray.

The quantity in parentheses on the right represents the product of $k_L$ and the interfacial area per unit volume of liquid, and is reported to vary little with tray design. The liquid-flow terms $L/\rho_L$ and $L'$ are based on active plate area and column width, respectively, but one is proportional to the other, so the two tend to cancel, along with $Z_L$, which is roughly equal to the column diameter. Thus for a fixed gas rate ($U_G$ or $F$), $k_L A$ evidently varies only because of a weak dependence of $Z_c$ on liquid-flow rate.

The data indicate that changes in gas velocity affect $k_L A$ and $k_G A$ differently. An increase in $F$ diminishes $k_G A/U_G$ slightly, as illustrated in Fig. 11.30 and by Eq. (11.11). On the other hand, $k_L A\rho_L/L$ responds in the opposite direction, as shown by Fig. 11.31. As indicated by Eq. (11.7), $Z_c$ decreases with increasing $F$, but the term $9.7F + 5.6$ in Eq. (11.12) provides a compensating factor, due to the increase of the interfacial area at the higher gas-flow rates. Again, variations in $L'$ affect $k_G RT A/U_G$ more than they affect $k_L A\rho_L/L$, as shown by Eqs. (11.11) and (11.12). Consequently, changes in tower diameter for constant $m$, $L_M/G_M$, and total-flow rate may change $k_L A$ and $k_G A$ in different ways, making a slight shift in the relative importance of gas- and liquid-phase resistances.

In many applications to distillation, where $mG_M/L_M$ is near unity for the key components, the gas-phase resistance will be the greater one. For slightly soluble substances, where $mG_M/L_M$ is large because $m$ is large, the liquid-phase resistance usually predominates, and $E_{OG}$ will be small, as illustrated by Fig. 11.29.

The degree of mixing of the liquid on a tray may have an important bearing on the Murphree efficiency, especially if the tray is more than a foot or two wide [35]. If lateral mixing is absent, the Murphree efficiency $E_{MG}$ for the whole tray may be greater than 100 percent, though the point efficiency $E_{OG}$ is always less than 100 percent. This has been shown by Lewis [60]; see also Chap. 9 in this connection. To express the effect quantitatively, one needs to know the eddy-diffusion

**Table 11.1   EFFECT OF LATERAL LIQUID MIXING ON A TRAY.***

Values tabulated are $E_{MG}/E_{OG}$

| $(mG_M/L_M)E_{OG}$ | Pe | | | | | | | | | |
| --- | --- | --- | --- | --- | --- | --- | --- | --- | --- | --- |
| | 0.5 | 1.0 | 2.0 | 3.0 | 5.0 | 10.0 | 20.0 | 50.0 | 100.0 | ∞ |
| 0.25 | 1.019 | 1.033 | 1.055 | 1.070 | 1.089 | 1.109 | 1.121 | 1.130 | 1.133 | 1.136 |
| 0.50 | 1.037 | 1.068 | 1.113 | 1.144 | 1.184 | 1.230 | 1.260 | 1.282 | 1.289 | 1.297 |
| 0.75 | 1.056 | 1.103 | 1.173 | 1.223 | 1.288 | 1.366 | 1.420 | 1.459 | 1.474 | 1.489 |
| 1.00 | 1.076 | 1.138 | 1.235 | 1.306 | 1.450 | 1.517 | 1.603 | 1.668 | 1.692 | 1.718 |
| 1.25 | 1.095 | 1.175 | 1.300 | 1.393 | 1.522 | 1.686 | 1.813 | 1.912 | 1.950 | 1.992 |
| 1.50 | 1.115 | 1.212 | 1.368 | 1.486 | 1.653 | 1.875 | 2.053 | 2.199 | 2.257 | 2.321 |
| 1.75 | 1.135 | 1.251 | 1.438 | 1.584 | 1.794 | 2.085 | 2.329 | 2.537 | 2.622 | 2.717 |
| 2.00 | 1.155 | 1.290 | 1.512 | 1.687 | 1.946 | 2.320 | 2.646 | 2.935 | 3.056 | 3.195 |
| 2.25 | 1.175 | 1.330 | 1.588 | 1.796 | 2.110 | 2.580 | 3.009 | 3.404 | 3.575 | 3.772 |
| 2.50 | 1.196 | 1.370 | 1.667 | 1.911 | 2.287 | 2.870 | 3.424 | 3.958 | 4.194 | 4.473 |
| 2.75 | 1.217 | 1.412 | 1.750 | 2.031 | 2.477 | 3.191 | 3.901 | 4.611 | 4.934 | 5.325 |
| 3.00 | 1.238 | 1.455 | 1.835 | 2.159 | 2.681 | 3.548 | 4.446 | 5.382 | 5.821 | 6.362 |
| 4.00 | 1.325 | 1.635 | 2.212 | 2.737 | 3.659 | 5.405 | 7.531 | 10.15 | 11.53 | 13.40 |
| 5.00 | 1.416 | 1.831 | 2.648 | 3.442 | 4.947 | 8.176 | 12.77 | 19.45 | 23.49 | 29.48 |
| 6.00 | 1.512 | 2.043 | 3.151 | 4.297 | 6.629 | 12.26 | 21.58 | 37.61 | 48.69 | 67.07 |
| 7.00 | 1.611 | 2.274 | 3.730 | 5.326 | 8.806 | 18.22 | 36.22 | 72.93 | 102.0 | 156.5 |
| 8.00 | 1.716 | 2.524 | 4.394 | 6.559 | 11.60 | 26.81 | 60.31 | 141.3 | 215.0 | 372.5 |
| 9.00 | 1.824 | 2.794 | 5.151 | 8.029 | 15.17 | 39.09 | 99.53 | 272.6 | 453.9 | 900.2 |
| 10.00 | 1.938 | 2.086 | 6.015 | 9.776 | 19.69 | 56.47 | 162.7 | 523.1 | 958.3 | 2202. |

*$U_F$ = total volumetric liquid flow rate divided by the product of $(1 - \varepsilon_F)$, $Z_F$, and tray width $B$. Pe = $U_F Z_L/D_E$ where $Z_L$ is the length of flow across the plate, and $D_E$ is the coefficient of eddy diffusion in the flowing froth.

coefficient $D_E$, based on the vertical cross section, $Z_F B$, of the froth flowing across the plate. With this information, the relation between $E_{MG}$ and $E_{OG}$ can be developed, as shown by Gerster et al. [38]. The result is similar to that described in Sec. 11.4 for axial dispersion in packed towers where the composition of one phase is essentially constant.

Table 11.1 lists calculated values of the ratio $E_{MG}/E_{OG}$ as a function of $(mG_M/L_M)E_{OG}$ and the Peclet number, $U_F Z_L/D_E$. Here $U_F$ is the horizontal velocity of the froth, which is the total volumetric liquid-flow rate divided by $(1 - \varepsilon_F)$, the froth height $Z_F$, and the tray width $B$.

Figure 11.32, from Gerster et al. [38], indicates the manner in which $D_E$ varies with liquid and gas flow rates, and with weir height, $W$. The data were obtained with trays of different widths, two values of $\rho_G$, and two weir heights. Calculated values of the ratio $E_{MG}/E_{OG}$ are illustrated in Fig. 11.33, for two weir heights and two gas densities. These data imply that trays 6 ft in diameter may be expected to give Murphree efficiencies $E_{MG}$ nearly as great as with no back-mixing ($D_E = 0$), except when the gas density and weir height are small.

Finally, the rate of liquid entrainment affects the effective Murphree efficiency $E_{MGE}$ at high gas velocities. The reduction in efficiency from $E_{MG}$ (no entrainment) to the effective value is given by an equation due to Colburn [13]:

$$E_{MGE} = \frac{E_{MG}}{1 + (\varepsilon_T G_M/L_M)E_{MG}} \tag{11.13}$$

This is exact if the change in mole fraction in the liquid from one tray to the next is the same for each pair of adjacent trays. Colburn concluded that the economic optimum value of $\varepsilon_T$ is $L_M/3G_M E_{MG}$, a relatively large value, which is not likely to be attained before the column floods for other reasons. From some detailed computations, Zenz [125] concluded that the optimum level of entrainment is about $0.175 (L_M/G_M)$, provided that stable operation is possible.

**EXAMPLE 11.4** Estimate the plate efficiency for the distillation of ethanol-water mixtures at atmospheric pressure to produce 80 mole percent alcohol and essentially pure water from a saturated liquid feed of 30 mole percent ethanol. The distilling tower is 5 ft in diameter. The trays are spaced 18 in vertically; each contains 3-in-diameter bubble caps on 4.5-in centers with 3-in weirs; 25 percent of the area of each tray is used for downcomers.

SOLUTION  A suitable basis for the design of the column, based on a McCabe-Thiele diagram, suggests the following: $L_M/G_M = 0.670$ and $m = 0.586$ at the top of the tower; $L_M/G_M = 1.55$ and $m = 8.9$ at the bottom. Most of the equilibrium stages will be needed at the top, where there is a close approach to the azeotropic composition of 84 mole percent alcohol. Only a few equilibrium trays will be required near the bottom where $mG_M/L_M = 5.7$.

The tray dimensions are such that 75 percent of the area is active with weirs 3.9 ft long. The average width of a tray is $B = 4.55$ ft, and the horizontal path for liquid flow between entrance and exit weirs is $Z_L = 3.23$ ft. The active tray area is 14.7 ft$^2$.

Choice of the vapor velocity in the tower (and the tower's capacity) is based on Eq. (11.4) and Fig. 11.20, using $K = 0.14$. Since the vapor velocities correlated by Souders and Brown [109]

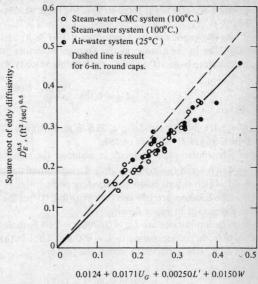

FIGURE 11.32

Experimental values of the eddy-diffusion coefficient in the froth. Here $U_G$ is the superficial gas velocity, in feet per second, $L'$ is the liquid flow as gallons per minute per foot of tray width, and the weir height $W$ is in inches.

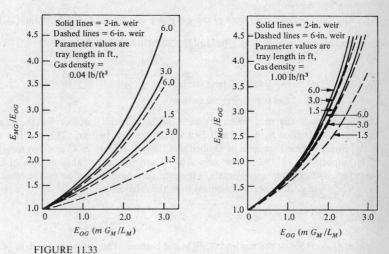

FIGURE 11.33

Calculated effect of lateral dispersion in the flowing froth on the ratio of the Murphree efficiency for the whole plate, $E_{MG}$, to the point efficiency, $E_{OG}$; for two weir heights and two gas densities.

were based on total tower cross section, the value based on active area must be $0.14(1.0/0.75) = 0.186$.

As the vapor density varies from $0.0367$ lb/ft$^3$ at the bottom of the column to $0.087$ at the top, due to the change in the average molecular weight from 18 to 40.4 and a change in temperature from 100 to 80°C, the limiting velocity will be that at the top, giving

$$U_G = 0.186 \sqrt{\frac{56.16 - 0.087}{0.087}} = 4.72 \text{ ft/s}$$

or $G = 1{,}478$ lb(h)(ft$^2$) and $G_M = 36.6$ lb mole/(h)(ft$^2$). The total vapor rate from the column will be $(36.6)(14.7) = 538$ lb mole/h.

The molar vapor velocity is assumed the same at the bottom of the column, giving $U_G = 4.98$ ft/s and $G = 659$ lb(h)(ft$^2$). Using bottom densities $K = 4.98\sqrt{0.0367/(62.3 - 0.0367)} = 0.121$, which is well below the flooding point.

The $F$ factors are 1.39 and 0.95 (ft/s)(lb/ft$^3$)$^{1/2}$ at the top and bottom, the difference arising from the change in vapor density.

The liquid rates are $L_M = 0.670 G_M = 24.5$ lb mole/(h)(ft$^2$) at the top and $1.55 G_M = 56.7$ at the bottom, corresponding to $L = 990$ and $1{,}021$ lb/(h)(ft$^2$), respectively. The corresponding liquid-flow rates horizontally across the tray are

$$L' = 990 \text{ lb/(h)(ft}^2) \times 14.7 \text{ ft}^2 \times \frac{1}{8.33 \times 0.9} \frac{1}{60} \frac{1}{4.55 \text{ ft}}$$

$$= 7.11 \text{ gal/(min)(ft)}$$

in the top and $L' = 6.60$ gal/(min)(ft) in the bottom.

The depth of clear liquid is calculated from Eq. (11.7). In the top trays it is $Z_c = 1.70$ in; for the bottom, 1.97 in. The depth of froth is found from Eq. (11.6) to be 8.96 in and 6.34 in at the top and bottom, respectively.

The mass-transfer resistance of the gas phase in a tray is found from Eq. (11.11), e.g.,

$$N_G = \frac{(0.776) + (0.116)(3) - (0.290)(1.39) + (0.0217 \times 7.11)}{\sqrt{1.04}}$$

$$= 0.858 \quad \text{near the top}$$
$$= 0.97 \quad \text{at the bottom}$$

The factor inside the root is an estimate of the Schmidt number for the vapor binary. Equation (11.12) gives $N_L = 23.2$ and 29.5 for the top and bottom. This involves an estimate of $D_L$, which is based on data of Dullien and Shemilt [25] at 25°C and the assumption that $D_L$ is proportional to $T/\mu$. These values of $N_L$ are above the range of data of Gerster and coworkers, and are only approximate. However, they have little effect on the result.

The point coefficients are obtained from the relations

$$E_{OG} = 1 - e^{-N_{OG}} \quad \text{and} \quad \frac{1}{N_{OG}} = \frac{1}{N_G} + \frac{m G_M}{L_M} \frac{1}{N_L}$$

whence $E_{OG} = 0.565$ at the top and 0.558 at the bottom. The terms on the right of the second equation represent the relative resistances of gas and liquid phases, and it follows that the gas phase accounts for 96.8 percent of the overall mass-transfer resistance at the top and 84.1 percent near the bottom.

The effect of liquid mixing in the trays can be found from Table 11.1 after the eddy-diffusivity values are estimated from Fig. 11.32. We find $D_E = 0.024$ and $0.025$ ft$^2$/s, respectively. The Peclet numbers are computed from the tray length and the froth velocity. We obtain $U_F = 0.11$ and $0.09$ ft/s and Pe $= 14.8$ and $11.6$, respectively. From Table 11.1, $E_{MV}/E_{OG} = 1.245$ and $4.1$, so that $E_{MG} = 0.703$ and $2.29$, for the top and bottom trays, respectively. The larger increase in efficiency for the bottom trays is because of the much greater $mG_M/L_M$ there. A small buildup of liquid concentration has a big effect on efficiency when the equilibrium line is very steep.

To estimate the entrainment effect, we use the Eq. (11.9) which indicates values of $\varepsilon_T = 0.095$ and $0.018$ lb liquid/lb gas, respectively. Using Colburn's equation [Eq. (11.13)], we see that the effective efficiencies are reduced but little, because of entrainment, to $E_{MGE} = 0.639$ and $2.23$, respectively. Since many more trays will be needed near the top of the column than at the bottom, an average efficiency for the whole column will be nearer the lower value. The O'Connell correlation, Fig. 11.28, shows that the whole-column value should be about 65 percent. ////

## 11.6  AGITATED VESSELS AND BUBBLE COLUMNS

As noted in Sec. 6.7, the coefficients $k_G$ and $k_L$ for mass transfer between a small gas bubble and a liquid are quite large. Swarms or clouds of bubbles provide enormous surface per unit volume of gas-liquid dispersion, and large values of $k_L a$ are obtained in such systems. This is the basis for the design of compact and inexpensive gas-liquid contacting equipment.

Figure 11.34 illustrates two types of contactors employing gas-liquid dispersions: the agitated (or "aerated") vessel and the bubble column. The former is widely used in the chemical industry for batch oxidations, chlorinations, and hydrogenations, with chemical reaction occurring in the liquid phase. It is also used widely for aerobic fermentation. (Papers constituting a symposium on mixing and oxygen absorption in fermenters were published in the February 1966 issue of *Biotechnology and Bioengineering.* See also Ref. 32.) Bubble columns are usually designed with a large ratio of height to diameter, and are agitated only by the sparged gas. They can be operated with no liquid flow or with either countercurrent or cocurrent flow of liquid and gas.

**The agitated vessel** is equipped with a motor-driven stirrer and a gas sparger. Four verticle baffles attached to the wall are standard; these increase turbulence and minimize swirling and vortex formation at the liquid surface. Diameters range from a few inches to several feet, and the impeller may be a flat paddle, marine propeller, or a turbine of the type suggested by Fig. 11.34a. The sparger may be a pipe ring with many holes or a single tube delivering the gas under the impeller near its tip. Typically, the impeller is located one-third of the distance from the bottom to the liquid level, and the ratio $H_i/T_i$ is unity.

The superficial gas velocity in these devices is low, usually being in the range of 0.01 to 0.25 ft/s, and limited by foaming. The gas is dispersed in the form of small bubbles by the impeller, the initial bubble size being determined primarily by

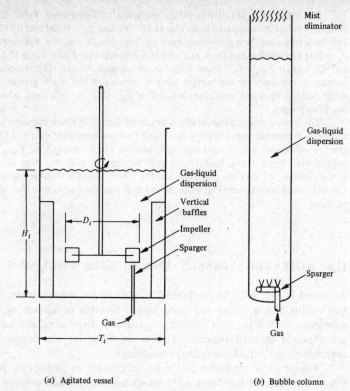

(a) Agitated vessel      (b) Bubble column

FIGURE 11.34
Sketches illustrating construction of agitated vessel and bubble column, used for gas-liquid contacting.

the impeller tip speed. At moderate to high impeller tip speeds (500 to 3,000 ft/min), the design of the sparger has almost no effect on bubble size. Mean bubble diameters are in the range of 0.1 to 0.5 cm in aqueous solutions of low viscosity, but they decrease with increase in viscosity. The bubbles tend to circulate with the liquid but rise due to buoyancy and tend to coalesce. Gas is drawn into the liquid from above by the vortex formed at the surface, especially when baffles are not used. This surface aeration by effluent gas contributes to the gas holdup, and causes some back-mixing of the gas. At low impeller speeds there is little back-mixing of bubbles, but at high speeds the distribution of bubbles tends to become uniform throughout the liquid. This fact complicates the analysis of experimental data on rates of mass transfer, since a log-mean driving force would be appropriate with no back-mixing of gas and no surface aeration, whereas with thorough mixing the potential corresponding to the composition of the effluent gas should be used to calculate $k_L a$.

Unfortunately many investigators fail to state what mean driving force was used. For typical vessels operated at moderate to high impeller tip speeds, conservative design calculations should probably employ the driving force at the gas outlet. The problem does not arise, of course, when a pure gas is used.

Gas holdup $\varepsilon$, defined as the volume-fraction gas in the dispersion, increases with agitation and gas flow but reaches a maximum of about 0.3 to 0.4. It is independent of vessel diameter but at low agitation increases with tip speed and power input per unit volume. The holdup attained is important, since the bubble diameter $d_b$ and the coefficient $k_L$ do not vary widely, and $k_L a$ depends primarily on $a$ [107]. For spherical bubbles, $a = 6\varepsilon/d_b$. Dissolved nonreacting electrolytes reduce coalescence, and $k_L a$ may be much greater in such solutions than in pure water [93, 67, 123, 99]. This phenomenon is attributed to "electrical effects" at the gas-liquid interface, which have not been well explained. The gas-side resistance is usually negligible in comparison with that in the liquid [70].

In principle, the interfacial area of gas-liquid contact can be found by measuring the rate of gas absorption by a liquid with which it reacts chemically at a high rate. The absorption of oxygen by aqueous $Na_2SO_3$, catalyzed by copper or cobalt ions, is commonly used in such tests. If the reaction is rapid, the rate of absorption is independent of the factors affecting the $k_L^\circ$ obtained with only physical absorption, as shown in Chap. 8. The rate then depends only on the speed of the homogeneous chemical reaction and the diffusivity. The rate of absorption is proportional to the interfacial area. See Sec. 8.15.

de Waal and Okeson [20] give an excellent description of the method. Westerterp, Van Dierendonck, and de Kraa [121] report extensive tests using air and sulfite solution containing a copper salt, with a series of agitated vessels having diameters from 14 to 90 cm. Though their study has been criticized by Linek and Mayrhoferova [61] and Reith and Beek [86], they demonstrate the enormous interfacial areas attainable at high impeller speed and gas holdup, and show how the overall absorption rate can be expected to vary with vessel geometry. A critical review of the sulfite oxidation method of measuring interfacial areas in aeration equipment is provided by Benedek, Bennett, and Ho [4a]. These authors state that their paper "reviews the endless round of mistakes committed" in the application of the technique and recommend that the technique not be used.

The fast reaction of oxygen with aluminum alkyls has been used by Bossier et al. [6a] to determine interfacial areas in agitated vessels containing organic liquids. These results indicate contact areas to be several times as large as in aqueous systems.

The discussions in the sections which follow will be concerned with physical absorption without chemical reaction. The symbols $k_L$ and $k_L a$ here replace $k_L^\circ$ and $k_L a^\circ$ used in Chap. 8.

Little information of use in predicting $k_L a$ is available in handbooks, but there are several good reviews [113, 107, 90, 10, 9]; that by Sideman, Hortacsu, and Fulton [107] is excellent. These provide the basis for very approximate estimates of $k_L a$, which, however, may be all that is needed for some purposes. Evidently $k_L$ does not vary widely with variations of $d_b$, $H_i$, $P/V$ (power input/unit volume), or $U_s$ (the superficial gas velocity). For the small bubbles produced in well-agitated vessels ($d_b < 0.25$ cm), Calderbank [9] suggests the following correlation for bubbles in aqueous electrolytes:

$$k_L = 0.31\left(\frac{\Delta\rho\mu_c g}{\rho_c^2}\right)^{1/3} (Sc)^{-2/3} \tag{11.14}$$

where $\mu_c$ and $\rho_c$ = viscosity and density of the liquid

$\Delta\rho$ = density difference between gas and liquid ($\approx \rho_c$)

$g$ = local acceleration due to gravity

$Sc$ = Schmidt number for transport in the liquid.

In dilute aqueous systems at room temperature, $k_L$ in bubble swarms falls roughly in the range 0.008 to 0.05 cm/s. The principal variable determining $k_L a$, as noted earlier, is the interfacial area $a$.

Westerterp et al. [121] found that they could correlate their extensive data for $a$ with turbine impellers by

$$\frac{aH_i}{1-\varepsilon} = 0.79\mu_c(n-n_0)D_i\sqrt{\frac{\rho_c T_i}{\sigma}} \tag{11.15}$$

in which $\mu_c$ must be expressed in centipoises. The impeller speed $n$ is in revolutions per second; $\sigma$ is the interfacial tension (g/s$^2$) and $H_i$ and $D_i$ are in centimeters. The constant $n_0$ is given by the dimensionless expression

$$n_0 = \left(\frac{1.22}{D_i} + 1.25\frac{T_i}{D_i^2}\right)\left(\frac{\sigma g}{\rho_c}\right)^{1/4} \tag{11.16}$$

Similar equations are given for four-bladed paddle agitators.

Calderbank suggests an equation for $a$ which involves both $P/V$ and the terminal velocity of rise of a single bubble and is difficult to use. An approximate value of $a$ may be obtained from Eq. (11.15) by taking $1 - \varepsilon$ to be 0.8 (maximum $\varepsilon$ is about 0.4 in clean solutions). However, Eqs. (11.15) and (11.16) give $a/(1 - \varepsilon)$, which is the interfacial area per unit volume of *liquid*, and a value of $\varepsilon$ is not then needed for design purposes (see Example 11.5). The data of Westerterp et al. indicate that $a/(1 - \varepsilon)$ is insensitive to gas-flow rate, though $a$ and $\varepsilon$ are not.

Power input to stirred vessels can be estimated by use of the data and correlations published by Rushton et al. [95]. Calderbank [9b], quoting Michel and Miller [71], shows a graph illustrating the marked decrease of power with increase in gas holdup.

---

EXAMPLE 11.5 *Oxygen Absorption in a Stirred Vessel*  A baffled cylindrical vessel equipped with a turbine impeller is to be used as a fermenter. The vessel is 3.0 ft in diameter filled to an unstirred depth of 3.0 ft, and the impeller diameter is 1.0 ft. The liquid is an aqueous solution having a density of 1.1, a surface tension of 65 dyn/cm, and a viscosity of 1.5 cP. The impeller speed will be 360 rpm with operation at 25°C and 1.0 atm. In the early stages of the fermentation, the microorganisms rapidly consume the dissolved oxygen, so the concentration of oxygen in the liquid phase is maintained near zero. The superficial velocity of the pure oxygen supplied will be 0.04 ft/s.

Estimate the fermentation rate to be expected, expressed as pound moles of oxygen per hour. What is the oxygen efficiency: the ratio of oxygen absorbed to oxygen supplied?

SOLUTION  From the data given, $H_i = 91.5$ cm, $T_i = 91.5$ cm, $D_i = 30.5$ cm, $n = 360/60 = 6$, $\rho_c = 1.1$ g/cm$^3$, $\sigma = 65$ g/s$^2$, $g = 981$ cm/s$^2$, and $\mu_c = 1.5$ cP. Substituting in Eq. (11.16),

$$n_0 = \left(\frac{1.22}{30.5} + \frac{1.25}{30.5}\frac{91.5}{30.5}\right)\left(\frac{65 \times 981}{1.1}\right)^{1/4} = 2.53 \text{ rps}$$

Using this value of $n_0$ in Eq. (11.15),

$$\frac{a}{1-\varepsilon} = \frac{0.79 \times 1.5}{91.5}(6 - 2.53)30.5\left(\frac{1.1 \times 91.5}{65}\right)^{1/2} = 1.71 \text{ cm}^2/(\text{cm}^3 \text{ liquid})$$

Then, using Eq. (11.14) to estimate $k_L$, with $D_L = 2.41 \times 10^{-5}$ cm$^2$/s,

$$k_L = 0.31\left(\frac{1.1 \times 0.015 \times 981}{1.1^2}\right)^{1/3}\left(\frac{0.015 \times 10^5}{1.1 \times 2.41}\right)^{-2/3} = 0.011 \text{ cm/s}$$

The volume of the dispersion is $(\pi/4)[91.5^3/(1 - \varepsilon)] = [602,000/(1 - \varepsilon)]$ cm$^3$, and $k_L a = 0.011 \times 1.71/(1 - \varepsilon)$. Using data from Perry for the Henry-law coefficient for oxygen in water at 25°C, $c_i = c^* = 1.27 \times 10^{-6}$ g moles/cm$^3$. The total rate of absorption is $k_L a V_d c_i$, where $V_d$ is the volume of the dispersion. This is

$$0.011 \times 1.71(1 - \varepsilon)\frac{602,000}{(1 - \varepsilon)}1.27 \times 10^{-6} \times 3,600 = 51.7 \text{ g moles/h} \quad \text{or} \quad 0.114 \text{ lb moles/h}$$

The gas-feed rate is $0.04 \times 3,600 \times 30.5 \times 91.5^2 \times (\pi/4)[273/(22,400 \times 298)] = 1,180$ g moles/h so 4.4 percent of the oxygen will be absorbed.

An estimate of the bubble size may be made by taking $1 - \varepsilon$ to be about 0.8, whence $a = 1.71 \times 0.8 = 1.37$ cm$^2$/(cm$^3$ dispersion). If the bubbles are spherical and uniform in size, then $a/\varepsilon = 6/d_b$, and $d_b = 0.88$ cm. This is somewhat larger than the range for which Eq. (11.14) applies, so the absorption rate can be expected to be appreciably greater than calculated above.

The power required can be estimated from the correlations proposed by Ruston, Costich, and Everett [95], described in Chapter 19 of Perry, 4th ed. The Reynolds number is

$$Re = \frac{D_i^2 n \rho_L}{\mu} = \frac{30.5^2 \times 6 \times 1.1}{0.015} = 409,000$$

whence the power number is 6.5 $(= Pg_c/\rho n^3 D_i^5)$

$$P_w = \frac{6.5 \times 62.3 \times 1.1 \times 6^3 \times 1^5}{32.2} = 2,990 \text{ ft-lb/s}$$

or 5.4 hp. This is for a nonaerated vessel; the aeration can be expected to reduce this by half, or to about 2.7 hp during aeration.

It is to be emphasized that the calculated absorption rate is no better than a good estimate, useful primarily in choosing equipment. (Fig. 6 of Ref. 32 indicates $k_L a$ to be about $0.07$ s$^{-1}$ for a similar fermenter and the same power input—some three times the value estimated above.) ////

*Bubble columns* are widely used as air-agitated fermenters, sometimes several feet in diameter and more than 50 ft tall. They cannot handle the high gas-flow rates attainable in packed columns, but they are attractive for use as chemical reactors, especially those involving the use of a finely divided solid catalyst in suspension in the liquid ("slurry reactors"). They may be operated batchwise with respect to the liquid, or with either cocurrent or counterflow of gas and liquid. The gas pressure drop is high for tall columns, and the compression of expensive sterilized air represents an important cost in deep-tank fermentation. They are relatively inexpensive and easily cleaned, however, and the bubble swarms provide enormous surface per unit volume for gas-liquid contacting.

Relatively shallow pools with porous-plate gas spargers are used to aerate sewage. Bubble-cap and sieve-plate columns might be described as bubble columns, but they usually carry only a few inches of liquid on each tray.

As in agitated vessels, $k_L$ varies little with operating conditions; $k_L a$ depends primarily on $a$, which varies widely. Liquid back-mixing is rapid, the liquid maintaining a nearly uniform concentration top to bottom, even with columns having a height-to-diameter ratio of 12 or more [123]. There is little back-mixing of the bubbles, however, except at the higher gas rates. Since the free cross section for gas flow is reduced by the presence of the liquid, the average rate of rise of the bubble swarm is greater than the superficial velocity. Moreover, in large columns the bubbles tend to collect and rise near the axis, with downflow of relatively bubble-free liquid near the column wall. This happens even though the gas is introduced uniformly over the cross section of the column at the bottom. The upflow of liquid near the axis carries the bubbles upward at velocities greater than velocity of free rise. In small-diameter columns the bubble swarm fills the column quite uniformly. Dissolved surface-active agents cause foaming at the top, setting a practical limit on gas flow, but $k_L a$ increases as about the inverse 0.62 power of the interfacial tension [1].

Published information on mass transfer in bubble columns is not extensive, and most of the experimental studies have been carried out with columns 6 in or less in diameter. The early work of Shulman and Molstad [104] describes the performance of small columns and provides good photographs of the bubble swarms. Using a porous plate sparger in 2- and 4-in columns, with downflow of liquid at rates as high as 40,000 lb/(h)(ft$^2$), they report (HTU)$_{OL}$ for $CO_2$ desorption from water to range from 0.5 to 2.5 ft, increasing linearly with $L$ but being essentially independent of gas-flow rate at the higher values of $G[> 35$ to $55$ lb/(h)(ft$^2$)]. Braulick, Fair, and Lerner [7] oxidized sulfite in 3-, 4-, and 6-in bubble columns at gas rates as high as 300 lb/(h)(ft$^2$), or 1.1 ft/s, using a single orifice sparger.

Data on gas absorption in larger columns have been reported by Sharma and Mashelkar [99], who used a variety of different liquids in columns up to 38.5 cm in diameter and 300 cm in height. They report $k_L a$ to increase as the 0.7 power of the superficial velocity of the gas (up to 1.2 ft/s) and to be essentially independent of height-to-diameter ratio in the range 3 to 12. (As noted above, Shulman and Molstad found no effect of $G$ in the range 35 to $55 < G < 100$.)

Akita and Yoshida [1] have recently reported extensive studies of gas absorption in bubble columns 6 to 24 in in diameter with aerated liquid heights up to 10 ft, using a single 0.5-cm orifice sparger. Figures 11.35 and 11.36 illustrate the nature

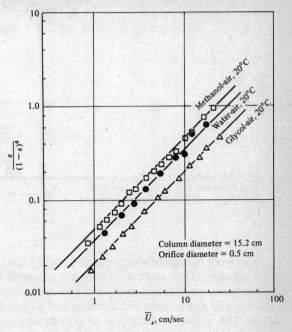

**FIGURE 11.35**
Gas holdup correlation for methanol-air, water-air, and glycol-air. Data of Akita and Yoshida [1].

of the data obtained. Equations involving dimensionless groups are derived to represent the results, one being an empirical correlation for the gas holdup $\varepsilon$ and another a correlation for $k_L a$. The value of $\varepsilon$ must be obtained from the first for use in the second to estimate $k_L a$. In simplified form the equations are

$$\frac{\varepsilon}{(1-\varepsilon)^4} = 0.744 \times 10^{-5} v_L^{-1/6} U_s \left(\frac{\sigma}{\rho_L}\right)^{-1/8} \tag{11.17}$$

(but use 0.935 in place of 0.744 for solutions of electrolytes)

$$k_L a(\text{second}^{-1}) = 363 D_L^{1/2} v_L^{-0.12} \left(\frac{\sigma}{\rho_L}\right)^{-0.62} D_T^{0.17} \varepsilon^{1.1} \tag{11.18}$$

where $v_L$ and $\rho_L$ = kinematic viscosity and density of the liquid
$\qquad D_T$ = column diameter
$\qquad D_L$ = molecular diffusion coefficient for the solute in the liquid
Neither equation is dimensionless, and cgs units must be employed. Though clumsy to use, these equations can provide useful estimates for design purposes.

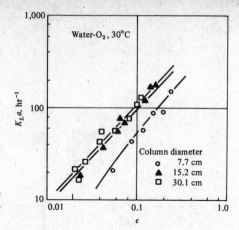

**FIGURE 11.36**
Bubble-column data on $k_L a$ vs. gas hold-up of $O_2$–water. From Akita and Yoshida [1].

This work has recently been updated [1a] by further studies on sizable square bubble columns. Data and correlations are presented for bubble-size distribution, interfacial area, and oxygen absorption by water, methanol, and aqueous solutions of glycerol and glycol. Both single orifices and porous plates were used for gas dispersion. This article would appear to provide an excellent source of data for the design of bubble columns to be used for oxygen absorption in clean nonelectrolyte solutions.

Back-mixing in a bubble column is difficult to prevent, and it is usually so great that only one or two equilibrium stages can be provided in a column of reasonable height. This difficulty can be partially avoided by the use of packing, so that the bubble column becomes a flooded packed column. Sahay and Sharma [95a] report data on such devices, obtained with several $\frac{3}{8}$- to 2-in rings and saddles in 10-, 20-, and 38.5-cm columns. Larger values of both $k_L a$ and interfacial area than in empty bubble columns are reported.

The power input for liquid agitation is a subject of economic importance in fermentation; agitation is needed to prevent clumping of mycelia and maintain free access of the microorganisms to the oxygenated liquid. This power can be large in deep gas-agitated fermenters and can be estimated for comparison with that in mechanically agitated vessels.

Assume operation near ambient temperature and neglect cooling due to vaporization of water. Assume further that the exothermic heat is lost through the walls of the vessel. Then the operation is adiabatic, and the whole device is the equivalent of a Joule-Thomson porous plug or "cracked valve." The enthalpies of inlet and effluent gas streams are the same, since the gas delivers no "shaft work" on the surroundings outside the vessel.

The gas bubbles formed at the bottom expand essentially reversibly as they rise, doing work on the liquid. This work energy is delivered to the liquid, causing it to be mixed and agitated. This same energy is then dissipated by viscous shear as heat, which is readily transferred back to the gas. The expanding gas tends to cool but is restored to its initial temperature by heat transfer from the liquid.

Work energy is required for gas flow through the sparger, but this is relatively small, and most of the kinetic energy of the gas jets is probably dissipated as heat. The agitation power is then [57]

$$W = \frac{C_1 V_1^{\ 2}}{2g_c} + \frac{RT_1}{M} \ln \frac{P_1}{P_2} \qquad \text{ft-lb/lb gas} \qquad (11.19)$$

where $R = 1,544$ (ft-lb)/(lb mole)($°$R)

$\kappa$ = ratio of heat capacity at constant pressure to that at constant volume ($= 1.41$ for air)

$M$ = molecular weight of gas

$T_1$ = temperature of gas forming the initial bubbles

$V_1$ = velocity of gas jets at vena contracta

$P_1$ = pressure at base of bubble column (inside)

$P_2$ = pressure of gas leaving column

$g_c = 32.2$

$C_1$ = fraction of the kinetic energy of the gas jet delivered as work energy to the liquid

The factor $C_1$ is probably no greater than 0.5. Following the treatment of orifices by McCabe and Smith [64], $V_1^{\ 2}/2g_c$ is given approximately by

$$\frac{V_1^{\ 2}}{2g_c} = \frac{\gamma^2}{\rho_0}(P_0 - P_1)\left(\frac{P_1}{P_0}\right)^2 \qquad (11.20)$$

where

$$\gamma \approx 1 - \frac{0.41}{\kappa}\left(\frac{P_0 - P_1}{P_0}\right) \qquad (11.21)$$

in which $P_0$ is the pressure at which the gas is supplied to the sparger.

The above applies when the molal flow rate of gas is constant and the vessel is isothermal, i.e., when the fraction absorbed is negligibly small or absorption is offset by pickup of water and $CO_2$. Sparging of pure ammonia gas into water would accomplish very much less agitation in a bubble column.

**EXAMPLE 11.6**  A large fermentation vessel is filled with water to a depth of 68 ft, before aeration. Air is supplied at the bottom through spargers consisting of orifices. The fraction absorbed is small and the air leaves at the top at 1.0 atm pressure. Operation is at $80°$F, with a gas flow corresponding to 0.3 ft/s based on 1.0 atm and $80°$F. The pressure drop through the sparger orifices is 10 psi. Estimate the agitation power, expressed as horsepower per 1,000 gallons.

SOLUTION  For air, $\kappa = 1.41$, $M = 29$. The gas density at the top is 0.0736 lb/ft$^3$. The pressures are $P_2 = 14.7 \times 144 = 2,120$ lb/ft$^2$; $P_1 = 2,120 + 68 \times 2,120/34 = 6,360$ lb/ft$^2$; $P_0 =$

$6,360 + 10 \times 144 = 7,800$ lb/ft$^2$.  Substituting in Eq. (11.21),

$$\gamma = 1 - \frac{0.41}{1.41} \frac{1,440}{7,800} = 0.946$$

and from Eqs. (11.20) and (11.21), with $C_1 = 0.5$,

$$W = 0.5 \frac{0.946^2}{0.271} 1,440 \left(\frac{6,360}{7,800}\right)^2 + \frac{1,544 \times 540}{29} \ln 3$$

$$= 1,581 + 31,585$$

$$= 33,166 \text{ ft-lb/lb gas}$$

or

$$\frac{0.3 \times 0.0736 \times 33,166 \times 1,000}{550 \times 68 \times 7.48} = 2.62 \text{ hp/1,000 gal}$$

To compress the air reversibly and adiabatically from 1.0 atm for supply to the sparger would require

$$\frac{\kappa}{\kappa - 1} \text{NRT}\left[\left(\frac{p_1}{p_0}\right)^{\frac{\kappa - 1}{\kappa}} - 1\right] \quad \text{ft-lb for } N \text{ moles}$$

or

$$\frac{1.41}{0.41} \frac{0.3 \times 0.0736}{29} \frac{1,544 \times 540}{550} \left[\left(\frac{7,800}{2,120}\right)^{0.29} - 1\right] \frac{1,000}{68 \times 7.48} = 3.6 \text{ hp/1,000 gal}$$

The actual power required is substantially greater than this, because the compressor efficiency is perhaps only 70 percent and the efficiency of the compressor drive is only 92 percent. Almost all this energy input is removed in the compressor aftercooler and does not go into the fermenter. In the case of a mechanically agitated mixing vessel used as a fermenter, the power turns up as heat in the liquid and sometimes has to be removed by relatively expensive refrigeration—requiring more power.     ////

# REFERENCES

1  AKITA, K., and F. YOSHIDA: *Ind. Eng. Chem. Process Des. Dev.*, **12**: 76 (1973).

1a  AKITA, K., and F. YOSHIDA: *Ind. Eng. Chem. Process Des. Dev.*, **13**: 84 (1974).

2  ARNOLD, D. S., C. A. PLANK, and E. M. SCHOENBORN: *Chem. Eng. Progr.*, **48**: 633 (1952).

3  ASHROF, F. A., T. L. CUBBAGE, and R. L. HUNTINGTON: *Ind. Eng. Chem.*, **26**: 1068 (1934).

4  ATTERIG, P. T., E. J. LEMIEUX, W. C. SCHREINER, and R. A. SUNDBACK: *AIChE J.*, **2**: 3 (1956).

4a  BENEDEK, A., G. F. BENNETT, and K. W. A. ITO: Paper presented at the Pittsburgh meeting of the American Institute of Chemical Engineers, June 1974.

5  BIBAUD, R. E., and R. E. TREYBAL: *AIChE J.*, **12**: 472 (1966).

6  BOND, J., and M. B. DONALD: *Chem. Eng. Sci.*, **6**: 287 (1957).

6a  BOSSIER, J. A., III, R. E. FARRITOR, G. A. HUGHMARK, and J. T. F. KAO: *AIChE J.*, **19**: 1065 (1973).

7  BRAULICK, W. J., J. R. FAIR, and B. J. LERNER: *AIChE J.*, **11**: 73 (1965).

8  BRITTAN, M. I., and E. T. WOODBURN: *AIChE J.*, **12**: 541 (1966).

9   CALDERBANK, P. H., (a) *The Chemical Engineer*, no. 212, CE 209, October 1967; (b) chap. 1 in V. W. Uhl and J. B. Gray (eds.), "Mixing: Theory and Practice," vol. 2. Academic, New York, 1967; (c) chap. 5 in N. Blakebrough (ed.), "Biochemical and Biological Eng. Science," Academic, New York, 1967.

10  CALDERBANK, P. H., and M. B. MOO-YOUNG: *Chem. Eng. Sci.*, **16:** 39 (1961).

11  CANNON, M. R.: *Oil Gas J.*, **58:** 68 (1956).

12  CHARPENTIER, J. C., C. PROST, and P. LEGOFF: *Chem. Eng. Sci.*, **24:** 177 (1969).

13  COLBURN, A. P.: *Ind. Eng. Chem.*, **28:** 526 (1936).

14  COOPER, C. M., R. J. CHRISTL, and L. C. PEERY: *Trans. AIChE*, **37:** 979 (1941).

15  COOPER, C. M., G. A. FERNSTROM, and S. A. MILLER: *Ind. Eng. Chem.*, **36:** 504 (1944).

16  CORNELL, D., W. G. KNAPP, and J. R. FAIR: *Chem. Eng. Progr.*, **56:** 68 (July 1960).

17  CRAWFORD, J. W., and C. R. WILKE: *Chem. Eng. Progr.*, **47:** 423 (1951).

18  DANCKWERTS, P. V.: *Chem. Eng. Sci.*, **2:** 1 (1953).

19  DAVIES, J. A.: *Ind. Eng. Chem.*, **39:** 774 (1947).

20  DE WAAL, K. J. A., and J. C. OKESON: *Chem. Eng. Sci.*, **21:** 559 (1966).

21  DE WAAL, K. J. A., and A. C. VAN MAMEREN: *Trans. Inst. Chem. Eng. (London), AIChE Inst. Chem. Eng. Joint Meeting, London*, paper 6.8, June 1965.

22  DODDS, W. S., L. F. STUTZMAN, B. J. SOLLAMI, and R. J. MCCARTER: *AIChE J.*, **6:** 390 (1960).

23  DOUGLAS, W. J. M., *Aerotech. Ind. Rev.*, Summer 1963.

24  DRICKAMER, H. G., and J. R. BRADFORD, *Trans. AIChE*, **39:** 319 (1943).

25  DULLIEN, F. A. L., and L. W. SHEMILT: *Can. J. Chem. Eng.*, **39:** 242 (1961).

26  DUNN, W. E., T. VERMEULEN, C. R. WILKE, and T. T. WORD: *Univ. of Calif. Radiation Lab. Rep. UCRL 10394* (1962).

27  EPSTEIN, N: *Can. J. Chem. Eng.*, 210 (October 1958).

28  ERGUN, S.: *Chem. Eng. Progr.*, **48:** 89 (1952).

29  FAIR, J. R., and R. L. MATTHEWS: *Petrol. Refiner*, **37:** 157 (April 1958).

30  FELLINGER, L.: Sc.D. thesis in chemical engineering, M.I.T., 1941.

31  FOSS, A. S., J. A. GERSTER, and R. L. PIGFORD: *AIChE J.*, **4:** 249 (1958).

32  FUCHS, R., D. D. Y. RYU, and A. E. HUMPHREY: *Ind. Eng. Chem. Process. Des. Dev.*, **10:** 190 (1971).

33  FURZER, I. A., *Chem. Eng. Sci.*, **28:** 296 (1973).

34  FURZER, I. A., and R. W. MICHELL: *AIChE J.*, **16:** 380 (1970).

35  GAUTREAUX, M. F., and H. E. O'CONNELL: *Chem. Eng. Progr.*, **51:** 232 (1955).

36  GEDDES, R. L.: *Trans. AIChE*, **42:** 79 (1946).

37  GERSTER, J. A., W. E. BONNETT, and I. HESS: *Chem. Eng. Progr.*, **47:** 523, 621 (1951).

38  GERSTER, J. A., A. B. HILL, and N. N. HOCHGRAF, and D. G. ROBINSON: Efficiencies in Distillation Columns, final report, Research Committee, American Institute of Chemical Engineers, New York, 1958.

39  GIANNETTO, A., V. SPECCHIA, and G. BALDI: *AIChE J.*, **19:** 916 (1973).

39a GIANETTO, A., V. SPECCHIA, and G. BALDI: *AIChE J.*, **19:** 929 (1973).

40  GOOD, A. J., M. H. HUTCHINSON, and W. C. ROUSSEAU: *Ind. Eng. Chem.*, **34:** 1445 (1942).

41  GRESCOVITCH, E. J.: *Ind. Eng. Chem. Process. Des. Dev.*, **11:** 81 (1972).

42  HARTLAND, S., and J. C. MECKLENBURGH: *Chem. Eng. Sci.*, **21:** 1209 (1966).

43  HIBY, J. W.: *Inst. Chem. Engrs. Symp. Interaction between Fluids Solids (London)*, 1962.

44  HOCHMAN, J. M., and E. EFFRON: *Ind. Eng. Chem. Fundam.*, **8:** 63 (1969).

45  HOLBROOK, G. E., and E. M. BAKER: *Trans. AIChE*, **30:** 520 (1934).

46  HORN, F. J. M.: *Ind. Eng. Chem. Process. Des. Dev.*, **6:** 30 (1967).

47  HUANG, C-J., and J. R. HODSON: *Petrol. Refiner*, **37:** 104 (February 1958).

48 HUGHMARK, G. A., and H. E. O'CONNELL: *Chem. Eng. Progr.*, **53:** 127 (1957).

49 HUNT, C. D'A., D. N. HANSON, and C. R. WILKE: *AIChE J.*, **1:** 441 (1955).

49a HUTTON, B. E. T., L. S. LEUNG, P. C. BROOKS, and D. J. NICKLIN: *Chem. Eng. Sci.*, **29:** 493 (1974).

50 JENSEN, V. G., and G. V. JEFFRIES, "Mathematical Methods in Chemical Engineering," Academic, New York, 1963.

51 JONES, J. B., and C. PYLE: *Chem. Eng. Progr.*, **51:** 424 (1955).

52 JUVEKAR, V. A., and M. M. SHARMA: *Chem. Eng. Sci.*, **28:** 825 (1973).

53 KEMP, H. S., and C. PYLE: *Chem. Eng. Progr.*, **45:** 435 (1949).

53a KERKHOF, P. J. A. M., and H. A. C. THIJSSEN: *Chem. Eng. Sci.*, **29:** 1427 (1974).

54 KLINKENBERG, A.: *Chem. Eng. Sci.*, **23:** 92 (1968).

55 KOHL, A. L., and F. C. RIESENFELD, "Gas Purification," McGraw-Hill, New York, 1960.

56 LARKINS, R. P., R. R. WHITE, and D. W. JEFFREY: *AIChE J.*, **7:** 231 (1961).

57 LEHRER, L. H.: *Ind. Eng. Chem. Process. Des. Dev.*, **7:** 226 (1968).

58 LEMONDE, H.: *Anales de Phys.*, **9:** 539 (1938).

59 LEVA, M.: "Tower Packings and Packed Tower Design," U. S. Stoneware Co., Akron, Ohio, 1953.

60 LEWIS, W. K., JR.: *Ind. Eng. Chem.*, **28:** 399 (1936).

61 LINEK, V., and J. MAYRHOFEROVA: *Chem. Eng. Sci.*, **24:** 481 (1969).

62 LOBO, W. E., L. FRIEND, F. HASHMALL, and F. ZENZ: *Trans. AIChE*, **41:** 693 (1945).

63 LYNCH, E. J., and C. R. WILKE: *AIChE J.*, **1:** 9 (1955).

64 MCCABE, W. L., and J. C. SMITH: "Unit Operations of Chemical Engineering," McGraw-Hill, New York, 1956.

65 MCMULLEN, A. K., T. MIYAUCHI, and T. VERMEULEN: *Univ of Calif. Radiation Lab.*, *Rep. UCRL-3911-suppl.*, (1958).

66 MANNING, E., S. MARPLE, and G. P. HINDS: *Ind. Eng. Chem.*, **49:** 2051 (1957).

67 MARUCCI, G., and L. NICODEMO: *Chem. Eng. Sci.*, **22:** 1257 (1967).

68 MAYFIELD, F. D., W. I. CHURCH, A. C. GREEN, D. C. LEE, JR., and R. W. RASMUSSEN: *Ind. Eng. Chem.*, **44:** 2238 (1952).

69 MAYO, F., T. G. HUNTER, and A. W. NASH: *J. Soc. Chem. Ind.* (*London*), **44:** 375T (November 15, 1935).

70 MEHTA, V. D., and M. M. SHARMA: *Chem. Eng. Sci.*, **21:** 361 (1966).

71 MICHEL, B. J., and S. A. MILLER: *AIChE J.*, **8:** 262 (1962).

72 MIYAUCHI, T., and T. VERMEULEN: *Ind. Eng. Chem. Fundam.*, **2:** 113 (1963).

73 MOHUNTA, D. M., and G. S. LADDHA: *Chem. Eng. Sci.*, **20:** 1069 (1965).

74 MORELLO, V. S., and N. POFFENBERGER: *Ind. Eng. Chem.*, **42:** 1021 (1950).

75 NEWTON, W. M., J. W. MASON, T. B. METCALFE, and C. O. SUMMERS: *Petrol. Refiner*, 141 (October 1952).

76 NIELSEN, C. H. (ED.): "Distillation in Practice," Reinhold, New York, 1956.

77 NORMAN, W. S.: "Absorption, Distillation, and Cooling Towers," Wiley, New York, 1961.

78 NORTON CO.: *Chem. Proc. Prod. Div.*, *Bull. S-32*, 1969.

79 O'CONNELL, H. E.: *Trans. AIChE*, **42:** 741 (1946).

80 ONDA, K., H. TAKEUCHI, and Y. OKUMOTO: *J. Chem. Eng.* (*Japan*), **1:** 56 (1968).

81 OTHMER, D. F., R. C. KOLLMAN, and R. E. WHITE: *Ind. Eng. Chem.*, **36:** 963 (1944).

82 PEAVY, C. C., and E. M. BAKER: *Ind. Eng. Chem.*, **29:** 1056 (1937).

83 PERRY, J. H. (ED.): "Chemical Engineers' Handbook," 4th ed., McGraw-Hill, New York, 1963.

83a PURANIK, S. S., and A. VOGELPOHL: *Chem. Eng. Sci.*, **29:** 501 (1974).

84 RAMACHANDRAN, P. A., and M. M. SHARMA: *Chem. Eng. Sci.*, **24:** 1681 (1969).

85 REISS, L. P.: *Ind. Eng. Chem. Process. Des. Dev.*, **6:** 486 (1967).

86  REITH, T., and W. J. BEEK: *Chem. Eng. Sci.,* **28:** 1331 (1973).

87  RENKIN, A.: *Chem. Eng. Sci.,* **27:** 1925 (1972).

88  Research Committee, American Institute of Chemical Engineers, "Tray Efficiencies in Distillation Columns," third annual progress report, June, 30, 1955.

89  Research Committee, American Institute of Chemical Engineers, "Bubble Tray Design Manual," New York, 1958.

90  RESNICK, W., and B. GAL-OR, "Advances in Chemical Engineering," vol. 7, p. 296, Academic, New York, 1968.

91  REYNIER, J. P., and J. C. CHARPENTIER: *Chem. Eng. Sci.,* **26:** 1781 (1971).

92  ROBINSON, C. S., and E. R. GILLILAND: "Elements of Fractional Distillation," McGraw-Hill, New York, 1950.

93  ROBINSON, C. W., and C. R. WILKE: *Chemica, 1970 Proceedings,* p. 65, 1971; *AIChE J.,* **20:** 285 (1974).

94  ROD, V.: *Brit. Chem. Eng.,* **9:** 300 (1964).

95  RUSHTON, J. H., E. W. COSTICH, and H. J. EVERETT: *Chem. Eng. Progr.,* **46:** 395, 467 (1950).

95a SAHAY, B. N., and M. M. SHARMA: *Chem. Eng. Sci.,* **28:** 2245 (1973).

96  SATER, V. E., and O. LEVENSPIEL: *Ind. Eng. Chem. Fundam.,* **5:** 86 (1966).

96a SATTERFIELD, C. N.: *AIChE J.,* **21:** 209 (1975).

97  SCHRODT, V. N.: *Ind. Eng. Chem.,* **59**(6): 58 (1967).

98  SELLERS, E. S., and D. R. AUGOOD: *Trans. Inst. Chem. Eng.* (*London*), **34:** 53 (1956).

99  SHARMA, M. M., and R. A. MASHELKAR: *Inst. Chem. Eng.* (*London*), *Symp. Ser.*; Montreal Symp. Mass Transfer with Chemical Reaction, 1968.

99a SHENDE, B. W., and M. M. SHARMA, *Chem. Eng. Sci.,* **29:** 1763 (1974).

100  SHERWOOD, T. K.: *Trans. AIChE.,* **36:** 177 (1940).

101  SHERWOOD, T. K., and F. A. L. HOLLOWAY: *Trans. AIChE,* **36:** 39 (1940).

102  SHERWOOD, T. K., and F. J. JENNY: *Ind. Eng. Chem.,* **27:** 265 (1935).

103  SHERWOOD, T. K., G. H. SHIPLEY, and F. A. L. HOLLOWAY: *Ind. Eng. Chem.,* **30:** 765 (1938).

104  SHULMAN, H. L., and M. C. MOLSTAD: *Ind. Eng. Chem.,* **42:** 1058 (1950).

105  SHULMAN, H. L., C. F. ULLRICH, A. Z. PROULX, and J. O. ZIMMERMAN: *AIChE J.,* **1:** 253 (1955).

106  SHULMAN, H. L., C. F. ULLRICH, and N. WELLS: *AIChE J.,* **1:** 247 (1955).

107  SIDEMAN, S. O., HORTACSU, and J. W. FULTON: *Ind. Eng. Chem.,* **58**(7): 32 (1966).

107a SIMKIN, D. J., C. P. STRAND, and R. B. OLNEY: *Chem. Eng. Progr.,* **50:** 565 (1954).

108  SLEICHER, C. A., JR.: *AIChE J.,* **5:** 145 (1959).

109  SOUDERS, M., JR., and G. G. BROWN: *Ind. Eng. Chem.,* **26:** 98 (1934).

109a SPECCHIA, V., S. SICARDI, and A. GIANETTO: *AIChE J.,* **20:** 646 (1974).

110  TELLER, A. J., S. A. MILLER, and E. G. SCHEIBEL: "Liquid-Gas Systems," in J. H. Perry (ed.), "Chemical Engineers' Handbook," 4th ed., chap. 18, McGraw-Hill, New York, 1963.

111  TICHY, J.: *Chem. Eng. Sci.,* **28:** 655 (1973).

112  TIMMERHAUS, K. D., D. H. WEITZEL, and T. M. FLYNN: *Chem. Eng. Progr.,* **54:** 35 (June 1958).

112a UFFORD, R. C., and J. J. PERONA: *AIChE J.,* **19:** 1223 (1973).

113  VALENTIN, F. H. H.: "Absorption in Gas-Liquid Dispersions," E. and F. N. Spon Ltd., London, 1967.

114  VERMEULEN, T., J. S. MOON, A. HENNICO, and T. MIYAUCHI: *Chem. Eng. Progr.,* **62:** 95 (1966).

115  VIVIAN, J. E., and C. J. KING: *AIChE J.,* **10:** 221 (1964).

116  WALTER, J. F., and T. K. SHERWOOD: *Ind. Eng. Chem.,* **33:** 493 (1941).

117  WATSON, J. S., and H. D. COCHRAN, JR.: *Ind. Eng. Chem. Process. Des. Dev.,* **10:** 83 (1971).

118  WENZEL, L. A.: *Chem. Eng. Progr.,* **53:** 272 (1957).

119 WEEKMAN, V. W., JR., and J. E. MYERS: *AIChE J.*, **10:** 951 (1964).

120 WEHNER, J. R., and R. H. WILHELM: *Chem. Eng. Sci.*, **6:** 89 (1956).

121 WESTERTERP, K. R., L. L. VAN DIERENDONCK, and J. R. DE KRAA: *Chem. Eng. Sci.*, **18:** 157 (1963).

121a WOODBURN, E. T.: *AIChE J.*, **20:** 1003 (1974).

122 WOZNIAK, M., and K. ØSTERGAARD: *Chem. Eng. Sci.*, **28:** 167 (1973).

123 YOSHIDA, F., and K. AKITA: *AIChE J.*, **11:** 9 (1965).

124 ZENZ, F.: *Chem. Eng. Progr.*, **43:** 415 (1947).

125 ZENZ, F.: *Petrol. Refiner*, **36:** 179 (1957).

126 ZENZ, F.: *Chem. Eng.*, **79:** 120 (November 13, 1972).

127 ZUIDERWEG, F. J., and A. HARMENS: *Chem. Eng. Sci.*, **9:** 89 (1958).

## PROBLEMS

**11.1** It is proposed to use a column packed with 1-in Raschig rings to recover ethanol from 1,000 ft³/min (40°C, 1 atm) of a dry air–alcohol vapor stream containing 0.142 mole fraction ethanol. The gas effluent will be at 1.0 psig. The liquid effluent will be distilled to recover alcohol and the column will be fed with water containing 0.3 percent alcohol by weight. Alcohol recovery from the gas is to be 98 percent.

(a) Assume operation to be *isothermal* with a ratio $mG_M/L_M$ (molal units) of 0.7 at the dilute end. Fix the mass flow of gas to correspond to 70 percent of flooding at the rich end. Make suitable allowance for the vaporization of water from the aqueous stream. What should be the diameter and height of the packed section?

(b) If operation is *adiabatic*, how tall would the packed section have to be providing the gas- and liquid-flow rates were the same as determined in part *a*?

*Data* Partial pressures of alcohol and water over their mixtures at 40°C are given in the International Critical Tables, vol. III, p. 290. Molecular diffusion coefficients for alcohol in aqueous solution and viscosity in centipoises at 18°C:

| Volume percent alcohol | 0 | 1 | 29 | 59 | 89 | 99 | 100 |
|---|---|---|---|---|---|---|---|
| $10^5 D$, cm²/s | | 1.09 | 0.62 | 0.32 | 0.82 | 1.10 | |
| $\mu$, cP | 1.045 | 1.10 | 2.62 | 3.00 | 1.86 | 1.30 | 1.24 |

The specific gravity of the aqueous solution at 40°C is nearly linear in concentration from 1.0 for water to 0.939 at 0.207 mole fraction alcohol.

**11.2** The alcohol recovery from air described in Prob. 11.1 might alternatively be accomplished using a bubble-cap plate column. If the liquid- and gas-flow rates are maintained at the values determined in Prob. 11.1, what would be the required number of theoretical and actual plates?

Assume the column diameter to be 2.5 ft, with plates carrying 1.5-in bubble caps on 2.5-in centers. The trays will be single-pass crossflow, with a segmental weir 18 in wide and 2 in high. The downcomer will also be 18 in wide extending to within 1.5 in of the plate below. Tray spacing will be 18 in.

**11.3** A gas mixture containing 0.40, 0.45, and 0.15 mole fractions methane, ethane, and ethylene is scrubbed with a nonvolatile absorber oil in a column of four theoretical plates operating at 25 atm and 100°F. The lean oil fed at the rate of 0.9 moles per mole "wet" gas contains 0.98 mole fraction oil and 0.015 and 0.005 mole fractions ethane and ethylene. The absorber oil has an average molecular weight of 200. Approximate values of $K(= Y/X)$ at the operating conditions are methane, 7.9; ethane, 1.6; ethylene, 2.3.

Determine the composition of the treated gas.

**11.4** An oil company plans to produce naphthalene by the hydrodealkylation of methyl naphthalenes contained in the high-boiling fraction obtained by distillation of the gasolines produced in a platformer. A high-temperature catalytic hydrogenation process carried out at 1,000 psia will be employed.

Hydrogen for the purpose will be supplied from the reformer, but it must be upgraded. It is proposed to accomplish this by operating a countercurrent packed absorber at 1,000 psia, using a 40° API kerosene as absorbent. The reformer gas supplied contains 80 mole percent hydrogen, the rest being essentially pure methane. The oil preferentially absorbs methane and so increases the hydrogen concentration of the gas. The oil leaving the absorber will be flashed to 50 psia to remove absorbed methane and hydrogen and then returned to the absorber at 1,000 psia.

For a gas supply of 15 million ft³/day (60°F, 1 atm), the proposed oil rate is 42,000 bbl/day.

(a) Assume that with this oil rate equilibrium will be attained at the rich end. What will be the composition of the enriched gas?

(b) The reformer hydrogen is insufficient for the needs of the naphthalene plant, so makeup hydrogen must be supplied at 35 cents per 1,000 ft³ (60°F, 1 atm). What is the annual cost of the makeup hydrogen equivalent to that lost in the flash gas? The flash gas goes to fuel, with a credit of 10 cents per 1,000 ft³.

(c) Suggest a modification of the simple flow sheet, designed to reduce the hydrogen loss (retaining the absorber).

*Data and assumptions*

Absorber oil: 40° API (specific gravity 0.825, viscosity at 75°F 1.63 cP, molecular weight 160.

One barrel = 42 US gal

Operating temperature: assume 75°F throughout

Pressures: absorber, 1,000 psia, flash drum, 50 psia

Equilibrium $K$ values:

| Pressure, psia | 50 | 300 | 1,000 |
|---|---|---|---|
| $K(= Y/X)$ methane | 88 | 16.75 | 5.03 |
| $K$ hydrogen | 440 | 113.3 | 34 |

Assume the flash gas to contain only hydrogen and methane (none of the oil). The suggested oil rate of 42,000 bbl/day is calculated on a methane- and hydrogen-free basis. Oil and gas withdrawn from the flash drum are in equilibrium.

**11.5** An organic solvent is to be removed from an airstream to reduce atmospheric pollution as the air is discharged. Two packed towers are to be employed, each packed with 1-in Berl saddles. The gas will be scrubbed with water, in which the organic vapor is moderately soluble.

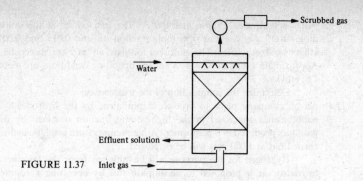

FIGURE 11.37

Both towers will be fed with fresh water and the aqueous streams discharged from the bottom of each. The gas to be treated passes up through the first tower, countercurrent to the water stream, and *down* through the second tower, cocurrent to the water. The organic content of the air is to be reduced from 1.0 to 0.05 mole percent in air.

If both towers are to be the same height, how many feet of packing should be used in each?

*Data* Water rate $= 10,800$ lb/(h)(ft$^2$) in each tower; gas feed rate $= 590$ lb/(h)(ft$^2$). Assume isothermal operation at 77°F. Molecular weight of the gas stream may be taken as constant at 29.5. The gas solubility at the operating pressure is given by $Y = 30X$, where $Y$ and $X$ are mole fraction organic in gas and liquid, respectively. Assume the height of an overall liquid-phase transfer unit to be the same for cocurrent as for counter-current operation. The molecular diffusion coefficient of the organic solute at low concentrations in water at 77°F is $1.4 \times 10^{-5}$ cm$^2$/s.

11.6 A plant uses a rather makeshift packed tower to remove ammonia from a waste air–ammonia gas mixture, with water as absorbent. The arrangement is illustrated diagrammatically in Fig. 11.37.

The installation operates satisfactorily over a period of months, giving the following performance:

Total pressure below packing, 753.7 mm Hg

Total pressure above packing, 740 mm Hg

Mole percent $NH_3$ in gas feed, 2.0

Effluent gas rate, 32,100 ft$^3$/min (68°, 1.0 atm)

Water rate, 69,000 lb/h

Effluent gas composition, 0.186 mole percent $NH_3$

Effluent solution, 2.23 weight percent $NH_3$

Operation isothermal at 68°F, negligible water evaporation

After some months of steady operation, the performance apparently changed rather suddenly. The effluent gas concentration dropped to 0.06 mole percent ammonia yet the absorber was clearly not performing properly. The metered effluent gas rate, all temperatures, the water rate and quality, the inlet gas composition, and the blower-motor power were found to be exactly the same as before. The new liquid effluent, however, is 1.86 weight percent ammonia.

What appears to be the nature and *location* of the difficulty?

Assume the equilibrium partial pressure of $NH_3$ to be given by $p = 6.0c$, where $p$ is in mm Hg and $c$ is in lb $NH_3$ per 100 lb water.

Assume all the test data to be completely reliable ($\pm 1\%$).

11.7  A packed column is used to remove ammonia from the gas purged from an ammonia plant in order to reduce the effluent to 0.10 mole percent $NH_3$. The gas to be treated contains 2.0 mole percent $NH_3$, 30 percent $N_2$, 2 percent argon, and 66 percent $H_2$. It will be scrubbed with water in a countercurrent absorber packed with 1-in Berl saddles, using an inlet-gas rate of 800 lb/(h)(ft²) and a pure-water rate 1.6 times the theoretical minimum.

(a)  How many equilibrium stages ("theoretical plates") are required?

(b)  How many "overall gas transfer units"?

(c)  How many feet of packing?

Notes  The heat of solution and the cooling due to water vaporization tend to offset—assume isothermal operation at 68°F. At 68°F the equilibrium partial pressure of $NH_3$ over an aqueous solution is given by $p = 6.03c$, where $p$ is in mm Hg and $c$ is lb $NH_3/100$ lb water. Neglect the small absorption of argon, $H_2$, $N_2$. The partial pressure of the inert gases may be taken as constant at 760 mm Hg (the total pressure drops slightly due to column pressure drop as the mole fraction inert gases rises due to absorption).

11.8  Several recent publications have claimed that periodic flow can result in improved performance of staged countercurrent absorbers, extraction equipment, distilling columns, and the like. The alleged advantage may be used to increase stage efficiency or throughput or both.

There are various ways to operate "periodic flow." In a plate column, for example, the gas flow would be stopped while the liquid was drained from the bottom plate, the liquid on the next plate moved to the bottom plate, and so on up the column. The top plate (in the case of the ammonia absorber) would then be filled with water. There would be no liquid flow during the gas "blow."

In order to test this concept, it is proposed to calculate the performance of a four-plate absorber operated in this way and compare the results with what might be expected using continuous countercurrent operation.

The gas described in Prob. 11.7 will be scrubbed with water, using the same time-averaged gas- and water-flow rates. Estimate the ammonia loss in the effluent gas, as a percentage of that in the purge gas fed, for both types of operation.

Data  Time required to drain liquid from plate to plate and fill the top plate with water is 6 min. Liquid level on each plate is 12 in. The Murphree efficiency of each plate is 100 percent. Ammonia holdup in the gas is negligible. The liquid on each plate is thoroughly mixed by the gas flow.

11.9  A gas containing 86 mole percent $CH_4$, 9 percent $C_3H_6$, and 5 percent $n$-$C_4H_{10}$ is treated in a tower operating under a pressure of 45 psig. The absorber oil is completely stripped and has a molecular weight of 180. The average temperature is 100°F. The tower contains 10 perfect plates, and 90 percent of the butane is absorbed.

(a)  What is the oil-gas ratio required, expressed as pounds of oil per 1,000 ft³ of rich gas at 60°F and 1 atm?

(b)  What percentage of the propane is removed from the gas?

(c)  Make a plot of butane and propane recovery vs. the number of plates from the top of the tower.

11.10 A water-gas stream from a petrochemical plant contains small quantities of aromatics which are to be recovered by absorption in a nonvolatile oil. The countercurrent absorber will operate at 25°C and 1.0 atm, with 20 percent more oil than required for complete absorption of paraxylene.

The feed-gas composition and the vapor pressures of the several constituents are as follows.

| | Mole percent in feed | Vapor pressure at 25°C, mm Hg |
|---|---|---|
| Benzene | 0.70 | 97 |
| Toluene | 0.30 | 28 |
| $o$-xylene | 0.303 | 6.5 |
| $m$-xylene | 0.206 | 8.0 |
| $p$-xylene | 0.305 | 8.6 |
| Nitrogen | 98.186 | |

(a) What is the oil rate, as moles per 100 moles gas fed?

(b) What percent of the toluene is absorbed?

*Notes* The gas is so dilute that the total gas flow may be taken as constant through the absorber. The absorber has a large number of equilibrium plates. The oil solvent is completely nonvolatile and contains no aromatics as fed. Nitrogen may be assumed to be insoluble in the liquids. Dalton's and Raoult's law apply under the operating conditions (equilibrium mole fraction in gas = mole fraction in liquid times ratio of vapor pressure to total pressure).

11.11 A 30-plate absorber is to be used to remove light hydrocarbons from a natural gas. The gas contains 11.6 mole percent $CO_2$, 7.2 percent $CH_4$, 1.0 percent $C_2H_4$, 9.3 percent $C_2H_6$, 1.2 percent $C_3H_6$, 4.7 percent $C_3H_6$, 0.9 percent $n$-$C_4H_{10}$, 0.6 percent $i$-$C_4H_{10}$, and 0.5 pentanes. The scrubbing oil is a completely denuded nonvolatile hydrocarbon having a molecular weight of 160, specific gravity 0.8, and specific heat 0.60.

Both gas and solvent oil will be supplied at 80°F and 20 atm. The liquid rate will be such as to absorb 90 percent of the ethane in the feed gas. The rich oil leaving the absorber will be flashed to 1 atm, distilled, cooled, and recycled.

What will be the composition of the treated gas?

*Data and assumptions* As a rough approximation, the heat of condensation and solution of each of the constituents may be taken as 6,000 Btu/lb mole.

The overall plate efficiency for propane is 50 percent and may be assumed to vary with $K^{-1/2}$ for the other components.

$K$ values at 20 atm for the several constituents are given below.

| | K, 80°F | K, 100°F |
|---|---|---|
| $CO_2$ | 1.2 | 1.5 |
| $CH_4$ | 8.9 | 9.8 |
| $C_2H_4$ | 2.45 | 2.8 |
| $C_2H_6$ | 1.65 | 2.0 |
| $C_3H_6$ | 0.70 | 0.83 |
| $C_3H_8$ | 0.61 | 0.76 |
| $n$-$C_4$ | 0.24 | 0.28 |
| $i$-$C_4$ | 0.29 | 0.36 |
| $C_5$ | 0.068 | 0.097 |

Pressure drop for gas flow may be neglected. The lean gas leaves the column at 80°F. The column operates adiabatically.

11.12   Natural gas is to be treated to remove $CO_2$ by the use of a 30-plate countercurrent absorber. The gas is available at 50 atm and, for simplicity, may be considered to contain 80 mole percent methane and 20 mole percent $CO_2$. The absorber operates isothermally at 80°F with an oil solvent having a molecular weight of 160 and a specific gravity of 0.80. The oil is essentially nonvolatile. The gas flow is to be 800,000 SCF (60°F, 1 atm) per 24-h day.

(a)   It is planned to throttle the effluent oil to 1.0 atm in a gas-liquid separator and vent the gas to a flare (waste). The oil is then recycled to the absorber. If the solute-free oil rate is 50 moles per 100 moles feed gas, what will be the methane loss, as percent of the methane fed?

(b)   In order to reduce the methane loss in the vented gas, it is proposed to flash the effluent oil in two stages. The oil is first flashed in a separator at 10 atm and the flash gas compressed and mixed with the gas feed to the absorber. The liquid from this first separator will then pass to a second separator at 1.0 atm and the gas vented as before. What will be the methane loss as percent of that in the natural gas feed?

Data   The K values are as follows (80°F).

| $P$, atm | 1 | 10 | 50 |
|---|---|---|---|
| $CH_4$ | 170 | 18.7 | 3.70 |
| $CO_2$ | 24 | 2.50 | 0.52 |

The separators are assumed to produce gas and liquid in equilibrium at 80°F.

11.13   A flooded packed column or packed bubble column is a gas-liquid contacting device having application where gas absorption and liquid-phase reaction must be carried on simultaneously.

One such device is 12 in in diameter and 8 ft tall, packed with $\frac{1}{2}$-in porcelain Raschig rings. It is fed with pure water at the top and pure $CO_2$ gas at the bottom. Flow is countercurrent and steady-state at 25°C, and at atmospheric pressure, with the liquid level being maintained just above the top of the packing. The superficial water velocity is 0.005 ft/s and the superficial gas velocity at the inlet is 0.05 ft/s.

Using the data given below (together with data from the literature, as needed), calculate the fractional approach to saturation of the $CO_2$ concentration in the outlet water.

Data   Void fraction dry column = 0.63; under the stated operation conditions the gas holdup is 12 percent and the liquid holdup 88 percent of the voids. The mass-transfer coefficient $k_L a$ is 0.001 s$^{-1}$, based on total packed volume. $CO_2$ solubility in water at 25°C, 1.0 atm is 0.00061 mole fraction. (Neglect pressure variation due to static head.)

11.14   (a)   Derive an expression for the effect of plate-to-plate entrainment of liquid on the performance of a countercurrent perforated-plate absorber.

In addition to the customary symbols, let $N$ = total number of plates; $n$ = plate number counting from the top; $E$ = liquid entrained from each plate to the plate above, moles/h; $Y_n$ = mole fraction solute in gas leaving the $n$th plate, $X_n$ = mole fraction solute on and leaving the $n$th plate.

Assume steady countercurrent flow, a single soluble solute, ideal plates, dilute gas and liquid streams, isothermal operation, no entrainment to bottom plate or from top plate, and equilibrium represented by $m = Y/X =$ constant. The desired expression is of the form $Y_F = f(Y_1, X_0, m, G_M, L_M, E)$, where $Y_F =$ mole fraction solute in gas fed.

(b) Show that the result reduces to the Kremser equation for $E = 0$.

(c) To test the effect of entrainment on performance, calculate the percent absorption for the case of an absorber having four ideal plates, with $L_M/mG_M = 1$, $E/L_M = 0.1$, $X_0 = 0$. Compare with the corresponding result for the same case with no entrainment.

11.15 In a laboratory experiment, liquid $\alpha$-methylstyrene was hydrogenated to form cumene by bubbling hydrogen through the liquid containing suspended fine particles of palladium black. The hydrogen entered the bottom of a vertical cylindrical reactor and the large excess used passed out at the top. The reaction was carried out at 27.8°C and 1.0 atm, and was found to be irreversible and first-order in hydrogen (g $H_2$ dissolved/$cm^3$ liquid). Liquid was not fed or removed during the test. The hydrogen bubbles kept the liquid well stirred.

In one test with 1.0 g palladium per liter, the hydrogenation rate was 0.40 g mole/(h)(l). In a duplicate test using 0.5 g palladium per liter, the rate was 0.25 g mole/(h)(l).

Estimate the rate expected by the same procedure, using the same gas rate, but with 10 g palladium per liter suspended in the liquid.

**INDEX**